The Anthropocene: Politik—Economics—Society—Science

Volume 37

Series Editor

Hans Günter Brauch, Peace Research and European Security Studies (AFES-PRESS), Mosbach, Baden-Württemberg, Germany

http://www.afes-press-books.de/html/APESS.htm
http://www.afes-press-books.de/html/APESS_37.htm

Jean Chrysostome K. Kiyala · Norman Chivasa
Editors

Climate Change and Socio-political Violence in Sub-Saharan Africa in the Anthropocene

Perspectives from Peace Ecology and Sustainable Development

With a Foreword by Professor Sibusiso Moyo and a Preface by Professor Ajume H. Wingo

INTERNATIONAL CENTRE OF NONVIOLENCE
DUT
DURBAN UNIVERSITY OF TECHNOLOGY
INYUVESI YASETHEKWINI YEZOBUCHWEPHESHE
Springer

Editors
Jean Chrysostome K. Kiyala
International Centre of Nonviolence
Durban University of Technology
Durban, South Africa

Norman Chivasa
Durban University of Technology
Durban, South Africa

Acknowledgement: The cover photo was provided by © Jean Chrysostome K. Kiyala with an exclusive license ro Springer Nature Switzerland. More on this book is at: http://www.afes-press-books.de/html/APESS_37.htm

ISSN 2367-4024 ISSN 2367-4032 (electronic)
The Anthropocene: Politik—Economics—Society—Science
ISBN 978-3-031-48374-5 ISBN 978-3-031-48375-2 (eBook)
https://doi.org/10.1007/978-3-031-48375-2

Copyediting: PD Dr. Hans Günter Brauch, AFES-PRESS e.V., Mosbach, Germany.
Copy and Language Editors: Ms. Elizabeth Norton-Amor, Dr. Jacqui Baumgardt, Ms. Pholile Maseko Zengele, Ms. Mmakgosi (Gadi) Malatsi, and Ms. Sury Bisetty.
Proofreading: By language editors.

This Springer imprint is published by the registered company Springer Nature Switzerland AG
The registered company address is: Gewerbestrasse 11, 6330 Cham, Switzerland

Paper in this product is recyclable.

This volume is dedicated to:

scientists who strive to provide evidenced-based frameworks under which human development and sustainable development can be pursued and achieved without causing a further degradation of the Earth Life's supporting systems;

scholars whose reflections have helped develop the scientific knowledge bridge between peace studies and environmental studies at which crossroads stands peace ecology; institutions that support peace studies and militate against bellicist mentality and is ideological assumptions;

victims of wildfires, fire fighters whose heroism, selflessness and care were proven when rescuing human lives, wildlife, and other forms of life in Australia and other parts of the Globe ravaged by a conflagration; and other parts of the Globe;

innocent lives shattered by the brutalities and inhumanities of wars and socio-political violence around the world; innocent orphans, widows, widowers and others who have survived environmental disasters and multiple types of violence worldwide;

all Climate Change's Prize laureates and those whose efforts towards ensuring sustainable development are pursued under the scientifically accepted conditions, particularly Prof. William Lee Steffen (25 June 1947–29 January 2023)—a pundit of Earth Science, a giant and climate science pioneer the World of Academia lost in 2023.

Foreword

Prof. Sibusiso Moyo, Stellenbosch University

Over the last decade, our world has seen geopolitical conflicts increase. For higher education institutions and society at large, these have a direct impact on partnerships, how and whom we engage with and affect the most vulnerable in our communities, especially women and children. The role that higher education institutions can play remains a question of interest in resolving conflict. The changing nature of contemporary conflicts and the scale of multidimensional violence that the globe is confronted with today question the ability of humans to be stewards of both creation and humanity. Modern theories of conflict resolution and peace research are evolving to construct a more comprehensive and integrated approach to peacebuilding that takes into consideration the geography of conflicts, various typologies of social and political violence that have become endemic and a world pervaded by the militaristic narratives and assumptions that peace must be imposed. Democratic and liberal peace theories espouse the logic of peace master plan to be imposed in nations and regions plagued by large-scale intra-national, regional and international violence. This waring predominant mentality still prevails as the world witness in prime time the genocide in Rwanda, the Arab Spring, the two Congo World Wars and recently Russia's war in Ukraine. Despite the unprecedented death tolls ensued from the 1994 Rwanda estimated at close to 900,000 lives lost, armed conflicts in the Democratic Republic of the Congo (DRC) (1996 to date) that have claimed millions of lives and the devastating war taken by Russia to Ukraine, world military powers are already engaging in an armed race and mass production of ammunitions. The depletion of arms pile stocks in the West is indicative of the powerlessness of violence to and military peace to achieve peace.

While military arsenals are thought of being the means of dealing to conventional socio-political violence, environmental conflicts, natural disasters and climate change-induced conflicts are begging for new means of fighting these challenges that have further expose humanity's fragility and vulnerability. Climate wars require new arsenals that are far from the military warehouses and nuclear arsenal. What is needed for the contemporary world and what is the desired future for humans, biodiversity and the Earth systems life support? There is no better condensed work to provide adequate insights, theoretical frameworks and pragmatic strategies to respond to these existential challenges than the content of this volume. It delves significantly on issues that threaten global peace and security and map out a series of strategies to save the Earth planet, humanity and biodiversity.

The editors and contributors to this book can only be commended for their deepest understanding of peacebuilding concepts and their actuation via a pluridisciplinary body of knowledge where environmental studies and peace studies intersect and span sufficient environmental peacebuilding literature and practice and the conventional frameworks. It is imperative to envision a holistic peacebuilding model and conceptualise sustainable development through the lens of peace ecology—an interdisciplinary paradigm of peace search to build positive peace during a period where climate change issues have come to the fore. Environmental justice, ecojustice and conventional justice are merged in efforts to ensure that structural violence is addressed to secure durable peace.

This volume is conceptually, theoretically and pragmatically all-inclusive of climate change violence, socio-political violence in Sub-Saharan Africa and the bridged peace ecology and sustainable development; this is uncontestably the agenda and recipe for durable and lasting peace on the African continent, and a contributing response to the risks posed by global warming and climate change. The book not only outlines the threats of climate change to food security, peace and stability, but simultaneously lays down realistic plans to mitigate those risks. The book expands beyond providing a linear exploration of direct violence such as military threats to peace and the impact of war on children to underscore the pathways to peace, security and sustainable development that stretch across structural problems that endanger the well-being of people in Sub-Saharan Africa. These include inequalities, poverty, unemployment, exploitation and child abuse, etc. These social, economic and political Upheavals are dealt with by creating viable structures such as developing social capital, fostering social climate and fostering social goods to meet humanity's aspirations towards happiness, harmony and welfare.

Justice and peacebuilding are tied, and the linkages between the two are extensively expounded in the book. Starting with justice for wartime injustices such as the use of child soldiers, the preservation of ones' culture and reputation by concealing grave human rights violence induced by rape cases, justice for institutional abuses of state power via clientelism and state capture, demonstration of the impact of restorative justice to redress injustice and empower children to become instruments of change, the book highlights the important role of civil society in peacebuilding.

Looking forward, the editors acknowledge the power of humanity in the face of adversities and the desire to overcome natural disasters, build peace as that had been the case—ending World Wars, prevailing the COVID-19 pandemic, seeking justice for victims or human rights' violations and environmental crimes. The future we desire is attainable when humanity becomes fully conscious of inter-human stewardship and stewardship of creation to mitigating anthropogenic causes of climate change, global warming and wage peace instead of waging war.

Finally, this exceptionally condensed volume presents distilled scholarly thoughts that make it a cutting-edge multi- and interdisciplinary scientific production to be a "must read". I recommend it as a personal and institutional tool of the modern peacebuilding paradigm.

Cape Town, South Africa
July 2023

Prof. Sibusiso Moyo
Office of the DVC Research
Innovation and Postgraduate Studies
Division of Applied Mathematics
Stellenbosch University
Matieland, South Africa
smoyo@sun.ac.za
https://www.sun.ac.za/english/research-innovation/Pages/default.aspx

Prof. Sibusiso Moyo holds a PhD in Mathematics from the University of Natal, Durban and a Masters (with distinction) in Tertiary Education Management from the LH Martin Institute, University of Melbourne Australia. As a scholar, she has published widely in the Mathematical Sciences with a focus on differential equations and optimization problems in international peer reviewed journals. She has also successfully supervised postgraduate students and continues mentoring and supervising at Master's and Doctoral level. She served in various capacities and levels at the Durban University of Technology (DUT) as Associate Director and Professor in the Department of Mathematics, Statistics and Physics and Director Research and Postgraduate Support up to mid-2017. She is currently an Associate Editor for the Journal of Higher Education Policy and Management published by Taylor and Francis and has served as Guest Editor of the Mathematical Methods in the Applied Sciences Journal (published by John Wiley & Sons) and Journal of Engineering Mathematics (published by Springer). She is also board member for Agenda Feminist Media (which hosts the Agenda journal published by Taylor & Francis). At DUT, Professor Moyo was responsible for driving the research, innovation and engagement agenda including supporting student entrepreneurship and innovation through the newly established DUT Center for Entrepreneurship and Innovation. She was part of the DUT Founding team that established two start up entrepreneurial desks which acted as student incubators both in the Midlands and Durban campuses established in 2018. In 2019 and 2021, she was awarded the special national Entrepreneurial Development in Higher Education (EDHE) award to a Deputy Vice-Chancellor responsible for facilitating remarkable institutional support for entrepreneurship development in terms of student entrepreneurship, entrepreneurship in academia and in growing an entrepreneurial university. She currently serves as Chairperson of the Board of Directors for CASME (Center for the Advancement of Science and Mathematics Education) and is board member on the boards of Innovate Durban, the Durban Chamber of Commerce and Industry, Advisory Board for the Future Professors Programme and innobiz. She is Honorary Dean of the School of Innovation and Entrepreneurship, Tianjin Vocational Institute, Tianjin, China from 2021 to 2023; and was co-Dean of the Confucius Institute from 2018 to August 2022. Her current position is Deputy Vice-Chancellor Research, Innovation and Postgraduate Studies at Stellenbosch University at Durban University of Technology.

Preface

Prof. Ajume H. Wingo, University of Colorado Boulder

Ngonnso, the woman who founded the state of Nso in the Northwest Region of present-day Cameroon more than 600 years ago, identified the principal source of human insecurity as our essential equality: since most people are equally powerful, absent some overarching authority, we each pose a potential threat to each other. We are, in short, our own worst enemies.

The most obvious evidence of the truth of that insight comes from the all-too-common instances of violence against ordinary people caught up in conflicts such as that in Sudan between General Abdel Fattah al-Burhan and General Mohamed Hamden or that roiling in the Northwest and Southwest Regions of Cameroon between the state and separatist groups.

But such violence is but one example of the threats we pose to each other. As Ngonnso recognised, humans pose a threat not just to each other's physical lives, but also to each other's spiritual lives. The death of that spirit is a kind of "death in life" that is the most regrettable and dreaded of all. Indeed, while humans' physical lives are of immense value among the Nso, even greater is the spiritual component of our lives that elevate us beyond being merely physical or biological creatures. From that perspective, the awful history of humans' violence on other humans is but a part of the story of humanity's self-destructive nature. More significant than even that is humanity's penchant for denying, destroying, and frustrating humans' ability to control their own lives and to have a share of control over the states and communities of which they are parts.

Nso people have a word for that ability. It is *vitavi*, which is translated as vitality. Vitavi is the precondition for individuals to realise full and fulfilling spiritual life ("Life in life")—lives in which they have autonomy and the ability to control their own lives. *Vitavi* as vitality is the control that an individual should have over her life and over creating the political world in which she is a part.

Now think of Africa. Separated from the West in so many ways—in space, in history, in power and wealth—there seems to be no greater distance between the two than in terms of this primitive control over oneself and one's life. Indeed, what youth across Africa are fighting and dying for is to have just this type of say over how their lives should go. This difference between the West and Africa does not reflect intrinsic differences in their peoples, but broad differences in individual determination to have that primitive control and audacity to ensure through the creation of institutions, constitutions and cultures that each individual has control over her life and the life of the organisation or state governing her. Europe, North America and the various outposts of Western liberal democracy today have produced a governing culture that to a remarkable (and enviable) degree gives individuals the experience of control in countless ways—and the expectation and confidence that they should enjoy that same kind of control in the political sphere.

Africans, however, seem to have nothing like this experience. When they look at their history, they see little except themselves being ruled by someone or something else—slave traders, colonial masters, domestic dictators, the cruel hands of poverty, war, disease and environmental degradation. And so, when they think about how things could be otherwise, they have little of their own to draw upon when they try to imagine what this better life of *vitavi*, vitality, power, freedom or control would be like.

There is piece of good news out of Africa. An inspiring new crop of African scholars, including the very authors of this volume, are leaving behind the simple linear African narratives of colonialism, imperialism and slavery that have hitherto bounded the intellect of African scholars and compelling them to look beyond received and antiquated narratives of Africa of their erstwhile colonial masters, both for the causes and the solutions to their problems. These young crops of intellectual leaders are doing something unprecedented—they are looking deep into themselves, their localities, their rich, complex African histories and asking themselves where Africans went wrong. In response to this haunting question, they are creating fitting analyses, diagnoses, prognoses, complex methodologies, theories and solutions. In so doing, they are creating a human world for themselves and the rest of the world including Western democracies and their outposts.

The authors who have come together in this volume emanate from different disciplines and academies from all over the continental Africa and beyond, bringing with them wealth of unprecedented and original intellectual resources to tackle the fundamental problem of peace and security in Africa in a new way. They are looking at the prospect of peace under the banner of our earth's ecology that is undergoing unprecedented human-induced climate changes that is not just causing serious, perhaps irreversible, damage to the natural environment, but will pose enormous challenges to our human social and political institutions.

The human need for peace and security is what it is that gave birth to the idea of organised associations we refer to as governments. At bottom, governments or states are made of concentrated zero or negative sum violence force—force of the kind that Thomas Hobbes talks about in his *Leviathan*, force that, as Max Weber argued, was to be wielded solely by the state. The *vitavi*, vitality or power (and by extension political power) that Ngonnso talks about, is a positive sum good that all can have at the same time and that each citizens need to protect herself against the excesses of the state use of violence and for use in participating in decisions that governs her life.

This crop of authors of this volume are speaking a new intellectual language which to an outsider to this outstanding new African scholarship may appear like a cacophonous melee. But that is just the face of this volume. Beneath that façade is a remarkably collaborative, coherent, and a systematic realm of thoughts that capture the way the indigenous people of Nso might understand, diagnose and address such an overwhelming all-encompassing problem like our present Anthropocene. In that communal intellectual process, each scholar put his/her own thoughtful views, informed by his or her discipline, his or her own insight, expertise and experience, into the communal pot to form a gumbo, a patchwork, a Senegalese coat of many colours forging enduring solutions in the service of each and all of humanity. And this is the deep African spiritual element engaged in the process of dealing with problems like our new Anthropocene that invite all to collaborate each with her own insight, input and/or output no matter how scanty or how grand.

This volume is about something greater than meets the eye. It is about a new understanding of peace in the face of violent conflicts. Western scholars are adept at theorising about war and have developed a rich intellectual repertoire on the subject matter, complete with moral guides on conduct in war. These very scholars are, however, often at a loss when it comes to peace. Their scanty vocabulary on peace and their focus on managing conflict give the impression that war has been here, and it is here to stay, and we just have to learn to live with it. These African scholars are giving us a fresh view on peace by positioning peace at the intersection of conventional understandings of peace among groups of human beings sharing political spheres and unconventional understandings of peace as it relates to human committing with our environment. This new view is one that is giving life to our environment and to our planet by revealing the extent to which we are obligated to husband, nurture, preserve and conserve it. It warns us (to paraphrase the due process clause of the US Constitution) that one cannot take the life of persons or that of our environment without due process of the law.

From this perspective, the environment is one with our lives, a fact that leads us to dispense with simplistic tradeoffs that pit human welfare against the welfare of the environment and come to treat the environment as continuous with ourselves, and so deserving of care, love and empathy. The *sotto voce* question just beneath the surface of all that is contained in this volume is this: can we have peace that does not emanate from conflict? The contributors to this volume's answer is a categorical yes with a caveat that such a peace must enliven our ecology and treat our environment as we would ourselves.

The authors of this collection have presented a sobering list of challenges posed by climate change and the destruction of the environment—climate change refugees, conflicts, food and environmental insecurity. That list provides further evidence for the truth of Ngonnso's observation of our capacity for self-destruction. But these inspiring authors also give us reason to take on those problems with hope. Our capacity for self-destruction, after all, must be matched by our capacity to preserve ourselves, and to produce things of peaceful power like the redefinition of our environment, life, conflict and the embrace of the oneness of our lives with the lives of our environment. This volume forces intellectuals all over the world, including the Western world, to look at old problems and solutions with new eyes and to appreciate the bursting bustling intellectual energy on the subject of peace coming out of Africa.

Boulder, CO, USA
July 2023

Prof. Ajume H. Wingo
Philosophy Department
University of Colorado Boulder
Boulder, CO, USA
ajume.wingo@colorado.edu

Associate Professor Ajume H. Wingo (University of Colorado Boulder. Ph.D. at Wisconsin, 1997) is a Prince from Nso in the Northwest Province of Cameroon. He is the Director of Law and Philosophy Program at the Philosophy Department, University of Colorado-Boulder. He attended Cameroon College of Arts, Science and Technology (CCAST). Bambili, where he studied history, economics and geography. He also attended the University of Yaounde, Cameroon, where he studied law at the Faculty of Law and Economics. He obtained his BA from the University of California Berkeley and an MA (1995) and Ph.D. (1997) from the University of Wisconsin Madison. He was a fellow at the Institute on Race and Social Division, Boston University; a fellow at the W. E. B. Du Bois Institute at Harvard; a Visiting Assistant Professor at Clark University and Emerson College; and an Assistant and Associate Professor at the University of Massachusetts, Boston. He has published widely on liberal democratic philosophy and politics, particularly on institutional building in places where there are non-liberal democratic or illegitimate political institutions. He has also published on Civic Education, power and political power, African Politics, African Art and Aesthetics. His book *Veil Politics in Liberal Democratic States* is published by Cambridge University Press in the Cambridge Studies in Philosophy and Public Policy series. He is currently working on two book manuscripts. The first one is tentatively entitled "In the Shade of Power: The Power to the Powerless" and the second with Michael Kruse entitled "The Citizen of Africa." The book is about how Africans can move beyond where their history has put them and begin to make their own future and secure their own political freedom.

Acknowledgements

We express our utmost gratitude to all the institutions and individuals who responded to our call to publish a cutting-edge volume that examines the impacts of socio-political violence and climate change on sustainable development while setting forth peace ecology as the underlying theoretical framework to develop new pathways of socio-economic development.

A scholarly endeavour inspired by the length and breadth of "Peace Ecology" covered under this volume's title *Climate Change and Socio-political Violence in Sub-Saharan Africa in the Anthropocene: Perspectives from Peace Ecology and Sustainable Development*, and the interlink between, on the one hand, conventional conflicts, violence and peacebuilding; and on the other hand, environmental conflicts, climate war and environmental peacebuilding, would not have been thinkable without the foundation laid by pundits and pioneers in peacebuilding and Earth Sciences and Climate Science, namely Prof. Paul Josef Crutzen (3 December 1933–28 January 2021), Prof. Johan Galtung (24 October 1930), Prof. William Lee Steffen (25 June 1947–29 January 2023), Prof. John Paul Lederach (17 April 1955) and many others whose works are recognised in the volume. Many thanks are due to Adj. Prof. (ret.) Hans Günter Brauch who initiates us to reflect and produce high-standard outputs in Peace Ecology. His invaluable guidance and insights are acknowledged here.

We express our heartfelt gratitude to Prof. Geoff T. Harris (Adjunct Professor, Faculty of Management Sciences, Durban University of Technology, DUT) whose former Ph.D. students have enormously contributed to this volume. Their zeal and enthusiasm to maintain a network of peace researchers in the African continent are discernible in this volume. That said, their efforts would have remained limited if they had not been supported by conventional peacebuilding and environmental peacemaking's scholars worldwide, and their outputs.

Our sincere appreciation is expressed to Prof. Sibusiso Moyo, the Deputy Vice-Chancellor, Innovation and Postgraduate Studies at Stellenbosch University and Prof. Ajume H. Wingo (University of Colorado Boulder), for their substantive contributions to the prelims of this volume; to Prof. Fulufhelo Netswera (The Executive Dean: Faculty of Management Sciences, DUT), to Prof. Melanie Lourens (The

Deputy Dean: Faculty of Management Sciences, DUT) and to Prof. Alain Jean-Charles Tschudin (The Director: International Centre of Nonviolence (ICON)), for their considerable support towards bringing this contributed volume to completion.

We remain profoundly grateful to over 160 scholars around the scientific world who responded to the invitations of reviewing over 31 chapters manuscripts and providing critical and invaluable comments to increase the scientific standards of this volume; they all contributed towards peer-review processes that helped in developing and producing this state-of-the-art scholarly book on peace ecology and sustainable development.

That said, we also thank all the authors whose passion was recognised and noted both at the editorial review and post-peer-review stages via their responsive attitude and meticulous revisions of their manuscripts.

Their zeal and passion to develop a peaceful continent in the Anthropocene are key drivers of this contributed volume. Thus, we acknowledge the efforts of the following colleagues who contributed to this volume via their anonymous insightful and constructive peer-review comments:

- Andrew Enaifoghe, Ph.D., Lecturer, University of Zululand, South Africa
- Andrew Okem, Ph.D., Greater Accra Region, Ghana
- Antony Egeru, Professor, RUFORUM, Makerere University, Uganda
- Ashton Murwira, Ph.D., Lecturer, Africa University, Zimbabwe
- Ashton Murwira, Ph.D., University of Zimbabwe
- Clayton Vhumbunu, Ph.D., Lecturer, University of Namibia
- Crispin C. Hemson, Lecturer, International Centre of Nonviolence, Durban University of Technology
- Daniel Nkosinathi, Ph.D., Lecturer, Tshwane University, South Africa
- Darlington Tshuma, Ph.D., Lecturer, University of South Africa
- David Makwerere, Ph.D., Lecturer, Bindura University, Zimbabwe
- Dorcas Ettang, Ph.D., Lecturer, University of KwaZulu-Natal, South Africa
- Dostin Lakika, Ph.D., Lecturer, University of the Witwatersrand, South Africa
- Edknowledge Mandikwaza, Ph.D., Lecturer, University of Zimbabwe
- Emmanuel Emem-Obong Agbenyeku, Ph.D., Lecturer, University of Johannesburg
- Farai Mukanya, Ph.D., Senior Lecturer of Peace and Leadership at Zimbabwe Open University
- Geoff T. Harris, Adjunct Professor, Faculty of Management Sciences, Durban University of Technology
- Ignatius K. Mukunto, Ph.D., Lecturer, Copperbelt University, Zambia
- Jeffrey Kurebwa, Ph.D., Lecturer, Bindura University of Science Education, Zimbabwe
- Jonathan Omuchesi, Ph.D., The Catholic University of Eastern Africa, Kenya
- Jude Cocodia, Ph.D., Associate Professor Niger Delta University, Fellow, International Peace Research Association, Research Fellow, University of the Free State, South Africa
- Morakinyo Akintolu, Ph.D., Lecturer, University of South Africa

- Mphathisi Ndlovu, Ph.D., Lecturer, National University of Science and Technology, Zimbabwe
- Mtshali Thabisile, Ph.D., Lecturer, University of Johannesburg, South Africa
- Mulanda 'Jimmy' Juma, Ph.D., Africa Peacebuilding Institute, Bujumbura, Burundi
- Noel Kiiza Kansiime, Ph.D., Dean, Faculty of Business Economics and Governance, Bishop Stuart University, Mbarara, Uganda
- Ndakaitei Makwanise, Ph.D., Lecturer, National University of Science and Technology
- Oripha Chimwara, Lecturer, University of Zimbabwe
- Phmuna Hamunakwadi, Ph.D., (affiliations not available)
- Rogers Manungo, Ph.D., Lecturer, Great Zimbabwe University, Zimbabwe
- Sejma Aydin, Professor, Department of Economics and Management, International University of Sarajevo
- Simbarashe Gukurume, Ph.D., Lecturer, Sol Plaatje University, South Africa
- Stanley Onwubu, Ph.D., Durban University of Technology
- Tafadzwa C. Makonese, Ph.D., Lecturer, University of Witwatersrand, South Africa
- Theo Mbazumutima, Ph.D., Director of Rema Aid Agency
- Thomas Nyagah, Ph.D., Saferworld, Kenya
- Thomas Otieno Juma (D.Phil.), Department Chair (HOD)—Humanities and Social Sciences, University of Kabianga, University of Kabianga, Kericho
- Tinashe Rukuni, Ph.D., Commissioner and Chairperson—Conflict Prevention and Non-recurrence Thematic Committee
- National Peace and Reconciliation Commission, Zimbabwe
- Tonderai Mabuto, Ph.D., The Aurum Institute—A Leading Health Impact Organisation
- Toyin Adetiba, Professor, University of Zululand, South Africa
- Vengesai Chimininge, Ph.D., Lecturer, Zimbabwe Open University, Zimbabwe
- Wilkister M. Shanyisa, Ph.D., Lecturer, Catholic University of Eastern Africa—Kenya

We acknowledge efforts of the following colleagues who contributed to this volume via their anonymous insightful and constructive peer-review comments.

We thank all language editors for their professionalism, particularly Ms. Elizabeth Norton-Amor, Dr. Jacqui Baumgardt, Ms. Pholile Maseko Zengele, Ms. Mmakgosi (Gadi) Malatsi and Ms. Sury Bisetty; and we finally express utmost gratitude to the Springer team at Chennai for the partnership we have built over the years—striving to publish research and reflections aimed at humanity's and Planet Earth's life.

Durban, South Africa
July 2023

Jean Chrysostome K. Kiyala
Norman Chivasa

Contents

Abbreviations and Acronyms

ACRC	African Charter of the Rights of the Child
ACRWC	African Charter on the Rights and Welfare of the Child
AFDL	Alliances des Forces Démocratiques Pour in Libération du Congo-Zaire
AJOL	African Journals Online
AR	Action Research
ASAL	Arid and semi-arid land
ASALs	Arid and semi-arid lands
ASM	Artisanal and small-scale mining
AU	African Union
AVP	Alternatives to Violence Project
BBSDP	Black Business Supplier Development Programme
BCC	Bulawayo City Council
BEE	Black Economic Empowerment
BICE	Basel Institute of Commons and Economics
CAADP	Comprehensive Africa Agriculture Development Programme
CAAFAG	Children Associated with Armed Groups and Armed Forces
CAR	Central African Republic
CAS	Complex adaptive systems
CBD	Central Business District
CC	Climate change
CCA	Climate change adaptation
CCC	Citizens Coalition for Change
CCDR	Country Climate and Development Report
CCJP	Catholic Commission for Justice and Peace
CCMT	Centre for Conflict Management and Transformation
CDC	Centre for Disease Control
CFSVA	Comprehensive Food Security and Vulnerability Assessment
CI	Confidence interval
CMIPP	Coupled Model Intercomparison Project Phase
COVID-19	Coronavirus disease of 2019

CPF	Community Policing Forum
CRC	Convention on the Right of the Child
CRS	Congressional Research Service
CSA	Climate Smart Agriculture
CSD	Commission on Sustainable Development
CSDC	Compendium of The Social Doctrine of The Church
CSIR	Council for Scientific and Industrial Research
CSO	Civil Society Organisation
CSOs	Civil Society Organisations
CST	Catholic Social Teaching
DDPA	Durban Declaration and Programme of Action
DDR	Disarmament, Demobilisation and Reintegration
DDT	Dichlorodiphenyltrichloroethane
DLF	Darfur Liberation Front
DoJCD	Department of Justice and Constitutional Development
DPT	Democratic Peace Theory
DRC	Democratic Republic of Congo
DRR	Disaster risk reduction
EA	Eastern Africa
EAC	East African Community
EITI	Extractive Industries Transparency Initiative
EPWP	Expanded Public Works Programme
ERRP	Economic Reconstruction and Recovery Plan
ESAB	Employment Services Amendment Bill
EU	European Union
EWS	Early warning systems
FAO	Food and Agriculture Organization
FARC	Revolutionary Armed Forces of Colombia
FARDC	Armed Forces of the Democratic Republic of Congo
FBOs	Faith-based organisations
FDGs	Focus group discussions
FEWS NET	Famine Early Warning Systems Network
FMLC	Freedom to Make Life Choices
GBM	Green Belt Movement
GBV	Gender-based violence
GCMs	General circulation models
GDP	Gross domestic product
GECAMINES	La Génerale des Carrières et des Mines
GHG	Greenhouse gas
GHGs	Greenhouse gases
GKI	Greening Kenya Initiative
GLR	Great Lakes region
GMST	Global mean surface temperature
GNU	Government of National Unity
GPA	Global Political Agreement

HBM	Health belief model
HCBD	Harare Central Business District
HCC	Harare City Council
HDC	Higher Degrees Committee
HECNW	Humanitarian and Environmental Consequences of Nuclear Weapons
HHS	Health and Human Services
HIV	Human immunodeficiency virus
HLE	Health Life Expectancy
HS	Happiness Score
ICC	International Criminal Court
ICD	International Crimes Division
ICL	Imperial College London
ICMM	International Council on Mining and Metals
ICRC	International Committee for the Red Cross
IDPs	Internally Displaced Persons
IGAD	Inter-Governmental Authority for Development
IGBP	International Geosphere Biosphere Programme
IHL	International Humanitarian Law
IHRL	International Human Rights Law
ILO	International Labour Organization
IMCs	Inter-Ministerial Committees
IMF	International Monetary Fund
IOM	International Organization for Migration
IPAP	Industrial Policy Action Plan
IPCC	Intergovernmental Panel on Climate Change
IPI	International Peace Institute
IPV	Intimacy Partner Violence
IRIS	Incident Registration Information System
IYWD	Institute for Young Women Development
JCPS	Justice, Crime Preventions and Security Cluster
KCC	Kamoto Copper Company
KFSSG	Kenya Food Security Steering Group
KNBS	Kenya National Bureau of Statistics
KZN	KwaZulu-Natal
LHR	Lawyers for Human Rights
LO	Life orientation
LPCs	Local Peace Committees
LRA	Lord's Resistance Army
LRC	Legal Resources Centre
LSA	League of Arab States
MDC	Movement for Democratic change
MDGs	Millennium Development Goals
MIBA	Société Minière de Bakwanga
MoU	Memorandum of Understanding

MRM	Moral Regeneration Movement
MTSFs	Medium-Term Strategic Frameworks
NAP	National Action Plan
NAPs/NASs	National Adaptation Plans/Strategies
NATJOINTS	National Joint Operational and Intelligence Structure
NATO	North Atlantic Treaty Organisation
NCA	Norwegian Church Aid
NCPF	National Cyber Security Policy Framework
NDC	Nationally Determined Contribution
NDP	National Development Plan
NEPAD	New Partnership for Africa's Development
NFSD	National Framework for Sustainable Development
NGOs	Non-governmental organisations
NID	Net import dependence
NIP	National Infrastructure Plan
NPRC	National Peace and Reconciliation Commission
NT	New Testament
NVAZ	National Vendors Association of Zimbabwe
NWoW	New Way of Working
OECD	Organisation for Economic Co-operation and Development
OHS	Occupational Health and Safety
ONHRI	Organ for National Healing, Reconciliation and Integration
OPAC	Optional Protocol to the Convention on the Rights of the Child
OR	Odds ratios
OSF	Open Society Foundations
OT	Old Testament
PB	Planetary Boundary
PB	Planetary boundaries
PDL	Datum line
PDU	Project Development Unit
PEPUDA	Promotion of Equality and Prevention of Unfair Discrimination Act
PES	Payment for Services
PESTEL	Political, Economic, Socio-cultural, Technological, Environment and Legal
PoC	Protection of Civilian
PPC	Provincial Peace Committees
PPE	Personal Protective Equipment
PRM	Population, Refugees and Migration
PSTDs	Post-traumatic stress disorders
RADP	Recapitalization and Development Programme
RBZ	Reserve Bank of Zimbabwe
RCD	Rassemblement Congolais pour La Démocratie—Goma
RECs	Regional Economic Communities

REDD	Reduction of Emissions from Deforestation and Land Degradation
REIPPPP	Renewable Energy Independent Power Producer Procurement Programme
SA	South Africa
SADC	Southern African Development Community
SAHRC	South African Human Rights Commission
SAPO	South African Property Owners Association
SAPS	South African Police Service
SCC	Sudan Council of Churches
SDG	Sustainable Development Goal
SDGs	Sustainable Development Goals
SEJA	Socio-Economic Justice for All
SG	Secretary-General
SGB	School Governing Body
SIPRI	Stockholm International Peace Research Institute
SIPs	Strategic Integrated Projects
SPLM/A	Sudan People's Liberation Movement/Army
SPSS	Statistical Package for Social Sciences
SS	Social support
SSA	Sub-Saharan Africa
TDM	Trade Data Monitor
TPNW	Treaty on the Prohibition of Nuclear Weapons
TWC	Third-World countries
UN	United Nations
UN OSAA	United Nations Office of the Special Adviser on Africa
UNAIDS	United Nations Programme on HIV/AID
UNCSD	United Nations Conference on Sustainable Development
UNDP	United Nations Development Programme
UNECA	United Nations Economic Commission for Africa
UNFCCC	UN Framework Convention on Climate Change
UNFPA	The United Nations Population Fund
UNHCR	United Nations High Commissioner for Refugees
UNICEF	United Nations Children's Fund
UNMISS	United Nations Mission in South Sudan
UNSD	United Nations Statistics Division
URJ	Ubuntu Restorative Justice
US	United States
USA	United States of America
VISET	Vendors Initiative for Social and Economic Transformation
Warsaw Pact	Warsaw Military Organisation
WBG	World Bank Group
WFP	World Food Programme
WHO	World Health Organisation
WHR	World Happiness Report

WHS	World Humanitarian Summit
WMAs	Water Management Areas
WNLA	Witwatersrand Native Labour Association
WSCM	World Social Capital Monitor
ZANU PF	Zimbabwe African National Union Patriotic Front
ZCC	Zimbabwe Council of Churches
ZELA	Zimbabwe Environmental Law Association
ZEP	Zimbabwe Exemption Permits
ZIMCET	Zimbabwe Civic Education Trust
ZNA	Zimbabwe National Army
ZPP	Zimbabwe Peace Project
ZRP	Zimbabwe Republic Police

Part I
Threats of Climate Change on Peace, Security and Sustainable Development

Chapter 1
Introduction to Climate Change, Violence, and Sustainable Development in the Anthropocene

Norman Chivasa and Jean Chrysostome K. Kiyala

Abstract This volume features contributions that address climate change-induced or the related peace, security, conflict, and development challenges in sub-Saharan Africa. Given that climate change constitutes one of the threats to peace and security, we argue that peace and security are no longer grounded in traditional peacebuilding theories that have evolved from liberal to democratic peace theories—rather, they have become critical questions and areas of peace studies and environmental studies, under the umbrella of peace ecology—because the need for peace, even if it may be delayed, arises when conflicts have ensued. Thus, our arguments in producing this volume were to underscore the effects of peace ecology—which is positioned at the intersection of peace studies and environmental studies, as conceptualised by Kyrou (Int J Peace Stud 73–92, 2007). Our aim was also to explore the theoretical contribution of peace ecology in understanding and bridging environmental peace and conventional peace; integrating environmental questions, and the factors that trigger socio-political and environmental conflicts and violence. We also aimed to raise an awareness of the contributions of the "peace ecology paradigm" in ensuring an understanding of violence, climate change and sustainable development issues, and their implications for global peace and security. Last, we designed this volume in such a way that it would generate a pluridisciplinary body of knowledge that draws from pluralistic methodological approaches that the various contributors in this volume have explored and employed—from their diverse fields of studies and expertise. The authors of some chapters in this volume discuss various mechanisms that are susceptible of reducing the Earth systems' vulnerabiliy to mitigate Anthropogenic

Dr Norman Chivasa holds a Master's and PhD in Conflict Resolution from the university of KwaZulu Natal and a PhD in Public Administration: Peace Studies from Durban University of Technology. He served as a research fellow at both. He is currently a lecturer in the Department of Peace, Security and Society at the University of Zimbabwe. Email: normanchivasa@gmail.com; ORCID: https://orcid.org/0000-0001-5953-7855

Dr Jean Chrysostome K. Kiyala holds a PhD in Public Administration: Peace Studies from DUT; he is a Senior Lecturer at the International Centre of Nonviolence (DUT), Associate Professor and visiting lecturer at the Université de Bandundu and the Université Evangélique en Afrique (Democratic Republic of Congo); Emails: JeanK@dut.ac.za; kljeanchrysostome@gmail.com; ORCID: https://orcid.org/0000-0002-9587-0230

J. C. K. Kiyala and N. Chivasa (eds.), *Climate Change and Socio-political Violence in Sub-Saharan Africa in the Anthropocene*, The Anthropocene: Politik—Economics—Society—Science 37,
https://doi.org/10.1007/978-3-031-48375-2_1

climate change and global warming, and their effects on humanity, the ecosystem and biodiversity, and provide various models of peacebuilding to avert the ongoing socio-political violence in sub-Saharan Africa, and in the Ukraine. Thus, we contend that the planetary boundaries framework remains the realistic framework within which human and sustainable development can be pursued and attained.

Keywords Anthropocene · Climate change · Climate wars · Conflict · Environmental justice · Justice · Peace · Security · Peace ecology · Environmental peacebuilding · Environmental peacemaking · Earth systems · Sub-Sharan Africa · Sustainable development · Violence · Wars

1.1 Background

Human activity has always been, from time immemorial, a primary source of environmental changes. The Anthropocene epoch refers to changes in the environment, due to human activity. Discourses around the Anthropocene are thought to have started around the 1960s, but other discourses around this issue seemingly started around the period of the industrial revolution, when carbon dioxide was largely emitted. According to some scholars, discourses around the Anthropocene began during the period of the rise of farming civilisation (about 11,000 years ago), when human societies supplanted natural vegetations, which led to the extinction of species and disrupting biochemical cycles (Lewis/Maslin, 2015). According to recent discourses, the period "is marked by a major expansion in human population, large changes in natural processes" (Lewis/Maslin, 2015: 176). Damage to the Earth's ecosystems, caused by human activity, has led to considerable increase in the frequency and intensity of climatic and weather extremes on land and water, such as torrential rains, droughts, and fires. These weather extremes not only had a direct impact on human and nonhuman mortality—but also have an indirect and long-term significance, due to an increase in food instability, water scarcity, livestock loss, and/or forced migration. Thus, climate change, and the related extreme weather events has a significant impact on everyday human existence, as well as on armed conflict and peace (Maximilian, 2023). For this reason, peace and security are no longer grounded in conventional peacebuilding that has evolved from liberal and democratic peace theories—on the contrary, they have become questions and areas of critical reflections under peace studies and environmental scientific discipline There is no doubt that the world today rests on negative peace, as articulated by Galtung (1969). The need for peace subsequently arises in times of conflicts, even though this may be delayed by the propensity for violence, fighting, or pronounced protracted hostilities. Therefore, our rationale for compiling this volume was to bring insights in respect of the effects of peace ecology, as it poses itself where peace studies and environmental studies' links crisscross, as Kyrou (2007) maintains. To provide clarify to the underlying epistemological and ontological frames of various themes explored in this volume, we set forth and upfront "Peace Ecology" as the main paradigmatic frame of reference to guide this volume's overarching argument.

Drawn from Kyrou (2007), "Peace Ecology asserts the underpinning values of environmental peacemaking: the use of the environment in building bridges of collaboration between parties in conflict" (p. 88). The theoretical contribution of peace ecology can be described as the bridging of environmental peace and conventional peace; it embraces environmental questions and factors that trigger socio-political violence that need redressing, and a fundamental response, which we believe, peace ecology can provide. To expound further, Kyrou (2007: 88) avers:

> Peace Ecology is built on key concepts such as bioregionalism, place, sustainability and interconnectedness and leads to a new definition for environmental peacemaking, and a new methodological approach to identifying and measuring resources for the transformation of conflict and violence located at the intersection of peace and ecology.

The scholar's assertion summarises the overall quest for seeking insights on violence, climate change and sustainable development, and their implications for global peace and security "Peace ecology" continues to evolve as a scientific model of peace studies, with an objective of building bridges between peace research and environmental studies (Brauch, 2021). It is ontologically, a composite discipline that requires a pluralistic methodological approach and a multimethodological paradigm to explain various aspects of conflicts, violence, and peace theories, as well as integrate them into environmental studies or vice-versa. This volume reflects that worldview, which contributors from diverse fields explored and employed to generate a more comprehensive and integrative understanding of peace, and the mechanisms to achieve it. Social sciences, applied sciences, ethics, political sciences, human sciences, theology, law, international relations, insights from geography, and geochemistry bring insights for a complete knowledge of our critical time, with its challenges to forge the pathways of lasting peace and sustainable development in every respect.

It is becoming evident that humanity lives in fear, and that there is a great sense of nervosity in the face of environmental disasters and scientific warnings in relation to the ongoing deterioration of the ecosystem. To this end, humans continue to be the main agents of climate change shifts and atmospheric conditions' supernatural deregulation. Such predicaments are real and worrisome, as world leaders are divided over climate action. The promises world leaders made to embark on the adjusted pursuit of sustainable development goals that consider the fragility and vulnerability of the climatic condition, as set by the UN's Earth Summit in 1992, and the Rio+20 in 2012 (Clémençon, 2012; United Nations, 2012) have not materialised. This is owing to world leaders' lack of commitment in taking bold decisions that uphold the Paris Agreement to contain the temperature increase to 1.5 °C above the preindustrial degrees. Article 2(1)(a) of the Paris Agreement states, in this regard that:

> Holding the increase in the global average temperature to well below 2°C above pre-industrial levels and pursuing efforts to limit the temperature increase to 1.5°C above pre-industrial levels, recognising that this would significantly reduce the risks and impacts of climate change (United Nations, 2015a)

The inception of the UN 2030 Sustainable Development Goals in 2015—"Transforming our World" (Papas, 2017; United Nations, 2015b), is yet to demonstrate the ability of world leaders to take a decisive action to mitigate the effects of climate change. Failure to do so will exacerbate the Earth's vulnerability; where climate change and global warming would engender new environmental conflicts, beside what is already happening (Ahmed, 2010; Molvaer, 1991). Climate wars (Dyer, 2010) will become more devastating, and the world will become the epicentre of environmental or ecological refugees. This will require new approaches outside the conventional approaches of disputes resolution to deal with such emerging conflicts (Biswas/Quiroz, 1996; Westra, 2009). These are the challenges that confront humanity today.

Apart from the multifaceted ecological crises we have alluded to above, traditional conflicts, wars, violence, and mass destruction continue to affect the international community. Europe, the Middle East, and Africa are no exceptions. International security is at stake, as the Russia-Ukraine war takes a dangerous turn, which does not exclude ominous escalation. Consequently, there are threats of conflicts between China and Taiwan (Goldstein 2015; Kastner, 2015). Adding to that, the prevailing tense military and political climate in the Korean Peninsula (Mastro, 2018; Mazarr et al., 2018); conflict in the Indo-Pacific Region (Chandra/Ghoshal, 2018), in the Middle East (Milton-Edwards/Hinchcliffe, 2007), and in sub-Saharan Africa (Buhaug et al., 2015; Fjelde/von Uexkull, 2012) are indeed concerning. Finally, the Great Lakes Region, particularly armed conflicts in the Democratic Republic of Congo (DRC) (Liwanga, 2015; Maystadt et al., 2014) typifies the extent of the socio-political violence in the aftermath of the two "World Wars of Africa" (1996–2008) (Kiyala, 2018).

In the sustainable development front, recent developments on peace building have seen a pragmatic turn from peace building to the sustainable peace framework at international level. This follows the adoption of a landmark framework, referred to as "sustaining peace" by the United Nations Security Council and the General Assembly in 2016 (de Coning, 2016, 2018; International Peace Institute (IPI), 2017). In part, the pragmatic turn is attributed to a myriad of versions to the definition of peace building. There is, to this end, no single definition to which scholars, practitioners, and agencies subscribe (de Coning, 2016). This conceptual variation of peace building is perceived to have ripple effects on the design and implementation of peace building initiatives.

Another factor is the linear cause-effect problem-solving model that underpins peace building interventions, where the failure of an intervention is usually blamed on the design and the methods of implementation. Further, conflict and the transition from violent conflict to peace, are considered the exclusive entry points of peace building, which led to the perception that peace building is an extension of both conflict resolution and conflict transformation processes (IPI, 2017), rendering it more of a reactive than proactive process.

This volume highlights three hypothetical elements that are central to peace building, as we perceive them, albeit the foregoing contestations: these are, first, at community level, where peace building has been achieved, and individual and group insecurities, owing to hunger, poverty and conflict are done away with, where people carry on with their normal day-to-day duties without any threats. Second, where peace building has been achieved, individuals and groups can work, with some engaging in income-generating activities to sustain or improve their livelihoods without any disruptions. Third, where peace building has been achieved, communities and nations create initiatives and institutions that support and sustain harmonious interactions between individuals and groups. Together, these three dimensions suggest that peace building is an on-going process and an indispensable fulcrum of human survival and is still relevant in contemporary communities. This book focuses on the sustaining peace framework as its thrust.

In contrast to peace building, the exclusive entry point for the sustainable peace framework is identifying those "attributes and assets that have sustained social cohesion, inclusive development [and other] the factors that together contribute to a peaceful society" (IPI, 2017: 2). The sustainable peace framework is strength-based, rather than needs-based. It fosters and sustains peace more than over-reliance on external interventions. It is based on the premise that culture specific resources exist already in a particular context, which must be tapped into, enhanced, and continue to be exploited. Although the sustainable peace framework differs from peace building in terms of their entry points, there are however, some overlaps, which we now focus on.

One major similarity between sustainable peace and the post-liberal peace building is that the latter feeds on local culture, traditions, and the needs of local people within their context. It moves beyond the criticism of liberal peace building and acknowledges that there are varieties of forms of peace in different contexts.

The post-liberal peace building framework articulates that the peace building process must engage with the everyday life of the local people, and that this culminates into the concept of everyday forms of peace (Mac Ginty, 2013). By its very nature, post-liberal peace building is more participatory, locally owned, and self-sustaining (Mac Ginty, 2014; Richmond, 2009; Wolff, 2015). The self-sustaining nature of post-liberal peace building lies in the fact that peace is not found in institutions, but in everyday interactions, informal relationships of individuals, and groups (Richmond, 2009). Mac Ginty (2014) contends that everyday forms of peace are "social practices that individuals and collectives use to navigate their passage through a deeply divided society" (549). The point is that every society or community devise culture specific methods and practices of promoting and sustaining peace among its members, which cannot be achieved without addressing humanity's survival, which is at stake because of climate change. Not only are contemporary societies at risk of annihilation, but the overall life of the Earth Planet is facing unprecedented Anthropogenic challenges that affect natural systems that support the Earth's life, biodiversity, and human life.

Inasmuch as climate change endangers global security and stability, it is imperative to reflect on innovative strategies that aim, not only to assess Anthropogenic consequence of atmospheric transformations and global warming on future world stability—we also need to think up and set forth effective plans to mitigate the evident risks of climate-destabilising effects on the future of international peace. These concerns justify the increasing interest in research, and the dissemination of outputs related to climate adaptation and resilience (Krampe/Mobjörk, 2018; Simpson et al.,

2019) and conflict-sensitive climate change adaptation (Yanda/Bronkhorst, 2011; Bob et al., 2024; Kiyala, 2022a).

For instance, a study conducted by Simpson et al. (2019) assessed the feasibility and effectiveness of adaptation courses of action, deemed crucial to orient the African continental responses to climate change in ways that minimise risks. The "Global Adaptation Mapping Initiative" was utilised along "a stocktake of adaptation-related responses to climate change", based on the peer-review of scholarly works, published between 2013 and 2020—the inquiry revealed 827 recorded actions towards climate adaptation in Africa (Simpson et. al., 2019: 1). The overall outcomes of that research suggest that climate adaptation is feasible and effective across the continent: 24 adaptation options emerged. Out of these, six dimensions of feasibility (including "economic, environmental, social, institutional, technological, and evidence of effectiveness") were contemplated for each option, and the authors note:

> Over half (51%) of all adaptation actions were reported in the food sector where sustainable water management (SWM) was the most reported option. The fewest actions were reported for cities (5%). The majority of actions (53%) were recorded in just six countries: Ghana, Ethiopia, Kenya, Tanzania, Nigeria and South Africa. Encouragingly, effectiveness was assessed as medium or high for 95% of adaptation options (Simpson et. al., 2019: 1)

The depletion of water grids across the continent, drying dams, and the scarcity of rains are worrisome, as they impact food security and the well-being of the people in the regions that are already affected, such as the Horn of Africa, South Africa, the Sahel Region, and other regions. The global water crisis may be attributed to climate change, but the responsibility to manage water resources rests with individual world governments. Water supplies across sub-Saharan Africa are becoming exhausted. This, in turn, contributes to food insecurity and sanitation crisis.

While the outcome of this investigation appears impressive, Simpson et al. (2019: 1) also note that "no options had high feasibility on any other dimension". To this end, technological and institutional factors have been identified as stumbling blocks to the implementation of those options. To this end, Simpson et al. (2019) has elaborated on the feasibility and effectiveness of adaptation options. Similar studies on climate adaptation (Filho et al., 2023; Alves et al., 2020) are indicative of the critical reality of global warming that challenges world stability. Thus, concerted efforts are required to prevent dramatic consequences of climate change on human life, biodiversity, and food security. Innovative approaches to deal with climate change demand robust climate policies that do not only rely on global processes, but consider national vulnerability, local geography and socio-economic development agenda, national adaptation strategies, and climate risks (Alves et al., 2020). These elements should be integrated with the globally recognised and standardised climate change policies, and the "National Adaptation Plans/Strategies (NAPs/NASs) that include mitigation measures and evaluation mechanisms" (Alves et al., 2020: 190). A comprehensive model should, thus, consider the decentralisation of policies from state-centred approaches that give more power to governments to implement adaptation schemes, amidst cooperation with other stakeholders, such as non-governmental organisations (NGOs), as Alvez and colleagues (2020) suggest.

Considering the reasons mentioned above, and the moderate feasibility of adaptation, as articulated by Simpson and colleagues, novel strategies should be implemented and governments across the continent take actions that acknowledge the gaps in environmental feasibility, and bring insights to the limits of adaptation, while increasing the level of adaptation for options with law feasibility and effectiveness.

1.2 Methods

This volume has 28 chapters, which are outputs of research that employed methodological approaches, varying from literary research, empirical, case studies, participatory research, which were relevant and adequate to yield credible findings that we are presenting in this volume. The aim of the contributors was to address peace, security, and development challenges in the face of environmental disasters and scientific warnings about the ongoing deterioration of the ecosystem. The main challenge is that humanity continues to be the main agents of climate change shifts, and atmospheric conditions' supernatural deregulation. Accordingly, the contributions in this volume provide, not only solutions, but are in line with science-based perspective of Sustainable Development Goals in the Anthropocene. They open further discussions on innovative ways of abating environmental disasters and scientific warnings about the ongoing deterioration of the ecosystem, socio-political violence, and sustainable development challenges.

1.3 Content and Structure

This book is divided into four major parts, which include uneven numbers of chapters in each, arranged according to the main themes developed and discussed. The introduction (Chapter one) and Chapter two provide a thematic and conceptual orientation to fundamental issues of concerns that are related to contemporary environmental and ecological issues, socio-political violence, the challenges posed by COVID-19, environmental conflicts and climate wars, justice, as well as peace and sustainable development in the Anthropocene. The contributors debate key issues vital to the survival of the Earth Planet, based on theories and ongoing research. At the end, some perspectives on justice and peace come to the fore, as the discussion unfolds, with the aim of pursuing sustainable development that guarantees the satisfaction of the needs of the present and future generations.

1.3.1 Introduction to Climate Change, Violence and Sustainable Development in the Anthropocene

Jean Chrysostome Kiyala and Norman Chivasa explored mechanisms to reduce the Earth's susceptibility, including climate change and global warming, which brings new environmental conflicts, at the micro and macro level. According to these scholars, these dynamics are becoming evident at a time when terrorism, conventional wars, violence, and conflicts of all kinds are still wreaking havoc in our world. It is noteworthy that even though practitioners, the international community, civic and political organisations, and governments continue to develop policies and offer programmes to address ecological crises, the threat posed by climate change to peace, security, and sustainable development does not appear to be abated. The authors, therefore, drew on secondary data on the Anthropocene, climate change, violence, peace, war, and security, and focus on sub-Saharan Africa. It examines contextual and novel perspectives to address the problems associated with climate change, violence, and their different manifestations, and remove barriers to sustainable development in sub-Saharan Africa and beyond.

Part I: Threats of Climate Change on Peace, Security, and Sustainable Development

This first section investigates the extent to which climate change threatens global peace, security, and development initiatives. It examines the origins and evolution of initiatives aimed at mitigating the negative effects of climate change at both the micro and macro levels. While these initiatives are not a panacea for the climate change problem, they provide insights into how modern communities can help mitigate the effects of climate change on human survival (Sects. 1.3.2–1.3.7).

1.3.2 Climate Change and Climate Wars in the Anthropocene

Chrysostome Kiyala employed documentary analysis to explore the nexus between climate change and sustainable development in the Anthropocene. He examined the development of the new geological epoch, referred to as the Anthropocene, which characterises its adverse impacts on human activities in relation to the well-being of the sum of natural systems that support the Earth Planet's health. For him, humanity, science, and technologies should not transgress without putting the whole Earth Planet at risks of further degradation and peril. This should be so because sustainable development, local, national, regional, and international stability rely on humanity's care and determination to refrain from further violating the green limits of the planetary boundaries. According to Chrys Kiyala, it is urgent and non-negotiable to heed scientific warnings regarding the planetary boundaries framework to pursue a meaningful and non-detrimental sustainable and human development. Such aspirations and expectations would, going forth, be possible if a responsible stewardship of the Earth guides human development.

1.3.3 The Impacts of Climate Change on Water Resources in the Anthropocene: Mitigation and Adaptation Strategies in Southern Africa

Rookmoney Thakur employed documentary analysis to examine the effects of climate change on water supply, as well as mitigation and adaptation strategies adopted in Southern Africa. The scholar argues that the depletion of water resources in sub-Saharan Africa (SSA), due to climate change, has various negative consequences, including increased food insecurity, transborder disputes, and increasing health problems, among other socio-economic factors. To minimise the vulnerability of communities in Southern Africa, Rookmoney suggests a rethinking of strategies aimed at mitigating the impacts of climate change on water resources, and further recommends an integrated approach, such as community-based disaster risk reduction (DRR) and climate change adaptation (CCA) programmes to improve the efficacy of aid, and minimise the uncertainty of communities of the nations in question.

1.3.4 Drought, the War in Europe and Its Impact on Food Security in Sub-Saharan Africa, East Africa

Katcho Karume, Jean Mondo Mubalama, and Chrysostome Kiyala examined secondary data to investigate the effects of drought, the European War, and food insecurity in East Africa. According to them, climate change has significantly impacted people's subsistence means, evidenced by a decline in crop yield and production. On the other hand, the prices of imported food and non-food commodities have increased since the launch of the Russia-Ukraine war. The scholars recommended, in this regard, that policymakers and development agencies prioritise formulating and implementing policies and programmes that promote farm level adaptation strategies, and that farmers across the region engage in climate smart agriculture (CSA), along other price regulating factors, such as subsidies on staple food crops and fertilisers. They further recommended that East Africa reduces the rate at which it imports food. This can be achieved if local productions or substituting some of the import products with locally available materials is encouraged. Social protection programmes in East Africa must account for potential increases in seasonal price variability to enhance household resilience to weather shocks.

1.3.5 Climate Change and Its Implications on Food Security in the Great Lakes Region

Jackson Ishara, Rehema Matendo, Jean Chrysostome K. Kiyala, and Katcho Karume used secondary data to examine climate change and its consequences on food security in the Great Lakes region. The scholars are of the view that changes in temperature, rainfall, and the prevailing extreme events significantly impact agriculture and livestock, which are the main pillars of smallholder farming systems and, therefore, essential sources of food. This vulnerability is exacerbated by multiple biophysical, political, and socio-economic constraints in the region; thus, limiting its adaptive capacity. They recommended that research be conducted and development actors, including farmers themselves, invest massively in identifying strategies to mitigate the situation.

1.3.6 Climate Change, Food Insecurity, Peace and Sustainable Development in East Africa: A Case Study of South Sudan, Sudan, Ethiopia and Kenya

Moses John evaluated secondary data to explore climate change, food hunger, peace, and sustainable development in South Sudan, Sudan, Ethiopia, and Kenya. He argues in this regard that while the extent and intensity of conflicts in sub-Saharan Africa has decreased in the 1990s, the East African region registered an increase of about 30 state-based conflicts in 2019. The scholars, therefore, recommends an early response to these problems to avert a humanitarian catastrophe. This brings to the fore the need for academic research and for books to be produced to create awareness and sensitise the relevant stakeholders to act early to save lives.

1.3.7 Gender, an Ecological Challenge to Peace in the Anthropocene: An African Perspective

Wilkister M. Shanyisa investigated the effect of gender in societal environmental peace, using secondary data. In her view, how society perceives men and women determines the specific roles of these two genders. The scholar argues that society is structured in a way that the existing power asymmetries continues to reinforce and perpetuate the gender divide between men and women in relation to their roles in ecosystem management. In part, gender injustice is attributed to the fact that efforts and policies to incorporate and integrate women in environmental issues remain superficial. Shanyisa argues in this regard, that even though gender mainstreaming and gender analysis may be the options available to ensure the realisation of the substantive, active participation, and representation of women from local to international levels in ecosystem management, collaborative efforts, community capacity building, commitment and the political will of governments remain integral to addressing this critical issue.

Part II: Addressing Direct Violence and Building Peace

This section explores the nature, causes, extent, and consequences of direct violence. In so doing, it investigates how such forms of violence are addressed at the micro and macro levels. (Sects. 1.3.8–1.3.10).

1.3.8 Community-Based Early Warning Mechanisms to Address Machete Violence in the Artisanal Mining Areas in Zimbabwe

Lawrence Mhandara evaluated secondary data to examine initiatives to counter the reoccurring threat of the machete violence in Zimbabwe's artisanal gold mining areas. For him, the community early warning approach has, hitherto, received less attention, and in some instances, been overshadowed by curative approaches that the security services are in favour of. Mhandara observed, in this regard, that despite the systemic and political challenges associated with community preventive initiatives, local and people-centred approaches can revolutionise the machete violence prevention. The scholar is of the view that local peacebuilidng actors—police, civil society, local leadership, as well as the youth and women—are spurred to develop the early warning systems within the affected communities by mounting tools to detect threats, reduce their impact, and mitigate the consequences of the violence.

1.3.9 Mapping the Conflict Between Street Vendors and the Harare City Council: Towards Building Peace in Zimbabwe

Washington Mazorodze investigated the conflict between street vendors and the Harare City Council in Zimbabwe, using empirical data. He observed that street vending has become the most common activity, not only in Harare, but throughout Zimbabwe's main city centres. For him, running battles between urban authorities and street vendors are common without any solution in sight. He believes that although urban authorities have their own council police, they often enlist the services of the Zimbabwe Republic Police (ZRP), and sometimes, the Zimbabwe National Army (ZNA) to help them remove illegal vendors from the streets. Mazorodze notes in this regard that street vendors, propelled by the need to survive, have vowed to remain in the streets, and spark, in the process, a complex, intractable conflict with so many dimensions, involving many actors. Mazorodze argues further that identifying parties in the conflict, the issues, the context, relationships, and the dynamics of the conflict is one way of addressing the conflict between street vendors and the Harare City Council authorities.

1.3.10 Youths in War in a Post-Colonial State and the Revolving Use of Child Soldiers: Perspectives from the Democratic Republic of Congo

Daniel Mlambo examined secondary data to investigate the employment of child soldiers and its implications for the Democratic Republic of the Congo's (DRC) economic growth potential, internal and regional stability, and Southern Africa's security architecture. Some regions have witnessed major internal conflicts and economic declines that have threatened socio-economic development, human security, sovereignty, the territoriality of states, and the stability and legitimacy of political regimes. Since attaining freedom from its erstwhile colonial ruler, Belgium, the DRC has been hampered by internal instability of many forms and sizes (primarily driven by the country's rich natural resource base), many rebel groups, and tribalism. Resources scarcity in the DRC has seen the formation of many rebel groups, which contributed to instability in the DRC. Other motives included the absence of the rule of law, competition for land, weak state capacity, and external influence.

Part III: Peace, Security, and Sustainable Development

This section assesses interventions aimed at achieving peace, security, and long-term development at both the micro and macro levels. Peacebuilding actors at the local and policy levels are doing everything in their power to address challenges to peace, security, and development. However, the effectiveness/ineffectiveness of these interventions is determined by a variety of factors, including social cohesion, partnerships between various actors or insufficient resources, institutional failure, and systemic injustice (Sects. 1.3.11–1.3.19).

1.3.11 Rethinking Peace: Some Reflections on the Imaginary Nature of the Contemporary Peace Model

Norman Chivasa reflects on the efficacy of imaginary peace in establishing good mental images that outlive conflict-ravaged nations, using secondary data. For him, imaginary peace motivates people to go beyond the boundaries of the existing conflict and violence by giving them something to strive for. As a method, rather than a goal, imaginary peace gives societies options on how to modify or improve their crisis situations. The scholar argues in this regard that peace is not only limited to the experienced reality of peaceful co-existence, tolerance, and the nonviolent resolution of conflict—it is also an invention and creative process of our imagination. Chivasa concludes that societies must exploit the imaginary peace by having dreams and visions of a peaceful future to outlive crisis situations such as violent conflicts. The strength of imaginary peace is that it can generate a commitment to achieve a better and peaceful future, even in the face of violence and conflict-ravaged communities.

1.3.12 Inequality, Unemployment and Poverty in the COVID-19 Era in the Eastern Cape Province, South Africa

Tafadzwa Maramura, Peter Makaye, and Torque Mude examined secondary data to evaluate how the three challenges inherited from the apartheid administration in colonial South Africa—namely poverty, inequality, and unemployment, worsened the country's burden at the height of the COVID-19 pandemic. They argue in this regard that the unprecedented economic implications of the COVID-19 pandemic have widened the inequality, unemployment, and poverty gaps in Eastern Cape. For them, the inequality, unemployment, and poverty nexus have particularly afflicted the young population, as evidenced by the high rates of crime, drug abuse, and teenage pregnancies during the COVID-19 era.

1.3.13 Sustainable Development in South Africa in the Anthropocene in the Post-COVID-19 Era

Jean Chrysostome K. Kiyala and Diaku K. Dianzenza employed documentary analysis to investigate South Africa's challenges and opportunities in respect of overcoming the developmental and economic setbacks ensuing from the global health crisis, and the Anthropogenic effects of climate change. They are of the view that South Africa can achieve sustainable development. For them, the path to reach this requires: (1) reinvigorating the health infrastructure to withstand global challenges associated with COVID-19, (2) setting forth robust strategies and policies that uplift its citizens' wellbeing to prevent violent riots that destroy the economy, (3) revitalising its struggling economy and creating jobs, (4) transitioning from coal-dependent industry, economy and trade to an eco-friendly economy that grows from cleaner and renewable energy resources, (5) dealing with brain drain push factors, (6) pursuing the Policy Coherence for Development Framework, supported by the political buy-in to align the National Development Plan's (NDP) priorities with the 2030 SDGs Agenda.

1.3.14 Civil Society-Community Partnership for Sustainable Livelihoods and Development in Ward 7 of Chivi District (Masvingo Province), Zimbabwe

Washington Mazorodze employed empirical data to analyse the impact of civil society- community partnerships on the livelihoods of the people of Chivi District, Ward 7. The scholar is of the view that a strong partnership between civil society organisations and rural communities can serve as an important element in promoting and achieving the much-cherished sustainable development goals (ending poverty and hunger), thereby making communities resilient and peaceful.

1.3.15 Peace, Security and Sustainable Development Goals in Zimbabwe

Torque Mude examines the options available in the pursuit of sustainable peace and human security for the attainment of SDGs in Zimbabwe. He is of the view that while negative peace prevails in Zimbabwe, the trajectory of human security in the country leaves much to be desired. The scholar argues further that with the recurring violent protests, post-election violence, inter-party and intra-party latent conflicts, human rights abuses; and most of all, the military rule that emerged with the military ouster of Zimbabwe's late president, Robert Mugabe, sustaining peace as a panacea for achieving SDGs in Zimbabwe hangs in the balance, and could remain a phantom on the horizons if "guns are not silenced", and human security is not prioritised. Torque concludes that democratisation, reconciliation, transitional justice, poverty eradication, the protection of the planetary biosphere, upon which human survival depends, and sustaining peace are *sine qua non* to attaining SDGs in Zimbabwe.

1.3.16 The Nexus Approach in Relation to Climate Action and Planetary Integrity: Delimiting SDG13 Relative to Accelerating Sustainable Development in Africa

Alain Jean-Paul Charles Tschudin analysed indicator data from the UN Department of Evaluation and Statistical Analysis (DESA) to review Africa's progress on all SDGs relative to the results of a survey conducted with participants from African member states. Tschudin focused exclusively on Climate Action (SDG13), which, in the light of climate change and degradation, has been increasingly located at the front and centre of the sustainability debate. According to him, human displacement, migration, and the related risks of refugee crises, human insecurity, and the related conflicts, are all attributed to climate change. The scholar proposes, in this regard, that greater attention be given to SDG13, relative to the other SDGs, located within a bigger-picture nexus approach. This may increase our awareness, which, in turn, may inform policy development and implementation to mitigate the negative consequences, due to failure to consider climate action relative to conflict, violence, and war.

1.3.17 Social Climate, Social Capital and Social Good in Sub-Saharan African Countries: Implications for Sustainable Development

Aliye F. Mataraci and Ozge Buyukdagli focused on Social Goods for Sustainable Development in the Anthropocene. The scholars analysed the results of the two important reports measuring social goods—namely the World Happiness Report (WHR), and the World Social Capital Monitor (WSCM). The WHR, in this regard, aims to measure the perception of happiness of people from countries around the globe. It has been conducted since 2006 and focuses on six factors: gross domestic product (GDP), social support, health life expectancy, freedom to make life choices, as well as the perception of corruption and generosity. The WSCM is a worldwide open access Social Capital Survey, which seeks to ensure that citizens have a better understanding of the social climate in their countries. The findings of these surveys, in respect of the selected sub-Saharan African countries, provided a better understanding of the social climate in the region, involving the impact of the Coronavirus disease (COVID-19), which is considered the greatest health crisis in the past 100 years. The analysis of the three concepts demonstrates that they can enhance world happiness and sustainable development. These are social climate, social capital, and social goods; which are tied to the pursuit of sustainable development, and enhance, and are sensitive to climate change, which calls for a global response to enhance the decreasing world happiness.

1.3.18 Occupational Health and Safety of Artisanal Miners in Kolwezi, the Democratic Republic of Congo: A Problem for Goals 3 and 8 of the Sustainable Development Agenda

Martha Chadyiwa, Lufuluabo E. Maweja, Thokozani Mbonane, Shalin Bidassey-Manilal, Charlotte Mokoatle, Claris Siyamayambo and Bheki Magunga conducted a cross-sectional study among 546 workers, using simple random sampling from five different artisanal mines namely – Kasulo, Mutoshi, Kamilombe, Uck Drain, Tilwezembe, and Intermediaires. The scholars argue in this regard that in recent years artisanal mining has become more prevalent in comparison to several other mining activities that have been in existence in the City of Kolwezi, in the Democratic Republic of Congo (DRC). The findings of their study revealed a significant relationship between socio-demographic characteristics and the possession of professional miners' cards, and that artisanal miners with professional miners' cards showed a better understanding of OHS and the DRC safety and mining regulations. To this end, artisanal mining knowledge bank can play a role in promoting safer and more sustainable practices, which can contribute to achieving SDGs 3 and 8.

1.3.19 Russia's War in Ukraine: Impacts on Ecology, Energy Market, Food Security, and Sustainable Development

Oyebamiji Sunday Israel, Ashe Muesiri Obero, and Kiyala Jean Chrysostome explored the possibility of reimagining new geopolitics and geostrategy, and how this can ensure the realisation of security, as well as economic and political goals in Europe. To this end, the international community can no longer afford to shy away from the reverberating effect of the Russian invasion of the Ukraine and the Ukrainian fighting to expunge the Russians from their territories. The scholars argue in this regard that the supremacy of Russia (formerly Soviet Union) in the Eastern world, and its hegemony in the Warsaw Pact until the collapse of socialism in the 1990s, has created a national pride that the country is determined to sustain, even through warfare, particularly as its rival military camp, NATO, is rapidly expanding its horizon. Further, the scholars posited that apart from the loss of lives, war is affecting a host of strategic issues, such as the environment, food security, energy supply, global and sustainable development at the regional and global level. They concluded that the failure of a permanent Security Council member to veto a peace resolution, where such a member turns an aggressor, calls for reforms of the UN Charter to ensure accountability and sustainability that will enhance sustainable development.

Part IV: Justice and Peacebuilding

One of the threats that has infiltrated human social interactions is social injustice, which has permeated our societies, particularly our institutions, and the social fabric. Despite these setbacks, individuals and communities are working tirelessly to combat social injustice through peacebuilding efforts. Individuals' and communities' reactions to social injustice, on the other hand, are influenced by their cultural contexts. This section investigates issues relating to justice and peacebuilding initiatives at the macro and micro levels (Sects. 1.3.20–1.4).

1.3.20 *Child Soldiering Justice and Peacebuilding: A Dialectic Analysis of the Conventional Criminal Justice System and African Indigenous Jurisprudence*

Jean Chrysostome K. Kiyala examined secondary data to assess a dialectic argument, drawn from literary research to argue for a justice model that takes cultural context, traditional values, and African jurisprudence into account, which can serve as an interplay of restorative retributivism and restorative sanctions against the logic of the mainstream conventional criminal justice system. According to Kiyala, the lack of justice endangers security and peace processes in war-affected communities. He is of the view that justice processes involving children accused of war-related crimes and other violations of international humanitarian and human rights laws, because of their participation in hostilities, often end in contested outcomes, because of the child soldiers' paradoxical portrayals as victims, perpetrators, and heroes. Kiyala decries that the ICC's refusal to consider the Ugandan traditional mechanisms of accountability in the sentencing of Dominic Ongwen, the abducted child soldier of the Lord's Resistance Army, who perpetrated war crimes and crimes against humanity, has raised serious concerns about the ability of traditional or indigenous jurisprudence in handling justice for former child soldiers. To this end, he commends some countries in sub-Saharan Africa that have made headways in holding child soldiers accountable and attaining justice for victims through traditional and restorative approaches. Their experiences can pave a way for a global model of community-based approach of post-conflict peacebuilding with former child soldiers.

1.3.21 *Honour Culture and Rape Denialism in the Southern Sudanese Culture: A Prospective Response from Ubuntu Restorative Justice*

Josephine Kimanu Mauwa and Jean Chrysostome K. Kiyala investigated rape denialism and the honour culture, while exploring the potential of the *Ubuntu* restorative justice approach to help manage and redress rape-related injustices and human rights violations in South Sudan. According to them, denialism of rape cases, owing to the South Sudanese honour culture, which intends to protect the reputation of societies affected by such injustice in wartime, deprives survivors of their human rights and dignity. Such a lack of acknowledgment of the suffering of rape survivors is reinforced by political leaders and traditional/tribal authorities, and the extensive monitoring of the affected women and girls who have, since the eruption of the civil war in 2013, been deprived of opportunities to narrate their ordeals. The scholars note with concern that the predicament of the criminal justice system that perpetuates rape injustices can be overcome by the *Ubuntu* restorative justice practices, which promote dialogue and acknowledges the rights of humans and ensures dignity, prevent retaliation, encourage forgiveness, and the empowerment of victims—giving them a voice, redressing injustices they are subjected to, and bringing restoration after the ordeals they suffered in times of war.

1.3.22 *Effects of State Capturing, Truth and Reconciliation Commissions in Zimbabwe on Sustainable Development*

Torque Mude explored ways to address, in the pursuit of sustainable peace, the trajectory of the state capture of the truth and reconciliation commissions in Zimbabwe. The scholar argues in this regard that since attaining independence in 1980, the country has experienced successive episodes of political violence and human rights abuses. In his view, the state has been responsible for unleashing violence on political opponents. What is worrisome is that state-led efforts, purported to redress the past injustices, have largely been politicised to uncover the truth, compel collective forgetting, and deny atrocities. Mude recommends that endogenous strategies be mapped out to address the state capture of transitional justice and reconciliation processes towards sustaining peace in Zimbabwe.

1.3.23 *Handling Indiscipline Involving Children at Home and Schools: A Restorative Justice Perspective to Unlearn Violent Behaviours in Zimbabwe*

To address the indiscipline involving children in homes and at schools, Ntombizakhe Moyo-Nyoni and Dorothy Moyo propose a restorative justice perspective that would ensure that Zimbabweans unlearn violent behaviours. The scholars argue in this regard that when adults (parents and teachers) in homes and at schools use violence to enforce discipline in children, the latter gravitate towards being violent, and will display this violent behaviour as adults. This, in turn, defeats the efforts to bring sustainable peace. Moyo-Nyoni and Moyo are of the view that restorative justice approaches to disciplining children within the endogenous systems of the Ndebele society in Bulawayo can lead to the adoption of nonviolent practices of curbing indiscipline at family and community level. Thus, the scholars concluded that parents and school authorities should be empowered and encouraged to cultivate restorative justice strategies to redress indiscipline.

1.3.24 Schools as Agents of Change: Impacts of Restorative Circles Mbambangwe School in KwaZulu-Natal, South Africa

Sibusisiwe Mlambo and Jean Chrysostome K. Kiyala explored strategies that can be employed to curb school violence in Mbambangwe High School. According to these scholars, restorative justice and conflict transformation mechanisms are promising ways of addressing school discipline issues; and, thus, avert violent behaviours among learners. The scholars argue that non-violent conflicts resolution strategies can settle inter-learners and learners-teachers disputes constructively, and bring restoration to the affected parties. Issues of trust between parents and learners, open channels of communication between parents and their children, and encouraging a good working relationship between learners, parents, and teachers are paramount to building peace and transforming the school environment—thus, creating a peaceful and conducive learning environment. Restorative justice and the spirit of *Ubuntu* are key principles transforming schools to peaceful environments.

1.3.25 Indaba and Violence Prevention in Educational Institutions: Perspectives from South Africa

Jean Chrysostome K. Kiyala, Primrose Sibusisiwe Mlambo and Crispin M. C. Hemson examined the role of Indaba as a violence prevention mechanism in educational institution in south Africa. From their perspective, violence in educational institutions is a widespread phenomenon which often results in injury, loss of life, and sustained negative impact on the mental, psychological and emotional health of survivors. They observed that educational institutions rely upon punitive justice and restorative sanctions as a means of deterrence. They concluded that a synergic humanistic and communitarian relationship between *Indaba,* restorative justice, peace education and *Ubuntu* can leverage peaceful resolution of conflicts in educational institutions, promote reconciliation and prevent recurring violence.

1.3.26 Peace-Building Processes to Address Xenophobia in South Africa

Clayton Hazvinei Vhumbunu examined secondary data to reflect on the accomplishments, problems, and prospects of combating xenophobia through peace-building activities in South Africa. The scholar is of the view that the Government of South Africa, in partnership with various stakeholders, has been implementing different peace-building activities, aimed at preventing the recurrence, escalation, and continuation of xenophobic attacks, yet xenophobic continues unabated. Vhumbunu suggests, in this regard, that new interventions be sought, and that the existing peace-building initiatives be adjusted to reduce the incidences of xenophobia in South Africa.

1.3.27 State and Civil Society-Led Peace Committees in Zimbabwe

David Makwerere examined secondary data to assess the value of Zimbabwe's Local Peace Committees (LPCs). The scholar contrasts LPCs, established by the National Peace and Reconciliation Commission (NPRC) with peace committees, established by various civil society organisations throughout the country that operate at the provincial level in all administrative provinces of the country. According to him, state-led LPCs have not had the much needed transformative effect, because of how they are set up. For instance, senior government officials and traditional leaders, most of whom are accused of being partisan, dominate these LPCs. On the other hand, the CSO-led LPCs have also failed to deliver the goods because their working relationship with government is very constrained, as they are often, accused of sinister agendas. Makwerere notes with concern that although there is enough room for collaboration between state-led and CSO-led CSOs, these actors have failed to effectively work together for the good of the people. He recommends that peacebuilding actors in Zimbabwe consider, on a serious note, work together to enhance their effectiveness.

1.3.28 Personal Identity and Peacebuilding: A Critical Reflection

Daniel Erdmann discusses the complex setting of a person's identity, and how this interferes with the process of peace building. A wide range of potential traits is highlighted and pointed out, to the effect that an individual's background supports the building of a specific personal's identity. During the dialog facilitation, the scholar focused on a wide range of potential diversity that ought to lead to peace building in the long run. During the facilitation, he challenged traditional, mostly academic approaches towards peace building, to address the urgent need to equip people at all levels of society to deal with conflict management that finds its place beyond strategies, methods, structures, and models.

1.4 Looking Towards the Future

Most of the contributors in this volume are of the view that the current climate change, peace, security, and development challenges confronting our global village call for a comprehensive, synergic, strategic, and international response that considers the vulnerability of individual countries, regions, and the global world. The linkages between climate change and violence are evident and impact sustainable development. Peace ecology, on the other end, reveals itself as a dynamic and versatile instrument of violent conflicts mitigation and peacebuilding that creates and sustains the conditions for durable development in the Anthropocene.

The authors unanimously agree that to ensure a better future for generations to come, governments, civil society, businesses, the international community, individuals, and social and political groups must all take the responsibility for the actual disintegrating status of the Earth's systems, and make honest and informed resolution to reduce Anthropogenic causes of climate change, prevent wars, object to the militaristic logic of conflict resolution, and always opt for the path of inter-nations and international peace.

At the heart of this debate is the need for all societies to adopt peace education that embodies ecojustice, and ecofriendly behaviours and foster peaceful social ideals to control their destiny without jeopardising the welfare of future generations. The desire to mitigate violence, social injustice, and environmental disasters pervades all the chapters in this volume, which is produced at a critical juncture of human history when terrorism, conventional wars, violence, and conflicts of all kinds are still wreaking havoc in our society. To this end, threatening challenges posed by climate change and global warming cannot be ignored. Even though practitioners, the international community, civic and political organisations, as well as governments continue to formulate policies and programmes that address ecological crises, the threat posed by climate change to peace, food security, regional and global stability, and sustainable development does not appear to be abating.

Finally, it is noteworthy to acknowledge humanity's resilience in the face of past adversities such as natural calamities, the challenge to maintain international peace after world wars, and fighting global pandemics, including the COVID-19 pandemic. To address the problems of climate change, violence, and its different manifestations, and remove barriers to sustainable development in sub Saharan Africa and beyond, this volume still celebrates humanity's ability to prevail and rise above contemporary challenges that preoccupy the global human family.

References

Ahmed, F. (2010). Approaches to and tools for managing environmental conflicts in coastal zones in Africa: Challenges and prospects in relation to Integrated Coastal Zone Management (ICZM). *African Journal on Conflict Resolution*, 10, (2), 31–47.

Alves, F., Leal Filho, W., Casaleiro, P., Nagy, G. J., Diaz, H., Al-Amin, A. Q., . . . Klavins, M. (2020). Climate change policies and agendas: Facing implementation challenges and guiding responses. *Environmental Science & Policy, 104*, 190–198. https://doi.org/10.1016/j.envsci.2019.12.001

Biswas, A. K. & Quiroz, C. T. (1996). Environmental impacts of refugees: A case study. *Impact Assessment, 14*(1), 21–39.

Brauch, H. G. (2021). Peace ecology in the Anthropocene. In Ú. Oswald Spring & H. G. Brauch (Eds.), *Decolonising Conflicts, Security, Peace, Gender, Environment and Development in the Anthropocene* (pp. 51–185). Cham: Springer. https://doi.org/10.1007/978-3-030-62316-6_2

Buhaug, H., Benjaminsen, T. A., Sjaastad, E. & Theisen, O. M. (2015). Climate variability, food production shocks, and violent conflict in Sub-Saharan Africa. *Environmental Research Letters, 10*(12), 1-11.

Chandra, S. & Ghoshal, B. (2018). *The Indo-Pacific axis:Peace and prosperity or conflict?* London: Routledge.

Clémençon, R. (2012). Welcome to the Anthropocene: Rio+ 20 and the meaning of sustainable development. *The Journal of Environment & Development, 21*(3), 311–338.

de Coning, C. (2016). From peacebuilding to sustaining peace: implications of complexity for resilience and sustainability. *Resilience, 4*(3), 166–181. https://doi.org/10.1080/21693293.2016.115377

de Coning, C. (2018). Adaptive peacebuilding. *International Affairs, 94*(2), 301–317.

Dyer, G. (2010). *Climate wars: The fight for survival as the world overheats*. Oxford: Oneworld Publications.

Filho, W. L., Wolf, F., Totin, E., Zvobgo, L., Simpson, N. P., Musiyiwa, K., . . . Efitre11, J. (2023). Is indigenous knowledge serving climate adaptation? Evidence from various African regions. https://doi.org/10.1111/dpr.12664

Fjelde, H. & von Uexkull, N. (2012). Climate triggers: Rainfall anomalies, vulnerability and communal conflict in sub-Saharan Africa. *Political Geography, 31*(7), 444–453.

Galtung, J. (1969). Violence, peace, and peace research. *Journal of Peace Research, 6*(3), 167–191.

Goldstein, S. M. (2015). *China and Taiwan*: John Wiley & Sons. https://unfccc.int/sites/default/files/english_paris_agreement.pdf. Accessed 23 February 2023.

International Peace Institute (IPI). (2017). *Sustaining peace: What does it mean in practice?* Retrieved https://www.ipinst.org/wp-content/uploads/2017/04/1704_Sustaining-Peace-final.pdf. Accessed 15 January 2023.

Kastner, S. L. (2015). Is the Taiwan strait still a flash point? Rethinking the prospects for armed conflict between China and Taiwan. *International Security,* 40(3), 54–92.

Krampe, F. & Mobjörk, M. (2018). Responding to climate-related security risks: Reviewing regional organizations in Asia and Africa. *Current Climate Change Reports, 4*(4), 330–337. https://doi.org/10.1007/s40641-018-0118-x

Krampe, F., Van De Goor, L., Barnhoorn, A., Smith, E., & Smith, D. (2020). Water Security and Governance in the Horn of Africa. *Stockholm International Peace Research Institute SIPRI, SIPRI Policy (Paper No. 54).* https://www.diva-portal.org/smash/get/diva2:1424710/FULLTEXT01.pdf. Accessed 22 March 2023.

Kyrou, C. N. (2007). Peace ecology: An emerging paradigm in peace studies. *International Journal of Peace Studies*, 73–92.

Lewis, S. L. & Maslin, M. A. (2015). Defining the Anthropocene. *Nature, 159*, 171–180

Liwanga, R.-C. (2015). The meaning of Gross Violation of Human Rights: A Focus on International Tribunals' Decisions over the DRC conflicts. *Denv. J. Int'l L. & Pol'y, 44*, 67.

Mac Ginty, R. (2013). Indicators+: A proposal for everyday peace indicators. *Evaluation and Program Planning, 36*(1), 56-63. https://doi.org/10.1016/j.evalprogplan.201.207.001

Mac Ginty, R. (2014). Everyday peace: Bottom-up and local agency in conflict-affected societies. *Security Dialogue, 45*(6), 548–564. https://doi.org/10.1177/096701061455089

Mastro, O. S. (2018). Conflict and chaos on the Korean Peninsula: Can China's military help secure North Korea's nuclear weapons? *International Security, 43*(2), 84–116.

Maximilian, L. (2023). Peace needs to embrace the Anthropocene, peace review: *A Journal of Social Justice, 35*(1), 166–174. https://doi.org/10.1080/10402659.2023.2174374
Maystadt, J.-F., De Luca, G., Sekeris, P. G. & Ulimwengu, J. (2014). Mineral resources and conflicts in DRC: A case of ecological fallacy? *Oxford Economic Papers, 66*(3), 721–749.
Mazarr, M. J., Gentile, G., Madden, D., Pettyjohn, S. L. & Crane, Y. K. (2018). *The Korean Peninsula: Three Dangerous Scenarios*: JSTOR.
Milton-Edwards, B. & Hinchcliffe, P. (2007). *Conflicts in the Middle East since 1945*. London: Routledge Taylor & Francis Group.
Molvaer, R. K. (1991). Environmentally induced conflicts? A discussion based on studies from the Horn of Africa. *Bulletin of Peace Proposals, 22*(2), 175–188.
Papas, M. (2017). The 2030 Sustainable Development Agenda and the Paris Climate Agreement—taking urgent action to combat climate change: how is Australia likely to fare? *Asia Pacific Journal of Environmental Law, 20*(1), 94–114.
Pius Yanda, P. & Bronkhorst, S. (2011). Climate change and conflict: Conflict-sensitive Climate Change Adaptation in Africa. *Policy & Practice Brief, 014 (November 2011)*. https://media.africaportal.org/documents/policy_practice14.pdf. Accessed 5 October 2021.
Richmond, O.P. (2009). A post-liberal peace: Eirenism and the everyday. *Review of International Studies, 35*(Suppl.), 557–580. https://doi.org/10.1017/S0260210509008651
Simpson, N.P., Shearing, C.D., & Dupont, B. (2019). Climate gating: A case study of emerging responses to Anthropocene Risks. *Climate Risk Management, 26*, 100196. https://doi.org/10.1016/j.crm.2019.100196
United Nations. (2012). The Future we want: Outcome document of the United Nations Conference on Sustainable Development, Rio de Janeiro, Brazil, 20–22 June 2012. https://sustainabledevelopment.un.org/content/documents/733FutureWeWant.pdf. Accessed 8 August 2023.
United Nations. (2015a). The Paris Agreement. https://unfccc.int/sites/default/files/english_paris_agreement.pdf. Accessed 3 July 2023.
United Nations. (2015b). Transforming our world: the 2030 Agenda for Sustainable Development. Resolution adopted by the General Assembly on 25 Sep 2015 [without reference to a Main Committee (A/70/L.1)] 70/1. Transforming our world: the 2030 Agenda for Sustainable Development. https://www.un.org/en/development/desa/population/migration/generalassembly/docs/globalcompact/A_RES_70_1_E.pdf. Accessed 23 January 2022.
Westra, L. (2009). *Environmental justice and the rights of ecological refugees* (1st ed.). London: Routledge.
Wolff, J. (2015). Beyond the liberal peace: Latin American inspirations for post-liberal peacebuilding. *Peacebuilding, 3*(3), 279–296.

Chapter 2
Climate Change and Climate Wars in the Anthropocene

Jean Chrysostome K. Kiyala

Abstract The nexus of climate change and sustainable development is increasingly pervading debates in the Anthropocene, along with environment conflicts and climate wars. Past narratives on global warming show both activism and denialism; however, science is crystal clear that climate change and global warming result from the impacts of human activities on atmospheric conditions. While several inquiries on the present state of the planet underline how inaction on earlier warnings about the threats posed to the earth by carbon dioxide and greenhouse gas emissions, few studies are dedicated to uncovering the potential of environmental peacebuilding to prevent further degradation of the structures that support life on earth. This chapter pinpoints two major frameworks for humanity's prosperity and that of future generations: (1) planetary boundaries framework and (2) environmental peacebuilding theory. The latter consists of (i) understanding and addressing environmental conflicts and violence; (ii) defining, conceptualising and contextualising climate wars, (iii) exploring and implementing environmental peacemaking processes; (iv) utilising the tenets of environmental justice and social justice as means to respond to climate-induced conflicts; and (v) deepening the notion of care, protection and stewardship of the environment. Based on literary research design and methods, the chapter found that sustainable development and prosperity of future generations are underscored by humans' ability to rise about selfish lucrative interests, denialism about scientific evidence of climate change's root causes, developing conscientious stewards of the earth, and bridging environmental peacebuilding theory and sustainable development.

Keywords Anthropocene · Climate wars · Climate change · Common good · Earth systems · Environmental peace building · Environmental peacemaking · Environmental justice · Social justice · Sustainable development

---

Dr Jean Chrysostome K. Kiyala holds a PhD in Public Administration: Peace Studies from DUT; he is a Senior Lecturer at the International Centre of Nonviolence (DUT), Associate Professor and visiting lecturer at the Université de Bandundu and the Université Evangélique en Afrique (Democratic Republic of Congo). Emails: JeanK@dut.ac.za; kljeanchrysostome@gmail.com; ORCID: https://orcid.org/0000-0002-9587-0230.

J. C. K. Kiyala and N. Chivasa (eds.), *Climate Change and Socio-political Violence in Sub-Saharan Africa in the Anthropocene*, The Anthropocene: Politik—Economics—Society—Science 37,
https://doi.org/10.1007/978-3-031-48375-2_2

2.1 Introduction

> Greenhouse gas emissions keep growing. Global temperatures keep rising. And our planet is fast approaching tipping points that will make climate chaos irreversible, he said. We are on a highway to climate hell with our foot still on the accelerator.
>
> UN chief Guterres tells COP27 on 7 Nov 2022

Climate change-induced-disasters and conflicts, along with socio-political violence have become an existential threat to humanity, biodiversity and the ecosystem, to the extent of causing unprecedented losses of life across the globe. Humanity is recognised for its longstanding of resilience against numerous adversities such as environmental catastrophes, health crises, and large-scale violence throughout history. But the outbreak of Covid-19, floods, wildfires around the world, tsunami hazards, civil wars in Sub-Saharan Africa, particularly in the Great Lakes Region, the Middle East, the Russia-Ukraine War, the ramifications of the 7th October 2023 Hamas assault on Israel and the unproportionate retaliation by the Israel Defense Forces (IDF) and its plausible genocidal actions against the Palestinians, have shown the limitations of humanity to prevent, speedily halt and avoid those nefarious and deadly military campaigns. The level of humans' resilience to fight those catastrophes or tribulations and bounce back continues to be tested. Humanity's vulnerability and paralysis in face of Covid-19 tragedy, natural calamities and ongoing war in Ukraine, the Gaza strip, the eastern Democratic Republic of Congo (DRC), and other socio-political violence should be acknowledged. The limits of resilience have been tested, and mankind is recovering from the Covid-19 pandemic albeit its frailty and emerging challenges of war.

The threats posed to the survival of humanity, the ecosystem and biodiversity have seen the emergence of a new discipline and peace conceptual framework, namely peace ecology (Brauch, 2016; Brisman, 2016) or environmental peacebuilding (Carius, 2006; Hardt/Scheffran, 2019; Kiyala, 2022a) or environmental peacemaking (Kyrou, 2007). Both peacebuilding paradigms are essential to mitigate the detrimental effects of conflicts on the environment. Conflicts and wars that are not triggered by climatic causes and those that have direct roots in environmental issues damage the land, seas, biodiversity, flora and fauna, whether conventional or unconventional means deployed to curb further disasters and tragedies. This is an obvious and tangible reality.

The Anthropocene is a term used to describe a new geological epoch characterised by a significant human influence on the environment and Earth's natural regulating systems. This has led to notable disruptions in the earth's atmospheric conditions, marking a departure from previous eras. Scholars trace the Anthropocene back to the end of the Holocene which is:

> the name given to the last 11,700 years of the Earth's history – the time since the end of the last major glacial epoch, or "ice age." Since then, there have been small-scale climate shifts – notably the "Little Ice Age" between about 1200 and 1700 A.D. – but in general, the Holocene has been a relatively warm period in between ice ages (Smith, 2011: n.p.)

Research on most reconstructions of the Holocene reveals that the world population has expanded exponentially since 2000 years ago; furthermore, there was an acceleration of population growth in the past 500 years as a result of progress made in disease control (Ruddiman, 2013: 50). Understanding the interconnectedness of population growth, the Anthropocene and sustainable development is crucial. The rapid increase in population has compelled humanity to develop technologies to meet the growing demands for food, housing and transportation, which in turn have spurred further industrial advancements and innovations in agriculture, transportation, clothing and food production. While all these human needs are critical, the lack of responsible and conscientious stewardship of the environment has exacerbated the vulnerability of the ecosystems and biodiversity (Kiyala, 2022a). In addition, climate stewardship is multileveled institutions also have a role to play. For instance, the UN Framework Convention on Climate Change (UNFCCC), sets out the basic legal framework and principles for international climate change cooperation with the aim of stabilising atmospheric concentrations of greenhouse gases (GHGs) to avoid "dangerous anthropogenic interference with the climate system" (International Institute for Sustainable Development, 1992–2021: n.p.). The UNFCCC consists of 198 countries that ratified the Convention which entered into force on 21 March 1994; those countries are called "Parties to the Convention" and have made the commitment to prevent "'dangerous' human interference with the climate system" (UNFCCC, 2022: n.p.).

This chapter attempts to establish how environmental peacebuilding is expected to enhance sustainable development at the intersection of climate change and climate wars; it further supports the idea that the pathways of human development and sustainable development rest upon a collective consciousness (ordinary citizens, politicians, scientists, environmental activists and non-governmental organisations (NGOs)) and their resolve to thrive within the planetary boundaries (PBs) framework. The specific aims of this framework are articulated are as follows:

(1) to demonstrate how wars and humans' influences exerted on the planet (via carbon dioxide and coal combustion and pollution created by greenhouse gas emissions and burning of fossil fuels) are persistently destroying humanity itself, the ecosystem and biodiversity;
(2) to underline the harmful effects of anthropogenic climatic perturbations and the possibility of pursuing sustainable development within PBs;
(3) to expand the awareness on the climate change and global warming and the extent to which human development and sustainable development can continue to be pursued – within the limits defined by scientific evidence such as the PBs framework;
(4) to look into climate transformations and its byproduct that is extremely consequential for the survival of Earth Planet;
(5) to develop an argument for a long-term environmental peacebuilding strategy that will help humanity thrive while refraining from further damaging the Earth systems, which comprise geosphere (earth itself, including rocks, minerals, and landforms of the earth—surface and interior), lithosphere (land), biosphere

(living creatures), cryosphere (frozen waters), hydrosphere (water), and atmosphere (air). Their interaction creates a familiar environment that supports humanity's sustenance and biodiversity;

(6) to develop strategies that make sustainable development viable and achievable in the Age of climate change and global warming, based on the conditions of human and environmental vulnerability, resilience capacity and prospective adaptation success, which should be context-fitting.

This argument focuses on humanity's ability to work towards achieving comprehensive peacebuilding, encompassing both conventional and environmental aspects. It emphasises the use of scientific evidence, particularly the PB, to guide efforts in addressing climate change-related conflicts and implementing climate-sensitive adaptation measures. The goal is to mitigate the impacts of human activities and global warming on the Earth's ecosystems and promote sustainable development.

Addressing the complexities of climate change requires bridging conventional peace and peace ecology; this stance is developed and supported by the findings from this literary research, which relies on a large range of electronic and printed versions of sources (Correa/Owens, 2010), and multifarious references drawn from Google Scholar, ResearchGate, internet sources, dictionaries, encyclopaedias and those sourced out from the library. A dialectic compound of theoretical frameworks is employed within the literary research design to guide this chapter, mainly peace ecology and sustainable development. The dialectic reasoning here refers to the paradoxical and contradictory argument that examines the threats posed by anthropogenic climatic changes and the pursuit of sustainable development dictated by humanity's needs. This leads to reconciling the prospects of feasible human and sustainable development while preserving Planet Earth's life supporting systems. The argument is developed in nine points that sometime overlap and end with some perspectives of environmental peacebuilding as the fundamental basis of durable human development and the survival of the earth, including the ecosystem and biodiversity.

This chapter is structured as follows: the introduction; a discussion on the Anthropocene; the relationship between the earth and the common good in the Anthropocene; the impact of ongoing climatic transformation of the ecosystem; and the rise in environmental conflicts and climate wars. These point to the consequences of climate change that requires new perspective to address them; those include examining the effects of war on the environment and considering PBs as new perspectives in the struggle to sustain the health of The Earth. Finally, exploring the interactions between environmental peacemaking, environmental justice and social justice can serve as the pathways to sustainable development in the age of climate change.

2.2 The Anthropocene

The end of the Holocene coincides with the start of the Anthropocene – the new epoch where humans started subduing natural forces and biodiversity on a larger scale by domestication of animals, control of flora and increase of production as the world population increased. Thus, the industrial revolution and the advancement of science and technology have incrementally become the new risks to the survival of humanity and biodiversity.

The increasing world population as it approximated 8 billion in 2023, with China with China and India both at 1.4 billion on the top of the list of world population density with the projection of these figures to rise to 9 billion in 2037 and 10 billion in 2058 according to Worldometer (2023). Sustainable development is essential in view of the global population increase and the means of producing food, medication, clothing and how that affects the environment. There is an urgent need to adopt new attitudes and technologies that are ecofriendly – to responsibly attain sustainable development goals. Development is needed to feed, clothe, house, educate and empower contemporary human communities, and cure contemporary pandemics, albeit not at the expense of the ecosystem and biodiversity' wellbeing.

There is a consensus among the great majority of scientists that human activity is the factor that causes "global warming" and its corollaries such as the destruction of some species' natural habitat, pollution, the ongoing extinction of flora and fauna, and there are projections according to which 20% of plant and animal species on earth could be extinct within the next 25 years (Smith, 2011). The loss of biodiversity caused by rapid population growth is a significant concern in scientific, political, economic, and development circles. This led to the concept of the "Anthropocene" introduced by Nobel Prize scientist Paul Josef Crutzen. The Anthropocene refers to the current geological epoch where human activities have disrupted the natural regulation of atmospheric conditions, contrasting with the previous Holocene era. This shift places a heavy responsibility on humans to make decisions regarding climate factors. By examining the observable changes in weather, including temperature increase, wind, clouds and precipitations resulting from human activities, we can analyse the impacts on the Earth's system (Crutzen/Brauch, 2016). Crutzen (2016: 23) observes: "It seems appropriate to assign the term 'Anthropocene' to the present, in many ways human-dominated, geological epoch, supplementing the Holocene – the warm period of the past 10–12 millennia". The escalation of human impacts on the environment, including carbon dioxide emissions will affect the planet for millennia, according to Crutzen (2016). Crutzen/Brauch (2016: 1) presented as first indicators of the Anthropocene,

> [...] the global accumulation of human population and urbanisation, the consumption of fossil fuels and water, the increase of synthetic chemicals, overfishing and the loss of biodiversity most notably in tropical regions, the massive increase of greenhouse gases and other harmful gaseous substances (esp. CO_2, CH_4, NO and SO_2).

Earlier, we attributed the causes of some contemporary natural disasters to human interference with nature at a large scale. Crutzen and Brauch (2016: 1) present as additional evidence of the Anthropocene Epoch, "the anthropogenic influences in lakes and the increase of natural disasters as well as the risk of man-made disasters, in particular the 'Nuclear Winter'. The nuclear winter is defined as:

> the environmental devastation that certain scientists contend would probably result from the hundreds of nuclear explosions in a nuclear war. The damaging effects of the light, heat, blast and radiation caused by nuclear explosions had long been known to scientists, but

such explosions' indirect effects on the environment remained largely ignored for decades ("Nuclear winter", 2023).

The awareness of such annihilation of humanity and biodiversity via a "nuclear winter" needs to be taken seriously to avert such decimation of humankind and biodiversity via a "nuclear winter". It follows that a better understanding of the Anthropocene is crucial to prevent the catastrophic consequences of the man-made age. Even on the account of the definition framed by Robock (2010), nuclear winter is a global and concerning problem. He describes the concept of nuclear winter as "the term for a theory describing the climatic effects of nuclear" (p. 418), and argues: "The only way to be sure to prevent the climatic effects of nuclear war is to rid the world of nuclear weapons." (Robock, 2010: 418). This is so important, because reaching long-term prosperity and pursuing the wellbeing of humankind and biodiversity can reach a dead end, and poverty eradication goals can stop being a dream should nuclear war erupts, accidentally or wilfully. These are eventualities that cannot be excluded as observed by Wallis (1982). Furthermore, reaching SDGs and attaing the dream of "the future we want" remain problematic and challenging if anthropogenic factors of the worsening atmospheric conditions and their implications of the Earth's life support systems are not addressed. Such failure will defeat the process of achieving the SDGs (Griggs et al., 2014).

The beginning of the new geological epoch when the influence of humans puts a strain on the natural functioning of Planet Earth is subject to extensive debate among scientists. However, two theories explain this, namely the industrial view which contends that great impacts of human activities on earth systems appeared in the early industrial era around 1950s, whereas the Anthropocene view acknowledges that this began thousands of years earlier (Ruddiman, 2013). He presents three indicators that shows the transition from the Holocene to the age of human influences on the planet on a large scale. These indices are: "forest clearance (and related use), emissions of greenhouse gases (CO_2 and CH_4), and effects on global temperature" (Ruddiman, 2013: 45). He also observes that the significant impact of human activity on earth started with industrialisation, arguing that:

Because reliable, systematic land-use surveys are rare prior to 1950, most reconstructions for early-industrial centuries and prior millennia are hind casts that assume humans have used roughly the same amount of land per person for 7000 years. But this assumption is incorrect (Ruddiman, 2013: 45).

For Crutzen (2016: 23),

The Anthropocene could be said to have started in the latter part of the eighteenth century, when analyses of air trapped in polar ice showed the beginning of growing global concentrations of carbon dioxide and methane. This date also happens to coincide with James Watt's design of the steam engine in 1784.

The argument is based on historical data obtained from archaeological databases, which highlight a significantly higher per-capita land usage during the pre-industrial era compared to recent centuries. This increased land usage is noticeable in early deforestation practices, which resulted in substantial greenhouse gas emissions and global temperature rise, surpassing those proposed within the industrial paradigm (Ruddiman, 2013). Additionally, scientific evidence acknowledges that the Anthropocene began with the industrial era when atmospheric concentrations of CO_2 and CH_4 started to rise exponentially by 1850 (Ruddiman, 2013). The work of Crutzen/Stoermer (2021) supports this claim by identifying significant changes in atmospheric conditions since the industrial era, which are linked to land use, rapid population growth (with over 85% of the current world population increasing to 7 billion since 1950), and the combustion of fossil fuels as a catalyst.

However, the ongoing human-induced disruption of natural atmospheric regulation continues to contribute to the degradation of earth's systems. This disruption is evident in phenomena such as global warming, rising sea levels, droughts, floods, and other natural disasters. It creates a paradox between managing abundant precipitation, leading to floods, and the consequences of scarce precipitation, such as droughts and sandstorms. In response, it is crucial for humanity to acknowledge the reality of the Anthropocene and reassess sustainable pathways for human development and economic growth. Numerous initiatives have been launched by the international community, research institutions and civil society organisations to address this critical challenge and ensure the earth's survival. These include the commitment made by the UNFCCC (United Nations, 2006), the UN Climate Change Summit and Rio+20 (Blanc et al., 2012; Clémençon, 2012), the Kyoto Protocol which was adopted in December 1997 to boost the effectiveness of the 1992 UNFCCC (International Institute for Sustainable Development, 1992–2021) and the Paris Agreement signed in December 2015 that "requires all parties to determine, plan, and regularly report on the nationally determined contribution that it undertakes to mitigate climate change" (International Institute for Sustainable Development, 1992–2021: n.d.). All these undertakings are parts of the global endeavour to maintain global warming at 1.5° C of the Holocene, pre-industrial revolution; and calls for human agency, responsibility and stewardship (Kiyala 2022a). Humanity's responsibility for creation stems from the relation between earth and common good. The next section looks into that rapport.

2.2.1 *Earth-Common Good Relation in the Anthropocene*

Faith-based organisations, as one of the major components of civil society, have been in the forefront of climate action, justice and people development. Since the time of Pope Paul VI, the Catholic Social Teaching has advocated "the development of peoples" and that the economy of the world serves the whole of mankind and not just a minority, according to the Encyclical Letter titled *Populorum progression* that was issued on 26 Mar 1967 (Papa Paulo VI 1967). The Catholic Social Teaching also promotes social justice theory that advances the common good which is relevant to the problem of public health; it also encourages equity in access to health resources (Cahill, 2021), and not excluding environmental issues that business enterprises must consider to be: "a community of solidarity, that is not closed within its own company interests. It must move in the direction of a "social ecology" of work and contribute to the common good also by protecting the natural environment"[1] (Pontifical Council for Justice and Peace, 2004: 148).

According to The Compendium of The Social Doctrine of The Church (CSDC), the common good does not entail merely the quantity of subjects' particular goods of a social entity; it "belongs to everyone and to each person, it is and remains 'common', because it is indivisible and because only together is it possible to attain it, increase it and safeguard its effectiveness, with regard also to the future"[2] (Pontifical Council for Justice and Peace, 2004: 72). Therefore, the connection between the common good and ecology necessitates individuals and communities to behave in morally acceptable ways, seeking what is good. By doing so, the collective moral actions of a society contribute to the realisation of the common good[3] (Pontifical Council for Justice and Peace, 2004). Understanding the inherent nature and essence of the common good, including our shared Earth and its resources, demands both individual and collective responsibility. Neglecting to take moral actions to preserve this common good makes us complicit in the adverse consequences of unethical and immoral treatment of the Earth's system.

No 164 of the CSDC defines the "common good" as:

> The principle of the common good, to which every aspect of social life must be related if it is to attain its fullest meaning, stems from the dignity, unity and equality of all people. According to its primary and broadly accepted sense, the common good indicates "the sum total of social conditions which allow people, either as groups or as individuals, to reach their fulfilment more fully and more easily" (Pontifical Council for Justice and Peace, 2004: 72)

The requirements of the common good vary according to the social conditions of each historical period and are closely linked to the respect and holistic promotion of human persons and their fundamental rights. These demands necessitate a dedication to peace, the effective functioning of governmental institutions, robust judicial systems, environmental care and protection, and the provision of essential human needs. The pursuit and preservation of the common good should address the existential basic needs of all humanity, as well as those of future generations (Pontifical Council for Justice and Peace, 2004). These include "food, housing, work, education and access to culture, transportation, basic health care, the freedom of communication and expression, and the protection of religious freedom [...]"[4] (Pontifical Council for Justice and Peace, 2004: 73). Nor must one forget the contribution that every nation is duty-bound to make towards a true worldwide cooperation for the common good of the whole of humanity and for future generations also [...].[5] The conditions of humanity also in our geological time recognised as the Anthropocene

---

[1] CSDC #340.

[2] CSDC #164.

[3] CSDC #164.

[4] CSDC #166.

[5] CSDC #166.

calls for adequate moral actions that are relevant and equal to addressing the enormity of climate challenges we are confronted with. It cannot be business as usual–to allow government-funded industries that relentlessly ignore the signs of the time and continue to make record profits out of fossil fuel and greenhouse gases that are destroying the planet's beauty and its inhabitants, both human and non-human.

The CSDC warns against business that seek record profits by disregarding the welfare of all, which it calls "the civilisation of 'consumption' or 'consumerism'"[6] (Pontifical Council for Justice and Peace, 2004). The connection between planetary boundaries (PBs) and the common good is a fascinating aspect that scholars explore, establishing a growing relationship within the field of environmental studies. For instance, Kosoy et al. (2012:76) remark: "Protection of planetary life-support systems is clearly a new category of scientifically defined common goods that demands a new kind of governance response, as articulated in Earth Charter Principle 5". Chapter 5 is articulated as follows: "Protect and restore the integrity of Earth's ecological systems, with special concern for biological diversity and the natural processes that sustain life" (Earth Charter International, 2000). This principle is consistent with the fundamentals of many societies, religions and cultures that acknowledge the "sacred nature of the Earth", namely the People of Colour in the USA (Environmental Working Group, 2023), major world religions (Pope Francis, 2015; The Ecumenical Patriarchate of Constantinople, 2022). This echoes the terms of advocacy for earth's protection by Kosoy et al. (2012: 76) who note: "the protection of earth's vitality, diversity and beauty is a sacred trust". These arguments provide a justification for treating the Earth and its inhabitants with fairness and recognising their interconnectedness. This extends beyond environmental justice to encompass social justice, highlighting the interplay between ecotheology and ecojustice. Understanding the interconnection between social justice and environmental justice becomes particularly relevant in the context of climate wars and climate change.

Environmental justice has a legal dimension as well as a non-adjudicative aspect which ties it with social justice and calls for collective responsibility to observe and preserve the sacredness of creation, the environment and safety of non-human creation (Kiyala, 2022a; Pope Francis, 2015; The Ecumenical Patriarchate of Constantinople, 2022). In line with the discussion on the interdependence of the earth and the concept of the common good, as argued by Kosoy et al. (2012), the realisation of a green economy, sustainable development, and the thriving of the Earth necessitates policy adaptations and a reevaluation of the economic growth model and sustainable development within the context of the Anthropocene. The next section deals with those complex issues and the paradox of sustainable economic prosperity and high life quality while the systems upon which such developmental agenda is built is at stake and science keeps warning of its crumbling state.

[6] CSDC #334.

2.2.2 Climatic Transformations

Kaufman et al. (2020:1) analysed temperature across the Holocene, applying five different statistical methods to reconstruct the global mean surface temperature (GMST) of the past 12,000 years (Holocene). They found "an extensive new multi-proxy database of paleo-temperature time series (Temperature 12k) enables a more robust analysis of GMST and associated uncertainties than was previously available". It appears that the reconstructed temperature levels of "Holocene global thermal maximum, GMST cooled at an average rate −0.08 °C per 1 000 years (−0.24,−0.05)". This serves as proof of the observed global-scale climate change, which has emerged as a significant existential threat to humanity and the survival of biodiversity. It is crucial for both humans and the scientific community to monitor these changes and develop strategies aimed at preserving the temperature levels experienced during the Holocene period, preventing any increase beyond 0.5 °C.

Climate change has transitioned from being a mere abstract concept to a tangible reality, as evident in the substantial transformations observed in atmospheric conditions today. These transformations manifest in various forms, including desertification, increased sea levels, floods, sandstorms, droughts, and more. During the 1970s and 1980s, global environmental threats, such as acid rain and ozone depletion, began to emerge and display their effects. However, industries responsible for fossil fuel and carbon pollution started to challenge the scientific evidence regarding the alarming climate alterations (Mann, 2021). Frederick Seitz, a renown physicist and prominent figure of anthropogenic global warming denialism, enlisted other scientists to dispute the evidence of human-induced atmospheric changes in the mid-1980s. He argued: "[T]he scientific facts indicate that all the temperature changes observed in the last 100 years were largely natural changes and were not caused by carbon dioxide produced in human activities."[7] (The Heartlander Institute, 2007). Despite this, President George H.W. Bush took action by signing the Clean Air Act, which mandated the reduction of sulphur emissions from coal-fired power plants (Mann, 2021). However, the increasing awareness of these issues faced resistance from industries and billionaires who profited from the production and commercialisation of fossil fuels, including the primary greenhouse gas, carbon dioxide (CO_2), emitted through human activities.

Amidst activism and denialism of human activities as the vector of climate change, the international community seems engaged in dealing with the emerging anthropogenic dangers on our planet. It is becoming more and more evident that climate change is taking an important space on world political and scientific agendas, discourse and debates. From the United Nations Conference on Sustainable Development held in Rio de Janeiro in 2012 to the Cop 27 Climate Change Conference held at the Sharm el-Sheikh in Egypt in 2022, climate change featured prominently on the agendas of these important international forums. For instance, at the Rio+20,

[7] Statement by Seitz quoted in a April 2007 interview with Frontline cited in The Heartlander, December 1, 2001. Archived September 29, 2012. Archive.is URL: https://archive.is/fUxPU (Accessed 3 February 2024).

participants were adamant about the urgent need to take action that would yield promising results. It was stated in Document A/RES/66/288 of the United Nations, paragraph 17 what follows:

> We recognise the importance of the three Rio conventions for advancing sustainable development and in this regard we urge all parties to fully implement their commitments under the UNFCCC, the Convention on Biological Diversity and the United Nations Convention to Combat Desertification in Those Countries Experiencing Serious Drought and/or Desertification, Particularly in Africa, in accordance with their respective principles and provisions, as well as to take effective and concrete actions and measures at all levels and to enhance international cooperation (United Nations, 2012a).

Rio+20 brought together 45,000 participants from various sectors of life all over the world to lay out the vision of the desired future for humanity, which was framed in terms of addressing global challenges in the context of sustainable development, such as "economic, social, and environmental" (Clémençon, 2012: 311). Since Rio+20 and its document titled "*The Future we Want*", there has been little to no evident commitment by world leaders to implement the *UNFCCC and the United Nations Framework Convention on Biodiversity Rio 1992* that showed the path for sustainable development (Clémençon, 2012). During the Rio+20 conference, countries made a collective pledge to work towards achieving the Sustainable Development Goals (SDGs), which were based on the Millennium Development Goals. This commitment was driven by extensive research demonstrating the significant impact of human activities on the Earth's life-supporting system. The magnitude of these pressures is such that ongoing global environmental transformations pose a serious threat to the long-term success of poverty eradication efforts, thereby endangering future prosperity (Griggs et al., 2014). However, the Rio-20's ambitions have not materialised, and the fate of the Earth systems hangs in the balance. According to Clémençon (2012) George W Bush was among the world leaders that received Rio+20 with hostility; however, Brazil was praised as its leadership showed skills in negotiations to avert fruitless discussions on climate action. As climate change intensifies in the Anthropocene era, its unpredictable and far-reaching consequences pose a significant threat to the survival of humanity. In response, certain nations are implementing measures to mitigate the risks of insecurity, displacement, and potential conflicts arising from climate change-induced conditions (Dyer, 2010 cited in Clémençon, 2012: 334). Dyer (2010) explains his perspectives by suggesting that scarcity of water resources will create tensions and disputes in already affected regions such as North Africa, the Middle East, Central Asia and China where attempts to re-route water trajectories from one country to another may trigger violent conflicts.

Additionally, according to Dyer (2010), the depletion of fish stocks is expected to result in significant price increases, while a large number of climate refugees from Sub-Saharan Africa may be compelled to migrate northward. These developments have raised concerns and placed a burden on The Intergovernmental Panel on Climate (IPCC). It is crucial to investigate the escalating dangers posed by climate change, as it could potentially give rise to a new form of warfare known as environmental warfare, as discussed by Dyer (2010). This suggests that we may be approaching an era of climate wars.

According to the IPCC Climate Report 2022 summary drawn from Working Group III Contribution to the Sixth Assessment Report of the Intergovernmental Panel on Climate Change, under Section C (C1) on System Transformations to Limit Global Warming, we are set to pass the 1.5°C threshold by 2040. It states:

> Global GHG emissions are projected to peak between 2020 and at the latest before 2025 in global modelled pathways that limit warming to 1.5 °C (>50%) with no or limited overshoot and in those that limit warming to 2 °C (>67%) and assume immediate action... In both types of modelled pathways, rapid and deep GHG emissions reductions follow throughout 2030, 2040 and 2050 (high confidence) (IPCC, 2022b).

To better understand the increasing concerns over climate change, it is necessary to go back to the post-industrial epoch which coincided to the Anthropocene Age in which we find ourselves. This will help humanity to mitigate the risks of climate alterations-induced violence.

Climate change research also aims to produce findings that can assist the global community in developing strategies to enhance climate resilience and mitigate the risks of full disintegration of Planet Earth. The study of the Lake Chad Region in Sub-Saharan Africa, located in the Sahelian Region of West Africa, is particularly significant as it serves as a model for exploring innovative approaches to address the negative impacts of climate change. As the need for effective measures to combat the adverse effects of climate change becomes increasingly crucial, examining this region can provide valuable insights. A study conducted by Pham-Duc et al. (2020), using a multi-satellite approach, shows that the current hydrology of the Lake Chad under actual climate change condition is resilient and has been so over the last two decades though there was, to some extent, a reduction of its northern pool. These outputs are significant in the sense that they could help towards the efforts to further enhance climate change adaptation strategies to address the unpredictability of the current Lake Chad hydrography whose stability is dependent on its ground water supply and the precipitation of 1990 (Pham-Duc et al., 2020). Human-induced climate change has resulted in extensive negative outcomes and the associated loss and damage to both people and the natural environment. However, there have been initiatives aimed at adaptation to minimise vulnerability and enhance climate resilience in sectors and regions that are particularly susceptible. These efforts focus on areas with a higher concentration of vulnerable populations and systems that are disproportionately affected by the impacts of climate change. (IPCC, 2022c). The IPCC's Summary for Policymakers on *Observed Impacts from Climate Change B (1)(1)* states:

> Widespread, pervasive impacts to ecosystems, people, settlements, and infrastructure have resulted from observed increases in the frequency and intensity of climate and weather extremes, including hot extremes on land and in the ocean, heavy precipitation events, drought and fire weather (high confidence). Increasingly since AR5, these observed impacts have been attributed to human-induced climate change particularly through increased frequency and severity of extreme events. These include increased heat-related human mortality (medium confidence), warm-water coral bleaching and mortality (high confidence), and increased drought-related tree mortality (high confidence). Observed increases in areas burned by wildfires have been attributed to human-induced climate change in some regions (medium to high confidence). Adverse impacts from tropical cyclones, with related losses and damages, have increased due to sea level rise and the increase in heavy precipitation

> (medium confidence). Impacts in natural and human systems from slow-onset processes such as ocean acidification, sea level rise or regional decreases in precipitation have also been attributed to human-induced climate change (high confidence) (IPCC, 2022c: 9)

The report also recognised, in Section B (1)(3) of the Summary for Policymakers, the certainty that climate change, including increased frequency and intensity of extreme events, has negatively impacted food and water security, impeding progress towards SDGs (IPCC, 2022c). It is well-established that sustainable development is closely linked to climatic conditions and how resources are managed to achieve the desired "*Future We Want*" (United Nations, 2005, 2012b). The extent to which humanity values the environment and incorporates scientific knowledge on the "nuclear winter" and PBs into climate and development policies, as well as political and economic decisions, will significantly influence the outcome. Ensuring global social, political, and economic responsibility is crucial for maintaining climate resilience.

Despite overwhelming scientific evidence on climate change and ozone layer depletion, there has been a significant challenge in the form of denialism. This predicament has been fuelled by disinformation campaigns carried out by certain scientists, multinational corporations, businesses, and politicians, who seek to preserve the current state of affairs regarding climate change. According to Mann (2021: 3), these disinformation campaigns emerged in the late 1980s, with ExxonMobil playing a prominent role by investing billions of dollars to undermine and discredit the scientific evidence linking climate change to human activities, particularly the burning of fossil fuels. Influential billionaires such as the Koch brothers, the Mercers, and the Scaifes were also involved in distorting scientific findings related to the anthropogenic causes of global warming, findings that were even confirmed by ExxonMobil's own scientists: "[…] the impacts of continued fossil fuel use could lead to 'devastating' climate-change impacts" (Mann, 2021: 3).

Among politicians who deny the existence of climate change are the conservative white right (McCright/Dunlap, 2011) and Donald J Trump is labelled as the 'denier-in-chief' (De Pryck/Gemenne, 2017). When the reality of climate change is thwarted by multinational companies, politicians and scientists, the future of the planet faces a threat that needs to be confronted in multiple fronts. One of such fronts is debunking erroneous narratives and disinformation campaigns about climate change, which have ventured to obstruct the efforts intended to preserve the planet (Mann, 2021). Over and above that, Mann (2021: 5) emphasises: "Our civilisation can be saved, but only if we learn to recognise the current tactics of the enemy—that is, the forces of inaction—and how to combat them". It is evident that there is ongoing climate war whose aim is to defeat the enemy of the planet we alluded to earlier whose mission is to detract people from the truth about climate change and its pernicious after-effects on humanity and non-human creation. Another front of the fight to save the planet is undertaken by faith-based organisations who stigmatise lifestyles of contemporary societies and unbridled consumerist attitudes (Deane-Drummond, 2016; Kiyala, 2022b; Pope Francis, 2015). Religious platforms are also used to convey the message of peace with the environment and protection of Mother Earth and non-human creation (Chryssavgis, 2014; Patriarch Bartholomew, 2021). It is time to make peace with our environment.

Climate pernicious risk mitigation requires concerted efforts that should incorporate environmental protection activism; climate relief advocacy and funding of low-carbon infrastructure projects, especially in less-developed economies (Ehlers et al., 2022); and mega-investment by the UNFCCC to pursue its ultimate objective, notably:

> to stabilise greenhouse gas concentrations "at a level that would prevent dangerous anthropogenic (human-induced) interference with the climate system". It states that "such a level should be achieved within a time-frame sufficient to allow ecosystems to adapt naturally to climate change, to ensure that food production is not threatened, and to enable economic development to proceed in a sustainable manner (UNFCCC, 2022).

The World Bank Group (WBG) maintains its position as the leading provider of climate action financing in developing nations. However, the vast disparity between military expenditures and the budget allocated to combat climate change is cause for concern. It is alarming to consider and reconcile these two realities: the immense funds allocated to weaponry that harms the environment, devastates human populations, and obstructs economic growth and sustainable development, juxtaposed with the insufficient funds directed towards meaningful climate action to protect our planet.

It is also acknowledged that private climate financing can play a crucial role in a time when emerging markets and developing economies are trying to restrain greenhouse gas emissions and contain climate change while struggling with its effects, but they must conjointly "invest at least $1 trillion in energy infrastructure by 2030 and $3 trillion to $6 trillion across all sectors per year by 2050 to mitigate climate change by substantially reducing greenhouse gas emissions" (Ehlers et al., 2022). Even that is still insufficient to win the battle against climate change. In addition, a further $140 billion to $300 billion is needed yearly by 2030 for investment in adaptation mechanisms to address the physical consequences of climate change, notably the rising seas level and intensified droughts (Ehlers et al., 2022). This estimate could increase to between $520 billion and $1.75 trillion annually beyond 2050; it will depend on how effective the climate mitigation measures are (Ehlers et al., 2022). Thus,

- boosting private climate speedily is key to climate stability, and essential solutions to that encompass: "adequate pricing of climate risks, innovative financing instruments, broadening the investor base, expanding the involvement of multilateral development banks and development finance institutions, and strengthening climate information" (Ehlers et al., 2022: n.p.).
- private sustainable finance in emerging market and developing economies is quite encouraging, because it rose to "a record $250 billion last year. But private finance must at least double by 2030, at a time when investable low-carbon infrastructure projects are often in short supply and funding of the fossil fuel industry has soared since the Paris Agreement" (Ehlers et al., 2022: n.p.).

The sum of facts presented by Ehlers et al. (2022) is a clear indication of how the International Monetary Fund (IMF) views the route to climate resilience that relies significantly on the synergic financial commitment of private economies and world major financial institutions. Climate change can be controlled only if there is political will and determination for action, individual conviction of the detrimental effects of atmospheric degradation, activism for a green economy, and concerted efforts by various stakeholders involved in producing and consuming products that destroy the environment. Acknowledging our collective negligence in safeguarding the essential systems that sustain the Earth, assuming responsibility for polluting the air, land, and water, it becomes imperative for the entire human family to cultivate a shared awareness and commitment to become stewards of the environment. This calls for embracing the principles of ecopeace or environmental peacebuilding as we strive to restore harmony with the natural world. The following section discusses this perspective in-depth; it shows its implications and promised to better human life, and other lives on the planet.

2.2.3 Environmental Conflicts and Violence

Conflicts exist, are widespread and stem from a variety of causes such as disputes over exploitation of natural resources, scarcity of shared resources (water, forests, etc.), disputes over arable lands, the struggle by natives to protect resources against migrants who attempt to settle on their territories. These conflicts occur on a small scale when it involves local communities, or on a large scale when local organisations and civil societies at national and international level rise against the damage inflicted upon the environment. Navas et al. (2018: 649) have written on conflicts that arise when local people (individuals and organisations) protest and denounce environmental damage leading to ecological conflicts. Their study was based on a multidimensional approach on 95 environmental conflicts in central America that were recorded in the EJAtlas database; these led to violence – ranging from "direct, structural, cultural, slow, and ecological forms of violence" (p. 649). These diverse forms of direct conflicts have the potential to impact not only environmental defenders but also the communities and their immediate environments. This interconnectedness becomes particularly evident in regions and periods marked by conflicts and violence, where the consequences extend beyond individuals and encompass the complex web of relationships within specific places and timeframes (Navas et al., 2018).

It emerges from research, existing scholarship and reports of the IPCC that climate change is driving conflicts and could generate new warfare that future generations will have to deal with. The IPCC's summary for policymakers clearly defines the parameters of climate-induced conflicts. According to Section B.1.7 of the report, climate change plays a role in exacerbating humanitarian crises, particularly in areas where there is an intersection of climate hazards and high vulnerability. These regions experience increasing displacement of people due to climate-related factors such as extreme weather conditions. Moreover, Africa, Central and South America have witnessed a rise in severe food insecurity and malnutrition attributed to flood and drought occurrences (IPCC, 2022c). Furthermore, Section B.5.3 reports that extreme weather and climatic conditions are driving "the redistribution of marine fish stocks, increasing risk of transboundary management conflicts among fisheries users, and negatively affecting equitable distribution of food provisioning services as fish stocks shift from lower to higher latitude regions" (IPCC, 2022c: 19).

Environmental conflicts are a common problem and call into question security at local, national, regional and global levels (Bob/Bronkhorst, 2010). The key issues related to those conflicts include people, their environment, climate change-related issues, including a range of ecosystems services, interactions and interrelationships between people and their environments, environmental conservation, water quality and availability and air quality (Bob/Bronkhorst, 2010). One aspect of environmental conflicts that is experienced by some parts of the Globe is related to the displacement of people who are facing a hostile environment as a result of global warming and climate change, namely the rise of ocean levels, floods and sandstorms. Bob and Bronkhorst (2010: 27) assert: "[...] a disconcerting trend is the migration levels associated with environmental and other conflicts that often result in existing or new conflicts emerging in receiving areas". That is one of the concerning issues in the Anthropocene and humanity should be prepared to continuously deal with such predicaments. Simultaneously, researchers continue to be concerned about the crises at hand, and they propose alternative approaches to enhance the preparedness of countries and regions affected by environmental conflicts and violence. In a similar vein, Navas et al. (2018: 658) have put forward a multidimensional violence approach aimed at understanding environmental conflicts more effectively, thus providing valuable insights for analysis. Such approaches comprehend "'slow', structural, cultural, and ecological forms of violence, and not only direct quick episodes of physical violence) as a tool" (p. 658). Configuring any strategy aimed at counteracting conflicts and violence tied to environmental issues in individual countries and different parts of the Globe will be counterproductive if the nature, contours of violence and conflicts drivers are not profoundly understood. For the benefit of policy making, Sakaguchi et al. (2017: 641) point out that such studies based on particular country-cases appear to lead in a fruitful direction as they provide "country-specific findings that might inform policies aimed at dampening the violence-inducing effects of certain climate change processes" (p. 641).

Climate science research and reports on the Earth's supporting systems are expanding due to vital challenges we are confronted with as atmospheric perturbations continue to drive global warming and climate change. This section discussed conflicts and violence occurring worldwide. The next section overviews the extent to which climate conflicts and violence could either escalate to climate wars – when people and countries conform each other to secure resources for the present and future generations, or climate as a frontline to defend and protect the Earth.

2.2.4 Climate Wars

The concept of climate wars is ambivalent. On the one hand, climate war would be the escalation of environmental conflicts where disputing communities and countries take up arms against each other where resources become scarce and the populations in respective societies fight for their survival. On the other hand, climate wars denote a non-military engagement where available means such as science and technology, lobbying, advocacy and activism are used to create a global awareness of the present and future catastrophic state of the systems that keep Earth's life regulating network functional; that is the byproduct and anthropogenic global warming and climate change. The former conceptualisation of war refers to the primary definition of war which is associated with "armed fighting between two or more countries or groups, or a war that is fought over a long period and only ends when one side has neither the soldiers and equipment nor the determination left to continue fighting" according to the Cambridge English Dictionary. The latter refers to "any situation in which there is strong competition between opposing sides or a great fight against something harmful" – definition provided by Cambridge dictionary.

The first interpretation of climate war is intrinsically related to conflict and violence arising from disputes over environmental resources. Within the scope of this chapter, the potential for climate wars (Dyer, 2010) to intensify is evident. As a result, the world will witness a surge in environmental or ecological refugees, necessitating novel approaches beyond traditional dispute resolution methods to address these emerging conflicts (Biswas/Quiroz, 1996; Westra, 2009). These are the challenges that confront humanity today. For example, the civil war in Darfur Region of Sudan originated from conflict induced by climate change which affected the regularity of rainfall in many parts of Darfur making the space unsuitable for livestock and coercing herdsmen to move southwards as nomads (Welzer, 2012). In addition, "[...] the drought produced large numbers of internally displaced persons, who were accommodated in newly built camps; as many as 80,000 starving people were on the move trying to reach one" (Welzer, 2012: 68). The conflict escalated between the northern nomads and populations in the South Sudan because of a change of policy regarding water and land which used to be handled at conciliation meetings chaired by a third party that came into existence following the 1989 push that saw General Al-Bashir rise to power (Welzer, 2012). The conflict took on military proportions as a result of interventions by non-government-backed militias leading to fighting between government troops and rebel organisations, particularly the Darfur Liberation Front (Welzer, 2012).

A review of mixed studies and empirical analyses conducted by Sakaguchi et al. (2017) substantiates a robust ear evidence between climate change and violence; however, based on their empirical findings, these scholars dispute the importance of climate change as the driver of conflicts (Fig. 2.1). It follows that seeking and designing policies on climate change adaptation and / or mitigation to help abate the likelihood of violence becomes a challenge, they note: "As a result [of their empirical analyses], the settings and types of violence likely to arise from climate

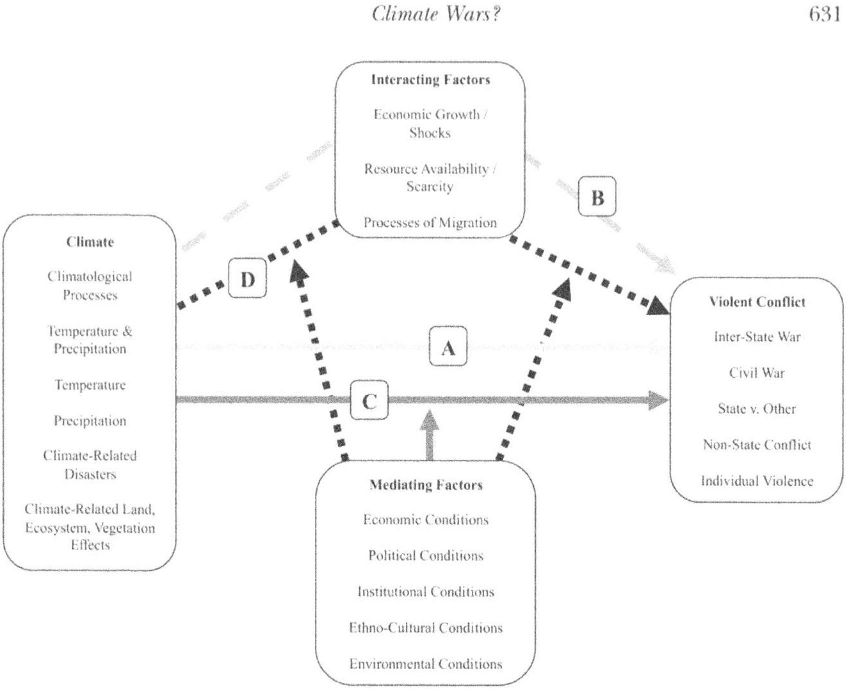

Fig. 2.1 Causal pathways linking climate change to violence. *Source* Sakaguchi et al. (2017: 631)

change were neither adequately disaggregated nor were the findings corroborated by using multiple data source" (Sakaguchi et al., 2017: 624).

Figure 2.1 gives a big picture of interacting and mediating factors that link climate change to violence; the authors excluded environmental scarcity (Sakaguchi et al., 2017: 625), which has been sustained by other studies as a driver of conflict and violence. Hauge and Ellingsen (1998: 299) point out:

> [...] factors like deforestation, land degradation, and scarce supply of freshwater, alone and in combination with high population density, increase the risk of domestic conflict, such as level economic development and type of political regime. The latter variable, however, prove more decisive that environmental scarcity in predicting the incidence of armed domestic conflict.

Economic and political circumstances are recognised as factors that shape and trigger the actual violence where environmental scarcities and also the scarcity of renewable resources prevail as was the case of South Africa, and affirmed by Percival and Homer-Dixon (1998: 279): "Environmental scarcity emerges within a political, social, economic, and ecological context and interacts with many of these contextual factors to contribute to violence".

It is evident that "human" protective instinct to use force to defend territory and resources and control them will remain a preoccupation of governments and the international community, especially in regions and countries with scarce resources or under ecological threats exacerbated by global warming and climate change.

The second conception of climate war is evident in the face of ecocide which is debated in the literature (Mehta/Merz, 2015; Welzer, 2012). This nature of war was described by António Guterres in emotive terms as a fight when addressing world leaders at the opening of the COP27 UN climate summit in Egypt in November 2022:

> This UN Climate Conference is a reminder that the answer is in our hands. And the clock is ticking. We are in the fight of our lives. And we are losing. Greenhouse gas emissions keep growing. Global temperatures keep rising. And our planet is fast approaching tipping points that will make climate chaos irreversible. We are on a highway to climate hell with our foot still on the accelerator (Guterres, 2022).

Indifference to ongoing destruction of human habitat and biodiversity habitat renders one an accomplice to such crimes. Aquatic species' habitats are impacted by pH pollutants resulting from human activities. As a consequence, humans have developed a higher tolerance for pH levels ranging from 4 to 11, experiencing minimal gastrointestinal irritation when consuming drinking water within this pH range (Fondriest Environmental, 2013). An observation about Easter Island in South America where Polynesians settled around AD 900, was that "twenty-one species of plant … have since disappeared from the island, the tallest, which could grow as high as 30 metres" (Welzer, 2012: 51).

The fight to win the planet started much earlier, in the 1960s when Rachel Carson led the campaign against the use of Dichlorodiphenyltrichloroethane (DDT) – a deadly insecticide used in agriculture which was catastrophic to the environment. Mann (2021) writes about the character assassination that Carson was victim of, because she detailed the way DDT was decimating populations of bald eagles, among other birds, by dissolving their eggs and killing the embryos within. Among other lethal effects of DTT were its "accumulation in food webs, soils, and rivers, creating an increasingly dire threat to wildlife – and ultimately, humans. Eventually, the United States banned DDT, but not until 1972" (Mann, 2021: 11). Carlson's campaign helped to salvage the remnants of bird species that survived DDT era.

Climate war refers to a non-military struggle aimed at restoring the natural processes that regulate the Earth system, encompassing atmospheric and climatic conditions, marine and terrestrial ecosystems, as well as biodiversity. However, it extends beyond these ecological aspects to encompass safeguarding humanity itself, which faces threats from historical instances of genocide, ethnic cleansing, terrorism and boundary disputes among some countries. These challenges, vividly described by Welzer (2012), underscore the need to address the broader implications of climate change on human survival. The interconnectedness between human suffering caused by armed conflicts and natural disasters in the context of climate war is apparent. Humanity occupies a central position within the intricate web of relationships among life, the environment and development. Excluding the human species from the discourse on climate, the environment and sustainable development would render the debate incomplete. Humans serve as the vital catalyst in any discussion concerning the future of the Earth system, as they possess a predominant role in regulating and transforming the norms that sustain life on earth. Thus, climate war is legitimately humanity's fight for its survival and that of non-human creation.

2.2.5 *Impacts of War on the Environment*

According to the Stockholm International Peace Research Institute (SIPRI), world military expenditure in 2021 exceeded the two trillion US dollar mark for the first time (Liang et al. 2023). Table 2.1 presents regional and national military expenditure data for the period of 2021, and highlights the trends over the decade between 2012 and 2021, based on the updated SIPRI Military Expenditure Database.

Global military spending in 2021 was 0.7% higher than in 2020 and 12% higher than in 2012. World military expenditure seen since 2015 was not affected by the economic effects of the Covid-19 pandemic; on the contrary, after the significant economic recovery across the globe in 2021, military spending globally "as a share of world gross domestic product (GDP) – the global military burden – reached 2.2 per cent, down from 2.3 per cent in 2020" (Liang et al., 2023: 1). With the Russia-Ukraine war causing the depletion of military capabilities, NATO chief, Jens Stoltenberg, called on members to ramp up ammunition production for Ukraine which had drained western military stockpiles. This means that military expenditure is expected to

Table 2.1 The 10 countries with the highest military expenditure in 2021

Rank			Spending ($ b.), 2021	2020–21	2020–21	Spending as a share of GDP (%)[b]		World share (%), 2021
2021	2020[a]	Country		2020–21	2012–21	2021	2012	
1	1	United States	801	*−1.4*	−6.1	*3.5*	*4.5*	*38*
2	2	China	[293]	*4.7*	*72*	*[1.7]*	*[1.7]*	[14]
3	3	India	76.6	*0.9*	*33*	*2.7*	*2.6*	*3.6*
4	6	United Kingdom	68.4	*3.0*	*3.7*	2.2	*2.4*	*3.2*
5	5	Russia	65.9	*2.9*	*11*	*4.1*	*3.7*	*3.1*
Subtotal top 5			1305	..	..	..	..	*62*
6	8	France	56.6	*1.5*	*13*	*1.9*	*1.9*	2.7
7	7	Germany	56.0	*−1.4*	*24*	*1.3*	*1.2*	2.7
8	4	Saudi Arabia	[55.6]	*−17*	*−15*	*[6.6]*	[7.7]	*[2.6]*
9	9	Japan	54.1	*7.3*	*18*	*1.1*	*1.0*	2.6
10	10	South Korea	50.2	*4.7*	*43*	*2.8*	*2.5*	2.4

SIPRI. *Source* Liang et al. (2023: 2)

increase exponentially over the next few years. The death toll will rise significantly and greater destruction of the environment is expected. Chancellor Olaf Scholz also announced the creation of a €100 billion (£85bn) fund to be set up to boost the strength of the country's armed forces over the coming years and support Ukraine's fight against Russian occupation of its territories (Connolly, 2022). The increase in military expenditure echoed across world powers. The militarist logic now prevails over the pacifism of the post-World Wars.

Paradoxically, in September 2022, the WBG provided a record $31.7 billion for fiscal year 2022, aimed to assist countries to address climate change (The World Bank, 2022). This was an increase of 19% from the $26.6 billion previously provided in the previous fiscal year and represents all-time high in funding (The World Bank, 2022).

The earth's environment has suffered significant consequences from wars, including the Russia-Ukraine conflict and others. These conflicts have resulted in devastating effects, with a high loss of life and ecological catastrophes. The impacts include the destruction of infrastructure through burning; damage to land caused by naval mines and landmines; extensive degradation of land due to destroyed tanks and crushed combat aircraft; the sinking of warships and other vessels; and the devastation of flora and fauna caused by artillery fire and the craters left by missiles and bombs. Furthermore, unplanned mass graves take up land where it is necessary to bury both civilians and military personnel. From the same perspective, Bob and Bronkhorst (2010: 27) note: "Wars and other types of violence (such as invasions and destruction of property) can denude the natural resource base in terms of quality and/or quantity". The effects of wars and conflicts on the environment are not limited to the damage caused to nature, but they deepen man's vulnerability because of scarcity of vital resources, food insecurity, the destruction of livelihoods and domestic animals; they create pollution, cause diseases, engender poverty and make children orphans and vulnerable to human trafficking. The impact of conflicts and wars on environment is immeasurable and is a factor that slows sustainable development.

That is more real as we assist and look powerlessly at the human and environmental catastrophe in Ukraine, the unfolding genocide of the Palestinians in Gaza, and the destruction of their homes by the IDF; finally, the increasing death toll in the war-torn eastern DRC. Russia and Ukraine continue to bear a significant human cost in this extremely brutal and atrocious conflict, which has far-reaching implications for food security, energy resources, international security and sustainable development. Liadze et al. (2022: 11) observe:

> The war in Ukraine represents not only a humanitarian tragedy but also a challenge to policymakers after the COVID-19 pandemic, and a massive cost for the global economy. Using the NiGEM model, we quantify these as 1% of global GDP in 2022, or about $1.5 trillion valued at Purchasing Power Parity exchange rates (PPP) compared with our GDP forecast made at the beginning of 2022.

While acknowledging that the Russia–Ukraine war was not induced by environmental conflict, elements of the environment are intrinsically tied to the conflict, namely the land of Ukraine, and the historical proximity of the two countries. Rawtani et al. (2022: 1) point out: "The war has serious negative consequences on people and the planet. The damage to industrial and commercial infrastructure can contaminate water sources, which can be hazardous for human and ecosystem health". The magnitude and diversity of the armed conflicts and related risks are complex and have significant implications for social, political and economic stability in various contexts and locations. Thus, military conflicts can threaten both nature and social security.

Science is crystal clear about the damaging impacts of the deployment of nuclear weapons on any life-bearing space (Rothschild, 2012; Westing, 1981). Noxious radioactive discharge renders the land and seas inhabitable. According to Rothschild (2012), a U.S. military-sponsored ecological research conducted during World War II to keep track of the release of radioactive wastewater into waterways from plutonium production caused a public outcry against nuclear testing that resulted in the radioactive fallout at the Nevada and Marshall Island test sites, especially following the Fukuryu Maru (Lucky Dragon) accident in 1954. Furthermore, the accident polluted neighbouring inhabited islands with radioactive byproduct which led to a significant influx of funding for environmental science by the Atomic Energy Commission (Rothschild, 2012). A number of biologists who led this research into nuclear after-effects and effluent at the Atomic Energy Commission raised concerns regarding radioactive contamination in the environment "from the long-term, cumulative effects of nuclear waste disposal, the use of atomic bombs for construction projects, and the potential ecological devastation wrought by nuclear war" (Rothschild, 2012: 492). These studies further helped towards developing a new subfield of 'radioecology' including research concerned with problems of environmental pollution at large scale (Rothschild, 2012).

In a similar perspective, a study by Bolton and Minor (2021: 81) on the use and testing of nuclear weapons emphasised transnational and disastrous humanitarian and environmental fallouts. These scholars note:

> Legacies of more than 2000 nuclear detonations in the territories of 15 states persist today, with serious implications for human rights and sustainable development. There is an inadequate global policy architecture for addressing the humanitarian and environmental consequences of nuclear weapons. However, the Treaty on the Prohibition of Nuclear Weapons (TPNW), adopted by 122 states at the UN in 2017, established obligations to assist victims of nuclear weapons and testing and remediate contaminated environments.

The TPNW became a legally binding instrument that interdicted the use of nuclear weapons – a process that aimed at the total elimination of such weapons (United Nations Office for Disarmament Affairs, 2021). It was adopted by the United Nations in July 2017, and it entered into force in January 2021 (United Nations Office for Disarmament Affairs, 2021).

Considering such catastrophic humanitarian and ecological radioactive after-effects, other international policymaking bodies have also decreed global action to address such concerns (Bolton/Mino, 2021). These scholars provided considerable insights on *Addressing the Humanitarian and Environmental Consequences of Nuclear Weapons* (HECNW); they provided a review of global evidence of the impact of past nuclear activity in various countries and some of the familiar and potential occurring "consequences of the blast, heat and radioactive energy released by past nuclear weapons detonations" (Bolton/Minor, 2021: 81).

We are currently in the middle of an epic battle that started with the aggression of Ukraine by Russia. It is interesting that António Guterres links the conventional war waged by Russia on Ukraine and climate change; he states:

> The war in Ukraine, other conflicts, have caused so much bloodshed and violence and have had dramatic impacts all over the world. But we cannot accept that our attention is not focused on climate change. We must of course work together to support peace efforts and end the tremendous suffering. But climate change is on a different timeline, and a different scale (Guterres, 2022: n.p.).

Research has provided insights to policymakers and urges them to pursue development policy around the TPNW and the collection of additional information about HECNW, and to enable efforts to create international "humanitarian, human rights and sustainable development policy assisting communities affected by nuclear weapons" (Bolton/Minor, 2021: 81). The most telling deadly cases of nuclear detonations that created a humanitarian and environmental tragedy were illustrated during World War II in 1945; they were deployed by the USA in the atomic bombings of Hiroshima and Nagasaki, Japan (Bolton/Minor, 2021). The death toll provoked by those nuclear weapons "caused an estimated 140,000 deaths in Hiroshima and 74,000 in Nagasaki by the end of 1945, along with increased rates of cancer and chronic disease among the survivors" (Bilfinger, 1945/2015; D'Olier, 1946; Douple et al., 2011; Groves, 1946/2008; Honda, 2002 cited in Bolton/Minor, 2021: 84).

For instance, according to these scholars:

> Between 1945 and 2017, more than 2000 nuclear test explosions were carried out at locations within what are today territories controlled by 15 states. Some of the locations of nuclear tests were in territories whose independence has since been recognised by the international community (Bolton/Minor, 2021: 84).

Despite the obvious and horrific consequences of nuclear weapons, and the interdict set by the TPNW, the arms race has shown no slowing down in terms of producing, testing and threats of deploying nuclear warheads. Although, as noted by Bolton and Minor (2021), people know the deadly harm resulting from the atomic bombings, very few are aware of how much nuclear weapons programmes have impacted many other places and populations. Thus, the production, testing, the transportation and deployment of nuclear weapons, including the disposal waste from such processes have harmed millions of people in the last seven decades and have caused perpetual environmental damage (Bolton/Minor, 2021). It is evident that the earth has been under constant violation of its welfare even before the industrialisation era, and much earlier in the age of World War II in 1945. It has further been under the burden of carbon dioxide (CO_2) emissions that originate from the combustion of fossil fuels and cement manufacture, including "carbon dioxide produced during consumption of solid, liquid, and gas fuels as well as gas flaring" (Eurostat, 2023a).

I do not consider it a mere coincidence that COP27 was held in the Egyptian coastal city of Sharm el-Sheikh turned to weigh in the cost of Russia–Ukraine was on the environment. This Russian–Ukraine war, the Hamas-IDF war in Gaza, the

Rwanda-backed M23 Rebels fighting the DRC in its eastern provinces show no sign of abating, and are being waged during the era of climate change, while the world is still grappling with the ecological and human repercussions of the Covid-19 pandemic. The brutal and inhumane nature of these hostilities are unfolding in real-time and are witnessed through the lens of modern communication technologies. The loss of civilian and military lives is unjustifiable and purposeless, just as the crimes committed against our planet are. It is now imperative to pursue "pacts" as has been done throughout history, where wars have been brought to an end through treaties and agreements. The world should heed the General-Secretary's appeal:

> That is why at the beginning of COP27, I am calling for a historic Pact between developed and emerging economies – a Climate Solidarity Pact. A Pact in which all countries make an extra effort to reduce emissions this decade in line with the 1.5-degree goal. A Pact in which wealthier countries and International Financial Institutions provide financial and technical assistance to help emerging economies speed their own renewable energy transition. A Pact to end dependence on fossil fuels and the building of new coal plants – phasing out coal in OECD countries by 2030 and everywhere else by 2040. A Pact that will provide universal, affordable, sustainable energy for all. A Pact in which developed and emerging economies unite around a common strategy and combine capacities and resources for the benefit of humankind. The two largest economies – the United States and China – have a particular responsibility to join efforts to make this Pact a reality. This is our only hope of meeting our climate goals. Humanity has a choice: cooperate or perish. It is either a Climate Solidarity Pact – or a Collective Suicide Pact (Guterres, 2022).

It is imperative to establish a pact to bring an end to bloodshed in Ukraine, in Gaza and the eastern DRC. War represents a squandering of valuable resources and energy, leading to the unnecessary destruction of innocent lives; it shows devastating consequence for the environment. Wars breed misery and inflict irreparable damage upon human life and the broader ecosystems that support all forms of life. The section that follows briefly presents the planetary boundaries (PBs) concept or framework, and then links it to climate change, environmental conflicts, food security and sustainable development. Thanks to scientific developments, a framework has been developed within which sustainable development and a good quality of life can be pursued and achieved. The PBs framework is explored and discussed here.

2.3 Planetary Boundaries Framework

In 2009, Johan Rockström, former director of the Stockholm Resilience Centre, led a group of 28 internationally renowned scientists in the process of identifying the nine procedures that regulate the stability and resilience of the Earth's natural regulating networks (Stockholm Resilience Centre, 2022). These scientists suggested "quantitative planetary boundaries" within which humanity could carry on with development and attain prosperity for future generations (Stockholm Resilience Centre, 2022). On a similar vein, Steffen et al. (2015: 736) define PBs as "scientifically based levels of human perturbation of the earth system (ES) beyond which ES functioning may be substantially altered". Scientists also warn that crossing these thresholds will increase the risk of creating far-reaching unanticipated or irreversible environmental changes (Stockholm Resilience Centre, 2022).

Science has found nine PBs as illustrated in Fig. 2.2. Since their identification, PBs have seen increased interest within the scientific community and policymakers. PBs have been studied in relation to agricultural development (Campbell et al., 2017); in climate change studies (Kiyala, 2022a); and in human development and sustainable development studies (O'Neill et al., 2018; Rockström et al., 2009; Steffen et al., 2015), among others. Other scientists concur with this observation, such as Gleeson et al. (2020: 223) who assert that the PBs framework "proposes quantified guardrails to human modification of global environmental processes that regulate the stability of the planet and has been considered in sustainability science, governance, and corporate management". These perspectives make PBs all the more important in the management of humanity's destiny, preserving the functioning Earth systems, and guaranteeing prosperity and safety of future generations. Embracing the prospects of pursuing quality and human life and sustainable development, a study conducted by O'Neill et al. (2018) concurs that human growth and progress that enhances life quality is achievable within PBs and without disrupting Earth's natural regulating processes, despite substantial challenges. This emerges from a study that used indicators designed to measure the possibility of attaining fair and less risky development space by quantifying the resource consumption associated with meeting humans' basic needs, and that was compared to downscaling PBs for about 150 nations (O'Neill et al., 2018). The outcomes of this inquiry suggest that it is hard for countries to meet the basic vital demands of the populations at a level that is globally sustainable; however, "physical needs such as nutrition, sanitation, access to electricity, and the elimination of extreme poverty could likely be met for all people without transgressing planetary boundaries" (O'Neill et al., 2018: 1).

Regrettably, scientific research indicates that humanity has surpassed five out of the nine PBs, which are regarded as the safe limits for maintaining a conducive environment for sustainable development and the wellbeing of future generations. Any breach of these PBs poses a significant risk of destabilising the Earth system's Holocene state, which has provided the foundation for the development of modern societies (Steffen et al., 2015). The five boundaries as per Figure 2.1 that have become critical for sustainable development are notably: climate change, land-system change, biosphere integrity loss, novel entities and biogeochemical flows. These need to be preserved in order to avoid a total collapse of Earth systems (Rockström, 2019; Steffen et al., 2015).

The nine PBs encompass:

(1) Biodiversity integrity loss, manifest in the extinction of many species of our diverse flora and fauna. Besides being endangered, according to the Zoological Society of London, there has been a decline by 68% of mammals, birds, and reptiles on average between 1970 and 2016 (Fransen, 2021).
(2) Altered biogeochemical cycles, caused by the modernisation of agricultural methods to supply enough food to cater the needs of the growing world population. However, the use of fertilisers has increased the amount of many gases

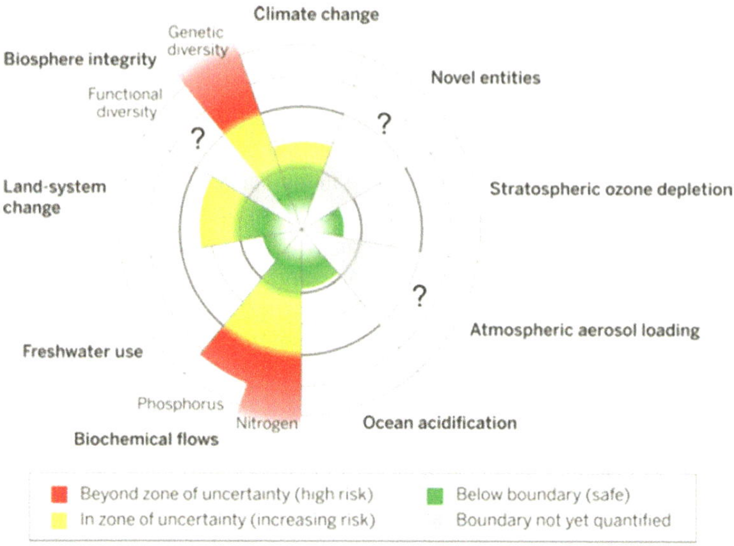

Fig. 2.2 Planatery boundaries. *Source* Steffen et al. (2015: 6)

being released into the atmosphere. The industrialisation of the means to produce fertilisers and their use in agriculture has further increased greenhouse gases (Fransen, 2021).

(3) Land-system change is discernible in the conversion of forests, wetlands and grasslands into agricultural land to feed the world's population. This practice is more than simply deforestation, because humans destroy the forest to create more available land for cultivation, instead of reforestation (Fransen, 2021).

(4) Novel entities: These are explained as "as new chemicals, new forms of existing substances, and changed life forms that can have undesirable geophysical and biological consequences" (Fransen, 2021). They result from profound human-driven changes introduced into the environment and their effects are long-term and discernible: "They reduce fertility and increase the potential of permanent genetic damage in the generations to come. They've already been linked to the decrease in bird populations and hampered development in marine animals" (Fransen, 2021: n.p.). Novel entities are critical to eco-conservation.

(5) Ozone layer; Fransen (2021: n.p) avers: "The ozone layer is responsible for filtering out the harmful ultraviolet rays from the sun". Unfortunately, during the 1980s, the rise in chemical concentrations resulted in the depletion of the ozone layer, creating a hole. As the layer becomes thinner, significant amounts of ultraviolet rays penetrate the earth's atmosphere, posing a subsequent threat to both marine and terrestrial biological systems (Fransen, 2021). There is, however, hope that ozone layer boundary will be protected through the actions undertaken as a consequence of the 1987 Montreal Protocol of 1987 (Fransen, 2021).

Crutzen (2016) highlights that numerous toxic substances released into the environment have profoundly damaging effects, despite not being inherently toxic. One example is the chlorofluorocarbons responsible for the Antarctic ozone hole. Since these are now regulated, the planet was fortunate that a catastrophic event did not occur. If chlorine had chemically behaved like bromine, the ozone hole would have become a global and year-round phenomenon, rather than being limited to the Antarctic spring (Crutzen, 2016). It is evident that humanity will continue to exert a significant impact on the environment for many centuries, unless events such as global catastrophes, meteorite impacts, world wars or pandemics occur (Crutzen, 2016). It is the responsibility of humans and the scientific community to monitor and prevent the worst-case scenario.

(6) Atmospheric aerosol loading: Fransen (2021) contends: The aerosols can formulate clouds that can cool and warm the Earth outside of standard natural cycles. Aerosols especially tend to affect the monsoon systems in tropical areas. The presence of aerosols also affects the reflection and absorption of solar rays in the environment (Fransen, 2021).

(7) Freshwater use: Research shows that by 2050 nearly half a billion people will struggle with access to water. It appears that freshwater resources are negatively impacted by their disproportionate usage by industries. In addition, "human modification of water bodies has a catastrophic impact on the environment, from changes in river flows and irregular vapour movements on a global scale" (Fransen, 2021: n.p.).

Water is a vital need for human and biodiversity; it is subdivided into different categories according to their source, use, colour and means of conservation. Freshwater is extracted "from lakes, rivers, the ground and man-made reservoirs for drinking, agriculture, recreation, and industrial processes" (University of California Regents, 2023). The growth of the world's population over the last century has disproportionately increased global water use, and as consequence, availability of freshwater has been reduced in many regions, both for human consumption and for ecosystems (University of California Regents, 2023). Besides the use of freshwater for human and development needs, it important to also note the effects of global warming on the constancy and severity of droughts in some regions, because warming alters atmospheric circulation patterns and the water cycle (University of California Regents, 2023).

It is important to understand the complexities of overusing freshwater and its ecological implication and why future generations may struggle and use this resource's scarcity to justify conflict and violence. Ros (2023: n.p.) notes:

> Freshwater is vital for life, supporting ecosystems and human civilizations. We use freshwater in many aspects of daily life including food production, power generation, manufacturing, and sanitation. However, it is becoming increasingly threatened. Although the Earth's surface is two-thirds water, less than three percent is fresh. Pollution in the world's rivers, streams, and lakes renders water undrinkable and harms local wildlife. Overuse, triggered in part by population growth, causes water shortages and dries up lakes, rivers, or streams before they reach the sea or downstream neighbors, sometimes causing conflict.

Later studies on the freshwater planetary boundary (PB) have suggested a subdivision of this boundary in order to better quantify its current distribution. For instance, Wang-Erlandsson et al. (2022: n.p.) recommends making freshwater green water "terrestrial precipitation, evaporation and soil moisture" and blue water "rivers, lakes, reservoirs and renewable ground water stores", and creating sub-boundaries around each, namely a green water sub-PB and a blue water sub-PB (Gleeson et al., 2020; Wang-Erlandsson et al., 2022). Green water is critical to ES dynamics but is sadly disturbed by human pressures from continental to planetary levels (Wang-Erlandsson et al., 2022). Thus, the violation of freshwater PB like that of other boundaries affects the well-being of the ecosystem and biodiversity (Wang-Erlandsson et al., 2022). Apparently, the "freshwater use" PB is close to being crossed while the green water sub-PB has already been crossed (Wang-Erlandsson et al., 2022). In the view of Gleeson et al. (2020: 223),

> The current freshwater use planetary boundary has been set at 4000 km^3/year blue water consumption, the lower limit of a 4000–6000 km^3/year range hat is considered a danger zone, as it takes us too close to the risk of blue and green water induced thresholds that could have deleterious or even catastrophic impacts on the ES.

So, freshwater, considered as a PB should be used responsibly because it is all-important to life in many aspects, including for food production, power generation, sanitation and manufacturing; in addition, it supports "ecosystems and human civilisations" (Ros, 2023). Unfortunately, freshwater is under threat by human use, pollution by man-made pH and climate change. Polluted water is not only hazardous to biodiversity, but also to human life and wildlife. So, the PB of freshwater must be preserved, and not further endangered.

(8) Ocean acidification: It is reported that about a quarter of CO_2 which reaches the atmosphere ends in into oceans; it forms carbonic acid in water, which subsequently shrinks the pH of surface water (Fransen, 2021). In consonance with the Water Science School (2019), pH stands for a measure of the correlative amount of free hydrogen and hydroxyl ions in the water.

Figure 2.3 shows various levels of pH. pH is a measure of how acidic/basic water is. pH ranges between 0 and 14, with 7 being neutral pHs of less than 7 shows acidity; considering that a pH of greater than 7 corresponds to a base (Water Science School, 2019). So, "water that has more free hydrogen ions is acidic, whereas water that has more free hydroxyl ions is basic" (Water Science School, 2019). An inadequate level of pH in water can be detrimental to biodiversity and metal lifespans, because "pH levels outside of 6.5–9.5 can damage and corrode pipes and other systems, further increasing heavy metal toxicity" (Fondriest Environmental, 2013).

According to the Fondriest Environmental (2013: n.p.), the ideal pH levels for fish range between 6.5 and 9.0. Any scale outside of the optimal ranges could cause stress and death of aquatic organisms. It is important to note: "As pH levels move away from this range (up or down), it can stress animal systems and reduce hatching and survival rates. The further outside of the optimum pH ranges a value is, the higher the mortality rates" (Fondriest Environmental, 2013: n.p.). Inconsistent degrees of pH can affect

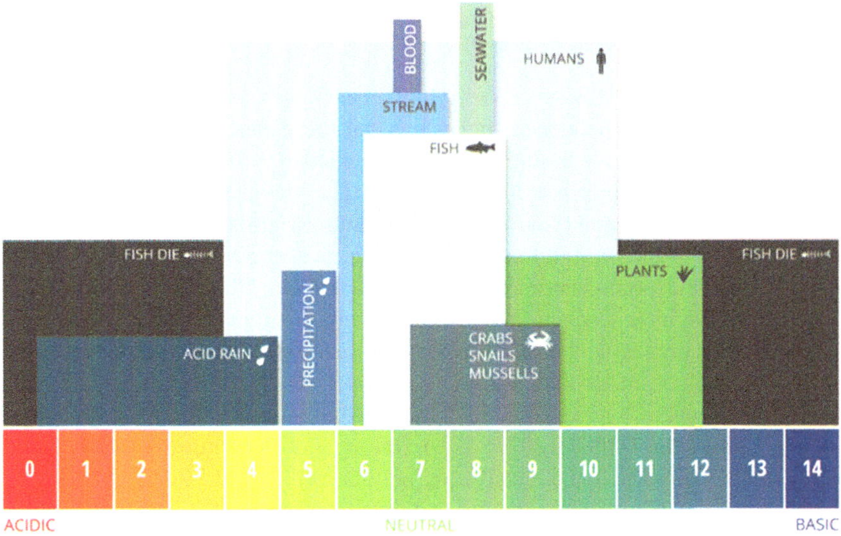

Fig. 2.3 Aquatic pH levels. *Source* Fondriest Environmental (2013). "pH of Water." Fundamentals of Environmental Measurements. https://www.fondriest.com/environmental-measurements/parameters/water-quality/ph/ (Accessed 28 Jan 2023) [Permission: **Fondriest Environmental, CC by 2.0**]

life under water and biodiversity. For instance, when the pH of water goes below 5.0 or increases above 9.6, the harmful consequences became manifest; such undesirable effects caused by acidification become more revealed in saltwater fish because of their adaptation to a higher pH (Fondriest Environmental, 2013). However, "when pH is below optimal levels, fish become susceptible to fungal infections and other physical damage" (Fondriest Environmental, 2013). Consumption of such affected fish can become a health hazard. The rise of acidity has damaging after-effects on the global marine environment.

So, ocean acidification has anthropogenic causes; it is produced by an influx of dissolved carbon dioxide which increases simultaneously with atmospheric CO_2 levels which subsequently reduces the pH of water; it is also generated by pollution (Fondriest Environmental, 2013).

(9) *Climate change:* Climate change is marked by the concentration of carbon dioxide (CO_2) in the air which used to be 280 ppm (parts per million) in the pre-industrial era, but it has gone up as far as 400 ppm (Fransen, 2021). The Working Group III Contribution to the Sixth Assessment Report B.1.2 of the IPCC points out:

> Climate change has caused substantial damages, and increasingly irreversible losses, in terrestrial, freshwater and coastal and open ocean marine ecosystems […] The extent and magnitude of climate change impacts are larger than estimated in previous assessments […] Widespread deterioration of ecosystem structure and function, resilience and natural adaptive

> capacity, as well as shifts in seasonal timing have occurred due to climate change [...], with adverse socioeconomic consequences (IPCC, 2022c: 9)

In the 1970s, James F. Black, ExxonMobil Fuel Giant's scientist alerted the world by providing scientific evidence suggesting that the planet was warming into the danger zone that requires measures to be implemented to curb planet warming caused by fossil fuel and carbon pollution (Mann, 2021). However, his factual findings on warming were thwarted and contested by interest groups and the public relation campaign of ExxonMobil so no preventive measures were taken to avert the looming crisis that is confronting the planet today (Mann, 2021). Scientists continue to warn that we have reached the tipping point in the climate change age to take action that will prevent global warming from exceeding 1.5 °C above pre-industrial levels. The truth of matter that there is – "a greater than 50% likelihood that global warming will reach or exceed 1.5 °C in the near-term, even for the very low greenhouse gas emissions" (IPCC, 2022a, b: 8). Halting global warming requires nothing less than a demonstrated and unwavering political will, accompanied by concerted efforts from scientists to highlight the ongoing deterioration of atmospheric conditions and the progressive disintegration of the Earth systems. Urgent and decisive actions are necessary today, without any delay, from multinational corporations, businesses and industries. It is crucial to accelerate the development and implementation of innovative technologies that promote green energy, while phasing out carbon and greenhouse-gas-dependent economies and fostering sustainable development. The questions that asked among world political leaders, business world, faith-based organisations (NGOs), environmental activists and the scientific world are:

- Which planet should we leave to posterity?
- What are the pathways to "Our Common Future"?
- What is "The Future We Want" and sustainable development we envisage when it is reported that an estimate of close to 800,000 people succumb annually as a result of inhaling air polluted by harmful aerosols (Fransen, 2021)?

This is why, once again, knowing PBs is critical for both the survival of humanity and biodiversity which sustain humans' life.

The Future We Want requires adapting development strategies and technologies to climate change realities and observing evidenced-based findings on PBs within which human development can progress and be attained without further increasing the multifaced dangers that the earth is facing. Earth Sciences continue supplying facts related to the operating limits with the nine boundaries. For instance, agriculture is viewed as one of the major contributors to atmospheric transformation and destabilisers of the Earth systems on a global scale, which should be practised with care regarding the nine PBs, or "safe limits", namely: "land-system change, freshwater use, biogeochemical flows, biosphere integrity, climate change, ocean acidification, stratospheric ozone depletion, atmospheric aerosol loading, and introduction of novel entities" (Campbell et al., 2017). The benefits of knowing PBs are enormously important to pursue sustainable development as contended by Steffen et al. (2015: 736): "There is an urgent need for a new paradigm that integrates the continued development of human societies and the maintenance of the ES in a resilient and accommodating state".

It is evident that humanity has made significant progress in various domains, including health, information technology, robotics applied to various developmental projects such as medical surgeries, automobile and aeronautical engineering, and more. Additionally, advancements have been made in inventing new technologies in agriculture, transportation, road infrastructure, armaments and pharmaceuticals. These developments aim to meet the needs of the growing world population and drive overall progress and innovation. However, "the perceived trade-offs between socioeconomic development and global environmental sustainability, for example between energy use and climate change caused by greenhouse gas emissions or land-use change for food production and biodiversity loss" (Griggs et al., 2014), remain an area of serious concern that could exacerbate the vulnerability of ES. Despite the promising agendas and commitments made by global political leaders, businesses (United Nations, 2005, 2012b), and civil society (Kiyala, 2022a, 2022b) to foster new paths of development aligned with the goals of the Paris Agreement, aimed at limiting the temperature increase to 1.5° C above pre-industrial levels, the actualisation of these ambitions and aspirations is falling behind, while the deterioration of the earth continues without respite. Hence, it is crucial to envision and strive for humanity's prosperity within the boundaries of sustainability—where our concerns and interests lie in preserving the environment for the wellbeing of future generations.

The overarching aims of the SDGs, as agreed by nations at Rio+20, can be summarised as poverty elimination, sustainable lifestyles for all and a stable resilient planetary life-support system. However, it is challenging to define, create and agree on SDGs that meet these overarching aims while resolving potential interactions between sectoral goals. For example, some approaches of in=creasing food security may come at a significant cost to the global climate system, in turn putting food security itself at risk in the long-term (Griggs et al., 2014).

The concept of sustainable development does imply limits – not absolute limits but limitations imposed by the present state of technology and social organisation on environmental resources and by the ability of the biosphere to absorb the effects of human activities (United Nations, 2005). The science of PBs is one of the strategies that science has considered as realistic to keep the impending environmental catastrophe under control while simultaneously pursuing human development.

2.4 Peace Ecology

Environmental peacemaking or peace ecology has emerged as a parallel disciple with conventional peacebuilding in peace studies. Kyrou (2007: 73) argues that this new paradigm stands at the intersection of peace studies and environmental studies and is described as "a theoretical framework, broad and integrative enough to allow a full understanding, functionally as well as philosophically, of the inherent capacities of the environment to inform and sustain peace". For Amster (2015), this entails creating a new model that connects the "essence of peace" and nonviolence with the principles of ecology and the fundamentals of environmentalism; it is interested in issues like "food justice, water sharing, climate change, peace zones and the free economy" (p. 9). Amster (2015: 9) asserts that peace ecology, in essence, contemplates the ways that the very environmental systems that usually trigger conflicts; these stem from "resources depletion, anthropogenic climate change, food and water shortages… [and] can also become profound opportunities for peaceful engagement". Peace can be driven by shared environmental crises such as hurricanes and earthquakes; these motivate people to work together, because a collective response is more needed than competition in these circumstances. In addition, he avers: "when people in a community works together to produce food or conserve shared resources, they can simultaneously build bonds of trust and mutual respect that serve to both forestall conflicts and help resolve them nonviolently when they arise" (Amster 2015: 9).

Thus, environmental peacemaking, environmental justice, social justice and ecojustice are interwoven; they are relevant and essential to ensure the survival of humanity. It is only by stopping the downward path to "climate hell" that real peace will prevail on Earth. At Sharm el-Sheikh, António Guterres, The General-Secretary of the UN, warned world leaders about the perspectives of winning or losing the global climate fight in this crucial decade. He said: "We are on a highway to climate hell with our foot still on the accelerator" (Guterres, 2022). The General-Secretary warns about the potential for humanity to perish alongside biodiversity is not excluded in the years to come, based on scientifically substantiated facts regarding climate change and global warming. This outcome may not be immediate, but it remains a distinct possibility as our understanding of the long-term impacts continues to grow (IPCC, 2022b).

Peace as a harmonious atmosphere of social interaction goes beyond social relations to embrace harmony with the environment, because the ecosystem entails the body of processes that regulates life in the *oikos*, which, in Greek, means "home". Environmental peacemaking is concerned with creating or bringing harmony and order in the home we inhabit. The implications of the interlink between peace and the environment are that the home should be clean, free from noise and pollution, free from mess and violence. To elaborate more on this, Patriarch Bartholomew associates climate change with social justice (The Ecumenical Patriarchate of Constantinople, 2022: n.p.), which he explains as follows:

> The word 'ecology' contains the prefix 'eco', which derives from the Greek word oikos, signifying 'home' or 'dwelling'. How unfortunate, then and indeed how selfish it is that we have reduced its meaning and restricted its application. This world is indeed our home. Yet it is also the home of everyone, just as it is the home of every animal creature and of every form of life created by God. It is a sign of arrogance to presume that we human beings alone inhabit this world. Moreover, it is a sign of arrogance to imagine that only the present generation enjoys its resources.

Care for the environment is intrinsically connected to social justice that calls for collective stewardship because the Earth is our home. The Patriarch adds: "Indigenous peoples throughout the world are the stewards and guardians not only of the forests and the seas, as well as of a vast store of knowledge about the natural world, which they regard as the "library of life". (The Ecumenical Patriarchate of Constantinople, 2022: n.p.). We need to maintain peace in our "common home"; that peace cannot exist if the home is on fire, as stated by the climate activist, Thunberg (2019). The fire must be extinguished, because no one will think of alternatives when their house is on fire. Since it is our only home, there is no alternative to run to. Thus, it is also a common responsibility to put out the fire set by humanity itself. This is fairness, justice and peacemaking because, as the Green Patriarch Bartholomew stated, we share this home with other creatures who are not responsible for the anthropogenic climate change and global warming.

Peace and security are no longer grounded in conventional peacebuilding that has evolved from liberal and democratic peace; they have become questions and areas of study in environmental peacemaking or peace ecology as well – for, today, the world rests on negative peace, in the sense provided by Galtung (1969). It is becoming evident that humanity lives in fear, and there is a great sense of nervosity in the face of environmental disasters and scientific warnings about the ongoing deterioration of the ecosystem, and humans continue to be the main agents of climate change shifts and atmospheric conditions' supernatural deregulation. The fear of violent conflicts increasing in the future is predicted by scientists who ground their admonitions on the current consequences of climatic and atmospheric perturbations (Sakaguchi et al., 2017) as climate wars and climate-induced conflicts on a large scale loom as described earlier (Dyer, 2010; Welzer, 2012). Such predicaments are real and worrisome as world leaders remain divided over climate action.

Apart from the multifaceted ecological crises we have alluded to above, traditional conflicts, wars, violence and mass destruction continue to affect the international community. Europe, the Middle East and Africa are no exception. International security is at stake: the ominous escalation of the Russia–Ukraine war; the threats by conflicts between China and Taiwan (Goldstein, 2015; Kastner, 2015), the tense military and political climate in the Korean Peninsula (Mastro, 2018; Mazarr et al., 2018); the conflict in the Indo-Pacific Region (Chandra/Ghoshal, 2018), conflicts in the Middle East (Milton-Edwards/Hinchcliffe, 2007), conflicts in Sub-Saharan Africa (Buhaug et al. 2015; Fjelde/von Uexkull, 2012); and in the Southern African Development Community Region, particularly wars and lack of accountability for human rights violations in the Democratic Republic of Congo (Liwanga, 2015; Maystadt et al., 2014).

Ecological peacemaking is still possible despite the anthropogenic warming of the Earth, the so-called ecocide in progress (Welzer, 2012). There has been even a suggestion by Mehta/Merz (2015) that the International Criminal Court should criminalise environmental crime in the same way it has done with genocide, war crimes and crimes against humanity.

Environmental crises and conflicts that have surged in recent years; that according to Ide et al., (2022: 1) rest and will rely upon innovative research streams that look into the environment, peace and conflict. Such studies encompass "environmental change and human security, climate change and armed conflict, environmental peacebuilding, political ecology, securitisation of the environment, and decolonising environmental security" (Ide et al., 2022: 1). Climate has become a new front of war for peace and global security. Social, political, economic and environmental instability expected to be created by climate change and global warming require an international consciousness to develop a multidisciplinary model of environmental peace. The approach should involve implementing legal measures within international criminal legislation to hold accountable individuals who deliberately cause harm to ecosystems and charging them with ecocide (Mehta/Merz, 2015). Additionally, it is important to adopt scientifically supported strategies that promote sustainability, in line with the PBs framework. Furthermore, in countries where a normative legal framework for addressing environmental conflicts is lacking, using indigenous alternative dispute resolution methods can be valuable. Climate refugees caused by sandstorms, floods and other adverse effects of global warming will need an all-encompassing worldwide solidarity and novel geostrategic approaches. As we address the immediate human-induced causes of environmental conflicts, it is crucial to enhance environmental peacebuilding by incorporating evolving challenges and new insights into the state of the Earth systems. This will empower policymakers to effectively engage with governments, encouraging them to prioritise the protection of PBs while ensuring the wellbeing and prosperity of future generations.

Figure 2.4 shows the interlocking relations between peace, environment and sustainable development. The three factors influence each other. Sustainable development relies on the strong interdependency between climate change and peace; this leads to understanding the nexus of development and climatic alterations though a multidisciplinary lens. Subsequently, concepts such as ecocide,environmental peacemaking and environmental justice compound the pathways of humanity's progress that does not harm the ecosystems and biodiversity. The following section provides useful perspectives on the environmental ambit of human development that should not stray outside the PBs and ecojustice. Environmental peacebuilding requires environmental justice that is advocated by scholars (Mehta/Merz, 2015) and religious leaders (Pope Francis, 2015; The Ecumenical Patriarchate of Constantinople, 2022).

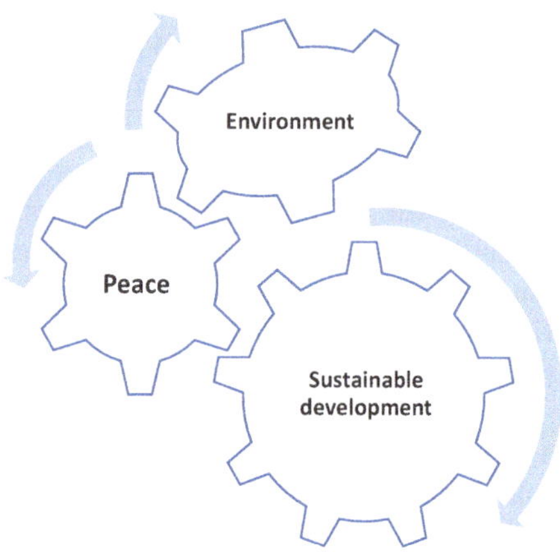

Fig. 2.4 Link between peace, environment and sustainable development *Source* The Author

2.5 Environmental Peacemaking and Environmental Justice

Environmental peacemaking is linked to environmental justice, social justice and ecojustice because these notions appear to overlap with one another. To explain the links among them, Kyrou (2007: 73) starts with the premise that "ontological, axiological, epistemological, rhetorical and methodological overlaps; interdependencies; conflicts and; potential synergism from their 'interaction'". This calls for a theoretical framework that is broad enough and sufficiently integrative to allow for "a full understanding, functionally as well as philosophically, of the inherent capacities of the environment to inform and sustain peace". He calls the novel paradigm, which is inspired by the notion of environmental peacemaking, "Peace Ecology". Hence, although the terms environmental peacemaking, environmental peacebuilding, and peace ecology often overlap and are used interchangeably, we endorse peace ecology as a theoretical framework within the fields of peace studies and environmental studies. This framework offers a more comprehensive understanding and encompasses both the interconnected and holistic aspects of how environmental issues can impact or promote peace (Kyrou, 2007).

As a logical consequence of human existential aspirations—without spirtualising too much – humanity always longs for peace after experiencing the horror of wars it creates. This has been validated over the centuries by peace treaties (Lesaffer, 2012); and the international peace mantras of "no more war", "no more deployment of nuclear arsenal", "never again war! never again war" (Pope Paul VI, 1965); "no more war, war never again" (Fox, 2015).

Before the General Assembly of the Organisation of United Nations, on Monday, 4 Oct 1965, His Holiness Pope Paul VI uttered his anguish in the following terms:

> It is enough to recall that the blood of millions, countless unheard-of sufferings, useless massacres and frightening ruins have sanctioned the agreement that unites you with an oath that ought to change the future history of the world: never again war, never again war! It is peace, peace, that has to guide the destiny of the nations of all mankind! (Pope Paul VI, 1965: 4)

Over five decades after the call made by the Roman Pontiff to end human bloodshed, a new needless, extremely brutal and ferocious war broke out between Russia and Ukraine beyond any reasonable imagination. On the UN Facebook page, António Guterres, the United Nations Secretary-General, reckoned: "The United Nations was born out of war to end war. Today, that objective was not achieved. But we must never give up. We must give peace another chance".[8] The Russian invasion of Ukraine and the massing of disproportionate force against civilians and civilian infrastructure cannot be justified; neither can the ecological impacts of this aggressive and murderous war (Rawtani et al., 2022) or its economic cost (Liadze et al., 2022).

[8] António Guterres, the United Nations Secretary-General. Facebook message on 25 Feb 2022, https://www.facebook.com/unitednations/posts/the-united-nations-was-born-out-of-war-to-end-war-today-that-objective-was-not-a/10160314880145820/ (Accessed 15 Feb 2023).

However/Nevertheless, it appeared that this war was only starting, though it was already one year old when this chapter was written. Following a Security Council meeting on Ukraine, Secretary-General António Guterres continued giving hope saying: "the principles and values of the UN Charter will prevail".[9]

Historians have for a very long time recognised that "the string of peace treaties that runs from Westphalia (1648) to Versailles (1919) formed the backbone of Europe's international constitution" (Lesaffer, 2012:), and later, there were Versailles and the other Parisian peace treaties signed at the end of World War I (1919–20) and World War II between European allies of Germany (Paris, 1947) (Lesaffer, 2012). As peace treaties ended historical major world wars such as those between the sixteenth and eighteenth centuries in Europe, environmental peace treaties are an option to end the war waged on the environment, and conventional wars in the Ukraine and other parts of the world that are having catastrophic effects on humans and the environment.

Against the backdrop of the assertions on the need for both civilian peace and environmental peace, the international mantra today should be "never again war to decimate humanity! Never again way to decimate the ecosystems and biodiversity". The remedy our contemporary world needs is, in the short term, to end wars around the globe and to limit the combustion of fossil fuels and carbon dioxide and GRH emissions, carbon pollution and planet warming.

In this fight for global peace and security, the focus needs to be on (1) empowering the entire human family with inalienable knowledge and understanding of environmental conflicts and violence; (2) demonstrating the horrific and deadly consequences of climate wars and conventional warfare on humans and the environment; (3) developing a global consciousness on the necessity to pursue environmental peacemaking; (4) reimagining environmental policies and the means to attain environment justice and social justice in the Anthropocene. These issues are discussed in the following sections.

2.6 Environmental Justice and Social Justice

According to Ben Crump Law (2023: n.p.), "[t]he principles of environmental justice focus on the core beliefs and structures around protecting the environment and giving people control over what happens to their communities". Environmental justice, as a theory associated with environmental peace, is much broader and demands an international commitment which quantifies and determines the extent of damages caused to individual communities and their environment, which, in most cases, occur unevenly. During the People of Colour Environment Leadership Summit in 1991, convened with the purpose of initiating a collective movement among people of colour to safeguard their lands from any form of harm, participants reaffirmed their deep spiritual connection to the "sacredness of Mother Earth". In this gathering, they

---

[9] Ibid.

extensively discussed and formally embraced the 17 Principles of Environmental Justice (Environmental Working Group, 2023).

Environmental justice is the logical response to the amplitude of suffering, pain and misery inflicted on humans and the wilful and persistent production of pollutants and toxicants that, over time, have proved catastrophic to the environment, biodiversity and ecosystems. Some of the tragic consequences amount to crimes against humanity and crimes against our planet as articulated by Mann (2021). He emphasises that humans have been entrusted with the stewardship of a magnificent treasure he refers to as a "Goldilocks planet". Positioned perfectly within the solar system, this planet provides ideal conditions for human existence and life, including a suitable temperature and air quality. As custodians, we bear the utmost responsibility to reject any changes to this planet that would endanger human and other forms of life. It is disheartening to witness large corporations consistently prioritising record profits over the wellbeing of the Earth and its supporting systems (Mann, 2021).

The underpinning source of the ongoing crisis in the relationship between humans and the environment, according to the biblical message and Catholic Church's Magisterium, resides in "man's pretension of exercising unconditional dominion over things, heedless of any moral considerations which, on the contrary, must distinguish all human activity"[10] (Pontifical Council for Justice and Peace, 2004: 201). Throughout history and influenced by cultural factors, humans have demonstrated a proclivity for heedless and inconsiderate exploitation of the Earth's resources. This behaviour stems from humanity's capacity to engage in innovative and transformative interventions (Pontifical Council for Justice and Peace, 2004: 201). The CSDC No 461 asserts:

> The aspect of the conquest and exploitation of resources has become predominant and invasive, and today it has even reached the point of threatening the environment's hospitable aspect: the environment as 'resource' risks threatening the environment as 'home'. Because of the powerful means of transformation offered by technological civilisation, it sometimes seems that the balance between man and the environment has reached a critical point (Pontifical Council for Justice and Peace, 2004: 201).

The world is experiencing the erosion of that stewardship that should define the relationship between humans and the environment. Big businesses and industries appear to be in an unrestrained race for unbridled pursuit of profits while willfully harming the environment. There have been catastrophic effects of human conduct as explained earlier and throughout this chapter with the production and use of fossil fuels, carbon and greenhouse emissions. Such pursuit of profits that threatens the Earth is beyond unacceptable and unethical. Exposure to acid rain and human-induced pH pollution poses a threat to human health. It is not only humans who are affected; aquatic organisms also experience increased mortality due to radioactive waste disposal during nuclear tests, as noted by Rothschild (2012). Additionally, human ecology plays a role in the unequal distribution of health issues resulting from environmental causes, as highlighted by Mohai et al. (2009). Processes of environmental justice should be cognisant of this fact. This is the rationale for environmental

[10] Compendium of the Social Teaching of the Church (CSDC) #461.

justice, which for Ben Crump Law (2023: n.p.), is "the concept that everyone should have a say in how the environment is managed and protects their communities. The idea of environmental justice grew out of the consistent and widespread damage caused by the decisions of a few people that affected large communities". For that reason, Environmental Justice Law has been elaborated "to protect disadvantaged communities". According to Ben Crump Law (2023: n.p.):

> Environmental justice law is a set of laws enacted to protect the environment and ensure that every group has representation. In recent decades, legal action has changed how states protect the rights of people, involving them in these decisions. In many cases, it is the direct result of individuals or groups taking legal action against larger entities.

Environmental justice for People of Colour in the USA entails a recognition of the violations of their rights, including the destruction and misuse of their lands. It involves reestablishing the spiritual interdependence between individuals and the sacredness of Mother Earth. It calls for respect and celebration of their languages, cultures, and beliefs about the natural world, as well as recognising the healing power of nature and their communities. Furthermore, it seeks to achieve liberation that has been denied for over 500 years of colonisation, during which People of Colour experienced political and economic oppression resulting in the poisoning of their communities and lands, and even the perpetration of genocide against their peoples (Environmental Working Group, 2023). Table 2.2 presents a summary of the 17 principles of environmental justice redesigned from Environmental Working Group (2023).

The principles of environmental justice can be summarised in four main concepts: (1) increase ecological protection and safety within disadvantaged communities; (2) expand cultural awareness and address potential language barriers; (3) promote environmental education; (4) provide additional opportunities for everyone to participate in the decision-making process, regardless of race or socioeconomic status (Ben Crump Law, 2023: n.p.). These four concepts defining the principles of environmental justice overlap with the religious and social perspectives of environmental justice that are conveyed by religious leaders such as the Green Patriarch Bartholomew (Chryssavgis, 2014; Patriarch Bartholomew 2021; The Ecumenical Patriarchate of (Constantinople, 2022) and the Roman Pontiff, His Holiness Pope Francis (Deane-Drummond, 2016; Ferrara, 2019; Pope Francis, 2015).

Patriarch Bartholomew, who is often referred to as the Green Patriarch, advocates for treating the environment with the same care and respect as we treat our fellow humans. He emphasises that sanctions for causing harm to the environment should be comparable to those imposed on individuals who harm others. He states:

> If human beings were to treat one another's personal property the way they treat the natural environment, we would view that behaviour as anti-social and illegal. We would expect legal sanctions and even compensation [...] The way we respond to the natural environment is directly reflects the way we treat human beings. The willingness to exploit the environment is revealed in the willingness to permit avoidable human suffering. So the survival of the natural environment is also the survival of ourselves. When will we understand that a crime against nature is a crime against ourselves and sin against God? (The Ecumenical Patriarchate of Constantinople, 2022: n.p.)

Table 2.2 Principles of environmental justice

1.	Environmental justice affirms the sacredness of Mother Earth, ecological unity and the interdependence of all species, and the right to be free from ecological destruction.
2.	Environmental justice demands that public policy be based on mutual respect and justice for all peoples, free from any form of discrimination or bias.
3.	Environmental justice mandates the right to ethical, balanced and responsible uses of land and renewable resources in the interest of a sustainable planet for humans and other living things.
4.	Environmental justice calls for universal protection from nuclear testing, extraction, production and disposal of toxic/hazardous wastes and poisons and nuclear testing that threaten the fundamental right to clean air, land, water and food.
5.	Environmental justice affirms the fundamental right to political, economic, cultural and environmental self-determination of all peoples.
6.	Environmental justice demands the cessation of the production of all toxins, hazardous wastes, and radioactive materials, and that all past and current producers be held strictly accountable to the people for detoxification and the containment at the point of production.
7.	Environmental justice demands the right to participate as equal partners at every level of decision-making including needs assessment, planning, implementation, enforcement and evaluation.
8.	Environmental justice affirms the right of all workers to a safe and healthy work environment, without being forced to choose between an unsafe livelihood and unemployment. It also affirms the right of those who work at home to be free from environmental hazards.
9.	Environmental justice protects the right of victims of environmental injustice to receive full compensation and reparations for damages as well as quality health care.
10.	Environmental justice considers governmental acts of environmental injustice a violation of international law, the Universal Declaration On Human Rights, and the United Nations Convention on Genocide.
11.	Environmental justice must recognise a special legal and natural relationship of Native Peoples to the U.S. government through treaties, agreements, compacts and covenants affirming sovereignty and self-determination.
12.	Environmental justice affirms the need for urban and rural ecological policies to clean up and rebuild our cities and rural areas in balance with nature, honouring the cultural integrity of all our communities, and providing fair access for all to the full range of resources.
13.	Environmental justice calls for the strict enforcement of principles of informed consent, and a halt to the testing of experimental reproductive and medical procedures and vaccinations on people of colour.
14.	Environmental justice opposes the destructive operations of multinational corporations.
15.	Environmental justice opposes military occupation, repression and exploitation of lands, peoples and cultures and other life forms.

(continued)

Table 2.2 (continued)

16.	Environmental justice calls for the education of present and future generations which emphasises social and environmental issues, based on our experience and an appreciation of our diverse cultural perspectives.
17.	Environmental justice requires that we, as individuals, make personal and consumer choices to consume as little of Mother Earth's resources and to produce as little waste as possible; and make the conscious decision to challenge and reprioritise our lifestyles to insure the health of the natural world for present and future generations.

Source Adapted from Environmental Working Group (2023: n.p.)

The Patriarch may not be suggesting the prosecution of people whose lifestyles becomes a burden to the environment, or scientists who derail actions to preserve the ecosystems by presenting false narratives on climate change (Mann, 2021), or politicians venturing in denialism about the existential threats posed by climate change (De Pryck/Gemenne, 2017; McCright/Dunlap, 2011), but he is trying to drive home the gravity of harm inflicted on the ecosystem which must stop. Mohai et al. (2009), in discussing the scope and dynamics of climate-induced-suffering inflicted on humanity and the ecosystems, found that environmental justice needs to be understood and attained within "the globalisation of the environmental justice movement, discourse and issues, as well as with some policy implications of finding and understanding environmental justice" (p. 405). This is evident, because of the widening scope and numbers of entities involved in environmental justice which is evolving as a pluridisciplinary model of justice. The section that follows expands on the rationale and framework of environmental justice viewed through the lens of earth-common good relations in the Anthropocene.

2.7 Sustainable Development in the Age of Climate Change

Sustainability has become one of the most debated concepts and theories since the UN's Earth Summit also known as the Rio Earth Summit that was held in 1992. The United Nations Conference on Sustainable Development commonly referred to as "Rio+20", which took place in June 2012 ended with slogans such as *The Future We Want* (United Nations, 2005, 2012b). Viewing the future from the perspectives of climate change, new ambitions and driving thoughts have emerged such as *Safe Future for Humanity* (Rockström 2019), or even *Care for Our* Common Earth (Pope Francis, 2015). These depictions of humanity's dreams and highest aspirations reveal the profound yearnings that humans possess, despite the global human community's failure to pursue a development path that safeguards the resilience of Earth systems and the biodiversity that supports all forms of life, including flora, fauna and marine life. The reckless and negligent exploitation of the environment puts these vital components of the Earth at risk. Rio+20 in 2012 envisaged a green development that would not imperils the Earth at risk (Clémençon, 2012; United Nations, 2012b).

Carbon dioxide and fossil fuels, the biggest energy sources used in development projects, are under severe scrutiny. World leaders have expressed their hopes to take action to avert a total collapse of the system that support life on the planet. Before discussing SDGs and the impact of climate change on the pursuit of this agenda, it is worth framing the parameters within which sustainable development is understood which we develop from several scientists such as Griggs et al. (2014: 2) who note:

> Alongside development-only targets and environment-only targets, these integrated targets would ensure that synergies are maximised and trade-offs are managed in the implementation of SDGs. Given the scale of humanity's impact on the planet, they argued that long-term sustainable development needs to be conceptualised in terms of an economy and society sustained within Earth's life-support system.

From the same perspective, humanity's aspirations conveyed by Rio+20 Summit's commitments by world leaders to reimagine a path of sustainable development that is concerned with the survival of the Earth depends heavily on how we pursue the SDGs. Sustainable development, according to The Report of the World Commission on Environment and Development: Our Common Future, Article 3 of Sustainable Development, paragraph 27, entails meeting "the needs of the present without compromising the ability of future generations to meet their own needs" (United Nations, 2005). Griggs et al. (2014) support the Brundtland Commission's 1987 definition which entails a type of progress that caters for the needs of current and future generations (Brundtland Commission 1987 clause 1 of Section IV Conclusions), which they link to the Anthropocene. These scholars state that "development that meets the needs of the present while safeguarding Earth's life-support system, on which the welfare of current and future generations depends" (Griggs et al., 2014:2). In similar vein, Steffen et al. (2015) envisage sustainable development with safer limits of operations within the PBs. For Fransen (2021: n.p.), PBs represent "the thresholds within which humanity can survive, develop and thrive for generations to come". Transgressing these may put the earth in an irreparable state.

The promises made by world leaders to embark on the gradual pursuit of sustainable development goals that take into account the concerning climatic condition's fragility and vulnerability as set by the UN's Earth Summit in 1992 and the Rio+20 in 2012 (Clémençon, 2012; United Nations, 2012b) has not materialised. There is a lack of commitment by world leaders to take bold decisions that uphold the Paris Agreement to contain the temperature increase to 1.5 °C above the pre industrial degrees. Article 2(1)(a) of the Paris Agreement states:

> Holding the increase in the global average temperature to well below 2 °C above pre-industrial levels and pursuing efforts to limit the temperature increase to 1.5 °C above pre-industrial levels, recognising that this would significantly reduce the risks and impacts of climate change (United Nations, 2015a).

Despite the 2015 inception of the UN 2030 sustainable development—"Transforming our World"—(Papas, 2017; United Nations, 2015c) the ability of world leaders to take decisive climate action has yet to be demonstrated. Failure to do so will further exacerbate the earth's vulnerability; climate change and global warming would engender new environmental conflicts beside what is already occurring (Ahmed, 2010; Molvaer, 1991). The enthusiasm about Rio+20 seems to have been muted as the world faces new challenges of climate wars, climate change and the Russian invasion of Ukraine which has exacerbated the already strained energy production. Unavailability of green energy sources and the lagging processes of transition from a carbon-dependent economy to a green economy seems to justify the surge in scepticism about carbon emissions reduction (Blanc et al., 2012; Clémençon, 2012). In a similar perspective, Seddon et al. (2016: 1) note: "Meeting the ever-increasing needs of the earth's human population without excessively reducing biological diversity is one of the greatest challenges facing humanity, suggesting that new approaches to biodiversity conservation are required". These scientists advocate for the incorporation of economic assessment models into biodiversity conservation efforts that embrace interdependent and complementary approaches, rather than solely replacing traditional approaches. They propose expanding beyond conventional biodiversity metrics, such as species richness, to include metrics based on traits and phylogeny (Seddon et al., 2016). Sustainable development is related to prosperity, food security and health (Lake et al., 2012) and the future of the planet (United Nations 2012b, 2015b).

Kosoy et al. (2012) advocate for economic growth that aligns with the human values articulated in the Earth Charter and operates within the boundaries of the Earth's life-supporting systems. They emphasise the importance of high-performance technology and sustainable consumption practices that do not exceed the limits of production. Additionally, they stress the need to restore the environment and acknowledge the strain placed on Earth's life-supporting structures in the Anthropocene era.

2.8 Conclusion and Recommendations

This chapter aimed to demonstrate the impacts of climate wars and climate change on the environment and the extent to which knowledge and practice of environmental peacebuilding can enhance sustainable development in the Anthropocene. To give a more comprehensive landscape of issues affecting the environment, an extensive discussion on the impact of wars, including World War II and the Russian invasion of Ukraine, revealed the tragic consequences of conventional and unconventional (nuclear) warfare on humans and the environment. War kills humans and other forms of life and destroys the environment, even leaving spaces uninhabitable. This was the case of Nagasaki and Hiroshima where atomic bombs were deployed in 1945. Beside the byproducts of conventional warfare on the surface of Earth, there is collateral damage to life under water where anti-warship mines are planted. Landmines, missile craters, wreckages of burnt tanks and other military equipment scattered on the earth surface render such spaces uninhabitable, hostile and dangerous to human life. It is evident that peace is the only way to save both humans and their natural environment.

The concept of the Anthropocene was briefly examined to provide the current context in which humanity exists, acknowledging that each generation faces unique challenges in its specific era. In this epoch shaped by human activities, where our actions significantly impact atmospheric conditions and give rise to severe consequences like climate change and global warming, it is crucial for individuals and societies to be aware of this reality. This awareness should serve as a catalyst for exploring new paths of human development and sustainable economic growth. However, the presence of climate change and global warming denialism hinders and postpones the necessary strategic response to mitigate the alarming decline of the Earth's life-supporting processes. The concept of PBs framework was debated, and it was shown that excessive discharge of carbon dioxide into the atmosphere, the combustion of fossil fuels and greenhouse gas emissions have put five of the PBs under severe constraints. This situation has been addressed by several reports of the IPCC and UN climate summits which engaged the world in reducing global warming at 1.5° C below pre-industrial levels to ensure that the world prospers, and future generations thrive.

Environmental peacebuilding was developed in five overlapping points which provide a roadmap to preventing climate wars and climate change in the Anthropocene. These encompass:

- Developing a collective consciousness about the noxious effects of anthropogenic climate change and long-term consequences, and the prospects of increased environmental conflicts and violence – by tackling anthropogenic causes of such conflicts and violence, namely reducing greenhouse gas emissions, phasing out coal-dependent economies and moving towards green energy and a green economy.
- Climate wars will last as long as the conditions that precipitate them persist, notably, scarcity of resources, human rights and right to life. Therefore, the prevention of future occurrences requires the development of climate resilience and conflict-sensitive adaptation models.
- The promotion of environmental peacebuilding is essential, advocating for an innovative peace paradigm that uses nonviolent strategies to encourage communities engaged in resource-related conflicts to view scarcity not as a catalyst for violent confrontation, but as an opportunity to unite and address the challenges that impact their wellbeing and prosperity.
- The link between environmental justice and social justice is useful to rethink the essence of and purpose of the earth and the sum of structures and systems that support it. Justice arises from the understanding of the earth as a gift to humanity and the habitat of all other forms of life and the inanimate part of creation. Such knowledge should prompt attitudes, behaviours and actions that are imbued with goodness and morality so that human activities do not undermine other lives.
- The intrinsic connection between earth and the common good reveals the need to handle the earth with care and responsibility. All it contains should be considered common goods intended to serve humanity and not be exploited and abused for mere lucrative purposes.

- Finally, scholars and scientists have concurred with scientific evidence based on PBs and the IPCC reports that define the parameters of sustainable development. This is still possible to achieve, but the limits set for viable economic growth, human and sustainable development should be pursued without further burdening the already strained earth systems. Thus, reduction of greenhouse gases, carbon dioxide and fossil fuel emissions are crucial/critical to the survival of humanity, the ecosystem and biodiversity.

In the final analysis and by way of conclusion, several perspectives on sustainable development overlap. In order for humanity to progress and ensure the wellbeing of future generations, it is imperative to embrace pathways that align with the boundaries set by scientific knowledge. This entails developing strategies that foster viable and attainable sustainable development in the context of climate change and global warming. Considering the vulnerabilities of both humans and the environment, as well as the capacity for climate resilience and adaptive measures, we must adopt a conscientious approach to stewarding nature. This involves implementing conflict-sensitive and risk-mitigating approaches to address climate wars and climate change, recognising the integral role of environmental peacebuilding in achieving the desired future for our shared home.

References

Ahmed, F. (2010). Approaches to and tools for managing environmental conflicts in coastal zones in Africa: Challenges and prospects in relation to Integrated Coastal Zone Management (ICZM). *African Journal on Conflict Resolution, 10*(2), 31–47.

Amster, R. (2015). *Peace ecology*. New York: Routledge. https://doi.org/10.4324/9781315632940.

Ben Crump Law. (2023). What Are the Principles of Environmental Justice? https://bencrump.com/environmental-justice-lawyer/what-are-the-principles-of-environmental-justice/. Accessed 15 February 2023.

Biswas, A. K., & Quiroz, C. T. (1996). Environmental impacts of refugees: a case study. *Impact Assessment,* 14(1), 21–39.

Blanc, D. L., Liu, W., O'Connor, D., & Zubcevic, I. (2012). Rio+20 working papers. Issue 1: Development cooperation in the light of sustainable development and the SDGs: Preliminary exploration of the issues Division for Sustainable Development, UNDESA. https://sustainabledevelopment.un.org/index.php?page=view&type=400&nr=761&menu=35. Accessed 16 February 2023.

Bob, U., & Bronkhorst, S. (2010). Environmental conflicts: Key issues and management implications. *African Journal on Conflict Resolution, 10*(2), 9–30.

Bolton, M. B., & Minor, E. (2021). Addressing the ongoing humanitarian and environmental consequences of nuclear weapons: an introductory review. *Global Policy, 12*(1), 81–99. https://doi.org/10.1111/1758-5899.12892.

Brauch, H. G. (2016). Sustainable Peace in the Anthropocene: Towards Political Geoecology and Peace Ecology. In H. Brauch, Ú. O. Spring, J. Grin & J. Scheffran (Eds.), *Handbook on Sustainability Transition and Sustainable Peace* (pp. 187–236). Cham: Springer https://doi.org/10.1007/978-3-319-43884-9_9.

Brisman, A. (2016). Peace ecology. *Contemporary Justice Review, 13*(3), 419–421. https://doi.org/10.1080/10282580.2016.1200886.

Buhaug, H., Benjaminsen, T. A., Sjaastad, E., & Theisen, O. M. (2015). Climate variability, food production shocks, and violent conflict in Sub-Saharan Africa. *Environmental Research Letters, 10*(12), 125015.

Cahill, L. S. (2021). Social justice and the common good: Improving the Catholic social teaching framework. *Journal of Moral Theology, 1*(CTEWC Book Series 1), 106–118.

Campbell, B. M., Beare, D. J., Bennett, E. M., Hall-Spencer, J. M., Ingram, J. S., Jaramillo, F., . . . & Shindell, D. (2017). Agriculture production as a major driver of the Earth system exceeding planetary boundaries. *Ecology and Society, 22*(4), 2–12. https://doi.org/10.5751/ES-09595-220408.

Carius, A. (2006). Environmental peacebuilding. *Environmental Cooperation as an Instrument for Crisis Prevention and Peacebuilding. Conditions for Success. Adelphi Report, 3*(07).

Chandra, S., & Ghoshal, B. (2018). *The Indo-Pacific axis: peace and prosperity or conflict?* London: Routledge.

Chryssavgis, J. (2014). The Green Patriarch: Ecumenical Patriarch Bartholomew and the Protection of the Environment. https://www.orthodoxcouncil.org/-the-green-patriarch-.html. Accessed 19 February 2024.

Clémençon, R. (2012). Welcome to the Anthropocene: Rio+ 20 and the meaning of sustainable development. *The Journal of Environment & Development, 21*(3), 311–338.

Connolly, K. (2022). Germany to set up €100bn fund to boost its military strength. *The Guardian Mail, Sun 27 Feb 2022 13.45 GMT*. https://www.theguardian.com/world/2022/feb/27/germany-set-up-fund-boost-military-strength-ukraine-putin. Accessed 27 February 2023.

Correa, D. D. S., & Owens, W. R. (2010). *The handbook to literary research* (2nd ed.). London: Routledge.

Crutzen, P. J. (2016). Geology of mankind. *Paul J. Crutzen: A pioneer on atmospheric chemistry and climate change in the Anthropocene*, 211–215. https://doi.org/10.1038/415023a.

Crutzen, P. J., & Brauch, H. G. (2016). *Paul J. Crutzen: a pioneer on atmospheric chemistry and climate change in the Anthropocene* (Vol. 50). AG Switzerland: Springer. https://doi.org/10.1007/978-3-319-27460-7.

Crutzen, P. J., & Stoermer, E. F. (2021). The 'Anthropocene'(2000). In *Paul J. Crutzen and the Anthropocene: A New Epoch in Earth's History* (pp. 19–21). Cham: Springer.

De Pryck, K., & Gemenne, F. (2017). The denier-in-chief: Climate change, science and the election of Donald J. Trump. *Law and Critique, 28*, 119–126.

Deane-Drummond, C. (2016). Pope Francis: Priest and Prophet in the Anthropocene. *Environmental Humanities, 8*(2), 256–262. https://doi.org/10.1215/22011919-3664369.

Dyer, G. (2010). *Climate wars: The fight for survival as the world overheats*. Oxford: Oneworld Publications.

Earth Charter International. (2000). The Earth Charter: A declaration of fundamental principles for building a just, sustainable and peacefulglobal society in the 21st century. https://www.kit.nl/wpcontent/uploads/2018/08/1401_ECI_Brochure2007_Webcolor.pdf. Accessed 19 February 2024.

Ehlers, T., Gardes-Landolfini, C., Natalucci, F., & Ananthakrishnan, P. (2022). How to scale up private climate finance in emerging economies: Scaling up private capital is crucial to finance vital low-carbon infrastructure projects, particularly in less developed economies. International Monetray Fund. https://www.imf.org/en/Blogs/Articles/2022/10/07/how-to-scale-up-private-climate-finance-in-emerging-economies#:~:text=Estimates. Accessed 17 February 2023.

Environmental Working Group. (2023). 17 Principles of Environmental Justice. https://www.ewg.org/news-insights/news/17-principles-environmental-justice. Accessed 15 February 2023.

Eurostat. (2023a). Glossary:Carbon dioxide emissions. https://ec.europa.eu/eurostat/statistics-explained/index.php?title=Glossary:Carbon_dioxide_emissions. Accessed 1 Feb 2023.

Eurostat. (2023b). Glossary:Climate change. https://ec.europa.eu/eurostat/statistics-explained/index.php?title=Glossary:Climate_change. Accessed 1 February 2023.

Ferrara, P. (2019). Sustainable International Relations. Pope Francis' Encyclical Laudato Si'and the Planetary Implications of "Integral Ecology". *Religions*, 10(466), 1–20. https://doi.org/10.3390/rel10080466.

Fjelde, H., & von Uexkull, N. (2012). Climate triggers: Rainfall anomalies, vulnerability and communal conflict in sub-Saharan Africa. *Political Geography, 31*(7), 444–453.
Fondriest Environmental. (2013). "pH of Water". Fundamentals of Environmental Measurements. 19 Nov. 2013. https://www.fondriest.com/environmental-measurements/parameters/water-quality/ph/. Accessed 19 February 2024.
Fox, T.C. (2015) Paul VI, 50 years ago: 'No more war, war never again'. National Catholic Reporter. Oct 4, 2015. https://www.ncronline.org/blogs/ncr-today/paul-vi-50-years-ago-no-more-war-war-never-again. Accessed 15 February 2023.
Fransen, B. (2021). What Are Planetary Boundaries, And Why Are They Significant? https://www.ecomatcher.com/what-are-planetary-boundaries-and-why-are-they-significant/. Accessed 28 January 2023.
Galtung, J. (1969). Violence, peace, and peace research. *Journal of peace research, 6*(3), 167–191.
Gleeson, T., Wang-Erlandsson, L., Zipper, S. C., Porkka, M., Jaramillo, F., Gerten, D., . . . Gordon, L. J. (2020). The water planetary boundary: interrogation and revision. *One Earth, 2*(3), 223–234. https://doi.org/10.1016/j.oneear.2020.02.009.
Goldstein, S.M. (2015). China and Taiwan. New York: John Wiley & Sons.
Griggs, D., Smith, M. S., Rockström, J., Öhman, M. C., Gaffney, O., Glaser, G., . . . Shyamsundar, P. (2014). An integrated framework for sustainable development goals. *Ecology and society, 19*(4). https://doi.org/10.5751/ES-07082-190449.
Guterres, A. (2022). Secretary-General's remarks to High-Level opening of COP27. https://www.un.org/sg/en/content/sg/speeches/2022-11-07/secretary-generals-remarks-high-level-opening-of-cop27. Accessed 26 February 2023.
Hardt, J. N., & Scheffran, J. (2019). Environmental peacebuilding and climate change: peace and conflict studies at the edge of transformation. *Policy Brief, 68*(Dec 2019), 1–20.
Hauge, W., & Ellingsen, T. (1998). Beyond environmental scarcity: Causal pathways to conflict. *Journal of peace research,* 35(3), 299–317.
Ide, T., Johnson, M. F., Barnett, J., Krampe, F., Le Billon, P., Maertens, L., . . . Vélez-Torres, I. (2022). The Future of Environmental Peace and Conflict Research. *Environmental Politics*, 1–27. https://doi.org/10.1080/09644016.2022.2156174.
International Institute for Sustainable Development. (1992–2021). UN Framework Convention on Climate Change – UNFCCC. Earth Negotiation Bulletin. https://enb.iisd.org/negotiations/un-framework-convention-climate-change-unfccc?field_start_date_time_value=&field_end_date_time_value=&field_participants_target_id=&page=7 Accessed 19 February 2024.
IPCC. (2022a). Climate Change 2022: Impacts, Adaptation and Vulnerability. Summary for Policymakers, Technical Summary and Frequently Asked Questions. https://www.ipcc.ch/report/ar6/wg2/downloads/report/IPCC_AR6_WGII_SummaryVolume.pdf. Accessed 24 January 2022.
IPCC. (2022b). Climate Change 2022: Mitigation of Climate Change. Working Group III Contribution to the Sixth Assessment Report of the Intergovernmental Panel on Climate Change. Summary for Policymakers. https://www.ipcc.ch/report/ar6/wg3/. Accessed 8 August 2023.
IPCC. (2022c). Summary for Policymakers. In H.-O. Pörtner, D.C.E. Roberts, S. Poloczanska, K. Mintenbeck, M. Tignor, A. Alegría, M. Craig, S. Langsdorf, S. Löschke, V. Möller & A. Okem A (Eds.), *Climate Change 2022: Impacts, Adaptation and Vulnerability. Contribution of Working Group II to the Sixth Assessment Report of the Intergovernmental Panel on Climate Change* (pp. 3–33). Cambridge: Cambridge University Press. https://doi.org/10.1017/9781009325844.001.
Kastner, S. L. (2015). Is the Taiwan strait still a flash point? Rethinking the prospects for armed conflict between China and Taiwan. *International Security, 40*(3), 54–92.
Kaufman, D., McKay, N., Routson, C., Erb, M., Dätwyler, C., Sommer, P. S., . . . & Davis, B. (2020). Holocene global mean surface temperature, a multi-method reconstruction approach. *Scientific data, 7*(1), 201. https://doi.org/10.1038/s41597-020-0530-7.

Kiyala, J.C.K. (2022a). Civil society, Climate Change, Security Risks Management and Peacebuilding in the Anthropocene. In J.C.K. Kiyala & G.T. Harris (Eds.), *Civil Society and Peacebuilding in Sub-Saharan Africa in the Anthropocene: An overview* (pp. 67–93). Cham: Springer Nature Switzerland. https://doi.org/10.1007/978-3-030-95179-5_3.

Kiyala, J.C.K. (2022b). Faith-Based Organisations and Environmental Peace: Impact of and Its Relevance in the Anthropocene. In J.C.K. Kiyala & G.T. Harris (Eds.), *Civil Society and Peacebuilding in Sub-Saharan Africa in the Anthropocene* (pp. 95–125). Cham: Springer Nature Switzerland. https://doi.org/10.1007/978-3-030-95179-5_4.

Kosoy, N., Brown, P.G., Bosselmann, K., Duraiappah, A., Mackey, B., Martinez-Alier, J., . . . & Thomson, R. (2012). Pillars for a flourishing Earth: planetary boundaries, economic growth delusion and green economy. *Current Opinion in Environmental Sustainability*, 4(1), 74–79. https://doi.org/10.1016/j.cosust.2012.02.002.

Kyrou, C. N. (2007). Peace ecology: An emerging paradigm in peace studies. *International Journal of Peace Studies*, 73–92.

Lake, I. R., Hooper, L., Abdelhamid, A., Bentham, G., Boxall, A. B., Draper, A., . . . Nichols, G. (2012). Climate change and food security: health impacts in developed countries. *Environmental health perspectives, 120*(11), 1520–1526.

Lesaffer, R. (2012). Peace treaties and the formation of international law. In B. Fassbender & A. Peeter (Eds.), *The Oxford Handbook of the History of International Law* (pp. 71–94). Oxford: Oxford University Press.

Liadze, I., Macchiarelli, C., Mortimer-Lee, P., & Sanchez Juanino, P. (2022). Economic costs of the Russia-Ukraine war. *The World Economy*. https://doi.org/10.1111/twec.13336.

Liang, X., Silva, D. L. d., Tian, N., Béraud-Sudreau, L., & Marksteiner, A. (2023). Trends in World Military Expenditure, 2021. *Stockholm International Peace Research Institute (SIPRI), Apr, 2022*. https://www.sipri.org/sites/default/files/2022-04/fs_2204_milex_2021_0.pdf. Accessed 17 February 2013.

Liwanga, R.-C. (2015). The Meaning of Gross Violation of Human Rights: A Focus on International Tribunals' Decisions over the DRC Conflicts. *Denv. J. Int'l L. & Pol'y, 44*, 67.

Mann, M.E. (2021). *The new climate war: The fight to take back our planet*. New York: Public Affairs.

Mastro, O.S. (2018). Conflict and Chaos on the Korean Peninsula: Can China's Military Help Secure North Korea's Nuclear Weapons? *International Security, 43*(2), 84–116.

Maystadt, J.-F., De Luca, G., Sekeris, P.G., & Ulimwengu, J. (2014). Mineral resources and conflicts in DRC: a case of ecological fallacy? *Oxford Economic Papers, 66*(3), 721–749.

Mazarr, M.J., Gentile G., Madden D, Pettyjohn S.L., & Crane, Y.K. (2018) The Korean Peninsula: Three dangerous scenarios. https://www.jstor.org/stable/pdf/resrep19919.pdf. Accessed 24 July 2023.

McCright, A.M., & Dunlap, R.E. (2011). Cool dudes: The denial of climate change among conservative white males in the United States. *Global environmental change, 21*(4), 1163–1172. https://doi.org/10.1016/j.gloenvcha.2011.06.003.

Mehta, S., & Merz, P. (2015). Ecocide–a new crime against peace? *Environmental Law Review, 17*(1), 3–7. https://doi.org/10.1177/1461452914564730.

Milton-Edwards, B., & Hinchcliffe, P. (2007). *Conflicts in the Middle East since 1945*: London: Routledge Taylor & Francis Group.

Mohai, P., Pellow, D., & Roberts, J.T. (2009). Environmental justice. *Annual review of environment and resources, 34*, 405–430. https://doi.org/10.1146/annurev-environ-082508-094348.

Molvaer, R.K. (1991). Environmentally induced conflicts? A discussion based on studies from the Horn of Africa. *Bulletin of Peace Proposals, 22*(2), 175–188.

Navas, G., Mingorria, S., & Aguilar-González, B. (2018). Violence in environmental conflicts: the need for a multidimensional approach. *Sustainability Science, 13*(3), 649–660. https://doi.org/10.1007/s11625-018-0551-8.

Nuclear winter. (2023). In The Editors of Encyclopaedia Britannica (Ed.).

O'Neill, D.W., Fanning, A.L., Lamb, W.F., & Steinberger, J.K. Kiyala (2018). A good life for all within planetary boundaries. *Nature sustainability, 1*(2), 88–95. https://doi.org/10.1038/s41893-018-0021-4.

Papa Paulo VI. (1967). Populorum progressio. *Revista de la Facultad de Derecho y Ciencias Políticas* (41–43), 3–26.

Papas, M. (2017). The 2030 Sustainable Development Agenda and the Paris Climate Agreement—taking urgent action to combat climate change: how is Australia likely to fare? *Asia Pacific Journal of Environmental Law, 20*(1), 94–114.

Patriarch Bartholomew (2021) Message of His All Holiness, Patriarch Bartholomew, for the day of prayer for the protection of the Environment (01/09/2016) Laudato Si Movement 2021 https://orthodoxia.be/en/2020/08/31/message-of-his-all-holiness-patriarch-bartholomew-for-the-day-of-prayer-for-the-protection-of-the-environment-01-09-2020/. Accessed 19 February 2024.

Percival, V., & Homer-Dixon, T. (1998). Environmental scarcity and violent conflict: the case of South Africa. *Journal of Peace research, 35*(3), 279–298.

Pham-Duc, B., Sylvestre, F., Papa, F., Frappart, F., Bouchez, C., & Crétaux, J.-F. (2020). The Lake Chad hydrology under current climate change. *Scientific reports, 10*(1), 1–10.

Pontifical Council for Justice and Peace (2004) Compendium of The Social Doctrine of The Church. To His Holiness Pope John Paul II, Master of Social Doctrine and Evangelical Witness To Justice and Peace. Rome: Libreria Editrice Vaticana. https://www.vatican.va/roman_curia/pontifical_councils/justpeace/documents/rc_pc_justpeace_doc_20060526_compendio-dott-soc_en.html. Accessed 10 February 2023.

Pope Francis (Cartographer) (2015) Encyclical Letter Laudato Si' of The Holy Father Francis on Care for Our Common Home [Online]. https://www.vatican.va/content/dam/francesco/pdf/encyclicals/documents/papa-francesco_20150524_enciclica-laudato-si_en.pdf. Accessed 14 January 2023.

Pope Paul VI (1965) Address of The Holy Father Paul VI to the United Nations Organization. https://www.vatican.va/content/paul-vi/en/speeches/1965/documents/hf_p-vi_spe_19651004_united-nations.html. Accessed 15 February 2023.

Rawtani, D., Gupta, G., Khatri, N., Rao, P.K., & Hussain, C.M. (2022). Environmental damages due to war in Ukraine: A perspective. *Science of The Total Environment, 850*, 157932. https://doi.org/10.1016/j.scitotenv.2022.157932.

Robock, A. (2010). Nuclear winter. *Wiley Interdisciplinary Reviews: Climate Change, 1*(3), 418–427. https://doi.org/10.1002/wcc.45.

Rockström, J. (Producer) (2019) Safe Future for Humanity on Earth. This is a Session at Re:publica19. The Thirteen Edition of the Conference (some call it a festival) took place on May 6, 7 and at Station Berlin. https://www.youtube.com/watch?v=ZmZ3Hp7rPo4. Accessed 8 Ausgust 2023.

Rockström, J., Steffen, W., Noone, K., Persson, Å., Chapin III, F. S., Lambin, E., . . . Schellnhuber, H. J. (2009). Planetary boundaries: exploring the safe operating space for humanity. *Ecology and society, 14*(2). https://doi.org/10.1126/science.1259855. originally published online 15 January 2015.

Ros, J.P. (1996–2023) Freshwater. https://www.nationalgeographic.org/education/freshwater/#:~:text=Freshwater. Accessed 10 February 2023.

Rothschild, R. (2012). Environmental Awareness in the Atomic Age: Radioecologists and nuclear technology. *Historical Studies in the Natural Sciences, 43*(4), 492–530.

Ruddiman, W. F. (2013). The anthropocene. *Annual Review of Earth and Planetary Sciences, 41*, 45–68. https://doi.org/10.1146/annurev-earth-050212-123944.

Sakaguchi, K., Varughese, A., & Auld, G. (2017). Climate wars? A systematic review of empirical analyses on the links between climate change and violent conflict. *International Studies Review, 19*(4), 622–645.

Seddon, N., Mace, G.M., Naeem, S., Tobias, J.A., Pigot, A.L., Cavanagh, R., . . . Walpole, M. (2016). Biodiversity in the Anthropocene: prospects and policy. *Proceedings of the Royal Society B: Biological Sciences, 283*(1844), 20162094. https://doi.org/10.1098/rspb.2016.2094.
Smith, D. (2011). *The Holocene Epoch*. https://ucmp.berkeley.edu/quaternary/holocene.php. Accessed 19 February 2024.
Steffen, W., Richardson, K., Rockström, J., Cornell, S. E., Fetzer, I., Bennett, E. M., . . . De Wit, C. A. (2015). Planetary boundaries: Guiding human development on a changing planet. *Science, 347*(6223), 736–747.
Steger, M. B. (2001). Peacebuilding and nonviolence: Gandhi's perspective on power. *Peace, conflict, and violence: Peace psychology for the 21st century*, 314–323.
Stockholm Resilience Centre. (2022). Planetary boundaries. https://www.stockholmresilience.org/research/planetary-boundaries.html. Accessed 30 January 2023.
The Ecumenical Patriarchate of Constantinople. (2022). Environmental justice and peace: Quotes by His All-Holiness Ecumenical Patriarch Bartholomew. https://www.orthodoxcouncil.org/-the-green-patriarch-.html. Accessed 15 March 2022.
The Heartland Institute. (2007). Do people cause global warming? The Heartland Institute's newsletter/blog "The Heartlander", 1 December 2001, Archived 29 September 2012. Retrieved 1 January 2024 from https://archive.is/fUxPU. Accessed 3 February 2024.
The World Bank. (2022). World Bank Group delivers record $31.7 billion in climate finance in fiscal year 2022. https://www.worldbank.org/en/news/press-release/2022/09/07/world-bank-group-delivers-record-31-7-billion-in-climate-finance-in-fiscal-year-2022. Accessed 17 February 2023.
Thunberg, G. (2019) Our house is on fire. The Guardian. https://www.theguardian.com/books/2020/mar/04/our-house-is-on-fire-review. Accessed 31 July 2021.
United Nations. (2005). Report of the World Commission on Environment and Development: *Our Common Future*. Transmitted to the General Assembly as an Annex to document A/42/427 - Development and International Co-operation: Environment. http://www.un-documents.net/wced-ocf.htm. Accessed 23 January 2023.
United Nations. (2006). *United Nations framework convention on climate change: Handbook*. UNFCCC, Bonn.
United Nations. (2012a). Future we want - Outcome document. https://sustainabledevelopment.un.org/rio20/futurewewant. Accessed 13 January 2023.
United Nations. (2012b). The future we want: Outcome document of the United Nations Conference on Sustainable Development, Rio de Janeiro, Brazil, 20–22 June 2012. https://sustainabledevelopment.un.org/content/documents/733FutureWeWant.pdf. Accessed 13 January 2023.
United Nations. (2015a). The Paris Agreement. https://unfccc.int/sites/default/files/english_paris_agreement.pdf. Accessed 23 March 2023.
United Nations. (2015b). Transforming our world: The 2030 agenda for sustainable development. A/RES/70/1. https://sustainabledevelopment.un.org/content/documents/21252030%20Agenda%20for%20Sustainable%20Development%20web.pdf. Accessed 8 January 2022.
United Nations. (2015c). Transforming our world: The 2030 Agenda for Sustainable Development. Resolution adopted by the General Assembly on 25 Sep 2015 [without reference to a Main Committee (A/70/L.1)] 70/1. https://www.un.org/en/development/desa/population/migration/generalassembly/docs/globalcompact/A_RES_70_1_E.pdf. Accessed 20 April 2021.
United Nations. (2022). What is the United Nations framework convention on climate change? https://unfccc.int/process-and-meetings/what-is-the-united-nations-framework-convention-on-climate-change. Accessed 20 February 2023.
United Nations Framework Convention on Climate Change. (2022). What is the United Nations Framework Convention on Climate Change? https://unfccc.int/process-and-meetings/what-is-the-united-nations-framework-convention-on-climate-change. Accessed 20 February 2023.
United Nations Office for Disarmament Affairs (2021) Treaty on the prohibition of nuclear weapons. https://www.un.org/disarmament/wmd/nuclear/tpnw/. Accessed 23 February 2023.

University of California Regents. (2023). Freshwater use. https://ugc.berkeley.edu/background-content/freshwater-use/. Accessed 10 February 2023.
Wallis, J. (1982). *Waging peace: a handbook for the struggle to abolish nuclear weapons.* San Francisco & Sydney: Harper & Rwo, Publishers, Inc.
Wang-Erlandsson, L., Tobian, A., van der Ent, R. J., Fetzer, I., te Wierik, S., Porkka, M., . . . Singh, C. (2022). A planetary boundary for green water. *Nature Reviews Earth & Environment, 3*(6), 380–392.
Water Science School. (2019). pH and Water. https://www.usgs.gov/special-topics/water-science-school/science/ph-and-water. Accessed 10 February 2023.
Welzer, H. (2012). *Climate Wars: what people will be killed for in the Twenty-First century.* Cambridge: Polity Press.
Westing, A. H. (1981). Environmental impact of nuclear warfare. *Environmental Conservation, 8*(4), 269–273.
Westra, L. (2009). *Environmental justice and the rights of ecological refugees* (1st Ed.). London: Routledge https://doi.org/10.4324/9781849770088.
Worldometer (2023) World population. https://www.worldometers.info/world-population/. Accessed 24 January 2023.

Chapter 3
The Impacts of Climate Change on Water Resources in the Anthropocene: Mitigation and Adaptation Strategies in Southern Africa

Rookmoney Thakur

Abstract This chapter explores the consequences of climate change on water resources in Southern Africa during the Anthropocene era, focusing on mitigation and adaptation strategies. The region's vulnerable water systems are significantly affected by the changing climate, resulting in increased variability, extreme weather events, and water scarcity. These impacts exacerbate existing water challenges, such as limited access, inadequate infrastructure, and inefficient management practices. Consequently, both human and natural systems face severe socio-economic and ecological consequences. The depletion of water resources in Sub-Saharan Africa (SSA) as a result of climate change has a variety of negative implications, including increased food insecurity, transborder disputes, and an increase in health problems, among other socio-economic factors. This reality appears to be exacerbated by the increasing pressure on anthropogenic causes and consequences of global warming in the Anthropocene. To minimise the vulnerability of communities in Southern Africa, it is vital to rethink strategies that aim at mitigating the impacts of climate change on water resources. Thus, climate change adaptation is paramount and vital. This requires support and resources to diversify the livelihoods of the people in the Southern African region, particularly Angola, Botswana, Zimbabwe, South Africa, and Mozambique, as these countries are the most susceptible to the negative impacts of climate change. The purpose of the current review is to assess viable alternative solutions to customise interventions for these regions. A review of the literature highlights various strategies that can be employed to minimise local community vulnerability and build resilience. As many of the impacts of these strategies are subjective in nature, this article recommends an integrated approach, such as community-based disaster risk reduction (DRR) and climate change adaptation (CCA) programs, to

Dr Rookmoney Thakur is a Postdoctoral Fellow at the International Centre of Nonviolence, Faculty of Management Sciences, Durban University of Technology. Her area of research includes water sustainability and conservation through behavioural change. Email: RookmoneyT@dut.ac.za; http://orcid.org/0000-0003-0102-3151

J. C. K. Kiyala and N. Chivasa (eds.), *Climate Change and Socio-political Violence in Sub-Saharan Africa in the Anthropocene*, The Anthropocene: Politik—Economics—Society—Science 37,
https://doi.org/10.1007/978-3-031-48375-2_3

improve the efficacy of aid and minimise the uncertainty experienced by communities in the nations mentioned here.

Keywords Anthropogenic · Sub-Saharan Africa · Water resources · Climate change adaptation

3.1 Introduction

In today's world, it is widely acknowledged that climate change is a reality. Even in the most hopeful scenarios of concerted worldwide action to decrease future emissions of greenhouse gases into the atmosphere, an age of fast and accelerating climate change is unavoidable. For example, it is predicated that through the course of the twenty-first century, temperatures in sub-Saharan Africa will rise faster than those throughout the world, with an increase in average surface air temperatures of 1.4–5.8 °C by the year 2100 (Masson-Delmotte et al., 2021). While mean land surface warming in southern Africa is projected to exceed the global mean land surface temperature increase in all seasons, temperature projections over west Africa for the end of the twenty-first century from both the Coupled Model Intercomparison Project Phase 3 (CMIP 3) General Circulation Models (GCMs) and CMIP 5 GCMs range between 3 and 6 °C above the late twentieth-century baseline (ibid). In such conditions, the stress and lack of water in these areas grows wider. As such, water will be the resource most impacted by climate change as a result of a decline in both quantity and quality (Bai et al., 2019).

African countries are most likely to experience the most severe effects of climate change due to their geographic location and poorer communities. They have limited institutional and technological capacity to adapt to rapid changes in the environment. More so, they also have greater reliance on climate-sensitive renewable natural resource sectors like agriculture and water (Benevolenza/DeRigne, 2019). Up to 250 million Africans are predicted to endure water stress as a result of climate change by 2020 (Leal-Filho et al., 2022). In the latter half of the 20th century, there were already reports of a decline in late austral summer precipitation over southern Africa's western areas, from Namibia to Angola and towards the Congo (van Schalkwyk et al., 2022). Furthermore, it was found that the Sub-Saharan area of Africa is the most susceptible to the consequences of a changing climate (Ringler et al., 2010). This is due to the region's relatively low adaptive capacity, which is linked to severe poverty and the inadequate infrastructure necessary to either mitigate or adapt to the consequences of climate change (Ringler et al., 2010). The ability of the people living in Sub-Saharan Africa (SSA) to meet their requirements for water, food, and energy is highly dependent on the amount of precipitation that the region receives (Ofori et al., 2021). For instance, the region's food supply is very vulnerable to changes in the pattern of precipitation because the majority of the region's agricultural production depends on rainwater harvesting (Ofori et al., 2021).

The current review assumes that SSA is more affected by climate change than any other sub-region in the world because of its reliance on crop-livestock and agro-pastoral production systems that are highly vulnerable to climate change, as well as the poor capacity of SSA nations and populations for adaptation (Barnard et al. 2014; Mensah 2014; Kotir 2011; Nelson et al. 2009). The low adaptive capacities of these countries and communities are associated with increasing water scarcity and land degradation, among other externalities (Wetende et al., 2018). As such, identifying and understanding viable adaptation solutions and alternatives in such contexts is urgent—it would enable targeted interventions.

3.2 Climate Change and Its Impact on Water Management

3.2.1 Background

The effects of climate change are being seen and will continue to be felt through the medium of water, notably in Africa, where global warming threatens water resources (Ziervogel et al., 2010). Anthropogenic and biological forces disrupt the water cycle's natural flow. Human activities that affect the water cycle include afforestation, deforestation, agriculture, amongst others. A rise in greenhouse gas emissions from human activity might affect cloud formation (Antonarakis, 2018). According to research, water shortage, water pollution, and global climate change threaten human water supplies (Stolberg et al., 2003). It can therefore be inferred that human actions will further worsen climate change, especially in areas facing water shortages.

When the climate is unpredictable, managing water resources may be challenging. In the most of sub-Saharan Africa, adapting to the effects of climate change requires sustainable water management strategies. Sustainable water management, however, demands responsive and adaptive institutions. Hence, when formulating solutions to climate change that aim to decrease risk, it is vital to consider the practicability and efficiency of potential adaption strategies. This chapter conducted a narrative review of the viability of adaption alternatives for countries within SSA, such Angola, Botswana, Zimbabwe, South Africa and Mozambique. Keywords, such as climate adaptation"; "climate risk management"; "water sustainability in sub-Saharan Africa" were used to search for the relevant literature on the widely used academic search engines, such as Google Scholar and Science Direct. The review was restricted to Angola, Botswana, Zimbabwe, South Africa, and Mozambique, as these countries are the most susceptible to the negative impacts of climate change.

3.2.2 *Sub-Saharan African Region*

It is anticipated that the threat of drought would intensify over most of SSA all throughout twenty-first century as a result of climate change. It would appear that fluctuations in sea surface temperatures, notably in the tropics, as well as associated changes in air circulation and precipitation are the primary drivers of droughts. Since the late 1960s, the effects of droughts have left SSA with long-lasting and harmful consequences. Between 1973 and 1984, almost the whole African continent had lower levels of precipitation, which contributed to the severity of the droughts that occurred during that time. During the same time period, the Horn of Africa, the Sahel, and Southern Africa had the worst drought scenarios, which affected millions of people (Gan et al., 2016). In addition to the Sahel region of western and northern Africa, Southern Africa also frequently experiences droughts. The quality of the water that is available has also been impacted by climate change. As a result of the warming of the global climate in recent decades, significant deviations have been seen in the world's oceans, lakes, and rivers. These changes have had ramifications for the world's freshwater ecosystems, including shifts in the amounts of salt and nutrients in the water (Miller/Hutchins, 2017). According to the aforementioned study, an increase in temperature of 1.5–2.0 °C will have a detrimental influence on fisheries and aquaculture in West African lakes. Changes in major biological parameters including temperature, salinity, ocean currents, and the strength of upwelling might have a significant impact on the abundance, distribution, and availability of fishing resources. These changes are likely to be induced by climate change.

The absence of effective governmental oversight, and consequently, the crisis in governance, may frequently be linked back to the root cause of a looming crisis in water supply (Bakker, 2010). According to the report that was compiled and distributed by the IPCC in the year 2001, the population that was at danger of facing water shortages rapidly rose as temperatures continued to rise over the second half of the century. In a similar line, study carried out by Parry et al. (2001) that assessed climate models shown that the consequences in dry and semi-arid regions are expected to be bigger than the averages reported worldwide.

Other significant factors that make SSA nations particularly vulnerable to climate change is the process of desertification, falling run-off from water catchments, declining soil fertility, dependency on subsistence farming, prevalence of AIDS and vector-borne illnesses, limited governance systems, and rapid population expansion (Mustapha, 2012). Even though Africa contributes very little to global greenhouse gas emissions, climate change is already having an impact on the majority of SSA countries' water supply. 1.6% of the world's greenhouse gas emissions come from SSA (Hassan/Tularam, 2018). Therefore, shedding greater insight on the relationship between water resources in Africa and climate change will help SSA in adopting sustainable water resource management.

Scientists and politicians have realised in the past decade that climate change affects water, land, and food resources in complex ways. This predicament demands for a nexus approach that considers the interconnectedness, synergy, and trade-offs between climate change and water, land, and food resources, which were once handled separately. Studies have endeavoured to demonstrate the climate change and WLF security nexus in Africa, but efforts to realise it in policies are less prominent in SSA.

In recent years, climate change and variability have been seen as the main cause for concern, particularly when connected to economic growth and human livelihoods (Sintayehu, 2018). Indeed, the planning process for economic growth may be significantly impacted by climate change (Hughes, 2017), which may also have an impact on community livelihoods and disturb people's ability to go about their daily lives as usual (Artur/Hilhorst, 2012). Following this, efforts have been made globally to comprehend how people and communities would be impacted by the predicted patterns in climate change (Cavicchioli, 2019).

From a continental perspective, Africa is expected to warm 3–4 °C more than the world average during the next century (Brooks et al., 2020). The variations in precipitation projected on the African continent by the end of the 21st century will be unique across the different areas (Niang et al., 2014). The West African area will have little or no change in heavy precipitation, but Eastern and Central Africa will see an increase (Niang et al., 2014; Girvetz et al., 2019). For Southern and Northern Africa, overall precipitation is expected to decrease, but Southern Africa's heavy precipitation is expected to increase (Brooks et al., 2020).

The 2004 International Dialogue on Water and Climate showed that SSA will face rising water stress. About 25% of the African population confronts water stress, whereas 69% enjoys relative water excess (Kadigi, 2019). In the SSA nations, 1100 million people depend on questionable sources of water, resulting in 5 million deaths each year (UNICEF World Health Organization, 2002).

Therefore, regions that are already experiencing water stress now, such asAfrica, will see their position become even more dire as a result of climate change.

For instance, fewer than 1.6% of global greenhouse gas emissions come from sub-Saharan Africa (Spore, 2008). Therefore, better understanding the relationship between climate change and Africa's water supplies will shed light on the problem and point the way toward sustainable water management in the region's Southern and Eastern regions. Some of the potential effects of climate change on SSA's water resources are summarised here. In the past, floodplains were favourable locations for establishing communities, cultivating land, and transporting goods and people. These once-dry floodplains have become podiums prone to flooding in the contemporary era.

Most of the Southern African Development Community (SSA) is at risk from flooding, as is much of Africa. Floods are most common in the East, South, and Central regions of the continent, and then move westward to West Africa (Osbahr et al., 2008). Urama et al. (2010) report that Burkina Faso, Chad, Ethiopia, Ghana, Senegal, Sudan, Togo, Kenya, Liberia, Mali, Niger, Rwanda, Nigeria, and Uganda are among the most hit nations in sub-Saharan Africa due to floods. On the same level as these two countries are Togo and Kenya, with Liberia rounding out the top 5.

Numerous lives have been lost, and hundreds of thousands of people have been forced to leave their homes because of the floods. This has repercussions for human settlements, agricultural land, and other ecological services. Similarly, between 1971 and 2001, floods accounted for 26% of all natural catastrophes in Africa (Nagarajan, 2010), with deadly results. In East Africa, El Nio was responsible for floods that year, causing around $1.8 billion in damage to Kenya's infrastructure and property.

The Mozambican flood of 2000 killed 800 people, affected about 2 million (of whom about 1 million needed food), displaced 329,000, and devastated vast areas of cropland, among other things. The yearly economic growth rate also dropped from 10 to 4% due to the flood.

In East Africa, the worst single flood occurrences in Southern and Eastern Africa have occurred, including one in 1997 that killed 2311 people in Somalia and another in 1999 that injured 1.8 million people in the Sudan. East Africa was the scene of each of these occurrences (Urama et al., 2010). Apparently the flooding in SSA is the worst it's seen in decades in certain areas. The path of the floodwaters is an arc that stretches from Kenya to Mauritania. It is estimated that at least 1.5 million people have been impacted at this stage (Ribeiro et al., 2009).

3.3 Climate Change Adaptation

Climate Change Adaptation (CCA) strategies in Africa are on the increase. A number of strategies have seen the light in aid of reducing disaster risk and adapting to climate change (Busayo/Kalumba, 2021). The DRR and CCA processes include the mapping of location and extent of known and potential hazards, vulnerable communities and environments, and opportunities that may exist to manage these risks. However, the mapping of often fast-changing urban and rural spaces in a standardised manner presents challenges that relate to processes, scales of data capture, level of detail recorded, software and compatibility related to data formats and net-works, human resources skills and understanding, as well as differences in approaches to the nature in which the map-ping processes are executed and spatial data is managed.

3.3.1 Angola

In light of the fact that Angola is extremely susceptible to the effects of climate change, there are signs that the country is beginning to place a greater emphasis on both the study of climate change and the preparation of strategies to adapt to its effects. However, there is a pressing need to increase awareness of the crucial connection that exists between climate change, urban land-use planning, and the administration of water resources (Cain, 2017).

If urban design is backed with credible climatic risk and flood data that can be delivered to water managers, it can significantly minimise the susceptibility of communities to the effects of climate variability and water-based natural catastrophes. Involving communities in planning strategies that prevent settlements from being located in vulnerable areas and developing corrective actions for settlements that are already situated in these areas is one way to improve resilience (Luetkemeier/Liehr, 2019). Another way to improve resilience is to involve communities in planning strategies processes that will prevent settlements from being located in vulnerable areas (Limones et al., 2020).

Research conducted by Development Workshop and presented to the Ministry of Water and Energy (MINEA22) revealed that the Water for All program failed to put in place sustainable water systems in many communities across the country (Cain, 2017). These systems were intended to guarantee true 'access' to water over the course of the long term. Only 48% of the pumping and water treatment systems that were installed were operational at any given time. This study also discovered that the local capacity for operation and maintenance of water systems is very low, that there is almost no collection of user fees, and that both local administrations and communities have a limited sense of local ownership of the water systems and, as a result, do not feel any responsibility for maintaining them. These studies came to a number of important conclusions, one of the most important being that of inadequate "water governance." One of the main findings of these studies was that technological solutions were rarely discussed with the communities, and they typically do not respond to the preferences of users.

Adaption is one of the five pillars of the climate change strategy in Angola (Luetkemeier/Liehr, 2019). The programmes on agriculture, coastal management and water resources mainly focus on early warning signs related to flood and drought management. The programme on health focuses on analysis of risk of epidemics, early warning and public awareness. The programme on infrastructure includes actions on risk mapping, particularly in relation to floods. The Civil Protection is systematically mentioned as a partner for the actions related to early warning.

3.3.2 Botswana

Climate change alone does not just generate vulnerability; it also has an impact when it interacts with other environmental, social, and economic factors. The severity of the strain on key GDP drivers including agriculture, water, wildlife, and ecosystems is an indication of how susceptible Botswana is to the consequences of climate change on all facets of its social and economic life (Bhaga et al., 2020). Botswana's susceptibility may very well be deduced from the severity of the stress since all these components are climate-dependent. It is projected that the current level of stress will rise due to the increasing unpredictability of precipitation, which has negative effects on regions' water resources and quality and increases the danger of floods. Extreme temperatures are becoming more common, and in certain regions, new environmental effects and changes in the state of people's health are also starting to be noticed (Nkemelang et al., 2018).

Droughts have been seen to become more frequent and severe, which is worsening poverty rates, especially among local producers who rely on rainwater for their livelihoods. The majority of the people are small-scale farmers who are unable to cultivate crops and animals in the traditional form of food production. The number of individuals who depend on social government grants operated by the government with the intention of eradicating poverty has increased as a result of the above-mentioned effect of climate change (Bhaga et al., 2020).

Additionally, agricultural production has been greatly impacted by both a lack of rainfall and changing patterns of rainfall (Mogomotsi et al., 2020). This has led to a decline in fertile farming land as a consequence of the unpredictable nature of rainfall, which has led to a drop in the total amount of land that has been cultivated. It was noted that whenever it does rain in Botswana, rather than having 'soft' showers that allow for conditioned soil moisture, the nation currently sees torrential rains that flush away soil surface and seedlings, which consequently results in low crop yield. This is a concern. As such, more resources are required for mitigating the effects of the harsh rainfall.

In response to the unavoidable economic impacts of climate change, the Government of Botswana saw the need to treat climate change adaptation as a development issue and incorporate it into its development planning process. As a result, this National Adaptation Plan (NAP) Framework was created to handle adaptation holistically and completely. NAP proposes feasible institutional arrangements based on existing governmental institutions that will improve effort coordination through vertical and horizontal integration throughout NAP process implementation (Mogomotsi et al., 2020).

Therefore, the process of developing Botswana's NAP includes a significant focus on the mobilisation of financial and other resources. The Government of Bermuda has, over the course of many years, committed a sizeable portion of its yearly budget to the process of adapting to the effects of climate change, such as droughts and disease outbreaks (Cavicchiol, 2019). Drought relief projects, water transfer networks, and health-related responses to disease outbreaks and prevention have received national support. Since the government considers adaptation as a development issue, the NAP Framework makes it apparent that ongoing state investment is needed.

3.3.3 *Zimbabwe*

The severe lack of available water is a huge challenge to the expansion of the economy in Zimbabwe, a nation that is predominately semi-arid with limited water resources (Chingombe/Musarandega, 2021). In addition, it has become an emotionally charged topic. Over the past decade, the local and international media have shown a wide array of issues relating to water, such as a lack of supply, malfunctions in wastewater services, and declining groundwater levels. These issues have all been at the forefront of the public's consciousness as the erratic weather patterns have had a detrimental effect on water supplies in both rural and urban areas, food security, energy generation, and industrial productivity. Damage to property, livelihoods, and lives can be inflicted by flooding, whether it be caused by cyclones or repeated floods. In addition, the country has had to deal with outbreaks of cholera, typhoid, and other diseases that are spread by water, and the contamination of both surface water and ground water has been a significant problem (Chanza/Gundu-Jakarasi, 2020). More so, the majority of people in Zimbabwe are employed in agriculture, which is rain-fed and supports around 80 percent of the country's population.

In order for there to be economic revival, it is imperative that water resources be developed and managed in a sustainable manner. This is due to the intimate link that exists between water and the key economic sectors that the Zimbabwe has. In response to the effects of climate change, the Government and the World Bank Group are working together to address climate-related risks and the impact those risks will have on the supply of water in the country (Dodman/Mitlin, 2015). This is in the form of adopting sustainable water development and management methods as part of the country's National Climate Change Response Strategy. This is aimed to allow the country to better adapt to the effects of climate change.

3.3.4 South Africa

Rainfall in South Africa is very unpredictable in terms of its spatial distribution, and it is impossible to predict with any degree of accuracy whether it will fall in one year or another (Serumaga-Zake, 2020). Water is therefore viewed as a resource that restricts growth in this country, and a change in the amount of water available could have a big impact on most of the economic sectors. One of the things that makes South Africa's water resources vulnerable is seasonal and interannual variations in rainfall, which are accentuated by high runoff and evaporation rates. The use of groundwater is another aspect that makes this vulnerability more likely. As a direct result of the drought, eleven of the country's nineteen Water Management Areas (WMAs) experienced water shortages in 2004.

Climate change is expected to have an impact on the hydrological resources that are now accessible in Southern Africa, putting additional strain on the flexibility of future water supply. Despite the fact that there are numerous uncertainties surrounding rainfall projections, the majority of models suggest that rainfall will decrease across the western part of southern Africa in the coming decades. When combined with an increase in temperature, this predicts an overall drying trend that will harm both agricultural production and water supplies (Garrote, 2017).

Furthermore, the Apartheid's legacy has left South Africa with the massive job of providing sustainable water delivery services to its inhabitants. The bulk of people who do not have access to basic domestic water services live in rural areas. According to the 2001 census, 84% of South Africans have piped water, with 32% receiving it straight into their houses.

3.3.5 Mozambique

Mozambique is bordered by Tanzania, eSwatini, Malawi, Zambia, Zimbabwe, and South Africa. It faces east toward the island of Madagascar and has a coastline that is 2500 kilometres long. By 2020, there will be over 31 million people living and working in the country, which means that over two thirds of the population will do so (Cabral et al., 2017). It has a richness of arable land, water, and energy, three deep seaports, and a rather sizable prospective labour pool, in addition to recently found natural gas and mineral treasures offshore. In Mozambique, where it still has a significant impact on the nation, agriculture is one of the most significant economic sectors.

However, the country is prone to experiencing large levels of climate change and often occurring, extremely severe weather events such as droughts, floods, and tropical cyclones. In the last 20 years, the frequency of extreme weather occurrences including storms, floods, and droughts has increased across the country. These natural calamities have frequently had a catastrophic effect on the country's socioeconomic system (Moorthy et al., 2018). It is predicted that temperatures will rise, precipitation patterns will alter, and extreme weather conditions like droughts would happen more frequently and be more intense. These climatic pressures might make both climate strains and non-climate stresses worse, which could have an impact on development objectives including resource management, public health, economic growth, and food security. It's possible that environmental stresses will have an autonomous impact.

The most frequent natural disasters, occurring every three to four years on average, are droughts. This is due to the fact that the majority of the population, particularly the poor, resides in rural areas and is reliant on agriculture that is irrigated by rain. This is a significant barrier to the nation's economy's ability to continue expanding and growing. The agricultural industry and the nation's infrastructure in Mozambique's deltas are always at risk from flooding, which is even worse when tropical storms are present. Mozambique has already begun to make financial investments to improve the effectiveness of its early warning systems and better protect its citizens from the destruction that potential natural disasters may wreak. Adaptation strategies are now being implemented across a number of industries, with a particular focus on coastal areas and the prevention of erosion. These industries include farming, aquaculture, power, the environment, and water (Broto et al., 2015).

The Mozambican government and foreign funders have launched projects to identify the most urgent vulnerability and adaptation concerns and to incorporate this knowledge into planning for both overall development and specific sectors in light of the rising challenges associated with climate change. With the help of these activities, the most urgent issues relating to vulnerability and adaptability will be identified, and this information will be used to inform strategic planning for the product market and the development process. However, despite this, there are still a number of obstacles to be overcome, including the creation of sufficient early warning systems, the improvement of planning capacities at the national and local levels, and the elimination of capacity and financial restrictions that impede the implementation of adaptation-related actions.

The National Climate Change Strategy seeks to make Mozambicans less vulnerable to climate change and to improve their quality of life (Broto et al., 2015). It suggests strategies for disaster risk reduction and climate change adaptation while also emphasising mitigation by concentrating on low carbon development.

3.4 Conclusion

African countries are becoming more aware of the urgent need for choices for addressing climate change that are based on facts. Africa features numerous populations and ecosystems that are among the most susceptible to climate change, confronts above-average increases in exposure to climate risks. In several African nations, climate variability and change, particularly catastrophic occurrences, have already had a detrimental effect on economic development, agricultural production, water security, and human health. In order to support planning for adaptation to climate change, information on the viability of a wide range of adaptation solutions across many sectors is required.

Because adaptation plans have not yet been formed to plan for the consequences that are predicted in the future, global warming is not yet considered to be a serious actual threat to the decrease of the current available water resources. Despite the climate unpredictability that is taking place, the mechanisms and regulations that are currently in place for water management have been created to make certain that the current supply of water is sufficient to meet the expanding demand.

Effective long-term plans are necessary, nevertheless, to ensure that the demand for water is satiated even during times of reduced availability. This study's framework provides a tool that may be used to screen for workable long-term approaches that also achieve the goals of local development. These plans must be put into action using a range of resources, notably financial and human capital and political support. Both the local government and the national government should explore policies and initiatives which would tackle the low concentrations of human and financial resources at the municipal level.

The idea of incorporating and instituting adaptability into development operations has to be carefully considered by planners in South Africa as a result of climate change. It shouldn't ever be an afterthought or an addition. There is currently very little proof for this. It is advised that long-term strategies be developed to satisfy local development's needs even while tackling the problems of managing water resources. This will guarantee that, notwithstanding the uncertainty of future climate forecasts, local governments have the adaptive resilience required to guarantee that the people they serve have enough clean water to satisfy their developmental requirements. This will guarantee that local authorities have the adaptable resilience needed to guarantee the people they serve have enough clean water.

Therefore, initiatives for combating climate change may play a vital role in fostering the sustainable growth of the economy and society. It is obvious that strategies that address both the country's development goals and climatic variability and change are in the national interest, particularly if they also aim to reduce poverty and create jobs.

References

Antonarakis, A., (2018). Linking carbon and water cycles with forests. *Geography, 103*(1), 4–11.

Bai, Y., Ochuodho, T.O., & Yang, J., (2019). Impact of land use and climate change on water-related ecosystem services in Kentucky, USA. *Ecological Indicators, 102*, 51–64.

Benevolenza, M.A., & DeRigne, L., (2019). The impact of climate change and natural disasters on vulnerable populations: A systematic review of literature. *Journal of Human Behavior in the Social Environment, 29*(2), 266–281.

Bhaga, T.D., Dube, T., Shekede, M.D., & Shoko, C. (2020). Impacts of climate variability and drought on surface water resources in Sub-Saharan Africa using remote sensing: A review. *Remote Sensing, 12*(24), p.4184.

Brooks, N., Clarke, J., Ngaruiya, G.W., Wangui, E.E. (2020). African heritage in a changing climate. *Azania: Archaeological Research in Africa, 55*(3), 297–328.

Broto, V.C., Boyd, E., & Ensor, J. (2015). Participatory urban planning for climate change adaptation in coastal cities: lessons from a pilot experience in Maputo, Mozambique. *Current Opinion in Environmental Sustainability, 13*, 11–18.

Busayo, E.T., & Kalumba, A.M. (2021). Recommendations for linking climate change adaptation and disaster risk reduction in urban coastal zones: Lessons from East London, South Africa. *Ocean & Coastal Management, 203*, p.105454.

Cabral, P., Augusto, G., Akande, A., Costa, A., Amade, N., Niquisse, S., Atumane, A., Cuna, A., Kazemi, K., Mlucasse, R., & Santha, R. (2017). Assessing Mozambique's exposure to coastal climate hazards and erosion. *International journal of disaster risk reduction, 23*, 45–52.

Cain, A. (2017). *Water resource management under a changing climate in Angola's coastal settlements*. International Institute for Environment and Development.

Cavicchioli, R., Ripple, W.J., Timmis, K.N., Azam, F., Bakken, L.R., Baylis, M., Behrenfeld, M.J., Boetius, A., Boyd, P.W., Classen, A.T., & Crowther, T.W. (2019). Scientists' warning to humanity: microorganisms and climate change. *Nature Reviews Microbiology, 17*(9), 569–586.

Chanza, N., & Gundu-Jakarasi, V. (2020). Deciphering the climate change conundrum in Zimbabwe: an exposition. *Global Warming and Climate Change.*

Chingombe, W., & Musarandega, H. (2021). Understanding the logic of climate change adaptation: Unpacking barriers to climate change adaptation by smallholder farmers in Chimanimani District, Zimbabwe. *Sustainability, 13*(7), p.3773.

Dodman, D., & Mitlin, D. (2015). The national and local politics of climate change adaptation in Zimbabwe. *Climate and Development, 7*(3), 223–234.

Gan, T.Y., Ito, M., Hülsmann, S., Qin, X., Lu, X.X., Liong, S.Y., Rutschman, P., Disse, M., & Koivusalo, H. (2016). Possible climate change/variability and human impacts, vulnerability of drought-prone regions, water resources and capacity building for Africa. *Hydrological Sciences Journal, 61*(7), 1209–1226.

Garrote, L. (2017). Managing water resources to adapt to climate change: facing uncertainty and scarcity in a changing context. *Water Resources Management, 31*(10), 2951–2963.

Girvetz, E., Ramirez-Villegas, J., Claessens, L., Lamanna, C., Navarro-Racines, C., Nowak, A., Thornton, P., & Rosenstock, T.S. (2019). Future climate projections in Africa: where are we headed?. In *The climate-smart agriculture papers* (pp. 15–27). Cham: Springer.

Hassan, O.M., & Tularam, G.A. (2018). The effects of climate change on rural-urban migration in Sub-Saharan Africa (SSA)—the cases of democratic Republic of Congo, Kenya and Niger. *Applications in water systems management and modeling.*

Hughes, S. (2017). The politics of urban climate change policy: Toward a research agenda. *Urban Affairs Review, 53*(2), 362–380.

Kadigi, R.M., Tesfay, G., Bizoza, A., Zibadou, G., & Zilberman, D. (2019). Irrigation and water use efficiency in Sub-Saharan Africa. *Gates Open Res, 3*(587), p.587.

Leal Filho, W., Totin, E., Franke, J.A., Andrew, S.M., Abubakar, I.R., Azadi, H., Nunn, P.D., Ouweneel, B., Williams, P.A., Simpson, N.P., & Team, T.G.A.M.I. (2022). Understanding responses to climate-related water scarcity in Africa. *Science of the Total Environment, 806*, p.150420.

Limones, N., Marzo-Artigas, J., Wijnen, M. and Serrat-Capdevila, A. (2020). Evaluating drought risk in data-scarce contexts. The case of southern Angola. *Journal of Water and Climate Change, 11*(S1), 44–67.

Luetkemeier, R. and Liehr, S. (2019). Integrated responses to drought risk in Namibia and Angola. *Water Solut, 3*, 56–60.

Masson-Delmotte, V., Zhai, P., Pirani, A., Connors, S.L., Péan, C., Berger, S., Caud, N., Chen, Y., Goldfarb, L., Gomis, M.I., & Huang, M. (2021). Climate change 2021: the physical science basis. *Contribution of working group I to the sixth assessment report of the intergovernmental panel on climate change*, 2.

Mbuli, C.S., Fonjong, L.N., & Fletcher, A.J. (2021). Climate change and small farmers' vulnerability to food insecurity in Cameroon. *Sustainability, 13*(3), 1523.

Miller, J.D., & Hutchins, M. (2017). The impacts of urbanisation and climate change on urban flooding and urban water quality: A review of the evidence concerning the United Kingdom. *Journal of Hydrology: Regional Studies, 12*, 345–362.

Mogomotsi, P.K., Sekelemani, A., & Mogomotsi, G.E. (2020). Climate change adaptation strategies of small-scale farmers in Ngamiland East, Botswana. *Climatic Change, 159*(3), 441–460.

Moorthy, R., Benny, G., Gill, S.S. (2018). Disaster communication in managing vulnerabilities. *Jurnal Komunikasi: Malaysian Journal of Communication, 34*(2), 51–66.

Niang, I., Ruppel, O.C., Abdrabo, M.A., Essel, A., Lennard, C., Padgham, J., & Urquhart, P. (2014). Africa. Climate change 2014: impacts, adaptation, and vulnerability. Part B: regional aspects. Contribution of working Group II to the fifth assessment report of the intergovernmental panel on climate change. *Climate Change 2014: Impacts, Adaptation and Vulnerability: Part B: Regional Aspects: Working Group II Contribution to the Fifth Assessment Report of the Intergovernmental Panel on Climate Change*, 1199–1266.

Nkemelang, T., New, M., & Zaroug, M. (2018). Temperature and precipitation extremes under current, 1.5 C and 2.0 C global warming above pre-industrial levels over Botswana, and implications for climate change vulnerability. *Environmental Research Letters, 13*(6), p.065016.

Ofori, S.A., Cobbina, S.J., & Obiri, S. (2021). Climate change, land, water, and food security: Perspectives from Sub-Saharan Africa. *Frontiers in Sustainable Food Systems, 5*, p.680924.

Price, R. (2020). The linkages between population change and climate change in Africa.

Ringler, C., Zhu, T., Cai, X., Koo, J., & Wang, D. (2010). Climate change impacts on food security in sub-Saharan Africa. *Insights from Comprehensive Climate Change Scenarios, 2.*

Serumaga-Zake, E. (2020). *The impact of rainfall variability on subsistence farmers in the North West Province, South Africa* (Doctoral dissertation, North-West University (South Africa)).

Sintayehu, D.W. (2018). Impact of climate change on biodiversity and associated key ecosystem services in Africa: a systematic review. *Ecosystem health and sustainability, 4*(9), 225–239.

UNICEF World Health Organization. (2002*). Children in the New Millennium: Environmental Impact on Health.* https://books.google.co.za/books?hl=en&lr=&id=UWdzf2d5I7EC&oi=fnd&pg=PA1&dq=UNICEF+World+Health+Organization+2002).+&ots=Vn6is5tkmw&sig=SGH5QGCw_BaMu3GHQTsW8LlRyT4&redir_esc=y#v=onepage&q=UNICEF%20World%20Health%20Organization%202002 Accessed 9 August 2023.

van Schalkwyk, L., Blamey, R.C., Dyson, L.L., & Reason, C.J.C. (2022). A climatology of drylines in the interior of subtropical southern Africa. *Journal of Climate*, 1–46.

van Weezel, S. (2020). Local warming and violent armed conflict in Africa. *World Development*, *126*, p.104708.

Wetende, E., Olago, D., & Ogara, W. (2018). Perceptions of climate change variability and adaptation strategies on smallholder dairy farming systems: Insights from Siaya Sub-County of Western Kenya. *Environmental development*, *27*, 14–25.

Ziervogel, G., Johnston, P., Matthew, M., & Mukheibir, P. (2010). Using climate information for supporting climate change adaptation in water resource management in South Africa. *Climatic Change*, *103*(3), 537–554.

Chapter 4
Drought, the War in Europe and Its Impacts on Food Insecurity in Sub-Saharan Africa, East Africa

Katcho Karume, Jean Mubalama Mondo, and Jean Chrysostome K. Kiyala

Abstract Conflict, drought, and dependency on food import are leading concerns for food security in Eastern Africa (EA). This region is vulnerable to drought and the adverse effects of international security, due to its heavy reliance on rainfed agriculture and food imports. This chapter reviews evidence of the adverse effects of drought and the war in Europe on food security in EA. The literature has established that climate change, in general, and specifically drought, has already impacted people subsistence means significantly, evidenced by a decline in crop yield and food production. It has also indicated a price increase of imported food and non-food commodities since the launch of the Russia-Ukraine war. These factors affect all components of food security, which include, but are not limited to food availability, food accessibility, food utilisation, and food stability and, hence, increase the risk of hunger and violent conflicts in the region. The chapter has confirmed that there is general consensus that EA is among the regions that are most vulnerable to drought (climate change), as well as global security. The chapter suggests further that policymakers and development agencies should focus on formulating and implementing policies and programmes that would ensure that farmers across the region promote farm level adaptation strategies, as well as climate-smart agriculture (CSA), and climate-smart

Prof Katcho Karume is Senior Lecturer at the Faculty of Agriculture and Environmental Sciences, Université Evangélique en Afrique, Bukavu, Democratic Republic of Congo, and Research Fellow at the Centre de Recherche en Géothermie, Bukavu, Democratic Republic of Congo. Email: kkatcho@yahoo.com

Corresponding author: Dr Jean Mubalama Mondo is a lecturer and researcher in Sustainable Agriculture, Food Security, Seed Systems, and Agricultural Innovation Adoption from the Université Evangélique en Afrique (UEA), Bukavu, Democratic Republic of Congo. E-mail: mondo.mubalama@yahoo.fr; ORCID: https://orcid.org/0000-0001-7197-6551, ResearchGate: https://www.researchgate.net/profile/Jean-Mondo-Mubalama

Dr Jean Chrysostome K. Kiyala holds a PhD in Public Administration: Peace Studies from DUT; he is a Senior Lecturer at the International Centre of Nonviolence (DUT), Associate Professor and visiting lecturer at the Université de Bandundu and the Université Evangélique en Afrique (Democratic Republic of Congo); Emails: JeanK@dut.ac.za; kljeanchrysostome@gmail.com; ORCID: https://orcid.org/0000-0002-9587-0230

J. C. K. Kiyala and N. Chivasa (eds.), *Climate Change and Socio-political Violence in Sub-Saharan Africa in the Anthropocene*, The Anthropocene: Politik—Economics—Society—Science 37,
https://doi.org/10.1007/978-3-031-48375-2_4

livelihoods (CSL). This should be done alongside other price regulating factors such as subsidies on staple food crops and fertilisers. It is also crucial that the region reduces its level of food imports by encouraging local production or substituting some of the import products by locally available materials. In this regard, social protection programmes must account for the potential increases in seasonal price variability to enhance household resilience to weather shocks.

Keywords Drought · Rainfed agriculture · Climate change · Climate-smart agriculture · Food security · Substitution · Food import

4.1 Introduction

Food security refers to when all people, always have physical and economic access to sufficient, safe, and nutritious food to meet their dietary needs and food preferences for an active healthy lifestyle. Achieving food security has been a major challenge for most African countries. Natural disasters such as drought and flood, demographic explosion, as well as corrupt leadership and violent conflicts are deepening food demand and supply gaps (Dodo, 2020; Giller, 2020; Anderson et al., 2021; Kakpo et al., 2022). Climate change, COVID 19, and Russia's invasion of the Ukraine added pressure on food security in Africa in general, and in sub-Saharan Africa (SSA) in particular. To this end, sub-Saharan Africa is the most vulnerable region to climate change, as its agriculture is mainly rainfed, and the poverty among most populations do not facilitate adaptation strategies, and access to stable food markets is limited by high transaction costs (Boliko, 2019; Kakpo et al., 2022; Karume et al., 2022). Food pricing is one of the key factors that determine access to food. An increase in food prices reduces, in this regard, purchasing power and, therefore, limits the quantity of food that can be purchased. Consequently, there is a large proportion of undernourished and food insecure populations in sub-Saharan Africa.

For instance, more than 123 million (12%) of the population in SSA are suffering from chronic hunger and food insecurity (International Monetary Fund (IMF), 2022a). Due to the recent pandemic crisis, price inflation, and the shortage of cereals and fuels, attributed to the Russia-Ukraine war, at least 28 million of these people became acutely food insecure over the past two years (IMF, 2022a). Because of its projected twofold population increase, SSA will be the region with the highest food security risk by 2050 (Sánchez/Namhata, 2019).

As it is, the 2030 United Nations' Sustainable Development Goal 2 (SDG2) on food security will be difficult to attain, amid the "new normal" of frequent and recurring droughts, floods, cyclones, and rising temperatures and sea levels. For instance, one-third of global droughts occur in SSA. Currently, EA (including Ethiopia and Kenya) is experiencing one of the most severe droughts in recent history, while Angola is undergoing the fifth consecutive year of drought. As a result, drought exacerbates existing price seasonality, which, in turn, can amplify negative impacts on households. Understanding how weather shocks affect rural sub-Saharan African households (through production and market price changes) is crucial in designing sustainable and effective programmes to increase the region's resilience (Kakpo et al., 2022).

There is ample empirical evidence of the links between weather shocks and economic, social, and political outcomes such as reduced crop yields, household income, household food consumption, reduced livestock weight gain and milk production, increased infant mortality rate and reduced life expectancy, as well as frequent conflicts, owing to internal and external migration (Anderson et al., 2021; Kakpo et al., 2022). Sánchez and Namhata (2019) established a link between political instability in SSA and the four dimensions of food security and found that the likelihood of protest for the 2005–2015 period was associated with low cereal production, low access to hygiene facilities, and high domestic food price variability.

The Food and Agriculture Organization (FAO) reports (2005–2018) show that the EA region has the largest proportion of people contending with food insecurity—that is, one in three people (Boliko, 2019) are affected by food insecurity. Droughts in the EA region tend to be more frequent, longer, and more severe, as the overall precipitation and water storage decline abruptly (Haile et al., 2019). The main causes of the changing pattern of droughts include climate variability and the Anthropogenic effects. Droughts have far reaching impacts on human beings, the environment, water resources, and agriculture (Haile et al., 2019). Thus, this chapter reviews the impact of global conflicts and drought on food security and the economy of EA countries, the coping strategies adopted by regional governments and their effectiveness, and how the prospects of food security can be sustained.

This chapter consists of eight major sections. The introduction provides the background of the problem, and maps out important features of the discussion. Subsequent sections focus on food prices inflation in Eastern Africa, as well as fuel and transport price inflation. The chapter further substantiate the effects of drought, particularly in four countries, namely Uganda, Ethiopia, the Democratic Republic of Congo (DRC), and Kenya. Following this, the chapter makes some remarks on climate shocks and food security, and appraises the attempts made by Uganda, Ethiopia, the Democratic Republic of Congo, and Kenya to respond to drought-driven food insecurity, and establish the effectiveness and limitations of their interventions. Next, we discuss Climate-Smart Agriculture (CSA) practices in Eastern Africa and how they are funded, before drawing conclusions in respect of the increasing governmental investments on the agricultural sector, to respond to climate change, reduce the costs of production, and stimulate the local production of food, instead of increasing imports.

4.2 Food Inflation in Eastern Africa

World food prices have increased by 52% since 2019. The East African countries have also been affected by the price increases. As highlighted in the afore going section, this inflation was driven by the global pandemic and shortages of food, attributed to the war in the Ukraine (IMF, 2022b). This situation is exacerbated by the fact that many households in EA are poor and spend large shares of their incomes on food consumption, and that many poor farmers are the net buyers of staple food. Table 4.1 provides the mean inflation for food commodities in the selected EA countries.

A typical consumption basket in the region is broadly made of two types of staples: cereals and starchy roots (IMF, 2022b). On average, cereals—such as maize, rice, and wheat—and starchy roots—such as cassava, sweet potatoes, and potatoes—account for almost two thirds of the daily caloric intake. The other third of daily food consumption mainly comprises vegetable oils, sugar, fruits, meat, and pulses. The five most consumed staples in SSA (including EA) are maize, rice, cassava, wheat, and palm oil, which account to ~ 54% of the average daily per capita caloric intake (IMF, 2022b). Within the region, maize—19.7% of the daily per capita caloric intake—is the top staple in Eastern and Southern Africa, while rice—11.4% of the daily per capita caloric intake—is widely consumed in Western Africa (IMF, 2022b). The consumption of cassava, wheat, and palm oil—which accounts for around 10.5, 8.3, and 3.5% of the daily per capita caloric intake respectively—seems more homogeneous across the region, albeit with some differences between certain countries (IMF, 2022b).

In SSA, where food makes up nearly 40% of the consumption basket, the local relative prices of staples surged by an average 19.3% in 2020–2021. This was the sharpest increase in the prices of staple food since the global financial crisis (IMF, 2022b). The recent steep change in the local costs of staple foods was broad-based, albeit of varying magnitudes. The largest real price increases were seen in relation to cassava (83.2%), wheat (18.6%), and maize (11.3%). The average price changes for rice and palm oil have been more muted (IMF, 2022b). Factoring in the consumption shares led to an 8.5% real increase in the cost of a typical food consumption basket in the region. However, most of these staple foods are imported, making food security vulnerable to global conflicts and other environmental shocks (Table 4.1). To this end, countries in SSA are mostly vulnerable to global food price shocks, as the estimated pass-through from global to local food prices for highly imported staples is close to

Table 4.1 Food inflation in eastern Africa during the Russia-Ukraine crisis

Country	Reference	Food inflation (%)
Burundi	September 2022	26.3
Kenya	October 2022	15.8
Rwanda	September 2022	41.2
Tanzania	September 2022	8.3
Uganda	September 2022	18.4
DRC	May 2022	6.1
Ethiopia	September 2022	31.0

Source Uganda Bureau of Statistics (2022), WFP (2022a), KIPPRA (2022), Kenya National Bureau of Statistics (2022). DRC = Democratic Republic of Congo)

Table 4.2 Net import dependence on the top five staples in selected east African countries

Country	Net import dependence (NID)				
	Maize	Rice	Cassava	Wheat	Palm oil
Ethiopia	0.7	78.9	0.0	21.1	100
Kenya	6.3	85.5	0.1	84.6	100
Rwanda	8.1	36.2	14.7	88.5	100
Tanzania	−1.7	31.5	−0.3	95.4	100
Uganda	−3.4	7.1	−3.2	93.5	99

Net import dependence (NID) = 100 × (imports − exports)/(production + imports − exports)
Source 2019 (IMF, 2022b)

unity (IMF, 2022b). Moreover, higher domestic production—resulted in lower net import dependence—and lower consumption shares correspond to lower food price inflation (Table 4.2).

The Ukrainian war has unleashed a "three-headed hydra" of food, energy, and finance shortages in Africa, further threatening the position of vulnerable Africans, and putting dozens of countries at a risk of default (Burrier, 2022). Even before Russia's invasion of the Ukraine, food assistance officials had warned that 2022 was going to be a terrible year. Other than COVID-19, severe weather and the surging food and fuel prices have pushed global hunger to "terrifying levels" (WFP, 2023). Ukraine which used to be one of the world's "breadbaskets", supplying key food exports such as wheat, barley, and sunflower oil to feed 400 million people worldwide (Burrier, 2022) is affected by the ongoing war with Russia. Over 40% of Africa's wheat supply is from Russia and the Ukraine. Yet, the Ukraine's key ports—like Odessa, through which 98% of grain exports typically pass—are being illegally blockaded by Russia. Some 20 million tons of wheat are stuck in the Ukraine, unable to reach those who desperately need it (Burrier, 2022).

4.3 Fuel and Transport Inflation in Eastern Africa

African economies are not only contending with higher prices and food unavailability— but rising energy prices as well. While Africa is home to significant oil reserves, a lack of refining capacity means that oil must be exported and then imported as refined fuels (Burrier, 2022). This has disrupted the entire continent, with drivers lining up at the pump, and airline flights halted. Moreover, health systems continue to be under strain, owing to the COVID-19 pandemic, with government resources being depleting. Burrier (2022) commented that the Covid-19 crisis was much bigger than that which followed the global financial crisis; everything was stacked against the low- and middle-income countries, he wrote.

Dalheimer et al. (2021) found that unlike the United States (US), or global corn markets, African food (corn) markets are much less sensitive to the impact of oil-specific demand shocks. Instead, disruptions in the global oil supply will likely lead to an increase in food prices in several markets. Countries affected by oil-supply shocks have neither strategic or natural oil reserves to buffer import shortages, nor efficient oil distribution systems that translate into food prices through higher transportation costs (Dalheimer et al., 2021). The scholars also found that a large share of corn price surges in 2011 and 2012 are attributed to oil-supply shortages, attributed to the Libyan revolution, and the oil embargo against Iran (Dalheimer et al., 2021). Conversely, the shale oil boom in the US, and the oil production expansion in the Middle East exerted downward pressures on corn prices in three African countries in 2014/15. Finally, political events related to the oil market, for instance the tensions between the US and Iran, or the recent oil price war between Saudi Arabia and Russia, can raise or lower corn prices by up to 25% (Dalheimer et al., 2021). Massive fuel inflation has been observed in all eastern African countries, due to the Russia-Ukraine crisis, which, in turn, exacerbated export food inflation (Table 4.3). The situation was worsened further by the currency devaluation against the US dollar during the crisis period.

Transportation costs for both farm inputs and marketable output are driven by fuel prices, which, in turn, are derivatives of crude oil (Dalheimer et al., 2021). In this regard, transport costs have been shown to be a particularly relevant transmission channel in SSA countries, where production and marketing are largely decentralised, and transport absorbs a relatively large share of production and marketing costs, compared to what happens in other parts of the world (Dillon/Barrett, 2016). Serra and Zilberman (2013) evaluated a large body of literature to empirically examines the relationship between crude oil and food prices. There is a broad consensus in the literature that disturbances in the energy markets impact food markets, and increasingly so after the emergence of biofuels (Serra/Zilberman, 2013; Dalheimer et al., 2021).

Table 4.3 Fuel inflation and currency depreciation in eastern Africa during Russia-Ukraine crisis

Country	Reference	Fuel inflation (%)	Currency depreciation (%)
Burundi	May 2022	36.0	– 4
Kenya	September 2022	11.1	– 9
Rwanda	June 2022	9.9	– 3
Tanzania	May 2022	17.1	– 3
Uganda	June 2022	50.0	– 8
DRC	April 2022	4.8	– 1
Ethiopia	September 2022	20.0	– 14

Source National Bureaus of Statistics of the different countries; WFP Regional Bureau for Eastern Africa (2022). Currency depreciation is evaluated against US dollar

4.4 Case Studies of the Selected EA Countries

No country is immune to the global food crisis. Indeed, food prices worldwide were ~ 23% higher in 2022 than they were in 2021. However, SSA is the most hit hard, where food takes ~ 40% of household budgets (IMF, 2022b). We are using the case studies of some East African countries; namely Kenya, Uganda, Ethiopia, and the DRC to elucidate on the influence of drought and global conflicts on food security, and how local governments have tackled these conflicts. The data pertaining to the case studies are online.

4.4.1 Uganda

Food prices in Uganda increased by 18.40% in September 2022, over the same month in the previous year (Uganda Bureau of Statistics, 2022). For instance, the consumption of bakery products, made from wheat, has continued to surge during the last decade or so, due to population growth, urbanisation, increasing standard of living, and disposable income, linked to economic growth. In Uganda for example, the domestic consumption of wheat products increased from 131,000 metric tonnes in 2010 to 480,000 metric tonnes in 2020. Most of the wheat consumed is imported – in this regard, 55% and 19% of the wheat is imported from the Ukraine and Russia respectively (Uganda Bureau of Statistics, 2022). Thus, the escalating war in these two countries has led to increased prices in wheat. To this end, wheat prices have more than doubled from 250 USD a metric tonne in December 2021 to 515 USD in June 2022. Similar trends were observed in Tanzania, the DRC, and other East African countries. The increasing cost of wheat puts a strain on foreign exchange earnings and contributes to food insecurity and poverty in these countries (Irakiza et al., 2021). This is destabilising the region's economy, and consequently, food security, as prices of wheat-based bakery products have soared, while imports undermine foreign reserves.

4.4.2 Ethiopia

There has been evidence of a relationship between conflict and food security in Ethiopia. For instance, hunger is tightening its grip on more than 20 million Ethiopians who are contending with conflict in the north, and drought in the south. The UN food relief agency (UN News, 2022) released a warning of dwindling food and nutrition support from July 2022. To this end, the World Food Programme (WFP) added that: "The combination of conflict and drought have caused inflation to soar", noting that as of April 2022, the Food Price Index in Ethiopia would be up by 43%, compared to the same month the previous year. Meanwhile, the prices of vegetable oil and cereals would be up by over 89 and 37% respectively, year-on-year (UN News, 2022).

According to the WFP, 19 months of civil war in Ethiopia have left more than 13 million people in the north requiring humanitarian food assistance, mainly in conflict-affected zones in Afar, Amhara, and Tigray regions (UN News, 2022). Food and humanitarian supplies had been flowing into the Tigray region over two months in 2022 (April to June), after the Government had announced a humanitarian truce. While the WFP had delivered over 100,000 tonnes of food from 1 April 2022, enough to feed 5.9 million people for a month – fuel deliveries had not kept pace (UN News, 2022). Less than half of the 2 million litres of fuel needed were delivered in the region in recent weeks. The WFP was, finally, able to meet the food needs of over 800,000 people in Tigray and had just completed its most recent deliveries of emergency food rations to 1.3 million people in Afar and in Amhara (UN News, 2022).

According to the WFP, conflicts and food shortages led to precarious health in conflict zones (Tigray). In this regard, over 20% of children under the age of 5, and half of the pregnant and breastfeeding women were malnourished. Meanwhile, 32% of parents in Afar zone 4 – one of five administrative zones in the region – and 16% in Amhara reported that over the previous three months in 2022 (February to May), malnutrition drove their children under the age of 5 to health centres (UN News, 2022). A recent WFP assessment of conflict-affected zones in both regions cited this as well. It is also estimated that in southern and south-eastern Ethiopia, 7.4 million people wake up hungry every day, as the country grapples with the effects of the fourth failed consecutive rainy season (UN News, 2022).

This food demand and supply unbalance has been aggravated by the Russia-Ukraine crisis. At the same time, the ripple effect of the war in the Ukraine is set to exacerbate Ethiopia's food security crisis. With over three-quarters of the WFP and government wheat – a country staple – coming from the Ukraine or Russia, Ethiopia is in a precarious situation, as the cost of food, as well as of fertilisers, will go up, beyond the means of millions of Ethiopian farmers (UN News, 2022).

4.4.3 DR Congo

Rising prices, driven by the Russia-Ukraine crisis, have stoked fear of devastating social turmoil in eastern DRC. Economic growth in the region has been hobbled by geographical remoteness and decades of violence (Africa News, 2022). Lately, Congolese who go at the pump have met with the same disappointment. Inexorably soaring fuel prices in the DRC are the latest consequence of the war in Ukraine (Africa News, 2022). The conflict is stoking fears and anger, especially among residents in the north-eastern part of the DRC. To this end, the authorities are unable to regulate rising prices. The rising prices not only affect fuel, commodities on the markets stalls get prohibitively expensive as well. A 50 kg bag of sugar, which previously cost around 43US$ went up to 60U$ (Africa News, 2022). This was also the case with cooking oil, rice, or maize, which were worth double of their price in the few weeks prior in 2022, according to a report released in March 2022 (Africa News, 2022). The chaotic situation, reported by Africa News at the time, is still prevalent in 2023, as the erratic weather conditions worsened the food security and increased the food inflation crises.

This inflation can be explained by the fact that the DRC relies heavily on imports. A dependence that could destabilise the country's food security according to economists (Africa News, 2022). In addition, the Congolese government has less invested in food production and agricultural research (Karume et al., 2022).

4.4.4 *Kenya*

Kenya's annual food price increase, the largest component in the inflation basket, quickened to 15.8% in October 2022, from 15.5% in the previous month. As of June 2022, the food inflation rate had more than doubled in the past year to 13.8% from 6.32% in June 2021, as shown in Fig. 4.1 (KIPPRA, 2022). This was the highest increase since June 2017. Some of the items in the food basket driving food inflation included wheat, sugar, cooking oil, maize flour, milk, potatoes, onion, and carrot. For instance, the inflation rate was the biggest for sugar (36.3%), cooking oil (33.6%), potatoes (32.6%), maize flour (33.5%), and beans (24.4%) (Kenya National Bureau of Statistics, 2022). The highest average price change was recorded for maize flour at 67%, as shown in Table 4.4 (KIPPRA, 2022).

According to an analysis conducted by the Kenya Institute for Public Policy Research and Analysis (2022), factors contributing to the rising food prices in Kenya are:

Fig. 4.1 Trend in food and overall inflation in Kenya (2021/2022). *Source* Kenya National Bureau of Statistics (2022), and KIPPRA (2022)

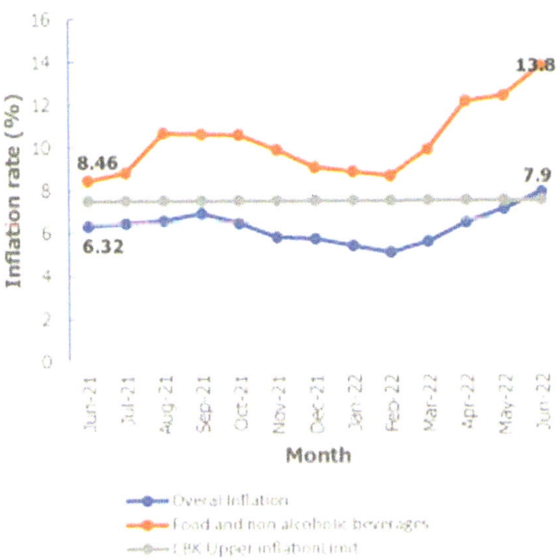

Table 4.4 Inflation of key food items in Kenya (2021/2022)

Commodity	Average price (Ksh)		Change	
	June 2021	June 2022	Kshs	%
Fresh packeted cow milk (500 ml)	49.3	58.4	9	19
Irish potatoes (1 kg)	69.9	84.8	15	21
Cooking oil (1 l)	255.8	388.0	132	52
Carrots (1 kg)	73.4	93.2	20	27
Wheat flour (2 kg)	129.6	186.9	57	44
White bread (400 g)	54.1	60.8	7	12
Onion (1 kg)	110.8	134.6	24	21
Maize flour (2 kg)	120.0	200.0	80	67
Beans (1 kg)	129.0	136.6	8	6

Data source Kenya National Bureau of Statistics (2022) and KIPPRA (2022)

- **Importation of food products**

Kenya relies on imports to meet its growing food demand. The country's import dependency ratio on food is 14.5% (KIPPRA, 2022). The latest data from the Kenya National Bureau of Statistics (KNBS) shows that the food import bill rose by 21%, shifting the overall monetary value from Sh128.06 billion in 2020 to Sh155.42 billion in 2022. Food imports put Kenya in a vulnerable position in respect of economic shocks in the global market. The global supply shocks that have contributed to the current rise in food prices in Kenya are the COVID-19 pandemic, the Russia-Ukraine war, and Indonesia's restriction of the export of palm oil (KIPPRA, 2022). The war between Russia and the Ukraine that began on 24 February 2022 has affected the global wheat supply chain. Kenya sources 44% of its wheat from Russia and the Ukraine. To this end, the sanctions imposed on Russia led to an increase in wheat importation costs, with the landing cost in Kenya increasing to US$500 per tonne from US$400 (KIPPRA, 2022). The war between Russia and the Ukraine has also led to a surge in fuel prices in Kenya, significantly increasing production and transportation costs, and thus, affecting the distribution of food commodities. Also, Indonesia's export restriction of palm oil in the period February 2022 to May 2022 contributed to the rising prices of cooking oil, as Indonesia is Kenya's largest supplier of palm oil (KIPPRA, 2022).

- **Weather conditions**

Since agriculture in Kenya is largely rainfed, farming is mainly dependent on favourable climatic conditions for good harvests (KIPPRA, 2022). In 2021, erratic rains yielded low harvests throughout the country. The situation was set to worsen in 2022, due to delayed long rainy season. Meanwhile, the historical drought led to increasing food prices. Examples are the sharp spikes in food inflation during the drought periods of 2011 and 2017. This trend continued in 2021/22 (KIPPRA, 2022).

- **Cost of production**

The rising cost of farm inputs such as fertilisers pushed production cost higher and, in the process, food prices went up. Kenya imports fertilisers from the global market, and is, thus, vulnerable to the price inflation in the domestic market. The price of fertilisers in the global market has been rising more than twofold, from 2019 to 2022 (KIPPRA, 2022), leading to increasing fertiliser prices, making things difficult for farmers who depend on the import to practice agriculture.

- **Depreciation of the Kenyan shillings against the US dollar**

The Kenyan shilling has depreciated by 8.8% in the period June 2021 to June 2022. This is likely to affect the price of imported food (KIPPRA, 2022). For instance, the relative strength of a country's currency is a major driver of food pricing, as it affects the costs of imported food items. According to Okou et al. (2022), a 1% depreciation in real effective exchange rates increases the price of highly imported staples by an average 0.3%.

- **Market imperfections**

Farmers tend to hoard maize, in anticipation of better prices. According to the Ministry of Agriculture, Kenyan maize farmers export their produce to Southern Sudan, Tanzania, and Uganda, due to high prices offered in those countries, leading to a maize deficit at the local markets in Kenya. The scarcity of maize has led to an increase in the cost of maize flour (KIPPRA, 2022).

- **Changes in sub-Saharan Africa trade**

During the 2021/22 season, Kenya was the largest maize importer in the region, accounting for 21% of the region's maize imports of 3.4 million tonnes. This is according to the data from the International Grains Council (KIPPRA, 2022). Tanzania and Zambia are the leading maize suppliers to Kenya. However, Tanzania, the biggest exporter in the region, is unlikely to fulfil this role this season, because its maize production is forecast to fall by 16% year-on-year to 5.9 million tonnes (KIPPRA, 2022). This is due to episodes of droughts at the beginning of the rainy season, combined with fall armyworm infestations, as well as reduced fertiliser usage in some regions, owing to prohibitively higher prices. The low production rates, coupled with firmer domestic consumption, means that Tanzania could have less maize for the export markets (KIPPRA, 2022).

Besides, South Africa's limited participation in the Kenyan maize market is arguably, hindered by regulations, rather than just the price and consumer preferences. Kenya continues to maintain an import ban on genetically engineered products (KIPPRA, 2022). This means that South Africa is less likely to supply maize to Kenya, as over 80% of its maize is genetically engineered. Indications are that Kenya is changing its longstanding policy, due to inflation pressure. Regulatory agencies have, recently, completed all trials for the approval of biotechnology maize. But any decision would still need to be approved by Kenya's cabinet (KIPPRA, 2022). Changing the policy on genetically modified (GM) maize would also prompt the US,

Mexico, Argentina, and Brazil to expand their market shares in the country (along with South Africa), as there are generally, few white maize producing countries in the world (white maize is the preferred staple grain among Kenyan consumers) (KIPPRA, 2022).

4.5 Climatic Shocks (Drought) and Food Security in EA

Climate shocks, mainly droughts, are recurrent in EA (Table 4.5). A drought in the Horn of Africa has left roughly 22 million hungry across Djibouti, Somalia, Ethiopia, and Kenya in 2022, after four consecutive failed rainy seasons (WFP, 2022a). The literature reviewed confirmed that a 1 °C temperature increase in developing countries is associated with a 3% reduction in agricultural output, leading to a 1.3% decline in growth (Haile et al., 2019). It has been projected that crop yields in SSA would decline by 5–17% by 2050, especially of key staples. Notably, rising temperatures and rainfall volatility are key contributors to the shrinking growing seasons and arable land. Consequently, productivity is reduced, owing to overuse of arable land—impeding total factor productivity in agriculture (IMF, 2022b). Besides, rising temperatures and water levels force insects and weed seeds to migrate into SSA. The 2019–2020 locust infestations in Ethiopia, Kenya, and Somalia affected 1.25 million hectares of land, and the infestation response increased the region's financing needs by about $70 million (IMF, 2022b).

East African regions have experienced droughts for a millennium (Kiros, 1991), which, have often, manifested as endemic events in the region (Lyon, 2014; Muller, 2014). In recent decades, the region experienced several prevailing dire drought events (Haile et al., 2019).

There are frequent drought occurrences in EA—however, these have been difficult to forecast, because of various natural and Anthropogenic factors, along with inefficient forecasting capacities. As a result, drought has had a severe impact on the environment, as well as the socio-economic welfare of societies in the region

Table 4.5 Drought events for 1900–2017 in EA countries

Country	Number of drought episodes	Number of people affected (in millions)	Number of deaths
Burundi	6	3.1	126
DRC	–	–	–
Ethiopia	15	77.1	402,367
Kenya	15	50.2	196
Rwanda	6	4.2	237
Tanzania	10	12.7	0.0
Uganda	9	5.0	194

Source Haile et al. (2019)

(Haile et al., 2019). Climate change, land use change, as well as socio-political and institutional transitions in EA have caused recurrent droughts. Drought frequency in EA has doubled—from once every six years to once every three years since 2005 (Haile et al., 2019). In some countries such as Somalia, Kenya, and Ethiopia, drought has been one of the main reasons for socio-economic instabilities (O'Loughlin et al., 2012; Ayana et al., 2016). For example, in Somalia alone, a wide range of food insecurity situations arose, and around 250,000 deaths were registered during the 2010–11 droughts (Hillbruner/Moloney, 2012; Lyon/Vigaud, 2017). Drought is also becoming a threat to food supplies in EA (Meier et al., 2007; Muller, 2014; Zhao/Dai, 2015; Haile et al., 2019; WFP, 2022a).

This insidious phenomenon results in malnutrition, epidemics, displacement of populations, food insecurity, famines, and death (Hillbruner/Moloney, 2012; Muller, 2014; Lyon/Vigaud, 2017; Haile et al., 2019). Drought is, therefore, considered a strategic enemy in EA—this, thus, calls for acute mitigation measures.

In parallel, past environmental, economic, political, and nomadic complexities have exposed the region to extremely high levels of poverty and made it vulnerable to natural and Anthropogenic shocks (instabilities, floods, and drought) (O'Loughlin et al., 2012; Shiferaw et al., 2014; Ayana et al., 2016; Haile et al., 2019; WFP, 2022a).

According to the Famine Early Warning Systems Network (FEWS NET, 2022a) in Kenya, the long rainy season going from March to May 2022 marked the fourth below-average season across most of eastern Kenya. The continuous decline in livestock productivity and crop production has resulted in below-average food availability. Livestock deaths due to drought and crop failure continue to constrain household food availability and income, driving crisis and emergency outcomes in the pastoral areas, as well as Stressed and Crisis outcomes in the marginal agricultural areas (FEWS NET, 2022a). In May, the Kenya Food Security Steering Group (KFSSG) conducted an Integrated Food Security Phase Classification (IPC) acute food insecurity and acute malnutrition projection update for May and June 2022, determining that 4.1 million Kenyans in the Arid and Semi-Arid Lands (ASALs) are acutely food insecure. This estimate is around 32% higher than the initial projection from the Short Rains IPC assessment of February 2022 (FEWS NET, 2022a). The acute malnutrition analysis reported that about 942,500 children under the age of five, and 134,270 pregnant and lactating mothers are acutely malnourished, and likely require treatment. The rise in acute malnutrition is likely driven by the increasing food insecurity, along with reduced milk consumption, increased morbidity, and poor child-care practices (FEWS NET, 2022a).

The high prices for staple food in Kenya are constraining household purchasing power and food access in both the urban and rural areas. Staple food prices currently range from 22 to 63% above average for maize, and 12 to 44% above average for beans. The high prices are attributed to the below-average production in 2021/2022, the expected below-average production for 2022/2023, the high market demand, due to low household stocks, high inflation, as well as bottlenecks at the Kenya-Tanzania border, following the enforcement of Tanzania's export permits (FEWS NET, 2022a). On 18 July 2022, the Kenyan Ministry of Agriculture allocated a 60% subsidy to a two-kilogram maize flour packet, lowering the price from 250 to 100

KES up to 15 August 2022. The below-average and poorly distributed long rains across the marginal agricultural areas resulted in the below-average planted acreage and moisture stress, with crop failure in Taita Taveta, Makueni, Kitui, Tharaka, and Embu (Mbeere). The reduced cropping activities resulted in the below-average food availability and income from crop sales and casual wage labour opportunities, driving Stressed outcomes across the marginal areas, with area-level crisis outcomes in Kitui, Makueni, Tharaka Nithi, and Meru North (FEWS NET, 2022a).

Below-average forage and water resources in the pastoral areas continue to drive fair to poor livestock body conditions and high levels of migration, resulting in below-average milk production and sale values. With livestock being in distant grazing areas, the provision of milk is particularly low for women, children, and the elderly (FEWS NET, 2022a). Milk consumption is 50–90% below the three-year average consumption across pastoral areas. Additionally, the below-normal livestock body conditions and high maize prices have reduced the goat-to-maize terms-of-trade by 28–42% below the five-year average (FEWS NET, 2022a). Although humanitarian assistance likely supported crisis outcomes in northern and north-eastern pastoral livelihood zones in June 2022, the continued loss of livestock, the below-average herd sizes, and falling household purchasing power are likely to lead to increasingly widespread emergency outcomes (FEWS NET, 2022a). A scale-up of food and nutrition assistance is required, at least through January 2023 to prevent further increases in levels of acute malnutrition and avert an increase in hunger-related mortality (FEWS NET, 2022a).

Such an alarming situation as reported in Kenya mirrors the situation in other countries in EA during the same period. For instance, by November 2022, cumulative rainfall in Uganda had totalled only ~ 60–90% of the normal average, and has been erratic, resulting in the below-average crop production. Besides, most poor rural households stored less than normal, and accessed below-normal seasonal income from crop sales. Consequently, the prices of staples such as sorghum, maize, and beans generally increased to 30–60% higher, compared to the same time last year, and significantly higher than the five-year average (FEWS NET, 2022b). The adverse effects of drought in Uganda have, since 2022, worsened because of the Ukraine-Russia war that affected the prices of import foods and farming inputs such as fertilisers and fuel. For instance, rising food prices in Uganda are driven largely by the significant above-average fuel prices, and the resultant high transportation costs, below-average domestic stocks following consecutive below-average harvests, and above-average regional demand for Ugandan exports (FEWS NET, 2022b). Such high prices for staple food, and the below-average seasonal income-earning, continue to limit food access in many poor households, making them vulnerable to food security crisis.

In Burundi, ~ 1.04 million people were experiencing high levels of acute food insecurity (Integrated Food Security Phase Classification phases 3 and 4) at the end of 2021, due to the combined effects of climate-related shocks and poor access to land and production (WFP, 2022b). The recurring climate shocks affect productivity and income and exacerbate malnutrition in the country. For instance, in 2020 52% of children under the age of 5 had stunted growth rate, with slightly higher rates for boys, at 55.5%, compared to 48.9% for girls. Acute malnutrition affects 6.1% of children under 5 (7.2% for boys and 4.9% for girls), with some districts close to or exceeding the 10% 'serious' threshold. Close to a quarter of adolescent girls and 18% of women are underweight (WFP, 2022b). It is noteworthy that the food insecurity prevalent in most of the regions in Burundi is due to a combined effect of the COVID 19 pandemics, the Ukraine-Russia war, and the erratic below-average rainfalls. For instance, in some livelihood zones, increased food prices, a below-average 2022 Season B harvest, below-average labour demand and wages, as well as atypical limitations on cross-border trade related to COVID-19 tests fees and migratory labour have led to reduced food access for poor and very poor households facing Stressed conditions (IPC Phase 2) outcomes (FEWS NET, 2022c).

4.6 Governmental Attempts, Efficiency, and Limits in Controlling Drought and Warfare Effects

Some EA governments such as Kenya, have waived all or part of the levies on food imports, a measure aiming at reducing the price of food commodities (Mburu/Ondieki, 2022). Besides grappling with growing food insecurity and the devastating effects of drought, Kenya has 'effectively' lifted its decade-old ban on the open cultivation of genetically modified (GM) crops (Wasike, 2022). For instance, Kenya has, since November 2012, been banned from engaging in the "open cultivation of genetically modified crops and the importation of food crops and animal feeds produced through biotechnology innovations" (Wasike, 2022). The meeting set to lift bans on GM maize also focused on various measures that could be implemented to boost Kenya's food security. To this end, 23 of the country's 47 counties struggle because of the devastating effect of drought (Wasike, 2022). Proposals presented included reducing Kenya's reliance on rain-fed agriculture by increasing irrigation, planting diverse and drought-resistant crops, and implementing early warning and response mechanisms (Wasike, 2022). According to a recent report by Famine Early Warning Systems Network (2022), Kenya is confronted by rising "acute food insecurity due to the impacts of drought on multiple below-average crop and livestock production seasons and high inflation" (Wasike, 2022).

Other measures that governments implemented to curb food inflation include offering subsidies to farmers (Mburu/Ondieki, 2022; KIPPRA, 2022; ITA, 2022). The fertiliser subsidies were allocated to shield the farmers from the high costs, and consequently, protect consumers from the high food prices. Subsidies, however, are unsustainable, and have been proven to distort the market with a negative outcome (KIPPRA, 2022). The other measure that governments implemented was to waiver levies on the imports of staples to curb inflation. However, food price inflation has not been abated, despite the measures implemented (KIPPRA, 2022). To this end, governments put in place irrigation interventions to reduce reliance on rain-fed agriculture (KIPPRA, 2022). This has, however, not yielded the desired outcomes. For instance, only 1.7% of the land in Kenya is under irrigation (KIPPRA, 2022). The existing irrigation schemes, however, have failed to realise their potential, contributing only 3% to Kenya's gross domestic product (GDP). Policies targeted to reduce post-harvest losses by half by 2025 have been implemented but are yet to produce results. There is, therefore, a need for more interventions to shield consumers from the rising food prices (KIPPRA, 2022). In Uganda, the strategies to tackle the adverse effects of war and climate on food inflation included import waiver and subsidies on critical products such as fuel, cooking oil, etcetera. The shortage of maize in Katanga and Kasaï regions in the DRC has propelled the government to negotiate its supply from Zambia and South Africa in April 2023, and to subsidise producers and waive the importation costs.

4.7 Finance for Climate Smart Agriculture (CSA) Practices in EA

Climate-Smart Agriculture (CSA) is one of the innovative approaches for sustainably increasing agricultural productivity, improving livelihoods and farmers' incomes, while at the same time, enhancing resilience and contributing to climate change mitigation (Karume et al., 2022). Evidence exists on how CSA practices could help mitigate climate shocks and, thus, strengthen food security and build farmers' resilience in SSA (Lipper et al., 2014; FANRPAN/Earth System Governance Project, 2017; Waaswa et al., 2022; Karume et al., 2022). Agricultural practices and technologies, and their knowledge, transfer, and adoption are fundamental in increasing productivity, rural incomes and growth, and subsequently, alleviating poverty, as well as adapting to climate change (Thornton et al., 2018; Karume et al., 2022). The government of EA countries should develop policy frameworks that are conducive that can effectively incentivise CSA. A strong emphasis must be on building the capacity of extension workers, producers, and other stakeholders in the use of CSA technologies, and improving infrastructures (Karume et al., 2022). University curricula should, therefore, be revised accordingly, to train a generation of CSA sensitive human resources. Climate Smart Agriculture practices are commonly clustered under five general themes: agronomy, agroforestry, livestock and aquaculture, post-harvest management, and energy systems (Rosenstock et al., 2016; Karume et al., 2022). Rosenstock et al. (2016) provide a checklist of 73 common CSA practices in developing countries, based on the literature from FAO CSA sourcebook and IPCC.

The effective implementation of such CSA practices can help reduce the rate at which food is imported and increase resilience to climate shocks, and thus, achieve stability in prices at local markets and increased food availability among the populations. Generally, barriers to the wide use/dissemination of CSA practices seem to be linked to low governmental investment, a lack of technical supervision, and weakness of extension services and farmer-support structures (Karume et al., 2022).

A faulty agricultural extension service is due to several causes: namely the shortage of qualified staff, lack of facilities and mobility means, low budget, dominance by old staff, and more importantly, a limited awareness of the extension officers on CSA practices and technologies (Karume et al., 2022). Consequently, such services have not yielded the desired outcomes of scaling-up innovative technologies at the village level.

4.8 Conclusions and Recommendations

Countries in EA can achieve sustainable food security if they promote farming practices that offer resilience to climate shocks, which can increase productivity per unit area, and limit the countries' reliance on imports (KIPPRA, 2022). Ways to achieve these objectives (according to KIPPRA, 2022) involve:

4.8.1 Increasing Governmental Investments on the Agricultural Sector

Countries in EA require heavy investment in agriculture to move from food deficiency to sufficiency. Being food sufficient will protect the public from the effects of global shocks, which precipitate food imports. For instance, budget allocation for agriculture in EA is way below the required 10% average recommended by the Comprehensive Africa Agriculture Development Programme (CAADP). The increased allocation will enable more investment in storage facilities that can help reduce post-harvest losses. Increased allocation can also result in more investment on irrigation schemes and other CSA practices that could enhance agricultural productivity. Investments could be used to provide more extension services to farmers to improve their productivity. The ratio of extension service officers to farmers in EA is far below the 1:400 that FAO recommends.

4.8.2 Responding to Climate Change

Most countries in EA, except the DRC, drafted CSA strategies (Karume et al., 2022) to promote high value agriculture. However, this has not translated entirely to CSA practices among farmers. To this end, farmers are not yet fully aware of some of the CSA practices that can engender resilience against climate change and limit over-reliance on rain-fed agriculture. Sectorial authorities may consider promoting CSA as a solution by localising the content to enable its fast adoption by farmers. These practices include supplying farmers with drought resistant crop varieties, training farmers on how to use more efficient production technologies: for instance, hydroponics and aquaponics, training farmers on best disease and pest management practices, providing targeted training extension services for agriculture and food-industry professionals, and promoting balcony farming in urban dwellings for a more sustainable farming and food system.

4.8.3 Reducing the Cost of Production

The reliance on fertiliser importation has exposed the EA to the adverse effects of inflation in the global markets. The governments must, therefore, invest in fertiliser plants to promote locally made fertilisers to shield farmers from the erratic price changes in the international markets. The governments have tried to shield farmers from high fertiliser prices by making subsidies available to them; however, the cost is still high. Besides, governments of the EA countries may consider tax exemptions on fertiliser to make it more affordable and accessible to farmers. These measures will help reduce the high food prices EA has been experiencing in the recent past.

4.8.4 Encouraging the Substitution of Import Food Products with the Locally Available

This proved to be effective in reducing wheat demands, as bakery products are now incorporating cassava flour to substitute wheat flour (Irakiza et al., 2021). Such a strategy was successful in the DRC after the government has encouraged the use of cassava in bakery products to cope with wheat flour price inflation and unavailability during the COVID-19 pandemic and the Ukraine-Russia conflict.

References

Africa News. (2022). DRC: War in Ukraine sparks food-security fears. https://www.africanews.com/2022/03/16/drc-war-in-ukraine-sparks-food-security-fears. Accessed 26 June 2023.

Anderson, W., Taylor, C., McDermid, S., Ilboudo-Nébié, E., Seager, R., Schlenker, W., Cottier, F., de Sherbinin, A., Mendeloff, D., & Markey, K. (2021). Violent conflict exacerbated drought-related food insecurity between 2009 and 2019 in sub-Saharan Africa. *Nature Food*, *2*(8), 603–615.

Ayana, E.K., Ceccato, P., Fisher, J.R., & DeFries, R. (2016). Examining the relationship between environmental factors and conflict in pastoralist areas of East Africa. *Science of The Total Environment*, *557*, 601–611.

Boliko, M.C. (2019). FAO and the situation of food security and nutrition in the world. *Journal of Nutritional Science and Vitaminology*, *65*(Supplement), S4–S8.

Burrier, E.A. (2022). *In Africa, Putin's War on Ukraine Drives Food, Fuel and Finance Crises*. https://www.usip.org/publications/2022/06/africa-putins-war-ukraine-drives-food-fuel-and-finance-crises. Accessed 4 August 2023.

Dalheimer, B., Herwartz, H. & Lange, A. (2021). The threat of oil market turmoils to food price stability in Sub-Saharan Africa. *Energy Economics, 93*, 105029.

Dillon, B. M. & Barrett, C. B. (2016). Global oil prices and local food prices: Evidence from East Africa. *American Journal of Agricultural Economics, 98*(1), 154–171.

Dodo, M.K. (2020). Understanding Africa's food security challenges. In B. Mahmoud (Ed.), *Food security in Africa* (pp. 1–17). London: IntechOpen. https://doi.org/10.5772/intechopen.91773

FANRPAN & Earth System Governance Project. (2017). Climate-Smart Agriculture in the DRC. *JSTOR*, 1–8. https://www.jstor.org/stable/resrep16466. Accessed 4 August 2023.

FEWS NET. (2022a). Kenya Food Security Outlook, June 2022 to January 2023. https://fews.net/east-africa/kenya/food-security-outlook/june-2022. Accessed 10 June 2023

FEWS NET. (2022b). Uganda Key Message Update: Rising prices continue to constrain food access as recent rainfall deficits threaten crop progress. https://reliefweb.int/report/uganda/uganda-key-message-update-rising-prices-continue-constrain-food-access-recent-rainfall-deficits-threaten-crop-progress. Accessed 10 March 2023.

FEWS NET. (2022c). Delayed season A rainfall will delay the harvest, leading to an extended lean season. https://fews.net/east-africa/burundi/food-security-outlook/october-2022. Accesed 23 June 2023.

Giller, K.E. (2020). The food security conundrum of sub-Saharan Africa. *Global Food Security, 26*, 100431.

Haile, G.G., Tang, Q., Sun, S., Huang, Z., Zhang, X. & Liu, X. (2019). Droughts in East Africa: Causes, impacts and resilience. *Earth-Science Reviews, 193*, 146–161.

Hillbruner, C. & Moloney, G. (2012). When early warning is not enough—Lessons learned from the 2011 Somalia Famine. *Global Food Security 1*, 20–28

IMF. (2022a). Climate Change and SSA's intensified food insecurity. Washington DC.: IMF Departmental Paper.

IMF. (2022b). Staple Food prices in sub-Saharan Africa: An empirical assessment. Washington DC.: IMF Departmental Paper.

Irakiza, P.N., Chuma, G.B., Lyoba, T.Z., Mweze, M.A., Mondo, J.M., Zihalirwa, P.K., Mapatano, S., Balezi, A. Z. & Mushagalusa, G.N. (2021). Fortification with mushroom flour (*Pleurotus ostreatus* (Jacq.) P. Kumm) and substitution of wheat flour by cassava flour in bread-making: Nutritional and technical implications in eastern DR Congo. *Agriculture & Food Security, 10*(1), 1–11.

ITA. (2022). Kenya – Country Commercial Guide: Agribusiness. https://www.trade.gov/country-commercial-guides/kenya-agribusiness. Accessed 3 March 2023.

Kakpo, A., Mills, B.F. & Brunelin, S. (2022). Weather shocks and food price seasonality in sub-Saharan Africa: Evidence from Niger. *Food Policy, 112*, 102347.

Karume, K., Mondo, J. M., Chuma, G. B., Ibanda, A., Bagula, E. M., Aleke, A. L., Ndjadi, S., Ndusha, B., Ciza, P. A., & Cizungu, N. C. (2022). Current Practices and Prospects of Climate-Smart Agriculture in Democratic Republic of Congo: A Review. *Land, 11*(10), 1850

Kenya National Bureau of Statistics (2022). Consumer price indices and inflation rates for November 2022. Kenya National Bureau of Statistics, Nairobi, Kenya. https://www.knbs.or.ke/. Accessed 24 June 2023.

KIPPRA. (2022). Four Ways to Address the Rising Food Prices in Kenya. https://kippra.or.ke/four-ways-to-address-the-rising-food-prices-in-kenya/

Kiros, F.G., 1991. Economic consequences of drought, crop failure and famine in Ethiopia, 1973–1986. *Ambio, 20*(5),183–185.

Lipper, L., Thornton, P., Campbell, B. M., Baedeker, T., Braimoh, A., Bwalya, M., Caron, P., Cattaneo, A., Garrity, D., Henry, K. & Hottle, R. (2014). Climate-smart agriculture for food security. *Nature Climate Change, 4*(12), 1068–1072.

Lyon, B., & Vigaud, N. (2017). Unraveling East Africa's Climate Paradox. In Simon Wang, S.-Y., Yoon, J.-H., Funk, C.C. & Gillies, R. R. (Eds), *Climate Extremes: Patterns and Mechanisms*, 265–281.

Lyon, B. (2014). Seasonal drought in the Greater Horn of Africa and its recent increase during the March-May long rains. *Journal of Climate*, *27*(21), 7953–7975.

Mburu, P., & Ondieki, E. (2022). Maize prices set to rise after lapse of tax waiver on imports. https://nation.africa/kenya/business/maize-prices-set-to-rise-after-lapse-of-tax-waiver-on-imports-3970254. Accessed 25 June 2023.

Meier, P., Bond, D., & Bond, J. (2007). Environmental influences on pastoral conflict in East Africa. *Political Geography*, *26*(6), 716–735.

Muller, J.C.Y., (2014). Adapting to climate change and addressing drought–learning from the Red Cross Red Crescent experiences in East Africa. *Weather and Climate Extremes*, *3*, 31–36.

Okou, C., Spray, J., & Unsal, D.F. (2022). Africa food prices are soaring amid high import reliance. United States of America, Washington DC.: International Monetary Fund. https://www.imf.org/en/Blogs/Articles/2022/09/26/africa-food-prices-are-soaring-amid-high-import-reliance. Accessed 25 March 2023

O'Loughlin, J., Witmer, F.D.W., Linke, A.M., Laing, A., Gettelman, A. & Dudhia, J. 2012. Climate variability and conflict risk in East Africa, 1990–2009. *Proceedings of the National Academy of Sciences*, 109(45), 18344–18349. https://doi.org/https://doi.org/10.1073/pnas.1205130109

Rosenstock, T.S., Lamanna, C., Chesterman, S., Bell, P., Arslan, A., Richards, M., Rioux, J., Akinleye, A.O., Champalle, C., Cheng, Z., Corner-Dolloff, C., Dohn, J., English, W., Eyrich, A.S., Girvetz, E.H., Kerr, A., Lizarazo, M., Madalinska, A., McFatridge, S., Morris, K.S., Namoi, N., Poultouchidou, N., Ravina da Silva, M., Rayess, S., Ström, H., Tully, K.L., & Zhou, W. (2016). The scientific basis of climate-smart agriculture: A systematic review protocol. CCAFS Working Paper no. 138. Copenhagen, Denmark: CGIAR Research Program on Climate Change, Agriculture and Food Security (CCAFS). Available online at: www.ccafs.cgiar.org. Accesed 8 August 2023.

Sánchez, A. & Namhata, C. (2019). What feeds protest participation in sub-Saharan Africa? An empirical analysis. *Global Food Security*, *23*, 74–84.

Serra, T., & Zilberman, D. (2013). Biofuel-related price transmission literature: A Review. *Energy Economics*, *37*, 141–151.

Shiferaw, B., Tesfaye, K., Kassie, M., Abate, T., Prasanna, B. M. & Menkir, A. (2014). Managing vulnerability to drought and enhancing livelihood resilience in sub-Saharan Africa: Technological, institutional and policy options. *Weather and Climate Extremes*, *3*, 67–79.

Thornton, P.K., Whitbread, A., Baedeker, T., Cairns, J., Claessens, L., Baethgen, W., ... & Keating, B. (2018). A framework for priority-setting in climate smart agriculture research. *Agricultural Systems*, *167*, 161–175.

Uganda Bureau of Statistics. (2022). Consumer price indices and inflation rates. Press release. Kampala, Uganda, 5p. https://www.ubos.org/wp-content/uploads/. Accessed 23 May 2023.

UN News. (2022). Conflict, drought, dwindling food support, threatens lives of 20 million in Ethiopia. https://news.un.org/en/story/2022/06/1121132. Accessed 23 June 2023.

Waaswa, A., Oywaya Nkurumwa, A., Mwangi Kibe, A., & Ngeno Kipkemoi, J. (2022). Climate-Smart agriculture and potato production in Kenya: Review of the determinants of practice. *Climate and Development*, *14*(1), 75–90.

Wasike, A. (2022). Facing food crisis, Kenya lifts ban on genetically modified crops. https://www.aa.com.tr/en/health/facing-food-crisis-kenya-lifts-ban-on-genetically-modified-crops/2702427. Accesed 21 June 2023.

WFP. (2021). *Global Report on Food Crises – 2021*. Washington, DC: World Food Programme. https://www.wfp.org/publications/global-report-food-crises-2021. Accessed 20 May 2023.

WFP. (2022a). Horn of Africa: Extreme drought deepens hunger in a region facing conflict. Washington, DC, USA: World Food Programme. https://www.un.org/africarenewal/magazine/. Accessed 20 May 2023.

WFP. (2022b). Burundi interim country strategic plan (2022–2024). First regular session, Rome, 28 February–2 March 2022. https://reliefweb.int/report/burundi/burundi-annual-country-report-2022-country-strategic-plan-2022-2024. Accessed 4 August 2023.

WFP. (2023). A global food crisis. https://www.wfp.org/global-hunger-crisis. Accessed 26 June 2023.

WFP Regional Bureau for Eastern Africa. (2022). Eastern Africa market and trade update. https://reliefweb.int/report/burundi/eastern-africa-market-and-trade-update-october-2022. Accessed 20 June 2023.

Zhao, T. & Dai, A. (2015). The magnitude and causes of global drought changes in the twenty-first century under a low–moderate emissions scenario. *Journal of Climate*, *28*(11), 4490–4512.

Chapter 5
Climate Change and Its Implications on Food Security in the Great Lakes Region

Jackson Ishara, Ayorinde Ogunyiola, Rehema Matendo, Jean Chrysostome K. Kiyala, and Katcho Karume

Abstract Climate change (CC) is an existential threat globally, and the Great Lakes Region (GLR) is not immune to the impacts of these extreme events. Erratic weather and extreme events such as rainfall and droughts have serious impacts on agriculture, a major income pillar for African smallholder farmers, and an essential source of food. Climate vulnerability in the GLR is exacerbated by multiple biophysical, political, and socio-economic constraints, thus limiting adaptive capacity within the GLR. Climate vulnerabilities in the GLR suggest that research and development actors, including farmers themselves, must take action to invest massively in mechanisms that can mitigate the situation. In light of the emerging literature, this chapter documents the most recent literature on climate change and its impact on food security, including the impact on food resources, food availability, food accessibility,

Corresponding author: Mr Jackson Ishara is a lecturer at the Department of Food Science and Technology, Université Evangélique en Afrique, Bukavu, D.R. Congo; and at the Department of Food Science and Technology, Jomo Kenyatta University of Agriculture and Technology, Nairobi, Kenya; Email: jackishara17@uea.ac.cd; jack.ishara@outlook.com; https://orcid.org/0000-0003-4789-4040

Dr Ayorinde Ogunyiola is aresearcher in the Department of Forestry & Natural Resources, Purdue University, West Lafayette, Indiana, USA and an Assistant Professor in the Department of Political Science and Sociology, Murray State University, KY, USA. Email: aogunyiola@murraystate.edu

Mrs Rehema Matendo is senior lecturer at the Department of Food Science and Technology, Université Evangélique en Afrique, Bukavu, D.R. Congo; and at the Department of Food Science and Nutrition, Jomo Kenyatta University of Agriculture and Technology, Nairobi, Kenya. Email: vecordcongoesther@gmail.com

Dr Jean Chrysostome K. Kiyala holds a PhD in Public Administration: Peace Studies from DUT; he is a Senior Lecturer at the International Centre of Nonviolence (DUT), Associate Professor and visiting lecturer at the Université de Bandundu and the Université Evangélique en Afrique (Democratic Republic of Congo); Emails: JeanK@dut.ac.za; kljeanchrysostome@gmail.com; ORCID: https://orcid.org/0000-0002-9587-0230

Professor Katcho Karume lectures at the Faculty of Agricultural and Environmental Studies, at the Université Evangélique en Afrique, P.O. Box 3323-Bukavu, D.R. Congo, and researcher at the Centre de Recherche en Géothermie, Bukavu, Democratic Republic of Congo; Emai: kkatcho@yahoo.com

J. C. K. Kiyala and N. Chivasa (eds.), *Climate Change and Socio-political Violence in Sub-Saharan Africa in the Anthropocene*, The Anthropocene: Politik—Economics—Society—Science 37,
https://doi.org/10.1007/978-3-031-48375-2_5

food utilisation and stability of food products, as well as overall food productivity, water sources and arable land in the GLR and proposes adaptation strategies based on climate-smart agriculture (CSA) to alleviate the impacts of the ongoing CC.

Keywords Climate change · Food security · Great Lakes Region · Climate-smart agriculture

5.1 Introduction

Climate change has received special attention for decades because of the threat it poses to our planet's ecosystems and human societies, with consequences for food security. Climate change is attributed, on the one hand, to anthropogenic activities that modify the composition of the global atmosphere and, on the other hand, to the natural variability of the climate (Werksman, 1994). The Intergovernmental Panel on Climate Change (IPCC) reported an average temperature increase at the global surface between 1.7 and 4.8 °C until the end of the twenty-first century caused by four scenarios, including population and economic growth as well as carbon use (Pörtner et al., 2022). Extreme weather events and natural disasters, such as floods, droughts, hurricanes, tsunamis, increased frequency of heavy precipitation, prolonged periods of drought, water acidification, and potential sea level rise, are among other effects of climate change that may affect fisheries and livestock production, as food production is linked to all of these different factors (Misiou/Koutsoumanis, 2022).

The erratic change in weather is adversely affecting communities already vulnerable to poverty and weak institutional support and can thus cause devastating consequences for food security and livelihoods in some African countries and communities, including the GLR (Akrofi, 2012; Devereux, 2007; Misselhorn, 2005). Moreover, extreme droughts are already hampering people's ability to grow food and raise livestock to maintain their food security and wellbeing (Negatu et al., 2010; Songok et al., 2011), in addition to multiple biophysical, political and socio-economic constraints that make the region too vulnerable by limiting its adaptive capacity (Davidson et al., 2003; Reid/Vogel, 2006). Especially since it is estimated that the livelihoods of 70% of Africans depend on rain-fed agriculture (Challinor et al., 2007; World Bank, 2009).

Although there is extensive empirical research on the impact of climate change on different aspects of agriculture, including fisheries and livestock production, there is a paucity of research on the effect of climate change on food security in the GLR (Change, 2020). Against this background, the implications of climate change on food security in the GLR are documented.

Thus, this chapter documents the implications of climate change on food security, including the impact on food resources, their availability, accessibility, utilisation and stability, as well as the overall productivity of food, water sources and arable lands, while paying particular attention to some mechanisms to mitigate the situation and adaptation strategies to deal with the scourge of climate change. Understanding the relationships between climate, food and livelihoods is not only a scientific imperative but also a necessity for guiding practical initiatives to improve livelihoods and food security. Such a perspective is based on the climate-smart agriculture (CSA) framework, which constitutes the main argument of this chapter.

CSA is defined by the World Bank as "an integrated approach to managing landscapes—cropland, livestock, forests and fisheries—that addresses the interlinked challenges of food security and accelerating climate change". CSA has the following three objectives, according to Lipper et al. (2014):

(1) to increase productivity in terms of sufficient quality food to ameliorate nutrition security and increase incomes, particularly for 75% of the world's poor population living in rural areas and who mainly rely on agriculture for their livelihoods.
(2) To strengthen resilience by reducing vulnerability to drought, pests, diseases, and other risks associated with climate factors and shocks and boosting the capacity to adapt and grow amidst longer-term stresses such as shortened seasons and erratic weather patterns.
(3) To minimise emissions by pursuing lower emissions food production, avoid deforestation from agriculture and identify ways to absorb carbon out of the atmosphere.

Lipper et al. (2014) averred that "climate-smart agriculture (CSA) is an approach for transforming and reorienting agricultural systems to support food security under the new realities of climate change" (p. 1068). In a similar vein, Azadi et al. (2021) remarked that "climate-smart agriculture (CSA) is seeking to overcome the food security problem and develop rural livelihoods while minimising negative impacts on the environment" (p. 1). The above descriptions of CSA overlap and focus on transformative, adaptative and innovative approaches to mitigate the negative impacts of climate change in order to ensure food productivity and quality are not affected by environmental factors, especially climate change.

While major international institutions actually argue on the compulsory need to transition to CSA to ensure there is sufficient supply for a projected world population of 50.9 billion by 2050, Taylor, (2018) has proposed an alternative 'climate-wise' framework to foreground the inherently political dimensions of food and agriculture in an era of climatic change.

Understanding the CSA framework is critical to enhancing productivity and limiting climate-driven food insecurity (Gardezi et al., 2022; Ogunyiola et al., 2022). We look at the Great Lakes Region case and suggest some ways to adopt and apply CSA to ensure that the agricultural sector in the region produces sufficient food of high quality, minimises climate vulnerability, boosts resilience and reduces emissions that occur in the process of producing food of high standards.

This chapter consists of 11 main sections, namely, the introduction, which lays out what the chapter intends to contribute to knowledge, and a general background of climate change and its impact on food security, food availability and overall food productivity. In subsequent sections, the impact of water on land resources, climate change and food security are discussed, and how existing mechanisms are shaping food production as the world faces the challenge of climate change is demonstrated. The last section before the conclusion argues for the adoption of CSA to address the adverse effects of climate change and ensure the Great Lakes Region produces sufficient and quality food, particularly for rural populations.

5.2 General Background and Climate Change Status

Climate change is one of the major threats of the current century because of its negative impacts on components of global food security. The mitigation of climate change is imperative to meet the future food demands of the growing population. Agricultural productivity in general and food crops (wheat, maize and rice) in particular are already threatened by climate change (Lobell et al., 2011; Syed et al., 2022). These impacts do not spare livestock and fisheries, which are also important for food security.

In the coming decades, we are likely to see disasters if nothing is done to limit these climatic variations, such as low and high temperature stresses, changes in precipitation frequency and intensity, and other disasters such as drought, salinity, and rising sea levels (Raza et al., 2021; Stablein et al., 2022; Ullah et al., 2021). Some uncertainties have been associated with climate fluctuations for decades and will likely continue to hinder food security goals unless appropriate research-based strategies are implemented (Heal/Millner, 2014).

The adverse effects of climate change are exacerbated by human activities, as they have intensified the warming process of the atmosphere, oceans and land by substantially and unequivocally increasing well-mixed greenhouse gas (GHG) concentrations since 1750, which has been followed by a continuous increase in the major gases, approaching annual averages of 410 ppm for CO_2, 1866 ppb for CH_4, and 332 ppb for N_2O in the atmosphere until 2019 (Myhre et al., 2017). The global average surface temperature between 2001 and 2020 was 0.99 °C higher than in the period 1850–1900. Furthermore, the global average surface temperature was 1.09 °C higher in the period 2011–2020 than in the period 1850–1900, with faster warming at the land surface (nearly 1.59 °C) than in the ocean (nearly 0.88 °C) (Misiou/Koutsoumanis, 2022). The projected relative magnitude of total human-induced global mean surface warming between 1850–1900 and 2010–2019 ranges from 0.8 to 1.3 °C, with a more precise estimate of 1.07 °C. The mixing of different GHGs was predicted to confer a warming of 1.0–2.0 °C, other human factors shared a mean temperature change of 0.8 °C, natural factors changed the global mean surface temperature ranging from − 0.1 to + 0.1 °C, while internal variabilities changed the surface temperature ranging from − 0.2 to + 0.2 °C (Zhou, 2021).

It is assumed that human activities have contributed to the observed precipitation shift patterns since the twentieth century and are most likely subject to near-surface ocean salinity (Hatfield/Prueger, 2015). Human activities are expected to be the primary drivers of global glacier retreat since the 1990s and the decrease in Arctic sea ice area between 1979 and 2019 (Wang et al., 2019). No significant trend change was observed in the Antarctic Sea ice area during 1979–2020 due to conflicting regional trends and huge internal variabilities, while the decrease in Northern Hemisphere spring snow cover was observed after the 1950s. Although little evidence exists on the contribution of human activities to the Greenland ice sheet surface melt observed in recent years, there is reason to emphasise their impact on Antarctic ice losses (Aschwanden et al., 2019; Moon et al., 2018).

The earth's surface temperature has been warming more rapidly since the early nineteenth century (Hegerl et al., 2019; Rohde et al., 2013). Global average land surface, sea, and ocean temperatures have increased by 0.8–1.5 °C since the beginning of the nineteenth century (Solomon et al., 2007), and most of this temperature increase was observed after 1975 (Hansen et al., 2006; Pierrehumbert, 2000). The various factors involved in climate change are predicted to likely lead to the incidence of more severe extreme events such as droughts, floods, and heat waves (Schiermeier, 2018; Solomon et al., 2007).

Global warming patterns are not expected to be uniform across the globe, with arid and oceanic regions at greater risk from the consequent impacts of global warming and extreme events (Solomon et al., 2007; Spinoni et al., 2021). Coastal livelihoods will be further threatened by flooding and salinity due to the anticipated climate change-induced rise in sea level. There are uncertainties in precipitation patterns, especially in tropical regions, due to the inability of climate models to present the hydrological cycle more accurately (Lorenz et al., 2016).

Atmospheric concentrations of CO_2 and CH_4 and N_2O are at their highest levels in 2 million years and 0.8 million years, respectively. Indications of observed changes in extreme events such as heat waves and frequent and erratic precipitation are, in most cases, the consequences of human influence. In contrast to extreme hot events that have become more intense and frequent across many global regions since the 1950s, cold waves have become relatively less severe and less frequent (Zheng et al., 2023), where human-induced climate change is one of the main drivers of these extreme events (Bell et al., 2018). In addition, the frequency and intensity of severe precipitation events have shown an increasing trend after the 1950s in different regions of the world due to many natural and anthropogenic factors (Ayugi et al., 2022; Jiménez-Donaire et al., 2020). An increase in precipitation after the 1980s resulted from increased atmospheric GHG levels and decadal to multi-decadal internal modulations (Johnson et al., 2020). It has been indicated that human-induced climate change generally causes heavy precipitation events associated with tropical cyclones. Since the 1950s, erratic and uncontrolled human activities have increased the risks of compound extreme events coexisting, including the increased frequency of simultaneous global heat waves and droughts, wildfires in various parts of all continents and increased flooding (Coscarelli et al., 2021; Ye et al., 2019).

The earth has experienced various cooling and warming fluctuations in the past, long before human activities. Projections and analyses have shown that the warming of the earth that has occurred mainly since the last century is occurring at a rapid rate that cannot be interpreted solely by natural causes. Cyclical changes in the earth's orbit and tilt can also cause climate change, as can volcanic eruptions that result in CO_2 release, as well as emissions of various aerosols, including volcanic dust or ash, and sulphur dioxide (Reikard, 2019). On the other hand, volcanic aerosols can even restrict a significant percentage of solar radiation and cause a cooling effect that can last for years. The total amount of solar radiation reaching the Earth's surface varies little, as the energy released by the sun differs by only 1.3 W/m^2, which is associated with darker areas on the sun called sunspots (Adnan et al., 2020; Hussain et al., 2018).

Climate change is considered a universal truth around the world, with consequent untoward impacts on agricultural production, water resources, biodiversity, human livelihoods, forestry systems, and socio-economic components (Muluneh, 2020), while forestry systems are a huge source of food such as edible insects (Ishara et al., 2023; Ishara et al., 2022a) and mushrooms (Ishara et al., 2022b) in most rural communities. Developing and underdeveloped countries, including the GLR, will suffer negative impacts from anticipated global climate change due to lower adaptive capacity. The most vulnerable are ordinary citizens, especially the poor at the regional level (Barnett et al., 2015). Uncontrolled urbanisation and the industrial revolution have increased GHG emissions that have caused a massive intensification of global warming and have a prolonged capacity (even for decades) to accelerate the warming process. The likely challenges of undernourishment, malnutrition and food insecurity are the most significant impacts of climate change. Extreme events associated with climate change include food insecurity resulting from the negative impact on livelihoods and social emotions (Iniguez-Gallardo et al., 2021), land use, reliability and quantity of available irrigation water and other agricultural resources (Davidson, 2018).

5.3 Global Climate Change and Impact on Food Resources

In Sub-Saharan Africa (SSA) and in the GLR in particular, 95% of agricultural activities are rain-dependent. Current and future climate change could have serious consequences not only for food security but also for the environment (Mupangwa et al., 2021), as well as changes in the length of the growing season with a propensity to decrease the length of the growing season (Change, 2014). The impact of climate change on agriculture is not limited to crop production but also affects livestock and other agriculture-related activities, as climate variables such as air, temperature, humidity, wind speed, and other climatic factors directly influence livestock performance such as livestock growth, milk production, wool production, and reproduction, while also influencing the quantity and quality of feed (pasture, forage, and grain) and the occurrence and spread of livestock diseases and parasites (Gezie, 2019).

Low temperature stress causes serious threats to agricultural production worldwide (Ferrante et al., 2021; Zhang et al., 2022). This loss is up to 100% in the winter for cereals. Photosynthesis is the essential source to optimise metabolic and energy transformation processes in plants, but it is highly sensitive to abiotic stresses (Shah, 2020), such as drought and high and low temperature stresses (Zhang et al., 2022).

Therefore, there is a need to improve plant metabolism, antioxidant mechanisms and osmolyte metabolism, and to modulate physiochemical attributes to combat biotic and abiotic stresses (Shah, 2020).

In addition, due to climate change, the rapid increase in the frequency and intensity of extreme weather disasters, including floods, droughts and wildfires, is having a devastating impact on food security and livelihoods. This rapid increase in natural disasters requires substantial disaster risk reduction management measures, potential policies, and intensified approaches to building resilience to the adverse effects of climate change that can ensure a sustainable and productive future (Sloggy et al., 2021). Compared to agriculture, industry, trade, and tourism, the agricultural sector alone suffers an immeasurable share of nearly 63% of the impact of natural disasters, and resource-poor countries are most at risk (Ali et al., 2017). Globally, between 2008 and 2018, the agriculture sector faced an economic loss of US$108 billion in terms of damage or reduction in crop and livestock production (Fao, 2018). This damage can be particularly disastrous for the livelihoods of communities practising subsistence agriculture with limited resources. Emerging crops and livestock pests and diseases have also become major disasters hindering food security and may cause large economic losses in the near future for crops, agroforestry, and livestock, ultimately jeopardising food security. Empirically, these impacts of climate change and natural disasters vary at different stages.

With respect to agriculture, the direct and positive impacts of natural disasters are actually quickly identifiable, such as typhoons improving the supply and availability of water for agriculture (Israel/Briones, 2012). Additionally, individually, windstorms, typhoons, and floods can limit the physical movement of farms, supply infrastructure, supply roads, and cause death or injury to the workforce involved in agriculture. Therefore, direct and negative variables can have indirect and negative impacts on agriculture and cause substantial economic losses (Kumar et al., 2022). In addition, these climate change disasters can also lead to the accumulation of several wastes, water pollution, deformation of land topography, reduced rainfall, decreased soil fertility, and increased saltwater intrusion (Akhtar et al., 2021; Shrivastava/Kumar, 2015).

All of these phenomena can indirectly reduce the sustainability of land and water resources, reduce ecosystem services, and put human health and safety at risk. Considering all disasters, the potential direct and indirect negative impacts on agriculture and agroforestry threaten food security, most obviously in areas already vulnerable to climate change and natural disasters. Although there is strong evidence of the impact of climate change on nutrition and mortality, there are inevitably uncertainties due to a limited understanding of the extent of climate change impacts on the components of food security (Shoaib et al., 2021; Springmann et al., 2016). Therefore, examining the mechanisms of climate change impacts on all components of food security and livelihoods is necessary to fill research gaps critical to achieving regional food security goals. Climate change has decreased consumable calories in the daily diet over the past few decades (Ray et al., 2019), stimulating household food insecurity issues as it impacts all determinants of food security and factors affecting

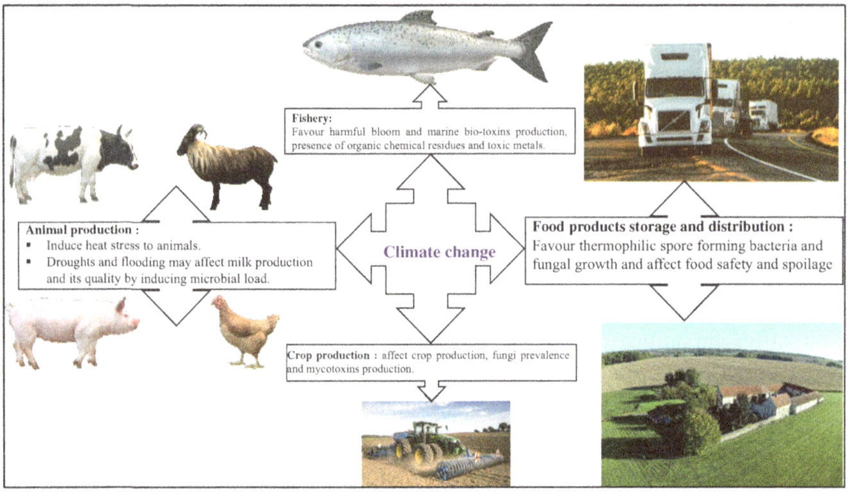

Fig. 5.1 Overview of climate change impact on food production and safety. *Source* The Authors

food production, access, availability, and stability (Mekonnen et al., 2021). Climate change influences all components of food production and security, as depicted in Fig. 5.1.

5.4 Impact of Climate Change on Food Availability and Access

Food availability is when people have enough food to meet their dietary needs, which also covers the supply side of the food chain. This is determined by food production, available technologies, inventories, supply chain efficiency, and trade policies at the national and international levels (Ali/Erenstein, 2017). As a result, the magnitude of change in agricultural production will vary from global to regional, country, and community scales due to alternative and contrasting projections of agricultural models. There is a strong and well-reasoned pattern of climate change impacts on food production and availability, with the most severe effects occurring in areas with the highest prevalence of undernourishment and food insecurity problems. Declines in the productivities of staple crops such as wheat, maize, and sorghum will be robust, but the analysis was inconclusive and showed conflicting results for other major crops such as rice and sugarcane (Knox et al., 2012).

Food accessibility can be defined as a household's ability to obtain a sufficient amount of quality food to meet its dietary needs (Masipa, 2017). Food access is largely based on household income, capabilities, and entitlements (Gartaula et al., 2017; Gupta et al., 2019). Food access issues are studied through two main channels: first, from top-down models that address and link major fluctuations to household responses and thus adaptation outcomes. Second, through communities and households, attempting to assess the impacts of climate variability from the bottom up. The production of certain food products, such as crops, fisheries, livestock, and forestry, is being transformed due to the adverse effects of climate change at the local, regional, and global levels, ultimately impacting the production of biomass, including fiber, feed, food, and fuel. The same is true for land use change, which impacts global food access by altering the geography of the food system that involves crops, livestock, and forestry and impacts farm-level incomes (Peña-Lévano et al., 2019). Therefore, micro- and macro level analyses for assessing the impacts of climate change on each component of food are needed to help achieve food security goals. Due to the impacts of climate change, some secondary structural consequences in a region or community may occur where there is a lack of property rights as well as a lack of protection of water and land resources that cause many food security problems, especially in developing and underdeveloped countries (Godfray et al., 2010).

Projections show that the impacts of climate change on food access will strain people's ability to purchase food. According to the protections of interdependent models on the prices of any agricultural and agroforestry food commodity and its trade in the near future under several climate, social, and economic change scenarios, the projected results show negative impacts on the food purchasing power of any population considered (Nelson et al., 2014). There are several indications that food prices will increase, suggesting uncertainties in the results projected by various macro- and micro-economic models (Nelson et al., 2014). Climate change significantly affects household purchasing power and access to food (Tol, 2009), while most likely affecting the geography of food production around the world and thus affecting prices, trade, and, consequently, access to food (Jones et al., 2017).

5.5 Climate Change and Food Utilisation as Well as Stability

Climate change also strongly influences the use of food to meet dietary and nutritional needs through factors such as water availability and sanitation facilities. There is a clear link between food use and climate change, as climate variability limits the availability of food items such as drinking water (Duran-Encalada et al., 2017). Hygiene problems during extreme weather events such as floods or droughts are more frequent in undeveloped areas as a result of unsound sanitation systems (Seneviratne et al., 2014). As a result, these sanitation problems can lead to gastric diseases, which can reduce the intake of essential nutrients (Lloyd et al., 2011).

The dimensions of food security across the food supply chain and direct or indirect health impacts due to climate change are two attributes that primarily influence food utilisation. Generally, high and low temperature stresses reduce the food security of needed commodities, particularly fruit and vegetables, due to increased rates of microbial activity under altered environmental conditions (Alegbeleye et al., 2022; Galstyan et al., 2022). Climate change influences the food system in dynamic ways, directly or indirectly, such as vector-borne diseases, high and low temperature stresses, and natural disasters including droughts and floods, thus affecting income, nutrition, safety, and childcare. Climate change indirectly impacts human health via the loss of jobs and critical livelihoods, migration and intervening public health services, which disproportionately influence indigenous communities and the already poor with negative consequences for food security (Ebi/Hess, 2020; Hayes et al., 2018).

Climate drives food price trends, both long- and short-term price variability, and thus influences food stability (Myers et al., 2017). Even small shocks in the food chain often increase food prices, whether on the demand or supply side (Haile et al., 2017). This situation of food instability is all too common in poor societies because it makes food supply and demand more unpredictable, thus increasing food volatility (Haile et al., 2017). This fluctuation in demand jeopardises food stability, especially when political programs and policies interfere, for example, with different types of subsidies for crops, livestock, or the forestry industry (Wheeler et al., 2021). These types of policy transitions have been implemented in some developed countries not only to mitigate climate impacts but also to adapt to them. It has been noticed that food crises of varying degrees have emerged due to the lack of short- and long-term adaptation and mitigation strategies to climate change and extreme events, the declining productivity of crops, livestock, fisheries, and agroforestry, and the absence of sound policies. In addition, this situation of food crises has been aggravated by a lack of policy implementation, a lack of expertise due to restrictions in developed countries, limited access to lucrative markets, and a lack of price regulation systems (Haile et al., 2017). As a result of climate change, concerns about food destabilisation continue to grow (Roy/Haider, 2018).

Economic and political factors that may cause food insecurity for indigenous populations are secondary risks to food destabilisation (Berazneva/Lee, 2013). This complex aggregate of constraints and risks can potentially turn into a ruinous system for food security. There is a direct and indirect link between food security and the ecosystem. Pressure on ecosystems is intensified by climate change and extreme events (Berazneva/Lee, 2013). A fluctuation in temperature and frequency of extreme events due to climate change negatively impacts biodiversity and also alters relationships between and within communities (Mach et al., 2016). Climate variability causes various types of damage to soil, water quality, and livestock, as well as damage to human health and ultimately the ecosystem. Thus, climate change also threatens the social and economic components of the food chain. Marginalised and resource-poor communities are most at risk (Panpakdee/Limnirankul, 2018).

5.6 Climate Change and Overall Food Productivity

Agriculture is the main source for meeting food needs, next to seafood (Pradeepkiran, 2019). Climate projections have indicated with some certainty that global food production will decline due to uneven and uncertain climate variability, despite the intrinsic shortcomings of climate-crop modeling (Martinich/Crimmins, 2019). Recent climate assessments with some uncertainties by the IPCC have shown that the average productivity of major staple crops (rice, maize and wheat) would decrease by 310% with a 1 °C increase in temperature (Hatfield/Dold, 2018). In addition, several studies have highlighted that increased CO_2 concentrations in the air could allow some crops to produce more harvestable products, particularly C3 plants (Ramalho et al., 2018). Climate change will impact livestock and fisheries production through natural resource contention, feed quality and quantity, livestock disease, heat and cold stress, and biodiversity loss, while knowing that demand for fisheries and livestock products is expected to increase by 100% by mid-century (Rojas-Downing et al., 2017). Forest productivity would not be spared, particularly if vulnerable but ecologically critical species were lost due to the physiological effects of climate stress (Kramer et al., 2020). Poultry and dairy products are significantly impacted in terms of quality and quantity when temperatures exceed the optimal range (Rojas-Downing et al., 2017).

There would be a decline in potential fish production (510%) by 2050, specifically in tropical marine ecosystems (Lam et al., 2016), followed by expected changes in fish distribution due to increased warming, fluctuations in nutrient availability, and changes in pH (Hollowed et al., 2013). This decline in fish production could likely lead to essential mineral and zinc deficiency in 845 million people and vitamin B-12 and fatty acid deficiency in 1.4 billion people (Fao, 2018). A systematic analysis of nearly 5000 fish farms worldwide reported that 68% of global fish production units have fallen below their potential biomass yield, and are projected to increase up to 88% biomass loss by 2050 (Gaines et al., 2018). Underdeveloped communities, including the GLR community, are most at risk due to their limited resources to access alternative solutions. In addition, climate change would increase the prevalence of various new pests and diseases (Fones/Gurr, 2017).

By influencing the food supply chain, climate change impacts the use, quality, and diversification of food. Ensuring the hygienic use of quality and diversified foods will be even more challenging as climate change is likely to induce intense and frequent occurrences of extreme events such as drought, floods, cold stress, and heat stress (Raza et al., 2021; Roy/Haider, 2018). At the same time, the use of quality, sanitised, and diverse foods will be more difficult for resource-poor populations due to negative impacts on livestock, crops, and forests (Hammond et al., 2018). Especially as climate change and extreme events limit the availability and use of sanitised food (Wu et al., 2016). Projections show that extreme events and natural disasters will be more frequent and intense, with a negative impact on the availability of diverse and sanitised foods. This would increase health and nutritional constraints (Hoekstra et al., 2018).

Table 5.1 Impact of climate change on overall food production

Climatic components	Livestock	Food and feed crops	Labor and capital
Precipitation variation	1. Drinking water shortage 2. Pests and diseases	1. Reduced yield 2. Reduced food and food quality 3. Shifts in production systems	1. Shifts in human health 2. Migration 3. Reduced yield 4. Conflicts among and within a community
Temperature	1. Heat stress • Reduction in livestock productivity • Enhanced mortality 2. Pests and diseases • Reduced resistance against pests • Biodiversity loss	1. Reduced yield 2. Reduced food and food quality 3. Shifts in production systems	
CO_2		1. Partial or total stomata closure 2. Shifts in production systems	

Source Biglari et al. (2019), Connolly-Boutin/Smit (2016), and Thornton et al. (2009)

While several studies estimate that insects, pests, and weeds are responsible for about 2540% of the decline in productivity of major food crops (Deutsch et al., 2018), and fungal attacks alone reduce hygienic food availability by 8.5% worldwide (Udomkun et al., 2017), climate change, primarily global warming, increases the survival rate of insects, pests, and weeds (Deutsch et al., 2018). Unimproved native crop varieties are highly susceptible to pests (Bebber, 2019). Climate change and extreme weather events affect agricultural production by providing nooks and crannies for the increased establishment of weeds, pests, and diseases (Powell/Reinhard, 2016). Climate change is expected to increase CO_2 concentrations, which would make herbicides and pesticides less effective in controlling weeds and pests (Ramesh et al., 2017). Table 5.1 summarises the impacts of climate change on global food production.

5.7 Impact on Water and Land Resources

Groundwater is primarily the only safe source of drinking water for many rural parts of the GRL. Although anthropogenic activities adversely affect freshwater ecosystems in the GLR, climate change also contributes to their degradation (MacDonald et al., 2012). Most of sub-Saharan Africa in particular and the GLR in general have low permeability and minor aquifers, with a few larger aquifer systems located only in the Congo, parts of Angola, and southern Nigeria (MacDonald et al., 2012). The impacts of climate change on freshwater ecosystems in the GLR are evidenced by the rise in lake surface water temperatures (Change, 2014). The water capacity of some large water systems, such as lakes and rivers, in the GLR is expected to decrease in response to decreased precipitation and increased evaporation (Bates et al., 2008). However, studies have suggested that climate change in the GLR will have an overall modest effect on future water scarcity compared to other anthropogenic activities (Change, 2014). The projected changes in water resources in response to climate change and other factors are different when considering the sub-regions of SSA.

Modeling the impact of climate change on resources in the GLR is limited by significant uncertainties in the models used with reference to future rainfall in the region. The impact of climate change on total runoff is expected to be minimal under the A2 scenario in the Congo Basin (Change, 2014). The occurrence of drought is a serious environmental issue in some sub-regions of the GLR, as it is a component of the natural climate variability on the African continent, with fairly high intensities on monthly, annual, decadal, or centennial scales. It has been shown how climate change, aerosol emissions, land use practices, and resulting land–atmosphere interactions lead to drought-inducing mechanisms (Bhaga et al., 2020). Furthermore, it is important to note that drought prediction is a slow and difficult process due to the overriding complexities related to its causes, temporal scale, extent, and timing at different spatial and temporal scales (Ndehedehe et al., 2020).

Changes in land surface characteristics such as albedo, evapotranspiration, and primary production contribute to climate change at different geographic scales and to the dynamics of atmospheric carbon concentrations. Note that in SSA in general and in the GLR, most households rely on land-based activities as a means of livelihood (Antwi-Agyei/Stringer 2021), which puts pressure on forest areas, leading to a decrease in the extent of natural vegetation while making some animal and plant species vulnerable (Sintayehu 2018). Although the primary driver of these changes is land use/land cover change caused by anthropogenic activities, climate change has important additional and interactive effects (Change, 2014). These impacts are commonly intensified by rural households' dependence on agricultural activities that are significantly climate dependent (Antwi-Agyei/Stringer, 2021). Poor smallholder farmers in SSA in general and in the GLR in particular depend on land to access necessary inputs, and with population growth, strong pressure is exerted, exacerbating land degradation due to climate change (Deen-Swarray et al., 2020).

5.8 Climate Change and Food Security

All continents are under water stress, especially North Africa and Western, Central, and South Asia, and the Food and Agriculture Organization of the United Nations (FAO) sees no progress on the sustainable development goals (SDGs), especially as efforts for sustainable agriculture, as well as long-term management of land and ocean resources have not been successful (WHO, 2019). In addition, more than 820 million people are hungry worldwide, about 60% of all livestock breeds are threatened with extinction, and little is being done to conserve the DNA of the most threatened plant species. In addition, one-third of marine stocks are currently overexploited. While some studies have indicated how water depletion in freshwater ecosystems and wetlands is impacting agricultural production, fisheries, and livestock in SSA in general and the GLR in particular (Léauthaud et al., 2013), others have highlighted the impact of inappropriate agricultural practices on land and water resources (Olisah et al., 2020).

The FAO has indicated that in developing countries, including the GLR, climate change will result in an increase in drylands and water-stressed lands, followed by a decrease in rain-fed cropland and, consequently, a decrease in cereal production (Fao, 2018). If left unchecked, this would put pressure on limited freshwater and groundwater resources to meet the increased demand for food and water resources (Bonsor et al., 2010). Climate change could result in a reduction of 50% or more in potential cereal production by 2080 in the Northern and Eastern countries of the African region, including the GLR, in response to future water scarcity events. While countries in the Western part of the Southern African region are expected to experience a 1550% decrease in agricultural productivity, countries such as the Democratic Republic of the Congo (DRC) would experience an increase of 25% or more (Nhemachena et al., 2020).

Climate change-induced depletion of freshwater soil in ecosystems has caused a reduction in fishing and livestock activities in some SSA countries including the GLR (Allison et al., 2009). Of the 132 countries that are vulnerable to the impacts of climate change on fisheries, the GLR is ranked first (Change, 2014). Projected changes in temperature and precipitation will also affect the quantity and quality of pasture biomass and drinking water needed to raise livestock. This would worsen during prolonged dry seasons (Descheemaeker et al., 2016). SSA experienced rapid deforestation of approximately 3.4 million hectares per year between 2000 and 2010 (Mapulanga/Naito, 2019). This deforestation reduces carbon storage and sequestration capacity, leading to increased atmospheric CO_2 concentration and higher rates of warming events, with impacts on freshwater ecosystems by causing reduced infiltration, increased sedimentation affecting water quality, increased soil erosion, and increased evaporation from nearby surface water ecosystems (Mapulanga/Naito, 2019). This negatively impacts cropping and livestock activities and further limits the amount of water available for domestic uses, resulting in food and water insecurity.

Increased runoff from farms reaching nearby freshwater ecosystems poses a risk of water pollution, affecting water quality and aquatic biodiversity (Change, 2014).

In mountainous regions of the GLR, warming could lead crop pests to expand their range into areas limited by cold temperatures. For example, in mountainous regions where Arabica coffee is produced, warming trends could cause the coffee berry borer (*Hypothenemus hampei*) to become a serious threat in the areas of Uganda, Rwanda, and Burundi that are involved in coffee production (Jaramillo et al., 2011). Increasing temperatures may also increase the risk of altitudinal range expansion of the highly destructive burrowing nematode *Radopholus similis* in regions that produce bananas (Nicholls et al., 2008). Another study by Zacarias (2020) predicted that climatic conditions would favor a strong spread of the fall armyworm (*Spodoptera frugiperda*), with potential increases in SSA, particularly in the GLR. The inappropriate handling of these chemicals by farmers and other users to deal with the problem of pests, weeds, and diseases is responsible for the high level of their presence in freshwater and terrestrial ecosystems, posing a threat to human health and the ecosystem as a whole (Olisah et al., 2020).

Crop and livestock productivity can decline suddenly or gradually due to rising temperatures, drought stress, or increased CO_2 concentrations. Changes in seasonal rainfall patterns and the occurrence of more severe precipitation events would lead not only to delays in planting but also in harvesting. Several studies suggest that global warming and climate variability have a very negative influence on food and feed production and food security worldwide (Rosenzweig et al., 2017). Climate variability often leads to droughts, decreases in crop yields, and even famine in areas where food security is not assured (Iizumi/Ramankutty, 2016). With changes in temperature, crop pests may survive or even reproduce with greater incidence, migrate to higher latitudes, and invade other regions, thus impacting crop growth (Castex et al., 2018).

The FAO convened a panel of experts to review the state of knowledge on the links between climate change and global warming, the risks posed by pests and diseases to plants and animals, and their potential effects on human health and food security (Change, 2020). These experts came to a most striking conclusion, highlighting the presence of particular risks with regard to the impact of diseases and pests directly induced by climate change. Global warming can alter phytoplankton, zooplankton, algae, and microalgae, and cause drastic changes in the geographic distribution of native species by forcing them to migrate to more favorable areas (Yanik/Aslan, 2018). The effects of increasing greenhouse gas emissions threaten coastal and marine ecosystems through changes in ocean temperatures, affecting ocean currents, weather patterns and sea level. In the case of extreme temperatures and ocean acidification, coral reefs are at risk with repercussions for many fisheries, altering the quantity and quality of phytoplankton and zooplankton, and dramatically affecting the entire food chain.

Microbial insects, pests and other vector populations are affected by climate change and, consequently, by the prevalence of microorganisms such as fungi and viruses. Abiotic factors, such as air pollution, nutrient deficiencies and temperature extremes, will affect soil quality, plant health and crop productivity (Change, 2020). Because crop growth capacity is highly correlated with increasing temperature and drought, drought-induced disruptions to the abundance and diversity of the soil microbial population can limit the ability to sustain plant growth (Bahram et al., 2018).

In an attempt to combat these increased infections due to climate change, more antimicrobial treatments may be required, leaving residues of veterinary drugs in foods of animal origin, which in turn may raise food safety concerns (Tirado et al., 2010). Given that antimicrobial efficacy is limited in many sectors, antimicrobial resistance challenges may be exacerbated by climate change (Change, 2020). Several studies on the impacts of climate change point to impacts on primary food production and consequences for food security, food safety, and nutrition, primarily addressing the effects of pathogens on human health. Given that, for many regions, it is estimated that food availability will be reduced as a result of a 2 °C temperature increase, ensuring food security is more critical than ever (Hoegh-Guldberg et al., 2019). As in the case of pathogenic bacteria, climatic factors such as temperature, humidity and heavy precipitation are expected to affect spoilage microorganisms and increase the risk of spoilage.

5.9 Some Mechanisms to Mitigate the Situation and Adaptation Strategies to the Ongoing Climate Change

The greatest technological challenge is to detect, attribute, and understand the effects of climate change on crops in order to define accurate prediction models for the future (Iizumi/Ramankutty, 2016). Addressing the effects of climate change requires a deeper understanding of its effects on crops by filling some of the gaps in knowledge at the global scale, coupled with a combination of indigenous and local knowledge and technological advances (Chandio et al., 2021). These strategies may be more effective when complemented by genomic innovations. Some U.S. and European countries have already diversified their livelihoods by harvesting rainwater or using alternative livelihoods for fishing to cope with climate change, and thus limit a variety of factors that are not only environmental and technological, but also political and market-related (Debaeke et al., 2017). Several diversification strategies can be adopted to achieve this, including the use of more profitable crops to increase income, the spread of income from traditional crops to shorten the time between harvests, the development of effective government policies, the imposition of restrictions to favor specific crops or facilitate access to improved vegetables or more efficient farming systems, and the reduction of vulnerability and adverse effects on markets, policies and the environment.

Approximately 760,000 km^2 of arable land is affected by salinity and acidity, which can greatly reduce food production. This requires avoiding soil degradation at all costs as long as climatic conditions still allow their use as arable land and developing efficient technologies to facilitate the adaptation of high latitude soils to crops and new cultivation techniques that allow the use of nutrient-poor soils, or no soil at all. To mitigate the effects of heat stress, animal productivity in hot environments must be supported through physical environmental modifications, nutritional measures, and the development of breeds that are more tolerant to heat stress (Collier et al., 2019). To increase the resilience of crops and livestock to extreme climates, several strategies have been used including cultivar selection based on drought and temperature tolerance, load and irrigation adjustment, canopy pruning, and the use of particle films and shading (Mir et al., 2022). For livestock, selection of better breeds; improving nutrition during periods of high heat load, using sunshades, evaporative cooling, or mechanical ventilation, minimising damage to rangeland and pasture by opting for rotational grazing, optimising forage stock management, and reducing herd size during dry periods, followed by rational water management, as well as facilitating access to information, developing risk management programs, regional awareness, extension, and education for farmers (Osei-Amponsah et al., 2019).

As for pests and diseases, consideration may be given to the use of biological agents to control pests or the introduction of pest- and disease-resistant crops and animal varieties. Although grain legumes have various sources of insect resistance, widespread adoption of insect-resistant crops would be a great asset (Van Lenteren et al., 2020). Grain legume pests can be effectively managed by manipulating the crop and its environment through intercropping, the use of natural essential oils, and biopesticides. CRISPR-based gene drives have been proposed as a way to reduce or eliminate insect-borne diseases, control invasive species, and even reverse insecticide resistance in pests (Champer et al., 2017). In the context of fisheries, adaptation and mitigation strategies are based on the particular characteristics and interactions of fisheries and aquaculture within the framework of feasible policy instruments to protect the livelihoods of fishing communities and food security. However, each country needs its own adaptation and mitigation strategy to protect itself while reducing greenhouse gas emissions (Change, 2020). Another strategy would be to promote options that facilitate carbon sequestration, like enriching the soil with organic matter to boost agricultural production and purify surface and groundwater (Autret et al., 2016; Tribouillois et al., 2018).

5.10 Climate-Smart Agriculture Perspectives in the Great Lakes Region

Being an urgent global challenge that demands immediate action, climate change effects are particularly evident in regions like the Great Lakes, where agriculture plays a vital role in the economy and food security. To ensure sustainable development, innovative approaches such as climate-smart agriculture (CSA) have gained significant attention. This section explores various perspectives on climate-smart agriculture in the Great Lakes region and highlights its potential to address the climate crisis while promoting agricultural resilience, productivity, and environmental sustainability (Lipper et al., 2014). In order to ensure climate-smart agriculture in the GLR, the following key elements must be effectively addressed, including understanding climate-smart agriculture, adaptation strategies, mitigation and sustainable practices, technological innovations, policy and institutional support, and collaboration and stakeholder engagement.

5.10.1 Understanding Climate-Smart Agriculture

Climate-smart agriculture is an integrated approach that seeks to address climate change challenges while ensuring food security and sustainable development. It comprises three main pillars: (a) increasing agricultural productivity and incomes; (b) adapting and building resilience to climate change; and (c) mitigating greenhouse gas emissions. CSA recognises that agriculture must adapt to changing climatic conditions and simultaneously contribute to reducing its own environmental impact (Azadi et al., 2021).

5.10.2 Adaptation Strategies

The Great Lakes region experiences changing weather patterns, including increased rainfall variability, heatwaves, and extreme weather events. The CSA offers several adaptation strategies to mitigate these challenges. First, farmers can adopt climate-resilient crop varieties and diversify their crops to enhance their resilience to changing conditions. Furthermore, implementing sustainable water management techniques, such as efficient irrigation systems, can mitigate the impacts of water scarcity. Conservation agriculture practices, such as no-till farming and agroforestry, help improve soil health, reduce erosion, and enhance carbon sequestration (Watson et al., 2013).

5.10.3 Mitigation and Sustainable Practices

The CSA recognises that agriculture is a significant contributor to greenhouse gas emissions. Therefore, it emphasises the adoption of sustainable practices to mitigate these emissions. For instance, optimising fertiliser use, adopting precision agriculture techniques, and implementing anaerobic digestion systems for managing organic waste can reduce emissions significantly. Additionally, integrating livestock systems with crop production and promoting silvopastoral systems can enhance carbon sequestration and reduce methane emissions (Zurek et al., 2022).

5.10.4 Technological Innovations

Technological advancements play a crucial role in promoting CSA in the Great Lakes region. Remote sensing, Geographic Information Systems (GIS), and precision agriculture technologies allow farmers to monitor crop health, predict weather patterns, and optimise resource use. Additionally, innovative tools like mobile applications can provide real-time information and advisory services to farmers, empowering them to make informed decisions. Encouraging the adoption of such technologies can enhance agricultural productivity while minimising environmental impact (Kaack et al., 2022).

5.10.5 Policy and Institutional Support

Effective policy frameworks and institutional support are critical for promoting CSA in the Great Lakes region. Governments can provide incentives and financial support for farmers to adopt climate-smart practices. Subsidies for climate-resilient seed varieties, low-carbon technologies, and sustainable management practices can encourage widespread adoption. Moreover, promoting research and development, strengthening extension services, and facilitating knowledge exchange platforms can ensure capacity building and knowledge dissemination among farmers (Bergquist et al., 2022).

5.10.6 Collaboration and Stakeholder Engagement

Implementing CSA requires collaboration among various stakeholders, including farmers, government agencies, research institutions, and civil society organisations. Farmers' participation and engagement are particularly crucial in ensuring the successful implementation of CSA practices. Building strong partnerships, fostering knowledge-sharing networks, and involving local communities in decision-making processes can foster a sense of ownership and empower stakeholders to address climate challenges collectively (Susskind/Kim, 2022).

5.11 Conclusion

Climate change has unprecedented consequences for food security, compounded by several uncertainties, thus requiring well-defined measures and environmental policy in the GLR as elsewhere. Climatic factors including global warming, natural disasters, increased frequency of heavy rainfall, prolonged periods of drought, and potential sea level rise directly and indirectly hinder agriculture, fisheries and livestock in multiple ways, thus increasing the level of food insecurity, which is already too precarious. The impacts of climate change on food security at different stages of the food chain were examined, including the impact on food resources, food availability, food accessibility, food utilisation and the stability of food products, as well as overall food productivity, water sources and arable land. Furthermore, attention was paid to some mechanisms to mitigate the situation and adaptation strategies to cope with the scourge of climate change. Finally, CSA perspectives in the GLR hold immense potential to address the challenges posed by climate change while ensuring sustainable food production. By adopting adaptive strategies, sustainable practices, technological innovations, and supportive policies, CSA can enhance agricultural resilience, reduce greenhouse gas emissions, and promote sustainable development. Collaboration and stakeholder engagement are key to the successful implementation of CSA, ensuring that the Great Lakes region can adapt to climate change while ensuring food security and environmental sustainability for generations to come.

References

Adnan, S., Ullah, K., & Ahmed, R. (2020). Variability in meteorological parameters and their impact on evapotranspiration in a humid zone of Pakistan. *Meteorological Applications, 27*(1), e1859.

Akhtar, N., Syakir Ishak, M.I., Bhawani, S.A., & Umar, K. (2021). Various natural and anthropogenic factors responsible for water quality degradation: A review. *Water, 13*(19), 2660.

Akrofi, S., Price, L.L., & Struik, P.C. (2012). HIV and severity of seasonal household food-related coping behaviors in rural Ghana. *Ecology of Food and Nutrition, 51*(2), 148–175.

Alegbeleye, O., & Sant'Ana, A. S. (2022). Impact of temperature, soil type and compost amendment on the survival, growth and persistence of Listeria monocytogenes of non-environmental (food-source associated) origin in soil. *Science of The Total Environment, 843*, 157033.

Ali, A., & Erenstein, O. (2017). Assessing farmer use of climate change adaptation practices and impacts on food security and poverty in Pakistan. *Climate Risk Management, 16*, 183–194.

Ali, S., Liu, Y., Ishaq, M., Shah, T., Abdullah, Ilyas, A., & Din, I. U. (2017). Climate change and its impact on the yield of major food crops: Evidence from Pakistan. *Foods, 6*(6), 39.

Allison, E. H., Perry, A. L., Badjeck, M. C., Neil Adger, W., Brown, K., Conway, D., ... & Dulvy, N. K. (2009). Vulnerability of national economies to the impacts of climate change on fisheries. *Fish and Fisheries, 10*(2), 173–196.

Antwi-Agyei, P., & Stringer, L. C. (2021). Improving the effectiveness of agricultural extension services in supporting farmers to adapt to climate change: Insights from northeastern Ghana. *Climate Risk Management, 32*, 100304.

Aschwanden, A., Fahnestock, M. A., Truffer, M., Brinkerhoff, D. J., Hock, R., Khroulev, C., ... & Khan, S. A. (2019). Contribution of the Greenland Ice Sheet to sea level over the next millennium. *Science Advances, 5*(6), eaav9396.

Autret, B., Mary, B., Chenu, C., Balabane, M., Girardin, C., Bertrand, M., ... & Beaudoin, N. (2016). Alternative arable cropping systems: a key to increase soil organic carbon storage? Results from a 16 year field experiment. *Agriculture, Ecosystems & Environment, 232*, 150–164.

Ayugi, B., Eresanya, E. O., Onyango, A. O., Ogou, F. K., Okoro, E. C., Okoye, C. O., ... & Ongoma, V. (2022). Review of meteorological drought in Africa: historical trends, impacts, mitigation measures, and prospects. *Pure and Applied Geophysics, 179*(4), 1365–1386.

Azadi, H., Moghaddam, S. M., Burkart, S., Mahmoudi, H., Van Passel, S., Kurban, A., & Lopez-Carr, D. (2021). Rethinking resilient agriculture: From climate-smart agriculture to vulnerable-smart agriculture. *Journal of Cleaner Production, 319*, 128602.

Bahram, M., Hildebrand, F., Forslund, S. K., Anderson, J. L., Soudzilovskaia, N. A., Bodegom, P. M., ... & Bork, P. (2018). Structure and function of the global topsoil microbiome. *Nature, 560*(7717), 233–237.

Barnett, J., Evans, L. S., Gross, C., Kiem, A. S., Kingsford, R. T., Palutikof, J. P., ... & Smithers, S. G. (2015). From barriers to limits to climate change adaptation: path dependency and the speed of change. *Ecology and Society, 20*(3).

Bates, B., Kundzewicz, Z., & Wu, S. (2008). *Climate change and water*. Intergovernmental Panel on Climate Change Secretariat.

Bebber, D. P. (2019). Climate change effects on Black Sigatoka disease of banana. *Philosophical Transactions of the Royal Society B, 374*(1775), 20180269.

Bell, J. E., Brown, C. L., Conlon, K., Herring, S., Kunkel, K. E., Lawrimore, J., ... & Uejio, C. (2018). Changes in extreme events and the potential impacts on human health. *Journal of the Air & Waste Management Association, 68*(4), 265–287.

Berazneva, J., & Lee, D. R. (2013). Explaining the African food riots of 2007–2008: An empirical analysis. *Food Policy, 39*, 28–39.

Bergquist, M., Nilsson, A., Harring, N., & Jagers, S. C. (2022). Meta-analyses of fifteen determinants of public opinion about climate change taxes and laws. *Nature Climate Change, 12*(3), 235–240.

Bhaga, T. D., Dube, T., Shekede, M. D., & Shoko, C. (2020). Impacts of climate variability and drought on surface water resources in Sub-Saharan Africa using remote sensing: A review. *Remote Sensing, 12*(24), 4184.

Biglari, T., Maleksaeidi, H., Eskandari, F., & Jalali, M. (2019). Livestock insurance as a mechanism for household resilience of livestock herders to climate change: Evidence from Iran. *Land Use Policy, 87*, 104043.

Bonsor, H. C., MacDonald, A. M., & Calow, R. C. (2010). Potential impact of climate change on improved and unimproved water supplies in Africa. *RSC Issues in Environmental Science and Technology, 31*, 25–50.

Castex, V., Beniston, M., Calanca, P., Fleury, D., & Moreau, J. (2018). Pest management under climate change: The importance of understanding tritrophic relations. *Science of the Total Environment, 616*, 397–407.

Challinor, A., Wheeler, T., Garforth, C., Craufurd, P., & Kassam, A. (2007). Assessing the vulnerability of food crop systems in Africa to climate change. *Climatic Change, 83*, 381–399.

Champer, J., Reeves, R., Oh, S. Y., Liu, C., Liu, J., Clark, A. G., & Messer, P. W. (2017). Novel CRISPR/Cas9 gene drive constructs reveal insights into mechanisms of resistance allele formation and drive efficiency in genetically diverse populations. *PLoS Genetics, 13*(7), e1006796.

Chandio, A. A., Jiang, Y., Akram, W., Adeel, S., Irfan, M., & Jan, I. (2021). Addressing the effect of climate change in the framework of financial and technological development on cereal production in Pakistan. *Journal of Cleaner Production, 288*, 125637.

Change, I. C. (2014). Impacts, adaptation and vulnerability. *Part A: Global and sectoral aspects. Contribution of working group II to the fifth assessment report of the intergovernmental Panel on Climate Change*, 1132.

Change, F. C. (2020). Unpacking the Burden on Food Safety. *FAO—Food and Agriculture Organization of the United Nations: Rome, Italy.*

Collier, R. J., Baumgard, L. H., Zimbelman, R. B., & Xiao, Y. (2019). Heat stress: physiology of acclimation and adaptation. *Animal Frontiers, 9*(1), 12–19.

Connolly-Boutin, L., & Smit, B. (2016). Climate change, food security, and livelihoods in sub-Saharan Africa. *Regional Environmental Change, 16*, 385–399.

Coscarelli, R., Aguilar, E., Petrucci, O., Vicente-Serrano, S. M., & Zimbo, F. (2021). The potential role of climate indices to explain floods, mass-movement events and wildfires in southern Italy. *Climate, 9*(11), 156.

Davidson, D. J. (2018). Rethinking adaptation: Emotions, evolution, and climate change. *Nature and Culture*, *13*(3), 378–402.
Davidson, O., Halsnaes, K., Huq, S., Kok, M., Metz, B., Sokona, Y., & Verhagen, J. (2003). The development and climate nexus: the case of sub-Saharan Africa. *Climate Policy*, *3*(sup1), S97-S113.
Debaeke, P., Pellerin, S., & Scopel, E. (2017). Climate-smart cropping systems for temperate and tropical agriculture: mitigation, adaptation and trade-offs. *Cahiers Agricultures*, *26*(3), 34002.
Deen-Swarray, M., Odularu, G., & Adekunle, B. (2020). Introduction: nutrition, sustainable agriculture and climate change issues in Africa. *Nutrition, Sustainable Agriculture and Climate Change in Africa: Issues and Innovative Strategies*, 1–11.
Descheemaeker, K., Oosting, S. J., Homann-Kee Tui, S., Masikati, P., Falconnier, G. N., & Giller, K. E. (2016). Climate change adaptation and mitigation in smallholder crop–livestock systems in sub-Saharan Africa: a call for integrated impact assessments. *Regional Environmental Change*, *16*, 2331–2343.
Deutsch, C. A., Tewksbury, J. J., Tigchelaar, M., Battisti, D. S., Merrill, S. C., Huey, R. B., & Naylor, R. L. (2018). Increase in crop losses to insect pests in a warming climate. *Science*, *361*(6405), 916–919.
Devereux, S. (2007). The impact of droughts and floods on food security and policy options to alleviate negative effects. *Agricultural Economics*, *37*, 47–58.
Duran-Encalada, J. A., Paucar-Caceres, A., Bandala, E. R., & Wright, G. H. (2017). The impact of global climate change on water quantity and quality: A system dynamics approach to the US–Mexican transborder region. *European Journal of Operational Research*, *256*(2), 567–581.
Ebi, K. L., & Hess, J. J. (2020). Health Risks Due To Climate Change: Inequity In Causes And Consequences: Study examines health risks due to climate change. *Health Affairs*, *39*(12), 2056–2062.
Fao, F. (2018). The impact of disasters and crises on agriculture and food security. *Report*.
Food and Agriculture Organization of the United Nations (FAO). (2017). The impact of disasters and crises on agriculture and food security.
Ferrante, L., Barbosa, R. I., Duczmal, L., & Fearnside, P. M. (2021). Brazil's planned exploitation of Amazonian indigenous lands for commercial agriculture increases risk of new pandemics. *Regional Environmental Change*, *21*(3), 81.
Fones, H. N., & Gurr, S. J. (2017). NOXious gases and the unpredictability of emerging plant pathogens under climate change. *BMC Biology*, *15*(1), 1–9.
Gaines, S. D., Costello, C., Owashi, B., Mangin, T., Bone, J., Molinos, J. G., ... & Ovando, D. (2018). Improved fisheries management could offset many negative effects of climate change. *Science Advances*, 4(8), eaao1378.
Galstyan, H., Shiri, B., & Safaryan, T. (2022). Heatwaves in Southern Armenia in the context of climate change. *International Journal of Climatology*, *42*(6), 3431–3447.
Gardezi, M., Michael, S., Stock, R., Vij, S., Ogunyiola, A., & Ishtiaque, A. (2022). Prioritizing climate-smart agriculture: An organizational and temporal review. *Wiley Interdisciplinary Reviews: Climate Change*, *13*(2), e755.
Gartaula, H., Patel, K., Johnson, D., Devkota, R., Khadka, K., & Chaudhary, P. (2017). From food security to food wellbeing: examining food security through the lens of food wellbeing in Nepal's rapidly changing agrarian landscape. *Agriculture and Human Values*, *34*, 573–589.
Gezie, M. (2019). Farmer's response to climate change and variability in Ethiopia: A review. *Cogent Food & Agriculture*, 5 (1), 1613770.
Godfray, H. C. J., Beddington, J. R., Crute, I. R., Haddad, L., Lawrence, D., Muir, J. F., ... & Toulmin, C. (2010). Food security: the challenge of feeding 9 billion people. *Science*, *327*(5967), 812–818.
Gupta, A. K., Negi, M., Nandy, S., Alatalo, J. M., Singh, V., & Pandey, R. (2019). Assessing the vulnerability of socio-environmental systems to climate change along an altitude gradient in the Indian Himalayas. *Ecological Indicators*, *106*, 105512.

Haile, M. G., Wossen, T., Tesfaye, K., & von Braun, J. (2017). Impact of climate change, weather extremes, and price risk on global food supply. *Economics of Disasters and Climate Change, 1*, 55–75.

Hammond, T. T., Palme, R., & Lacey, E. A. (2018). Glucocorticoid–environment relationships align with responses to environmental change in two co-occurring congeners. *Ecological Applications*, 28(7), 1683–1693.

Hansen, J., Sato, M., Ruedy, R., Lo, K., Lea, D. W., & Medina-Elizade, M. (2006). Global temperature change. *Proceedings of the National Academy of Sciences, 103*(39), 14288–14293.

Hatfield, J. L., & Dold, C. (2018). Agroclimatology and wheat production: coping with climate change. *Frontiers in Plant Science, 9*, 224.

Hatfield, J. L., & Prueger, J. H. (2015). Challenge for future agriculture. *Crop wild relatives and climate change*, 24–43.

Hayes, K., Blashki, G., Wiseman, J., Burke, S., & Reifels, L. (2018). Climate change and mental health: risks, impacts and priority actions. *International Journal of Mental Health Systems, 12*(1), 1–12.

Heal, G., & Millner, A. (2014). Reflections: Uncertainty and decision making in climate change economics. *Review of Environmental Economics and Policy, 8*(1), 120–137.

Hegerl, G. C., Brönnimann, S., Cowan, T., Friedman, A. R., Hawkins, E., Iles, C., ... & Undorf, S. (2019). Causes of climate change over the historical record. *Environmental Research Letters, 14*(12), 123006.

Hoegh-Guldberg, O., Jacob, D., Taylor, M., Guillén Bolaños, T., Bindi, M., Brown, S., ... & Zhou, G. (2019). The human imperative of stabilizing global climate change at 1.5 C. *Science, 365*(6459), eaaw6974.

Hoekstra, A. Y., Buurman, J., & Van Ginkel, K. C. (2018). Urban water security: A review. *Environmental Research Letters, 13*(5), 053002.

Hollowed, A. B., Barange, M., Beamish, R. J., Brander, K., Cochrane, K., Drinkwater, K., ... & Yamanaka, Y. (2013). Projected impacts of climate change on marine fish and fisheries. *ICES Journal of Marine Science, 70*(5), 1023–1037.

Hussain, A., Ali, S., Rizwan, M., ur Rehman, M. Z., Javed, M. R., Imran, M., ... & Nazir, R. (2018). Zinc oxide nanoparticles alter the wheat physiological response and reduce the cadmium uptake by plants. *Environmental Pollution, 242*, 1518–1526.

Iizumi, T., & Ramankutty, N. (2016). Changes in yield variability of major crops for 1981–2010 explained by climate change. *Environmental Research Letters, 11*(3), 034003.

Iniguez-Gallardo, V., Lenti Boero, D., & Tzanopoulos, J. (2021). Climate change and emotions: analysis of people's emotional states in southern Ecuador. *Frontiers in Psychology, 12*, 644240.

Ishara, J., Ayagirwe, R., Karume, K., Mushagalusa, G. N., Bugeme, D., Niassy, S., ... & Kinyuru, J. (2022). Inventory reveals wide biodiversity of edible insects in the Eastern Democratic Republic of Congo. *Scientific Reports, 12*(1), 1576.

Ishara, J., Buzera, A., Mushagalusa, G. N., Hammam, A. R., Munga, J., Karanja, P., & Kinyuru, J. (2022). Nutraceutical potential of mushroom bioactive metabolites and their food functionality. *Journal of Food Biochemistry, 46*(1), e14025.

Ishara, J., Cokola, M. C., Buzera, A., Mmari, M., Bugeme, D., Niassy, S., ... & Kinyuru, J. (2023). Edible insect biodiversity and anthropo-entomophagy practices in Kalehe and Idjwi territories, DR Congo. *Journal of Ethnobiology and Ethnomedicine, 19*(1), 3.

Israel, D. C., & Briones, R. M. (2012). *Impacts of natural disasters on agriculture, food security, and natural resources and environment in the Philippines* (No. 2012–36). PIDS discussion paper series.

Jaramillo, J., Muchugu, E., Vega, F. E., Davis, A., Borgemeister, C., & Chabi-Olaye, A. (2011). Some like it hot: the influence and implications of climate change on coffee berry borer (Hypothenemus hampei) and coffee production in East Africa. *PloS one, 6*(9), e24528.

Jiménez-Donaire, M. D. P., Tarquis, A., & Giráldez, J. V. (2020). Evaluation of a combined drought indicator and its potential for agricultural drought prediction in southern Spain. *Natural Hazards and Earth System Sciences, 20*(1), 21–33.

Johnson, N. C., Amaya, D. J., Ding, Q., Kosaka, Y., Tokinaga, H., & Xie, S. P. (2020). Multi-decadal modulations of key metrics of global climate change. *Global and Planetary Change, 188*, 103149.

Jones, C., Hine, D. W., & Marks, A. D. (2017). The future is now: Reducing psychological distance to increase public engagement with climate change. *Risk Analysis, 37*(2), 331–341.

Kaack, L. H., Donti, P. L., Strubell, E., Kamiya, G., Creutzig, F., & Rolnick, D. (2022). Aligning artificial intelligence with climate change mitigation. *Nature Climate Change, 12*(6), 518–527.

Negatu, W., Iassen, A., & Kebede, A. (2011). A comparative analysis of vulnerability of pastoralists and agro-pastoralists to climate change: a case study in Yabello Woreda of Oromia Region, Ethiopia. *Ethiopian Journal of Development Research, 33*(1), 61–95.

Knox, J., Hess, T., Daccache, A., & Wheeler, T. (2012). Climate change impacts on crop productivity in Africa and South Asia. *Environmental Research Letters, 7*(3), 034032.

Kramer, R. D., Ishii, H. R., Carter, K. R., Miyazaki, Y., Cavaleri, M. A., Araki, M. G., ... & Hara, C. (2020). Predicting effects of climate change on productivity and persistence of forest trees. *Ecological Research, 35*(4), 562–574.

Kumar, L., Chhogyel, N., Gopalakrishnan, T., Hasan, M. K., Jayasinghe, S. L., Kariyawasam, C. S., ... & Ratnayake, S. (2022). Climate change and future of agri-food production. In *Future Foods* (pp. 49–79). Academic Press.

Lam, V. W., Cheung, W. W., Reygondeau, G., & Sumaila, U. R. (2016). Projected change in global fisheries revenues under climate change. *Scientific Reports, 6*(1), 32607.

Leauthaud, C., Belaud, G., Duvail, S., Moussa, R., Grünberger, O., & Albergel, J. (2013). Characterizing floods in the poorly gauged wetlands of the Tana River Delta, Kenya, using a water balance model and satellite data. *Hydrology and Earth System Sciences, 17*(8), 3059–3075.

Lipper, L., Thornton, P., Campbell, B. M., Baedeker, T., Braimoh, A., Bwalya, M., ... & Torquebiau, E. F. (2014). Climate-smart agriculture for food security. *Nature Climate Change, 4*(12), 1068–1072.

Lloyd, S. J., Kovats, R. S., & Chalabi, Z. (2011). Climate change, crop yields, and undernutrition: development of a model to quantify the impact of climate scenarios on child undernutrition. *Environmental Health Perspectives*, 119(12), 1817–1823.

Lobell, D. B., Schlenker, W., & Costa-Roberts, J. (2011). Climate trends and global crop production since 1980. *Science, 333*(6042), 616–620.

Lorenz, R., Argüeso, D., Donat, M. G., Pitman, A. J., van den Hurk, B., Berg, A., ... & Seneviratne, S. I. (2016). Influence of land-atmosphere feedbacks on temperature and precipitation extremes in the GLACE-CMIP5 ensemble. *Journal of Geophysical Research: Atmospheres, 121*(2), 607–623.

MacDonald, A. M., Bonsor, H. C., Dochartaigh, B. É. Ó., & Taylor, R. G. (2012). Quantitative maps of groundwater resources in Africa. *Environmental Research Letters, 7*(2), 024009.

Mach, K. J., Mastrandrea, M. D., Bilir, T. E., & Field, C. B. (2016). Understanding and responding to danger from climate change: the role of key risks in the IPCC AR5. *Climatic Change, 136*, 427–444.

Mapulanga, A. M., & Naito, H. (2019). Effect of deforestation on access to clean drinking water. *Proceedings of the National Academy of Sciences, 116*(17), 8249–8254.

Martinich, J., & Crimmins, A. (2019). Climate damages and adaptation potential across diverse sectors of the United States. *Nature Climate Change, 9*(5), 397–404.

Masipa, T. (2017). The impact of climate change on food security in South Africa: Current realities and challenges ahead. *Jàmbá: Journal of Disaster Risk Studies, 9*(1), 1–7.

Mekonnen, Adisu, Ayele Tessema, Zerhun Ganewo, and Ashenafi Haile. 2021. "Climate Change Impacts on Household Food Security and Farmers Adaptation Strategies." *Journal of Agriculture and Food Research* 6: 100197.

Mir, R. A., Bhat, B. A., Yousuf, H., Islam, S. T., Raza, A., Rizvi, M. A., ... & Zargar, S. M. (2022). Multidimensional role of silicon to activate resilient plant growth and to mitigate abiotic stress. *Frontiers in Plant Science, 13*, 819658.

Misiou, O., & Koutsoumanis, K. (2022). Climate change and its implications for food safety and spoilage. *Trends in Food Science & Technology, 126*, 142–152.
Misselhorn, A. A. (2005). What drives food insecurity in southern Africa? A meta-analysis of household economy studies. *Global Environmental Change, 15*(1), 33–43.
Moon, T., Ahlstrøm, A., Goelzer, H., Lipscomb, W., & Nowicki, S. (2018). Rising oceans guaranteed: Arctic land ice loss and sea level rise. *Current Climate Change Reports, 4*, 211–222.
Muluneh, A. (2020). Impact of climate change on soil water balance, maize production, and potential adaptation measures in the Rift Valley drylands of Ethiopia. *Journal of Arid Environments, 179*, 104195.
Mupangwa, W., Nyagumbo, I., Liben, F., Chipindu, L., Craufurd, P., & Mkuhlani, S. (2021). Maize yields from rotation and intercropping systems with different legumes under conservation agriculture in contrasting agro-ecologies. *Agriculture, Ecosystems & Environment, 306*, 107170.
Myers, B. J., Lynch, A. J., Bunnell, D. B., Chu, C., Falke, J. A., Kovach, R. P., ... & Paukert, C. P. (2017). Global synthesis of the documented and projected effects of climate change on inland fishes. *Reviews in Fish Biology and Fisheries, 27*, 339–361.
Myhre, G., Forster, P. M., Samset, B. H., Hodnebrog, Ø., Sillmann, J., Aalbergsjø, S. G., ... & Zwiers, F. (2017). PDRMIP: A precipitation driver and response model intercomparison project—Protocol and preliminary results. *Bulletin of the American Meteorological Society, 98*(6), 1185–1198.
Ndehedehe, C. E., Agutu, N. O., Ferreira, V. G., & Getirana, A. (2020). Evolutionary drought patterns over the Sahel and their teleconnections with low frequency climate oscillations. *Atmospheric Research, 233*, 104700.
Negatu, W., Iassen, A., & Kebede, A. (2011). A comparative analysis of vulnerability of pastoralists and agro-pastoralists to climate change: a case study in Yabello Woreda of Oromia Region, Ethiopia. *Ethiopian Journal of development research, 33*(1), 61–95.
Nelson, G. C., Van Der Mensbrugghe, D., Ahammad, H., Blanc, E., Calvin, K., Hasegawa, T., ... & Willenbockel, D. (2014). Agriculture and climate change in global scenarios: why don't the models agree. *Agricultural Economics, 45*(1), 85–101.
Nhemachena, C., Nhamo, L., Matchaya, G., Nhemachena, C. R., Muchara, B., Karuaihe, S. T., & Mpandeli, S. (2020). Climate change impacts on water and agriculture sectors in Southern Africa: Threats and opportunities for sustainable development. *Water, 12*(10), 2673.
Change, I. C. (2014). Impacts, adaptation and vulnerability. *Part A: global and sectoral aspects. Contribution of working group II to the fifth assessment report of the intergovernmental Panel on Climate Change, 1132.*
Nicholls, T., Norgrove, L., & Masters, G. (2008). *Innovative solutions to new invaders: managing agricultural pests, diseases and weeds under climate change* (No. 654-2016-44584).
Ogunyiola, A., Gardezi, M., & Vij, S. (2022). Smallholder farmers' engagement with climate smart agriculture in Africa: role of local knowledge and upscaling. *Climate Policy, 22*(4), 411–426.
Olisah, C., Okoh, O. O., & Okoh, A. I. (2020). Occurrence of organochlorine pesticide residues in biological and environmental matrices in Africa: A two-decade review. *Heliyon, 6*(3).
Osei-Amponsah, R., Chauhan, S. S., Leury, B. J., Cheng, L., Cullen, B., Clarke, I. J., & Dunshea, F. R. (2019). Genetic selection for thermotolerance in ruminants. *Animals, 9*(11), 948.
Panpakdee, C., & Limnirankul, B. (2018). Indicators for assessing social-ecological resilience: A case study of organic rice production in northern Thailand. *Kasetsart Journal of Social Sciences, 39*(3), 414–421.
Peña-Lévano, L. M., Taheripour, F., & Tyner, W. E. (2019). Climate change interactions with agriculture, forestry sequestration, and food security. *Environmental and Resource Economics, 74*, 653–675.
Pierrehumbert, R. T. (2000). Climate change and the tropical Pacific: The sleeping dragon wakes. *Proceedings of the National Academy of Sciences, 97*(4), 1355–1358.

Stablein, M. J., Cruz, J. G., Fidan, E. N., Talbot, J., Reed, S. P., Walters, R. S., ... & Rodríguez, L. F. (2022). Compound [ing] disasters in Puerto Rico: Pathways for virtual transdisciplinary collaboration to enhance community resilience. *Global Environmental Change, 76*, 102558.

Stuart-Smith, R., Otto, F. E., & Wetzer, T. (2022). Liability for Climate Change Impacts: the Role of Climate Attribution Science. *Rupert F Stuart-Smith, Friederike EL Otto & Thom Wetzer, Liability for Climate Change Impacts: the Role of Climate Attribution Science, in Elbert R De Jong et al (eds) Corporate Responsibility and Liability in Relation to Climate Change (Intersentia 2022).*

Pörtner, H. O., Roberts, D. C., Adams, H., Adler, C., Aldunce, P., Ali, E., ... & Ibrahim, Z. Z. (2022). *Climate change 2022: impacts, adaptation and vulnerability*. IPCC.

Powell, J. P., & Reinhard, S. (2016). Measuring the effects of extreme weather events on yields. *Weather and Climate Extremes, 12*, 69–79.

Pradeepkiran, J. A. (2019). Aquaculture role in global food security with nutritional value: a review. *Translational Animal Science, 3*(2), 903–910.

Ramalho, J. C., Pais, I. P., Leitão, A. E., Guerra, M., Reboredo, F. H., Máguas, C. M., ... & DaMatta, F. M. (2018). Can elevated air [CO_2] conditions mitigate the predicted warming impact on the quality of coffee bean?. *Frontiers in Plant Science, 9*, 287.

Ramesh, M., Palanikumar, K., & Reddy, K. H. (2017). Plant fibre based bio-composites: Sustainable and renewable green materials. *Renewable and Sustainable Energy Reviews, 79*, 558–584.

Ray, D. K., West, P. C., Clark, M., Gerber, J. S., Prishchepov, A. V., & Chatterjee, S. (2019). Climate change has likely already affected global food production. *PloS One, 14*(5), e0217148.

Raza, S., Zamanian, K., Ullah, S., Kuzyakov, Y., Virto, I., & Zhou, J. (2021). Inorganic carbon losses by soil acidification jeopardize global efforts on carbon sequestration and climate change mitigation. *Journal of Cleaner Production, 315*, 128036.

Reid, P., & Vogel, C. (2006). Living and responding to multiple stressors in South Africa—Glimpses from KwaZulu-Natal. *Global Environmental Change, 16*(2), 195–206.

Reikard, G. (2019). Volcanic emissions and air pollution: Forecasts from time series models. *Atmospheric Environment: X, 1*, 100001.

Rohde, R., Muller, R. A., Jacobsen, R., Muller, E., Perlmutter, S., Rosenfeld, A., ... & Wickham, C. (2013). A new estimate of the average Earth surface land temperature spanning 1753 to 2011. Geoinfor Geostat: An Overview 1: 1.

Rojas-Downing, M. M., Nejadhashemi, A. P., Harrigan, T., & Woznicki, S. A. (2017). Climate change and livestock: Impacts, adaptation, and mitigation. *Climate Risk Management, 16*, 145–163.

Rosenzweig, C., Arnell, N. W., Ebi, K. L., Lotze-Campen, H., Raes, F., Rapley, C., ... & Warszawski, L. (2017). Assessing inter-sectoral climate change risks: the role of ISIMIP. *Environmental Research Letters, 12*(1), 010301.

Roy, A., & Haider, M. Z. (2018). Stern review on the economics of climate change: implications for Bangladesh. *International Journal of Climate Change Strategies and Management, 11*(1), 100–117.

Schiermeier, Q. (2018). Climate as culprit. *Nature, 560*(7716), 20–22.

Seneviratne, S. I., Donat, M. G., Mueller, B., & Alexander, L. V. (2014). No pause in the increase of hot temperature extremes. *Nature Climate Change, 4*(3), 161–163.

Shah, T. (2020). Climate change and groundwater: India's opportunities for mitigation and adaptation. In *Water resources policies in South Asia* (pp. 213–243). Routledge India.

Shoaib, S. A., Khan, M. Z. K., Sultana, N., & Mahmood, T. H. (2021). Quantifying uncertainty in food security modeling. *Agriculture, 11*(1), 33.

Shrivastava, P., & Kumar, R. (2015). Soil salinity: A serious environmental issue and plant growth promoting bacteria as one of the tools for its alleviation. *Saudi Journal of Biological Sciences, 22*(2), 123–131.

Sintayehu, D. W. (2018). Impact of climate change on biodiversity and associated key ecosystem services in Africa: a systematic review. *Ecosystem Health and Sustainability, 4*(9), 225–239.

Sloggy, M. R., Suter, J. F., Rad, M. R., Manning, D. T., & Goemans, C. (2021). Changing climate, changing minds? The effects of natural disasters on public perceptions of climate change. *Climatic Change*, *168*, 1–26.

Solomon, S. (Ed.). (2007). *Climate change 2007-the physical science basis: Working group I contribution to the fourth assessment report of the IPCC* (Vol. 4). Cambridge university press.

Songok, C. K., Kipkorir, E. C., Mugalavai, E. M., Kwonyike, A. C., & Ng'weno, C. (2011). Improving the participation of agro-pastoralists in climate change adaptation and disaster risk reduction policy formulation: a case study from Keiyo district, Kenya. *Experiences of climate change adaptation in Africa*, 55–68.

Spinoni, J., Barbosa, P., Bucchignani, E., Cassano, J., Cavazos, T., Cescatti, A., ... & Dosio, A. (2021). Global exposure of population and land-use to meteorological droughts under different warming levels and SSPs: a CORDEX-based study. *International Journal of Climatology*, *41*(15), 6825–6853.

Springmann, M., Mason-D'Croz, D., Robinson, S., Garnett, T., Godfray, H. C. J., Gollin, D., ... & Scarborough, P. (2016). Global and regional health effects of future food production under climate change: a modelling study. *The Lancet*, *387*(10031), 1937–1946.

Susskind, L., & Kim, A. (2022). Building local capacity to adapt to climate change. *Climate Policy*, *22*(5), 593–606.

Syed, A., Raza, T., Bhatti, T. T., & Eash, N. S. (2022). Climate Impacts on the agricultural sector of Pakistan: Risks and solutions. *Environmental Challenges*, *6*, 100433.

Taylor, M. (2018). Climate-smart agriculture: what is it good for?. *The Journal of Peasant Studies*, *45*(1), 89–107.

Thornton, P. K., van de Steeg, J., Notenbaert, A., & Herrero, M. (2009). The impacts of climate change on livestock and livestock systems in developing countries: A review of what we know and what we need to know. *Agricultural Systems*, *101*(3), 113–127.

Tirado, M. C., Clarke, R., Jaykus, L. A., McQuatters-Gollop, A., & Frank, J. M. (2010). Climate change and food safety: A review. *Food Research International*, *43*(7), 1745–1765.

Tol, R. S. J. (2009). The economic effects of climate change. *Journal of economic perspectives*, 23(2), 29–51.

Tribouillois, H., Constantin, J., & Justes, E. (2018). Cover crops mitigate direct greenhouse gases balance but reduce drainage under climate change scenarios in temperate climate with dry summers. *Global Change Biology*, *24*(6), 2513–2529.

Udomkun, P., Wiredu, A. N., Nagle, M., Müller, J., Vanlauwe, B., & Bandyopadhyay, R. (2017). Innovative technologies to manage aflatoxins in foods and feeds and the profitability of application–A review. *Food Control*, *76*, 127–138.

Ullah, H., Fordham, D. A., & Nagelkerken, I. (2021). Climate change negates positive CO2 effects on marine species biomass and productivity by altering the strength and direction of trophic interactions. *Science of The Total Environment*, *801*, 149624.

Van Lenteren, J. C., Alomar, O., Ravensberg, W. J., & Urbaneja, A. (2020). Biological control agents for control of pests in greenhouses. *Integrated pest and disease management in greenhouse crops*, 409–439.

Wang, Y., Bi, H., Huang, H., Liu, Y., Liu, Y., Liang, X., ... & Zhang, Z. (2019). Satellite-observed trends in the Arctic sea ice concentration for the period 1979–2016. *Journal of Oceanology and Limnology*, *37*(1), 18–37.

Watson, J. E., Iwamura, T., & Butt, N. (2013). Mapping vulnerability and conservation adaptation strategies under climate change. *Nature Climate Change*, *3*(11), 989–994.

Werksman, J. (1994). The United Nations framework convention on climate change: The first conference of the parties opening in Berlin. *Global Environmental Change*, *4*(4), 339–340.

Wheeler, S. A., Nauges, C., & Zuo, A. (2021). How stable are Australian farmers' climate change risk perceptions? New evidence of the feedback loop between risk perceptions and behaviour. *Global Environmental Change*, *68*, 102274.

World Bank. (2009). Making development climate resilient: A World Bank strategy for sub-Saharan Africa.

World Health Organization. (2019). *The state of food security and nutrition in the world 2019: safeguarding against economic slowdowns and downturns* (Vol. 2019). Food & Agriculture Org.
REGION, A. (2009). Making Development Climate Resilient: A World Bank Strategy for Sub-Saharan Africa.
Wu, X., Lu, Y., Zhou, S., Chen, L., & Xu, B. (2016). Impact of climate change on human infectious diseases: Empirical evidence and human adaptation. *Environment International*, *86*, 14–23.
Yanik, T., & Aslan, I. (2018). Impact of global warming on aquatic animals. *Pakistan Journal of Zoology*, *50*(1).
Ye, L., Shi, K., Xin, Z., Wang, C., & Zhang, C. (2019). Compound droughts and heat waves in China. *Sustainability*, *11*(12), 3270.
Zacarias, D. A. (2020). Global bioclimatic suitability for the fall armyworm, Spodoptera frugiperda (Lepidoptera: Noctuidae), and potential co-occurrence with major host crops under climate change scenarios. *Climatic Change*, *161*(4), 555–566.
Zhang, H., Zhu, J., Gong, Z., & Zhu, J. K. (2022). Abiotic stress responses in plants. *Nature Reviews Genetics*, *23*(2), 104–119.
Zheng, F., Ren, H., Lin, R., & Zhu, J. (2023). Realistic ocean initial condition for stimulating the successful prediction of extreme cold events in the 2020/2021 winter. *Climate Dynamics*, *61*(1), 33–46.
Zhou, T. (2021). New physical science behind climate change: What does IPCC AR6 tell us?. *The Innovation*, *2*(4).
Zurek, M., Hebinck, A., & Selomane, O. (2022). Climate change and the urgency to transform food systems. *Science*, *376*(6600), 1416–1421.

Chapter 6
Climate Change, Food Insecurity, Peace and Sustainable Development in East Africa: Case Study of South Sudan, Sudan, Ethiopia and Kenya

Moses John

Abstract The relationship between climate change and food insecurity is apparent, as the former triggers the latter's consequences for individuals, groups and countries. Food insecurity and hunger persist in many East African countries, particularly South Sudan, Sudan, Ethiopia and Kenya, driven primarily by conflicts, wars and drought. While the extent and intensity of conflicts in Sub-Saharan Africa decreased in the 1990s, the East African region registered an increase of about 30 state-based conflicts in 2019. The following year, the number of state-based conflicts increased to 56, including eight wars. Consequently, many human and animal lives were lost, properties, including food stores, were burned, and millions of populations migrated amidst increased insecurity. With the ongoing War between Russian Federation and Ukraine taking global attention and resources, the international community has not responded promptly and adequately to the hunger crisis in the region. The quest for peace in the region remains elusive and more complex to realise as the 2030 Agenda for Sustainable Development is in grave jeopardy due to multiple factors, including COVID-19, climate change, and conflict impacts on all of the Sustainable Development Goals. Drawing on a broad range of literature, this chapter explores the conceptual basis of climate change and its implications for food insecurity and conflicts in the region. The chapter further discusses underpinning problems of climate, Anthropocene and food crisis. It concludes with an analysis of the roadmap of survival and policy recommendations to put the East African region on track to food security, peace and sustainable development.

Keywords Climate change · Food insecurity · Conflict · Peace · Sustainable development · South Sudan · Sudan · Ethiopia · Kenya

---

Dr Moses Monday John is a PhD graduate in Public Administration-Peace Studies from Durban University of Technology. He is an Associate Lecturer of Conflict Management at Graduate School of Public Service in University of Juba in South Sudan. Email: mosesjoa@gmail.com

---

J. C. K. Kiyala and N. Chivasa (eds.), *Climate Change and Socio-political Violence in Sub-Saharan Africa in the Anthropocene*, The Anthropocene: Politik—Economics—Society—Science 37,
https://doi.org/10.1007/978-3-031-48375-2_6

6.1 Introduction

The world food insecurity prevalence is highest in East Africa compared to other regions, with climate change and associated variability being important contributing factors (FAO, 2017; Smith et al., 2017; Drammeh et al., 2019; Kabubo-Mariara/Mulwa, 2019). According to FAO et al. (2019) and Thome et al. (2019), more than 50% of the African population experiences moderate or severe food insecurity. The highest level of food insecurity in 2018 was recorded in Africa, where the East Africa region had 63% (272 million), Southern Africa at 54% (35 million), West Africa at 48% (183 million), and North Africa at 30% (70 million) of the entire population were food insecure (FAO et al., 2019). This implies that most of the African population, particularly the East African population, do not have regular access to sufficient and nutritious food for a healthy and productive life. According to the Food and Agricultural Organization FAO (2017), the major attributes of the rise of food insecurity in East Africa are violent conflicts, climate adversity, and the global economic environment. I am of the opinion that, other underlying factors include global economic and political exploitative policies towards Africa and its resources. Throughout 2022, levels of food insecurity continued to increase in Eastern Africa, with 82 million people who have suffered from hunger.

In South Sudan, approximately 6.6 million people—or over half of South Sudanese have experienced high levels of acute food insecurity, particularly between Oct and Nov 2022. According to the latest Comprehensive Food Security and Vulnerability Assessment (CFSVA), over 15 million people—34% of the population, were food insecure in Sudan in the first quarter of 2022. Projections indicate that this may have increased by up to 18 million people (39% of the population) by the third quarter of 2022 (WFP Sudan Country Brief, Sep 2022). Meanwhile, according to FAO, WFP and IGAD (2022), an estimated 22.6 million people are suffering from food insecurity across Ethiopia (Relief Web, 2022). In comparison, an analysis carried out by the Kenya Food Security Steering Group (KFSSG) reported that the number of food insecure people in pastoral and marginal agricultural areas has risen from 3.1 million in Feb 2022 to over 4.1 million in May 2022, driven by the impacts of late on set below average rainfall, resource-based conflict, and above-average staple food prices (FEWS Net, 2022). Recent drought and little rainfall in Kenya and Ethiopia have caused crop failure and loss of livestock.

Food security has a variety of definitions. However, the World Food Summit adopted the following definition, "Food security, at the individual, household, national, regional and global levels [is achieved] when all people, at all times, have physical and economic access to sufficient, safe and nutritious food to meet their dietary needs and food preferences for an active and healthy life" (Thome et al., 2019). In contrast, according to the FAO, food insecurity exists when people do not have adequate physical, social or economic access to food as defined above (FAO, 1996). Food production, access and security is a complex system that involves the function of the demographic, social, physical, financial and policy decisions that determine how effectively households can utilise their resources to meet their food needs (Shah/Dulal, 2015; Sam et al., 2019). The same demographic, social, physical, and policy settings and climate vulnerability are fundamental determinants of food availability and access (Schmidhuber/Tubiello, 2007; Sam et al., 2019).

According to the United Nations, climate change refers to long-term shifts in temperatures and weather patterns due to natural (variation in the solar cycle) or human activities, such as burning fossil fuels like coal, oil and gas (UN, 2022; IPCC, 2007: 21). Burning fossil fuels generates greenhouse gas emissions that increase the sun's heat and raising temperatures on earth. Furthermore, human activities such as excessively uncontrolled tree logging, clearing land and forests for farming, charcoal making or other developmental purposes can also release carbon dioxide and affect climate variability. Besides, landfills for garbage are a major source of methane emissions. In contrast, the production and utilisation of energy, industry, transport, buildings, agriculture and land use are among carbon emitters. Consequently, greenhouse gas concentrations of our time are at their highest levels in 2 million years as emissions continue to rise. As a result, the Earth is now about 1.1 °C warmer than it was in the late 1800s. The last decade (2011–2020) was the warmest on record (UN, 2022).

The impacts of climate change in our time include but are not limited to prolonged droughts, water scarcity, severe fires, rising seawater levels, flooding, melting polar ice, catastrophic storms and loss of biodiversity. In 2019 and 2020, the largest desert locusts outbreak swarmed East Africa, damaging hundreds of thousands of cropland and over a million pastures (FAO, 2020). However, climate change is one of the factors escalating food insecurity across East Africa with adverse effects on macroeconomics, especially on economic growth, thus high levels of poverty and under-development. Successive shocks from Russia's invasion of Ukraine and the COVID-19 pandemic have worsened food insecurity in Sub-Saharan Africa by at least 30% since early 2020 (IMF, 2022).

Surveys, research and discussions on the worrying impacts of climate change on the food crisis and forging ideal strategies to adapt and mitigate the impacts of climate change are ongoing in several countries. The discussions often focus on halting conflicts and restoring peace as a precondition to sustainable development. This is because in conflict situations, civilians who provide labour for food production are often displaced, forced to join combatants or migrate to safer areas to avoid suffering. As argued by Helvey, almost everywhere, guerrilla warfare results in massive civilian suffering at the hands of all of the warring parties, who are often viewed as brutal and oppressive (Helvey, 2004, p. 34–92). There are more or less no conversations in East African countries on the negative contribution of the military, armed conflicts and wars on gas emissions and how to reduce its impacts on global warming. While the extent and intensity of conflicts in Sub-Saharan Africa decreased in the 1990s, the East African region registered an increase of about 30 state-based conflicts in 2019 (Fang et al., 2020, p. 2). The number of active conflicts in Africa increased to 56 in 2020, including nine new or restarted ones (Pettersson et al., 2020, p. 2). Conflicts, droughts and floods disrupt agricultural production and productivity, denying countries economic gains and hindering their efforts to achieve sustainable development goals.

The concept of sustainable development is not new, as it has invoked numerous intellectual debates over the decades. Many of these debates are situated around global environmental politics, which examine relationships between global socio-economic and political forces and environmental change (IUCN, 1980; Ojo, 2022). The politics of the environment also focus on the implications of local-global interactions for environmental management, as well as the implications of environmental change and environmental governance and the future of 'environmental' policy in times of earth system transformations and the recognition of the 'Anthropocene' as a new epoch in planetary history (Biermann, 2020). Despite diverging perspectives about the concept of sustainable development, there is some convergence around the notion that it is centred on "improving human security and wellbeing" through poverty elimination, ending hunger, provision of health care and, climate action, peace and justice among other SDG goals (UNDP, 2015).

The 2022 Sustainable Development Goals Report hails progress towards realising the 17 Goals but acknowledges multiple challenges. Although SDGs were supported by more than 50 international and regional agencies and governments in over 200 countries and areas, the report paints a particularly sobering picture, which reveals that the 2030 Agenda for Sustainable Development is in grave jeopardy due to multiple cascading and intersecting crises. Whereby COVID-19, climate change and conflict predominate. These crises and their complex interactions impact all 17 Goals, creating spin-off crises in food and nutrition, health, education, the environment, peace and security. To put the world on track to sustainability will require concerted action on a global scale (UNDP, 2022, p. 3). The East African communities are not the only region facing the dangers of climate change and food insecurity, but most importantly, the world is facing a confluence of crises that threaten the very survival of humanity. According to the UNDP, all of these crises—and ways to prevent and navigate them—are addressed holistically in the 17 sustainable development goals (SDGs 17). UNDP warns that if humankind ignores them, it is only at our peril (UNDP, 2022, p. 3).

Humankind and its human institutions protect the environment, produce and export foods to global markets, and promote sustainable development. Similarly, humankind has been widely recognised as a crucial force pushing the Earth's system into a new geological epoch called the Anthropocene. Humans impact the environment in a variety of ways. For instance, extracting natural resources, logging trees, and polluting air and waterways are some of the most damaging examples of industrial destruction. These activities can destroy some or all of an entire ecosystem, wiping out the plants and animals that call these ecosystems home. Thus, humans have become the most influential species on the planet, causing significant global warming and other changes to land, environment, water, organisms and the atmosphere. The Anthropocene is a new concept that gained ground in the early 2000s when the Nobel Prize laureate in chemistry (Crutzen, 2002; Benner et al., 2021) used it during a meeting of the International Geosphere-Biosphere Programme (IGBP) in Mexico when he claimed that due to human interventions in nature and the Earth System, humankind had entered a new geological epoch in the Earth's history (Brauch, 2022,

p. 18). Since then, 'Anthropocene' has been used extensively in the natural and social sciences to gauge human influence in society and our planet, Earth.

Earth is a dynamic planet made up of complex and inter-connected spheres—atmosphere (air), geosphere (land and solid earth), hydrosphere (water), cryosphere (ice), and biosphere (life)—which together provide life-sustaining systems for both humans and non-humans (Bosselmann, 2020, p. 20). Environmental degradation results when the planet's natural resources, such as air, water, ice and soil, are eroded, polluted or destroyed. Once complex and interconnected aspects of any system are damaged or changed, this has far-reaching consequences on the remaining parts. Consumption of Earth's resources is currently in overshoot, causing devastating flow-on effects, such as the deterioration of global ecosystems, habitat destruction and the extinction of wildlife (Bombaci et al., 2018; Wallace/Clarkson, 2019). Human consumption of the Earth's resources has accelerated air and water pollution rates, deforestation, soil degradation, biodiversity loss, and climate change (Change, 2018; McGregor, 2013; Moore/Rees, 2013).

The links between climate issues and sustainable development are manifold. A rich body of knowledge advocates an integrated approach to linking climate change adaptation and mitigation in the context of sustainable development (Bizikova et al. 2011, p. 273). Wilbanks and Sathaye (2007) further emphasise that the linkages between climate adaptation and mitigation are highly context-specific depending on the priorities guiding development. For example, environmental but, often referred to as "border conflicts", exist between the Turkana ethnic group of Kenya and the Toposa of South Sudan. Natural resource conflicts are fundamentally linked to the day-to-day livelihood of people and communities in accessing or controlling the resources and utilisation of the same in areas under competing interests or contestation.

This chapter has three objectives. First, it aims to explore the concepts of climate change and its implications for food insecurity, conflicts, peace and sustainable development. Second, the chapter attempts to rethink and analyse the problems of climate and food crisis in East Africa in light of the Anthropocene. The third is to discuss the roadmap and make policy recommendations for survival to put East African countries back on track for food security, peace and sustainable development. This chapter employs desk review methods of data collection, analysis and presentation. Historical analysis of related reports from development partners and academic articles was considered important to explain the threats of climate and food crisis and refocus resources and efforts towards realising peace and sustainable development in Africa. As the popular saying goes, "A problem identified is a problem half-way resolved". Thus, the rationale for this chapter is to provide updated reflections on the implications of climate change to food insecurity and conflicts to mobilise masses, governments and development partners to support food production as a foundation for peace and durable development. Context-specific literature helps people to understand and change attitudes and behaviour to avert a crisis and transform anthropogenic activities into potentials for peace and development. This chapter covers the following:

- An introduction with a focus on the theoretical framework.
- Rethinking the problems of climate change and food insecurity with attention to South Sudan, Sudan, Kenya and Ethiopia.
- A roadmap for survival from climate crisis to peace and sustainable development and conclusion.

6.2 Rethinking the Problems of Climate Change, Food Insecurity and Conflict

Conflicts are inevitable, but violence remains a matter of choice for individuals, groups and countries. The challenge is that most climate variability and drought lead to violent conflicts or subsequent wars, which trigger food crises and reverse the development wheel. As Albert argues, climate change and conflicts constitute fundamental threats to peace and sustainable development in Africa. They are responsible for the loss of human life, destruction of properties, mass displacement of people and diversion of resources meant for promoting sustainable development into arms purchase and funding expensive peacekeeping support operations (Albert, 2008, p. 31).

Despite the immense human and natural resources, the East African region remains economically poor and underdeveloped. The problem is that there are significant challenges in East African countries regarding the fair allocation of financial resources to government expenditure priorities. Income generated from natural resources such as oil, minerals, and taxes is barely directed towards improving agricultural productivity to limit food crises and poverty and deliver basic social and development services. Instead, income generated by governments is directed towards unproductive conflict management, the so-called improving security and other state-centric priorities, which are often mismanaged and not properly accounted for. Corruption and mismanagement of public resources have been a major problem in East African countries to provide social and development services such as provision of security, rule of law, quality education, health care, clean drinking water, affordable clean energy, construction of tarmac roads to connect agricultural areas with the markets, and more. According to the World Bank, corruption is one of the greatest economic and social development obstacles. It distorts the rule of law and weakens the institutional foundation on which economic growth depends (World Bank, 2013). Corruption has been a major driver towards poverty and recurrent violence in many fragile and conflict-prone countries.

Literature reveals sufficient evidence and the straightforward role and involvement of most if not all, African governments in African armed conflicts and wars, be it directly or indirectly, at all stages ranging from pre-conflict to conflict, its resolution and post-conflict situation (Kiplagat, 1997, p. 132). The role often varies from manipulating ethnic and militia groups to fight to achieve governments' political objectives, including using food as a weapon of War. For example, in its special report, the World Peace Foundation argued that; all pillars of livelihoods and food entitlement in Tigray have been destroyed, and the belligerents deliberately used food as a weapon of War (World Peace Foundation, 2021). Nevertheless, prior to the recent Tigray armed conflict, about 80% of the population in northern Ethiopia relied on agriculture as their primary source of food and income. Since November 2020, conflict surge in the Tigray region has constrained agricultural production, forcing the population to depend on humanitarian assistance following extreme food insecurity, partially due to missed harvests and crop destruction. Additionally, the conflict has disrupted the food system and market supply chain as access has become constrained, leading to rising prices and making basic food commodities inaccessible or unaffordable for many households (World Peace Foundation, 2021, p. 5).

There are challenges in implementing sustainable development goals (SDGs 17) as the agenda has no clear responsibility and accountability to individual governments and stakeholders. Some people, countries, and even regions perceive or misconstrue SDGs as the United Nations Development Programme (UNDP) project. East African countries have no clear obligation to ensure that income generated locally in each country is fairly allocated and regularly released to fund climate actions, food production and productivity. The African Union and all member countries should own and set clear milestones to support sustainable development goals at the highest levels. According to Strange and Bayley (2008, p. 24), sustainable development involves developing to benefit the widest range of sectors across borders and even between generations. Strong (1992) argues that it involves a process of deep and profound change in the political, social, economic, institutional and technological order. The concept is based on three interconnected pillars, including eight social, economic and environmental processes (Strange and Bayley, 2008; Rogers et al., 2008). Strange and Bayley (2008, p. 27) argue that a society with unrest, poverty and disease will not develop long-term. Therefore, there is a strong link between social wellbeing or stability and sustainable development.

As the populations in the East African region recover from the impacts of the COVID-19 pandemic, the unfolding crisis in Russia's invasion of Ukraine has posed yet another crisis to food security. Ukraine and the Russian Federation are large producers and exporters of key food commodities, fertiliser, minerals and energy. Together they are considered the world's breadbasket, supplying 30 and 20% of global wheat and maize exports, respectively, and 80% of global exports of sunflower seed products. At least 50 countries import 30% or more of their wheat from these two countries, with many African countries importing more than 50%. Ukraine and the Russian Federation are also leading exporters of fertilisers. The conflict has caused a steep and sudden reduction in grain exports, sunflower seeds and fertilisers. As a result, import-dependent countries are vulnerable to rising food costs and supply chain disruptions. Joint, coordinated activities and policy solutions are urgently needed to avert food shortages for the world's poorest people and reduce the impact of the conflict and lingering consequences of the pandemic on global food insecurity (FAO, 2022). Tables 6.1 and 6.2 reveal huge exports of selected commodities in thousands of metric tons from the Russian Federation and Ukraine in 2020 and 2021 (FAO, 2022).

From the analysis of exports of selected commodities (in thousands of metric tonnes) in the two tables above, it is clear that the Russian Federation and Ukraine are

Table 6.1 Russian Federation: exports of selected commodities (thousands of metric tonnes)

Commodity	Year	Jan	Feb	Mar	Apr	May	Jun	Jul	Aug	Sep	Oct	Nov	Dec	Total
Wheat	2020	2101	1576	3361	4765	605	214	2337	4671	5015	4643	4290	4867	38445
	2021	3083	4537	1150	802	458	2228	1882	5221	4581	2834	3190	2951	32917
Barley	2020	469	239	307	863	135	140	712	593	885	721	394	549	6007
	2021	223	404	777	368	550	77	505	564	553	292	410	433	5156
Maize	2020	352	335	548	877	248	87	250	156	88	140	333	389	3803
	2021	374	451	982	287	551	202	134	68	79	252	351	407	4138
Soybean	2020	93	108	210	98	76	47	138	122	100	99	123	174	1388
	2021	674	52	31	12	18	26	27	19	14	35	36	50	994
Rape	2020	49	61	24	28	23	12	29	77	99	138	97	77	714
	2021	14	26	33	30	24	36	19	23	15	14	14	29	277
Sunflower	2020	157	201	342	72	61	10	1	2	33	204	184	106	1373
	2021	6	3	8	3	2	2	3	1	4	1o	16	35	93
Sunflower oil	2020	283	289	455	437	359	276	300	329	107	180	291	357	3663
	2021	298	297	495	375	176	143	153	374	99	92	318	292	3112
Rapeseed oil	2020	57	46	50	40	65	32	38	39	82	84	92	63	688
	2021	56	41	53	68	66	53	50	64	83	91	85	92	802

Source Trade Data Monitor (TDM), FAO (2022, p. 9)

Table 6.2 Ukraine: exports of selected commodities (thousands of metric tonnes)

Commodity	Year	Jan	Feb	Mar	Apr	May	Jun	Jul	Aug	Sep	Oct	Nov	Dec	Total
Wheat	2020	924	681	1310	1200	1191	257	1239	3701	3710	2156	1191	495	18055
	2021	508	709	697	713	858	662	961	3613	4363	3415	2375	1174	20048
Barley	2020	152	141	309	339	152	190	839	1315	750	491	296	71	5045
	2021	120	61	131	25	23	64	1097	1658	1016	737	435	244	5611
Maize	2020	4543	3457	3529	3091	2379	1547	425	179	29	1842	3106	3824	27951
	2021	1996	2476	2620	2628	2245	1698	962	302	165	895	3792	4897	24676
Soybean	2020	333	176	122	103	53	35	9	1	57	344	301	255	1789
	2021	109	92	86	104	36	44	31	5	10	172	215	192	1096
Rape	2020	2	2	2	5	1	4	183	880	546	316	276	164	2381
	2021	11	3	13	3	1	0	52	772	879	635	234	57	2660
Sunflower	2020	5	4	4	9	10	2	3	2	8	69	38	34	188
	2021	12	20	2	2	4	1	0	1	3	8	23	4	80
Sunflower oil	2020	581	627	608	717	639	588	593	329	304	525	756	594	6861
	2021	482	484	381	391	502	325	328	202	277	434	639	690	5135
Rapeseed oil	2020	0	33	2	0	0	0	5	35	35	19	6	2	137
	2021	0	0	0	0	0	0	4	47	61	35	15	2	164

Source Trade Data Monitor (TDM), FAO (2022, p. 9)

important global food producers and suppliers in international markets. Thus, continuing the War between the two countries will no doubt disrupt the food production system and increase food prices and food insecurity.

6.3 South Sudan, Sudan, Ethiopia and Kenya

South Sudan, Sudan and Kenya, except Ethiopia, are part of the East African Community (EAC) and share international borders. The Republic of South Sudan is bordered to the east by Ethiopia, to the north by Sudan, to the West by the Central African Republic, to the southwest by the Democratic Republic of the Congo, to the south by Uganda and to the southeast by Kenya. The four countries are members of the Inter-Governmental Authority for Development (IGAD), a sub-regional body. The four countries have many climates and landscapes, including cool highlands, swamp areas, tropical rain forests and other features typical of an equatorial region. Furthermore, the countries possess diverse ecosystems and agro-ecological zones.

6.3.1 Sudan and South Sudan

Climate changes, successive droughts, population pressure and chronic food shortages are the major threats to livelihoods in the North Kordofan state. The drought in North and Western Sudan, North Kordofan and Darfur, Kassala State and some parts of the rain-fed areas in central Sudan influence agriculture, livestock, water resources and health. Historically, Sudan suffered a severe famine in the mid-1980s. Examination of the mean annual precipitation for some of Sudan's most severely affected regions revealed that the most severe famines of the past two decades (1984–85, 1973–74, 1965–66) were associated with the worst multiple-year droughts. The livestock prices decrease during drought periods, indicating the disposal of assets on a large and widespread scale. There has been a drought in some of the main grain-producing areas, such as Gadarif, and some more impoverished parts of the country. In other places, there has been a disastrous dry spell mid-production season, and water is becoming very scarce in parts of the West and East. Pastures across northern Sudan are reported to be very poor. Many ongoing national policy processes have parallel aims to climate change adaptation. Water harvesting techniques have been implemented for nine areas in several states (North Darfur, Nile, North Kordofan, and West Kordofan). These projects have increased community access to reliable water, increasing their capacity to cope with the impacts of reduced precipitation, increased temperature and drought, all of which have been integrated into the NAPA consultation process (NADA, 2008).

Sudan has been at War for most of its past and modern history.Since the withdrawal 321 of the British from Sudan on 1 January 1956, the country has struggled with two 322 major civil wars (1955–1972 and 1983–2005). The third to be a Civil War and an ongoing armed conflict erupted on 15 April 2023. The crisis in Darfur, Nuba Mountains, Southern Blue Nile, and Eastern Sudan remains a security concern to the government in Khartoum (McCutchen, 2015). The same conflicts were used as excuses to divert a lot of financial resources, which could have been used to sponsor agricultural projects in Gezira, Algadarif and other schemes to limit the food crisis and gain economic resources to improve livelihoods in Sudan. The first north–south civil War lasted until 1972 and briefly ended with the Addis Ababa Peace Agreement, which ushered in more than a decade of relative peace in Sudan. However, in 1983 the second civil War erupted in Sudan led by the Sudan People's Liberation Movement/ Army (SPLM/A) after the Khartoum regime of General Niemeri abrogated the 1972 agreement by re-dividing the Southern region into three regions (Equatoria, Bahr elGazal and Upper Nile) following the discovery of oil in Upper Nile region to weaken the unity of Southerners.

In Sudan during Bashir's regime, the popular Janjaweid militiamen were adequately sponsored and used by the government to fight against the Darfur rebels and the perceived civilian sympathisers. Many examples of conflict are intertwined with the effects of climate change, including the rainfall variability identified as an important factor in the Darfur conflict (Ki-Moon, 2007). The variability in rainfall and the accompanying droughts have forced pastoral communities to change routes and water holes, leading them to new and potentially unfriendly areas. People's fight for survival has resulted in new conflicts. A closer look at pastoralism, characterised by nomadic migration and the tradition of raiding, may be warranted. It is a common thread in many cases of conflict-driven or exacerbated by climate change (Willms/ Werner, 2009, p. 26). For instance, droughts in Darfur forced the Rizega't Pastoralists to seasonally migrate to Northern Bahr elGazal State of South Sudan in search of greener pastures and water. However, this annual movement crossing an international boundary often results in conflicts between the Rizeigat of Sudan and the Dinka Malual of South Sudan due to unresolved political, economic, and security issues between the two countries.

The vulnerability of South Sudan to climate changes and extreme weather variability is attributed to years of negligence and under-development by successive colonial and neo-colonial regimes in Sudan. For example, the beginning of severe flooding in the 1960s could have sent signals to the previous governments in Sudan to develop water resources management infrastructure in the Southern region. Instead, much effort and financial resources were directed towards accessing the vast reserves of freshwater resources in the Sudd Wetlands for cotton growth in Sudan and Egypt through the development of the Jongeli Canal project (Deng, 2022). Sufficient reliable sources of evidence suggest that the South Sudanese people have suffered from acute food insecurity, poverty and under-development, largely inherited from years of political and economic marginalisation of Sudan's southern region. This situation of hardship experienced under the subsequent Sudanese regimes was sufficient justification for the secession of South Sudan in July 2011. However, not much has changed since the independence, as financial resources are not being released to manage water resources infrastructure to boost agricultural productivity and production to enhance food security, generate income through exports and improve livelihoods. Instead, South Sudan imports food from neighbouring Uganda, Kenya, Sudan and even chicken from Brazil, respectively. Lack of allocation and actual release of financial resources to manage the impacts of flooding by the government, particularly between 2019 and 2022, resulted in submerging of villages in the Greater Regions of Upper Nile, Bahr el-Ghazal, and Equatoria and an uncounted number of human and animal lives lost, roads, bridges and buildings damaged due to floods.

Conflict, economic decline and soaring food prices have driven about 20.3 million people across Sudan (over 42 percent of the country's population) into high levels of acute food insecurity, classified in Integrated Food Security Phase Classification (IPC) Phase 3 or above (Crisis or worse) between July and September 2023. Of those, 14 million people (29 percent of the population) are under in IPC Phase 3, Crisis, and almost 6.3 million people (13 percent of the population) are experiencing worse conditions in IPC Phase 4, Emergency. The ongoing conflict in Sudan that erupted on 15 April 2023, between the Sudanese Armed Forces (SAF) and Rapid Support Forces (RSF) is deepening the humanitarian crisis and worsening the food security situations of the people in several urban and rural areas. The escalation of violence in Khartoum, Greater Darfur, and Greater Kordofan has triggered the large-scale displacement of 2.6 million people across all states of Sudan-forcing 757,230 people to flee to neighbouring countries (IPC, 2023).

After the independence of South Sudan, the government, in its 2019–20 national budget, allocated $2.3 million to the Ministry of Agriculture and Food Security to address food insecurity through climate change measures, investments in agriculture, and food relief to those in emergencies. The government has developed and adopted several climate change and environmental protection policies to improve food security for the country, including adapting the Paris Climate Agreement, the National Environment Policy 2015–2025, and other policy frameworks (Concern Worldwide, 2022). Although the government of South Sudan has demonstrated efforts to improve food security and tackle the climate crisis, these efforts are mostly on paper, lips-service and hardly backed with financial resources for project implementation. To bridge the gap of state failure to improve food security, individual senior government officials and citizens have since 2012 embarked on food production. While there have been marginal improvements in food security in some parts of the country, these efforts are largely the contribution of rural farmers and individual elites and citizens. The results are promising, with good harvests of fresh food crops, fruit and vegetables but are often interrupted by insecurity, which forces the rural population to migrate in search of food, safety and other resources.

South Sudan is among the richest countries in Africa, endowed with immense natural resources, including oil, fertile land, extensive hardwood forests, large amounts of fresh water and minerals, as well as a wide variety of livestock and fisheries (GOSS Ministry of Environment and Forestry, 2018, p. 9). Despite its agricultural potential, South Sudan's economy is largely dependent on oil revenue, which contributes about 90% of the country's Gross Domestic Product (GDP). Ironically, instead of the government using the oil money to invest in agricultural production and productivity to reduce food shortage, hunger, and the prevailing higher food prices, South Sudan remains a consumer country. This situation is often blamed on the armed conflicts and War, which commenced in December 2013 and continued in July 2016 despite the peace agreement signed in August 2015.

In addition to armed conflicts, South Sudan has been crushed by corruption, mismanagement of public resources and impunity; a case in point in this regard is the popular DORA SAGA that occurred in 2012, where President Salva Kiir accused 75 ministers and officials of having stolen $4 billion from state coffers and demanded the money to be returned. However, according to news reports, only 1.5% of this sum was effectively recovered (Aljazeera, 2014). Thus, the post-independent era has witnessed successive challenges that have resulted in protracted violent conflicts and increased corruption cases, which has impeded the ability of the government to deliver social and development services.

However, from 2012 to 2022 onwards, civil society and faith-based activists, youths and women groups across South Sudan have raised their voices against corruption, poor social and development service delivery. On 30 March 2020, hundreds of angry protesters, mostly young artists, women and men from different civic groups, seized Mobil Roundabout in Juba and protested the catastrophic car accident and poor health services, which led to the death of a famous female artist Tarisha Cee a day before. The protesters denounced corruption and poor working conditions for medical personnel, which have affected the delivery of health services. One protester claimed, "The Juba Teaching Hospital, the only teaching and referral hospital in the country, is just a building. It has no experienced doctors on duty, nor an accessible blood bank to save lives". Over 20 activists were arrested and detained by the South Sudan Police Service for participating in the protest but were released without charges on the same day (John, 2023). In August 2022, a handful of youth also staged a peaceful demonstration at the Konyokonyo market in Juba, protesting high food prices despite the local currency's appreciation against the hard currency. In 2020, civil society activists rolled-out a nationwide #Gurushwen campaign (Arabic for #WhereIsthe Money? which demanded transparency and accountability in managing public resources for service delivery, among other campaigns, increased government transparency in receipt of non-oil revenue collections.

Consequently, the National Revenue Authority (NRA) has, on several occasions, reported an increase in the collection of tax revenues since 2021 and 2022. Besides, the national government introduced the "oil for development" project, which connected Juba-Terekaka and Juba-Bor with tarmac roads. There is also commendable progress achieved by the government of South Sudan towards public finance management reforms. These is commendable progress being made by the government of South Sudan. However, the challenge remains with how the financial resources are distributed to priority areas, including mitigating climate crisis and food production.

Besides, South Sudanese nonviolent campaigns to achieve reforms in public financial management, CSOs, and faith-based organisations have been instrumental in the provision of agricultural inputs and services such as health care, shelter, education, nutrition, peacebuilding, conflict mitigation and environmental conservation (Civil Society Sustainability Index for South Sudan, 2019, p. 7). For example, after the signing of the Addis Ababa peace accord, the semi-autonomous regional government in Juba attempted Project Development Unit (PDU) of the World Bank, Norwegian Church Aid (NCA) and the Sudan Council of Churches (SCC) to improve the food production of rural farmers in few specific districts. However, it was short-lived with little impact on food security. As a result, rural farmers in South Sudan have perpetually faced numerous problems in their efforts to produce food. Some problems include climate variability in some ecological zones, inappropriate government policy, very low levels of agricultural technologies, and limited extension and research programmes (Odra, 2004, pp. 42–47).

In South Sudan, the search for scarce resources has been in the form of an invasion of a nearby community to raid cattle and other resources, or a migration to a new area to look for opportunities to bring new arrivals into competition with the host communities or land owners (Tiitmamer et al., 2018). Examples of migration to a new area due to climate-induced displacements include Jonglei Agro-pastoralists migrating to the Equatoria region, which has caused conflicts with farmers in the region since the 1990s. The scarcity of resources in rural areas due to drought and floods has forced people to migrate to towns, increasing socio-economic stresses that sometimes fan political upheavals. Major floods of the 20th century have also been associated with conflicts. There are several cases of conflict associated with floods and droughts worth mentioning. These include Pastoralist–Farmer Conflict in Equatoria and Dinka–Nuer Conflict, among others. While other factors have caused these conflicts, climate-induced stresses and shocks in the form of floods and droughts have significantly triggered or exacerbated the hostilities (Tiitmamer et al., 2018, p. 6).

One of the causes of the conflict between Dinka pastoralists and Equatoria farmers can be traced to the devastating flood of the 1960s, which destroyed homes, crops and pastures in the then Yirol Districts. Pastoralists, mostly from Bor areas, migrated to Equatoria, which has highlands with good pastures almost all year round. Tensions emerged as herds of Dinka cattle began to stray and devour the crops in Equatoria. The farmers' grievances fed into the political grievances of Dinka domination in the regional government in Juba and led to the Kokora, a campaign to decentralise Southern Sudan so that each group could go to their administrative region (Willems/Deng, 2015). As the conflict heightened, the Khartoum regime used the tensions within the Southern region to divide it into three administrative regions to fulfil its political agenda of controlling Southern Sudan to benefit from its oil and other natural resources. For some, it was perceived as a temporary conflict resolution measure to the Dinka pastoralists and Equatoria farmers' climate-induced conflicts.

6.3.2 Kenya

Approximately 80% of Kenya is arid and semi-arid, with pastoralism and agro-pastoralism being the dominant rural livelihood. According to Ellis/Freeman (2004), low household incomes in Kenyan rural areas are associated with low land and livestock holdings, high reliance on food crop agriculture, and low monetisation of the rural economy. However, there are also cases of land disputes or inheritance, declining security in rural areas, deteriorating access to proper agronomic advice and inputs, and predatory taxation imposed by county governments, often worsening these adverse conditions (Ellis/Freeman, 2004). While South Sudan, Sudan and Ethiopia have access to benefit from the River Nile water, Kenya suffers from water scarcity as a resource to boost the agro-pastoral industry and improve livelihood. Kenya heavily depends on rainfall, small and big rivers, springs and dams as water sources for life, including agro-pastoral industry. Most water sources are seasonal, and 43% of households attribute this seasonality primarily to drought and delayed rains as major challenges for food production (Speranza, 2010: 361).

The consequences of drought in rural areas affect the economic life in urban areas as income-earning opportunities remain low for most households due to scarcity. The COVID-19 pandemic exacerbated an already deteriorating situation of economic hardship for the average Kenyan. Recent research conducted to examine the experiences of Kenyans who migrated from rural areas to urban towns because of drought or water shortages shows that Kenyans who have relocated due to droughts and water shortages are more likely to be attacked outside of their home than those who have not reported moving for this reason. Migrants are often physically vulnerable compared to host communities (Linke et al., 2018).

Despite the challenges of climate change, agriculture is the backbone of the Kenyan economy, contributing to about 24% of the national Gross Domestic Product (GDP) directly and another 27% indirectly. The agriculture sector is the driver of Kenya's economy and the means of livelihood for most Kenyan people. The sector provides income to more than 80% of the population, employing over 40% of the total population and over 70% of the rural population. These milestones were achieved through the Green economy and the Green agriculture policy shifts. A Green economy focuses on improving human wellbeing. At the same time, Green agriculture is characterised by shifting both commercial and subsistence farming towards ecologically sound farming practices, such as efficient use of water, extensive use of organic and natural soil nutrients, optimal tillage, integrated pest control and agroforestry (UNEP, 2014: 5). Consequently, Kenya became the largest economy in the East African Community (EAC), with an estimated per capita GDP of $862 in 2012. The country has experienced recent stable economic growth, averaging 5.7% in 2013, and that is expected to remain the same for 2014, according to Kenya's National Bureau of Statistics (KNBS) (World Bank, 2014; Kimenyi/Kibe, 2014).

The Government of Kenya has taken substantial steps towards operationalising the green economy, including enacting the policy and legal framework promoting a green economy and green agriculture and integrating the same into its vision for 2030 (GoK, 2007). Kenya's key policies and programmes for a green economy include investments in renewable energy, promotion of resource-efficient and clean production, pollution control and waste management, environmental planning and governance and restoration of forest ecosystems. In addition, Kenya has developed a National Climate Change Response Strategy (NCCRS) and seeks to embrace a low-carbon development pathway that is inclusive and equitable and that contributes to Kenya's global competitiveness. Through the Greening Kenya Initiative (GKI), the government developed a database on green economy activities, which includes the manufacture of eco-friendly materials, tree planting, organic farming, fish farming, renewable energy, eco-labelling, solid waste management and environmental management (UNEP, 2015, p. 5).

It must be noted that the progress being made by the Kenyan government in implementing a green economy and green agriculture was largely a result of the mounting pressure from and the work of the civil society sector. Several civil society organisations such as the Green Belt Movement (GBM), co-founded by Wangari Maathai in 1977, have contributed to reducing the negative impacts of climate change through planting trees and nonviolent campaigning to limit environmental disasters, desertification, and deforestation. The GBM's successes in forest conservation, education, and women's economic empowerment have gained worldwide recognition. Dr Wangari Maathai, a Nobel Peace Laureate, a woman who advocated for investments in agriculture and worked for the environment, human rights and democracy issues, was also recognised not only as a Kenyan or African woman leader but globally (Ndura et al., 2011: 440).

Despite investment in agriculture which has placed Kenya as one of the growing economies in East Africa, the agricultural sector is facing major challenges, including stagnant or declining productivity levels, under-exploitation of land, inefficiencies in the supply chain due to limited storage capacity, lack of post-harvest services, poor access to input markets and low-value addition of most agriculture exports. The rate of industrialisation is low, and the current external account is in deficit. In addition, Kenya has a high poverty level, about 34–42% (World Bank, 2014), strong inequalities, and a relatively high unemployment rate, particularly for the youth. Environmentally, the country is trending towards overshooting its ecological capacity and depleting its natural resources and ecosystems.

6.3.3 Ethiopia

Food insecurity is one of the most crucial problems threatening millions of people in Ethiopia. The descriptive statistics results revealed that 60.5% of the households faced food insecurity, which was higher than the findings reported from Tigray (42.7%) and Amhara (43.8%) (Gebrehiwot, 2009). Beyond the change in climatic variables, the demographic factors influence the household's food security status. Using Mekonnen's FGDs result as a secondary source of information, "households with the large family size were found to be food insecure than the small family size." Family size was a negative correlation with food security. This could be because the increased population led to the fragmentation of farmlands, shifts in cropping patterns, acceleration of land use conflict and competition, and land degradation (Mekonnen et al., 2020).

Ethiopia experiences drought conditions every eight to ten years. Drought is dryness due to an acute shortage of water, which lasts for several months or years.

Drought considerably endangers food security. As a complex natural event, it stems from a lack of precipitation over a prolonged period, and its effect can only be witnessed slowly over a period. The country has been facing drought at a growing incidence throughout the past many decades. For instance, between 1984–1985, droughts affected the lives of more than two hundred thousand people and millions of livestock in Ethiopia (Liou/Mulualem 2019). Ethiopia's economy depends on rain-fed agriculture, which is vulnerable to climate change (Schmidt et al., 2017).

Despite the drought and other climate variabilities, Ethiopia's economy is dominated by the subsistence agricultural sector, which plays the dominant role in the country's economic development, accounting for about 42% of the gross domestic product (CSA, 2010). Moreover, the country is highly affected by climate change because; (a) about 80% of the population largely depends on rain-fed agriculture, (b) it is a low-income country, and (c) it has varied geographical locations with different magnitudes of climate impacts (Alemu/Mengistu, 2019). The nature-dependent agricultural sector associated with the country's geographical location, topography, and low adaptive capacity made the country highly vulnerable to the adverse effects of climate change (Negash, 2013). Thus, a large portion of the Ethiopian population is affected by chronic and transitory food insecurity, which is highly linked to severe, recurring food shortages and famine associated with recurrent drought (climate change) (Mota et al., 2019).

Ethiopia's rapid population growth of about 120 million, the second largest in Africa after Nigeria, with about 206 million in 2022, increases pressure on land resources, expanding environmental degradation and raising vulnerability to food shortages. Poverty, drought, and communal and political armed conflicts have caused Ethiopia's migration since the 1960s. Since the defeat of the Derg in 1991, Ethiopians have migrated to escape violence among some of the country's ethnic groups or to pursue economic opportunities amidst the growing problem of women and children trafficking for domestic work, and prostitution as a growing problem. Like Sudan and South Sudan, Ethiopia experienced worsening food insecurity due to multiple shocks affecting food availability and access, including the conflict in the northern Tigray Region, which began in November 2020. Consequently, in the Tigray region alone, 5.3 million people are estimated to be severely food insecure; the failure of the Mar–May 2022 'Gu-Genna' rains in southern pastoral areas of the southern Oromiya Region and southern Somali Region exacerbated drought conditions prevailing since late 2020, causing severe crop and livestock losses. The conflict situation further led to severe macroeconomic challenges, including insufficient foreign currency reserves and the continuous depreciation of the national currency; as a result, inflation is at very high levels, with the year-on-year food inflation rate estimated at 35.5% in July, one the highest of the last decade. These difficulties are exacerbated by the ripple effects of the Ukraine war, which triggered hikes in international prices of wheat, fuel and fertilisers (World Fact Book, 2022).

Ethiopia is also known for having border conflicts with its neighbours, particularly Eritrea, Somalia and the Sudan. Ethiopia's construction of a large dam (the Grand Ethiopian Renaissance Dam) on the Blue Nile since 2011 has triggered diplomatic tensions with Egypt and Sudan; as of 2020, four years of three-way talks between the three capitals over operating the dam and filling its reservoir had made little progress. Ethiopia began filling the dam in July 2020, which led to a military confrontation with Sudan over the al-Fashaga triangle's un-demarcated boundary with Ethiopia. Taking advantage of Ethiopia's civil War in Tigray, Khartoum deployed troops in December 2020 to reconquer al-Fashaga's historically disputed territory between the two countries.

6.4 A Roadmap for Survival from Climate Crisis to Peace and Sustainable Development

For most of the post-independent history of Africa, peace has remained elusive. Peace and sustainable development have largely remained complex concepts and more difficult to achieve than the Afro-optimists envisaged in the immediate post-independence period, owing to various domestic and external factors (Francis, 2008, p. 3). Africa's abundant natural, mineral, and human resources and the enormous wealth they produce have not been translated into poverty reduction, directed towards long-term economic growth or increased livelihood and welfare for the majority of Africans. The result was the relevance of poverty across the region endowed with enormous resources. This situation of "poverty amidst plenty of resources" and what some analysts described as "the natural resource curse or dead resources" is largely due to bad governance, i.e. corruption and mismanagement of natural resources by corrupt officials within the ruling elites, impunity, including state and non-state actors' weakness to hold corrupt officials accountable for their actions. Externally, with the end of the Cold World War, there has been renewed international interest in Africa's natural resources because of the threats to global energy resources and competition posed by China and India's new emerging economic powers. Thus, Africa's natural resources ended up being exploited to meet the needs of superpowers rather than used to develop the continent (Africa Development Bank, 2007, xv–xix). Therefore, any attempts to support peace and sustainable development initiatives in Africa should be de-colonised.

Superpowers should base their support on international organisations to tackle climate and food security crises entirely on "humanity and human security". Human security and wellbeing are central to peace and sustainable development goals. Human security has several forms and dimensions, including personal, community, economic, political and health (food and sanitation) wellbeing. There are diverse root causes that underpin different forms of insecurity. For example, personal insecurity is associated with physical and psychological violence in all its forms (Hope/Sparks, 2000). Community-level insecurity arises from inter-ethnic, religious and other identity-based tensions and crime and terrorism (Duvall et al., 1999). Economic insecurity is generally associated with persistent poverty, unemployment, and lack of access to credit and other economic opportunities (Western et al., 2012).

Similarly, food insecurity is linked to hunger, famine, and sudden increases in food prices (Barrett, 2010). Political insecurity is connected to political repression, human rights violations, and lack of rule of law and justice (Bates, 2005). Epidemics, malnutrition, poor sanitation, and lack of access to basic health care are often associated with health insecurity (Gama, 2016).

Despite the challenges of conflict, war, drought and limited progress towards the realisation of sustainable development goals in East African countries, there are also promising cases of success and resilience in the region. For example, despite the drought, resource-based conflicts and water scarcity in Kenya, the country has demonstrated their audacity not only to limit climate change and food crisis but also to gain income from the agricultural sector and employ plenty of people in farming, pastoral and tourism industries. This positive experience can be emulated by other countries too.

The governments in South Sudan, Sudan, Kenya and Ethiopia have collaborated with international development partners such as the FAO, UNDP and the UNEP, as well as the civil society sector to mitigate the impacts of climate change, food insecurity and conflicts. However, the responses lack proper coordination and coherent approaches to developing a sustainable food security policy for Africa. Notable responses are often isolated and specific. For example, the efforts of Lake Olbollosat area residents in the heart of Kenya, who slowed down deforestation on the mountain slopes, exacerbated by erratic rainfall caused by climate change — through reforestation. Similarly, the Konso of Ethiopia, living in one of Africa's harshest climates, have created inter-connected terracing systems that preserve fertile lands (Ground Up, 2007). In Sudan, communities living along the River Nile used the water to irrigate desert lands to produce vegetables, fruits and other stable food crops. Such individual responses are essential signs of resilience to assure food production in threatened areas. In a broader sense, the African response must be locally owned and reflect cooperation among African states and stakeholders to demonstrate socio-economic and political maturity. Such practical actions will also act as the basis of the argument to press upon the relevant international organisations the need for more financial and technical support to tackle the challenges of food insecurity. This way, an integrated and effective approach and response to food insecurity will be achieved with Africans shaping that effort.

Most importantly, the severity and magnitude of the challenges of climate change and food insecurity in East African countries demand shift, creative and sustained actions. In order to formulate and roll out an effective roadmap for survival from the climate change and food crisis, East African civil society organisations', governments of the four countries and international development partners should undertake the following policy recommendations seriously:

- **Civil Society Organisations**

First, *CSOs to coordinate, develop, launch and* jointly implement food security advocacy strategies for each of the four countries. The strategy should promote resilient food security systems for better production and environmental protection and ensure that the government, private sector and international development partners allocate adequate financial resources to improve climate change. *CSOs* and media should be financially and technically supported to promote green agriculture and economy, focusing on rural communities. CSOs continue their legitimate roles to shape the socio-economic and political landscape through nonviolent actions to resist corruption, mismanagement of resources and impunity in the region. Every effort should be made to halt intra and interstate-armed conflicts and wars within Africa. CSOs, media and researchers expose the use of food as a weapon of war as well as advocate for holding perpetrators accountable for their actions under international humanitarian law.

Civil society actors should improve preparedness through coherent coordination with local communities, government authorities and international development partners to better respond to environmental challenges, conflict early warning systems, and limit tree logging and other anthropogenic destructive human activities. They should use media, civic education and advisory services to the rural farmers and pastoralists to strengthen community resilience.

- **Governments of the four countries**

Second, governments of the four countries must note with great concern that despite the region's agricultural potentials, climate variabilities, conflicts, and other anthropogenic human activities have qualified the region as world food insecure. Therefore, the governments of South Sudan, Sudan, Kenya and Ethiopia should consider food insecurity, climate crisis and poverty as real threats to the region's security and territorial integrity. Thus, each government must undertake (or strengthen) immediate, creative and coordinated response to prioritise agriculture to grow food, create employment opportunities and generate income to deliver social and development services to the population. Agricultural production and productivity are fundamental determinants of the availability and the price of food and hence have a major impact on food security. The government must allocate and release adequate financial resources, provide policy guidance and support rural farmers and agricultural industries to produce enough food for local consumption and export the surplus to achieve economic growth, peace and sustainable development.

The four governments take advantage of growing revenues from oil, tourism, and other revenue resources to develop and boost the green economy (agriculture) and infrastructural development projects that put people, the environment and livelihoods at the forefront. The oil for infrastructure project employed by the government of South Sudan in 2020 should prioritise linking agricultural areas with markets to transport food to the local markets. For example, in 2021, the FAO declared Western Equatoria state as the leading food-producing state in South Sudan, yet the state's main road connecting the markets in Juba remained largely inaccessible during rainy seasons. The government's plan to tarmac the said road should be expedited. Meanwhile, the governments of Ethiopia and Kenya should use revenues generated from non-oil and mineral sources to consolidate a green economy and restore stability and peace.

The four countries and, indeed, the entire region should coordinate, regulate and support the migration of the population within and between one country to another in such a way as to ensure conflict sensitivity, respect and protection of local resources, culture and traditions. Reinforce policies to limit human activities such as irresponsible extraction of natural resources, indiscriminate tree logging, and polluting air and waterways by individuals, groups and multi-national industrial companies. These activities should be regulated to avoid ecosystem destruction, wiping out the lives of plants, humans and animals. The governments of South Sudan, Sudan, Ethiopia and Kenya, together with their development partners, should prioritise supporting the rural communities with adequate financial, technical and technological resources to improve crop varieties, irrigation, agroforestry, water-saving landscape and climate management.

- **International development partners**

Third, the international development partners such as the UNDP, FAO, UNEP, and others should, in coordination with the relevant government authorities and civil society organisations, build consensus over individual obligations and coordinate efforts on how best to implement the road map laid out in the Sustainable Development Goals (SDGs). There is a need for the stakeholders to address the root causes of increasing inequality and transform international financial and debt architecture to avoid leaving developing countries behind (UNDP, 2022). The international community ought to ensure that Russia and Ukraine respect and allow trade in food, fertilisers and oil to limit the gaps in world food insecurity. The War's impact on the food security of vulnerable groups necessitates timely and well-targeted social protection interventions to alleviate hardship and foster a fast recovery.

The international development partners should mobilise financial resources and local technical, human and material resources to jointly and effectively implement SDGs with clear short-term, medium-term, and long-term achievable benchmarks.

6.5 Conclusion

Although the reflections on the implications of climate change to food insecurity and conflicts discussed in this chapter may turn to present a worrisome future for the four countries, the threats can be reversed. There are always creative possible alternatives to every crisis. Fortunately, the civil society organisations, the governments of the four countries and the international development partners have demonstrated efforts towards tackling climate and food crises, managing conflicts realising sustainable development goals. Therefore, the stakeholders should double more practical efforts and achievable actions to produce more food to fight hunger, generate financial resources and improve climate conditions for the current and future generations.

References

Albert, I. (2008). Understanding peace in Africa. In Francis, D. (Ed.), *Peace and conflict in Africa* (pp. 31–45). London: Zed books.

Alemu, T., & Mengistu, A. (2019). Impacts of climate change on food security in Ethiopia: Adaptation and mitigation options: A review. In P. Castro, A.M. Azul, W. Leal Filho & U.M. Azeiteiro (Eds.), *Climate change-resilient agriculture and agroforestry* (pp. 397–412). Cham: Springer.

Bates, R. H. (2005). "Political Insecurity and State Failure in Contemporary Africa." CID Working Paper Series 2005.115, Harvard University, Cambridge, MA, January 2005. https://dash.harvard.edu/bitstream/handle/1/42482324/115.pdf?sequence=1. Accessed 4 August 2023.

Benner, S., Lax, G., Crutzen, P. J., Pöschl, U., Lelieveld, J., & Brauch, H. G. (2021). *Paul J. Crutzen and the Anthropocene: A New Epoch in Earth's History.* Cham: Springer.

Biermann, F. (2020). *The future of environmental policy in the Anthropocene time for a paradigm shift, Environmental Politics.* https://doi.org/10.1080/09644016.2020.1846958

Bizikova, L., Robinson, J., & Cohen, S. (2011). Linking climate change and sustainable development at the local level. *Climate Policy*, 7 (4), 271–277.

Bombaci, S., Pejchar, L., & Innes, J. (2018). Fenced sanctuaries deliver conservation benefits for most common and threatened native island birds in New Zealand. *Ecosphere*, 9(11), e02497.

Bosselmann, K. (2020). The role of trusteeship in earth governance. In *Ecological Integrity in Science and Law* (pp. 241–252). Cham: Springer.

Brauch, H. G. (2022). Peace ecology in the Anthropocene for Africa. In G.T. Harris & J.C.K. Kiyala (Eds.), *Civil Society and Peacebuilding in Sub-Saharan Africa in the Anthropocene: An Overview* (pp. 17–66). Cham: Springer. https://doi.org/10.1007/978-3-030-95179-5

Carter, T. R., Jones, R.N., Lu, X., Bhadwal, S., Conde, C., Mearns, L.O., O'Neill, B.C., Rounsevelt, M.D.A., Zurek, M.B. (2007). New assessment methods and the characterisation of future conditions. In M.L. Parry, O.F. Canziani, J.P. Palutikof, P.J. Van den Linden & C.E. Hanson (Eds.), *IPCC, 2007: Climate change 2007: impacts, adaptation and vulnerability: contribution of Working Group II to the fourth assessment report of the Intergovernmental Panel on Climate Change* (pp. 133–171). Cambridge: Cambridge University Press.

Central Statistics Agency (CSA). (2010). *Agricultural Sample Survey Report on Area and Production of Crops.* The Federal Democratic Republic of Ethiopia Central Statistics Agency.

Christopher, N. & Collins, M. (2012). Understanding Intractable Conflict in Sudan. *American Intelligence Journal*, 30(1), 23–26.

Colglazier, W. (2015). Sustainable development agenda: 2030. *Science*, 349(6252), 1048–1050.

Concern Worldwide. (2022). *Hunger in South Sudan: Understanding a decade of food crisis.* https://www.concernusa.org/story/hunger-in-south-sudan/. Accessed 4 August 2023.

Cruickshank, E. W., Schneeberger, K., & Smith, N. (2012). *A pocket guide to sustainable development governance.* Stakeholder Forum.

Crutzen, P, J. (2002). Geology of Mankind. *Nature*, 415,3 (January), 23.

Deng, P. (2022). *In report of the 6th Governors' Forum on Irrigation and Water Management.* Ministry of Irrigation and Water Management, Juba 21–30 November 2022.

Drammeh, W., Hamid, A.N., & Rohana, J.A. (2019). Determinants of household food insecurity and its association with child malnutrition in sub-Saharan Africa; A review of the literature. *Curr. Res. Nutrit. Food Sci.*, 7(3), 610–623.

Duvall, R. D., Weldes, J., Laffey, M., & Gusterson, H. (1999). *Cultures of insecurity: States, communities, and the production of danger.* Minneapolis: University of Minnesota Press.

Ellis, F. & Freeman, H.A. (2004). Rural livelihoods and poverty reduction strategies in four African countries. *Journal of Development Studies*, 40(4), 1–30.

Fang, X., Kothari, S., Mcloughlin, C. & Yunice, M. (2020). *The Economic Consequences of Conflicts in Sub-Saharan Africa.* A working paper published by the International Monetary Fund.

FAO, WFP and IGAD. (2022). *In Eastern Africa, over 50 million to face acute food insecurity in 2022.* https://reliefweb.int/report/ethiopia/eastern-africa-over-50-million-face-acute-food-insecurity-2022. Accessed 10 January 2023.

FAO. (1996). *Rome Declaration on World food security and World food summit plan of Action.* World food summit 13–17 November 1996. Rome: FAO.
FAO. (2017). *Regional Overview of Food Security and Nutrition in Africa 2017.* The food security and nutrition–conflict nexus: building resilience for food security, nutrition and peace. Accra.
FAO. (2019). *The State of Food Security and Nutrition in the World.* Safeguarding against economic slowdowns and downturns. Rome: FAO.
FAO. (2020). *Impacts of desert locust infestation on households' livelihoods and food security in Ethiopia.* Rome: FAO.
FAO. (2022). *The importance of Ukraine and the Russian Federation for global agricultural markets and the risks associated with the War in Ukraine.* Rome: FAO.
FEWS Net. (2022). *Kenya Food Security Outlook June 2022 to January 2023.* Food Emergency Warning System Network.
Francis, D. (2008). Understanding the context of peace and conflict in Africa. In *Peace and conflict in Africa.* London: Zed Books.
Gama, E. (2016). Health insecurity and social protection: Pathways, gaps, and their implications on health outcomes and poverty. *International Journal of Health Policy and Management*, 5(3), 183–187.
Gebrehiwot, T. (2009). Determinants of food security in rural households of the Tigray Region, 29 Tsegay.
Gleditsch, N.P. (2011). *Regional conflict and climate change.* Paper prepared for the Workshop on Research on Climate Change Impacts and Associated Economic Damages, 27–28 January in Washington DC, USA.
GoK (Government of Kenya). (2007). *Kenya Vision 2030.* Nairobi. Nairobi Government Printers. https://vision2030.go.ke/publication/kenya-vision-2030-popular-version/ Accessed 7 August 2023.
GOSS Ministry of Environment and Forestry. (2018). *Initial national communication to the United Nations framework convention on climate change.* Government of South Sudan, Ministry of Environment and Forestry.
Ground Up. (2007). Sustainable community development. A PELUM/PSAf publication promoting sustainable community development, 3 (19).
Helvey, R. L. (2004). *On strategic nonviolent conflict: Thinking about the fundamentals.* Albert Einstein Institute Boston. https://www.mkgandhi.org/ebks/nonviolent_conflict.pdf. Accessed 4 August 2023.
Hope, T., & Sparks, R. (2000). *Crime, risk and insecurity: Law and order in everyday life and political discourse.* London: Routledge.
Griggs, D. J., & Noguer, M. (2002). Climate change 2001: the scientific basis. Contribution of working group I to the third assessment report of the intergovernmental panel on climate change. *Weather, 57*(8), 267–269.
Houghton, J.T., Ding, Y, Griggs, D.J., Noguer, M., van der Linden, P.J. & Xiaosu, D. (eds). (2001). *Climate change 2001: the scientific basis.* Contributions of Working Group I to the Third Assessment.
IMF. (2022). *Climate change and chronic food insecurity in Sub-Saharan Africa.* International Monetary Fund. Intersections of Truth, Justice and Reconciliation in South Sudan.
IPC (2023). IPC Acute Food Insecurity Analysis Jun 2023-Feb 2024 Report. Integrated Food Security Phase Classification. Available: https://www.ipcinfo.org/fileadmin/user_upload/ipcinfo/docs/IPC_Sudan_Acute_Food_Insecurity_Jun2023_Feb2024_report.pdf. Accessed 13 Feb 2024.
IPCC. (2007). *Contribution of Working Group II to the Fourth Assessment Report of the Intergovernmental Panel on Climate Change.*
IUCN. (1980). *World Conservation Strategy: Living resource conservation for sustainable development.* International Union for Conservation of Nature and Natural Resources.

John, M. (2023). Unarmed civilian protection and nonviolence with attention to Sub-Saharan Africa. In J. Furnari & Kabaki (Eds.), *Unarmed civilian protection: A new paradigm for protection and human security* (pp. 115-125). Bristol: Bristol University Press.

Kabubo-Mariara, J., & Mulwa, R. (2019). *Adaptation to climate change and climate variability and its implications for household food security in Kenya*. Food security.

Kimenyi, M.S., & Kibe, J. (2014). *Africa's Powerhouse*. Brookings Institution. https://www.scirp.org/%28S%28351jmbntvnsjt1aadkposzje%29%29/reference/referencespapers.aspx?referenceid=3082431. Accessed 12 February 2023.

Kinda S.R., Badolo F., & Tajani, F. (2019). Does rainfall variability matter for food security in developing countries? *Cogent Economics & Finance*, 7(1):1640098.

Kiplagat, B. (1997). African Government-African Conflicts. In: *Out of conflict from War to peace in Africa*. Edited by Gunnar Sorbo and Peter Vale. Published by Motala, Sweden. Uppsala.

Linden, P.J, & Hanson, C.E. (Eds). (2007). *Climate change: impacts, adaptation and vulnerability*. Contribution of Working Group II to the Fourth Assessment Report of the Intergovernmental Panel on Climate Change. Cambridge University Press, Cambridge, pp 133–171.

Linke, A. M., Witmer, F. D., O'Loughlin, J., McCabe, J. T., & Tir, J. (2018). The consequences of relocating in response to drought: human mobility and conflict in contemporary Kenya. *Environmental Research Letters, 13*(9), 094014. https://doi.org/10.1088/1748-9326/aad8cc

Liou, Y.-A., & Mulualem, G. M. (2019). Spatio–temporal assessment of drought in Ethiopia and the impact of recent intense droughts. *Remote sensing, 11*(15), 1–19.

McCutchen, A. (2014). *The Sudan Revolutionary Front: Its Formation and Development*. Geneva: Small Arms Survey, 8.

McCutchen, (2015). The Sudan Revolutionary Front, 8. 6 Thomas P. Ofcansky, "*Historical setting", in Sudan: A Country Study,* ed. LaVerle Berry (Washington: Federal Research Division, 2015): 3–58

Mekonnen, A., Tessema, A., Ganewo, Z., & Haile, A. (2020). *Climate change impacts on household food security adaptation strategies in Southern Ethiopia*. https://onlinelibrary.wiley.com/doi/full/https://doi.org/10.1002/fes3.266. Accessed 11 March 2023.

NAPA. (2007). *National Adaptation Programme of Action*. Republic of Sudan, Ministry of Environment and Physical Development, Higher Council for Environment and Natural Resources, Khartoum.

Ndura, E., Meyer, M., & Atiri, J. (2011). *Peace, environment and social change: A conversation between Elombe Brath and Wangari Maathai in; Seeds bearing fruit*. Pan-African Peace Action for Twenty-first Century. Africa World Press, Eritrea.

Odra, J. (2004). Crop production and food security in; *Framework for rehabilitation and development of postwar South Sudan* edited by Odra, J, Deng, I and Nhiem, P. Published by University of Bahr el Ghazal.

Ojo, A. (2022). Human safety and security for sustainable and inclusive settlements. *Journal of Community Safety and Well-Being*, 7(4), 178–183.

Pettersson, T., Engstrom, G., Sollenberg, M., Davies, S., Hwach, N., Olberg, M., Deniz, A. & Hogbladh, S. (2020). *Organised violence 1989–2020, with a special emphasis on Syria*. Sage Publication.

Reliefweb. (2021). *Starving Tigray. How Armed Conflict and mass atrocities Have Destroyed an Ethiopian Region's Economy and Food Security and are threatening famine*. Report by World Peace Foundation via OCHA. https://reliefweb.int/report/ethiopia/starving-tigray-how-armed-conflict-and-mass-atrocities-have-destroyed-ethiopian. Accessed 4 August 2023.

Sam, A.S., Abbas, A., Surendran Padmaja, S., Kaechele, H., Kumar, R. & Müller, K. (2019). Linking food security with household's adaptive capacity and drought risk: implications for sustainable rural development. *Soc. Indic. Res.*,142(1), 363–385.

Schmidhuber, J. & Tubiello F.N. (2007). Global food security under climate change. *Proc. Natl. Acad. Sci.*;104(50), 19703–19708.

Schmidt, W., Peter Uhe, A., Kimutai, J., Otto, F., & Cullen, H. (2017). Climate and development knowledge network and world weather attribution initiative raising risk awareness. *Royal Netherlands Meteorological Institute: De Bilt, The Netherlands*, 2016–2017.

Shah, K.U. & Dulal, H.B. (2015). Household capacity to adapt to climate change and implications for food security in Trinidad and Tobago. *Reg. Environ. Change*,15(7), 1379–1391.

Smith, L.C., & Haddad, J.L. (2000). *International Food Policy Research Institute; Explaining Child Malnutrition in Developing Countries: A Cross-Country Analysis*. Vol. 111. Washington: International Food Policy Research Institute (IFPRI).

Smith, M.D., Rabbitt, M.P. & Coleman-Jensen, A. (2017). Who are the world's food insecure? New evidence from the food and agriculture organisation's food insecurity experience scale. *World Dev.*, 93, 402–412.

Speranza, C. (2010). Drought coping and adaptation strategies: Understanding adaptations to climate change in Agro-pastoral livestock production in Makueni District, Kenya. *European Journal of Development Research*, 22 (5), 623–642.

Strange, T.& Bayley, A. (2008). *OECD Insights Sustainable Development Linking Economy, Society, Environment.* OECD Report, Geneva.

Thome, K., Smith D.M., Daugherty K., Rada N., Christensen C. & Meade M. (2019). *International Food Security Assessment*. US Department of Agriculture, Economic Research Service; 2029, GFA-30. https://www.ers.usda.gov/webdocs/outlooks/93659/gfa-30.pdf?v=8164. Accessed 8 August 2023.

Tiitmamer, N., Mayai, A.T., & Mai, N.J.H. (2018). *Climate change and conflicts in South Sudan.* https://www.suddinstitute.org/assets/Publications/5af4422f2eed4_ClimateChangeAndConflictsInSouthSudan_Full.pdf. Accessed 5 February 2023.

UNDP. (2015). *Sustainable development goals*. Published by the UNDP.

UNDP. (2022). *The Sustainable Development Goals Report 2022*. Published by the UNDP. Available: https://annualreport.undp.org/ Accessed 22 December 2022.

UNEP (United Nations Environment Programme). (2014). *Green Economy Assessment Report: Kenya*. https://wedocs.unep.org/handle/20.500.11822/32899. Accessed 5 February 2023.

United Nations. (2022). *What is climate change*? https://www.un.org/en/climatechange/what-is-climate-change. Accessed 30 January 2023.

Wallace, K. J., & Clarkson, B. D. (2019). Urban forest restoration ecology: A review from Hamilton, New Zealand. *Journal of the Royal Society of New Zealand*, 49(3), 347–369.

Western, B., Bloome, D., Sosnaud, B., & Tach, L. (2012). Economic insecurity and social stratification. *Annual Review of Sociology*, 38, 341–359.

Wilbanks, T.J., & Sathaye, J. (2007). Integrating mitigation and adaptation as responses to climate change: a synthesis. *Mitigation and Adaptation Strategies for Global Change*, 12, 957–962.

Willems, R., & Deng, D. (2015). *The legacy of Kokora in South Sudan*. Briefing Paper.

Willms, D., & Werner, K. (2009). Climate change and conflict in Africa. In N. Sewankambo & H. Besada (Eds.), *Climate change in Africa. Adaption, mitigation and governance challenges* (pp. 26–49). Ontario: The Centre for International Governance Innovation.

World Bank. (2013). *Fraud and corruption: Frequently asked questions*. http://web.worldbank.org. Accessed 10 December 2022.

World Bank. (2014). *World Development Indicators*. http://data.worldbank.org/data-catalog/worlddevelopment-indicator. Accessed 5 February 2023.

World Factbook. (2022). *Countries-Ethiopia*. https://www.cia.gov/the-world-factbook/countries/ethiopia/. Accessed 15 February 2023.

Yagoub, Y.E., Li, Z., Musa, O.S., Anjum, M.N., Wang, F., & Bo, Z. (2017). Detection of drought cycles pattern in two countries (Sudan and South Sudan) by using Standardized Precipitation Index SPI. *Am J Environ Eng, 7*(4), 93–105.

Chapter 7
Gender, an Ecological Challenge to Peace in the Anthropocene: An African Perspective

Wilkister M. Shanyisa

Abstract Gender plays a critical role in shaping the world around us, but gender issues remain a challenge in Africa. The way men and women are situated in society describes the kinds of roles performed by these groups, especially in the African context. Existing power asymmetries lead to society being structured in a manner that reinforces and perpetuates the situation of men and women. Although men and women assume different roles in the use, management, and conservation of the environment, women assume and play a critical role due to their close contact with the environment. Their daily activities expose them to the dangers of environmental health hazards and suffer the greatest brunt from climate change, thereby threatening their security. This chapter seeks to explore the role that gender plays in environmental peace from an African perspective. It is based on the analysis of available literature on gender, climate change, and the environment. It is inevitable that for a long-time, women in Africa have been negated in decision-making and in top positions regarding environmental matters. Efforts and policies to incorporate and integrate women in environmental issues have remained superficial. A feminist view is key to integrating women's participation in environmental peace. Detrimental effects felt by women can be dealt with, thus empowering them to cope with the adversities of the environment. In Africa, a gender lens may be the option to realise substantive, active participation, and representation of women in environmental peace, but the commitment and political will of governments remain integral in addressing this critical issue.

Keywords Gender · Climate change · Feminist view · Power asymmetries · Environmental peace

Dr Shanyisa Milimu Wilkister is currently a lecturer and Ag HoD Centre for Social Justice and Ethics at The Catholic University of Eastern Africa, Kenya. She is also gender consultant. Email: wmilimu@yahoo.com; ORCID ID: https://org.ocid/0000-0002-5476-4271

J. C. K. Kiyala and N. Chivasa (eds.), *Climate Change and Socio-political Violence in Sub-Saharan Africa in the Anthropocene*, The Anthropocene: Politik—Economics—Society—Science 37,
https://doi.org/10.1007/978-3-031-48375-2_7

7.1 Introduction

Gender continues to pose a challenge to environmental peace. Climate change–'*threat multiplier,*' for example–continues to pose a major challenge to world peace in the twenty-first century, especially in Africa. Its effects are cross-cutting but felt mostly by vulnerable groups with little recourse to protect themselves. Loss of lives, increased food insecurity, competition over scarce resources, inequality, and power asymmetries are some of the end results of climate change. In the worst-case scenario, climate change has undermined the developmental benefits made by a nation, especially its economic benefits. This is especially so where political, economic, and social factors, among others, come into play. The above factors present avenues for human vulnerabilities, which render humanity, especially women, less susceptible to the foreseen effects of oceanic acidification (Bob/Bronkhorst, 2014). African nations have greatly experienced the brunt of climate change. Developing countries are more prone to ecological crises since they rely overly on agriculture for their survival. Ecology on its own does not foster crises, but several factors combine to fuel them within the environment. Multiple pressures differ across nations, among which are persistent poverty, disparities in wealth distribution, competition and rivalry among communities, unemployment, and population growth, which result in political struggle and ecological crises in some places (Richard et al., 2003).

This assertion could be true considering the increasing poverty levels in the African continent. Africa has been deemed a capitalist society which, coupled with the effects of covid 19 pandemic, the poor continue to be pushed to the periphery as the rich continue to amass more wealth for themselves. This dichotomy makes life more difficult for poor people who rely on the environment.

The extraction of natural resources is a critical cause of the ecological crisis. Oil extraction, for instance, leads to massive militarisation, as evidenced in parts of the Middle East where the government invests so much in the armies, unlike human needs. In oil-producing countries, internal conflicts have led to violations of human rights and the erosion of human security. In such countries, carbon emissions into the air result in climate change, posing a threat to human safety across borders (Prugh et al., 2005). For example, African countries known for the extraction of resources are reportedly experiencing conflicts: Somalia, DRC, Niger, etc. The concept of security is about risk and vulnerability. The ecological crisis poses a high risk to human justice, making people more vulnerable (Barnett, 2001). In DRC, for example, natural resources continue to fuel conflict. Most of those worst affected are women and girls perceived as vulnerable and highly dependent on climate-related resources, land cultivation, and water lowering their adaptation mechanism. As this happens, the peace process is disrupted due to ensuing violent conflicts by humans, which make communities poorer, less intense, and unprepared to mitigate the challenges of climate change. This chapter adopts an analytical approach to addressing the subject matter.

7.2 Gender in the Anthropocene

There is growing evidence of how the concept of gender is a significant influencer on global ecology in the current Anthropocene. Debates around this concept take on different voices. The realisation has been about more than achieving environmental and social equality or equity but enhancing institutional efficiency to mitigate ongoing global ecological challenges. The influence of climate change and the associated security challenges assume a crucial gender perspective of how men and women affect and are affected by the risks associated with climate change.

Gender—*the socially constructed roles and responsibilities assigned to men and women by society* (Kabonesa, 2002; Milimu, 2020), plays a vital role in climate change mitigation. In most cases, it is negated among policy frameworks and policy-makers, especially where issues of global ecological challenge in the Anthropocene where issues are articulated. Gender inequalities remain inevitable across the globe, and both men and women feel the impact, but women and girls are more vulnerable and account for most of the world's poor. Women's close interaction with the environment to secure their livelihood exposes them to detrimental effects of the domain. Evidence continues to unfold of violence directed towards women environmental campaigners and advocates of environmental rights across the world. Recognising the consequences of climate change to both men and women and acknowledging that both are affected differently helps to reduce the risks and inequality between men and women, hence, allowing a fairground where they can equally participate, hence fostering peace. The existing power asymmetries/dichotomy (Pratto, 1999) within society confines women in the private sphere limiting their mobility to the public sphere, where important decisions regarding environmental security are made. This implies that Success can only be realised when both sexes, men, and women, assume an active role as players and actors in the ecology.

The way women and men relate to the environment differs in many ways. Evidence continues to unfold of the unique effects women face, their concerns about the environment, and why they continue to be environmental actors against environmental degradation, unlike men. Previous studies also exhibit existing gender differences in understanding environmental challenges and the importance given to ecological concerns and taking actions, including activism to curb environmental degradation by both sexes (McKinney, 2014). Further, the literature reveals that aspects of disparity in vulnerability and exposure to the effects of climate change across genders, emphasising that women are not inherently more at risk, but that connection between gender, power changes, socio-economic structures, and societal expectations lead to effects of climate change impacting differently on women (Djoudi et al., (2016). In addition, there are lost chances and avenues for action when women are not fully integrated into the policy and decision-making processes (Olsson et al., 2014).

Notably, women and men have a critical role in enhancing environmental justice. Women, in most cases, are perceived as major contributors to environmental injustice, but often, programs aimed at promoting environmental justice negate women's contributions. Acknowledging the efforts of women in enhancing environmental justice is critical for a durable solution to be realised. Besides women having a right to participate in decision making including environmental matters, participation also offers a unique chance to overcome the socially constructed and deep-rooted inequalities in society that negatively affect the environment (UN, 2002). The evidence further reveals that environmental justice can only be necessitated by environmental analysis, which entails the adoption of a gender-sensitive lens.

This allows for a transformed society where the right to participate in promoting environmental justice is embraced not as a chance but as a human right by all sexes. With this approach, existing complex relationships between men and women will be closely examined. The examination needs to consider the impact of existing power asymmetry for the realisation of women's human rights and empowerment.

There is a need to break the existing power asymmetries between men and women, which continue to cause ecological crises. Ecological peace entails a fair playground where the efforts and gains made by both men and women equally matter in policy development and implementation. The existing strata, as Marxist thinkers present them in the capitalist theory (Glabau, 2017), makes structural inequalities inevitable for different sexes and access to and control over resources limited to the apparently 'inferior' sex. Such inequalities pervade all spheres of life, including ecology. Overcoming the ecological crisis entails a need to urgently understand how intersecting structural inequalities continue to strengthen the male–female divide that paves the way for the ecological crisis. This realisation could help towards achieving a fair distribution, representation, and ecological justice in the Anthropocene. A feminist perspective needs to be adopted to help challenge the marginalisation of a specific sex from an equal position in the global society and challenge a system that endangers all human beings on planet Earth. A feminist perspective is key not only in realising gender equality and equity but also for the equity, efficiency, and achievement of ecological justice and addressing other Anthropocene challenges.

7.3 A Feminist View of Environmental Justice

A feminist view on gender and environment emerged from the recognition that women performed much work and closely interacted with the environment. Their work, however, remained absent in environmental decision-making, policies, and programs. A need to integrate both men and women in the environment to bring women and their nurturing skills and knowledge into decision-making and to protect the environment has begun to unfold. It was out of the understanding that the relationship between men and women in society is asymmetrical due to the Culture and the patriarchal nature of the society (Sjoberg, 2010; Glabau, 2017). Such asymmetries are inevitable in all areas of life, including the environment. This also explains why there exists power inequality. It is out of these asymmetries that led to the growth of feminist theories.

Carefully analysing contemporary developments in women's roles and changes in urban and regional structure and activity patterns reveals asymmetrical relations. The relationship led to feminist curiosity and urged to challenge the existing inequalities and ensure that women's nurturing skills and knowledge are integrated into environmental issues and other developmental programs. Dynamics in class relations relative to environmental alterations were carefully explained. The dynamics are key in understanding women's lives to environmental production and alterations. As noted by Glabau (2017), feminist views and ideas did not readily find acknowledgment and recognition, still, they created a platform for recognising women's existence and the critical role they could play in enhancing environmental justice. The argument for the contention arose from the overemphasis on women's issues, negating men, who are also key players in environmental justice, the same opportunity. This was contested and created conflicts in women's lives because the focus was mainly on women's concerns negating men the opportunity also to participate actively. The dynamics led to a new view of women shifting parameters on women's lives and their location, politically in the political arena (Mackenzie, 1984). This struggle led to the emergence of women's movements in the late 1960s and, thus, a feminist analysis. The feminist analysis saw women as a social group and made their roles and activities more visible not only in the private but also in the public setting.

Women started to enter public life, particularly the labor force, which led to the emergence of liberal feminists who started to criticise the private setting/family that women were confined to. They saw women suffering as emanating from confinement in the private sphere which was also a source of 'feminine 'roles/norms. They aimed at securing the social role of women as wage workers and to ensure equal rights for them as men in capitalist social relations (Friedan, 1963, in Mackenzie, 1982). This progress in one area created problems in another area due to their dual roles (private and public sphere). The focus shifted from the private to the public sphere, where women began to make known their skills and knowledge. Liberal feminists were not explicit in the explanations. Out of these limitations, a new feminism which held that women's roles are socially constructed and cannot be eliminated through alterations distinct from a private or public setting emerged. It saw women's position as being created and challenged by the functioning of society. Their activities, constitution, and change of the category, 'woman,' were seen to be embedded in the process of reproducing capitalist social relations and a component of such social reproduction.

Women's position can only be understood through a re-analysis of complex social change from the perspective of how it structures women's activities. The re-analysis calls for a holistic approach to examine women's life and work, the reason radical and socialist feminists emerged (Mackenzie, 1984).

Radical feminist aims to understand the struggles between male Power and the social institutions through which it is reproduced, for example, marriage. Radical feminists focus on gender, which fixes women's oppression at the psychological or biological level and isolates women's issues from a social analysis and action. It ignores other forms of oppression, like sex preference in society (Mackenzie, 1984).

Socialist feminists, on the other hand, focus on analysing institutions and social practices of capitalism and how it impacts women. They try to understand gender and class relations and how gender constitution translates to class formation and change. Socialist feminists see gender constitution as a process implicated within a complex of 'productive' and 'reproductive' relations and roles of women in society (Mackenzie, 1984). Although a socialist view aims at creating a just society where environmental justice is paramount, it is not clear how this view can realise its goal. Glabau (2017) pointed out that environmental justice presents a form of global capitalism where the institutions mandated with learning ecological justice have been marred with oppression, which impedes this goal.

Existing classes tend to be skewed toward the "winning team," which aims at enjoying the benefits of environmental justice while creating a dichotomy for the less privileged to enjoy the fruits of their hard work. In a capitalist world, the workers (majority) work hard to sustain the means of production owned by the 'haves' of the society (few) who own and control the means of production. In this case, men may be referred to as the 'haves' of society, while women are the workers (majority) who closely interact with the environment; the proceeds are owned and controlled by the 'haves' of the society (men). This dichotomy disadvantages the workers and the environment, thus resulting in the existing environmental injustice. Although other factors have been pointed out as key contributors to the dichotomy, capitalism can be argued to be a significant defining moment for environmental injustice, as noted by Moore (2015). A socialist feminist perspective will be vital in breaking the dichotomy and ensuring that the proceeds of ecological justice and not detrimental to any one sex but are enjoyed and protected by those directly and indirectly involved in promoting environmental justice.

As pointed out by Griffin (1978), Ecofeminists acknowledge that the social mentality that leads to the domination and oppression of women is linked to the social mentality that paves the way for the abuse of nature. Ecofeminists relate women to nature and men to Culture. Culture—a people's way of life is assumed to be superior as opposed to nature. This means that women are inferior and relate to nature, while men are superior and relate to Culture. Further, there has been an assumption that women are good nurtures and, therefore, have a role in healing the environment. The domination and oppression mentality emerges from the restricted emotionality to conform to the existing masculine roles/norms that call for men as controllers of the inferior sex. Conformity to such standards results in environmental injustice, which leads to ecological conflicts that can either be constructive or destructive, driving society towards negative or positive peace (Galtung, 1990). In his view, Galtung (1990) presents conflict as a multi-contextual approach to related interconnected and ever-changing structures. There is a need for practical conflict management skills and a conducive environment where such conflicts are resolved non-violently. The dichotomy between men and women in relation to environmental justice presents a conflict that can be resolved by adopting a multi-contextual approach with a focus on effective conflict management skills to break the dichotomy. An agender lens and a feminist view are critical in this process.

Feminist strands are key to understanding the relationship between gender and environment. Feminists may not be explicit in their analysis of the relationship between nature and society, but their explanations lay a basis for a general understanding of the relationship between human beings and nature. The understanding is that, due to the asymmetrical relation in the environment, efforts to realise environmental justice remain futile unless and until a paradigm shift is realised.

7.4 Gender and Climate Change

Climate change is perceived as a "rise in sea level, the risks associated with it, and changes in weather conditions with frequent occurrence of more intense and extreme events, loss of species and ecosystem, water pollution, among others" (Agostino/Lizarde, 2012). In the UPMUNC-UNEP twelfth conference, there were proposals to establish projects aiming at fighting the effects of climate change but only one project incorporated gender mainstreaming.

The way women and men relate to the environment differs in many ways. Evidence continues to unfold of the unique effects women face, their concerns about the environment, and why they continue to be environmental actors against environmental degradation, unlike men. Previous studies also exhibit existing gender differences in understanding environmental challenges and the importance given to ecological concerns and taking actions, including activism to curb environmental degradation by both sexes (McKinney, 2014).

Individual view of environmental matters originates from both cultural and structural settings of the society, which positions and confines women in the private sphere. This position impacts their gender division of labor, thus strengthening direct, powerful links to the ecosystem during their livelihood. It is to be noted that women's concern for environmental issues is hypothesised to originate in part from the past forces and cultural situations that view them as caregivers, subsistence providers, and collectors of resources needed by the household.

Often, women are in close interaction with the environment, and through this, they play a crucial role in mitigating climate change. Their efforts remain behind the scenes. Accounting for 70% of the world's 1.3 billion people living below the poverty line, women remain the poor of the poorest, and this situation is heightened by persisting gender disparities in society. They suffer all the challenges of society, including atrocities of climate change. They are more vulnerable, and as a result, climate change poses a specific challenge/risk to their security which heightens existing gender inequalities, further threatening their livelihood, health, and safety. This widens the disparity between women and men in society regarding their coping skills. Even though they have good coping strategies, they remain on the periphery of society in decision-making, policy formulation, and implementation regarding efforts to mitigate climate change. This is due to their physical location, social aspects, education factors, economic factors, and political will (GGCA, 2009).

Though they suffer significantly from the effects of climate change, this offers them a unique opportunity to challenge and change their gendered status in society. During disasters, women assume masculine roles in responding to the effects of climate change. They become the driving force in the community, educating disaster risk management skills (Dankleman, 2003, 2008). For example, in a CARE project in Bangladesh, women tended to prioritise adaptation strategies that could be implemented close to home, such as homestead gardening and duck rearing. In the project, which recruited female field officers, women comprised 58% of the total project participants. This has also been the case in some nations, including Kenya, which has been and continues to be affected by natural disasters.

Water, sanitation, and health challenges put an extra burden on women. This adds to the double burden of productive and reproductive labor, decreasing their coping mechanism in disaster and a general collapse of livelihood. Women's mobility is hampered during these atrocities due to socio-cultural norms. Besides, women are mostly confined to the private sphere, even during natural disasters. As a result, women's limited mobility limits their active participation in public life and access to information. This has decreased their bargaining power at the household and community levels (Patt et al., 2009). Where humanitarian assistance is needed, women remain passive and active participants. For example, in the Humanitarian assistance in disasters during the Haitian earthquake, it was evident that most women who were affected were at home, and efforts to support them did not recognise the role they played in the reconstruction of the society. This limits women's Power in society.

Besides, female disaster mortality rates compared to those of men remain high. This is due to the specific gender-constructed vulnerability of women built-in everyday socio-economic patterns. Further, women suffer the aftermath effects of disasters, an increase in domestic and sexual violence, and general abuse of human rights. Girls are forced to drop out of school to reduce household expenses by saving on school fees, to assist in the household chores, or because of pregnancy and early marriage, as evident in a discussion with one of my Somali woman friends. She acknowledged that Somalia is one of the most affected countries in the Horn of Africa with droughts and famines and is more prone to these abuses. Girls drop out of school to allow a boy child to pursue education as girls help to vend for the family. As this happens, most men migrate in search of food and, sometimes, jobs from neighboring countries. The irony is that they never return, even when they know they have left women and children behind. This increases the vulnerability of women and girl-child as they are forced to assume new masculine roles.

A visit to Arid and Semi-Arid (ASAL) Counties of Kenya presented a rather glaring impact of drought on human beings. The impact of drought has been real on women and girls, with most men and boys moving in search of greener pastures and water for their livestock. As this happens, women track over 20 km to fetch water for their animals and domestic use.

Personal interviews with these women and girls further revealed that human security remains compromised as the country continues to experience global warming. Women's and girls' rights will continue to be violated, and since they are confined in the private sphere, their bargaining power to claim and assert their rights is negated. Early marriage, female genital mutilation, and tracking for long distances in search of water, among others, are some of the critical issues affecting women and girls during environmental injustices.

Curbing climate change is an issue that requires a gender perspective, with gender mainstreaming and gender analysis as key strategies. This will help bridge the gap between men and women in society, allowing for equality hence peaceful co-existence. This calls for a holistic approach with community-mobilised efforts and a strong political will from the local to international level.

Climate change, as mentioned, is the "rise in sea level, rise in sea level, the risks associated with it, and changes in weather conditions with frequent occurrence of more intense and extreme events, loss of species and ecosystem, water pollution among others." As this happens, it impacts differently on different locations, ethnicity, different income earners, and different groups of people within society. Poor suffer more from the impacts of climate change. Women form a majority (70%) of the world's 1.3 billion poor people, and they are the most affected. They bear these consequences since they lack political, economic, social, educational, and cultural autonomy, which limits their voices from being heard and considered in decision-making matters of society (Agostino/Lizarde, 2012).

Women have, for a long time, been viewed as 'vulnerable' even with those agencies that aim for women's advancement. This view is problematic since it looks at women as passive recipients, not active participants who cannot solve their problems. To realise environmental justice between men and women in society, an urgent need to decolonise this perception and recognise the essential roles that women play in climate change. Climate justice implies an avenue for dialogue to realise a symbiotic relationship between the human person and the environment (Kyrou, 2007), where both men and women benefit. Although this will go a long way, notably, it will enhance the mitigation of climate injustice, thus creating peaceful coexistence in the Anthropocene.

Fossil fuel which has led to the emergence of a movement led by the Catholic church, namely the *Laudato Si'* movement, continues to advocate against the effects of climate change.

Laudato Si' is an encyclical of Pope Francis published in May 2015. It aims to care for the natural environment, all humanity, and the relationship between God, man, and mother earth. With its recent screening in Kenya at the Catholic University of Eastern Africa, *Laudati Si'* remains one of the global movements that call for reducing fossil fuel burning to mitigate climate change. In his letter, Pope Francis develops the aspect of environmental care by presenting the environmental challenge as an innate human predicament of love. He further presents the question of ecological injustice as an intrinsic human appeal for survival that compromises the future of humanity.

The environment should be perceived as an intimate human challenge that needs moments of self-reflection and critical examination. This view will help humanity rethink and re-discover its actions concerning the environment. It will also arouse the urge to revisit the lost environmental intimacy between the human person and mother earth. A paradigm shift focusing on positive transformation is key to viewing the environment as part and parcel of our human survival. A gender lens that integrates the gains made by both men and women is key to realising environmental intimacy. It allows for the full incorporation and integration of women in decision-making, policy formulation, and implementation of climate change programs and projects equally to men (Agostino/Lizarde, 2012). A paradigm shift is needed to change the view of men and women by society, and how men and women are impacted differently by climate change will be necessary. In addition, integrating women into the existing policies and programs is not the solution instead, it leads to the assumption of "add women and stir." This process calls for adopting a gender lens and using gender mainstreaming and gender analysis as strategies (Brody, 2011; Agostino/Lizarde, 2012).

7.5 Gender Roles in Climate Change Mitigation and Adaptation

In understanding the role that gender plays in climate change mitigation and adaptation, it is necessary to recognise men and women both as change agents, the different relationships that both have about the environment, and the role women play in mitigation and adaptation to climate change. It explains why gender is essential in climate change adaptation and mitigation.

Men and women are both agents of change (GGCA, 2009). There is also growing evidence that gender-segregated vulnerabilities to climate change are inevitable in the agricultural sector, where women dominate Caretta and Börjeson (2015)—societal structures, situations, and positions that women get exposed to lead to their vulnerability. For instance, the kinds of socialisation women get to make them feel weaker than men and thus passive recipients of support and assistance. Women should not be seen from this perspective. Rather, they should be seen as active participants capable of responding to the challenges of climate change.

Research, As noted by Mavisakalyan and Tarverdi (2019), reveals that women's representation and participation in different spheres of life determines how much they will make a difference, especially in environmental justice. For example, women and men relate differently to the environment. Due to their activities, women are ever interacting with the environment, but still, they have limited access to using and controlling resources. This implies that women and men use and manage natural resources differently. They are affected by environmental degradation and natural resource depletion. In Africa, women dominate the agricultural sector and work towards ensuring stability in food security in society. In most scenarios, they have been absent in policy formulation and implementation—normally dominated by men.

In adapting to and mitigating the effects of climate change, women organise themselves, and through these organisations, they can tackle these challenges. For instance, in some parts of Africa, Senegal, women have organised themselves and built canals to help prevent soil erosion—a significant threat to food security (GGCA, 2009). Notably, gender equality can promote practical actions to realise environmental justice (Andrijevic et al., 2020). Stumbling blocks, for example, the aspect of Culture which allows avenues for discrimination, should be tackled to allow women participation and representation, especially in the public arena, for effective adaptation and mitigation.

7.6 Gender in Forest Management and Conservation

Men and women in society play different roles in managing and conserving forests. These roles may be productive or reproductive, including planting, protecting, and caring for seedlings. Men, in most cases, engage in productive work which has an economic gain, like extraction of timber, and charcoal burning, while women, on the other hand, engage in reproductive and productive work: a collection of firewood, fodder for livestock, and herbal medicine, which increase family income; hence they do not gain any economic benefit from these activities (Aguilar et al., 2007 in GGCA, 2009; Andrijevic et al., 2020). For example, in my community, I have seen women help men extract timber and even burn charcoal on top of their role, but they rarely benefit from such activities.

Women have also played an important role in advocating for the protection and care of forests as individuals and as groups. This is in the realisation of the symbiotic relationship that exists between man and ecology. Man, inevitably benefits from ecology in several ways, including its source of food and medicine (Dankleman, 2008; Raimi et al., 2019). As individuals, the late Wangari Maathai from Kenya presents an example. She is one of the women who advocated for the protection and Conservation of the Mau Forest, among other forests destroyed by human activities.

Further, Wangari started a campaign encouraging schools, especially students, to plant trees to save the future. Such initiatives have enhanced ecological peace mitigating against climate change. This advocacy ogres well with the writings of Weber (1999), who emphasised the need to protect plant ecology as it plays a critical role in the life of humanity. It is true since, in the current world, the depletion of some traditional plants and trees, which acts as a buffer against global warming, has seen humanity suffer a great deal.

As groups, the Chipko movement of Himalaya, formed of village women activists, presents a case example of the difference women make in afforestation projects which improved landslides and provided fodder (GGCA, 2009).

All these activities indicate the role that gender plays in mitigating the effects of climate change and ensuring the reduction of emissions from deforestation and land degradation (REDD) and the need to incorporate and integrate them in local, national, regional, as well as international policymaking, design, formulation, and implementation of REDD. Concerning payment for services (PES), carbon emission, for instance, both men and women must benefit from equality. Property ownership, especially land, should be owned equally and promote the rights for both sexes, especially women, to access and control other valuable resources within society. This will provide them with a strong economic base to enable them to participate substantively in climate change, mitigation, and adaptation strategies hence curbing the effects of climate change. Further, education will help both men and women to adopt and use new technology in the Conservation and management of forests (GGCA, 2009; Milimu, 2020).

Since women are closely involved in environmental issues, they should be at the forefront in decision-making and policy formulation as regards climate justice.

They should be incorporated and integrated into all sectors, programs, and policy formulation, design, and implementation regarding climate change. This will allow solutions that are gender sensitive and hence counter the impacts of climate change, therefore, environmental security.

Through local, regional, and global adoption of a gender-sensitive National Adaptation Programme of Action and Disaster Risk Reduction plans, greater sensitisation will be realised and, hopefully, the achievement of a secure environment. Besides, reviewing other existing policies on mitigating climate change effects using gender mainstreaming and gender analysis as strategies is necessary. Thus, Success depends on individuals' support and willingness to realise that "change begins with me," governments, community capacity building, and mobilised efforts globally.

Reflection Questions on climate change from a gender perspective according to ***Ulrike Röhr, director of geneanet***.

- Are there gender differences in the perception of, and the response to, climate change?
- Who is causing climate change, by which activities, and for what purposes?
- How is the polluter-pays-principle considered in mitigation and adaptation policies and measures?
- Are there gender differences in what policies and measures are preferred, and if so, what are the reasons for these differences?
- Are women and men affected differently by the effects of climate change?
- What are the gender-specific impacts of climate change and its resulting environmental damage?
- Are there gender differences in negotiations and decisions on climate change policy?
- How and to what extent are women participating when it comes to working out and deciding about climate protection programs and measures?
- How do the results and programs impact gender relations, for example, climate policy guidelines and directives at the national, regional, and international levels?

7.7 Gender and Energy

Gender and energy are recently receiving much attention, especially internationally. Despite this much attention, gender equality is still viewed with suspicion and not widely understood, making women's concerns viewed as affirmative and not political issues. In policy and decision-making regarding energy, women often remain behind the scenes, assuming their concerns are unrelated to energy production and supply. It is noted that incorporating women's energy needs into development policies is a stairway towards achieving the SDGs: poverty eradication, education, environmental sustainability, health, and employment hence women empowerment. There have been international frameworks to help bring women on board, but challenges remain, especially in the implementation. Such frameworks include: The 1992 Earth Summit, which saw the need for women's advancement as a critical component of sustainable development; the 1995 Fourth World Conference on Women, which emphasised in Chapter 4 the role of women in the use and management of natural resources and called upon all the governments to ensure equality and support for women in accessing affordable and sustainable energy through a participatory needs assessment, the 2001 Ninth Session on the UN Commission on Sustainable Development (CSD9) urged all governments to address health issues resulting from the search for fuel like walking long distances and carrying heavy loads, and the 2002 World Summit on Sustainable Development which reinforced the Johannesburg Plan of Implementation on women's participation (Lambrou/Piana, 2006).

Although the realisation and adoption of some of such frameworks of which most countries, including Kenya, signatories, challenges persist. In most cases, women are affected, especially during droughts where they must track for long distances in search of fuel. Health and security challenges ensue, forcing them to become victims of circumstances.

Gender is essential in addressing energy issues since most people, especially women in rural areas, rely on agriculture for their livelihood. Lack of access to modern energy affects their farming, production, and household maintenance, making women over-rely on natural energy sources like fuel, wood, biomass, and agricultural wastes. This situation continues to deteriorate their health. The over-reliance, which is common in the south, is attributed to increasing poverty levels, especially in the rural areas; hence, the poor people remain on the periphery, and since women are the majority of the world's 1.3 billion poor people (70%), they suffer the most (UNDP, 2004). This scenario has been worsened by the inception of corona Virus pandemic, which saw a massive loss of employment. Although there have been attempts through campaigns to overcome overreliance on fuel, the situation continues to deteriorate. It is likely to worsen, especially in Kenya with the ongoing state of living and persistent drought.

Women in most countries, especially those in the south, resort to using fuel wood, agricultural waste, and biomass since the other sources of energy are not within their reach. The use of these natural sources of energy continues to affect their health. Further, the effect is felt on the environment, which continues to be depleted, leading to climate change. In most African countries, for example, women spend most of their time looking for firewood instead of men. They walk long distances to collect firewood and even return it home. This is due to Culture, which limits their mobility to the private sphere, further limiting their time to participate in other productive work within society hence limited bargaining power.

As women continue to be confined in the private sphere, their nurturing skills and knowledge, critical for ecological justice, remain untapped. The environmental justice process needs to integrate women in the planning and implementation to mitigate the challenge.

7.8 Challenges Facing Women in Enhancing Environmental Justice

Women continue to face numerous challenges in their efforts to enhance environmental justice.

7.8.1 Culture

Culture *is a people's way of life* – values, beliefs, perceptions, attitudes, and knowledge. It defines what society views as right and wrong based on given sexes. A culture of domination in most communities continues to push women to the periphery. This negates the opportunity to enjoy the gains and fully exploit their skills and knowledge, especially in the environment (Aghtaie/Gangoli, 2014). In most instances, Culture provides a plan and cultural restrictions for control and exploitation symbolised by unequal gendered relationships. Culturally rooted ideology masculinises men giving them more control and Power over women (Milimu, 2020). Some practices within society have been legitimised and continue to 'normalize' certain behaviors. Such ideologies define the relationship between men and women in different spheres of life, which either empowers or disempowers women (Francis, 2004; Aghtaie/Gangoli, 2014). Women have no control over and access to resources. In addition, men are socialized to be masculine and to believe in competition as a significant determinant of their end (Connell, 1994; Connell and Messerschmidt, 2005; Connell, 2013). Culture thus portrays women as *proletarians* and men as *bourgeoisie* who continue to own and control the means of production in the environment. Women relate more often with the environment, but the gains made are enjoyed more by men. This dichotomy can only be challenged when such inequalities are named and shamed to create a fair playground for both sexes where both work to protect and benefit from the environment.

7.8.2 Patriarchy

Patriarchy – *the belief in male dominance* has been seen as a key driver in perpetuating discrimination against women concerning critical aspects of life. It has been used as a theoretical tool and a cause of all abuses against women as it determines Power. It is also an approach to sustaining Power. Patriarchy – *deeply entrenched in all structures of society,* is the underlying force that makes the existing gender asymmetries continue thriving (Njuguna/Itegi, 2013). Through this system, men dominate, oppress, and exploit women – *a practice that is seen to be widely accepted to reinforce violence of all types in society* (Merry, 2006). The differential treatment society has accorded to men and women leads to inequalities between these sexes in all spheres, especially the environment. Experience reveals that Patriarchy bestows more Power on a man, which makes him a key decision-maker and that his decision is not to be challenged. He decides, for example, when the land must be utilized and who enjoys the benefits from the same. Due to their socialization, women accept and internalise the practice increasing their interaction with the environment and limiting the enjoyment of the gains made from the same. The gendered socialisation and internalisation of cultural norms continue to push women on the periphery making their work in the environment invisible. Women must be supported at all levels to make the gains and benefits from the atmosphere more visible to all. Challenging Patriarchy can be the only avenue to enhance the achievement of environmental justice.

7.8.3 Power

Power – *the ability to control* – is a key cause of violence against women in the environment. Shanyisa (2020) noted that Power is viewed in terms of the economic and social forces that exploit female labor and the female body. We live in a world that blends Power, powerlessness, privileges, and pain. Power determines the ability to make key decisions in all spheres of life, including the environment. Masculinisation of Power in most benefits men at the expense of women. Power should not be used as an excuse to disregard the gains made by women in realising environmental justice. Anderson 2021 noted the following on international women's Day;

> Environmental decision making – *decisions for the people and the planet* should be inclusive and involve all voices, including equal representatives from women. The planetary crisis – climate, nature, biodiversity loss, pollution, and waste – places a triple burden on women. According to UN Women, climate-related extreme events can increase violence against women and girls. And we all know that biodiversity loss places profound pressure on women and girls as land managers and resource users. And as food producers, women are frequently at more significant risk through endocrine-disrupting chemicals in pesticides, or persistent organic pollutants.

Anderson's speech raises numerous questions that need urgent answers. For example, to what extent are women included in environmental decision-making? Is the inclusion an "*add up and stir*" process, or is it a substantive process? To what extent are women enjoying the gains made from the environment? These questions need urgent responses if Power must be challenged and create a just environment where men's and women's efforts count.

All the above aspects are learned behavior and can easily be unlearned if environmental justice is enhanced where the gains benefit both men and women. Factors like religion and gender socialisation have also been pointed out as crucial causes of women's discrimination concerning environmental justice, which needs urgent attention.

7.9 Conclusion

Gender presents a global ecological challenge where the role of women in environmental justice is yet to be embraced. Women's close interaction with the environment offers an avenue that can enhance environmental justice. Their nurturing skills and knowledge are crucial in ecological planning and implementation. Even with this realisation, women still face challenges due to Culture, power asymmetries, inadequate empowerment, and inadequate economic base, which continues to confine them in the private sphere. Addressing these issues calls for a gender perspective with gender analysis and mainstreaming as critical strategies. There is a need to assess the existing policies and structures and their impact on men and women in society. Through a gender lens, women's hidden experiences and uniqueness are unveiled, thus, recognising their role in the use, management, and Conservation of energy.

Further, there is a need to provide alternative sources of energy that are environmentally friendly and within the reach of all the members of society, including impoverished people in rural areas. Political goodwill, mobilised efforts, and local and international community participation will also be crucial. This will make the concept of environmental justice "*everyone's business and everyone's responsibility*" and thus, promoting environmental intimacy as is the call by Pope Francis in his *Laudato Si'* Encyclical Letter of 2015.

References

Aghtaie, N., & Gangoli, G. (2014). Key issues. Understanding gender-based violence: National and international contexts.

Agostino, A., & Lizarde, R. (2012). Gender and climate justice. *Development, 55*(1), 90–95.

Aguilar, L. (2004). *A Gender Analysis of the Environment and Sustainable Development: Study Guide*. Costa Rica: University for Peace.

Anderson, I. (2021). UN Environmental Programme, International Women's Day.

Andrijevic, M., Crespo Cuaresma, J., Lissner, T., Thomas, A., & Schleussner, C.-F. (2020). Overcoming gender inequality for climate resilient development. *Nature Communications, 11*(1), 6261. https://doi.org/https://doi.org/10.1038/s41467-020-19856-w

Barnett, J. (2001). The Meaning of Environmental Security: Ecological Politics and Policy in the New Security Era (Chapter 5). New York: Zed Books

Bob, U., & Bronkhorst, S. (2014). *Conflict-sensitive adaptation to climate change in Africa*. BWV Verlag

Caretta, M.A. & Börjeson, L. (2015). Local gender contract and adaptive capacity in smallholder irrigation farming: a case study from the Kenyan drylands. *Gender Place Cult., 22*, 644–661

Cincotta, R.P., Engelman, R., & Anastasion, D. (2003). The security demographic: Population and civil conflict after the Cold War. Population Action International-Security Demographic, Washington DC

Connell, R.W. (1994). Psychoanalysis on masculinity. In H. Brod & M. Kaufman (Eds.), *Theorizing masculinities* (pp. 11–38). Thousand Oaks: SAGE.

Connell, R. (2013). *Gender and power: Society, the person and sexual politics*. John Wiley & Sons. Connell, R.W. (2014) Gender and power: Society, the person and sexual politics. Cambridge: John Wiley & Sons.

Connell, R.W., & Messerschmidt, J.W. (2005). Hegemonic masculinity: Rethinking the concept. *Gender & society*, *19*(6), 829–859

Dankelman, I. (2003). Gender, Environment, & Sustainable Development: Theoretical Trends, Emerging Issues, and challenges (pp. 5–7). Review Paper. New York: INSTRAW

Dankelman, I. E. M (2008) Gender, climate change and human security lessons from Bangladesh, Ghana and Senegal

Djoudi, H., Locatelli, B., Vaast, C., Asher, K., Brockhaus, M., & Basnett Sijapati, B. (2016). Beyond dichotomies: Gender and intersecting inequalities in climate change studies. *Ambio*, *45*, 248–262.

Francis, D. (2004). Culture, Power Asymmetries and Gender in Conflict Transformation. In Austin, A., Fischer, M., & Ropers, N. (Eds.) *Transforming Ethnopolitical Conflict. VS Verlag für Sozialwissenschaften,* Wiesbaden: Springer. https://doi.org/10.1007/978-3-663-05642-3_5

GGCA. (2009). *Training Manual on Gender and Climate Change. Module 4. Gender mainstreaming in adaptation efforts*. New York: UN

Galtung, J. (1990). Cultural violence. *Journal of peace research*, *27*(3), 291–305

Glabau, D. (2017). Feminists write the Anthropocene: three tales of possibility in Late Capitalism. Journal of Cultural Economy, *10*, 6, 541–548. https://doi.org/10.1080/17530350.2017.1350597

Griffin, S. (1978). *Women and Nature: The Roaring within Her*. New York: Harper & Row

Homer-Dixon, T. (1999). *Environment, Scarcity, and Violence*. Princeton: Princeton University Press.

Itegi, F. M., & Njuguna, F. W. (2013). Gender based violence in educational institutions and its impacts on girls' education: A comparative study of selected countries. *Research Journal in Organizational Psychology & Educational Studies*, *2*(5), 276–279.

Kabonesa, C. (2005). Gender Mainstreaming and Implications for Peace. In Dina Rodriguez (Ed.) *Gender and Peace Building in Africa* (pp. 13–22). Costa Rica: University for Peace

Kyrou, C.N. (2007) Peace ecology: An emerging paradigm in peace studies. *International Journal of Peace Studies*, 73–92

Lambrau, Y., & Piana, G. (2006). *Energy and gender issues in rural sustainable development*. Rome: Food and Agricultural Organization of the United Nations

Mackenzie, S. (1984). A socialist Feminist Perspective on Gender and Environment (p 182–186). [Original text from Antipode 16 (3), 3–10]. Canada: McGill -Queen Publisher, University Press

Mavisakalyan, A. & Tarverdi, Y. (2019). Gender and climate change: do female parliamentarians make difference? *Eur. J. Polit. Econ. 56*, 151–164

McKay, S.R., & Winter, D.D. (1998). The United Nations' Platform for Action: Critique and Implications. *Peace and conflict*, *4*(2), 167–178.

Merry, S.E. (2006). *Human rights and gender violence. Translating international law into local justice*. Chicago: University of Chicago

Moore, J.W. (2015). *Capitalism in the web of life: ecology and the accumulation of capital*. London: Verso Books

Olsson, L. M. et al (2014) Livelihoods and Poverty. In C.B. Field, V.R. Barros, D.J. & K.J. Dokken (Eds.) *Climate Change 2014: Impacts, Adaptation, and Vulnerability. Part A: Global and Sectoral Aspects. Contribution of Working Group II to the Fifth Assessment Report of the Intergovernmental Panel on Climate Change* (793–832). Cambridge: Canbridge University Press

Page, E., & Redclift, M. (2002). *Human Security and the Environment*. Cheltenham Glos UK: Edward Elgar Publishing Limited

Paolisso, M. (1995). *New Directions for the Study of Women and Environmental Degradation*. Washington, D.C.: International Center for Research on Women.

Patt, A.G., Dazé, A., & Suarez, P. (2009). *Gender and climate change vulnerability: What's the problem, what's the solution? In Distributional impacts of climate change and disasters*. Edward Elgar Publishing

Peterson, V., & Runyan, S. (1999). *Global Gender Issues: Dilemmas in World Politics* (2nd edition). University of Notre Dame: Westview Press.

Pratto, F., & Stewart, A.L. (2011). Social dominance theory. *The Encyclopedia of peace psychology*

Pratto, F. (1999). The puzzle of continuing group inequality: Piecing together psychological, social, and cultural forces in social dominance theory. *Advances in experimental social psychology, 31*, 191–263

Prugh, T., Flavin, C., & Sawin, J. L. (2017). Changing the oil economy. In *State of the world 2005* (pp. 126–147). Routledge.

Raimi, M.O., Suleiman, R.M., Odipe, O.E., Tolulope, S.J., Modupe, O., Olalekan, A.S., & Christianah, M. B. (2019). Women role in environmental conservation and development in Nigeria. *Ecol Conserv Sci*, *1*(2).

Milimu, W.S., & Kaye, S.B. (2020). *Peacebuilding Skills: A Gender-Based Violence Mitigation Strategy*. Mauritius: Lambert Academic Publishing.

Milimu, W.S., & Kaye, S.B. (2020). *Gender and Environmental Security, A training Manual*. Lambert Academic Publishing, Mauritius

Sjoberg, L. (2010). *Gender and International Security: Feminist Perspective*. New York: Routledge

Su, Y., Bisht, S., Wilkes, A., Pradhan, N.S., Zou, Y., Liu, S., & Hyde, K. (2017). Gendered responses to drought in Yunnan Province, China. *Mountain Research and Development*, *37*(1), 24–34.

UN. (2002). Gender Mainstreaming an Overview (pp. 5–7). Office of the Special Adviser on Gender Issues: Department of Economic and Social Affairs. New York: UN

UNDP/ENERGIA. (2004). Gender and Energy for Sustainable Development: a toolkit and resource guide (pp. 7–42). www.undp.org/energy/genenergykit/genderengtoolkit.pdf. Accessed 8 August 2023.

Part II
Addressing Direct Violence and Building Peace

Chapter 8
Community-Based Early Warning Mechanisms to Address Machete Violence in the Artisanal Mining Areas in Zimbabwe

Lawrence Mhandara

Abstract This chapter steers the debate towards the adoption of community early warning mechanisms in the efforts to counter the recurring threat of machete violence in the artisanal gold mining areas in Zimbabwe. The community early warning approach has hitherto received less attention and in some instances been overshadowed by the curative approaches desired by the security services. Despite the systemic and political challenges associated with community preventive initiatives, the article argues that local and people-centred approaches present the potential to revolutionise machete violence prevention. Local peacebuilidng actors – police, civil society, local leadership, youth and women – are spurred to develop the early warning systems within the affected communities by mounting tools to detect threats, reduce the impact and mitigate the consequences of the violence. This entails developing preventive systems that are able to assess, monitor, communicate and respond to the machete violence risk in a timely and coordinated manner. The import of the contribution lies not in seeking to prescribe a rigid way in which the community early warning mechanisms must work but to highlight an existential need by bringing to the fore a framework that can guide the next step of bringing affected communities and pracademics (practitioners and academics) to design and operationalise violence prevention in the artisanal gold mining areas in Zimbabwe.

Keywords Community peacebuilding · Early warning · Machete violence · Artisanal mining · Zimbabwe

---

Dr Lawrence Mhandara is a senior lecturer in the Department of Governance and Public Management at the University of Zimbabwe. Email: lmhandara@gmail.com

---

J. C. K. Kiyala and N. Chivasa (eds.), *Climate Change and Socio-political Violence in Sub-Saharan Africa in the Anthropocene*, The Anthropocene: Politik—Economics—Society—Science 37,
https://doi.org/10.1007/978-3-031-48375-2_8

8.1 Introduction

Instances of governments in developed and developing countries failing to prevent conflicts and atrocities have been noted, raising critical questions on the practice of conflict prevention (The Sentinel Project for Genocide Prevention, 2013, p. 1). This has provoked calls for the adoption of the most robust systems that exist in the natural disaster management field and the atrocity prevention community as models for conflict early warning systems. By understanding the process and drivers that lead to violence, preventive mechanisms such as community early warning systems can be established. The burden of the violence shows that out-group members, especially populations in the immediate locale of occurrence, are among the most affected in the conflict areas. For peacebuilidng actors, it is thus crucial to develop the capacity of the affected people by supporting them with tools to detect threats, reduce the impact and mitigate the consequences of the violence. This argument has also gained currency with the rise of vile forms of violence perpetrated by terrorists and other radical groups. No wonder the role of communities in preventing or responding to radical forms of violence is increasingly finding prominence within government strategies, nationally and internationally (Spalek/Weeks, 2017, p. 991; Subedi, 2017, p. 135). Constrained by legal frameworks, governments fighting terrorists and other criminal groups are increasingly partnering communities in the efforts to counter the groups.

The role of community based EWS has thus been hailed as a cornerstone of efforts to prevent and counter violent extremism (Grossman, 2018). 'The point to be made here is that the topics of radicalisation, terrorism, and countering violent extremism are no longer the sole purview of government. Communities are not only embroiled in the debate; they passively monitor the individuals around them and the events that affect their community' (Spalek/Weeks, 2017, p. 987). Although the importance of EWS is widely recognised, in many cases they are not adequately invested in by stakeholders who implement preventive or mitigating policies. At the same time, implementation of effective community-based partnerships has been nominal in most jurisdictions and hitherto underutilised or undervalued an approach to countering the risk of radical violence. Yet a preventive approach (detection and pre-emption), which could be cost-effective and function as a "threat minimiser" to machete violence, has received less attention from both academics and practitioners in Zimbabwe. Recent studies on machete violence have tended to focus on drivers, actors and consequences of the violence (Zimbabwe Environmental Law Association, 2020; Mkodzongi/Zano, 2020; International Crisis Group, 2020), and the preventive aspects are conspicuously missing in the literature.

This chapter attempts to address the existing gap on violence prevention in the artisanal gold mining areas in Zimbabwe and proposes the establishment of the EWSs in communities suffering from machete violence as a sustainable peacebuilding intervention. The EWSs are central to limiting the loss of lives and livelihoods as a result of hazards, natural or man-made (Cowan et al., 2014, p. 6). The same logic can be applied to the bane of machete violence in Zimbabwe. The chapter argues for the EWSs that evolve from centralised processes to becoming local knowledge and community-centred approaches. The findings of the study may be useful in the development of preventative tools to counter the emergent pattern of radicalisation and violent extremism by organised groups in artisanal mining sites in Zimbabwe. This intervention stresses the theoretical and empirical links between early warning mechanisms and sustainable peacebuilding. By doing so, as well as hoping to assist the security services and the communities, it is intended to stimulate academic debate on an understudied but important aspect of countering the antonym to peace. The contribution does not seek to achieve an ambitious aim of prescribing how the community mechanisms must work but to highlight an existential need by bringing to the fore a framework that can guide the next step of bringing affected communities and pracademics (practitioners and academics) to design and operationalise the mechanisms. Put simply, this piece intends to set the agenda for sustainable peacebuilding in violence-prone artisanal gold mining areas in Zimbabwe.

8.2 Informality and the *Mashurugwi* Phenomenon in Zimbabwe

The origins of artisanal mining in Zimbabwe are traced to the late 1990s when structural changes to the economy emerged. A significant development was the closure of industrial mines, leaving the unemployed workers to adapt to informal ways of survival. The jobless miners organised into nomadic groups looking for mineral deposits to explore. Rarely would the groups resort to violence. However, competition over mining spaces increased in the post-2000 period when formal employment astronomically dwindled (Government of Zimbabwe, 2016). This followed the changing political and economic fortunes in the late 1990s that were confronted by radical economic reform policies adopted by the government at the beginning of the millennium. The resultant period was defined by a constrained economic environment in which the formal industry receded, imposing a huge dent on livelihoods that depend on the formal economy. As economic instability left limited formal employment opportunities, informal artisanal mining became increasingly important as an income stream. Zimbabwe claimed its place among the countries in which a significant number of people turned to artisanal mining for economic survival (Spiegel, 2015). According to the Zimbabwe Environmental Law Association (ZELA) (2020) a significant number of the population's livelihoods are depended directly on artisanal and small-scale mining (ASM), making Zimbabwe a country with the fourth-highest rate of dependence on the sector in Africa. It is estimated that over one million people are directly employed in artisanal mining, with an estimated five million people less directly engaged in the activity (Mkodzongi/Zano, 2020).

As a general observation, mining is one of the remaining key pillars of the economy. In particular, gold is strategic to the country's development agenda. Gold is central to the economic revival plans of President Emmerson Mnangagwa. It was unpredicted that such a lucrative sector would incubate affluent instability incarnating in deadly machete violence that would undermine output. The violence that has attracted exasperation of the people in Zimbabwe is attributed to the *Mashurugwi* (violent groups from Shurugwi) or machete gangs. As Tinhu (2020) notes, these groups have a long history. One of the earliest incarnations of these gangs arose in the town of Shurugwi where groups of criminals on the run from the police sought refuge in the late 1980s. An empty mine provided a safe haven for these men due to its inaccessibility and difficult terrain. The gangs mined gold illegally and engaged in violence locally often using machetes. The members were labelled as the *Mashurugwi*, which became the umbrella term to refer to all the gold mining gangs that emerged elsewhere in Zimbabwe. Although not all these groups engaged in violence, and not all machete gangs come from Shurugwi, the terms *Mashurugwi* and machete gangs have become synonymous. The violence has become malignant in most gold-producing areas. Appositely, some of the eminent factors driving the violence are (Mkodzongi/Zano, 2020; ICG, 2020; ZELA, 2020):

- The de-dollarisation policy adopted by Mnangagwa's government in late 2018 propped the significance of gold. Criminals turned the ASM into a lucrative hunting ground for the elusive dollar.
- The mining title administration system is open to manipulation and mistakes leading to double allocations of mining claims.
- Gold prospecting technology has made it easier to illegally prospect and discover gold. Once gold is discovered, social media are used to spread the news easily attracting the gangs.
- Local business people, foreign nationals and political leaders have also been fingered in sponsoring rival groups in a way that ferments conflict while they control access to the rich gold sites and the illegal gold market (Tinhu, 2020; ICG, 2020; Mkodzongi/Zano, 2020).
- The ASM-related violence is linked to the illicit trade in gold and the Reserve Bank of Zimbabwe (RBZ) policy of buying gold on a no-questions-asked basis. Under the RBZ gold mobilisation strategy, anyone can sell to Fidelity Printers gold-buying centres.

Since 2018, the *Mashurugwi* have become bolder and more active. The groups perpetrate violence in the artisanal mining areas and abandoned gold mining sites such as Mukaradzi and Msasa in Mashonaland Central Province or once lucrative mines like Shamva, Ren and Jumbo (ICG, 2020, p. 7; Tinhu, 2020). The violence perpetrated by the gangs takes several forms of more or less organised violence, namely, (i) inter-gang fights over the pits and shafts promising the highest returns (ICG, 2020, p. (ii) gangs simply robbing artisanal miners and without extracting ore themselves (The Chronicle, 5 Dec 2019, p. (iii) gang members and armed artisanal miners fighting in bars and attacking communities surrounding the mining sites (The Herald, 22 Oct 2018); and gangs kidnapping and forcing victims to work at prolific production sites (ZELA, 2020).

The gangs have expanded in size and number, recruiting women and children while engaging in more violent activities, targeting both rivals and civilians in Kadoma,

Kwekwe, Battlefields, Mazowe and Bindura. Some machete-wielding groups have had the temerity to impose curfews in towns such as Mazowe and Shamva and the suburbs of Chitungwiza and Gweru. They have also attacked security personnel. In late 2017, the Zimbabwe Peace Project (ZPP) warned of escalating violence after fifteen died in clashes between gangs and private mine security forces (Zimbabwe Peace Project, 2017). In Jul 2019, *Mashurugwi* killed three soldiers, injured several more in Bindura and raided police stations in the town of Gokwe to free arrested members and murdered a police officer (Tinhu, 2020). It is difficult to determine the number of casualties resulting from these forms of violence, but the death toll runs into thousands. For instance, the Zimbabwe Peace Project recorded 105 casualties between August–October 2019 in just one mining area (Zimbabwe Peace Project, 2019). In early 2020, the police launched operation *Chikorokoza Ngachipere* (no to anarchy in the artisanal mining areas) to contain the gangs. The operation involved several deadly encounters between police and soldiers, on one hand, and machete gangs on the other (ICG, 2020).

By early 2020, the media spoke of a 'civil war' between machete gangs against the security forces. After the police operation was interrupted by COVID-19, there have been infrequent reports of gang clashes and gangs robbing artisanal and industrial miners. In spite of the violence in the artisanal mining areas apparently diminishing after the police operation and the COVID-19 travel restrictions that are making it difficult for anyone to freely journey, there is every reason to believe it will flare up again as movement restrictions are lifted, making it important to rethink the conflict prevention strategy.

8.3 The Evolution of Early Warning Systems (EWSs) and the Significance of the Community

The EWSs are used as a strategy in a variety of prevention arenas to warn individuals and communities of impending danger. The EWSs work through the provision of timely and effective information, through identified institutions or structures, which allows individuals exposed to danger to take action to avoid or reduce their risk and prepare for an effective response. A functional EWS consists of organised surveillance mechanisms or actions that collect and analyse information on a potential hazard in a given location, in order to trigger timely and coordinated responses (Cowan et al., 2014: 6). The EWSs have become an important tool for the practitioners of conflict prevention and peacebuilding since the 1990s when asymmetric warfare, most of it occurring within the borders of the nation state, escalated in the post–Cold War era in Africa, Eastern Europe, Latin America, and Asia (Subedi, 2017: 137). They provide communities, governments, non-governmental organisations (NGOs) and humanitarian actors with the information required to act effectively and efficiently. Over the years, EWS has evolved in terms of both methodologies and actors. Subedi (2017) and The Sentinel Project for the Prevention of Genocide (2013) identify three phases in this evolution leading to the present prominence of systems etched in communities.

The first generation of EWS depended entirely on external secondary sources of information, and the analysis was conducted outside of the conflict zones. Such analysis could predict some form of macro trends of conflict and violence, but it had notable limitations to forecast micro-dynamics of conflict and violence at national and local levels. The international community developed a wide variety of EWSs during the first two decades of the post-Cold War era. For example, in 1992, the former United Nations Secretary-General, Boutros Boutros-Ghali, published *'Agenda for Peace: Preventive Diplomacy, Peacemaking and Peacekeeping'* which launched a new commitment to developing a comprehensive and integrated global EWS (The Sentinel Project, 2013, p. 4).

The detachment of the UN-based EWS from the conflict theatres precipitated calls for a new practice. The result was the second generation of the EWS. This contrasted with the antecedent generation in that monitoring of conflict and collecting information was undertaken within countries and regions of conflict. However, the analysis of the information was still done outside of the conflict areas. Therefore, although the second generation of the EWS could better predict conflict trends both national and local because of the advantage of geographical proximity to the conflict, its impacts on local and national decision making with regard to conflict prevention were limited. It also involved top-down early warning systems that failed to take into account local social, political, economic, and security dynamics that would have impacted how conflict unfolded, and the possible response system that might be more effective to address the violence.

By contrast, the third generation of the EWS underscores a bottom-up as well as a localised approach to conflict monitoring, information collection and analysis. It involves local expertise and actors including civil society, youth, and women who can better detect indicators of potential conflict and violence. The idea of the third generation of the EWS rests on the assumption that local people who live in the violence prone areas have better access to the best and most reliable information and understanding of the dynamics of conflict and violence. The unprecedented use of the internet, communication technology, and social media in information sharing aids the easy application of the localised EWSs.

Under the first and second generations of the EWSs, there has been a reliance on the ability of the international and national security institutions to contain violence, or as is more often the case, investigate crimes once they have been committed with the goal of arresting those responsible (Garland, 2002 cited in Spalek/Weeks, 2017, p. 994). Once arrested, individuals are prosecuted and imprisoned for their crimes. This approach is classically considered the retributive justice model that relies on the cycle of arrest, conviction, and imprisonment. The basis of the retributive justice model is that individuals convicted of violence must be punished and thus experience the consequences of their actions. This leads to the reality that most of the resources of the security establishments are not organised for the prevention of violence but rather the investigation of violence already committed (Spalek/Weeks, 2017, p. 993).

However, this traditionally based reactive rather than the proactive model is politically and professionally defied in the changing environment of countering violence churned by radical groups. The new forms of violence necessitate that individuals be interdicted before they act, putting tremendous pressure on the security services to seek means to disrupt plots before they mature while simultaneously attempting to gather sufficient evidence for arrest and conviction. This approach is consistent with the third generation of EWS.

Literature on radicalisation and counter-terrorism shows a growing trend and emphasis placed on building social resilience to radical ideologies and groups. The resilience programs have targeted communities. In most Western countries – Britain, the USA, Australia and Canada – preventive strategies against radical violence have underscored the indispensable role of communities (Grossman, 2018; Cherney/Hartley, 2017; Huq, 2017; Levitt, 2017; Lambert/Patterson, 2017; Spalek/Weeks, 2017; Hewit, 2015). The role of community-based EWSs has thus increasingly been hailed as a cornerstone of efforts to prevent and counter violent extremism. The community approaches are part of a broader conceptualisation of preventing and countering all forms of violence as the strategic mobilisation of a range of soft power initiatives designed to pre-empt or complement more traditional security-oriented frameworks. This in turn reflects the current emphasis on international countering violent extremism policy and thinking on not only of "whole of government" but also "whole of society" efforts (Levitt, 2017) both to identify and act on early warning signs of radicalisation to violence at the micro-level of families and social networks, and also to address pre-emptively some of the enabling conditions that erode social cohesion and community resilience to violence. Grossman (2018) adds that the focus on the role of communities reflects a heightened understanding of the limits that are prevalent in more security-oriented logic and strategies for dealing with the risks of violence. The logical question that arises is: why are communities so important in fighting violence perpetrated by radical groups? To address this question, I exploit arguments posited in the counter-radicalisation and counter-terrorism literature. The reliance on this body of literature is informed by the objective view of machete violence as being perpetrated by radicalised individuals and groups who are ready to mete violence on other people to achieve their goals. Thus the proposed community approaches to prevent radicalism and terrorism can also be adapted to deal with a variant of radical violence in Zimbabwe. In other words, my view is that machete gangs and terrorists are competitive siblings in the same family of violence agitated by radical individuals and groups.

Traces of the first two generations of EWS have a visible footprint in Zimbabwe. As an active member of international organisations such as the UN, African Union (AU), Southern Africa Development Community (SADC) all of which have EWS, entails that the country to some extent depends on external secondary sources of information on local conflicts. The processing of such information is conducted outside of local conflicts that may arise. The UN, AU and SADC use the global, continental and sub-regional EWSs to predict macro trends of conflict and violence that have not proved to be useful when dealing with micro-level conflicts like *machete* violence. Zimbabwean communities cannot rely on these external systems that are predominantly made to respond to national level conflicts. The externally driven systems lack effectiveness because of lack of risk knowledge and coordination with the affected communities that are far placed away from the institutions of international diplomacy.

Given the obvious challenges associated with the typical first generation EWS associated with the inter-governmental organisations, such organisations have also developed mechanisms to conscript their local proxies to undertake 'in-country' information gathering and monitoring of conflicts. But the analysis of the information is still done outside of the conflict areas. For example, Zimbabwe is expected to collect data on conflict and convey the information for further analysis to the regional early warning centre. Decisions on how the SADC would respond to such a conflict entirely depends on the recommendations from the early warning centre (Anticipation Hub, 2022). This imposes a limitation on local decision making on conflict prevention. The reason being that most of the organisations and national institutions delegated to this role by the international organisations are mostly located outside the conflict areas. The hierarchical flow of information favours the top-down EWSs, dominated by state actors, especially the security apparatus, and agents of international organisations.

The current context shows a reliance on the ability of the international and national security institutions to contain violence, investigate crimes once they have been committed with the goal of arresting, prosecuting and incarceration. This has (re)produced a reactive EWS model that fails to protect the communities from violence because its structure is national in outlook. The new forms of violence like machete violence need a more proactive EWS that is fully constituted by locals affected by the conflicts that produce direct violence that directly impacts the locals concerned. The involvement of the locals (traditional leaders, youth, women and civil society organisations, security services at that level) presents opportunities for EWSs that generate context-relevant and specific risk knowledge, monitoring, dissemination and communication that can be converted into timeous response by the community itself or the national security institutions, where the communities' response capabilities are constrained.

Although existing studies on EWS in Zimbabwe mostly focus on natural disasters, it can be argued that there has been diffusion of practice with regard to artificial disasters early warning systems in terms of confining actors involved to predominantly state agents and international organisations. Early warning systems are focused on the natural disaster response and risk management legislative framework, guided by the Civil Protection Act (1989), resulting in systems that are top-down are exclusionary. Key actors in natural disaster systems have been the government, through the Civil Protection Unit, UN Agencies and the Meteorological Department, and with a few exceptions, non-governmental organisations. These systems are designed in such a way that they facilitate efficient access of information by locals. The information is collected and disseminated by these actors, placing the locals in the positions of consumers of processed information (Reliefweb, 2022). Private actors and the people excluded from the EWS, depriving the system of a people-centred and multi-hazard founded on shared knowledge (Khoza/Nhamo, 2021). Indeed, "current early warning communication is a one-way process, with those issuing warnings not fully aware of the needs and priorities of vulnerable communities, and therefore not responsive to their needs." (Gwimbi, 2007: 152).

The value that EWSs can add to preventing violence is certainly less debatable. Apart from the benefits and advantages, certain challenges and risks are also discernible. Delving into these challenges is critical as the first step to developing a more rigorous framework of EWSs to prevent violence. The following is a synopsis of the findings.

8.3.1 *Political Issues*

The political interpretation of violence occurring at the local level undermines the local agency of people directly affected by the violence, and also discredits and demotivates local people from engaging with outsiders. Early warning monitors themselves require formal recognition that ensures their protection. Thus, political willingness to work with local people including youth, and turning the will into relevant policies is important, not only to enhance local agency, but also for formal recognition of local EWSs. In the lack of formal recognition, local monitors have felt ignored, and are demotivated to engage in monitoring and protection of civilians. In addition, while EWS is designed to prevent the impact of conflict, violence, and security, those involved in EWS become the most vulnerable from the security point of view. For example, civilians implementing the EWS can be caught in the battle between armed groups and national security forces. In some cases, politicians benefitting from the violence may sabotage the local EWS.

8.3.2 *Lack of Skills*

Despite in-depth knowledge of local contexts and cultures, local monitors lack specialised skills and capacities to collect and report the information systematically. Most third generation EWSs, therefore, keenly emphasise the capacity-building of local actors, which can come in many forms, such as training on how to collect and process information in a targeted and unbiased manner. On many occasions, EW monitors need to work with a range of actors including local to national, as well as multinational and multilateral security personnel. Effective communication, as well as crisis communication, across the spectrum of actors involved is, therefore, another skill that local people might lack. A more critical systemic challenge, however, is how to link the EW system with effective early response mechanisms. The lack of an effective response mechanism at the local and national levels may cause the EWER mechanism to have little impact in averting the breakdown of the peace agreement.

Despite these shortcomings or challenges, there are ample reasons supporting the adoption of community-based prevention measures in the context of 'the whole of society' approaches to conflicts. From a conceptual standpoint, it can be said that governments and communities have similar goals when it comes to countering violence in so much as violence is never a tolerable alternative in any robust democratic society (Spalek/Weeks, 2017, p. 992). Additionally, because human nature is universal in that no one wants to feel vulnerable, the notion of security can be seen as universal at the governmental, community, and individual levels (Ibid). At the community level, the methods to achieve a sense of security can be both remarkably similar to government and at the same time significantly different. As an example, community resources do not possess the same traditional hard power resources that government has, such as the power of arrest. But given the attentive nature of communities, coupled with the fact that government actions alone have been unable to stem the tide of individuals adopting violence, the natural evolution of government is to partner with communities in countering violence (Spalek/Weeks, 2017; Lambert/Parsons, 2017, p. 1056; Huq, 2017). Most notably, communities do possess similar and arguably more effective soft power approaches (Weeks, 2013 cited in Spalek/Weeks, 2017, pp. 992–993). Not only do communities establish the boundaries of what is acceptable and what is not, they passively monitor that environment to ensure compliance. When individuals fall outside of defined communal boundaries they are seen as outliers and eyed with scrutiny. Those outside of the margins arguably become the socially excluded 'other' and strive to make sure that there are limited unchallenged spaces where individuals can be recruited, plan, or execute violent acts (Ibid). The value of the community in this instance is its latent power to forestall plans by gang leaders to recruit locals into their ranks.

The participation of communities opens the avenues for engagement while nurturing trust between the people and the law enforcement services. This can be realised through methods of consultation that help to enhance the perceived legitimacy of counter violence measures (Mueller-Johnson/Sherman, 2015 cited in Cherney/Hartley, 2017, p. 752). This logic dates back to a well-accepted adage that effective crime control relies upon support and cooperation from the general public and that if members of the public do not regard authorities and laws as legitimate, they are less likely to defer to those authorities and laws (Tyler, 2012; Cherney/Murphy, 2013). Forms of community engagement are an important way of minimising backlashes (Lambert/Parsons, 2017; Baker, 2012 cited in Cherney/Hartley, 2017, p. 752). Community approaches are central to the hearts and minds approach to deradicalisation and countering violence (Cherney/Hartley, 2017, p. 750). In fact, three overarching assumptions supporting the engagement of communities in preventing violence churned by radical groups can be distilled from literature (Cherney/Hartley, 2017, p. 752; Lambert/Parsons, 2017; The Sentinel Project for Genocide Prevention, 2013):

1. The first is the recognition that the security services have been overstretched by the frequency and ferocity of the violence. This indicates that sole engagement by the security institutions needs to be replaced by a multi-agency approach. A people-centred EWS capitalises on the knowledge, tools and systems within a community. The core idea is that for any EWS to be effective, the message from the "'top level' (e.g. government, research institutions) must reach the

populations who stand to be affected by the hazards being monitored. In addition, communities can contribute substantially to EWS from 'the bottom-up', in that they can raise initial warnings about changes in key indicators (e.g. rising tensions among groups, siting machetes in the area, use of inciting language, verbal confrontations, isolated fights). The assumption here is that communities are pillars of resilience, constituting essential building blocks in the key task of fostering community resilience to violent extremism, especially in the very early stages of prevention and resistance to the appeal of violent extremist ideology and rhetoric.

2. The second is a recognition that soft approaches to countering violence stirred by radical groups are missing in the affected areas, leaving the threat to emergency and curative responses by the security institutions. The involvement of communities is attributed to value construction in that communities play a vital role in influencing and shaping people's attitudes and responses to the world as they explore independence and autonomy, but where they may also encounter harmful influences through violent extremist rhetoric and propaganda. Community influencers can provide guidance, support and context for young people's questioning of and seeking in the world around them that is crucial in shaping and sustaining the values by which youth interpret and respond to broader social and cultural influences. In particular cultural contexts, community influencers such as women can draw on key cultural, religious and social narratives that enshrine the importance of women's guidance and wisdom on issues of how to behave and what values are better or worse to live by.
3. The third is the recognition that the communities or citizens around the sites of violence have also experienced the impact of hard approaches during the operations to counter the violence. An EWS that is people-centred is respectful and cognisant of the participation of communities in the development of an EWS that concerns them at the local level, minimising the impact of external interventions. Recognising this dimension has the potential to convert communities into 'the first line of defence' whereby communities assume the important role of detecting early signs of radicalisation, in particular subtle changes in behaviour, attitude, social networks, or emotional responses. In the context of machete violence in Zimbabwe, the gangs operate within communities and disconnecting them from the communities is a potentially highly effective approach to combating the violence.

8.4 Embracing the Community EWS Concept for Zimbabwe's Artisanal Mining Areas

Arguably, with machete violence recurring since 2018, the need for the EWSs in the gold-rich areas hosting artisanal mines to produce systematic and organised collection and analysis of open sources information on machete gangs and then link the information to formal and institutionalised response mechanisms in order to prevent violence before it occurs cannot be over-emphasised. The intent is to develop capacity for early warning within communities and use that locally-generated information in the security structures for an early response. This way, we anticipate that the proposed mechanisms will add value to the security services response plans. Thus, while early response by the security services is the goal of preventive action, early warning by the communities is the mechanism to achieve it. Since the effectiveness of the response mechanism hinges on the quality of early warning information, the proposed frameworks should be designed in such a way that guarantees quality inputs into the response systems. Thus, there should be significant efforts to place the onus on each of the at-risk community's ability to detect the gangs but with the full support of the communities to generate usable information. It is a collective responsibility of the government of Zimbabwe and the communities to take a stance against violent radical groups. There are four elements in people-centred EWSs that have a central role to play in increasing resilience of the machete violence-prone areas. The components, adapted from Cowan et al.'s (2014) work on natural disaster risk reduction, are essential and should be addressed to operationalise EWSs at the community level.

8.4.1 Risk Knowledge

This entails systematic collection of data and undertaking risk assessments. This element is addressed when the stakeholders are informed or have knowledge of: the existence of a risk and the arising vulnerabilities – social, economic, political, psychological, environmental and personal; the patterns and trends in the identified risks; and the ability to draw risk maps marking the actual and potential areas of risk (maps of all known artisanal mining sites drawn by the community stakeholders). To achieve a truly-people-centred EWS, risk assessment and mapping can be achieved through memory (historicisation) or through focus group discussions.

8.4.2 Monitoring

This element underscores the community's ability to develop risk monitoring and early warning services. This demands that the right conflict factors be monitored, scientific or evidence-based forecasts be made, and that accurate and timely warning be generated. Monitoring would entail physical visits to mining sites, shops and alcohol selling points to detect inter-group tensions, siting machetes and other dangerous weapons outside the working areas, use of inciting language, verbal confrontations, and isolated group fights. This will allow for an announcement of an accurate and timely looming danger to stakeholders.

8.4.3 Response Capability

The community should build response capabilities based on response plans that are regularly updated, realistic and tested and mirror local capacities. The response plans should be based on local knowledge and experiences. Response capability is also demonstrated when the people facing the risk are prepared and ready to react to early warnings. Escape routes to nearby safe houses such as police camps should be identified.

8.4.4 Communication

The dimension deals with communicating risk information and early warnings. The functionality of the element is gauged by the ability to transmit warnings that percolate all at-risk people and whether the risk and warnings are understood. The aspect of communication also entails that the warning information should be clear and usable. Use of local devices for communication such as word of mouth, runners, criers, drums, flags, bells, telephone, WhatsApp messages, live calls to local and national radio stations, and loud- speakers would be handy.

The recommended framework is based on the assumption that in each affected community, the requisite expertise needed to develop the four elements is available among local stakeholders in the crime prevention business. These include police officers, neighbourhood police, civil servants operating at local levels, teachers, and civil society or non-governmental organisations (NGOs) in the peacebuilding sector. To be sure, civil society or NGOs employ full-time practitioners who have a sole focus on peacebuilding in each of the areas affected by the violence. We acknowledge that in real life, not all the affected areas have similar expertise for the four elements. The deficient areas may tap from national CSOs who can assist the locals to build the required capacity at the inception level and undertake to transfer skills to sustain the system in the event that their expertise will be required in other locations. The key local actors for each element of the EWS are shown in Table 8.1.

A significant number of locals, who are the vulnerable, have an opportunity to be involved at all the stages of the EWS from situation analysis to responding to violence. Further, the involvement of a multiplicity of local actors in each of the elements create opportunities for inclusiveness and bestow a sense of community ownership. In turn, inclusivity and ownership encourages the locals to take responsibility in the implementation of the EWS. Meaningful participation is also possible given that decision-making on each of the four elements of the EWS is controlled by a cluster of local actors designated to take charge of each element.

Table 8.1 Key Elements and the local actors

Element	Key local actors
Risk knowledge	• Researchers and academics (peace, conflict and security issues) • CSOs (peace, conflict and security) • Security services • Community leaders, including traditional leaders • Civil Servants (teachers, local government officials)
Monitoring	• Youth groups • Women groups • Security services • CSOs
Response capability	• Community leaders, including traditional authorities • Youth groups • Security services • CSOs
Communication	• Schools • Pupils • Public transport operators • Community radio stations • Women • Youth

Source Author's own construction

8.5 Conclusion

Since 2018, violence caused by machete-wielding groups in Zimbabwe's ASM sector has wreaked havoc in the country, affecting peace and stability at community levels. Although addressing machete violence has been a principal aim of the government of Zimbabwe, much of its responses have been curative with the main focus on addressing the consequences of the violence (investigations, surveillance and policing). The threat of violent machete gangs however remains high with little sign of abatement. The desire to stem the tide and recurrence of machete violence around Zimbabwe's artisanal mining area resulted in this chapter which has highlighted and proposed the establishment of EWSs in communities suffering from the curse of machete violence as a sustainable peacebuilding intervention. The components of risk mapping, monitoring, warning communication and response capability are essential and should be addressed in each of the affected communities to operationalise the EWSs. In so doing, the article has not pursued an ambitious aim of prescribing how the community mechanisms must work but only highlighted an existential necessity by bringing to the fore a potential and pragmatic framework that can guide the next step of bringing communities, practitioners and academics to design and operationalise violence prevention in artisanal mining areas.

References

Cherney, A., & Hartley, J. (2017). Community engagement to tackle terrorism and violent extremism: Challenges, tensions and pitfalls. *Policing and Society*, 27(7), 750–763.

Cowan, Y., O'Brien., E & Rakotomalala-Rakotondrandria, N. (2014). Community-based early warning systems: Key practices for disaster risk reduction. https://www.fao.org/3/i3774e/i3774e.pdf. Accessed 18 February 2024.

Government of Zimbabwe. (2016). Employment creation potential analysis by sector. Labour and Economic Research Institute/International Labour Organisation: Harare.

Grossman, M. (2018) The role of families and civil society in detecting radicalisation and promoting disengagement from violent extremism. https://www.kas.de/documents/288143/288192/Terrorism_Grossman.pdf/6186a3c7-a10e-be16-05bb-7be3d5d09b42. Accessed 20 June 2021.

Gwimbi, P. (2007). The effectiveness of early warning systems for the reduction of flood disasters: Some experiences from cyclones-induced floods in Zimbabwe. *Journal of Sustainable Development in Africa*, 9(4), 152–169.

Hewitt, S. (2015, Jan 29). Canada should learn from UK: You can't arrest your way out of terrorism. The Globe and Mail. https://www.theglobeandmail.com/opinion/the-critical-counterterror-tool-of-skepticism/article22710643/ Accessed: 20 June 2021.

Huq, A. A. (2017). Community-led counterterrorism. *Studies in Conflict and Terrorism*, 40(12), 1038–1053.

International Crisis Group. (2020). All that glitters is not gold: Turmoil in Zimbabwe's mining sector. *Africa Report*, 294, 1–48.

Khoza, S., & Nhamo, G. (2021). Revisiting Zimbabwe's early warning systems in the light of tropical cyclone Idai. https://link.springer.com/chapter/https://doi.org/10.1007/978-3-030-74262-1_4. Accessed 14 May 2023.

Lambert, R., & Parsons, T. (2017). Community-based counter-terrorism policing: Recommendations for practitioners. *Studies in Conflict and Terrorism*, 40(12), 1054–1071.

Levitt, M. (2017). Recent trends in terrorism and counterterrorism: National practices in countering violent extremism. https://www.un.org/securitycouncil/ctc/sites/www.un.org.securitycouncil.ctc/files/8-statement-as-delivered-by-mr.-matthewlevitt-the-washington-institute.pdf.. Accessed 20 June 2021.

Mkodzongi, G., & Zano, V. (2020). The political economy of artisanal and small-scale gold mining in Zimbabwe: The problem of formalisation. *Resource Insight*, 19, 1–34.

Reliefweb. (2022). Towards a community focused Early Warning Systems for the Buzi-Pungwe and Save (BuPuSa) Transboundary River Basins project inception workshops held. https://reliefweb.int/report/zimbabwe/community-focusedearly-warning-system-buzi-pungwe-and-save-transboundaryriver.

Spalek, B., & Weeks, D. (2017). Community-based counterterrorism. *Studies in Conflict and Terrorism*, 40(12), 987–990.

Spiegel, J. S. (2015). Shifting formalisation policies and recentralising power: The case of Zimbabwe's artisanal gold mining sector. *Society & Natural Resources*, 28, 543–558.

Subedi, D.B. (2017). Early warning and response for preventing radicalization and violent extremism. *Peace Review*, 29(2), 135–143.

The Chronicle. (2019, Dec 5). Machete rampage in Inyathi: One person dead as 50 Mashurugwi attack. The Chronicle. https://www.chronicle.co.zw/machete-rampage-in-inyathi-one-person-dead-as-50-mashurugwi-attack/. Accessed 5 August 2023.

The Herald. (2018, Oct 22). Army call for ceasefire in machete attacks. https://www.herald.co.zw/army-calls-for-ceasefire-in-machete-attacks/. Accessed 21 June 2021.

The Sentinel Project for Genocide Prevention. (2013). Localized early warning systems: From natural disasters to human security. https://thesentinelproject.org/2013/08/13/localized-early-warning-systems-from-natural-disasters-to-human-security-leveraging-a-people-centered-protection-network/. Accessed 20 June 2021.

Tinhu, S. (2020). Why is Zimbabwe doing too little to tackle growing gang violence? https://africanarguments.org/2020/02/zimbabwe-gangs-mashurugwi-mnangagwa-chiwenga/. Accessed 20 June 2021.

Zimbabwe Environmental Law Association. (2020). Why is violence rising in Zimbabwe's small-scale mines. https://miningzimbabwe.com/why-violence-is-rising-in-zimbabwes-small-scale-mines/. Accessed 20 June 2021.

Zimbabwe Peace Project. (2017). Statement: Escalating Violence in Gold Panning Communities. https://zh-cn.facebook.com/Kubatana/posts/escalating-violence-in-gold-panning-communities-via-zpp-ngothe-zimbabwe-peace-pr/1575789579099102/. Accessed 20 June 2021.

Zimbabwe Peace Project. (2019). Monthly Monitoring Report, Sep 2019. https://reliefweb.int/report/zimbabwe/zpp-monthlymonitoring-report-september-2019.

Chapter 9
Mapping the Conflict Between Street Vendors and Harare City Council Towards Building Peace in Zimbabwe

Washington Mazorodze

Abstract Street vending has become the most common economic activity not only in Harare, but also throughout the Zimbabwe's main urban centres. Street vending has provided employment for the greater majority of Zimbabweans. It has provided a means of survival and contributed to the efforts to end poverty and hunger, which are sustainable development goals number 1 and 2 respectively. However, running battles between urban authorities and street vendors are common without any solution in sight. Although urban authorities have their own council police they often enlist the services of the Zimbabwe Republic Police (ZRP) and sometimes Zimbabwe National Army (ZNA) to help them remove illegal vendors from the streets. Pushed by the need to survive from vending, street vendors vowed to remain in the streets creating a complex intractable conflict with so many dimensions and actors. This chapter sought to map the conflict between Harare City Council (HCC) and street vendors using the conflict mapping analysis tool, which helps to identify the parties in conflict, the issues, the context, relationships and the dynamics of the conflict for the successful resolution of the conflict. A nuanced understanding of these dynamics stands a better chance of peaceful resolution of the conflict. This is premised on the idea that sustainable peace can only be attained if there is an exploration and understanding of the root causes of the conflict, the interests, wants and needs of the conflict parties. Again, addressing the root causes is key when resolving conflicts to avoid spoilers. It is only through the consideration of all stakeholders' interests that peace can be attained, in fulfilment of the sustainable development goal number 16 on peace, justice and strong institutions. The chapter adopted a qualitative research methodology because of the nature of the research objectives. Interviews, observation, journal and newspaper articles were used as data collection methods to map and analyse the conflict.

Dr Washington Mazorodze is a PhD graduate in Public Administration (Peace Studies) from the Durban University of Technology. He researches about domestic, regional and international politics, peace and conflict issues, human security, and governance. He is a Lecturer at the University of Zimbabwe, Department of Peace Security and Society. Email: washvvvone@gmail.com

---

J. C. K. Kiyala and N. Chivasa (eds.), *Climate Change and Socio-political Violence in Sub-Saharan Africa in the Anthropocene*, The Anthropocene: Politik—Economics—Society—Science 37,
https://doi.org/10.1007/978-3-031-48375-2_9

Keywords Street vendors · Urban authorities · Conflict · Violence

9.1 Introduction

In Zimbabwe's capital city, Harare, there are two main conflict parties which are fighting for the use of public space, namely street vendors and the Harare City Council (HCC). There are, however, other fringe players such as the vendor organisations and formal businesses, Zimbabwe Republic Police (ZRP), and Zimbabwe National Army (ZNA). This chapter sought to explore the relationships and dynamics that exist among a plethora of actors in street vending in Harare with a view to find a lasting solution to the conflict. The chapter highlights the importance of street vending for the survival of many families and the fulfilment of sustainable development goals number 1 (no poverty) and 2 (zero hunger). The conflict is a result of the closure of industry partly due to the effects of economic sanctions imposed by the Western countries against Zimbabwe. The sanctions were a response to the latter's land reform programme and human rights violations. The resultant economic hardships compelled many people to resort to vending as an alternative source of livelihoods (Staff Reporter, 2021). The closure of industries witnessed an influx of vendors into all the major cities in Zimbabwe seeking to make a living, a situation which caused congestion and filthiness in urban city centres around the country. Urban authorities and government reacted by forcibly evicting vendors from the streets in an attempt to maintain the cleaniliness of the urban centres, a move which generated an intractable conflict with vendors.

An interview with a city council of Harare representative revealed that street vendors caused filthy and congestion in the city centre because they operated from undesignated sites, hence the move by the Harare city council to remove street vendors from the streets.[1] Street vendors on the other hand argued that they derived their livelihoods from the streets and resisted attempts by the city of Harare to drive them from the streets. A street vendor interviewed by the researcher argued that removing them from the streets would worsen their economic conditions.[2] Therefore, the conflict between street vendors and Harare city council became an intractable conflict based on the needs of the conflict actors. Zimstats (2022: 6) notes that 88% of the employed population in Zimbabwe were informally employed. This means that the majority of employed people in Zimbabwe are in the informal sector, which includes street vending. Therefore, the efforts by the HCC to drive out vendors from the streets and the ensuing conflict affected the vendors' source of livelihoods with far-reaching implications on sustainable development goals number 1, 2 and 16 (zero hunger and no poverty as well as peace and justice). Despite being chased and harassed, street vendors have continued to operate from illegal selling points, further deepening hostilities.

---

[1] Interview with HCC member, November 2020.

[2] Interview with an illegal street vendor, Jan 2020.

The objectives of this chapter are to;

- Explore the actors involved in conflict between street vendors and Harare city council.
- Discuss the issues, interests, needs and fears among the conflict actors.
- Analyse the strategies currently deployed by the conflict actors to resolve their conflict.
- Recommend sustainable peacebuilding strategies among the conflict actors.

This chapter seeks to answer the following questions:

- Who are the actors in the conflict between street vendors and Harare city council?
- What are the issues, interests, needs and fears of the conflict actors?
- What strategies are the conflict actors using to resolve their conflicts?
- What peacebuilding strategies have the potential to sustainably resolve the conflict that can be adopted by the conflict actors?

The assumption in this chapter is that the use of force to resolve conflict is not sustainable. Thus, a strategic peacebuilding approach is advocated as the way forward for street vendors, city of Harare and government security services to sustainably address the root causes of the conflict.

The first section deals with the methodology employed in this chapter. The second section deals with the conceptual frameworks informing the analysis in this chapter, which are peacebuilding and sustainable development. Section three looks at mapping the conflict. Section four looks at street vendors in general while section five looks at legal street vendors and emerging issues. Section six looks at illegal street vendors and emerging issues. Section seven highlights pedestrians and their issues with street vendors. Section eight considers Harare city council and its issues with street vendors. Section nine discusses how vending has been politicised in Harare. Section ten discusses formal businesses and their issues with street vendors. Section eleven looks at vendor organisations and their concerns with city of Harare's conduct in relation to street vendors. Section twelve considers Zimbabwe' security forces and their involvement in the conflict. The last section is the conclusion.

9.2 Methodology

The purpose of this chapter is to apply the conflict mapping tool in order to explore actors, issues and themes in the conflict which on the surface appears to be between street vendors and HCC. It also seeks to understand the undertones influencing the actors and the power relationships between and among a plethora of actors in the conflict under review. The chapter is unique in the sense that it explores a variety of actors, interests, needs and wants of the various actors in the conflict in an attempt to finding a lasting peaceful resolution of the conflict which appears intractable. There is no any other work of a similar nature in Zimbabwe. This chapter made use of qualitative research methodology. A qualitative study approach was adopted in this chapter in order to explore the relationships, experiences and power dynamics among the various actors fighting for space in the Harare Central Business District (CBD). These phenomena were explored using different sources of data as a way of enhancing the reliability and validity of the findings (Yin, 1994: 94). Qualitative research methodology has the advantage that it understands the phenomenon from its original context using small samples. The data gathering techniques that were used include structured interviews, observations and secondary sources of information such as journal and newspaper articles. The interviews were conducted between 2019 and 2023 in Harare's CBD at Copacabana bus terminus where there is a high concentration of vendors.

In this chapter, the perspectives of those affected by the street vending phenomenon were taken into consideration. Interviews were conducted with a variety of actors to explore their experiences. This was done to understand different experiences of these numerous actors. The respondents were conveniently selected. A total number of six respondents were interviewed. These comprised one legal vendor, one illegal vendor, one formal business or shop operator, one council member, one pedestrian and the National Vendors Union of Zimbabwe President. The number of respondents interviewed seems small but this was done to make sure that the data collected was rich and in-depth. The research was conducted in Harare because it is the capital city of the country and this is where most of the vending and clashing between vendors and urban authorities takes place. The researcher further read and analysed relevant literature on street vending and conflict between street vendors and urban authorities not only in Zimbabwe but in other countries as well. It made use of books, journal and newspaper articles for content analysis which complemented data gathered from interviews and observation.

9.3 Conceptual Framework

9.3.1 Peacebuilding

Just like any other term, Peacebuilding is complex and evolving. The term was introduced by Galtung in the 1970s but only gained prominence in the 1990s when Boutros Boutros Ghali (1992: 5) defined it as action to identify and support structures which will tend to strengthen and solidify peace in order to avoid a relapse into conflict. When Boutros-Boutros Ghali defined peacebuilding, it was conceptualised as a post-conflict activity aimed at preventing recurrence of conflict. The term has undergone some changes and now encompasses a wide range of activities, not only limited to the post-conflict phase. Peacebuilding now aims to strengthen and solidify local social institutions so that society becomes resilient and self-sustainable. Peacebuilding assumes that violent conflicts are prevented, threats of violent conflicts are absent, marginalised groups are empowered and structural violence is addressed resulting in positive peace. Peacebuilding encompasses conflict prevention, management, resolution and transformation. Peacebuilding aims to transform the conflict system as a whole to prevent violent conflict from occurring in the future. Ettang et al. (2011) further notes that peacebuilding is a long-term and evolving process. In this chapter, the aim of mapping the conflict is to explore the actors, their interests, fears and needs. Mapping the conflict becomes a peacebuilding activity and addressing these interests, needs and fears prevents violent conflicts emerging between street vendors and HCC and its partners in the future which is the end goal of peacebuilding.

9.3.2 Sustainable Development

The concept of sustainable development became popular in the 1970s and 80s when economic growth put pressure on the environment which resulted in pollution, limited living space, poverty and diseases (Tomislav, 2018). Sustainable development is development that meets the needs of the present generation without compromising the ability of future generations to meet their own needs (WCED, 1987). It is a concept that relates to socio-economic development, distribution of resources to ensure quality life for all as well as concern for future generations. It focusses on a balance between three pillars of sustainability, environmental sustainability, social sustainability and economic sustainability. Environmental sustainability seeks to maintain the quality of the environment, social sustainability ensures human rights, cultural identity, preservation of cultural diversity, and economic sustainability ensures the required income and living standards. Since the establishment of the United Nations (UN) in 1945, it has been actively involved in various activities to promote sustainable development. Heads of States and Government and High Representatives launched sustainable development goals (SDGs) in September 2015. These SDGs were built upon the millennium development goals and sought to fulfil the latter's unfinished business. The two most important commitments which are key in this chapter which the world grappled with were to end poverty and hunger (SDGs number 1 and 2 respectively). The world boldly committed to end poverty and hunger in all their forms by 2030.[3]

The most vulnerable populations were going to be targeted and, in this discussion street vendors are vulnerable because of their dire economic situations. The majority of street vendors are former factory employees who were retrenched when the economic situation in Zimbabwe worsened since the beginning of the millennium. The mapping of the conflict between street vendors and HCC is meant to underscore the importance of vending in ending hunger and poverty. Thus the transformation of the conflict between street vendors and HCC can result in the recognition of vending as a key strategy to contribute to the fulfilment of the aforementioned SDGs.

[3] https://www.globalgoals.org/goals/.

9.4 Mapping the Conflict

Conflict mapping is one of the most common tools used to analyse conflicts and is used to identify the main actors in a conflict. The method produces a graphical presentation as shown below which enables practitioners to clarify the following issues: main conflict theme and issues, actors and their influence or power over the conflict, and relationship between the actors (Fisher, 2005). The conflict mapping diagram below helps to identify alliances between the actors, patterns of power, potential alliances for cooperation and possible entry points for non-violent intervention (FIg. 9.1).

9.4.1 Mapping Components

Circles indicate parties and power to the situation. The bigger the circle, the more power the conflict party has over the issue and vice versa.

1. A straight line indicates a link or a fairly close relationship between the actors.
2. Double lines represent an alliance between the actors.
3. A dotted line indicates an informal or intermittent link between the actors.
4. An arrow indicates the direction of influence or power.
5. Lines like lightning indicate discord or conflict.
6. Broken line arcs indicate shadow parties, those who have an influence but are not directly involved in the issue.

The following letters represent the parties in the conflict and their relationships.

A = Illegal Street vendors
B = Harare city council
C = Government of Zimbabwe

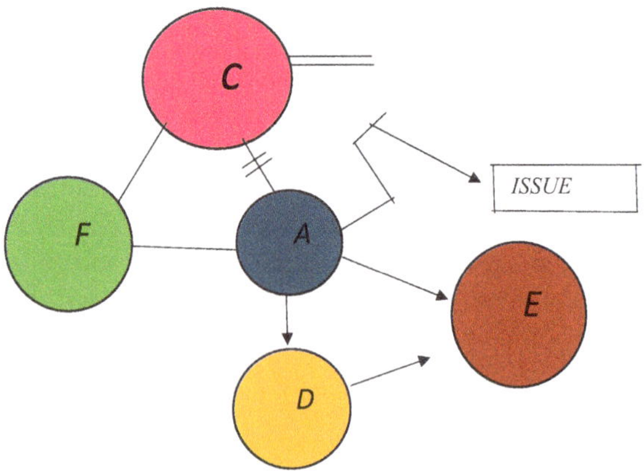

Fig. 9.1 Conflict map
Source Adopted from: Fischer: 23 (2005)

D = ZRP and ZNA
E = Formal Business and Legal Street vendors
F = Vendor Organisations

In the conflict mapping diagram above, there is discord between HCC (B) which is the major actor and street vendors (A) as shown by lines like lightning. There is an alliance between (C) and (B) which represent the Government of Zimbabwe through the Ministry of Local Government and Public Works and HCC respectively. The HCC acts on the directives that it receives from the Ministry to clean up the city and drive out vendors from the streets. The Ministry has influence over HCC. For example, in March 2020, the Ministry issued a directive to all urban authorities through a letter which stated that,

> at the nineth cabinet meeting, it was resolved that local authorities should take advantage of the national lockdown to clean up and renovate Small and Medium Enterprises and informal traders' workplaces so that these areas will be more conducive to operate from when business reopens" (Zimpricecheck.com).

Street vendors (A) have some relationship with formal shops (E) where (A) is allowed to operate on (E)'s doorsteps and pavements as (E) does not have the power to chase (A) away. There is a fairly close relationship between the Government of Zimbabwe (C) and vendor organisations (F) who often engage each other in discussions and dialogue on issues affecting street vendors. There is also a fairly close relationship between street vendors (A) and vendor organisations (F) who often work together in protecting (A) from evictions from the streets by the partnership between (B) and (D). There is also a close relationship between (B) and (D) as they partner to chase (A) away from the streets. The following section discusses various actors and, issues, perceptions and relationships.

9.4.2 Street Vendors

Street vendors are people who sell on the streets and have no permanent structures where they sell their goods from (Bhomik, 2005). Street vendors have occupied all street corners of Harare seeking to sell all kinds of goods in order to make a living due to mainly lack of alternative economic activities (Ndebele, 2017). There are no formal jobs due to the closure of industries as the country has been affected by an economic crisis which began at the beginning of the new millennium when relations with the Western countries nosedived owing to the country's land reform programme, among other factors (Njaya, 2014). It was estimated that there were close to 100,000 street vendors in Harare in 2016 (Njaya, 2016). Because of their sheer numbers, vending sites in the city centre are not enough to cater for the ever-growing number of street vendors leaving many operating as illegal street vendors. Two groups of vendors can therefore be identified.

9.4.2.1 Legal Street Vendors

These have been allocated space by the HCC from which to sell their goods in terms of the city by-laws statutory instrument 159 of 2014 that requires vendors to sell their goods at designated sites usually located at various bus terminus (Rank/Renkini) in the city centre (Harare Vendors By-Laws, 2014).

9.4.2.2 Context

Section 14(f) of the Harare Vendors By-Laws (2014) prohibits vendors from selling their goods at any place other than designated vending sites. Street vendors sell a variety of products ranging from vegetables to clothes. This group of street vendors usually have a good relationship with the HCC as they pay rates to the city council for space allocated to them. However, due to limited space the designated sites can only accommodate few vendors. Their relationship with city council is therefore transactional. It is dependent on their continued payment of rates to the city council and the continued rendering of reliable social services such as water and ablution facilities by HCC. The city council has the authority to remove them from their sites as and when they deem it necessary. Therefore, street vending provides street vendors with a means of decent livelihood in the absence of formal employment. The provision of informal employment helps to achieve SDGs 1 and 2. The city council also derives part of its revenue from taxing street vendors. However, in recent times, relations between legal vendors and Harare city council have deteriorated owing to reasons explained in the following section.

9.4.2.3 Emerging Issues

With the increasing number of illegal vendors in town, legal vendors have been affected as the volume of their sales have declined. Legal vendors' products are more expensive than illegal vendors due to the rates that they pay to the council and other recurrent expenditures. For example, the researcher observed that legal vendors were selling bananas at $1 (USD) for ten bananas whereas those selling illegally were selling the same quantity of bananas at $ 0.50 (USD). This negatively affected the relationship between legal vendors and HCC as the latter has been blamed for failing to deal with illegal vendors.[4] The relationship between legal street vendors and illegal street vendors themselves have also been strained as the former accuse the latter of taking business from them but the former have no authority and power to act against the latter as they are outnumbered.[5] The researcher noted that there are more illegal street vendors than legal vendors owing to the limited space that is allocated to vendors by the city council. Legal vendors appealed to HCC to remove illegal

[4] Interview with legal vendors in Harare CBD, June 2019, Harare.

[5] Ibid.

vendors who have occupied all street pavements in town arguing that the latter's presence on the streets have negatively affected legal street vendors' daily incomes. Resultantly, some of the legal vendors have moved from their designated sites and are now operating from illegal points such that it is now difficult to distinguish between illegal and legal vendors. They argued that illegal street vendors sell cheap goods as they do not incur expenses such as rates and rentals to HCC which have left them without business. The few who have remained at designated sites have since stopped paying rates to HCC. The legal vendors complained that they were failing to sell from their allocated points due to the influx of illegal vendors whom HCC failed to deal with.[6] This created another conflict between HCC and legal vendors as HCC continues to demand the payment of rates and is allegedly owed a lot of money by these vendors.[7]

9.4.3 *Illegal Street Vendors*

9.4.3.1 Context

The second group of vendors are the illegal street vendors who have invaded the city due to the lack of economic opportunities in the formal sector owing to the closure of companies and retrenchments which left many out of formal employment (Finscope, 2014; Mangudhla, 2021). These are in direct conflict with the HCC as they fight for space and operate from illegal sites where they do not pay rates to the council.

9.4.3.2 Emerging Issues

Illegal street vendors have occupied all spaces available in town and demand convenient vending space where it is strategic for them, and their clients should they be relocated from their current illegal locations.[8] HCC demands that these vendors should be regularised and relocated to vending spaces allocated by city council which are on the periphery of town as there were limited vending sites in the city centre. Illegal street vendors shunned these designated points created by HCC on the outskirts of town as the HCC aimed to decongest the city centre. Vendors claim that these points are not convenient to their clients which led them to continue occupying the streets where they believed there is more business.[9] City council, on the other hand, has called on vendors to abide by city by-laws and to desist from trading at undesignated sites as doing so resulted in congestion, dirty and increased crime levels. Efforts by

[6] Ibid.

[7] Interview with HCC member, November 2020, Harare.

[8] Interview with illegal vendor, Jan 2020, Harare.

[9] Ibid.

city council to regularise street vending through creating vending sites out of town have been scorned by vendors resulting in running battles.

Another issue which also force street vendors to operate illegally are the high rates charged by the city council. An interview with an illegal street vendor revealed that they did not pay for license discs because they were charged US$140 for a hawker license valid for a year and allocated a space out of town where there are no clients. For these reasons, the street vendors preferred to operate illegally in the city centre where there is a ready market without paying rates. These vendors operate against Hawkers and Street vendor's by-law 4(1) which requires them to work in the areas specified without which results in collision with the HCC. The relationship between illegal street vendors and HCC is therefore bad and characterised by constant running battles, confiscation of vendor's goods and arrests which sometimes turn violent as illegal street vendors try to resist. In some cases, HCC police members get beaten up by angry illegal street vendors resulting in injuries. This intractable conflict was also demonstrated by street vendors who accused the city council of causing poverty among them when they insist on chasing them from the streets where they earned their living from. One vendor stated that, "They [HCC] may chase us from the streets but where will we go? We will come back because this is where we earn our living. They want us to die of hunger by chasing us from the streets."[10]

Vendors do not believe the city council's narrative that it wants to maintain order and cleanliness in the city by designating areas for vending. Vendors believe that any place in the city is suitable for vending. On the other hand, HCC believes that the disorder in the city is caused by street vendors, hence its intention to drive out street vendors from the streets. The council wants street vendors to operate from designated selling points and those without designated sites should apply to get allocated space on the periphery of the town. The city council argues that not every place is suitable for vending and has therefore allocated limited space for vending. This lack of common ground between illegal street vendors and Harare city council fuelled conflict that is raging between the two actors over the utilisation of space in the city centre. The researcher also noted that illegal street vendors had conflicts with pedestrians which is discussed in the section below.

9.4.4 Pedestrians

9.4.4.1 Context

Pedestrians are ordinary people who move around in town for different purposes. Most pedestrians pass through town going to different workplaces. They encounter street vendors when they step on vendors' goods which are often displayed on pavements where pedestrians walk. As a result, conflicts occasionally erupt between street vendors and pedestrians. Street vendors occupied all pavements making it difficult for

[10] Ibid.

pedestrians to walk on these pavements. Many pedestrians have been at the receiving end of street vendors' tongue lashing or even beatings for stamping on their goods displayed on the pavements. This conflict is a result of negative perceptions of pedestrians held by street vendors who perceive pedestrians as people who undermine their trade as street vending has been traditionally looked down as a trade for people who do not qualify for formal employment. Illegal street vendors who were interviewed noted that pedestrians have a negative perception of vendors, and this makes them to stamp on their goods displayed on pavements deliberately. Interviewed street vendors noted that they were not given the respect that they deserved by pedestrians. They argued that pedestrians should respect their trade as any other trade because they were only trying to make a decent living from the streets and their presence on the streets was not a matter of choice but necessity.[11]

9.4.4.2 Emerging Issues with Vendors

On the other hand, pedestrians lamented what they called a disregard of the city by-laws by street vendors. They accused street vendors of not considering the inconveniences they were causing to pedestrians. They complained that street vendors made it difficult for them to walk as many times vendors occupied both sides of the pavements. Pedestrians considered street vendors as violent and inconsiderate people who sold their products at undesignated sites which often attracted conflict with other people like pedestrians.[12] This was confirmed by Mutami/Gambe (2015: 126) who noted that conflicts often erupt between street vendors and pedestrians after the latter stumble over vendors' displayed goods on the pavements and pedestrians are ordered to pay for the damaged products despite vendors operating illegally. Illegal street vendors actually acted as if they owned the spaces they occupied illegally. Pedestrians called upon the HCC to make sure that it regulates vendors as they were making it difficult for them to move freely in the city centre. This has also resulted in strained relations between pedestrians and Harare city council as the former accuse the city council of failing to deal with illegal street vendors who are blocking the pavements with their displayed goods and assault them in the event that they mistakenly step on their goods.

9.4.5 Harare City Council (HCC)

HCC operates under the mandate from the government of Zimbabwe through the Ministry of Local Government and Public Works. The Ministry issues directives to the HCC to make the city of Harare orderly and clean. The HCC is then responsible for the running of city affairs which include allocation of vending sites through its

---

[11] Ibid.

[12] Interview with one pedestrian, Mar 2021, Harare.

Department of Physical Planning. It is also responsible for making the city clean and orderly. Its vision is to make the city of Harare a sunshine city.

9.4.6 What Are the Issues with Street Vendors?

As an authority which is responsible for the administration and running of city of Harare, the HCC has clashed with street vendors whom it accused of creating disorder, congestion, dirty and increased crime levels in the city centre (Mutami/Gambe, 2015: 126). It has responded by forcefully removing illegal street vendors from the streets and sometimes in partnership with ZRP and ZNA which has resulted in bloody battles. Sometimes, the HCC carries its own operations. The city council has also designated some vending sites on the outskirts of the city centre which have not been occupied by street vendors who argued that the designated sites are not viable as clients are concentrated in the city centre. This resulted in violent clashes as HCC tried to enforce city by-laws whereas illegal street vendors sought to derive their livelihoods from selling on the streets. The researcher had the opportunity to interview a council official who likened their conflict with vendors to war. She retorted that,

> The problem with street vendors is that they think they are the only ones who have the right to use the streets. They decided to invade the streets without coming to us and we have responded heavy handedly because this is war. They are disregarding other city users and have become a law unto themselves. They have occupied every corner of the city, where do other city users conduct their business? Moreover, they do not pay anything to the city council. The majority of them are illegal vendors. We cannot accommodate all of them in the city centre.[13]

This council position is supported by a study that was carried out by Nyavava (2015) in which then HCC town clerk Tendai Mahachi was interviewed and said that in a bid to contain lawlessness in Harare CBD area, all vendors operating at undesignated sites in the CBD were ordered to relocate to the designated sites or risk being banned from trading in the capital. He also noted that the city council had agreed with vendors' representatives that there should be organised trading and order in the city. He had this to say,

> The CBD sites can only accommodate limited number of traders and the rest will be accommodated in the suburbs and outskirts of the city centre. The city council will continuously review provision of marketplaces with a view to increasing the number of trading sites. We want to empower our people as well as transform our economy so we will engage stakeholders at all levels so that we consult extensively before we move our people (Nyavava, 2015).

This is consistent with Mitullah's (2003) observation that the greatest challenge facing informal traders has to do with site of operation and the right to trading space as most of the spaces occupied by informal traders were illegal since the spaces have

[13] Interview with HCC member, November 2020, Harare.

not been set aside for this kind of business. Mitullah further noted that the major conflict occurs when street vendors are required to move from these illegal sites to those proposed to them by city council where they feel there is no business. This brings them in direct confrontation with city authorities. This has happened to the HCC and street vendors where the former designated areas for vending out of town. Street vendors shunned these sites citing lack of business. However, HCC maintained that if they occupy the designated sites, clients will follow them. No solution has been found yet as HCC continues to raid illegal street vendors. On the other hand, street vendors continue to occupy illegal vending sites in the city centre creating a vicious circle of violence.

9.4.7 The Politicisation of Vending in Harare

City council officials who were interviewed also added that the problem of street vending in Harare was being exacerbated by politics. The city's drive to cleanse the city of illegal street vending had always been over overruled by politics. They provided an example when the former first lady Grace Mugabe condemned police and council brutality against illegal street vendors at a rally in 2014. She encouraged illegal street vendors to continue selling their goods on the streets contrary to HCC policy that wanted the street vendors to regularise their business and sell at designated sites. This was supported by Hove et al. (2020) who quoted a Harare City Council official saying,

> The illegal street vending is becoming political. At one time we wanted to remove them (illegal street vendors) from the streets completely but then the first lady said no vendor will be removed from the streets and everyone started to chant the slogan "everyone belongs to mother", first lady Grace Mugabe and as Harare city council we were left with nothing to do (Hove et al., 2020: 77).

A council member also noted that the council and the government were to blame for the problem of illegal street vending that was prevalent in Harare. He noted that,

> This problem of illegal street vending has been allowed to go on for a long time until a time it was difficult to rein them [street vendors] in. Had the city council and government acted proactively to drive illegal vendors from the streets before they established themselves in the city centre, we would not be having difficulties in chasing them from the streets. But we allowed them to settle down and established semi-permanent positions which are illegal in the city centre making it difficult for us to drive them out.[14]

He added that HCC was justified in using violence to drive illegal vendors from the streets noting that street vendors ignored HCC call for them to regularise their trade.

Ndawana (2018) delved into the HCC-vendors dynamics when he argued that the politicisation of street vending has made it impossible to come up with a lasting

[14] Interview with HCC member, November 2020, Harare.

solution to the problem of street vending in Zimbabwe. He identified political expediency as the major reason why a lasting solution is elusive in handling the illegal street vending problem. There continues to be conflict between opposition run councils and national government dominated by the ruling party (ZANU PF). This conflict over the running of council affairs made it difficult to come up with lasting solutions to the problem of illegal vending. For instance, towards the 2013 elections, the then Minister of Local Government and Public Works directed all rural and urban councils to cancel debts owed to them by ratepayers. This affected the performance of councils, the majority of whom were run by the opposition as they lost huge sums of money that ran into billions of dollars. In the run up to the 2018 elections, Vice President Chiwenga also told people at a rally that they should boycott paying rates to councils if the service delivery is poor. This reflects the politicisation of the council business, in particular street vending. Politicisation is detrimental to resolving conflicts between street vendors and urban authorities.

Ndawana (2018) further writes that the ZANU-PF government has purposely used the illegal street vending activities in Harare to wrestle control of the city from the opposition MDC-T, now Citizens Coalition for Change (CCC). On the other hand, the opposition was bent on keeping stranglehold on Harare capitalising on the illegal street vending activities. As a result, the two parties (one in government and one running HCC) are ever opposed to finding sustainable solutions to illegal street vending in Harare, the main reason being the need for political gain from illegal street vendors by the political players.

In addition, large parts of open spaces in Harare have been occupied by ZANU-PF youth and followers. For example, the apportionment of vending sites at Mupedzanhamo and Mbare Musika is run and managed by ZANU-PF officials beyond the power of the HCC (Kriger, 2012). The ZANU-PF youth militia code named 'Chipangano' (a youth vigilante group) runs the Mupedzanhamo and Mbare Musika markets and often extorts money from desperate street vendors thereby usurping council authority to effectively conduct its work. Opposition members are not allocated space to trade at these marketplaces which is a well- known ZANU-PF territory. The Mupedzanhamo marketplace had to be closed due to power struggles between ZANU-PF and opposition CCC youth over the control of the place. Matenga (2021) wrote that the political fights between the opposition and ZANU-PF over Mupedzanhamo had resulted in delays to reopen the marketplace. This affected a lot of people who used to make a living through selling a variety of items at Mupedzanhamo marketplace. In July 2021, there were reports of youths from the two major political parties fighting over the control of the market, which left two people nursing gunshot wounds.

9.4.8 Formal Business Caught in The Middle

9.4.8.1 Context

Shop operators (formal business) have been caught up in this conflict as they have suffered the effects of illegal street vending. Illegal street vendors have invaded their pavements selling products that they also sell in their shops at much reduced prices since they are not paying rentals. Due to the economic crisis, people are forced to buy cheap products from illegal street vendors thereby putting shop operators at risk of liquidation since they can no longer record any significant sales. This situation has put shop operators on collision course with the HCC threatening to stop paying rates and taxes as they have been affected by illegal street vendors unless the HCC deals with the problem of illegal street vending at their doorsteps decisively.[15] Shop operators have failed to deal with street vendors on their own due to not lack of the authority to do that, but also the sheer number of street vendors who invade shop operators' pavements selling their goods.

The researcher interviewed representatives of established or formal businesses in the CBD who lamented that they have lost business because of street vendors. They blamed the conflict raging on between street vendors and HCC arguing that illegal street vendors have disregarded everyone else besides themselves for the use of space in the CBD. The HCC has failed to protect the formal shops against illegal vendors. Formal businesses complained what they termed harassment of their clients by street vendors who sell their goods on the pavement and sometimes at doorsteps of their shops. They intercept clients who would want to enter into formal shops and entice them to buy from them as they are cheaper than formal shops. Not only do they bar customers from entering formal shops, but they also harass them if they refuse to buy from them. Clients who refuse to buy from vendors are barred from entering formal shops by these vendors who sometimes become aggressive towards potential clients. As a result, shops have made losses and some have been forced to close. One shop operator had this to say,

> While we sympathise with street vendors as victims of the ailing economy, they should also understand that we are also victims of the same economy. Street vendors have the courage to sell at my doorstep goods found in my shop at prices much lower than mine without even negotiating with me. They have virtually taken over my business but I pay rates to city council every month while they do not pay anything. I am contemplating to defer payment of rates to council because I am not doing business here. I better close the shop and stay at home.[16]

This view was supported by the study carried out in Bulawayo by Ndiweni et al. (2014) who noted that many people were operating without licenses from Bulawayo city council (BCC). This is because unregistered vendors sold their goods at a lower price creating stiff competition with registered businesses. The prices of goods of the unregistered vendors were very flexible because they did not pay rentals to the

[15] Interview with shop operator, May 2020, Harare.

[16] Interview with shop operator, May 2020, Harare.

responsible authority. This created conflict between registered businesses and city authorities as the former felt short changed because they were expected to pay rentals, yet they were losing business to unregistered vendors who did not pay anything to the city council.

This position that formal businesses found themselves in was also confirmed by one council planning official who was cited by Hove et al. (2020: 79) saying that "formal shops are suffering loss and face closure because the vendors have brought unfair competition. They sell their goods at low prices because they do not pay rentals and tax". The foregoing quotation revealed that council is aware of the challenges that formal shops have encountered as a result of the influx of illegal street vendors and has been battling to chase vendors without much success.

Formal shops further noted that they are forced to close their shops almost on daily basis as street vendors engage in cat and mouse chase with council police in the streets. Street vendors running away from council police and Zimbabwe Republic Police (ZRP) seek refuge in their shops and on several occasions ZRP had thrown teargas inside their premises accusing them of sheltering street vendors. This had interrupted their business due to these running battles. As a result, the conflict between street vendors and HCC has negatively affected the formal business. Formal business wants this conflict to be resolved sooner as they have lost business. They want street vendors removed from their premises' pavements and be relocated to designated points.

9.4.8.2 Vendor Organisations

Two major vendor organisations have played an important role in the conflict between street vendors and HCC. These are the National Vendors Association of Zimbabwe (NAVUZ) and the Vendors Initiative for Social and Economic Transformation (VISET). The two organisations purport to represent vendors in all their spheres of life. VISET director Samuel Wadzanai support vending activities and was quoted saying,

> We are witnessing the wanton abuse of street traders at the hands of the Zimbabwe Republic Police and the army. Hundreds of our supporters have been rounded up and arrested…. Vendors are not criminals but ordinary people seeking livelihoods and must be allowed to feed their families, especially when the government has failed to provide formal jobs (Manayiti 2019).

The two organisations have argued that vendors have nowhere to go and vowed that they will not leave the streets until adequate vending sites with all facilities including ablution have been created. The two organisations occasionally consult lawyers with a view to approach the courts to seek an interdict to stop the continued attack on vendor's livelihoods. VISET argued through its director Wadzanai that forceful eviction of street vendors would not solve but exacerbate the problem. Wadzanai also revealed that his organisation is willing to sit down with all the stakeholders to find solutions and to work amicably with both formal sector and government.

Vendor organisations representatives blamed the HCC for failing to avail enough vending space to the vendors despite vending becoming the biggest employer in the country. They also criticised the brutal manner in which HCC was chasing vendors away from the CBD. They expressed concern over the lack of consultation by council when allocating vending space. For instance, they lamented the fact that vending sites designated by council out of city centre were not convenient and as a result vendors had no choice but to return to the city centre. A NAVUZ leader had this to say,

> We have seen in recent years the abuse of vendors by the Republic of Zimbabwe Police (ZRP), the council police who harass and victimise vendors for doing their work in the streets. The informal economy continues to sustain Zimbabwe with 7.4 billion dollars circulating in the informal economy annually. We however call for the increase in vending sites from the current position where we have 85,000 vendors competing for space in 1.185 vending sites in Harare city centre. There are also no sanitary facilities and safety provisions in designated areas. We therefore call upon the city of Harare to stop confiscating vendors' goods until they are allocated vending space.[17]

Another official from NAVUZ had this to say,

> Removing us from the streets is tantamount to saying go home and die. Our stance remains the same. As long as the economy does not change and government is not willing to chip in and give a conclusive and beneficial resolution to the matter our members will continue trading at sites that suite them. If council has agreed with associations as they have claimed, they should make sure that the registration process is done transparently to avoid the problem of corruption which some are engaging (Nyavava, 2015: 3).

In March 2016, there were a series of clashes between street vendors and municipal police in conjunction with ZRP which resulted in many street vendors losing their goods and some being arrested. NAVUZ released a press statement and said:

> We have learnt with utter disbelief that the Harare Municipality has engaged the services of the riot police who are assisting them during their raids in the Central Business District. Navuz has over the past days recorded several cases of vendors by riot police who have taken these raids as an opportunity to loot goods from vendors as some of the confiscated goods were never recovered by their rightful owners. We condemn in the strongest sense the militarisation of Zimbabwe *(The Zimbabwean 22 Sep 2016).*

Therefore, according to vendors' representatives, HCC needs to do the following things if they are to resolve the current conflict with street vendors.

- Consult vendors with regards to vending sites.
- Provide more vending sites at reasonable rates.
- Be more transparent in the allocation of vending sites.
- Stop confiscating vendors' goods until they allocate them vending spaces which are convenient to their clients and viable.[18]

---

[17] Interview with Navuz President, June 2019, Harare.

[18] Ibid.

9.4.9 The Security Forces' Involvement in the Conflict: Zimbabwe Republic Police and Zimbabwe National Army

Zimbabwe's security forces namely the Zimbabwe Republic Police and the Zimbabwe National Army have also been involved in this conflict though from the peripheral standpoint. They have been outsourced by the city council to help them to chase street vendors from the streets. The reason why the Zimbabwe Republic Police (ZRP) in particular is involved in this conflict is that although HCC has its own police department, it does not have arresting powers. Lack of arresting and prosecuting powers for the council police has made it very difficult for HCC to arrest and prosecute vendors. As a result, HCC has brought in ZRP to arrest and prosecute street vendors who refuse to abide by the city by-laws.[19] In this case, the Harare municipal police work together with the ZRP. HCC police only have the power to confiscate vendors' goods and hand them over to ZRP who then fine and prosecute the vendors. ZRP has also been involved in some raids on street vendors in joint operations with the city council and at times they incorporate ZNA in joint operations when the violence is likely to escalate. Frosty relations have therefore emerged between the security forces and street vendors with the latter enjoying support from human rights defenders who have lamented the heavy handedness of the security forces in the manner in which they brutalise vendors. Human rights organisations such as Lawyers for Human Rights, NAVUZ and other street vendors' organisations have accused HCC of militarising and securitising the streets (*The Zimbabwean* 22 September 2016). There have been isolated incidences where ZRP and HCC police members have been attacked by street vendors while on duty trying to drive vendors from the streets. These incidences have justified the engagement of ZNA to help the city council to enforce their by- laws against the marauding and vicious street vendors. It is therefore important to note that there are strong links between ZRP, ZNA and HCC who work jointly to maintain order in the streets. The security forces have created tense relations with street vendors and vendor organisations. This chapter asserts that the presence of the security services in the streets cannot resolve the conflict between street vendors and the city council. It argues that for sustainable peace to be realised, there should be dialogue between the vendors and city council. Using force to drive vendors off the streets has so far proven beyond doubt that it is futile.

---

[19] Interview with HCC member, November 2020, Harare.

9.5 Conclusion

This chapter used the conflict mapping analysis tool to map the conflict actors in the conflict, which on prima facie appears to be between street vendors and HCC. It was established from the discussion above that the conflict is both intractable and complex. It involves many interlinked players and issues. It is key for any peaceful resolution of conflict to explore the relationships, issues, influences and linkages between the various actors. Exploring these issues and dynamics enhances chances of resolving the conflict without leaving other actors' interests unaddressed. This may act as spoilers. Having explored these issues and actors, it is recommended that the conflict can be resolved peacefully through dialogue without resorting to violence as what has been happening until now. There is consensus among the various actors in this conflict that the use of force to resolve the conflict has failed to yield better results. Even the involvement of the security forces could not resolve the conflict. There is need for introspection and change of strategy in finding a common solution. What is needed is peaceful resolution guided by dialogue which should include all actors identified here. The HCC needs to create more vending sites in the central business district to accommodate more vendors. Vending must be allowed as long as the vendors are paying for the services. Vending has become the major employer in Zimbabwe, which is contributing to the attainment of SDGs number 1 and 2 of ending poverty and hunger respectively. When there is no poverty and hunger, this contributes to another sustainable development goal number 16, peace, justice and strong institutions, which is also a peacebuilding strategy. The allocation of vending sites should be transparent. Vendors too should be organised and clean their vending sites which in most cases are left filthy. They should also make sure that they are not blocking human and vehicular traffic. Lastly, the city council should consult and engage vendors and other stakeholders often if they intend to implement by-laws that have a bearing on these actors. Lasting peace between street vendors and HCC may seem unachievable, but it can be attained if there is dialogue among all the actors. The use of force to resolve conflicts is unsustainable and costly. Vending is a major source of employment and should be allowed. Favourable conditions should be created to allow vending to take place as it is key to achieving SDGs in the absence of formal employment.

References

Bhowmik, S.K. (2005). Street vendors in Asia: A review. *Economic and Political Weekly*, 40: 2256–2264.

Ettang, D., Maina, G., & d Razia, W. (2011). A regional approach to peacebuilding. Policy and Practice Brief (006). Durban, African Centre for the Constructive Resolution of Disputes.

Finscope. (2014). Consumer Survey Zimbabwe. Accessed 29 March 2019.

Fisher, S. (2005). *Working with conflict: Skills and Strategies for action*. New York. Zed Books.

Ghali, B.B. (1992). An agenda for peace: Preventive diplomacy, peacemaking and peacekeeping. United Nations. https://unrol.org/files/A_47_277.pdf. Accessed 25 Feb 2023.

Harare Vendors By-laws. (2014). Statutory Instrument 159 of 2014.

Hove, M., Ndawana, E., & Ndemera, W. S. (2020). Illegal street vending and national security in Harare, Zimbabwe. *Africa Review, 12*(1), 71–91.

Kriger, N. (2012). ZANU PF politics under Zimbabwe's 'Power-sharing' government'. *Journal of Contemporary African Studies* 30(1), 11–26. https://doi.org/10.1080/02589001.2012.644947.

Matenga, M. (2021). Zanu PF, MDC Alliance clash over Mupedzanhamo-Newsday Zimbabwe. *The Standard | Sep. 21, 2021*. https://www.newsday.co.zw/southerneye/2021/09/21/zanu-pf-mdcalliance-clash-over-mupedzanhamo. Accessed 18 February 2024.

Manayiti, O. (2019, Jan 28). Military crackdown: Vendors feel the heat. *Newsday*. https://www.newsday.co.zw/2019/01/military-crackdown-vendors-feel-the-heat/. Accessed 22 March 2020.

Matenga, M. (2021, Sep 21). ZANU PF, MDC Alliance clash over Mupedzanhamo. *Newsday*. https://www.newsday.co.zw/2021/09/zanu-pf-mdc-alliance-clash-over-mupedzanhamo/. Accessed 22 September 2021.

Mitullah, W. V. (2003). Street vending in African cities: A synthesis of empirical findings from Kenya, Cote D'ivoire, Ghana, Zimbabwe, Uganda, and South Africa. https://openknowledge.worldbank.org/bitstream/handle/10986/9211/WDR2005_0027.pdf?sequence=1/. Accessed 22 September 2021.

Mutami, C., & Gambe, T. R. (2015). Street multi-functionality and city order: The case of street vendors in Harare. *Journal of Economics and Sustainable Development, 6*(14), 124–129.

Ndawana, E. (2018). Sacrificing urban governance at the altar of political expediency: Illegal Street vending in Harare. *African Security Review, 27*(3–4), 253–277. https://doi.org/10.1080/10246029.2018.1550428.

Ndebele, H. (2017, Jan 20). Harare, Vendors enduring stand off as city battles disease outbreaks. Zimbabwe: Harare, Vendors' Enduring Stand-Off-all *Africa.com*. https://allafrica.com/stories/201701200389.html. Accessed 22 September 2021.

Ndiweni, N., Mashonganyika, S., Ncube, N., & Dube, N. (2014). Informal economy as a source of livelihood in Zimbabwean urban areas: The case of Bulawayo metropolitan province. *International Journal of Scientific and Research Publications, 4*(1), 1-8.

Njaya, T. (2014). Employment creation or sheer survival: Case of street vendors in the Harare Metropolitan in Zimbabwe. *Global Journal of Advanced Research*, 1(2), 264–274.

Njaya, T. (2016). An evaluation of income disparities between male and female street vendors of Harare in Zimbabwe. *Journal of Studies in Social Sciences and Humanities*. 2(3), 106–114.

Nyavava, K. (2015, Apr 7). Vendors ordered to relocate to designated sites. *Newsday*, p. 3.

Staff Reporter. (2021). Zimbabwe: Closure of foreign companies cleared way for local controlled economy - Muchingura, All Africa. https://allafrica.com/stories/202108020407.html. Accessed 7 December 2022.

The Global Goals. https://www.globalgoals.org/goals/ Accessed 25 February 2023.

The Zimbabwean. (2016, Sep 22). NAVUZ concerned over riot police brutality against vendors. *The Zimbabwean*. NAVUZ concerned over riot police brutality against vendors (thezimbabwean.co). Accessed 7 Dec 2022.

Tomislav, K. (2018). The concept of sustainable development: From its beginning to the contemporary issues. *Zagreb International Review of Economics and Business*, 21(1), 67–94.

Yin, R.K. (1994). *Case study research*. California: Sage Publications.

WCED, United Nations Commission on Environment and Development. (1987). Our Common Future. www.un-documents.net/our-common-future.pdf. Accessed 28 Feb 2023.

Zimpricecheck. (2020). Harare City Council Destroys market stalls at the behest of the national government. *Zimpricecheck*. https://zimpricecheck.com/market-intelligence/harare-city-council-destroys-market-stalls-at-the-behest-of-the-national-government. Accessed 30 April 2020.

Chapter 10
Youths in War in a Post-Colonial State and the Continuing Use of Child Soldiers: Perspectives from the Democratic Republic of Congo

Daniel N. Mlambo

Abstract Since most African states gained their independence from erstwhile colonial masters, some parts of Africa have remained unstable, driven by various factors, including dictatorship and political instability. Some regions have witnessed major internal conflicts and economic declines that have threatened socio-economic development, human security, sovereignty, the territoriality of states, and the stability and legitimacy of political regimes. Since its liberation from its erstwhile colonial ruler Belgium, the Democratic Republic of Congo (DRC) has been hampered by internal instability in many forms and sizes, primarily driven by the country's rich natural resource base, numerous rebel groups, and tribalism. The attraction of the country's resources has resulted in the formation of many rebel groups, which has made the country unstable. Other motives have included the absence of the rule of law, competition for land, weak state capacity, and external influence. The existence of these rebel groups has resulted in the creation of a new form of the army in child soldiers, especially in the unstable Kivu province. We can trace the emergence of child soldiers in Africa to the post-colonial predicament of globalisation driven by increasing inequalities, insecure livelihoods, a weakened and strained social fabric, a downturn in the capacity of households, and debilitated social norms and the erosion of traditional African values. In addition, it has been driven by the easy access to light weapons by armed rebels and, by extension, by children. Employing a qualitative research methodology supplemented by secondary data, this chapter delves into the use of child soldiers and what impact this has had on the country's economic growth prospects, internal and regional stability, and Southern Africa's security architecture. The chapter's objective is to reveal the dire consequences of instability, especially the use of child soldiers, for the development and economic growth prospects of African states and the DRC in particular.

Dr Daniel N. Mlambo is a Post-Doctoral Research Fellow at the Office of the Executive Dean of the Faculty of Humanities at the Tshwane University of Technology. He would like to acknowledge the National Institute for the Humanities and Social Sciences (NIHSS) for their financial support. Email: MlamboDN@tut.ac.za; ORCID: https://orcid.org/0000-0001-9898-1414

J. C. K. Kiyala and N. Chivasa (eds.), *Climate Change and Socio-political Violence in Sub-Saharan Africa in the Anthropocene*, The Anthropocene: Politik—Economics—Society—Science 37,
https://doi.org/10.1007/978-3-031-48375-2_10

Keywords DRC · Child soldiers · North kivu · Instability · Security · Socio-economic development.

10.1 Introduction

The DRC (with an estimated population of about 92.4 million people) shares its borders with nine other countries: Angola, Burundi, the Central African Republic (CAR), Congo-Brazzaville, Rwanda, Sudan, Tanzania, Uganda, and Zambia. The Kivu Province comprises three main cities: Goma, the provincial capital, and Beni and Butembo, with different territories and chiefdoms. The driving forces of conflict in North Kivu are as complex as the instability. North Kivu borders Rwanda and Uganda to the East, Maniema to the west, Orientale province to the north and North West, and South Kivu to the south. It also borders three Great Lakes, Lake Kivu to the south, Lake Edward to the East, and Lake Albert to the north. Mlambo and Adetiba (2020) noted that post the colonial era and despite various interventions by numerous bodies such as the United Nations (UN), the African Union (AU), the Southern African Development Community (SADC), and other African states to bring peace and stability to the DRC the country continues to face multiple conflict-related challenges that pose a severe threat to its delicate transition and damage its post-colonial security architecture. While the country is no longer involved in large-scale armed conflict, it remains a country and society in ongoing turmoil. Oyewole (2018) points out that violent conflicts are the bane of human social interactions. From the medieval ages, humankind has resorted to organised violence to resolve disputes driven by resources along social, political, economic, religious, military, and diplomatic lines. Consequently, the overwhelming effects of War caused by the ineffectiveness of the formal structure may render incapable or even destroy the informal protection structures for protecting youths, thus leaving them vulnerable to harm. In most cases, affected youths who find themselves in these positions respond by seeking alternative and effective structures that can provide both resources and protection, and the most prominent of these happen to be armed groups or rebels.

According to Neethling (2014), several non-state armed movements or militias have continued to pose a severe security threat to neighbouring states and the Great Lakes Region (GLR) over the last few decades. The involvement of Burundi, Rwanda, and Uganda in the past Congo Wars driven by shared history and populations, porous borders, access to mineral resources and further driven by the Rwandan and Ugandan government support of insurgents stimulated many conflicts in the region (Lalaba-hadur, 2017; Prunier, 2010; Mlambo, 2021). Ekyamba (2022) describes the Kivu region of the DRC as, for more than two decades, the epicenter of non-state armed groups and continuous violence. The constant presence of these groups has led to high death tolls and undermined development, resulting in severe socio-economic problems and humanitarian crises, which have undermined the capacity of state institutions and governance. Over the years, these groups have also been pivotal in recruiting, training, and using child soldiers to promote this instability further.

Including youths in warfare (especially in the eastern and northern parts of the country) continues to be one of the most problematic issues in the DRC. According to Kiyala (2018b), youths in warfare are generally referred to as *Kadogo,* (which translates to child soldiers or little ones in Swahili). Building from this, how can one link globalisation with the continuous conflict and use of child soldiers in the DRC? Firstly, globalisation refers to the notion of the world becoming a global village and brings several implications, some of which involve states becoming independent, self-determining, and self-reliant. Hence, the nexus of globalisation and child soldiers helps us understand how globalisation exposes youths to the particular concept of War they experience and engage in. Again, many elements drive child soldiering in the DRC and are stimulated by different interests, such as political, religious, socio–ethnic and economic factors. Similarly, media exposure, War scenes, border crossings, and the presence of foreign fighters are push and pull factors that further bring children closer to the realities of War.

A 'child' is defined as someone below 18 unless a majority is attained earlier under the law applicable to the child (UNICEF, 2007). However, Kiyala (2018a) noted some contradictions in the definition of child soldiers, mainly because professional child soldiering is supposed to be an adult job and not for minors. To further unpack this, a soldier is an individual that has undergone the required training to appropriately and robustly carry out duties in government service, which is sponsored by the state, under the defined rules and discipline of the armed governmental forces. Contrarily, when a specific rebel group trains a child soldier, s/he is inadequately trained and outfitted and, in most cases, operates under the influence of drugs. In such instances, these youth harass, loot and kill innocent civilians. From an Afrocentric perspective, the DRC is home to thousands of child soldiers that have been used to fight other rebel groups and government forces, destabilising the country's growth prospects post the colonial era. The presence of these armed groups is mainly due to long-standing grievances, unresolved post-colonial governance challenges, the need to fight for natural resources, and the limited success of peace efforts. Drawing from the above, this necessitates understanding the DRC's post-colonial instability and peace and security challenges, the increasing use of child soldiers in the country, and some of the underlying factors behind the failure of various peace efforts.

10.2 Fragmentation in a Post-Colonial State: A DRC Perspective

Over five decades since most African states attained their freedom from colonial rulers, Africa has continued to be called the dark continent, mainly because of its ongoing violence and human rights violations driven by dictatorship rule, non-state actors, terrorism, coups, and War crimes. Since the 1990s, there has been constant instability in the GLR, primarily stimulated by the fundamental shift in the domestic politics of core states and the diminishing Western engagement in the region. As

Vlassenroot (n.d) puts it, the DRC has often been presented as one of the most prominent examples of institutional collapse and state failure since its freedom from Belgian colonial rule. This scholar substantiates this by asserting that the DRC seems to have become a "forsaken black hole characterised by calamity, chaos, confusion, and a bizarre form of social cannibalism where security has become its prey." Jobbins (2010) noted three characteristics in connection with the continued instability in the DRC. Firstly, like most African states, the DRC is a developing country with abundant resources. Because of this, minority groups of wealthy individuals possess resources that could, if properly utilised, support community protection and response mechanisms. Secondly, the collapse of the Zairian state marked a shift in the breakdown of public order. This led to moving public provisions outside the public sphere to local communities and institutions. Thirdly, in various parts of the DRC, violence became identity-driven. Numerous militia groups were mobilised to protect local interests and played a very active and brutal role in the violence. Thus, such patterns of violence were often determined by local claims and grievances along ethnic and regional lines. While the DRC's instability stems from national and international forces, local factors have significantly shaped violence patterns and civilian suffering.

The presence of Rwandan refugees and the Congolese state's decay primarily drove the first Congo War (1996–1997). The second Congo War (1998–2003) started when the Rwandan government fell out with Laurent Kabila, whom it had backed until he became president. This led to establishment of the Congolese Rally for Democracy (CRD) against Kabila in 1998. This contributed to new waves of conflict and took a turn for the worse when Kabila asked all Rwandan troops to vacate the DRC in the same year. This resulted in one of the deadliest Wars in modern African history and caused the deaths of millions of individuals (Stearns, 2012a). Indeed, as Neethling (2014) noted, most states emerging from violent conflict carry a significant risk of sliding back into violence. As previously mentioned, since the 1990s, two major wars have been fought in the DRC.

The War between the DRC and its neighbours ended in 2003, but many armed groups are still operating. This has to do with a history of group rivalry, the quest for territorial control, and, importantly, the presence of mineral resources that have been exploited illegally over the years. According to N'Gambwa (2011), the instability in the DRC, political unrest, and minimal socio-economic progress led to a wave of emigration from the country, mainly to its neighbours.

Ekyamba (2022) observes that the historical origins of armed rebels in the Kivu region can be traced back to the eruption of the first Congo War in 1996, occasioned by the Alliances des Forces Démocratiques Pour la Liberation du Congo-Zaire (AFDL). Again, the second Congo War, involving the Rassemblement Congolais pour La Démocratie—Goma (RCD—Goma) faction in 1998, stimulated the establishment of localised armed self-defence groups in communities, which enabled new waves of armed groups in the Kivu region. As already alluded to, many other states surround the DRC, and instability in the country has vast regional implications; the impact of these spill beyond the country's borders and affect the entire region. The GLR has suffered immensely in terms of loss of human lives, continuous instability, and effects on the region's growth and economy (UNECA, 2015).

Scholars such as Monyae (2014) agree that the conflict in the 1990s in the GLR was driven by post-colonial nation and state-building issues, weak regional institutions for economic integration, resources, and environmental vulnerability. This was exacerbated by the Rwandan genocide, which brought instability and turbulence in the region and regional relations.

Over the last few decades, these groups have used incitement of violence to intimidate each other and the state and local citizens. Scholars (Newman, 2007; Whitaker, 2010; Borowski and Fuchs, 2012) argue that illicit transnational networks, predominantly militias groups, have been highly active in weak or so-called failed states. The DRC has not been immune to this. This is also driven by the fact that since the 1990s, African states have not only struggled against international terrorism but have also had to engage with local terrorism and militia groups. In addition, some African states (apart from grappling with different forms of violence) also have to contend with poverty, famine, and disease that further undermine human security. These have all been prominent in various recent security policy discussions in Africa.

What creates such acts of violence is that they are often aimed at a group, individual, or symbol that may not be directly linked to their real target. Instead of government ideology, militia groups and terrorists generally focus on a particular target, such as intimidating a particular community or undermining a specific political and economic system (Neethling, 2014). From the many items of literature by various scholars, one may deduce that there are innumerable reasons why individuals take up arms and make some states ungovernable. In the literature, grievance, and avarice are two of the most important explanations, specifically from an African perspective (Richards, 2014). What drives conflict in the northern parts of the DRC is that some non-state actors have legitimate grievances, but at the same time, they are responsible for extreme human rights violations. To clearly understand some of the driving motives of the conflict in the DRC, one can look at the following five elements:

1. State bureaucracy
2. National Boundaries as the Legacy of Colonialism
3. Marginalisation and disenfranchisement of communities
4. Ethnic consciousness and identities and
5. Corruption

Regarding the above, the author believes that points 2, 3, and 4 are predominantly at play in the Kivu region. This is because the post-Rwandan genocide has allowed many rebel groups from neighbouring Rwanda to occupy the DRC. They have used these bases to plan and execute their attacks illegally. Such activities were also stimulated by the Rwandan genocide in 1994, which saw many Hutu and Tutsi individuals escape or move to the DRC. In addition, the DRC is home to many ethnic groups (some from neighbouring states) that do not share the same vision, values, and culture and come from different communal backgrounds. All of these also fuel instability and threaten peace in the Kivu region. Since its independence, numerous heads of state of the DRC have singularly failed to meet what N'Gambwa (2011) calls the deep-seated political, economic, and social challenges facing the Congolese population. Over the years, the DRC has not fulfilled its fundamental duties, principally providing peace and security to the populace, protecting its borders, and fighting poverty. This is even though the DRC, over the years, has signed numerous peace agreements (see Table 10.1).

Table 10.1 The signing of various peace agreements between the DRC and other stakeholders

Agreement	Place and date[1]
Lusaka ceasefire agreement	Lusaka, Zambia, Jul 1999
Pretoria accord	Pretoria, South Africa, Jul 2002
Luanda agreement	Luanda, Angola, Sep 2002
Global and inclusive agreement on Transformation in the DRC	Pretoria, South Africa, Dec 2002
The final act	Sun City, South Africa, Apr 2003
The pact on security, stability, and development in the GLR	Dar es Salam, Tanzania, Nov 2004 and ratified in Nairobi, Kenya, in 2006.
The peace agreement between the government, the national congress for the defence of the people (CNDP), and other armed groups	Goma, DRC, Jan 2008
Peace agreement between the government, the national congress for the defence of the people	Goma, DRC, Jan 2008
The goma conference	Goma, Jan 2008
Operation Kimia 2	Kinshasa, Mar 2009
23 March agreement	Kinshasa, Jan 2009
Peace, security, and cooperation for the DRC and the region	Addis Ababa, Ethiopia, Feb 2013
Operation amani leo	Kinshasa, Jan 2020
Peace, security, and cooperation framework for the drc and the region and the deployment of the east african community (EAC) regional force:	Kinshasa, Feb 2022
The 10th high-level meeting of the peace, security, and cooperation framework for the DRC	Kinshasa, Feb 2022
Sun city agreement	Sun city, Rustenburg, South Africa, Apr 2022

Source Stearns (2012a, 2012b), Muraya and Ahere (2014), South African Government (2023), African Union (2023)

[1] In these agreements, various stakeholders were involved including, but not limited to, Angola, the DRC government, Namibia, Rwanda, Zimbabwe, Civil society, different political parties, various governments, Rebel groups, South Africa, Tanzania, South Sudan, CAR, Burundi, Uganda and the SADC.

The above table illustrates that for almost all stages in the life of the DRC, peace processes have been characterised by the involvement of internal and external actors who have been supportive but destructive in many cases. This is because some states and militia groups were directly or indirectly involved in the DRC conflict. Again, despite numerous ceasefire agreements (in addition to those in Table 10.1), over the years, and democratic elections which integrated former rebel groups, the instability continues to shape the DRC's post-colonial trajectory. Thus, despite many treaties signed along the way, the continual instability threatens to reverse the limited gains made in the DRC over the years.

10.3 Child Soldiers in the DRC: The Underlying Factors

In the last six decades, the phenomenon surrounding children with warfare has received significant attention in some parts of the world and Africa, which attests to this growing phenomenon (Hart, 2006). Youths who participate in warfare is a global phenomenon and historically derives from the era of World Wars at a time when many child soldiers participated in hostilities, some of which entailed the conscription of the Nazi juvenile soldiers known as the Hilterjugend (the Hitler Youth; Lewis, 2019). Child soldiers in warfare have served all over the globe. They have thus become integral parts of both organised military units and non-military violent organisations, including rebel and terrorist groups. The youth in these groups have various roles and responsibilities, such as sentries, spies, trench diggers, and porters. There is a child soldier element in all corners of the world.

For example, in Central and South America, since 1990, child soldiers have been involved in conflicts in Colombia, Ecuador, El Salvador, Guatemala, Mexico, Paraguay, and Peru. Colombia is the only country using child soldiers in this region, where youths serve in rebel groups, the Revolutionary Armed Forces of Colombia (FARC), and the National Liberation Army (ELN) organisations. In Europe, most child soldiers are drawn from the East in countries such as Kosovo and Macedonia (Ortiz, 2002). For instance, youths fought in the Kosovo Liberation Army (KLA) in the War against Serbia in 1997–1998. Others have joined Albanian rebel groups attempting to break away bits of territory from Serbia and Macedonia, serving in both the Liberation Army of Presevo, Medveda, and Bujanovac (UCPMB) and the Albanian National Liberation Army. In the Middle East, youths engaging in warfare can be found in Azerbaijan, Iran, Iraq, Lebanon, and Yemen. These youths serve in several radical Islamic groups.

From an African perspective, the continent is known as the epicenter of having youths in warfare, particularly in Sub-Saharan Africa (SSA), as is evident in African states such as Liberia, South Sudan, Burundi, Rwanda, Somalia, and the DRC. From a Rwandan perspective, thousands of youths seem to have participated in the 1994 genocide between the Hutus and the Tutsis. However, child soldiers became a prominent feature of the continent's most significant conflict. The DRC War began in Zaire in 1996 through a revolt by Laurent Kabila. His army comprised more than 10,000 youths who were soldiers. While some youths volunteer to join rebel groups for protection, some do so because of hunger and poverty, being orphans because of warfare, and the need to put bread on the table for their families. On the other hand, some are abducted, trained, and forced to be part of rebel groups. For instance, the Congolese Rally for Democracy – Goma (RCD – Goma) and the Rwandan Patriotic Army (RPA) are two Rwandan groups known to target schools for kidnapping youths to join their ranks.

Every child anywhere in the world is essential; however, in some parts of the world, these children are taught how to use guns and kill and destroy. Globally, innocent children being forced to take up arms in unstable states is a dreadful crime against humanity. To fully understand the effects on the development of children in the DRC, the DRC's economic, social, and developmental aspects need to be examined. In the eastern parts of the DRC, institutional instability has added to the presence of multiple armed groups and weak political leadership and fostered many children's increased susceptibility. From Matambura's (2017) assessment, in some African states, particularly the DRC, not only have children become the new target of violence and atrocities but have constantly been made perpetrators of the same. The use of child soldiers is not new but has escalated in the last few decades, particularly in Africa. During military coups in Africa (1970–1989), children were used as child soldiers, especially when separated from their family members (Falola/Adebayo, 2010).

Rich or developed regions generally do not use child soldiers in conflicts compared to poor or developing areas. This is driven by the fact that developed regions are relatively stable, unlike poor or developing areas where children have constantly pointed to poverty as their reason for joining rebel groups (Kakhuta-Banda, 2014). In some African states, such as Somalia, the CAR, South Sudan, and Sierra Leone, many youths are coping with exposure to violence and ongoing conflicts. Violence can severely affect children in the short and long run (particularly the latter). Zuilkowski et al. (2016) posit that violence can disrupt children's relationships across the spectrum of social ecology among families, religious groups, schools, and broader communal settings. Because not all children have access to such support, they most generally rely on some of these elements for their physical, emotional, and developmental needs. Thus, any form of violence disrupts their development and resilience.

From an African perspective, the emergence of child soldiers can be directly linked to the post-colonial African predicament and globalisation. Kiyala (2018b) maintains that this arises from numerous circumstances that create the use of child soldiering, and such causes have several pushes and pull factors. Push factors may include poverty, wars, political conflicts, famine and droughts, inadequate public service, religious and political persecution, absence of good governance, and unemployment.

Pull factors may be fame, security, greed, humanitarian protection, and communal respect. In like manner, this use of child soldiering has been driven by the ever-escalating nature of inequalities, insecure livelihoods, a weakened and strained social fabric, the failure of government, households, and communities to provide for the children, and debilitated social norms and values and the erosion of traditional African values. For decades, most African regions have been embroiled in ever-increasing internal conflicts and sectarian violence driven by political disagreements and the need to control natural resources. Elements such as civil wars, political intolerance, and coups have directly and indirectly contributed to many African states' poverty, malnutrition, and underdevelopment. Inevitably, this is further escalated by the new forms of warfare and the easy accessibility of light weapons to armed groups and militias, making them easily accessible to children (Mlambo et al., 2019).

Taking advantage of political and social instability, rebel groups have abducted young boys to fight in armed combats over the years. This has severely impacted children's continuous development and the possibility of a stable society. These children have been injured illegally and deprived of access to proper upbringing, education, adequate nutrition, and other necessities as a direct or indirect result of such instability. This is even though there are legal obligations to ensure child soldiers are assisted and integrated into society. However, the DRC government has failed to develop and implement a robust and effective national strategy to identify and protect them. Long-term policies for successfully integrating child soldiers must be integrally linked to post-conflict reconstruction programmes that firmly address the frequent cause and consequences of their recruitment and use. This is important because in Africa, whatever the number of child soldiers may be, the fact remains that children have become components of militia groups not only in the DRC but also in South Sudan, Uganda, and the CAR.

10.4 Theoretical Framework

North and South Kivu are one of the most vulnerable places where continuous conflict is at play, and this has been the case for many years. Some of this is also driven by the fact that since the Rwandan genocide, there has been almost no reconciliation between local communities, thus, leaving groups with growing prejudice towards each other. The little reconciliation that has been achieved has been chiefly driven by churches and Non–Governmental Organisations (NGOs) and not the government. Thus, the lack of reconciliation has fuelled support for insurgents, who, in most cases, are isolated and poor, and this has become the norm for many youths in the DRC (Stearns et al., 2012a). Many theories in the literature explain conflict, War, terrorism, and revolutions. Because of being shaped by practical events and processes, theories are born out of reality and are of assistance as they aid in constructing a more robust understanding of the fluctuating dynamics of the global arena. This chapter uses the human needs theory to understand better what stimulates conflict. Drawing on this, a theory is meant to guide the researcher/author in conceptualising the study; thus, without a theoretical setting, the study is consigned to an epistemological vacuity.

Conflict will likely emerge because of the increasing competing interests of states and social groups. Scholars have presented different reasons for the cause of conflict, ranging from unsatisfied basic needs to a clash of individual interests. While human needs, conflict, and peace are interrelated and affect all spheres of life, human needs theory proposes that all humans have specific basic universal needs. These needs may include but are not limited to distributive justice, safety and security, belongingness, self-esteem, personal fulfillment, identity, cultural security, and freedom. When these are not met, conflict is likely to occur. Danesh (2012) believes competition, structured inequality, revolutions, and instability drive modern-day conflict.

Thus, competition will likely occur when resources are too scarce to satisfy needs and wants. Structural inequality involves unequal power distribution that leads to social class revolutions. Again, instability arises directly from competition, limited resources, and unequal power distribution. The human needs theory is based on the notion that essentials are necessary for survival. These are not only food and shelter but include physical and nonphysical elements (Walsh, 2015). While the theory does not necessarily offer a comprehensive list of human needs, it is largely agreed that it refers to ontological conditions ranging from biological and physical requirements to identity and recognition (Azam/Rehman, 2018). However, according to Christie (1997), the theory offers insights into various peacebuilding processes that reduce direct and structured violence. It further assists in developing insights into why security needs and identity are, in most cases, proponents in conflicts. The theoretical setting links to child soldiering in that other than being kidnapped by other rebel groups, most youths join these rebel groups for protection and to sustain their livelihood, driven by the need for money and alleviating poverty for themselves and their families. In addition, the fierce competition between rebel groups along ethnic lines also stimulates conflicts, as in many cases in the DRC. Concisely, the theory outlines how the needs of an individual can, directly and indirectly, fuel conflict in both the short and long run.

10.5 Research Methodology

Research methods involve collecting and using data for a specific study in any discipline. Thomas (2003) maintains that no method is considered sacrosanct when undertaking research, especially in the social sciences. Thus, one can apply two main methods, namely qualitative and quantitative research, and there is a mixed method element between the two. However, when the research method becomes questionable, the entire research becomes questionable (Ndagubu/Hanyane, 2019). This chapter utilised a qualitative approach; when a researcher uses this method, s/he intends to construct a comprehensive viewpoint regarding human behaviour and its rationale.

10.6 Child Soldiering in the Post-Colonial DRC: What Needs to be Done?

The history of the DRC contains several sobering legacies from slavery, genocide, political instability, and mineral resources fuelled by conflict driven by domestic and neighbouring parties. Because of the nature of instability in the DRC, a sound and workable resolution of the conflict and, in particular, a decrease in child soldiers and limited instability will require implementing several measures. Lorch (2020) noted that tackling the use of child soldiers in the DRC is a complex phenomenon, just like the conflict itself, and needs both short and long-term solutions. As long as some parts of the continent continue to experience conflict, much-needed economic reforms will remain elusive. Thus, fighting child soldiering must be driven by the government (in this case, the DRC government), NGOs, and, in some instances, the private sector. This is even though in other African states, the practice and voice of NGOs are limited due to the government's tough stance on their practice. However, the government must drive all reforms in a particular state. This should be underpinned by the belief that youths should grow up with a sense of belonging, be vision orientated and know how to contribute to the DRC's long-term vision. This can only be achieved if the government introduces major reforms as outlined below. In line with some of the findings of this study, the researcher put forward some recommendations regarding the need for the following:

1. Development of a strong judiciary not only in the Kivu region but also throughout the DRC as so to build effective court presentation and justice.
2. Better awareness of and education on the role of conflict in developing states, neighbouring states, and regional integration.
3. Improvement of security mechanisms in unstable areas like north Kivu.
4. Better educational policies for youths and awareness of the effects, complications, and post-conflict restructuring of unstable states.
5. More robust monitoring and evaluation collaboration between agencies, states, Regional Economic Communities (RECs), and the private sector in dealing with child soldiers in unstable states like the DRC.

In his article titled "Africanizing Rebel Insurgency in a Post-colonial State and Ramifications for African Regional Integration: Insight(s) from the Democratic Republic of Congo", Mlambo (2021) brings to the fore the following measures to assist in bringing peace in unstable states such as the DRC:

1. The consolidation of real peace (physical and social security of the people affected by instability);
2. The creation of resilient social, political, and legal institutions that will sustain peace, economic growth, and development;
3. Enhanced protection of gender equality and civil and human rights;
4. The formulation of coherent regional programmes which are driven by solid policies aimed at limiting conflicts caused by ethnicity, resources, poverty and
5. The equitable distribution of natural resources.

The points above outline some reforms that the DRC, other African states, and Regional Economic Communities (RECs) may pursue to limit conflict in the DRC and youth joining rebel groups. In any state, reforms and policies in both the short and long run are essential for consolidating peace and driving domestic policies. Again, regional integration is essential in Africa, as the continent possesses weak states that must work collectively to stimulate economic growth and development. Hence, regional integration can be seen as a development stimulator that generates economic growth and political stability (Oloruntoba/Muchie, 2018). It serves as a critical driver for conflict resolution and post-conflict reconstruction. In a similar vein, it can assist in rebuilding the relations between the DRC and its neighbours.

To effectively tackle the DRCs ongoing conflict and poor governance, a clear and practical vision for the DRC's future must be articulated, implemented, monitored, and rigorously evaluated by the government and other internal and/or external stakeholders. The immediate need for social governance is vital in the DRC's post-colonial reconstruction. This should be driven in an open, accountable, equitable, and responsive manner to society's present and future needs. While the government is essential in stimulating peace negotiations, other important role players include NGO's, religious leaders, political parties, the media, and continental and international donors.

The DRC government should have a clear vision focusing on and guaranteeing socio-economic opportunities for the local populace, especially the most vulnerable youth. Post the colonial era, and ever since the escalation of instability in the DRC, youths have been susceptible to joining armed rebels because of poverty. As long as poverty persists and children remain vulnerable in the DRC, the risk of instability will continue, further hampering any socio-economic growth and development for the country in the short and long run. Recruiting and using child soldiers in modern-day DRC will not end until more sensible and realistic policies are implemented. These should be based on the need to ensure clarity of long-term peacebuilding strategies.

10.7 Conclusion

In any state, intrastate and interstate instability brings dire consequences. It can, among other things, halt regional activities such as the movement of individuals, the flow of resources, human capital development, and the destruction caused by physical and spillover effects. As we look to the future, what is worrisome is that there seems to be no end to the instability in some parts of the continent and the DRC. Most parts of the continent, especially in SSA, remain unstable and driven by continuous conflicts that lead to youths growing up with no vision about their lives and economic growth opportunities. Stability has remained elusive in the DRC despite numerous peace processes, and the culture of violence has continued to exist. Thus, finding a lasting solution to the DRC conflict will take many more years, considering that many peace accords have been agreed upon and signed but have yet to produce the required results. This chapter has shown that most of the instability in the DRC has been driven by the country's history post its independence era, caused by constant instability, the

Rwandan genocide, the greed for natural resources, and territorial control. Apart from the massive investment from continental role players in supporting peace and security in post-colonial DRC, the country is far from escaping its continuous and ongoing instability issues.

The continuous instability in the DRC has increased the number of rebels and child soldiers, crippled the country's economic growth, and severely affected the GLR. Successfully bringing peace and the rule of law to the DRC will be a huge challenge that will need domestic and regional measures over several years. If this fails to materialise, instability will continue and spill over to other parts of the DRC, such as South Kivu, Maniema, and Katanga. Moving forward and drawing on some of the recommendations provided, various role players must work collectively and effectively to steer the country in the right direction.

References

African Union. (2023). Communique of the 1140th meeting of the PSC held at the level of the Heads of State and Government on 17 Feb 2023, on the situation in the Eastern Democratic Republic of Congo: Peace, Security and Cooperation Framework for the DRC and the Region and deployment of the Eastern African Community (EAC) Regional force. https://reliefweb.int/report/democratic-republic-congo/communique-1140th-meeting-psc-held-level-heads-state-and-government-17-february-2023-situation-eastern-democratic-republic-congo-peace-security-and-cooperation-framework-drc-and-region. Accessed 7 February 2024.

Azam, A., & Rehman, S. (2018). Assessing Human Needs Theory: An Approach to Conflict Resolution. *The Nucleus*, 55(3), 128–132.

Borowski, H., & Fuchs, I. (2012). Weak States and Terrorist Organizations: A Proposed Model of Intervention. *Minn. J. Int'l L.*, *21*, 177.

Christie, D.J. (1997). Reducing Direct and Structured Violence: The human needs theory. Peace and Conflict. *Journal of Peace Psychology*, 3(4), 315–332.

Danesh, H.B. (2012). Human needs theory, conflict, and peace. *The Encyclopedia of peace psychology*, *4*, 63–67.

Daudu, I.A., & Shulika, L.S. (2019). Armed Conflict in Africa: Examining Sexual Violence as an Instrument of War. *Journal of African Union Studies*, 8(1), 51–71.

Ekyamba, I. S. (2022). Assessing the Challenges of Armed Groups in the Democratic Republic of Congo's Kivu Region. *International Journal of African Renaissance Studies-Multi-, Inter-and Transdisciplinarity*, 1–18.

Falola, T., & Oyebade, A.O. (2010). *Hot Spot: Sub-Saharan Africa: Sub-Saharan Africa*. ABC-CLIO.

Hart, J. (2006). The Politics of "Child Soldiers." *The Brown Journal of World Affairs*, *13*(1), 217–226.

Jobbins, M. (2010). Community-Driven Civilian Protection in the DRC: Preventing Violence and Mitigating Harm. https://www.brookings.edu/wp-content/uploads/2012/04/Case-Study-DRC-Oct-2010-Mike-Jobbins.pdf Accessed 12 February 2022.

Kakhuta-Banda, F.B. (2014). *The use of child soldiers in African armed conflicts: A comparative study of Angola and Mozambique* (Doctoral dissertation, University of the Witwatersrand, Faculty of Humanities, International Relations).

Kiyala, J.C.K. (2018a). Child soldiers and restorative justice: Participatory action research in the eastern Democratic Republic of Congo. Cham: Springer.

Kiyala, J.C.K. (2018b). The impact of globalization on child soldiering: challenges and opportunities in the context of the Democratic Republic of Congo. *International Social Science Journal*, *66*(221–222), 271–284.
Lalbahadur, A. (2017). Improving SADC Responses to bring peace to the DRC. Johannesburg, South African Institute of International Affairs Policy Brief.
Lewis, B.R. (2019). *Hitler Youth: the Hitlerjugend in War and peace 1933-45*. Amber Books Ltd.
Lorch, J. (2020). Terrorism in Bangladesh: Understanding a complex phenomenon. *Asian Survey*, *60*(4), 778–802.
Matambura, I. (2017). Child Soldiers Recruitment and Use in the Democratic Republic of Congo: An Ethical Perspective. https://scholarcommons.scu.edu/cgi/viewcontent.cgi?article=1003&context=jst_dissertations Accessed 1 Apr 2022.
McCarroll, T.N.G. (2014). Uganda and the Democratic Republic of Congo: Children at War. *Global Majority E-Journal*, 5(2), 4–17.
Mlambo, D.N., & Adetiba, T. C. (2020). Post-1994 South Africa's peacekeeping and military intervention in Southern Africa, reference from the Democratic Republic of Congo (DRC) and Lesotho. *Journal of Public Affairs*, *20*(1), e1984.
Mlambo, D.N. (2021). Africanizing Rebel Insurgency in a Post-Colonial State and Ramifications for African Regional Integration: Insight(s) from the Democratic Republic of Congo (DRC). *Journal of Social Political Studies*, 2(3), 205–224.
Mlambo, V.H., Mpanza, S., & Mlambo, D. N. (2019). Armed conflict and the increasing use of child soldiers in the Central African Republic, Democratic Republic of Congo, and South Sudan: implications for regional security. *Journal of Public Affairs*, *19*(2), e1896.
Monyae, M.M.D. (2014). Learning to Lead: South Africa's Role in Africa: Lesotho, Zimbabwe, Burundi and the Democratic Republic of Congo. PhD thesis. University of the Witwatersrand, Johannesburg.
Muraya, J., & Ahere, J. (2014). Perpetuation of instability in the Democratic Republic of Congo: When the Kivus sneeze, Kinshasa catches a cold. *ACCORD Occasional Paper*, *2014*(1), 1–46.
Ndaguba, E.A., & Hanyane, B. (2019). Ndaguba, E. A., & Hanyane, B. (2019). Conceptualizing community economic development strategic framework for poverty alleviation: The case of Raymond Mhlaba Local Municipality. *Journal of Public Affairs*, *19*(2), e1929.
Neethling, T. (2014). Rebel Movements in the DRC as Sub-National Terrorists and Need for Appropriate Counter Initiatives. *The Journal of Transdisciplinary Research in Southern Africa*, 10(3), 393–408.
Newman, E. (2007). Weak states, state failure, and terrorism. *Terrorism and political violence*, *19*(4), 463–488.
N'Gambwa, K. (Ed.). (2011). *Leadership, peace, stability, and prosperity in the DRC*. US Institute of Peace.
Oloruntoba, S.O., & Muchie, M. (2018). *Innovation, Regional Integration, and Development in Africa: Rethinking Theories, Institutions, and Policies*. Cham: Springer.
Ortiz, R.D. (2002). Insurgent strategies in the post-cold War: the case of the revolutionary armed forces of Colombia. *Studies in Conflict and Terrorism*, 25(2), 127–143.
Oyewole, T.C. (2018). The Proliferation of Child Soldiers in Africa: Causes, Consequences, and Controls.
Prunier, G. (2010). *Africa's World War, Congo, the Rwandan Genocide, and the making of a Continental Catastrophe*. New York, Oxford University Press.
Richards, J. (2014). Forced, coerced, and voluntary recruitment into rebel and militia groups in the Democratic Republic of Congo. *The Journal of Modern African Studies*, *52*(2), 301–326.
South African Government. (2023). President Ramaphosa to Participate in the 10th High–Level Meeting of the Peace, Security, and Cooperation Framework for the DRC. https://www.gov.za/news/media-statements/president-cyril-ramaphosa-participates-10th-high-level-meeting-peace-security. Accessed 7 February 2024.
Stearns, J. (2012a). *North Kivu: the background to conflict in North Kivu Province of eastern Congo*. Nairobi: Rift Valley Institute.

Stearns, J. (2012b). Mai-Mai Yakutumba: Resistance and Racketeering in Fizi, South Kivu.
Thomas, R.M. (2003). *Blending Qualitative and Quantitative Research Methods in Theses and Dissertations*. California, Thousand Oaks.
UNICEF. (2007). The Paris Principles, Principles and Guidelines on Children Associated with Armed Forces for Armed Groups. https://www.unicef.org/mali/media/1561/file/ParisPrinciples.pdf Accessed 22 Feb 2023.
Vlassenroot, K. (n.d). Armed Groups and Militias in Eastern DRC. https://www.diva-portal.org/smash/get/diva2:610652/FULLTEXT01.pdf Accessed 8 March 2022.
Walsh, D. (2015). How a Human Needs Theory Understanding of Conflict Enhances the Use of Consociationalism as a Conflict Resolution Mechanism: The Good Friday Agreement in Northern Ireland. *Ethnopolitics*, *15*(3), 285–302.
Whitaker, B.E. (2010). Compliance among weak states: Africa and the counter-terrorism regime. *Review of International Studies*, *36*(3), 639–662.
Zuilkowski, S.S., Collet, K., Jambai, M., Akinsulure-Smith, A.M., & Betancourt, T.S. (2016). Youth and Resilience in Post Conflict Setting. *Human Development*, 59(2/3), 64–80.

Part III
Peace, Security and Sustainable Development

Chapter 11
Rethinking Peace: Some Reflections on The Imaginary Nature of Contemporary Peace Model

Norman Chivasa

Abstract In this theoretical article, the imaginary nature of the contemporary peace model is explored. Imaginary peace takes people beyond the boundaries of existing conflict and violence, by giving people something to strive for. As a method rather than a goal, imaginary peace gives societies options on how to modify or improve their crisis situations. Using a qualitative research approach through documentary analysis, this article reflects on the use of the imaginary peace in creating positive mental images that outlive conflict-ravaged societies. The analysis reveals that peace is not only limited to an experienced reality of peaceful co-existence, tolerance and nonviolent resolution of conflict, but is also an invention and creative process that originates in our imagination. The article concludes that societies need to utilise the imaginary peace through dreams and visions of a peaceful future in order to outlive crisis situations such as violent conflict. The strength of imagery peace is that it has the potential to generate commitment to achieving a better and more peaceful future, even in the face of violence in conflict-ravaged communities.

Keywords Conflict · Crisis situations · Imaginary peace · Utopia · Violence

11.1 Introduction

This article reflects on the imaginary nature of peace as a social practice devised by societies to cope with existing challenges such as ecological crises, climate-induced conflicts and violence, in co-creating positive images and visions of a preferred future, in order to outlive present realities characterised by agony and violent conflict. These positive images have the potential to generate commitment to achieving a better and more peaceful future (Boulding, 2001). Lederach (2005) argues that people live

Dr Norman Chivasa holds a Master's and PhD in Conflict Resolution from the university of KwaZulu Natal. He is currently a lecturer in the Department of Peace, Security and Society at the University of Zimbabwe. Email: normanchivasa@gmail.com; ORCID: https://orcid.org/0000-0001-5953-7855

J. C. K. Kiyala and N. Chivasa (eds.), *Climate Change and Socio-political Violence in Sub-Saharan Africa in the Anthropocene*, The Anthropocene: Politik—Economics—Society—Science 37,
https://doi.org/10.1007/978-3-031-48375-2_11

in relation to future aspirations, displaying positive visions that are dramatically different from those generated in conflict-ravaged societies. Albeit this reality, there is a tendency to think that imagery involves daydreaming and wishful thinking, yet in peace discourses there is evidence that peace has some utopian elements embedded in it.[1] Boulding (2001) posits that peace is something that happens in people's minds; as such it invokes positive mental images that are futuristic – a yearning for something better than the current state of affairs, characterised by ocean acidification, damaged flora and fauna, and global warming, amongst other major challenges. The point is that imagery is one of the social practices devised by societies to cope with existing challenges.

The imaginary nature of peace is that, although peace is attainable, there is still a yearning for something better and more fulfilling in the future. This seems to be the argument by Horner (2013), when she argued that peace is *utopian*. Another helpful example of the imaginary nature of peace is that all human societies are currently making efforts to address physical violence, ecological crisis, and environmental disasters in our midst (which is negative peace in the sense proposed by Johan Galtung); nonetheless, in the same human societies there is still that yearning for a good society free from all forms of violence, and with environmentally friendly practices, gender equality, and climate justice (which is also positive peace in the sense proposed by Galtung) (McCandless, 2007).

Imaginary peace falls under the realm of utopias – popularly known as aspirations, dreams or visions of a better future. They can also refer to practices that envision a better future (Jensen, 2016). In real life, when people are faced with catastrophic situations, they embrace utopias in order to cope with existing or looming challenges, but also to align their life patterns to the perceived status quo or visions of a better life. In biblical theology, when Jews in the Old Testament and Christians in the New Testament were faced with catastrophic situations, utopianism manifested in various cultural forms, one of which is popularly known as apocalypses. Such utopias were embodied in dreams, visions and fairytales. As argued already, peace in itself is future oriented in that it projects visions of a better future in light of existing conditions and potentialities (Boulding, 2001). Given that violence has invaded our social fabric, and ecological crisis such as plastic pollution is at our doorstep, the imaginary nature of peace presents alternative options to counter such crises. The battle against environmental conflict and violence, as well as climate wars in our societies, is not yet over. Hence, there is the need for dreams and visions for environmental user-friendly practices and a violence-free world. Whether such visions or dreams will be achieved or we may not live to witness their fulfillment, is not the objective of this article. The reasoning behind it is that the imaginary nature of peace is one of the options to counter current climate wars and environmentally-induced conflicts and violence. By implication, thinking about peace using this frame of mind is not wishful thinking, but rather such imagery is directed towards a possible peaceful future. Peaceful images

[1] This article is in series with the 'How peace is visualised? Perspectives from Seke District in Zimbabwe' published by the *African Journal of Peace & Conflict Studies*. https://doi.org/10.31920/2634-3665/2022/v12n1a5.

help communities to outlive the present realities characterised by climate-induced conflicts and violence, food insecurity, and climate wars.

Building on the notion of the imaginary nature of peace by Horner (2013), who strongly believed that peace is *utopian*, this article reflects on the importance of positive imageries of peace in a world that is faced with violence and its various forms. Basing on Horner's model of the utopian nature of peace, the researcher was inspired to begin to think of peace using two categories, namely, *the already peace* and *the not yet peace*. These phrases are an extension of the Christian eschatological phrases *'the already'* and *'the not yet'* view of the kingdom propounded by George Eldon Ladd, in which the kingdom of God is construed as both a present and futuristic reality. In my theological class in 2004, where I first learnt about *the already* and *the not yet* kingdom model, these phrases were made in reference to the kingdom of God. The *already kingdom* represented the present, or what is experienced in the now, or the present positive elements which are a replica of what is to come, while *the not yet kingdom* was visionary, what might be, a futuristic or positive state of being, or eternal bliss that everyone looks forward to experiencing. Thus, *the already* and *not yet kingdom* meant that the kingdom of God was a present reality, but not in its fullness.

While this article is not a theological thesis, I have distanced myself from contextual and eschatological debates surrounding George Eldon Ladd's theological phrases, limiting myself to the terminology *'the already'* and *'not yet'* themselves, for the purpose of putting peace into two broad categories, which I propose here as the *already peace* and the *not yet peace*. The *already* and *not yet* phrases provide valuable insight into the concepts of peace in this article, namely that peaceful co-existence, getting along and good relationships make up peace in the present (the *already peace*), while the *not yet peace* far exceeds peace in the present and remains an ideal. While bombs are dropping in Ukraine, that is not to say there is no peace in the world or that peace is nonexistent, as we still have dreams and visions that one day Ukraine and other parts of the world under the threat of war and other forms of violence will live in peace. This aspect is central to this article.

As an extension of Horner's argument, what this researcher calls the *already peace* is peace in the community understood in everyday life as having food in the home, gender equality, community development, cordial harmony, or peaceful co-existence. The *not yet peace* consists of images of tomorrow, for example, that the conflict in Ukraine will finally be resolved, climate justice will be achieved, or that Africa will achieve its aspirations. In other words, the *not yet peace* is more visionary – what might be in the future.

In academic discourses, peace has conventionally been classified as negative and positive peace, and these concepts have been mutually influential in coming up with the *already peace* and *not yet peace* model proposed in this article. It is, therefore, important to say that the negative and positive peace paradigms (McCandless, 2007), which are currently the ultimate goal of all peace efforts, were also instrumental in the development of the *already peace* and *not yet peace* models proposed in this article. Negative peace is akin to the *already peace* model, while positive peace is akin to the *not yet peace* model. As explained already, all societies are making efforts to achieve *already peace* whether as negative peace (absence of physical violence, climate justice, gender equality) or *not already peace* as positive peace (what is idealised as a good society) (McCandless, 2007). The creation of peace initiatives across the globe are efforts aimed at achieving a good society where every household has sufficient food to eat, and conflicts are resolved without the use of violence. Improved livelihoods, climate action, environmental user-friendly practices, and a reduction in cases of violence are a classic example of efforts towards achieving positive peace (the *not yet-peace* model). Positive peace relates to values such as "calm, harmonious relationships, mutual understanding and compassion" (McCandless, 2007: 93). This type of peace can be both a present reality and an ideal form, depending on communities. For example, in one country residents might be experiencing calm and mutual understanding, evidenced by not experiencing ongoing conflict; on the other hand, in another society such as Ukraine, there is a yearning for calm and harmonious relationships – that becomes imaginary peace; as such, positive images may potentially generate commitment to achieving a better and more peaceful future. Following from the reasoning above, this article reflects on the use of the imaginary nature of peace in the contemporary world ravaged by conflict and violence.

11.2 Methodological Note

To reflect on the imaginary nature of peace, this article utilised the qualitative research approach through documentary analysis. The qualitative research approach seeks to understand phenomena within their social and cultural context. Specifically, the paper examined how peace has been understood and presented in its binary form. The documentary analysis involved peer-reviewed articles and grey literature. A literature search was conducted using a variety of accessible databases such as EBSCOHost, Google Scholar, Elsevier, SAGE, Science Direct, Taylor & Francis, Sabinet, and African Journals Online (AJOL).

During the searches, keywords such as 'peace,' "negative peace," "positive peace," and "imagining peace" each produced 50 articles when paired with "peacebuilding." Thirty-four articles published between 1999 and 2023 were chosen for thematic analysis based on predefined criteria that included imagining peace, visions of peace, and utopia. A multidisciplinary approach was purposefully taken by first drawing insights from history, psychology, sociology, cultural studies, and social anthropology, and which were then supplemented by insights from other disciplines such as political science, peace studies, law, and sociology, to enlighten not only the scope of the subject under discussion, but also the analysis of the already and not yet peace models.

11.3 Findings and Discussion

11.3.1 Peace as a Binary Concept

Peace is a binary concept which involves a present reality (*the already* peace) and as an ideal state (the *not yet* peace). The *already* and *not yet* peace are not the first peace constructs ever to feature in peace discourses in its binary form. Johan Galtung developed a pair of peace constructs: *negative* and *positive peace*. The former refers to the absence of direct violence whereas the latter refers to conditions that strive toward the attainment of freedom, equality, and justice (Galtung, 1964, 1996). Based on the negative and positive peace paradigms, peace has become a framework for academic, politics, policy, and practice around the world (McCandless, 2007; Tilahun, 2015).

Scholars either define peace with a focus on the present reality dimension of peace or its ideal form. Those who define peace as a present reality often define it as a value-laden concept. For example, Fisher et al. (2000: 11) note that "peace is more easily recognized by its absence." By implication, when people experience direct violence, such as fighting or killing, they can be very sure that peace has been ruptured. Fisher et al. stressed the connection between peace and violence. They argue that peace is ruptured when violence takes precedence. Meanwhile, Gervais defines peace as a value-laden concept (2004). Values are culturally conditioned; this, therefore, suggests that there is a myriad of experiences of peace.

Other scholars maintain a middle way between understanding peace as a present reality and an ideal form. For instance, Sponsel defines peace "as a relative condition involving dynamic processes that are life enhancing..." (1996: 98). First, it is apparent that Sponsel embraces the binary forms of peace when he classified it 'as a relative condition' and 'peace as a dynamic process.' The phrase 'peace as a relative condition' gives the impression that peace lies in the eyes of the beholder, and is characterised by values which individual people and groups desire and aspire to as positive and worth the effort (Webel/Galtung, 2007). Peace as a dynamic process suggests that peace is elusive because it exists in an ideal form. Lederach (1997: 20) contends that "peace is a dynamic social construct." This suggests that peace is relative and open ended. In support of this, Horner (2013: 371) posits that peace is not a closed but an open-ended phenomenon. For her, the open-endedness of peace means that it "is a signifier, a pointer or indicator which serves only to point people to something greater than the term (peace) itself" (p. 370). By implication, even though peace is attainable, it produces images that prompt people to yearn for something better.

Some scholars have overemphasised the ideal form of peace. For example, Levitas' (2007) and Horner's (2013) discussions about the concept of 'imaginary peace', to refer to peace as a future or projected reality. There are three significant characteristics of imaginary peace which are discussed here. First, imaginary peace elicits critical thought on societies' lived experiences of conflict and violence, implying that peace is conceivable. Second, visualising peace sends a message of hope to our conflict-torn societies, indicating that peace is still possible in the near future. Third, imaginary peace suggests that if people can dedicate themselves to ideas of a peaceful future, those aspirations will one day become a reality. Accordingly, the ideal form of peace is espoused by Levitas (2007) and Horner (2013), who claim that imaginary peace plays a significant role in shaping people's attitudes and behaviours toward a desirable future. Richmond (2005) contends that even in the binary paradigm of negative and positive peace, peace is conceived as both a present reality and an ideal. The binary nature of peace is due to the fact that people live in relation to future aspirations, and display positive visions that are dramatically different from those found in conflict-ravaged villages (Lederach, 2005). These dreams for the future are crucial for peace theory because visions of a world free from conflict and violence inspire hope, which outlives communities ravaged by these scourges.

11.3.2 Imaginary Peace Takes People Beyond the Boundaries of a Crisis

Boulding (2001: 569) posits that "positive images of the future act as releasers of social energy in the societies holding such images..." As a result, images of a peaceful future cannot simply be dismissed as pipe dreams (Levitas, 2007; Chitando, 2008) or escapism; rather, these should be considered as one method of adopting a state of peace. As this article suggests, imagination is important in evoking thoughts of a better future and so aiding in the advancement of the peace agenda. One of the key promises of imaginary peace appears to be its focus on a future free of violence, societies free of hatred and sectarianism, while also expressing optimism in the potential for peaceful communities (Moore, 2006). Lederach (2005) and Reber-Rider (2008) posit that adopting imaginative frameworks as a means of overcoming violence has promise, in that they arouse dreams and hope that transcend humanity's violent history. However, even though critics may accuse imaginary peace as creating castles in the sky, Reber-Rider (2008: 83) argues that imaginary peace is critical in that it "gives people something to strive for; it gives them an alternative to current situations and options for how to modify or improve them". From her perspective, "images of the future are not permanent buildings with specified truths...they are like temporary scaffolding, readily adjusted and replaced, and they do not need to be fixed" (Reber-Rider, 2008: 83). Imaginary peace helps in the creation of images of peace, fosters optimism, and rekindles hopes for a brighter future even in the face of conflict-ravaged situations (Lederach, 2005).

11.3.3 Peace Begins from the Heart

Adolf (2009) classified peace into the three categories. The first is inner peace—how individuals develop ways to become peaceful and stay at peace with themselves.

The second is social peace – how groups become peaceful and stay at peace within themselves. The third and final category is collective peace – how groups become peaceful and stay at peace with one another. The same author observes that some people, whether as individuals, or within groups or communities, are able to testify about their experiences of peace (Adolf, 2009). Adolf (2009: 4), on the other hand, recommends that collective peace can only be achieved via inner peace. The mantra "peace begins with me and with all of us" explains Adolf's (2009: 243) model that inner peace is a prerequisite for social peace, which in turn is a prerequisite for collective peace. We are more interested with inner peace as visions and dreams that all occur at the individual level.

Inner peace, which we can translate as *peace beginning from the heart*, informs us that the *already* peace is not something 'over there', but something that must exist here and now—a present reality. Inner peace is an appeal to individuals that they should not wait for other people to bring peace to them; peace is a reality that has to be experienced at an individual level (Chitando/Mukenge, 2015). *Peace begins from the heart* is another way of saying '*peace begins with me*', a slogan initiated in Zimbabwe prior to the 2013 general elections. Thus, the phrase *peace begins from* ***the heart*** and the slogan '*peace begins* ***with me***', as Chitando/Mukenge (2015: 19) put it, both highlight "the role of individual responsibility in creating peace." In that sense, the '***me***' and '***heart***' in both the slogan and the phrase respectively refer to the 'self' or the 'I', which is the individual.

11.3.4 Individuals and Groups as Creators and Inventors of Peace

There is general agreement in that individuals and groups must consciously work to realise peace because peace does not fall from heaven (Chitando/Mukenge, 2015). By implication, peace is not something that happens on its own; rather, it requires initiatives, commitment and sustenance. Thus, people need to work for peace in order to taste, realise, achieve, and sustain it. Without initiatives and participation by individuals and communities, people do not build peace at all (Maruta, 2008). In the process of working for peace, catastrophic situations can prompt people to dream and envision a peaceful future (the *not yet* peace model).

Chivasa/Mutswanga (2014: 123) note that "peace is a positive concept with a number of positive aspects that give it a particular meaning...with positive aspects... [such as] happiness, harmony, love, justice and freedom that any person and culture relatively assert and highly esteem." As a result, to say that peace is positive is to say that it is a desirable entity and therefore is not something undesirable. Again, what is positive and undesirable vary from culture to culture. Usually, when people desire something, they put some effort into it in order to achieve the desired outcome. For instance, if peace involves justice, love or happiness, it suggests that people must consciously work to realise any of these values because that is the desirable state of affairs. This puts Chitando/Mukenge (2015: 19) into context when they say, "peace does not fall from heaven: it begins with individuals living here and now." By implication, people have the responsibility to create peace in their villages—no one else except them and them alone. All societies equally have the responsibility to create the peace that they desire.

Creating peace is not a preserve of the elites – people at all levels of the social stratum have this responsibility. However, the challenge with peace is that it lies in the eyes of the beholder because people have different criteria for what constitutes peace. Even though societies have different criteria for peace, the important thing is that peace is a public good across cultures and societies. Maruta (2008) notes that the survival of any individual, group, household, or community/village from the scourge of conflict and violence, is linked to how each of these entities responds to the peacebuilding call. By implications, without initiatives and participation by individuals and communities to create peace initiatives, peace cannot come about by itself. Maruta (2008: 8) goes further to say, "...unless we do these things for ourselves, nobody will come and do them for us, and we will continue to wallow in self-annihilating conflict, marginalization and poverty". In a nutshell, individuals and groups that have created peace initiatives, embraced environmentally user-friendly practices, and promoted climate justice should continue doing such practices for themselves.

11.3.5 Threats to Peace

The issue about threats to peace is suggested in literature as scholars such as Sponsel (1996), Lederach (1997) and Webel/Galtung (2007) believe strongly that peace is a dynamic process, which suggests that peace is elusive. The dynamic nature of peace implies that a state of peace is likely to change from time to time due to two major factors suggested here: first, by its very nature peace exists in an ideal form, and this makes it dynamic and therefore elusive. Second, peace is linked to the everyday interactions and social practices between individuals and groups, which exist in a state of flux due to constantly changing environments. It is within this reasoning that a threat to peace is conceived as a situation with the potential to provoke hostile relations and to destabilise peace (Vark, 2009). Hohe/Nixon (2003) posit that in almost all societies peace is a reality which exists, but in a state of flux. Ferdinando (1999) shares the same view that peace is construed as fragile across communities. In some societies it is explicitly believed that peace has possible threats. In my local African (Shona) culture, peace is perceived to be a gift sanctioned by God and the ancestral spirits, in which case the generality of the people, in particular traditional leadership, serve as custodians of peace.

Studies by Gelfand (1969) showed that the moral values of the traditional African (Shona) society of Zimbabwe were defined by ancestral spirits who frowned on incest, adultery, seduction, and many more vices. Consequently, traditional Shona people believe that the spirit world is the source and referent for peace on earth (Shumba,

1983). They believe that peace involves social harmony, unity and solidarity between humans, the environment, the spirits and God (*Musikavanhu*—creator of mankind). Accordingly in the traditional Shona society of Zimbabwe, the spirits and God sanction peace on earth for the benefit of mankind (Ferdinando, 1999; Haverkort et al., 2003). Mankind has the responsibility to preserve and secure social harmony, order and tranquility through compliance with imperatives such as *'do not kill a person'* and *'do not beat or scold your mother/parent'* to cite but a few. If for any reason a person violates any of the imperatives, it is believed that such behaviour disrupts the sanctioned peace and therefore invokes spirits to act against the villain. Spirits can cause the villain to get sick, have bad luck or suffer in any other devastating manner, as a way to put the villain back into line, as well as to communicate that peace has been disrupted (Ferdinando, 1999). To ensure that social harmony and friendship is sustained between spirits and humankind, traditional African (Shona) people endeavour to keep the spirits in a favourable mood by avoiding violating any of the imperatives. Accordingly, the Shona people of Zimbabwe believe that social harmony and friendship between spirits and surviving family members, which is peace by another name, can be fragile from time to time. This fragility results from offences committed by some individual members or groups in the community. In the Shona society of Zimbabwe, threats to peace include, but are not limited to, offences such as murder, adultery, or incest, as they are believed to bring about the disruption of peace (Ferdinando, 1999). To restore the disrupted peace, some reparations have to be made by the villain through some rituals (Chivasa, 2019).

Similarly, amongst traditional East Timorese peace is perceived as "order [that is] a system of value circulation" that has been sanctioned by the ancestors (Hohe/Nixon, 2003: 17). This sanctioned peace is disturbed when a crime is committed because:

> A wrongdoing is perceived as a disturbance of the system of value circulation. If an individual commits a crime or a social conflict occurs, the flow of value is interrupted and an imbalance in the socio-cosmic system is the result (Hohe/Nixon, 2003: 17).

In that sense, when a member of the traditional East Timorese has committed murder, violated a marriage custom, or any other misdemeanor, the existing order (peace) in the community is construed as under threat. In the East Timorese context, the primary threats to peace are murder, adultery, or incest (Ferdinando, 1999). For traditional East Timorese people, murder is conceived as a threat in that when a person kills someone, peaceable values such as love, kindness, tolerance, or mercy are replaced with violence, creating a chaotic situation in the society (Hohe/Nixon, 2003: 18).

To this end, it can be logical to argue that in traditional societies, threats to peace are considered a reality. Threats to peace point the finger to violence as the chief culprit. Violence is both a threat and antithesis to peace in that, at its core, violence does not secure happiness, health, rest, comfort, leisure, relaxation, cooperation, mutuality, tranquility, security, or safety to name but a few. Violence threatens the wellbeing of people and specific cultural values, which peace seems to secure (Bonta, 1996:

405; Webel and Galtung, 2007: 5). As such, innovative mechanisms and initiatives are needed to help promote peace.

11.4 Conclusion

The *already* peace (negative peace in the sense put forward by Johan Galtung) is the kind of peace we have been familiar with over the years. Imaginary peace is a vision of peace, what might be in the future- our aspirations and future images of peace in our societies. It helps communities to outlive the present realities characterised by climate change catastrophes, environmental crisis, and violent conflict, towards imagining a preferred future. Simply put, these positive images have the potential to generate commitment to achieving a better and more peaceful future. This article advances the notion that, while human societies are hopeful for imaginary peace (positive peace in the sense put forward by Johan Galtung), they must come up with mechanisms to sustain the *already* peace – the peace present in our communities. The creation of peace initiatives is a classic example of efforts by individuals and societies to take responsibility for their own peace and to sustain it in the present. This article concludes by saying that the contemporary world has come up with a broader framework for expressing the *already peace* and imaginary peace, which is popularly known as peacebuilding. Peacebuilding embraces both the *already* peace and *not yet peace* models proposed in this article. Accordingly, the processes involving individuals and groups coming up with visions and dreams directed at a violence-free world, are the *not yet peace* model by another name. Visions of a village without conflict, a relationship without squabbles, a community without offenders, a village without disrupted livelihoods where everyone is happy and all people having their needs met, are imaginative frameworks. They serve as means to overcoming conflict and violence, ecological crises, climate wars, and gender inequality in our societies. These visions should be emulated because the goal of imaginary peace is to arouse a dream, hope, and human imagination that transcend current crises situations.

References

Adolf, A. (2009). Peace: A world history. Cambridge: Polity Press.

Bonta, B.D. (1996). Conflict resolution among peaceful societies: The culture of peacefulness. *Journal of Peace Research, 33*(4), 403–420.

Boulding, E. (2001). Peace culture and social action. *Peace Review,* 13(4), 567–570.

Chitando, A. (2008). Imagining a peaceful society: a vision of children's literature in a post-conflict Zimbabwe. Discussion paper 40, Uppsala: Nordiska Afrikainstitute.

Chivasa, N., & Mutswanga, P. (2014). An Analysis of conceptions of peace among people with disabilities in Zimbabwe: a case of Harare-Chitungwiza urban. *International Journal of Humanities Social Sciences and Education (IJHSSE)*, 1(11), 122–134.

Chivasa, N. (2019.) *Kuripa Ngozi* as a conflict resolution model in Shona communities of Zimbabwe: a conceptual analysis. *Critical African Studies*, 11(2), 159–175. https://doi.org/10.1080/21681392.2019.1657028.

Galtung, J. (1964). An editorial. *Journal of Peace Research*, 1(1), 1–4.

Galtung, J. (1996). *Peace by peaceful means: peace and conflict, development and civilization.* London: Sage Publications.

Gelfand, M. (1969). "Medical" arthritis in african practice. *Central African Journal of Medicine, 15*(6), 131–135.

Gervais, M. (2004). The Baha'i curriculum for peace education. *Journal of Peace Education, 1*(2), 205–224.

Ferdinando, K. (1999). The Triumph of Christ in African perspective: a study of demonology and redemption in the African context, Paternoster Press.

Fisher, S., Abdi, D., I., Ludin, J., Smith, R., Williams, S., & Williams, S. (2000). *Working with conflict: skills and strategies for action.* New York: Zed Books.

Haverkort, B., Millar, D., & Gonese, C. (2003). Knowledge and Belief Systems in sub-Saharan Africa. In B. Haverkort, K. van Hooft & W. Hiemstra (Eds.), *Ancient Roots, new Shoots: Endogenous Development in Practice* (pp. 127–152). London: Zed Books.

Hohe, T. & Nixon, R. (2003). *Reconciling Justice, 'Traditional' Law and State Judiciary in East Timor*. Washington D.C.: United States Institute of Peace.

Horner, K.L. (2013). Peace as an event, peace as utopia: a re-imagining of peace and its implications for peace education and development. *Discourse: Studies in the Cultural Politics of Education,* 34(3), 366–379.

Jensen, U.J., (2016). Utopianism. In ten Have, H. (Ed.) *Encyclopedia of Global Bioethics*. Cham: Springer. https://doi.org/10.1007/978-3-319-09483-0_428.

Lederach, J.P. (1997). *Building peace: sustainable reconciliation in divided societies*. Washington, D.C.: United States Institute of Peace.

Lederach, P.J. (2005). The moral imagination: the art and soul of building peace. Oxford: Oxford University Press.

Levitas, R. (2007). Looking for the blue: the necessity of utopia. *Journal of Political Ideologies,* 12(3), 289–306.

Maruta, S. (2008). Hands-on community peacebuilding, democracy and development in Zimbabwe: Proceedings of the Symposium on community peacebuilding, democracy and development, held in Epworth near Harare on 22 October.

McCandless, E. (2007). Synopses of major concepts. In E. McCandless, A. K. Bangura, M. E. King & E. Sall (Eds.), *Peace research for Africa: critical essays on methodology* (pp. 83–109). Addis Ababa: University for Peace.

Moore, M.M. E. (2006). Imagine Peace: Knowing the Real, Imagining the Impossible," In J. McDaniel & D. Bowman (Eds.), *Handbook of Process Theology* (pp. 201–216). St. Louis: Chalice.

Richmond, O.P. (2005). The transformation of peace. New York: Palgrave.

Reber-Rider, L. (2008). Building cultures of peace in the world: One peace center at a time. *International Journal on world peace*, 73–88.

Sponsel, L. (1996). The Natural history of peace: the positive view of human nature and its potential. In T. Gregor (Ed.) *A Natural History* (pp. 95–125) Vanderbilt University Press.

Tilahum, T. (2015). Johan Galtung's concept of positive and negative peace in the contemporary Ethiopia: an appraisal. International Journal of Political Science and Development, 3(6), 251–258.

Vark, R. (2009). Terrorism as a Threat to Peace. *Juridica Int'l, 16*, 216.

Webel, C., & Galtung, J. (2007). *Handbook of Peace and conflict studies*. New York: Routledge.

Chapter 12
Inequality, Unemployment, and Poverty in the COVID-19 Era in Eastern Cape Province, South Africa

Tafadzwa Maramura, Peter Makaye, and Torque Mude

Abstract Just like other Third World countries (3WCs), South Africa is faced with the triple threat of inequality, unemployment, and poverty, a burden which, though inherited from apartheid, has been intensified by the COVID-19 pandemic. Using the case of the Eastern Cape Province, which was one of the COVID-19 hotspots, this paper examines the challenges of inequality, unemployment, and poverty in low-income areas during a pandemic. Using methodological triangulation of purposively sampled secondary data and physical observations, this qualitative paper argues that more than two decades into democracy, socio-economic development is still a far cry from what it should be in South Africa, particularly in previously marginalised demographics. Against this background, this chapter maintains that the unprecedented economic implications of the COVID-19 pandemic have widened the inequality, unemployment, and poverty gaps in the Eastern Cape. The chapter further exposes how the accompanying by-products of the inequality, unemployment, and poverty nexus have particularly afflicted the young population, as evident in the high rates of crime, drug abuse and teenage pregnancies in the COVID-19 era.

Keywords Colonial legacy · Entrepreneurship · Inequality · Poverty · Sustainability · Unemployment

---

Corresponding author: Dr Clementine Maramura Tafadzwa holds a PhD in Public Management and Governance from the North-West University, a Master of Public Policy, Honours Public Policy (Cum Laude) and a B. Soc. Sci (Cum Laude) from the University, of Fort Hare. Email: tmaramura@gmail.com

Dr Peter Makaye is a Zimbabwean academic who holds a PhD in Development Studies from Nelson Mandela University in South Africa, Master of Arts Degree in Economic History and Bachelor of Arts Honours Degree in Economic History. Email: pmakaye2002@gmail.com

Dr Torque Mude (PhD) is a Postdoctoral Research Fellow at the Centre for Africa-China Studies of the University of Johannesburg. His research engages African politics, international relations, governance, peacebuilding, international law and development. Email: mudetorque@gmail.com. ORCID: https://orcid.org/0000-0002-9885-6637

J. C. K. Kiyala and N. Chivasa (eds.), *Climate Change and Socio-political Violence in Sub-Saharan Africa in the Anthropocene*, The Anthropocene: Politik—Economics—Society—Science 37,
https://doi.org/10.1007/978-3-031-48375-2_12

12.1 Introduction

This chapter analyses the intersection between inequality, unemployment, and poverty during the COVID-19 pandemic in a typically rural province. Undeniably, their interconnectedness is obvious even to the ordinary observer, considering that Sen (1973) notes how poverty has been identified consistently in association with inequality and unemployment. Despite receiving considerable attention at global and national levels, the challenges of inequality, unemployment, and poverty have a proven resilience at all levels. In some cases, such as that of South Africa, these have worsened since the advent of democracy (Klasen, 1997; Meldrum 2021). This makes it imperative to carry out more research to grasp the historical, structural, and cultural underpinnings of the challenges to generate successful ways of understanding and confronting them. Therefore, this study proffers practical recommendations and action points for policy planning as well as programming to ameliorate the poverty, inequality, and unemployment nexus in the Eastern Cape province of South Africa (the Eastern Cape). The importance of addressing the nexus of the triple challenges in South Africa cannot be over-emphasised, considering the socio-economic and development challenges they pose if left unresolved.

12.2 Contextualising the 'Triple Threat' Nexus in South Africa

The triple threat of inequality, unemployment, and poverty has been the focus of several studies (Resha, 2021; Bernstein, 2019; Chibba and Luiz, 2011). These challenges have preoccupied many researchers because they affect the lives of the majority as well as the potential that these threats must destabilise the country if not addressed. Bernstein (2019) notes that higher job levels are concentrated with more of Whites than Blacks, and that amongst these, there are more White males than females. This confirms that, in middle management, Blacks, and women in general, are disproportionately under-represented. However, when it comes to the semi-skilled and unskilled levels, most of the workforce is Blacks at 39.3 % males and 29 % females, followed by Coloureds (m—7.1%, f—6.8%), Indians (m—2.3 % and f—2%), and then whites (m—5.6% and f—6.7%) (Bernstein, 2019: 295). The implications of these statistics are that employment at lower levels is matched by lower levels of remuneration, hence the predominant poverty levels amongst the black population.

Leibbrandt et al. (2007) and Westaway (2012) traced the historical origins of poverty and inequality in South Africa and emphasised the racial footprint undergirding these challenges. They appropriately note that poverty and inequality in South Africa can be traced to colonial and apartheid policies that favoured white people at the expense of Blacks. Leibbrandt et al. (2007) also analysed the effect of state transfer incomes into households as well as the effect of increasing real wages on inequality. The latter was found to be quite significant given that while wages were increased for some, others such as the low-skilled failed to secure employment, which exacerbated the gap between these groups of people.

Despite the focus that has been given to these three variables, there is still room to conduct more research because currently, poverty, unemployment, and inequality studies have some limitations. For example, despite the voluminous literature on the triple challenges, a few scholars have analysed the inequality, unemployment, and poverty nexus in South Africa, especially in the context of the global coronavirus pandemic. This could be attributed to several factors. One could be that several previous studies (Resha, 2021; Chibba and Luiz, 2011; Sen, 1973) looked at each of these factors from different perspectives and before the devastating effect of the pandemic on societies. Resha (2021) concentrated on how women and youths have historically been absent from the labour market and the vulnerabilities that they have faced within the labour market. This, in part, can be attributed to their general lack of competences to manipulate information and communication technologies (ICT) (Asongu and Odhimbo, 2020). Sen (1973) analysed how poverty could be the outcome of inequality and unemployment.

In the poverty, inequality, and unemployment nexus, the legacy of apartheid looms large, and South Africa has been susceptible to external shocks such as the 2008 global financial crisis due to its interlinkages with western economies (Chibba and Luiz, 2011). Understanding this background is integral as it creates solutions to these societal challenges since they have a socio-economic, political, and administrative bearing. The worsening of these challenges means that there is still need not only for more studies but also for better action and policies targeted at this nexus. Moreover, we cannot ignore the structural challenges that underlie the poverty, inequality, and unemployment nexus.

Enaifoghe et al. (2021) argue that the failure to discuss the effects of poverty, unemployment, and inequality nexus on rural livelihoods in the context of Covid–19 could be attributed to the fact that COVID–19 is a new phenomenon that began in 2019, spreading from China to other parts of the globe in 2020. As this pandemic continued to unfold, its exposure of structural challenges in South Africa is an ongoing discourse receiving scholarly attention. In explaining the causes of the poverty, inequality, and unemployment nexus in South Africa, scholars partly attribute this to the enduring legacy of apartheid-era policies, as well as post-apartheid governments' public service delivery and poverty alleviation policies (Chibba and Luiz, 2011; Klasen 1997). While South Africa spends more than 4% of its GDP on social welfare, Chibba and Luiz, 2011) argue that social grants only deal with the symptoms of the problem while failing to address the root cause of poverty in South Africa. Clearly, it is arguably reasonable that these grants do not address the structural challenges of the poverty, inequality, and unemployment nexus because they are mostly invested in consumption and are thus an unsustainable solution to the nexus.

12.3 Methodological Parameters

This study employed document-based research through the grounded theory approach, whose principle is that theory is constructed from data. Thus, data for this study were obtained from reports of organs of state, reports from non-governmental institutions, newspaper articles, peer-reviewed journal articles, policy papers, and books. Using grounded theory, abstract classifications on the 'triple threat' nexus were acquired from secondary data sources and broken down into conceptual categories as the researchers inductively theorised these dynamics in the Eastern Cape. Hence, secondary data sources were analysed to generate themes, concepts, and inferences on the trajectory of inequality, unemployment, and poverty in the Eastern Cape during the COVID-19 pandemic.

Descriptively, this study analyses the nature of the inequality, unemployment, and poverty nexus in the Eastern Cape. de Lima (2011) argues that descriptive research is concerned with explaining the inherent characteristics of a phenomenon being studied. Descriptive research was considered appropriate for this study because the characteristics of the 'triple threat' in the Eastern Cape constitute the "what questions" in terms of the inequality, unemployment, and poverty nexus, which is the premise of this type of research. To this end, the review of secondary data sources was aimed at developing inferences regarding the nature of the phenomenon being studied. The inferences were then used to generate theoretical views on the manner and extent to which inequality, unemployment, and poverty are interrelated in their manifestation and are compounded by the COVID-19 crisis in the province. Examining and thematically analysing key documents on the trajectory of inequality, unemployment, and poverty in South Africa, and in the Eastern Cape in particular, inferences were developed on the nexus between these variables. This is against the background that an investigation on the dynamics of the inequality, unemployment, and poverty nexus in Eastern Cape warrants amplification. Thus, this chapter constitutes an epistemic contribution to knowledge regarding the dynamics mentioned above, given the novelty of COVID-19 and limited studies that reflect the triple threat in a global pandemic. The next section theorises the inequality, unemployment, and poverty nexus in the Eastern Cape province.

12.4 Theoretical Considerations

The nature, causes, and effects of poverty are the interest of most poverty studies, but in the current study, it is imperative to also proffer suggestions on how the poverty, inequality, and unemployment nexus can be reduced. Theories of poverty can be grouped into 3 broad categories, namely behavioural, structural, and political. First, behavioural theories concentrate on individual behaviours as being driven by behaviours and culture (Brady, 2018). Secondly, structural theories emphasise both demographic and labour market contexts, which affect behaviour and cause poverty.

Political theories contend that power and institutions formulate policies that cause poverty, and these two also moderate the relationship between behaviour and poverty (Brady, 2018: 1). In other words, political theories view poverty as the outcome of power relations and the institutionalisation of collective decisions.

Structural factors state that capitalism is responsible for unleashing conditions that cause poverty, inequality, and unemployment (Sameti et al., 2012). Arguably therefore, capitalism promotes the gap between the 'haves' and 'have nots', thus creating a platform for inequality. The economic context which includes economic growth and development, industrialisation, and deindustrialisation, as well as spatial and skills mismatches, also contributes to poverty (Brady, 2018). Poverty becomes endemic among the lower classes, and unequal employment opportunities leave a trace of widespread unemployment, hence the reflection on the poverty, inequality, and unemployment nexus. The functionalist perspective on poverty holds that poverty has social, economic, and political functions in society (Davis and Moore, 1945). Capitalism needs the poor, from which it receives labour for unsustainable meagre wages, thus perpetuating the vicious cycle of inequality. To the political elites, the poor masses are the necessary base for their survival as they continuously hold onto power. Considering these clarifications, social stratification, therefore, serves many functions in society.

Economic theories argue that there is a general failure by the labour market to address the poverty, inequality, and unemployment nexus because the labour market does not provide sustainable employment opportunities for the lowly skilled or unskilled segment of the population (Chibba and Luiz, 2011). The argument is that the labour market caters for those who are educated or skilled while excluding the lowly or unskilled, thus exposing the latter into a vicious cycle of poverty. As a result, the poverty, inequality, and unemployment conundrum is therefore perpetuated. This then highlights the need for a pro-poor strategy that addresses the challenges emanating from the discriminatory practices of the labour market. Notably, this study argues and recommends that central and local governments need to address the triple threat by making efforts to turn the pendulum towards discriminated segments of society.

In the South African context, the social exclusion theory also plays a critical role in amplifying the reflection on poverty, unemployment, and inequality in the country. Sameti et al. (2012: 51) reinforce that social exclusion is defined as a situation in which a citizen is deliberately excluded from participating in the normal activities of citizens, whether he/she desires to participate. Through social exclusion, citizens are declined active socio-economic participation, thus limiting their access to resources (Plagerson and Mthembu, 2019). Considering this enlightenment in the context of South Africa, this can be attributed to both colonialism and apartheid, where resources were limited to a certain minority racial group while excluding majority races. The 1913 Natives Land Act deliberately excluded Africans from full participation in productive activities and, according to Hall (2014), most of the challenges underpinning the triple threat challenges in South Africa can be easily traced back to this piece of legislation. In an unfair distribution of resources, the Group Areas Act of 1950 divided state lands into distinct residential zones separating Blacks and Whites.

Another socially excluding policy was the 1950 Population Registration Act, which, in retrospect, shaped the long-term trajectory of the development of South Africa. Clearly, the residual effects of pre-1994 social exclusion dynamics are being experienced currently in South Africa, given its status of the highest Gini coefficient in the world, making it the most unequal society. Moreover, COVID-19 intensified these nuances.

12.5 The Trajectory of the 'Triple Threat' in Eastern Cape, South Africa

Using the Eastern Cape as a case study, this study further examines the interconnectedness of the triple challenges of poverty, inequality, and unemployment amidst the COVID-19 pandemic. This is not to suggest that this nexus does not exist elsewhere in South Africa, but the argument is that the Eastern Cape is within a rural context, and the province also exposes the disparities of resource distribution among provinces post-1994. The Eastern Cape was purposively chosen to be a theatre of segregation during apartheid and thus still suffers from the residual effects of colonial neglect and social exclusion. To reiterate the impact of social exclusion, this study magnifies how the long-term effects of the Natives Land Act, 1913, are still very much alive in the Eastern Cape, where, according to (StasSA, 2012), household poverty lies at 12.7% while unemployment lies at 47.1%. Moreover, the Eastern Cape is also the most unequal province in South Africa (StatsSA, 2021). It can thus be argued that the COVID-19 crisis compounded and exposed the structural trajectory that has contributed to inequality and high unemployment levels of poverty in the Eastern Cape (Kollamparambil and Oyenubi, 2021).

South Africa has the highest levels of inequality by international standards, with the combination of inequality and unemployment being almost unique to the Southern African state (Tregenna and Tsela, 2008). The Eastern Cape dilemma leaves a lot to be desired. Having remained a considerably rural province compared to other South African provinces, StatsSA (2021) argues how the economic slowdown due to the COVID-19 pandemic compounded inequality, unemployment, and poverty levels in this least developed province of South Africa. With strict COVID-19 lockdown measures and fear of exposure to the deadly pandemic, business became a mammoth task, resulting in economic stagnation. Subsequently, Enaifoghe et al. (2021) affirm how the COVID-19 lockdown contributed to food insecurity by rendering people's income-generating projects and occupations insufficient. StatsSA (2021) further confirms that coupled with the socio-economic implications of COVID-19, 75% of Eastern Cape residents have economically suffered and are still economically suffering disproportionately to the rich. Such disproportionate implications of COVID-19 can only be attributed to the existence of economic class differences emanating from the legacy of apartheid segregation and social exclusion in the Eastern Cape.

Central to the doctrine of transformative constitutionalism, one key question pertains to the impact of South Africa's transformative constitutionalism in ameliorating socio-economic and political upheavals in the Eastern Cape. Having gained political independence in 1994, the South African government embarked on a series of policies and measures to address the economic imbalances that existed because of the segregating policies of apartheid. At the core of these policies and measures adopted by arguably one of the most democratic states in Africa is the doctrine of transformative constitutionalism, which is premised on the idea of political, economic, and socio-cultural rights as well as the equality of South Africans irrespective of race, gender, and religion among others (Rapatsa, 2014; Hailbronner, 2017). Nearly three decades since the attainment of independence and transition to democracy in South Africa, provinces such as the Eastern Cape are still characterised by the marginalisation of especially the Black majority (Kollamparambil and Oyenubi, 2021). With uneven distribution of resources, poor service delivery, and unequal distribution of economic opportunities still proliferating, unemployment and poverty are still prevalent in the province being studied. As a result, the COVID-19 pandemic further widened the inequality gap and impoverished the already suffering segments of the Eastern Cape population (Casale and Posel, 2021).

The spatial inequality emanating from the legacy of colonial policies of discrimination has contributed to the inherent dynamics of inequality whose concomitant implications have led to rampant unemployment and poverty, with the COVID-19 pandemic increasing their impact. Looking into history, this problem sufficiently reflects the effects the 1913 Land Act, *Bantustans,* and the apartheid organisation of provinces, all of which have had a major impact on the contemporary issues of inequality, unemployment, and poverty in present day South Africa. Sadly, the adverse impacts of post-apartheid policies in South Africa, perpetrated by skewed agricultural patterns and weaknesses in public administration, have also contributed to the chain of causation which has culminated in the 'triple threat' in Eastern Cape. To further argue, by contextualising, the land question in this province is far from being genuinely addressed, particularly considering how large parts of the Eastern Cape are made up of former homelands of Ciskei and Transkei. Hence, the current high unemployment and poverty rates in the province may be linked directly to this historical economic imbalance. With the emergence of the COVID-19 pandemic, whose containment measures deprived people of their economic activities, the undertones of the 'triple threat' became magnified in every angle.

12.6 Reducing the Implications of the 'Triple Threat' in Eastern Cape: Looking Forward

Having exposed the nexus between the 'triple threat' in the Eastern Cape Province, it suffices to proffer suggestions for reducing these challenges. To enable sustainable development for all, there is need to put into perspective the nexus of the triple challenges facing the men, youth, and women in the Eastern Cape. The nature, causes, and effects of the triple challenges of poverty, unemployment, and inequality in the Eastern Cape need to be understood well. We hold that the nexus of the triple challenges requires a clear grasp of not only its causes but also its manifestations, so that tentative remedies can be put on the table. In essence, this may disconnect causes emanating from the past and moot strategies for improvement. Hence, the relevance of this nexus sits at the epicentre of people's socio-economically affected livelihoods, a matter that warrants detailed investigation. Further empirical studies of the Eastern Cape are necessary as it is one of the provinces that still suffer from these challenges in a big way.

This study adds to the existing conceptual and theoretical literature on poverty, inequality, and unemployment in South Africa by examining the nexus of these triple challenges in the context of COVID-19 that glaringly exposed the structural issues underlying the Eastern Cape society. Further investigations on these issues ought to be conducted. The emphasis on the remedies also needs attention because there has never been a dearth of explanatory literature on the causes and effects, but the fact that challenges still abound justifies the need for more studies on the strategic methods needed to improve rural livelihoods in this province engulfed by the poverty, unemployment, and inequality nexus.

12.7 Conclusion

In conclusion, the dynamics of the inequality, unemployment, and poverty nexus for the Eastern Cape province of South Africa and similar contextually rural provinces knotted by the triple threat during the COVID-19 pandemic illustrate how the average rural citizen is suffers daily. Grasping the effects of these challenges and how they are perpetuating the poverty cycle and ultimately inhibiting sustainable development for future generations should then enable the stitching together of a comprehensive tailor-made approach that seeks to dissolve previous social exclusion, whose fruits are still visible in present day South Africa. It suffices to indicate that a lot needs to be done to address these structural inequalities towards transformation, inclusivity, and sustainable development.

References

Asongu, S., & Odhimbo, N.M. (2020). Inequality and gender inclusion: Minimum ICT policy thresholds for promoting female employment in Sub- Saharan Africa, *Telecommunications Policy*, 44. Elsevier; 101900.

Bernstein, C. (2019). Digitisation and thriving within the contested terrain of intersections of gender, race, education and in the South African context. In M. Coetzee (Ed.), *Thriving in digital workspaces: Emerging issues for research and practice* (pp. 285-307). Pretoria: Springer.

Brady, D. (2018). "Theories of the causes of poverty", Forthcoming *annual review of sociology*, University of California, Oct 14.

Casale, D., & Posel, D. (2021). Gender inequality and the Covid – 19 crisis: Evidence from a large national survey during South Africa's lockdown, *Research in social stratification and mobility*, 71. Elsevier. 100569.

Chibba, M., & Luiz, J.M. (2011). Poverty, inequality, and unemployment in South Africa: Context, issues, and the way forward. *Economic Papers*, 30(3), 307–315.

Davis, K., & Moore, W. (1945). Some principle of stratification. *American sociological review*, 10(2), 242–249.

de Lima, D.V.M. (2011). Research design: A contribution to the author. *Online Brazilian Journal of Nursing*, 20(2), 1–19.

Enaifoghe, A., Maramura, T.C., & Ndlovu, M. (2021). Critically Assessing the Effect of COVID-19 beyond the Pandemic: The Social and Economic Impacts in Sub-Saharan Africa. *J.Hum Ecol*, 74(1–3), 40–48.

Hailbronner, M. (2017). Transformative constitutionalism in South Africa: Not only in the Global South. *The American Journal of Comparative Law*, 65(3), 527–565.

Hall, R. (2014), The Legacies of the Natives Land Act of 1913, *Scriptura*, 113, University of the Western Cape.

Klasen, S. (1997). Poverty, Inequality and Deprivation in South Africa: An Analysis of the 1993 SALDRU Survey. *Social Indicators Research*, 41, 51–94.

Kollamparambil, U., & Oyenubi, A. (2021). Behavioural response to the Covid-19 pandemic in South Africa. *PLoS ONE*, 16(4), 1–19.

Leibbrandt, M., Woolard, I., & Woolard, C. (2007). Poverty and inequality in South Africa: Post-apartheid developments in the light of the long-run legacy, Prepared for the IPC0 DRCLAS workshop. Brasilia, 11–13 Jan.

Meldrum, A. (2021). Explainer: What caused South Africa's week of rioting? https://apnews.com/article/africa-business-biden-cabinet-south-africa. Accessed 5 August 2023.

Plagerson, S., & Mthembu, S. (2019). Poverty, inequality and social exclusion in South Africa: A systematic assessment of key policies, strategies, and flagship programs. Johannesburg: Centre for Social Development in Africa.

Rapatsa, M. (2014). Transformative constitutionalism in South Africa: 20 years of democracy. *Mediterranean Journal of Social Sciences*, 5(27), 887–895.

Resha, G. (2021). *Prioritising youth and women: How the digital economy can protect South Africa's most precarious*. Johannesburg: The Brenthurst Foundation.

Sameti, M, Esfahani, R.D., & Haghighi, H.K. (2012). Theories of poverty: A comparative analysis. *Kuwait chapter of Arabian journal of business and management review*, 1(6), 45–56.

Sen, A. (1973). Poverty, Inequality, and Unemployment: Some Conceptual Issues in Measurement. *Economic and political weekly*, 8(31/33), 1457–1464.

Statistics South Africa. (2021). Statistical Release P9110.1 - Quarterly Financial Statistics of Selected Municipalities, Mar 2021. http://www.statssa.gov.za/?page_id=1856&PPN=P9110.1&SCH=72947 Accessed 19 November 2021.

Tregenna, F., & Tsela, M. (2008). Inequality, unemployment, and poverty in South Africa. Report of the Trade and Industrial Policy Strategies.

Westaway, A. (2012). Rural poverty in the Eastern Cape Province: Legacy of apartheid or consequence of contemporary segregationism? *Development Southern Africa*, 29(1), 115–125.

Chapter 13
Sustainable Development in South Africa in the Anthropocene in the Post-Covid-19 Era

Jean Chrysostome K. Kiyala and Diaku K. Dianzenza

Abstract South Africa's development strategy aligns with the Millennium Development Goals (MDGs) signed by leaders of 189 countries, which focuses on fighting poverty, promoting education, protecting the environment, and fighting hunger. The MDGs were superseded by the United Nations Sustainable Development Goals (SDGs) Agenda. Unfortunately, the global call to transition from coal dependency to renewable energy, climate change-induced disasters and the outbreak of Covid-19 have increased uncertainties about achieving the SDGs domestically and internationally. This article investigates both South Africa's challenges and its opportunities with regard to overcoming the developmental and economic setbacks ensuing from the global health crisis and the anthropogenic effects of climate change. Designed as a case study, relying on a literary research method and an interpretive paradigm, the findings indicate that South Africa has the potential to achieve sustainable development. However, the path to reach this requires: (1) reinvigorating the health infrastructure to withstand Covid-19's global challenges, (2) setting forth robust strategies and policies that uplift its citizens' wellbeing to prevent violent riots that destroy the economy, (3) revitalising its struggling economy and creating jobs, (4) transitioning from coal-dependent industry, economy and trade to an eco-friendly economy that grows from cleaner and renewable energy resources, (5) dealing with brain drain push factors, (6) pursuing the Policy Coherence for Development Framework that is supported by political buy-in so as to align the National Development Plan's (NDP) priorities with the 2030 SDGs Agenda.

Dr Jean Chrysostome K. Kiyala is a Senior Lecturer at the International Centre of Nonviolence, Faculty of Management Sciences at Durban University of Technology (South Africa), Associate Professor and visiting lecturer at the University of Bandundu, Evangelical University in Africa (Democratic Republic of Congo). Emails: JeanK@dut.ac.za; kljeanchrysostome@gmail.com; ORCID: https://orcid.org/0000-0002-9587-0230

Dr Diaku Dianzenza Kunsikila holds a Degree of Doctor of Philosophy (PhD) in Management Sciences Specialising in Public Administration-Peacebuilding. Email: diakudianzenza@gmail.com

J. C. K. Kiyala and N. Chivasa (eds.), *Climate Change and Socio-political Violence in Sub-Saharan Africa in the Anthropocene*, The Anthropocene: Politik—Economics—Society—Science 37,
https://doi.org/10.1007/978-3-031-48375-2_13

Keywords Anthropocene · Climate change · Covid-19 · Medium-term strategic frameworks · Millennium development goals · National development plan · National framework for sustainable development · Policy coherence for development · South Africa · Sustainable development goals

13.1 Introduction

While science is unequivocal about anthropogenic climate change in the Anthropocene, the world is lacking common ground to respond adequately to the global ecological crisis of our time. This is coupled with Covid-19 outbreak, the worse health crisis the world has experienced since the 1918 influenza epidemic. SARS-CoV-2 spread from Wuhan in China in 2019, and this situation posed serious challenges to the achievement of the 2030 UN Sustainable Development Goals (SDGs) (Wang/Huang, 2021, p. 2) in several nations. To social unrest and Covid-19 lockdowns, climate change-induced floods have come as another factor to slow attainment of the SDGs. For instance, Bonorchis (2022) observes that South Africa has to budget $34 million to meet the need of repairing two roads damaged by the floods of Apr/May 2022. It is reported that the Table Mountain wildfire, floods and 'severe weather' would cost SA billions in 2021 (Omarjee, 2021). However, there is still a necessity to assess the impact of climate change in South Africa, and to incorporate that with research that can help develop mechanisms to support a climate-resilient development scheme (Ziervogel et al., 2014).

Regarding the spread of SARS-CoV-2, with the help of advanced research and technology there have been concerted strategic preventive and curative responses to address the scourge of the Covid-19 pandemic. According to the World Health Organization, as of 29 Jul 2022, there have been 572,239,451 confirmed cases of COVID-19 globally, with 6,390,401 deaths; and as of 26 Jul 2022, a total of 12 248 795 623 vaccine doses have been administered (WHO, 2022b). However, several questions have been raised amidst the multifaceted challenges of our time about the future of sustainable development. Despite some progress being made by pharmaceutical firms striving to develop drugs that can effectively combat Covid-19, the development of vaccines, imposing travel restrictions, and developing health protocols to mitigate the spread of this deadly virus have given some hope that people can return to work while observing local and international health practices to prevent new infections. However, South Africa still faces multiple challenges to achieve the 17 SDGs. Firstly, the country was heading to recession (Kanu, 2020; Odeku, 2020) in the period leading the outbreak of the Covid-19 international health crisis. Secondly, socio-political instability has caused great losses and the destruction of several businesses. Adding to that, it was observed:

> By Apr 2020, all major credit rating agencies had downgraded South Africa's sovereign credit status to noninvestment ('junk') status, as the COVID crisis was unfolding, further complicating economic recovery, and raising the cost of government borrowing to address the crisis (Marquard et al., 2021: 12).

Thirdly, the 'brain drain' phenomenon has caused South Africa to lose skilful professionals (physicians, engineers, and health personnel, etc.). Fourthly, the overwhelming dependency on coal and liquid gas as its main source of energy could put South Africa's economy on the back foot due to the increasing international pressure to phase out coal and fossil fuels—the main drivers of climate change. Lastly, the numerous and multifaceted disruptions inflicted on international trade and various sectors of development by the surge in numbers of Covid-19 infections remain an impediment to sustainable development.

While South Africa's Government had already developed the National Framework for Sustainable Development (NFSD) at the service of the National Development Plan's (NDP) implementation, it is however, necessary to assess the emerging opportunities and the challenges that may impede the pursuit of sustainable development. Thus, South Africa's compounded development plan of action built around the NDP 2030, the NFSD and the Medium-Term Strategic Frameworks (MTSFs) could pave the way to forge economic growth in the aftermath of the Covid-19 outbreak. South Africa is no exception to the scourge of this pandemic that has severely hit the world's most powerful nations. South Africa has been one of the countries in Africa most affected by the pandemic, and from 3 Jan 2020 to 29 Jul 2022, there were 4,004,201 confirmed cases of COVID-19 with 101,977 deaths, according to the WHO (2022a); as of 17 Jul 2022, a total of 37,025,629 vaccine doses had been administered.

It is undeniable that the transformations induced by Covid-19 demand that the world rethinks development policies and strategies (Leach et al., 2021). The emerging challenges in this global health crisis and its impact on the sustainability of development are framed in the following questions: (1) To which extent could scientific advice and experiential resilience help frame innovative policies in the context of rigid conditions of uncertainties, even in centre-periphery power relations? (2) How do economies operate in the global environment of the Covid-19 crisis which has unearthed new shortcomings in the conventional paradigms of doing trade and conducting national and international affairs? (3) To what extent can emerging forms of politics and solidarity help in revamping citizen-state relations in confronting a pandemic such as Covid-19[1] (Leach et al., 2021: 1). (4) How does one address the challenges of the brain drain and socio-political unrest that often deals a blow to domestic economy and creates a climate of distrust on the side of investors and external development founding institutions? (5) To what extent can transitioning from coal to renewable energy create sustainable jobs and reduce the burden of greenhouse emissions on the ecosystem in the Anthropocene?

This article attempts to address these questions, while focusing on three major development programmes initiated at the domestic level (namely, the NDP 2030, the NFSD and the MTSFs) in view of promoting policy coherence for implementation of sustainable development frameworks to attain the UN SDGs 2030. It further explores developmental opportunities such as the economic and environmental benefits of transitioning from a coal-dependent economy to renewable and cleaner energy

[1] Questions inspired by Leach et al. (2021).

resources; such a transition is susceptible of creating jobs and enhancing sustainable development. At the same time, embracing the competing cleaner energy resources will alleviate the burden on coal and greenhouse emissions on our environment; this action is pursuant to Goal 13 of the 2030 sustainable development agenda.

This article is subdivided into seven sections: (1) the introduction followed by (2) design, methods and materials, (3) the rationale of the study, (4) a discussion of the UN 2030 SDGs, (5) South Africa's Sustainable Development agenda, (6) resilience in the Covid-19 pandemic era, (7) recommendations and finally, (8) the conclusion.

13.2 Design and Methodological Approaches

Designed as a case study based on literary research (Correa/Owens, 2010), this article uses South Africa's context to explore and examine, in the light of the current international crisis related to the Covid-19 pandemic, what strategy and policy-framing approach should be utilised to achieve the SDGs. Most of data is qualitative and consists of literary sources (online and library) (Towheed, 2010). The qualitative content analytical method (Frey et al., 1999) was applied to the texts used as units of analysis, because we were interested in understanding the meaning conveyed by the data, we gathered to respond to the questions we asked above. This inquiry employs an interpretative paradigm—the content analytical method—to understand and interpret data collected. The paper argues that South Africa has the potential to attain sustainable development because of its infrastructure, its resilient economy, its vaccination scheme, and policy coherence adopted across government departments to meet the priorities of the NDP 2030 and the SDGs 2030. The limitations of the findings are explained by the changing nature of SARS-CoV-2 and the unpredictability of its mutations, and the future that lies ahead of nations—and even the future of the SDGs. Finally, different theoretical frameworks such as development ownership paradigms (NDP, NFSD, MTSFs) and PCDF will guide this inquiry by creating a ground for policy adaptability and adjustments to mitigate the total collapse of world economies and the achievement of SDGs in South Africa, and the rest of the world. These theoretical frameworks are not discussed individually, but rather incorporated in the arguments and integrated into the discussion of policies, priorities, and sustainable development.

13.3 Rationale

The rationale of this study is obvious: to improve development policies aimed at helping South Africa meet the SDGs. It is urgent to rethink policy coherence between local development priorities and the global development agenda proposed by the UN, in light of the devastating effects of the pandemic and the extent to which it has crippled powerful nations. This study is significant because the relation that is established between the COVID-19 crisis and the SDGs, and the linkages between the two have emerged as an ongoing field of research in sustainable development (Hörisch, 2021). This scholar notes: "The pandemic has been found to severely threaten the achievement of the SDGs, while opportunities concerning selected SDGs can also be found. In this regard, the author identifies patterns, concerning which types of SDGs opportunities or respectively, threats exist" (Hörisch, 2021: 877). In addition, research in sustainable development is useful in supporting efforts engaged in by the UN and member states to attain the 2030 SDGs; it helps define performance indicators to measure the degree of progress by individual countries towards meeting the SDGs.

Following on from this argument, research is useful to help define the urgencies and the sustainability of human activities, and to assess the pressure that those activities put on human natural resources and the natural environment, especially as human ecology becomes a burden on planet Earth (Leal Filho et al., 2018). Similarly, inquiries on sustainable development can no longer evade the demands for climate action (Goal 13); consequently, a series of novel strategies for development is required at this time. Any development endeavour should be mindful of future generations—a responsibility that relies upon science and think tanks. This view underscores the need for reconceptualising sustainable development in the Anthropocene, as argued by (Griggs et al., 2014); Lim et al. (2018). For instance Griggs et al. (2014) suggest an integrated developed model built upon the SDGs, but that also takes into account development needs and environmental considerations ; it entails a comprehensive framework of goals and related targets, which maintains that it is possible and necessary to develop integrated targets associated with the following: food production, generating energy, and supplying water, while pursuing ecosystem services goals as well (Griggs et al., 2014). In a similar vein, Carr et al. (2021: 1) aver: "Sustainable development requires knowledge of trade-offs and synergies between environmental and nonenvironmental goals and targets". Striking such a balance is essential to meet the needs of humanity's wellbeing today, without sacrificing future generations' wellbeing, and the wellbeing of our natural environment.

Lastly, development theory in a time when progress is struggling (Pieterse, 2010) requires innovative strategies and policy coherence (Cloete, 2018; Fourie, 2018; National Science and Technology Forum, 2018) in order to enhance integral sustainability. Those concerns make this study a very important one.

13.4 The Anthropocene

The concept of the Anthropocene was postulated by Paul Josef Crutzen (1933–2021), the co-laureate with F. Sherwood Rowland and Mario J. Molina of the Nobel Prize for Chemistry in 1995 for work on the ozone layer (Lelieveld, 2021). Lelieveld writes in his obituary of Crutzen:

> At a conference in Cuernavaca, Mexico, in 2000, Crutzen stood up and proclaimed that we live in the "Anthropocene". The term immediately caught on and stimulated discussion in many disciplines. The "age of humans" is now considered to have begun in the mid-twentieth century, as the exploitation of the planet's resources accelerated (Lelieveld, 2021: 29)

The Anthropocene is explained as a new geological epoch, during which human activity appears to affect the environment quite significantly, from the "Earth's major biogeochemical cycles to the evolution of life" (Lewis/Maslin, 2015: 172). Human actions have ended the Holocene Era to trigger the new geologic epoch that places homo sapiens, the only current member of the genus Homo, at the centre of this Epoch's name and ramifications.

In a similar vein, Zalasiewicz et al. (2010) support the notion that humankind's influence on nature is not new, speaking of the Anthropozoic Era, which, according to Stoppani (1873), represents the period when the first traces of human presence are seen on Earth (cited in Rull, 2020). The designation of a geological time with the "human" as the core noun, contrasts with the customary practice. The traditional determinants of a new epoch when dividing geological-scale time are normally "global-scale changes to Earth's status, driven by causes as varied as meteor strikes, the movement of continents and sustained volcanic eruptions" (Lewis/Maslin, 2015: 171).

However, for the Anthropocene, it is human activity that has globally become the dominant driver of the most contemporary environmental alterations (Billings, 2014; Lewis/Maslin, 2015), including other human-prompted repercussions on the Earth's geology and ecosystems, most notoriously anthropogenic climate change and related global warming. This latest determinant of geological time has grown an expanded interest in the Anthropocene in Earth systems science (Lepori, 2015), in Peace Ecology (Brauch, 2016; Brisman, 2016; Kyrou, 2007), eco-theology (Chryssavgis, 2014; Kiyala, 2022; Pope Francis, 2015), eco-justice (Mueller/Tippins, 2015; Tippins/Britton, 2015), etc.

Several studies suggest that the beginning of the Anthropocene coincides with the advent of the Industrial Revolution, due to the incremental and accelerating use of fossil fuel, twinned with rapid societal transformations that presaged an unusual, critical and unique occurrence in human history (Lewis/Maslin, 2015; Ruddiman, 2013; Steffen et al., 2011). Ruddiman (2013) situates the start of the Anthropocene in the Industrial Revolution (around 1850), a period marked by observable large-scale effects of human activity on planet Earth. According to the industrial view, most significant anthropogenic impacts on this planet have occurred since the early industrial era (~1850), while the early anthropogenic view dates such large effects thousands of years earlier (Ruddiman, 2013). The debate about the time scale of the Anthropocene is beyond the scope of this chapter, which focuses on the way in which the recent and ongoing deterioration of atmospheric conditions and climate change affect sustainable global development, especially locally, in South Africa. However, it is worth mentioning that the signs of human-induced change of the geological time that bear the stem "Anthropos" (human), are "forest clearance (and related land use), emissions of greenhouse gases (CO_2 and CH_4), and effects on global temperature" (Ruddiman, 2013).

Lepori (2015) describes the Anthropocene as revolutionary; he writes: "we live now in a geological epoch defined by Homo sapiens. That is to say, the human species has radically reshaped the Earth and its systems to the point that they no longer resemble those of the prior epoch, the Holocene" (Lepori, 2015: 103–104). However, his argument diverges from the generalisation of responsibility, while bringing to light the impact of political-economic structures in climate change. He argues: "[..] the Anthropocene discourse demonstrates the importance of critical theory—the revelation of power relations, struggle, and difference (of interests, outcomes) that are concealed on the surface of thought and practice—to the study of environmental politics" (Lepori, 2015: 107). This scholar advocates a better understanding of the Earth system and ecosystem boundaries to avoid pushing beyond these, since this has resulted in the actual environmental crisis. He commends the notability of "Earth system science" for "its attempt to know the entirety of the Earth 'system of systems,' and its depiction of humanity as a unit within the system" (Lepori, 2015: 108).

The works of scientists on planetary boundaries can be very impactful in mitigating further anthropogenic causes of climate change (Rockström et al., 2009; Steffen et al., 2015). The political-economic responsibility is vital to maintain planetary boundaries, especially in agricultural sectors (Campbell et al., 2017). This point leads to rethinking sustainable development in the Anthropocene, as most of the 17 Sustainable Development Goals (SDGs) are achievable mostly via the agricultural sector, which is paradoxically a multiplier factor of climate change.

The Anthropocene raises many questions about the ability of *homo sapiens* ("wise human") to inflict such tremendous damage to the environment, in the name of progress, and technological and scientific advancement. Tables 13.1 and 13.2 give an indication of how various human-driven activities have affected the ecosystem and threaten the survival of biodiversity.

The indices of what is attributed to humanity's consequential activities that are deemed responsible for atmospheric transformations, which have wounded the ozone layer, are scientifically demonstrable, causing climate change and global warming, and endangering the ecosystems and biodiverse systems. The itinerary of the Industrial Revolution stands as clear evidence of the diverse ramifications of compounded technological and industrial occurrences that are making the Anthropocene the most consequential Epoch of all geological times.

It is apparent that the relevance of research in the Anthropocene has grown incrementally because of the existential challenges that climate change is posing to ecosystems and biodiversity (Dudgeon, 2019), and this demands, more than ever before, novel adapted policies which aim to mitigate the degradation of ecosystems, support the resilience of biodiverse systems and develop structural frameworks to better manage and preserve biodiversity (Seddon et al., 2016).

The area concerned with the Anthropocene that is discussed in this chapter is sustainable development. The linkages between these two disciplines are demonstrated by several scholars such as Reid et al. (2017), Lim et al. (2018) and (Dudgeon, 2019). How do we attain sustainable development goals without further deepening the ongoing climate change crisis?

Table 13.1 Major transformations during the industrial era

Innovation	Environmental consequence
Mechanized agriculture	Farming in prairies and steppes
	Deep-soil erosion in some areas
Damming of rivers	Siltation of catchments behind dams
Mechanized earth-moving	Greater than all natural processes
	Ecosystem fragmentation leading to species loss
Industrial (nitrogen-rich) fertilizers	Runoff to coasts, algal blooms
Industrial power plants	Release of sulfates to atmosphere
	Acidification of lakes, streams, rivers
Use of fossil fuels	CO_2 emissions
	30% increase in ocean acidity
Use of CFCs for refrigeration[a]	Creation of ozone hole
Mechanized fishing methods	Depletion/collapse of primary fisheries
Mechanized irrigation	Human use of >50% of river water; pollution of many rivers
Mechanized pumps	Depletion of groundwater

[a]CFC, chlorofluorocarbon

Table 13.2 Major characteristics of the three industrial revolutions

	British industrial revolution	Second industrial revolution	ICT revolution
Period	1760–1850	1880–1913 (?)	1990 (?)–
Macroinventions (General purpose technologies)	Cotton spinning, water power, steam originc, railway, machine tools	Electricity, steel products, heavy chemicals, automobile, aircraft	Semiconductor, compeer, telecomrnunication equipment
Core input	Cotton, iron, coal	Steel, copper, oil, synthetic materials	Chips, knowledge (?)
Constellation of technical innovations	Water- and steam-powered mechanization of industry	Electrification and moiorization of industuy and transport	Computerizatiian of the entire economy
Organizational changes	Separation of the household and product ion	Giant firms, Fordism	Teams, networking, work al home
Firm organization	Factory	M-form	Project-based form
Internal organization of firms	Strict supervision, control and discipline	Delegation of decision-making rights to division	Empowerment, dclayering, teams with responsibility

Source Kapás (2007, p. 26)

13.5 UN 2030 Sustainable Development Goals Agenda

The SDGs evolved from the Millennium Development Goals (MDGs) signed by the leaders of 189 countries gathered at the United Nations headquarters in Sep 2000; this was a historical Declaration, in which they expressed their commitment to achieving a set of eight measurable goals, ranging from the eradication of extreme poverty and hunger to promoting gender equality and reducing child mortality, and growing a global development partnership. The MDGs consisted of the following eight goals: eradicate extreme poverty and hunger (Goal 1), achieve universal primary education (Goal 2), promote gender equality, and empower women (Goal 3), reduce child mortality (Goal 4), improve maternal health (Goal 5), combat HIV/AIDS, malaria and other diseases (Goal 6), ensure environmental sustainability (Goal 7), and develop a global partnership for development (Goal 8). It was the Organization for Economic Co-operation and Development (OECD), which, in 1996, as a result of the 21st Century Policy Paper initiated the 8 MDGs (Scheyvens et al., 2016); these were finally approved by country leaders at the Millennium Summit in 2000. The MDGs were the first incontrovertible growth partnership agenda between the industrialised world and emerging nations (McArthur/Rasmussen, 2018). As part of development process, the MDGs have been superseded by the 2030 SDGs.

The SDGs were officially introduced at the UN Conference on Sustainable Development in the city of Rio de Janeiro in 2012; these aim at alleviating poverty, protecting the planet, ensuring the enjoyment of peace and prosperity by the whole of humanity, and addressing the imperative ecological, political and financial challenges that confronted nations. The Rio summit succeeded in approving 17 SDGs that are to be achieved by 2030. The SDGs function as guidance charter and stimulus to help countries move towards workable growth (Le Blanc, 2015: 176–178). They stand as an overall strategy to encourage sustainability in most countries. On one hand, part of the task attributed to the 2030 Agenda was to pursue 17 goals and meet 169 targets based on 247 indicators—though only 160 of them are available (Goal Tracker South Africa, 2019). The monitoring process also targeted the 17 SDGs in the areas of financial, ecological and communal issues (Dlouhá/Pospíšilová, 2018). On the other hand, the SDGs function to promote sustainable economic growth; they are focused on industrial revolution to attain fruitful employment, and decent work for women and men by 2030. Sustainability is now comprehended as an all-inclusive methodology that targets the financial, societal, and biophysical setting instead of it being limited to a *green agenda* (Amba, 2010). Unfortunately, achieving the SDGs has been slow; this was pointed out by the UN Secretary General (United Nations, 2019a), although the UN State members demanded a swift implementation of this international development agenda (United Nations, 2019b).

We should point out that the execution of the SDGs is now challenged by the Covid-19 pandemic and a myriad of issues that affect the ecosystem which is degrading due to unconcerned and uncaring human activities in our "common home", the Earth (Patriarch Bartholomew, 2021; Pope Francis, 2015). Several studies support the necessity to envisage and frame development policies and priorities in a synergistic approach that acknowledges and safeguards the interrelatedness of the SDGs and targets.

Sustainable development policies should be based on thoroughly-studied and -evidenced impacts of some SDG targets on various sectors in order to adopt policies that are grounded in mutually beneficial outcomes for environmental and non-environmental goals (Carr et al., 2021). These scholars systematically reviewed the literature on the effects of the SDGs' targets on forest ecosystems; they picked out 63 targets that could potentially be beneficial (targets connected to social progress and wellbeing), harmful (target linked with energy and infrastructure) or show mixed impacts of impacts forest ecosystems (Carr et al., 2021). They argue: "Institutions seeking to help achieve one or more non-environmental SDG targets must remain aware of the implications of their actions for natural biological systems and resources (illustrated here in the case of forests)" (Carr et al., 2021: 9).

13.6 South Africa's New Development Plan 2030 and the UN SDGs 2030

South Africa has been working to ensure that there is policy coherence between the NDP and the SDGs, and the country initiated the five-year cycle Medium-Term Strategic Frameworks (MTSFs) as an instrument for the implementation of the NDP and to ensure policy coherence across government departments (PCSD) for this purpose (National Science and Technology Forum, 2018). To proactively evaluate how the SDGs have been implemented around the world, innovative materials have been used, such as the "Global Reporting Initiative", the "SDG compass" as well as the "SDG industry matrix" (Blasco et al., 2018). South Africa was able to report on 128 (64%) of the 199 reportable indicators (Statistics South Africa, 2019). In addition, the country reported on an additional 26 indicators that were not part of the original set of SDG indicators (Statistics South Africa, 2019). However, generally, reporting on SDG implementation has been problematic in several countries in Sub-Saharan Africa; this is an impediment to assessing the progress that individual countries have made on achieving each target goal.

To illustrate this situation, Erin and Bamigboye (2021) evaluated and analysed the reporting of SDGs in Africa, and the findings reveal that there is a very low disclosure of the indicators of each SDG target from a list of 80 firms, with the exception of South African firms. Several firms seem not concerned nor interested in tabling data on SDG activities; the same research survey indicates challenges with voluntary disclosure, non-existence of management commitment, lack of regulations to enforce such reporting, and there are also costs involved for low SDG disclosure by the selected African firms (Erin/Bamigboye, 2021). In a similar vein, Ali et al. (2018) observe that South African companies show very low interest in adopting UN SDGs. The lack of implementation of the UN goals on Quality Education (Goal 4),

Climate Action (Goal 13), and Life Below Water (Goal 14) in the BRICS countries (of which South Africa is one) is concerning; these do not feature prominently in the vision and mission statements of several companies in BRICS countries, according to Ali et al. (2018). These are critical goals that should feature in each nation's development priorities, especially inasmuch as climate action is concerned.

Evaluating the effects of SDGs should refer to various elements to prove the effectiveness of the ground-based outcomes, and it is important to highlight that different government partners and stakeholders, including the private sector, are involved in the SDGs via the NFSD that supports the implementation of NDP 2030. Therefore, among the emphasis considered to gauge the previous MDGs and the new SDGs is included an identification of 1–6 agenda goals from the MDGs and the 7–17 goals of the new SDGs (Environment Statistics Section of the United Nations Statistics Division (UNSD), 2014). Goal Tracker South Africa (2019) highlights some improvement as far as Goal 1 and Goal 4 of the SDGs are concerned, between 2002 and 2017; during this period, there were some improvements with regard to school-going learners aged 6–18 years: school attendance increased from 91.3 to 96.0%, and the difficulties in accessing text books decreased from 24.3 to 5.9%; the rise in school attendance was stimulated by the implementation of the "no-fee" policy and the percentage of learner beneficiaries of this policy increased from 0.3 to 64.8% by 2017. Learners attending public schools in the Limpopo, Free State and Eastern Cape Provinces were the major legatees of the "no-fee" policy (Goal Tracker South Africa, 2019).

However, Goal 3 still remains a great concern in South Africa, since a considerable percentage of the population relies on the public health system to attend to healthcare needs, coupled with low levels of access to medical aid across all population groups, but particularly among black Africans (Goal Tracker South Africa, 2019). Regarding Goal 1, the World Bank Group (2020) found that South Africa is "one of the most unequal countries in the world with Gini index at 63 in 2014/15"; inequalities are "high, persistent, and [have] increased since 1994" (World Bank Group, 2020); there is polarisation of high levels of income; there exist also high levels of chronic poverty, "a few high-income earners and a relatively small middle class" (World Bank Group, 2020). It is reported that South Africa has made progress in reducing poverty since 1994, but the progress achieved on poverty alleviation moved backwards between 2011 and 2015; this reversal has imperilled some of the gains made since 1994 (World Bank Group, 2020). The World Bank Group (2020: n.p.) remarks:

> Approximately 55.5 percent (30.3 million people) of the population is living in poverty at the national upper poverty line (~ZAR992) while a total of 13.8 million people (25 percent) are experiencing food poverty. Similarly, poverty measured at the international poverty lines of \$1.90 and \$3.20 per person per day (2011 PPP) is estimated at 18.9 percent and 37.6 percent in 2014/15, up from 16.6 percent and 35.9 percent in 2010/11, respectively.

Regarding the agricultural sector, through which Goal 2 can be reached, the regional statistics reveal that agriculture in Sub-Saharan Africa (SSA) creates about 20–60 % of the Gross Domestic Product (GDP) but also uses 54% of the working force in the region where 70% of countryside households depend on this sector, according to the OECD-FAO (2022). It appears from the above statistics that South Africa should work harder to meet the priorities of the NDP and Goal 1 of the UN Sustainable Development Agenda 2030; obviously this requires political will. The country has made some progress, but further development setbacks as a result of the Covid-19 pandemic could hinder human development (Dube/Nhamo, 2021). For instance, the tourism sector requires collaboration with diverse stakeholders, including academics (Dube/Nhamo, 2021), and the aviation sector similarly remains in need of revitalisation in the aftermath of the Covid-19 pandemic (Dube et al., 2021).

The key continental development scheme espoused by South Africa was the New Partnership for Africa's Development (NEPAD) spearheaded by former President Thabo Mbeki. Four major locally-developed programmes are the National Development Plan (NDP) 2030, The National Framework for Sustainable Development (NFSD), also referred to as the National Strategy for Sustainable Development and Action Plan (NSSD 1) 2011–2014 (South Africa Department of Forestry Fisheries and the Environment, 2011), and the MTSFs, which are described as the NDP's vehicle (Fourie, 2018). The 2014–2019 MTSF was initiated to ensure that the activities and planning of South African government departments aligned to the NDP (National Science and Technology Forum, 2018).

Before the SDGs, South Africa had identified nine challenges facing it (The Republic of South Africa National Planning Commission, 2011). The NPC sought to tackle poverty and inequality as a part of the journey to 2030 (Fig. 13.1) so it launched the NDP which was conceived as a guiding document for South Africa's ownership of developmental policies and priorities.

The NDP was adopted by the South African Government as a "long-term vision and plan for the country" and "the blueprint for tackling South Africa's challenges" (National Science and Technology Forum, 2018: 2). Nine priorities were listed in the NDP, along existing challenges facing the country (Figure 13.1). A close examination of the NDP shows that its goals are consistent with the SDGs. To achieve the NDP's goals and SDGs, South Africa relies upon its human capital and political will; this vision was elucidated by the Republic of South Africa National Planning Commission (2012: 14):

> South Africa can realise these goals by drawing on the energies of its people, growing an inclusive economy, building capabilities, enhancing the capacity of the state, and promoting leadership and partnership throughout society.

Notwithstanding national initiatives to attain SDG 1 and 2, the NDP is also challenged by the poor and uneven education system with disparities between poor and well-off communities; thus, all the other SDGs ought to feature in the NDP so that robust policies and strategies are put in place to achieve this. Besides, Goals 7 (Clean water and sanitation) and 13 (Climate action) remain the stumbling block for South Africa's growth, especially because the country depends heavily on fossil fuels such as coal which are the biggest drivers of climate change. A report by the South African Department of Mineral Resources and Energy predicts that: "About 77 percent of South Africa's primary energy needs are provided by coal. This is unlikely to change

Fig. 13.1 Nine challenges facing South Africa and nine NDP 2030's priorities. *Source* National Science and Technology Forum (2018: 3)

significantly in the next two decades owing to the relative lack of suitable alternatives to coal as an energy source" (South Africa Department of Mineral Resources and Energy, n.d). However, the NDP 2030 encourages the transition from fossil fuel sources of energy to a low-carbon economy (Statistics South Africa, 2020). Statistics South Africa (2020: n.p.) states:

> Full-time employment decreased by 568,000 or −6.2% year-on-year between June 2019 and June 2020. This was largely due to decreases in the following industries: trade (−150,000 or −7.3%), business services (−137,000 or −6.5%), construction (−102,000 or −19.2%), manufacturing (−88000 or −7.7%), community services (−41,000 or −1.7%), transport (−35,000 or −7.3%), mining (−12,000 or −2.6%) and electricity (−3000 or −5.0%).

As a result of the depressed economy, Moody's autonomous debt assessment of the country was reduced to junk status in March 2020, stating the lack of political constancy, and other aspects related to national debt and debt payments (Greyling et al., 2020). The South African government has several possible means to encourage SDGs, given its strong development agenda (NDP 2030 and the NFSD), and the continental development agenda, also known as NEPAD. It is noteworthy to come back to the NFSD, because it is connected to the SDGs.

Thus, South Africa has shown keen interest in following the path of cooperation and international integration in the pursuit of the UN SDGs, and this direction takes into consideration local realities (policies and priorities). Being aware of its leadership role at the continental and international level—as the Chairman of the ad hoc Group of 77 countries plus China who played a pivotal role in getting the 2030 SDGs approved —South Africa has a moral obligation to represent the views of partner countries who framed and adopted the UN 2030 SDGs (Fourie, 2018). In a similar way, South Africa is duty-bound to oversee its international commitments, and ensure it adheres to a programme that does not contradict its own national development agenda (Fourie, 2018). We agree with Fourie about the critical role that the MTSFs should play to foster policy coherence in order to stimulate integral national growth within the global itinerary of sustainable development provided by the United Nations via its 2030 SDGs Agenda. The framework that can potentially help in meeting the SDGs is the PCDF which is discussed in the next section.

13.6.1 Policy Coherence for Development Framework Prospects in South Africa

The PCDF aims, in the first instance, at "understanding the development dimensions of member country policies and their impacts on developing countries" (Fourie, 2018); therefore, it considers compromised and potential synergies across various areas of development such as "trade, investment, agriculture, health, education, the environment and development cooperation" (Fourie, 2018). This definition denotes the centre-periphery relation that exists between donors and recipients of funds where development policies are to be executed. The PCDF requires a mutual understanding between South Africa and its funding partners in terms of balancing the implementation of local policies and priorities with global development responses framed in the 2030 SDGs Agenda. The same applies to the challenges pre- and post-Covid-19 pandemic. Even prior to the SDGs, South Africa put in place the NFSD that is the vehicle for the NDP. Therefore, the MTSFs need to work beyond policy coherence across government departments to look at opportunities for policy coherence in the implementation and management of the NDP 2030 and the 2030 SDGs Agenda. Scholars like Cloete (2018) found that efforts to manage the implementation of national policies and monitor performance indicators via MTSFs, the NDP, and the NFSD are doubtfully efficient to optimise coherence among these frameworks. Thus, it is necessary to improve the content and the management of actual South African policy indicators in order to maximise policy governance outcomes to effectively measure the compliance of South Africa with the UN SDGs (Cloete, 2018). The usefulness and significance of policy coherence is sustained by Cloete (2018: 10) who observes: "[…], policy coherence and consistency is imperative to avoid policy programme contradictions and conflicts". Harmonising policies and integrating national developmental priorities into the global development agenda is essential to achieve economic growth domestically and keep the local agenda consistent with the international one. This symmetrical policy implementation is susceptible of unearthing imbalances that individual nations may be confronted with.

In the case of South Africa, it is noteworthy to highlight some important challenges that the NDP is coming up against in aligning itself with the UN 2030 SDGs. Firstly, the NDP preceded the 2030 SDGs: the former emphasises national priorities and focus of ownership and observance of national policies, while the latter calls for global responses to international challenges. In a sense, the SDGs transcend the NDP, without superseding it, but rather call for national implementation of the SDGs. As a matter of fact, the NDP and the SDGs are not mutually exclusive, nor are they incompatible with each other. However, national governments are challenged by the prospects of maintaining the entirety of national planning processes while at the same time orienting and aligning national development (Fourie, 2018). Considering the necessity of constructive coherence, national ownership of priorities is an essential factor of development effectiveness, as noted by Fourie (2018); this implies, in the context of South Africa, that coherence between the NDP and the SDGs gives South Africa the opportunity to call on elements of the 2030 Agenda and utilise its own development agenda (policies and priorities) as a point of departure for policy alignment within the global development discourse (Fourie, 2018).

For instance, to support policy coherence for a development paradigm, Cloete (2018: 10) remarks: "The OECD model of policy coherence is an effective guideline to optimise the implementation of sustainable development programmes". Fourie (2018) has discussed the consonance and discordance of the NDP with the SDGs; these are framed in terms of their chronological development, the scope of activities, and the timeframe for implementation and evaluating the results. However, South Africa is well-positioned to achieve the needed cooperation and harness the existing inter-departmental and cross-departments -policy coherence and to broaden this in the pursuit of the SDGs, which are achievable via the NDP and its MTSFs support structure. The Department of International Relations and Cooperation, Statistics South Africa and other structures have been identified to enhance CPSD in South Africa. These two departments, along with the Department of Labour, should consider and implement the International Labour Organisation's (ILO) "Guidelines for a just transition towards environmentally sustainable economies and societies for all", which, in "Guiding Principles" 13(d) states:

> Coherent policies across the economic, environmental, social, education/training and labour portfolios need to provide an enabling environment for enterprises, workers, investors and consumers to embrace and drive the transition towards environmentally sustainable and inclusive economies and societies (ILO, 2015: 6).

Policy coherence and alignment of the NDP 2030 with the 2030 UN SDGs' Agenda (National Science and Technology Forum, 2018; Republic of South Africa National Planning Commission, 2012) are paramount in all efforts intended to meet the SDGs and the NDP agenda.

13.6.2 National Framework for Sustainable Development (NFSD)

The NFSD was approved by the South African Cabinet in 2008. This endorsement was a sign of a new way of thinking that aimed at "promoting the effective stewardship of South Africa's natural, social and economic resources" (South Africa Department of Forestry Fisheries and the Environment, 2011: n.p.). This National Strategy for Sustainable Development and Action Plan is also known as NSSD 1 (2011–2014), which the Cabinet approved on 23 Nov 2011.

Table 13.3 presents the priorities of the South African NFSD and the National Strategy for Sustainable Development and Action Plan (NSSD 1) 2011–2014.

The NFSD's purpose is "to enunciate South Africa's national vision for sustainable development and indicate strategic interventions to re-orientate South Africa's development path in a more sustainable direction (South Africa Department of Forestry Fisheries and the Environment, 2011: n.p.). Unfortunately, it does not provide detailed strategies or actions, but rather suggests "a national vision, principles, trends, strategic priority areas and a set of implementation measures that will enable and guide the development of the national strategy and action plan" (South Africa Department of Forestry Fisheries and the Environment, 2011: n.p.). The strategic priority put forward by NSSD1 indicates that the South African government has conceded that indicator evaluation is important in enhancing monitoring and reporting systems for improved environmental performance (Montmasson-Clair and Plooy, 2012 cited in the (Republic of South Africa National Planning Commission, 2012).

Despite substantive efforts made by South Africa and its commitment to improve the wellbeing of its people by 2030, the country still has several hurdles to surmount in the areas of job creation, action for climate and life under water, transitioning to cleaner and renewable energy resources, etc. Furthermore, its current path of

Table 13.3 National strategy for sustainable development: strategic priorities

NFSD strategic priorities	Reformulated NSSD 1 strategic priorities
Priority 1: Enhancing systems for integrated planning and implementation	**Priority 1**: Enhancing systems for integrated planning and implementation
Priority 2: Sustaining our ecosystems and using natural resources efficiently	**Priority 2**: Sustaining our ecosystems and using natural resources efficiently
Priority 3: Economic development through investing in sustainable infrastructure	**Priority 3**: Towards a green economy
Priority 4: Creating sustainable human settlements	**Priority 4**: Building sustainable communities
Priority 5: Responding appropriately to emerging human development, economic and environmental challenges (including climate change, rising oil prices, globalisation and trade)	**Priority 5**: Responding effectively to climate change

Source The Republic of South Africa National Planning Commission (2012)

economic development relies primarily on maximising economic growth (determined by the GDP), in particular, via its mining sector and manufacturing and agricultural activities (South Africa Department of Environmental Affairs, 2011: 13). This economic growth model has resulted in an economy intensively relying on fossil-fuel energy, and subsequently, leading to the erosion of the resource base—making the situation clearly unsustainable (South Africa Department of Environmental Affairs, 2011: 13).

13.6.3 South Africa's Sustainable Development: Contributing Sectors

The advancement of technology and research opportunities in issues of sustainability and development have increased the understanding of the key expansion of opportunities in different sectors of investment. Coal, for instance, is identified as contributing to an increase in job opportunities, given its huge consumption in the country, the region, and the world. The CEIC Data (2021) reports that South African annual coal production was 248.339 million tonnes in Dec 2020; this was a decrease from the previous quantity of 258.360 million tonnes in Dec 2019; the updated yearly records average 223.537 million tonnes from Dec 1981 to Dec 2020, based on 40 observations; the data reached a height of 261.399 million tonnes in 2014 and a record low of 130.420 million tonnes in 1981; lastly, South African coal production data continues to be in active status according to CEIC[2] and remains reported by BP PLC. As stated by Walker (2022), South Africa's thermal coal exports declined by about 13% in 2021 due to logistical constraints, social unrest and the decline in demand by India. These records testify to the preponderance of coal in the South African economy, as reported also by Venage (2021: n.p.): "The diplomatic standoff between China and Australia has provided an unexpected boost for South African coal producers who are now once again shipping to the Chinese mainland".

Coal production is therefore one of the key sectors backing South Africa's economy, because almost 70% of the major energy demand, as well as approximately 90% of national power production in the country, is provided by this sector (World Coal Association, 2016). Thus, its direct impact as a multiplier consequence of energy manufacturing in South Africa offers many thousands of jobs for the people of the country, which indirectly decreases the dearth of jobs and helps the progress of the economy (Saint Akadiri et al., 2019). However, the aspect of dependency on coal and use of fossil fuels as the main energy supply for South African industries and those of its neighbours, and the impact of these resources on the degradation of the environment has not been sufficiently addressed.

[2] CEIC was founded in Hong Kong in 1992 by a team of expert economists and analysts, and today employs data analysts in multiple offices around the globe, with more than ten offices in Asia alone (https://www.ceicdata.com/en/about-us/introduction-ceic).

Phasing out coal and moving towards a renewable energy-oriented economy should be part of the Economic Reconstruction and Recovery Plan (ERRP), which aims at building a novel economy and unleashing South Africa's true potential; and the creation of a "sustainable, resilient and inclusive economy" (South Africa Government, 2021).

13.7 Challenges to Attaining the UN Sustainable Development Goals 2030

The South African government has its own development programmes that encourage and finance local initiatives, while external or international organisations play a crucial role in supporting macro-development projects, and knowledge transfer. We identified four main predicaments that could either hinder or slow the pace of achieving sustainable development. These are (1) the brain drain phenomenon (Crown Publications, 2022; Kalipeni et al., 2012; PricewaterhouseCoopers ('PwC'), 2017), (2) socio-political unrest, (3) economic and industrial dependency on coal and fossil fuels, and (4) the Covid-19 pandemic.

13.7.1 The Brain Drain

This concept has been used differently to mean the same thing from various perspectives. It refers to "human capital flight" (Sako, 2002), while El Saghir et al. (2020, p. 185) defines "brain drain" as "the migration of educated and skilled individuals from a less developed region or country to a more economically established one". For Iravani (2011), "Brain drain or human capital flight is a large emigration of individuals with technical skills or knowledge, normally due to conflict, lack of opportunity, political instability, or health risks" (p. 285). Adding to the preceding definitions, brain drain entails a movement of people who hold skills, or educated people from a less-developed country to a developed country (El Saghir et al., 2020). A study conducted by Tankwanchi et al. (2013) shows an increase in the migration trends among Sub-Saharan Africa-trained physicians from 2002 to 2011, although South Africa was the only country that presented a decrease of 8% in migration of its physicians to the US (about 2156). Another study showed that more than 150 doctors were migrating annually from South Africa to the US, UK, Canada and Australia (Kalipeni et al., 2012).

Crown Publications (2022) reports an estimated 95,000 white people have migrated out of South Africa since 2011; 304,112 white South Africans left the country between 1986 and 2000, with an additional 133,782 leaving between 2001 and 2005, and close to 112,046 between 2006 and 2011. This gives a total of 644,940 white people who had left the country over 30 years (Crown Publications, 2022). It is noteworthy to point out that many young white people who complete matric or university studies depart South Africa to further their studies (Crown Publications, 2022).

Despite the fact that the brain drain phenomenon may produce both losers and winners in the formation of human capital (Beine et al., 2008; Docquier/Rapoport, 2008), it also affects sustainable development, such as in the case of Zimbabwe and Macedonia (Dinkovski/Markovska-Simoska, 2018). For instance, Gwaradzimba/Shumba (2010) argue, in the context of Zimbabwe, that brain drain constitutes a paramount development constraint in terms of development opportunities and lost investment; it drains countries of origin of their human capital—an asset that requires enormous resources to generate and raise. Thus, the effects of the phenomenon on the development of a nation, and its socio-political wellbeing (Gwaradzimba/Shumba, 2010), need redress if a country intends to undertake the path of sustainable development. This is equally a concern for South Africa, whose various sectors are affected by the migration of skilled professionals, especially in the wake of Covid-19 pandemic; this represents a serious threat the country's economic recovery scheme (Mfeka, 2022).

It is encouraging that the South African government, via the department of Home Affairs, is working on new policy regarding international migration (Crown Publications, 2022)—a move that could minimise the flight of the labour force and skills that the country needs to meet the NDP 2030 and attain the 2030 SDGs. Therefore, we agree with Kalipeni et al. (2012) who observe: "[...] while Africa can do nothing about the pull factors, it can mitigate the push factors by creating conducive working environments to help retain its health care professionals" (p. 168). The same applies to South Africa, because, as observed by PricewaterhouseCoopers ("PwC") (2017), South Africa is a country where skills scarcity remains critical, and it is imperative to retain and develop talent. It follows that the brain drain could be tackled by providing better working conditions and creating opportunities to develop and maintain skilled professionals.

13.7.2 Socio-Political Unrest

Factors contributing to the collapse of a country's stability in relation to its socio-economic situation take into consideration numerous aspects, political, economic, socio-cultural, technological, environment and legal (PESTEL). Foreign investment to boost the economy of other nations, either business-to-business or country-to-country, investigates all PESTEL aspects to ensure that the investment of millions of dollars is not wasted in a country where PESTEL conditions are not met. Yüksel (2012) notes: "PESTEL analysis, as it stands, mainly provides a general idea about the macro environmental conditions and situation of a company". It has become evident in the context of globalisation that managers seek to establish environmental conditions of business where the need for expansion is obvious; and research has even developed studies such as "REGLO" analysis, which is "an improved version of PESTEL analysis which marks out globalisation and regionalisation forces that are a great part of today's external environment" (Menet, 2016). It follows that the business climate must be propitious to attract investors and inspire development funding institutions.

Unfortunately, 2021 was a difficult year, as social unrest and the violence of thousands of South African citizens who took the law into their own hands by engaging in brutal and public violence that culminated in widespread theft from shops in the KwaZulu-Natal and Gauteng provinces. Cele and Wilson (2021) report that the deadly riots that broke out in the aftermath of former President Jacob Zuma's arrest cost South Africa close to 50 billion rand ($3.4 billion) in lost output, with 150,000 jobs simultaneously in peril (estimates from the South African Property Owners Association (SAPO) cited in (Cele/Wilson, 2021).[3] In addition to 200 malls that were targeted, about 3000 shops were looted by rioters or protesters, while 200 banks and post office facilities were vandalised, according to a report by the Minister in the Presidency, Khumbudzo Ntshavheni (Cele/Wilson, 2021).

The July 2021 socio-political unrest dealt a severe blow to development structures, which SAPO reports as follows: 150,000 informal traders targeted by looters, 40,000 businesses affected by the July 2021 protestors, 1400 bank teller machines destroyed, 100 malls set alight and damaged, and stocks in stores worth about 1.5 billion rand lost in the eastern city of Durban alone (Cele/Wilson, 2021). The July 2021 riots have badly affected South Africa's struggling economy, as echoed by the International Monetary Fund (IMF), which released its findings on the economic consequences of social unrest, stating: "They bode ill for an economy embarking on a fragile recovery from the monumental 7% contraction of 2020" (Stoddard, 2021). The records of violent protests, torched schools, and setting heavy motor vehicles and tankers alight are not of the nature to invite investors; such brutalities are harmful to development and to the country's economy.

13.7.3 Transitioning from Coal and Natural Gases to Cleaner Energy Resources

South Africa's dependency on coal is a permanent danger to the country's economic growth as well as to the social living conditions of its citizens given the high cost of coal in the long term. South Africa is among the countries that is highly ranked internationally as a "major emitter of CO_2, which is 1% of the total emission of what is produced by the world due to its use of coal considered as a component of CO_2" (Shahbaz et al., 2013: 1453). On the same issue, Burton et al. (2019) have discussed the compatibility of transitioning from coal to cleaner energy resources in

[3] The South African Property Owners Association (SAPOA) indicated that the cost of the unrest is estimated to exceed R20bn in KZN and the overall impact on the national GDP will be R50 billion. Over 200 shopping malls were targeted, and 1787 retail stores were impacted and damaged. Source: https://pmg.org.za/committee-meeting/33438/ (12 Jan 2022).

the light of the Paris Agreement. Regarding coal they note that South Africa has had coal as an integral source of its energy system and economy for decades, but coal is becoming increasing uncompetitive and this state of affairs poses significant risks to the economy, social life and the environment. Of great importance, it is revealed that "77% of South Africa's energy needs are provided by coal, while 53% of what makes the reserve is used in electricity generation, 33% in petrochemical industries, 12% in metallurgical industries and 2% in domestic heating and cooking" (Shahbaz et al., 2013: 1453).

South Africa's dependency on coal's emissions and liquid gases is an established fact, as The Republic of South Africa National Planning Commission (2013) reports that South Africa's energy emissions overall stand at 80% that comes from energy supply Carbon Brief Ltd. (2018), particularly from coal-related combustion emissions, and this is due to the dependence on the fuel for energy. In addition, coal comprised 67% of the primary energy supply of South Africa in 2015 (DoE, 2015); of this 92% went towards electricity and approximately "25% of liquid fuels consumption originates from the conversion of coal into synthetic fuels at Sasol's Secunda plant" (Burton et al., 2019: 3). The need to transition in the transport sector is obvious—by phasing out internal combustion engines and moving to alternative technologies such as hybrid/electric vehicles to ensure that sustainable development is achieved (Marquard et al., 2021: 22).

To mitigate South Africa's main fossil carbon, it has been suggested that the country needs to consider changing its main dependency on coal in the power sector, a key recommendation that complies with the Paris Agreement (Pricewaterhouse-Coopers ("PwC"), 2017). To comply with the objective of the Paris Agreement, efforts should be made to phase out coal and liquid fuels; thus, renewable energy policy implementation to mitigate climate change will contribute to poverty alleviation, the creation of jobs and will help South Africa address socio-economic challenges and attain sustainable development (Burton et al., 2019). This is necessary to prevent the adverse effects of the transition from coal to cleaner energy that coal-dependent regions may suffer, also a result of international policy to fight climate change that could leave coal workers stranded (Burton et al., 2019). Such transition also requires dealing with the developmental challenges confronting the country, putting in place transition planning that considers "worker transition schemes, local economic resilience and the development of new sectors" (Burton et al., 2019: 1). In the likely event of South Africa embracing the Paris-adjusted scenario, it could phase out coal in the electricity sector and possibly in coal-to liquids approximately by 2040 and show a large reduction in coal use by 2050 (Burton et al., 2019).

The pathway to sustainable development beyond 2030 remains the competitive transition to renewable energy, because climate change is expected to impair economic growth, and that is why climate action features in the 2030 SDGs Agenda (Goal 13). Considering that coal in South Africa is under structural decline on economic grounds (McCall et al., 2019 cited in Burton et al. (2019: 5), the transition from a coal-dependent economy to renewable energy resources becomes imperative for South Africa's economic growth, and renewable energy production appears more cost-effective than coal production (Burton et al., 2019). This is confirmed by Creamer (2021), based on the estimations provided by The Council for Scientific and Industrial Research (CSIR), who states that the tariffs reached during the fifth bid window of the Renewable Energy Independent Power Producer Procurement Programme (REIPPPP) back its initial analysis demonstrating that renewable energy is the lowest-cost energy production option available for South Africa. Furthermore, The CSIR indicates that "the actual tariffs achieved under the REIPPPP since 2011 have fallen a dramatic 78% for new wind, from R1.90/kWh to 49.5c/kWh and 91% for solar PV, from R4.60/kWh to only 42.9c/kWh" (Creamer, 2021: n.p.).

Thus, the UN Department of Economic and Social Affairs urges the world to take urgent action to combat climate change and its impacts (United Nations, 2015b; United Nations Department of Economic and Social Affairs, 2022). As reported by The World Bank (2022), climate change could induce the migration of 216 million people within their own countries by 2050, and climate migration could cause further world instability, trigger conflicts and threaten global peace (Bronkhorst, 2011; Wilson, 2020). Therefore, climate change remains an important challenge that requires urgent action. In the opinion of (Gopalakrishnan et al., 2022: n.p.):

> If unchecked, climate change can push up to 132 million people into extreme poverty by 2030 according to World Bank estimates and most of the world's poorest people will live in situations characterized by fragility, conflict, and violence. Poverty is already intertwined with vulnerability to climate-related threats such as flooding and vector-borne diseases, making climate change a major impediment to alleviating extreme poverty.

This situation is even more serious and should impel to South Africa to adopt an immediate stance against greenhouse gas emission and pursue the salutary path of renewable and cleaner energy, a path being embraced by several countries.

13.7.4 Covid-19's Effects on Sustainable Development

The most unprecedented health crisis the world has experienced since the 1918 'Flu represents a significant economic setback both domestically and internationally. After the WHO declared Covid-19 as a worldwide threat, the South African government embarked on severe protective measures to contain, if not to flatten the spread of the virus; President Cyril Ramaphosa, on 23 Mar 2020, announced a countrywide lockdown of all activities as the first response to the COVID-19 crisis (Winck, 2020). The Covid-19-related crisis was labelled by the President as a "National Disaster" (Nordling, 2020). There have been several reports on the severe consequences of the pandemic on various sectors of the international economy (Lone/Ahmad, 2020). A short account of these factors has been compiled below as evidence. At the South African level, these cover the consequences of Covid-19 for different sectors, such as aviation (Dube et al., 2021), transport and trade (Marquard et al., 2021), catering and the accommodation industry (BusinessTech, 2021), and tourism (Bama/Nyikana, 2021; Rogerson/Rogerson, 2020), etc.

At the international level, the Covid-19 pandemic's socio-economic effects encompass: the fear of seeing about 40–60 million persons fall into acute poverty

(World Bank Group, 2020); the shrinking of the GDP per capita of 170 nations that might result in a 3.0 percent GDP reduction worldwide (Stoddard, 2021); fear of a new surge in Covid-19 infections which has been predicted to have further deteriorating effects on the global economy (Winck, 2020); COVID-19's cost to the financial world of about US$ 5.8 trillion US$ 8.8, or likely an international GDP of 6.4 percent to 9.7, while the drop in universal employment might reach between 158 and 242 million jobs (Asian Development Bank, 2022). International trade, that contributes tremendously to development, was brought to standstill. These statistics are of great concern for SDGs. World governments, together with international financial development agencies and economic international bodies, need strategic ways to handle this health crisis. Health specialists have also warned that the disease will have dire consequences in the countries with small and medium revenue and these will face challenges in dealing with the pandemic due to the dearth of health maintenance benefactors as well as medical infrastructure.

This deadly virus has plunged the world into uncertainty; because of this, novel thinking in designing development frameworks has become an imperative of development strategy. To expound further, Leach et al. (2021) suggest an approach that takes into consideration the prerequisites of development—that entails anticipating significant shocks that ensue from the crisis, building resilience capacity. This reflects a development paradigm shift as explained here: "Where mainstream approaches to development have been top down, rigid and orientated towards narrowly-defined economic goals, post-COVID-19 development must have a radically transformative, egalitarian and inclusive knowledge and politics at its core" (Leach et al., 2021: 1).

Ataguba (2020) makes an analogy between war and fighting the Covid-19 pandemic, containing its spread and minimising its economic impact; such war strategy necessitates "a multipronged approach and cooperation from all parties, including substantial reductions in uncertainty levels, panic levels and disease transmission, the primary channels through which the economic impact manifests" (p. 328). More significantly, the author argues that similar to times of war, when defence spending increases appreciably, African countries must consider the Covid-19 pandemic as a "war" to be fought and won and be ready to deal with the increase in public health spending (Ataguba, 2020). Two years after the onset of this pandemic, nations across the globe have substantially lifted stringent Covid-19 regulations and most health protocols as the virus shows a downward trend in infection rates. It is time to believe that global growth and sustainable development gaols can still be pursued and attained.

13.8 Resilience in the Covid-19 Pandemic Era

While acknowledging the far-reaching impact of the pandemic, which has touched every possible sector development, causing the poor and the vulnerable to bear the brunt of it, the pandemic is slowing the prospects of ending poverty and magnifying shared prosperity (Gopalakrishnan et al., 2022). However, the world is not beyond hope, because there have been some positive developments, as the global economy has shown signs of growth: trade is rebounding, food commodity prices are being stabilised, and remittances showed a robust recovery (Gopalakrishnan et al., 2022). Despite this optimism, newer Covid-19 variants and unequal distribution of and access to vaccines require more national and international responses (Gopalakrishnan et al., 2022). Simultaneously, since a number of countries are beginning to experience recovery, this becomes their opportunity to attain durable economic growth, but without further damaging the environment or exacerbating inequalities (Gopalakrishnan et al., 2022). These scholars note: "The Bank Group is helping countries chart a recovery that is green, resilient, and inclusive through achieving economic stability and growth, leveraging the digital revolution, making development greener and more sustainable, and investing in people" (Gopalakrishnan et al., 2022: n.p.).

As proof of such hope, several countries have lifted health restrictions and protocol, and South Africa has shown significant downtrends in infections rate. The Word Health Organisation (2023) reports:

> Globally, nearly 4.1 million new cases and 28,000 deaths were reported in the last 28 days (13 February to 12 March 2023), a decrease of 40% and 57%, respectively, compared to the previous 28 days…; however, there are significant regional differences including increases in some regions. As of 12 March 2023, over 760 million confirmed cases and over 6.8 million deaths have been reported globally (p.1).

Noting the variations in various regions, it is necessary to remain vigilant and keep testing and boosting individual immune system by taking recommended jabs and respecting health protocols pertaining to Covid-19 prevention. This is imperative, because the WHO reports that South Africa had the highest numbers of new cases (6828 new cases; "11.5 new cases per 100,000; −4%)" as per data collected between 13 February to 12 March 2023 in the African Region (World Health Organisation, 2023: 12). To harness resilience, the following might be considered in various sectors of the South African society: merging of various wellbeing systems complying with Centres for Disease Control (Van Nguyen et al., 2020); creating entrepreneurial resilience founded on preparations for developing situations and constriction of resources and coping strategies (Purnomo et al., 2021); mobilising collectively as one entity around UN SDGs to meet the objectives and ensure the safety, and access to potable water, as well as clean air that is central to several objectives (Mukherjee et al., 2020); and working towards SDGs that encourage Sustainable Accounting, Management and Policy (SAMP) might play a crucial role in building a post-Covid-19 era (Hörisch, 2021).

13.9 Conclusion

This article attempted to respond to questions inspired by Leach et al. (2021): to examine the extent to which scientific evidence can be utilised to frame policies in the context of severe uncertainties related to the development disruptions induced by the Covid-19 pandemic; to investigate how world economies function in the global environment of the Covid-19 health crisis, and uncover important shortcomings in a conventional paradigm of economic growth; and finally, to determine to what extent emerging forms of policy, such as the PCDF and domestic sustainable development models, could work synergically with a global development agenda to create productive conditions for economic growth and sustainable development. While the development priorities of a country remain the responsibility of the government, partnering with the world in an imperative of "mundialisation" is necessary. Partnership has been encouraged by the development agenda as stipulated in the SDGs (SDG 17), which recommends its key role within the country's political management, the private sector and civil society.

It emerged from our analyses that the prospects of South Africa keeping the NDP 2030's pace while pursuing the SDGs Agenda are promising, despite the real health challenges that confront the world. Firstly, because South Africa is committed to its ERRP and health protocols and measures put in place to flatten the Covid-19 infection rate, and the enhancement of testing and vaccination programmes in various places and circumstances. Secondly, the South African Government has a long legacy of overcoming adverse national calamities; the nation has the potential, infrastructure and resolve to turn its economy around as research continues around developing a national response to the Covid-19 pandemic and maintain its commitment and leadership on the continent and at international level. Thirdly, the country has shown significant signs of resilience, moving from hard lockdowns to easing restrictions on business and carrying out research on SARS-CoV-2, and discovering emerging variants such as SARS-CoV-2's Omicron variant. The opening of the vaccine manufacturing plant in Brackenfell by President Cyril Ramaphosa and the South African-born health technologies billionaire, Patrick Soon-Shiong was significant; this facility aims to produce vaccines to help stop the Covid-19 pandemic (Felix, 2022). Here we see the importance of maintaining skilled scientists to ensure that such development in the health sector and other sectors is achieved; the causes of the brain drain causes must be tackled.

Resilience remains, therefore, a key driver of improved economic growth and mitigation of Covid-19's effects, though the whole world remains on alert due to the mutating nature of SARS-CoV-2. As to the South African leadership and partners at all levels of the society, it is time to focus on new strategies and policies as the development agenda continues its journey, and maintain the PCDF as an instrument of balancing the priorities to reach the SDGs. Such policies should be cognisant of the multiple threats climate change is posing to South Africa and accelerate the transformation process from coal-dependent industry to an eco-friendly industrial revolution.

13.10 Recommendations

The successful attainment of the SDGs as well as the NDP in the context of the Covid-19-related global health crisis requires a compounded strategy that:

- Firstly, seeks to maintain multiple coherent policy-approaches across South Africa's government departments, and in alignment with the NDP 2030 and 2030 SDGs Agenda.
- Secondly, focuses on Goal 3 (good health and wellbeing) and Goal 13 (climate action) of the SDGs to ensure robust protocol and health policies are put in place and maintained to fight Covid-19.
- Thirdly, strengthens South Africa's commitment to the Economic Reconstruction and Recovery Plan (ERRP) which aims at building a novel economy and unleashing South Africa's true potential; the overarching aim of the ERRP is the creation of a "sustainable, resilient and inclusive economy" (South Africa Government, 2021).
- Fourthly, continues pursuing the Government's priorities, namely *energy security, an industrial base to create jobs, a mass public employment programme, infrastructure development, macro-economic interventions, a green economy, food security, and reviving the tourism sector* (South Africa Government, 2021).
- Ffifthly, accelerates the process of transition to cleaner energy resources to create more jobs and mitigate the impacts of climate change.
- Sixthly, implements the World Bank's series of recommendations such as (1) reducing greenhouse gas emissions and keeping the temperature goals of the Paris Agreement (Burton et al., 2019; The World Bank, 2022), pursuant to Article 2(1)(a) of the Paris Agreement which consisting in keeping "the global average temperature to well below 2°C above pre-industrial levels and pursuing efforts to limit the temperature increase to 1.5°C above pre-industrial levels, recognizing that this would significantly reduce the risks and impacts of climate change" (United Nations, 2015a).
- Lastly, balancing its national priorities and development action plan with the UN Global Development Agenda 2030.

It is important to emphasise that sustainable development cannot be achieved if the questions around climate change, socio-political unrest, insufficient health infrastructure, and the brain drain are not addressed. The success of the NDP 2030 and the SDGs relies upon a good health infrastructure, education, and retaining skilled professionals within the country; and all these strategies will be productive if there is political will to implement them.

References

Ali, S., Hussain, T., Zhang, G., Nurunnabi, M., & Li, B. (2018). The implementation of sustainable development goals in "BRICS" countries. *Sustainability, 10*(7), 2513.

Amba, K. (2010). The need for popular participation in Abuja: a Nigerian story of informal settlements. *Journal of Place Management and Development, 3*(2), 149–159. https://doi.org/10.1108/17538331011062694.

Asian Development Bank. (2022). COVID-19 Economic Impact Could Reach $8.8 Trillion Globally — New ADB Report. News Release | 15 May 2020. Available: https://www.adb.org/news/covid-19-economic-impact-could-reach-8-8-trillion-globally-new-adb-report. Accessed 7 August 2023.

Ataguba, J.E. (2020). COVID-19 pandemic, a war to be won: understanding its economic implications for Africa. *Applied Health Economics and Health Policy, 18*(3), 325–328.

Bama, H.K.N., & Nyikana, S. (2021). The effects of COVID-19 on future domestic travel intentions in South Africa: A stakeholder perspective. *African Journal of Hospitality, Tourism and Leisure, 10*(1), 179–193.

Beine, M., Docquier, F., & Rapoport, H. (2008). Brain drain and human capital formation in developing countries: winners and losers. *The Economic Journal, 118*(528), 631–652.

Billings, L. (2014). Brave New Epoch: A search for humankind's mark on the Earth. https://nautil.us/brave-newepoch-234749/. Accessed 16 February 2024.

Blasco, J. L., King, A., & Jayaram, S. (2018). How to report on the SDGs: What good looks like and why it matters. Madrid, Spain: KPMG.

Bonorchis, R. (2022). South Africa Budgets $34 Million for Flood Repairs to Two Roads. *Bloomberg (6 May 2022, 15:32 SAST).* https://www.bloomberg.com/news/articles/2022-05-06/south-africa-budgets-34-million-for-flood-repairs-to-two-roads. Accessed 8 August 2023.

Brauch, H.G. (2016). Sustainable Peace in the Anthropocene: Towards Political Geoecology and Peace Ecology. In H. Brauch, Ú.O. Spring, J. Grin & J. Scheffran (Eds.), *Handbook on Sustainability Transition and Sustainable Peace* (pp. 187–236). Cham: Springer.

Brisman, A. (2016). Peace ecology. *Contemporary Justice Review, 13*(3), 419–421. https://doi.org/10.1080/10282580.2016.1200886

Bronkhorst, S. (2011). *Climate change and conflict: lessons for conflict resolution from the southern Sahel of Sudan.* Umhlanga Rocks: ACCORD.

Burton, J., Marquard, A., & McCall, B. (2019). Socio-Economic Considerations for a Paris Agreement-Compatible Coal Transition in South Africa. https://lifeaftercoal.org.za/wp-content/uploads/2020/05/Burton-Marquard-McCall-JT_Paris_South-Africa-Final.pdf. Accessed 14 August 2023.

BusinessTech. (2021). New lockdown rules threaten businesses looking for Easter weekend boost in South Africa. https://businesstech.co.za/news/business/479417/newlockdown-rules-threaten-businesses-looking-for-easter-weekend-boost-in-south-africa/. Accessed 8 September 2023.

Campbell, B.M., Beare, D. J., Bennett, E. M., Hall-Spencer, J. M., Ingram, J. S., Jaramillo, F., . . . Shindell, D. (2017). Agriculture production as a major driver of the Earth system exceeding planetary boundaries. *Ecology and Society, 22*(4).

Carbon Brief Ltd. (2018). The Carbon Brief Profile: South Africa. https://www.carbonbrief.org/the-carbon-brief-profile-south-africa/. Accessed 12 March 2023.

Carr, J.A., Petrokofsky, G., Spracklen, D.V., Lewis, S.L., Roe, D., Trull, N., . . . Sallu, S.M. (2021). Anticipated impacts of achieving SDG targets on forests-a review. *Forest Policy and Economics, 126*(2021), 1–11. https://doi.org/10.1016/j.forpol.2021.102423.

CEIC Data. (2021). South Africa Coal Production 1981–2020. https://www.ceicdata.com/en/indicator/south-africa/coal-production. Accessed 5 August 2023.

Cele, S., & Wilson, L. (2021). South Africa Economy Set to Take $3.4 Billion Hit From Riots. *Bloomberg*. https://www.bloomberg.com/news/articles/2021-07-20/south-african-economy-set-to-take-3-4-billion-hit-from-riots. Accessed 5 August 2023.

Chryssavgis, J. (2014). The Green Patriarch: Ecumenical Patriarch Bartholomew and the Protection of the Environment. https://www.apostolicpilgrimage.org/the-green-patriarch.html. Accessed 5 August 2023.

Cloete, F. (2018). Policy indicator consistency and coherence in measuring public sector development programmes in South Africa. *African Evaluation Journal, 6*(2), 1–11. https://aejonline.org/index.php/aej/article/view/338/545. Accessed 7 August 2023.

Correa, D.D.S., & Owens, W.R. (2010). *The handbook to literary research* (2nd ed.). London: Routledge.

Creamer, T. (2021). CSIR says fall in renewables tariffs points to need for higher deployments. https://www.engineeringnews.co.za/article/csir-says-fall-in-renewables-tariffs-points-to-need-for-higher-deployments-2021-11-12. Accessed 5 August 2023.

Crown Publications. (2022). SA government wants to stop the brain drain. https://www.crown.co.za/crown-blog/general/item/234-challenges-and-opportunities-in-the-latest-brain-drain. Accessed 5 August 2023.

Dinkovski, V., & Markovska-Simoska, S. (2018). Brain drain as a function of sustainable development in the Republic of Macedonia. *Industrija, 46*(1), 173–188.

Dlouhá, J., & Pospíšilová, M. (2018). Education for Sustainable Development Goals in public debate: The importance of participatory research in reflecting and supporting the consultation process in developing a vision for Czech education. *Journal of Cleaner Production, 172*, 4314–4327.

Docquier, F., & Rapoport, H. (2008). Brain drain and human capital formation in developing countries: winners and losers. *The Economic Journal, 118*(528), 631–652.

Dube, K., & Nhamo, G. (2021). Sustainable development goals localisation in the tourism sector: Lessons from Grootbos private nature reserve, South Africa. *GeoJournal, 86*(5), 2191–2208.

Dube, K., Nhamo, G., & Chikodzi, D. (2021). COVID-19 pandemic and prospects for recovery of the global aviation industry. *Journal of Air Transport Management, 92*(1). https://doi.org/10.1016/j.jairtraman.2021.102022.

Dudgeon, D. (2019). Multiple threats imperil freshwater biodiversity in the Anthropocene. *Current Biology, 29*(19), R960–R967.

El Saghir, N.S., Anderson, B.O., Julie Gralow., Lopes, G., Shulman, L.N., Moukadem, H.A., Yu, P.P., & Hortobagyi, G. (2020). Impact of merit-based immigration policies on brain drain from low-and middle-income countries. *JCO global oncology, 6*(5 Feb 2020), 185–189.

Environment Statistics Section of the United Nations Statistics Division (UNSD). (2014). United Nations General Assembly's Open Working Group on Sustainable Development Goals, New York, 19 Jul 2014. *35*(Aug2014). https://unstats/nsd/ENVIRONMENT/newsletters.htm. Accessed 5 August 2023.

Erin, O.A., & Bamigboye, O. A. (2021). Evaluation and analysis of SDG reporting: evidence from Africa. *Journal of Accounting & Organizational Change. 18(3)*, 369–396. https://www.researchgate.net/profile/Olayinka-Erin/publication/351955976_Evaluation_and_Analysis_of_SDG_Reporting_Evidence_from_Africa/links/613d20e94e1df271062e9cb9/Evaluation-and-Analysis-of-SDG-Reporting-Evidence-from-Africa.pdf. Accessed 18 February 2024.

Felix, J. (2022). Ramaphosa, tech billionaire open new Covid-19 vaccine manufacturing laboratory. https://www.news24.com/news24/southafrica/news/ramaphosatech-billionaire-open-new-covid-19-vaccine-manufacturing-laboratory-20220119. Accessed 5 August 2023.

Fourie, W. (2018). Aligning South Africa's National Development Plan with the 2030 Agenda's Sustainable Development Goals: Guidelines from the policy coherence for development movement. *Sustainable Development, 26*(6), 765–771. https://doi.org/10.1002/sd.1745.

Frey, L.R., Botan, C.H., & Kreps, G.L. (1999). *Investigating Communication: An introduction to research methods* (2nd Ed.). Boston: Allyn and Bacon.

Goal Tracker South Africa. (2019). South Africa sees Improvement on Social Challenges. https://www.statssa.gov.za/?p=12731. Accessed 5 August 2023.

Gopalakrishnan, V., Wadhwa, D., Haddad, S., & Blake, P. (2022). 2021 Year in Review in 11 Charts: The Inequality Pandemic. https://www.worldbank.org/en/news/feature/2021/12/20/year-2021-in-review-the-inequality-pandemic. Accessed 5 August 2023.

Greyling, T., Rossouw, S., & Adhikari, T. (2020). *A tale of three countries: How did Covid-19 lockdown impact happiness?* https://www.econstor.eu/bitstream/10419/221748/1/GLO-DP-0584.pdf. Accessed 5 August 2023.

Griggs, D., Smith, M. S., Rockström, J., Öhman, M. C., Gaffney, O., Glaser, G., . . . Shyamsundar, P. (2014). An integrated framework for sustainable development goals. *Ecology and society, 19*(4). https://doi.org/10.5751/ES-07082-190449.

Gwaradzimba, E., & Shumba, A. (2010). The nature, extent and impact of the brain drain in Zimbabwe and South Africa. *Acta Academica, 42*(1), 209–241.

Hörisch, J. (2021). The relation of COVID-19 to the UN sustainable development goals: Implications for sustainability accounting, management and policy research. *Sustainability Accounting, Management and Policy Journal, 12* (5), 877–888. https://doi.org/10.1108/SAMPJ-08-2020-0277.

ILO. (2015). Guidelines for a just transition towards environmentally sustainable economies and societies for all. https://www.ilo.org/wcmsp5/groups/public/@ed_emp/@emp_ent/documents/publication/wcms_432859.pdf. Accessed 5 August 2023.

Iravani, M.R. (2011). Brain drain problem: A review. *International Journal of Business and Social Science, 2*(15), 284–289.

Kalipeni, E., Semu, L. L., & Mbilizi, M. A. (2012). The brain drain of health care professionals from sub-Saharan Africa: A geographic perspective. *Progress in Development Studies, 12*(2–3), 153–171. https://doi.org/10.1177/146499341101200305.

Kanu, I.A. (2020). COVID-19 and the economy: an African perspective. *Journal of African Studies and Sustainable Development, 3*(2). https://doi.org/10.13140/RG.2.2.18801.43362.

Kapás, J. (2007). *Industrial Revolutions and the Evolution of Firm Organization.* Paper presented at the Economic Growth, Development, and Institutions, Paper for the EAEPE 2007 Conference "Economic Growth, Development, and Institutions – Lessons for Policy and the Need for an Evolutionary Framework of Analysis", Nov 1–3, 2007, Porto.

Kiyala, J.C.K. (2022). Faith-based Organisations and Environmental Peace: Impact of Laudato Si' and its relevance in the Anthropocene in Africa. In J.C.K. Kiyala & G.T. Harris (Eds.), *Civil Society and Peacebuilding in Sub-Saharan Africa in the Anthropocene: An overview* (pp. 65–125). Cham: Springer Nature Switzerland.

Kyrou, C.N. (2007). Peace ecology: An emerging paradigm in peace studies. *International Journal of Peace Studies*, 73–92.

Le Blanc, D. (2015). Towards integration at last? The sustainable development goals as a network of targets. *Sustainable Development, 23*(3), 176–187. https://doi.org/10.1002/sd.1582.

Leach, M., MacGregor, H., Scoones, I., & Wilkinson, A. (2021). Post-pandemic transformations: How and why COVID-19 requires us to rethink development. *World Development, 138*(Feb 2021), 1–11. https://doi.org/10.1016/j.worlddev.2020.105233.

Leal Filho, W., Azeiteiro, U., Alves, F., Pace, P., Mifsud, M., Brandli, L., . . . Disterheft, A. (2018). Reinvigorating the sustainable development research agenda: the role of the sustainable development goals (SDG). *International Journal of Sustainable Development & World Ecology, 25*(2), 131–142.https://doi.org/10.1080/13504509.2017.1342103.

Lelieveld, J. (2021). Paul J. Crutzen (1933–2021): Ozone Nobel prizewinner who coined the term Anthropocene. *Nature, 591*(4 Mar 2021), 29. https://www.nature.com/articles/d41586-021-00479-0. Accessed 5 August 2023.

Lepori, M. (2015). There is no Anthropocene: Climate change, species-talk, and political economy. *Telos, 172*, 103–124. https://doi.org/10.3817/0915172103.

Lewis, S.L., & Maslin, M. A. (2015). Defining the anthropocene. *Nature, 519*(7542), 171–180. https://doi.org/10.1038/nature14258.

Lim, M.M., Jørgensen, P. S., & Wyborn, C. A. (2018). Reframing the sustainable development goals to achieve sustainable development in the anthropocene—A systems approach. *Ecology and Society, 23*(3). https://doi.org/10.5751/ES-10182-230322.

Lone, S.A., & Ahmad, A. (2020). COVID-19 pandemic–an African perspective. *Emerging microbes & infections, 9*(1), 1300–1308. https://doi.org/10.1080/22221751.2020.1775132.

Marquard, A., Merven, B., Hartley, F., McCall, B., Ahjum, F., Burton, J., . . . Charitar, D. (2021). *Technical Analysis to support the update of South Africa's First NDC's mitigation target ranges.* Energy Systems Research Group. Cape Town.

McArthur, J.W., & Rasmussen, K. (2018). Change of pace: Accelerations and advances during the Millennium Development Goal era. *World Development, 105*(May 2018), 132–143. https://doi.org/10.1016/j.worlddev.2017.12.030.

Menet, G. (2016). The importance of strategic management in international business: Expansion of the PESTEL method. *International Business and Global Economy, 35*(2), 261–270.

Mfeka, B. (2022). Brain drain is another threat to economic reforms and recovery. *Independent Online and affiliated companies (IOL).* https://www.iol.co.za/business-report/opinion/brain-drain-is-another-threat-to-economic-reforms-and-recovery-a19ac26c-7cbe-475e-b493-7f6c11b11d83. Accessed 5 August 2023.

Mueller, M.P., & Tippins, D.J. (2015). *Ecojustice, citizen science and youth activism.* Cham: Springer International Publishing.

Mukherjee, A., Babu, S.S., & Ghosh, S. (2020). Thinking about water and air to attain Sustainable Development Goals during times of COVID-19 Pandemic. *Journal of Earth System Science, 129*(1), 1–8. https://doi.org/10.1007/s12040-020-01475-0.

National Science and Technology Forum. (2018). The National Development Plan (NDP). http://www.nstf.org.za/wp-content/uploads/2018/04/All-The-NDP.pdf. Accessed 5 August 2023.

Nordling, L. (2020). 'A ticking time bomb': Scientists worry about coronavirus spread in Africa: South Africa declares "state of disaster" after virus is found to spread locally and cases rise to 61. https://www.science.org/content/article/ticking-time-bomb-scientists-worry-about-coronavirus-spread-africa. Accessed 6 August 2023.

Odeku, K.O. (2020). The Plight of Women Entrepreneurs during Covid-19 Pandemic Lockdown in South Africa. *Gender & Behaviour, 18*(3).

OECD-FAO. (2022). Agricu lture in Sub-Saharan Africa: Prospects and challenges for the next decade In *OECD-FAO Agricultural Outlook 2016-2025* (pp. 59–95). Paris: OECD Publishing.

Omarjee, L. (2021). Table Mountain wildfire, floods and 'severe weather' cost SA billions in 2021. https://www.news24.com/fin24/economy/table-mountain- wildfire-floods-and-severe-weather–cost-sa-billions-in-2021-20220126. Accessed 5 August 2023.

Patriarch Bartholomew. (2021). Message of His All Holiness, Patriarch Bartholomew, for the day of prayer for the protection of the Environment (01/09/2016). *Laudato Si Movement 2021.* https://orthodoxia.be/en/2020/08/31/message-of-his-all-holiness-patriarch-bartholomew-for-the-day-of-prayer-for-the-protection-of-the-environment-01-09-2020/. Accessed 5 August 2023.

Pieterse, J.N. (2010). *Development Theory Deconstructions/Reconstructions* (2nd ed.). Los Angeles: Sage.

Pope Francis. (2015). *Encyclical Letter Laudato Si' of The Holy Father Francis on Care for Our Common Home.* Rome Libreria Editrice Vaticana.

PricewaterhouseCoopers ("PwC"). (2017). Integrated report 2016. https://examples.integratedreporting.org/wp-content/uploads/2022/07/Truworths-Intnl_IntRep_2016.pdf. Accessed 5 August 2023.

Purnomo, B.R., Adiguna, R., Widodo, W., Suyatna, H., & Nusantoro, B.P. (2021). Entrepreneurial resilience during the Covid-19 pandemic: navigating survival, continuity and growth. *Journal of Entrepreneurship in Emerging Economies, 13*(4), 497–524.

Reid, A.J., Brooks, J.L., Dolgova, L., Laurich, B., Sullivan, B.G., Szekeres, P., . . . Cooke, S.J. (2017). Post-2015 Sustainable Development Goals still neglecting their environmental roots in the Anthropocene. *Environmental Science & Policy, 77*, 179–184. https://www.sciencedirect.com/science/article/abs/pii/S1462901116307900.

Republic of South Africa National Planning Commission. (2012). Our future - make it work. National Development Plan 2030: Executive Summary. https://www.gov.za/sites/default/files/Executive%20Summary-NDP%202030%20-%20Our%20future%20-%20make%20it%20work.pdf. Accessed 5 August 2023.

Rockström, J., Steffen, W., Noone, K., Persson, Å., Chapin III, F.S., Lambin, E., . . . Schellnhuber, H. J. (2009). Planetary boundaries: exploring the safe operating space for humanity. *Ecology and society, 14*(2). https://doi.org/10.1126/science.1259855. Originally published online Jan 15, 2015.

Rogerson, C.M., & Rogerson, J.M. (2020). COVID-19 tourism impacts in South Africa: government and industry responses. *GeoJournal of Tourism and Geosites, 31*(3), 1083–1091.

Ruddiman, W.F. (2013). The anthropocene. *Annual Review of Earth and Planetary Sciences, 41*, 45–68. https://doi.org/10.1146/annurev-earth-050212-123944.

Rull, V. (2020). The Anthropozoic era revisited. *Lethaia, 54*(3), 289–299.

Saint Akadiri, S., Bekun, F.V., & Sarkodie, S.A. (2019). Contemporaneous interaction between energy consumption, economic growth and environmental sustainability in South Africa: what drives what? *Science of the total environment, 686*, 468-475.

Sako, S. (2002). Brain drain and Africa's development: A reflection. *African Issues, 30*(1), 25–30.

Scheyvens, R., Banks, G., & Hughes, E. (2016). The private sector and the SDGs: The need to move beyond 'business as usual'. *Sustainable Development, 24*(6), 371–382.

Seddon, N., Mace, G.M., Naeem, S., Tobias, J.A., Pigot, A.L., Cavanagh, R., . . . Walpole, M. (2016). Biodiversity in the Anthropocene: prospects and policy. *Proceedings of the Royal Society B: Biological Sciences, 283*(1844), 20162094. https://doi.org/10.1098/rspb.2016.2094.

Shahbaz, M., Tiwari, A.K., & Nasir, M. (2013). The effects of financial development, economic growth, coal consumption and trade openness on CO2 emissions in South Africa. *Energy Policy, 61*(2013), 1452–1459. https://doi.org/10.1016/j.enpol.2013.07.006.

South Africa Department of Environmental Affairs. (2011). National Strategy for Sustainable Development and Action Plan (NSSD 1) 2011–2014. https://www.dffe.gov.za/sites/default/files/docs/sustainabledevelopment_actionplan_strategy.pdf. Accessed 5 August 2023.

South Africa Department of Forestry Fisheries and the Environment. (2011). National Strategy for Sustainable Development and Action Plan. https://www.dffe.gov.za/content/documents/strategic_docs/national_strategy_sustainable_development. Accessed 5 August 2023.

South Africa Department of Mineral Resources and Energy. (n.d). Coal resources. http://www.energy.gov.za/files/coal_frame.html. Accessed 5 August 2023.

South Africa Government. (2021). South African Economic Reconstruction and Recovery.

Statistics South Africa. (2019). Tracking South Africa's Social Development Goals. http://www.statssa.gov.za/?p=12813. Accessed 5 August 2013.

Statistics South Africa. (2020). Quarterly employment statistics (QES) June 2020. https://www.statssa.gov.za. Accessed 5 August 2023.

Steffen, W., Grinevald, J., Crutzen, P., & McNeill, J. (2011). The Anthropocene: conceptual and historical perspectives. *Philosophical Transactions of the Royal Society A: Mathematical, Physical and Engineering Sciences, 369*(1938), 842–867.

Steffen, W., Richardson, K., Rockström, J., Cornell, S. E., Fetzer, I., Bennett, E. M., . . . De Wit, C. A. (2015). Planetary boundaries: Guiding human development on a changing planet. *Science, 347*(6223), 736–747. https://doi.org/10.1126/science.1259855.

Stoddard, E. (2021). IMF study suggests civil unrest will have a significant economic impact on South Africa. *Daily Maverick*. https://www.dailymaverick.co.za/article/2021-07-20-imf-study-suggests-civil-unrest-will-have-a-significant-economic-impact-on-south-africa/. Accessed 5 August 2023.

Tankwanchi, A.B.S., Özden, Ç., & Vermund, S.H. (2013). Physician emigration from sub-Saharan Africa to the United States: analysis of the 2011 AMA physician masterfile. *PLoS medicine, 10*(9), e1001513.

The Republic of South Africa National Planning Commission. (2011). Diagnostic overview. https://1213www.gov.za/sites/default/files/gcis_document/201409/npcdiagnosticoverview1.pdf. Accessed 5 August 2023.

The Republic of South Africa National Planning Commission. (2013). National development plan vision 2030. https://www.gov.za/issues/national-development-plan-2030. Accessed 7 August 2023.

The World Bank. (2022). Climate Change Could Force 216 Million People to Migrate Within Their Own Countries by 2050. *Press Release No: 2022/12/CCG, Sep 13, 2021*. https://www.worldbank.org/en/news/press-release/2021/09/13/climate-change-could-force-216-million-people-to-migrate-within-their-own-countries-by-2050. Accessed 5 August 2023.

Tippins, D., & Britton, S.A. (2015). Ecojustice pedagogy. *Encyclopedia of Science Education. Springer, Dordrecht.*

Towheed, S. (2010). Tools and techniques for literary research: Using online and printed sources. In D.d.S. Correa & W.R. Owens (Eds.), *The Handbook to Literary Research* (2nd ed., pp. 9-35). London: Routledge.

United Nations. (2015a). The Paris Agreement. https://unfccc.int/sites/default/files/english_paris_agreement.pdf. Accessed 5 August 2023.

United Nations. (2015b). Transforming our world: the 2030 Agenda for Sustainable Development. Resolution adopted by the General Assembly on 25 Sep 2015 [without reference to a Main Committee (A/70/L.1)] 70/1. Transforming our world: the 2030 Agenda for Sustainable Development. https://www.un.org/en/development/desa/population/migration/generalassembly/docs/globalcompact/A_RES_70_1_E.pdf. Accessed 5 August 2023.

United Nations. (2019a). *Report of the Secretary-General on SDG Progress 2019: Special Edition*. https://sustainabledevelopment.un.org/content/documents/24978Report_of_the_SG_on_SDG_Progress_2019.pdf. Accessed 5 August 2023.

United Nations. (2019b). Resolution adopted by the General Assembly on 15 Oct 2019 [without reference to a Main Committee (A/74/L.2)] 74/4. Political declaration of the high-level political forum on sustainable development convened under the auspices of the General Assembly. https://digitallibrary.un.org/record/3833352?ln=en. Accessed 30 June 2023.

United Nations Department of Economic and Social Affairs. (2022). Goals 13: Take urgent action to combat climate change and its impacts. https://sdgs.un.org/goals/goal13. Accessed 5 August 2023.

Van Nguyen, H., Van Hoang, M., Dao, A.T.M., Nguyen, H.L., Van Nguyen, T., Nguyen, P.T., . . . Gilmour, S. (2020). An adaptive model of health system organization and responses helped Vietnam to successfully halt the Covid-19 pandemic: What lessons can be learned from a resource-constrained country. *The International journal of health planning and management, 35*(5), 988–992.https://doi.org/10.1002/hpm.3004.

Venage, G.d. (2021). Coal Remains Vital to South African Economy. https://www.coalage.com/features/coal-remains-vital-to-south-african-economy/. Accessed 5 August 2023.

Walker, L. (2022). S Africa coal exports decline 13% amid constraints. *Montel AS, 09 Sept 2021 12:27*. https://www.google.com/search?client=firefox-bd&q=S+Africa+coal+exports+decline+13%25+amid+constraints.+Montel+AS%2C+09+1253+Sept+2021+12%3A27. Accessed 5 August 2023.

Wang, Q., & Huang, R. (2021). The impact of COVID-19 pandemic on sustainable development goals–A survey. *Environmental Research, 202*(2021), 1–16.

World Health Organization. (2023). COVID-19 weekly epidemiological update, edition 134, 16 March 2023. https://apps.who.int/iris/handle/10665/366534 Accessed 1 July 2023.

WHO. (2022a). South Africa. https://covid19.who.int/region/afro/country/za. Accessed 24 June 2023.

WHO. (2022b). WHO Coronavirus. (COVID-19). Dashboard. https://covid19.who.int/. Accessed 20 June 2023.

Wilson, G. (2020). Climate Change as a" Threat to the Peace": Responding to Climate Change through the UN Collective Security System. *International Journal of Climate Change: Impacts & Responses, 12*(3).

Winck, B. (2020). The IMF says its forecast for the COVID-19 recession might now be too optimistic. https://www.weforum.org/agenda/2020/04/imf-economy-coronavirus-covid-19-recession/. Accessed 5 August 2023.

World Bank Group. (2020). South Africa. *Poverty & Equity Brief. Sub-Saharan Africa, Apr 2020.* https://databank.worldbank.org. Accessed 5 August 2023.

World Coal Association. (2016). Coal in the energy mix of South Africa. https://www.energy.gov.za/files/media/explained/south-african-coal-sector-report.pdf. Accessed 5 August 2023.

Yüksel, I. (2012). Developing a multi-criteria decision making model for PESTEL analysis. *International Journal of Business and Management, 7*(24), 52.

Zalasiewicz, J., Williams, M., Steffen, W., & Crutzen, P. (2010). The new world of the Anthropocene. *Environmental Science & Technology*, 44(7), 2228-2231. https://doi.org/10.1021/es903118j.

Ziervogel, G., New, M., Archer van Garderen, E., Midgley, G., Taylor, A., Hamann, R., . . . Warburton, M. (2014). Climate change impacts and adaptation in South Africa. *Wiley Interdisciplinary Reviews: Climate Change, 5*(5), 605–620. https://doi.org/10.1002/wcc.295.

Chapter 14
Civil Society Organisations and Community Partnership for Sustainable Livelihoods and Development in Ward 7 of Chivi District, Masvingo Province, Zimbabwe

Washington Mazorodze

Abstract A symbiotic partnership between civil society organisations and rural communities is an important element for poverty reduction and improvement of livelihoods. This type of partnership is not always appreciated in Zimbabwe as it is often perceived as a ploy for regime change. This article examined civil society-community partnership using ward 7 of Chivi district as a point of reference. This partnership involved a coalition of international non- governmental organisations consisting of CARE International, Christian CARE, Food and Agricultural Organisation, Heifer International, World Vision, Red Cross and communities in Chivi District ward 7 of Masvingo Province. Results showed that this partnership resulted in the improvement of community livelihoods through the provision of safe water by drilling solar powered boreholes and promoting gardening to alleviate food shortages in this drought prone district. The improvement of livelihoods in ward 7 culminated into the achievement of sustainable development goals number 1 and 2.

Keywords Civil society · Resilience · Livelihoods · Sustainable development · Hunger · Poverty · Zimbabwe

14.1 Introduction

Civil society is defined as the space outside family, market and the state (WEF, 2013). The concept of civil society has evolved since the term was introduced to the developing countries in the 1980s. Civil society consists of a variety of organised

---

Dr. Washington Mazorodze is a PhD graduate in Public Administration (Peace Studies) from the Durban University of Technology. He researches about domestic, regional and international politics, peace and conflict issues, human security, and governance. Email: washvvvone@gmail.com

---

J. C. K. Kiyala and N. Chivasa (eds.), *Climate Change and Socio-political Violence in Sub-Saharan Africa in the Anthropocene*, The Anthropocene: Politik—Economics—Society—Science 37,
https://doi.org/10.1007/978-3-031-48375-2_14

groups which include non-governmental organisations, communities, trade unions, social movements (VanDyck, 2017). They can operate at local, regional and international levels. Civil society organisations vary in size, structure and objectives. Civil society organisations provide service to the community, advocate for human rights, provide an oversight to the government and act as a watchdog on government's actions, among other roles. The term civil society has been mostly associated with non-governmental organisations. There is a tendency to use civil society and non-governmental organisations interchangeably. Non-governmental organisations are part of civil society but not all civil society are non-governmental organisations (Van Dyck, 2017) This article is based on the partnerships between international non-governmental organisations and the local communities in Chivi district, Masvingo Province in the improvement of the community's livelihoods. This was done through the introduction of an irrigation project and the drilling of the solar powered borehole which helped to reduce poverty and hunger which are placed at the centre of the world's sustainable development priorities. Since the government cannot fully realise the sustainable development goals alone, this necessitated collaboration and the involvement of international civil society organisations who possess the financial resources that the government lacks, given the scope and ambition of the commitments made by governments to achieve these goals by 2030. This chapter was written in response to the lack of serious academic attention given to the localisation of the sustainable development goals through international civil society organisations.

Even though there are tense relations between non-governmental organisations, particularly those that support political rights, and developing nations, the former plays a significant role in raising the living standards of those in the latter by introducing a range of developmental programs like poverty reduction and the provision of safe and clean water as a complement to the latter's efforts, which are frequently insufficient due to limited financial resources. The government of Zimbabwe implemented the Private Voluntary Organizations Amendment Bill 2021 in January 2023 as a result of the dubious link between human rights civil society organisations and the government. The first goal of the bill was to stop terrorist organisations from using non-governmental organisations as fronts or channels for financing terrorism. Secondly, the legislation aimed to prevent private nonprofit organisations from engaging in political lobbying (Moyo, 2023). It was believed that enacting the Bill would silence civil society groups that speak out about Zimbabwe's governance and human rights violations. According to some commentators, the passage of this law will make it impossible even for international civil society organisations engaged in humanitarian work to operate.

Despite the promulgation of the Private and Voluntary Organisations Bill into law by the government of Zimbabwe in January 2023, several international non-governmental organisations have been active in numerous development initiatives in both rural and urban regions in an attempt to improve the welfare of the communities. In this article, the local community and civil society organisations, including the Food and Agricultural Organization (FAO), Heifer, Red Cross, and Care International, launched the Zimari irrigation project, a popular irrigation project that includes six villages and 68 households (interview with community member, 28

August 2021). The project, which was started in 2020, gave the locals access to clean water for domestic use and irrigation, increasing their food security (eliminating hunger and poverty) and improving their way of life. Data collection and analysis employed a qualitative methodology. To determine how much civil society efforts have contributed to enhancing the standard of living for Chivi people, particularly in eradicating hunger and poverty in order to achieve sustainable development goals, interviews with community members were conducted. The study's qualitative nature made it appropriate for a thorough examination of the impact of the phenomena on the communities where the project is located, resulting in accurate and valid data collection and analysis. The article's focus is on the relationship between human security, sustainable livelihoods, and development and peace, and it includes an introduction, information on the project's history, a conceptual framework, a methodology, and a conclusion.

14.2 Background

The Chivi district is located 300 km south of Harare. Masvingo, the provincial capital, lies 65 km to the southwest of Chivi District Centre. It is situated in natural regions 4 and 5 (Mvumi et al. 1998). Due to the poor rainfall patterns in this region of the country, people in the Chivi district engage in subsistence farming and cultivate traditional crops like sorghum, millet, and groundnuts, among others, that are drought resistant. The majority of residents in Chivi did not have access to clean, safe water because of the region's ongoing drought. They drank dangerous river water that had been contaminated by human waste. In addition to having limited access to water, the area lacked sanitation, with most residents relieving themselves in bushes. There was an upsurge in diarrheal and water related diseases as a result of both a lack of sufficient rain and limited access to sanitation. The economic state of the nation, which made it challenging for the government to expand social spending to provide water services and sanitation to rural populations, did not assist. In order to supplement government efforts in providing social services that the government had failed to offer, international civil society organisations' intervention became essential and vital. As a result, the project was started by the aforementioned international organisations in collaboration with local leaders and members of the community to drill a solar-powered borehole and provide residents of Chivi District Ward 7 with safe, dependable cooking, drinking, and irrigation water.

Given that agriculture is Zimbabwe's mainstay of the economy, contributing to 18% of the country's (GDP) gross domestic product (GDP), the provision of clean water for both consumption and irrigation in Chivi District Ward 7 is consistent with the new government of Zimbabwe's vision 2030, which is intended to, among other things, improve agricultural productivity. The president declared that the new Zimbabwean government had started a number of initiatives to improve food security in the nation. These programs included building dams, restarting, and constructing irrigation systems. The government's vision 2030, which was driven by international civil society organisations in collaboration with local communities, was realised with the introduction of an irrigation system in Chivi district Ward 7. The Zimbabwean President also mentioned in his speech that Agenda 2063, which aimed to alleviate poverty in Africa, and the achievement of Sustainable Development Goals were the inspiration for Zimbabwe's Vision 2030. The President stated that Zimbabwe's Vision 2030 was a development strategy that aimed to leave no one behind (Office of the President and Cabinet 2019). It is a development strategy that is put into action through the national development strategy 1 (2021–2025), a five-year development plan. The development strategy aims to elevate Zimbabwe to middle-income status by 2030.

14.3 Conceptual Framework: Sustainable Livelihoods Framework

A holistic approach known as the sustainable livelihoods approach seeks to identify the root causes and aspects of poverty without condensing them into a small number of components. It also seeks to outline the connections between various dimensions of poverty, allowing for a more efficient operational prioritisation of action. By employing poverty indicators that they themselves identify, the Sustainable Livelihoods Framework intends to assist the disadvantaged groups in achieving sustainable livelihood improvements (Sustainable Livelihoods Support Office, 1999). According to Chambers and Conway (1992: 7–8), a livelihood consists of the skills, possessions (such as stores, resources, claims, and access), and activities necessary for a means of subsistence. A livelihood is considered sustainable if it can withstand stress and shocks and bounce back, maintain or improve its capabilities and assets, offer opportunities for future generations to live sustainably, and contribute positively to other livelihoods both locally and globally, as well as over the long and short terms.

The Sustainable Livelihoods concept is discussed in this chapter together with the peacebuilding idea. In order to prevent a relapse into violence, Boutros Boutros Ghali, the then-UN Secretary General, defined peacebuilding as efforts made to identify and support systems that tend to maintain and solidify peace (Ghali, 1992). According to his concept, peacebuilding was the opposite of preventative diplomacy, which entailed taking measures to avert crises, whereas peacebuilding involved measures intended to stop conflicts from recurring. Galtung coined the phrase "peacebuilding," but it wasn't until Boutros-Boutros Ghali popularised it in the 1990s that it gained widespread acceptance and use. The phrase "peacebuilding" has changed somewhat since Boutros Boutros Ghali made it popular in the 1990s. Instead of focusing solely on preventing conflict from recurring, it is now also about developing and bolstering neighborhood social structures, which help to build local resilience that enables societies to endure both internal tensions and external pressures and shocks. Galtung separated positive peace from negative peace, the latter which only involved the absence of military conflict. According to De Coning (2008), peacebuilding is primarily concerned with the consolidation of peace and entails addressing both the risk factors for violent conflict and the underlying causes of conflict that could jeopardise peace. The sustainable livelihoods framework and the sustainable development goals' focus on eradicating poverty and hunger are peacebuilding enhancing strategies that seek to assist communities in coping with and recovering from stress and shocks, thereby strengthening social cohesion and peace structures. It is well known that poverty is one of the causes of conflict. Threats to peace are removed in this manner.

A framework for long-term poverty alleviation is provided by the sustainable livelihoods framework. The framework for sustainable livelihoods differs from conventional approaches that attempted to combat poverty by identifying and meeting the needs of the poor without their input. According to conventional methods, those in need only received assistance in a passive manner (Mcleod, 2001). The local communities, who are not only the recipients of aid but also have local expertise, operate in conjunction with the external actors who typically have financial resources under the new framework for sustainable livelihoods. As a result, there is a tight connection between the sustainable livelihoods framework and peacebuilding because by reducing poverty with the active involvement of local communities, the sustainable livelihoods framework lays the foundation for peace consolidation. The goal of a sustainable livelihood framework is to build on local residents' existing resources (UNDP, 1999). In this instance, the local population had the manpower to produce bricks, irrigation canals, and other forms of capital that do not require a lot of financial resources. The framework for sustainable livelihoods brings together a variety of actors, including local government, civil society, and local communities, and emphasises the connections between micro-level livelihood systems and macro-level policies that have an impact on these livelihoods (UNDP, 1999). Several parties were involved in this article's discussion on sustainable livelihoods, including the Red Cross, Care International, FAO, Heifer, the Chivi Rural District Council, and local residents. Borehole drilling and irrigation activities in Chivi district's Ward 7 were made possible by financial and material support from international civil society organisations. As a representative of the central government, the Chivi Rural District Council provided direction for the execution of the international civil society intervention and a supportive working environment between civil society organisations and the local community.

The labour needed to drill, run, and maintain the borehole was provided by the local community. To emphasise the community's engagement in the planning and execution of the project as opposed to being passive recipients of aid, the local communities also supplied the irrigation project with a site that was convenient for every community member.

According to the Sustainable Livelihood Framework, people act in a vulnerable environment. The residents of Chivi District Ward 7 in this article are at risk of poverty because they consistently receive below-average rainfall, which causes crops to fail and, as a result, creates a food crisis. The Chivi tribes frequently relied on importing food from other regions with greater rainfall, but doing so was expensive for the majority of households, who could not afford to do so because of limited financial capacity. They also depended on international non-governmental organisations for food aid, with the World Food Programme (WFP) serving as the district's top provider. Unfortunately, it was unable to supply enough food for every household, leading to the introduction of the irrigation project to increase community production instead of providing food aid. According to the Sustainable Livelihood Framework, the local population has access to a variety of resources that help reduce poverty and are given meaning and value by the social, institutional, and organisational environment in which they are found (UNDP 1999). If local populations are given access to resources like boreholes that they are unable to get on their own, such as labour and land, they can reduce their reliance on foreign aid. The success of livelihood strategies is determined by the availability and accessibility of assets, services, and opportunities, which can be positively or negatively influenced by ecological conditions, social structures, and institutional processes. In this situation, the ecological parameters were favorable for borehole drilling because the water table was placed with enough water for the irrigation project. The project was embraced by the local social structures as an enabling platform for community members to address poverty and hunger, which happen to be Sustainable Development Goals 1 and 2. Government institutions were also welcomed and attentive to civil society's help to the local people in Ward 7.

14.4 UN Sustainable Development Goals

In 2014, the UNDP emphasised the importance of civil society in identifying crucial development objectives and presenting practical solutions and policy opportunities (UNDP 2014). Heads of State and Government, as well as High Representatives, launched sustainable development goals on September 25–27, 2015. These sustainable development goals were based on the millennium development goals and attempted to complete the latter's unfinished business. The two most critical commitments that the world faced with in this article were to abolish poverty and hunger. The world made a bold commitment to end poverty and hunger in all forms by 2030 (www.un.org/sustainabledevelopment/hunger). One of the objectives of the Sustainable Development Goals was to leave no one behind, which meant that development should benefit the poor, marginalised, vulnerable, and oppressed as well. The most vulnerable groups would be targeted, and in this case, international civil society organisations chose the ward 7 community in Chivi district because it is one of the locations that has undergone severe perennial droughts. The decision by international civil society organisations to partner with poor communities fulfills the commitments made by heads of states and governments to ensure the mobilisation of resources from a variety of sources in order to implement programs and policies to end poverty in all of its dimensions and strengthen the resilience of the poor by the year 2030. International civil society organisations working in the district not only targeted the poor and vulnerable for sustainable development, but they also listened to their issues and worked with the community to draw out locally based solutions. Chambers (1983) agreed, noting that the goal of localisation is to prioritise the last. In this scenario, multinational civil society organisations are working with disadvantaged and vulnerable communities to attain sustainable development goals by localising them.

Aside from poverty elimination, the second Sustainable Development Goal is based on the concept of zero hunger. The heads of states and governments pledged to abolish hunger and malnutrition by 2030 by ensuring that everyone has access to adequate and nutritious food at all times. This entails introducing sustainable agricultural production through international cooperation, land access, and market access. A commitment was made to double small-scale farmers' agricultural production by 2030 through secure and equal access to land, other productive resources and inputs, knowledge, financial services, markets, and opportunities for value addition and non-farm employment (www.un.org/sustainabledevelopment/hunger/). The launch of the Zimari irrigation project in Chivi District Ward 7 as part of a civil society-community collaboration highlighted the importance of zero hunger because the community is now producing from the irrigation project, thereby addressing hunger which used to be a perennial challenge in the area.

14.5 Research Methodology

A qualitative research approach was used to answer the study objectives outlined in this article in a systematic manner. Because qualitative research does not strive to generalise results, it uses small samples to better understand phenomenon in their original context. In this article, a small sample of respondents was conveniently chosen for interviews. Their perspectives on the irrigation project's influence on eradicating poverty and hunger represented the entire ward. Qualitative technique also aims to comprehend a specific study problem through the eyes of the local population. The research in this example is concerned with analysing the impact of community-civil society partnership programs on the local people of Chivi ward 7, hence a qualitative research approach was used. In this chapter, respondents were sampled using convenience sampling. This sampling strategy involves picking available and easily accessible respondents. The researcher questioned local residents who were physically available and accessible. This method of sampling has the advantage of saving time and interviewing persons who had also witnessed the phenomena.

According to Snape/Spencer (2003), qualitative research is a natural, explanatory technique concerned with understanding the significance that humans place on an event in their context. They drew out some crucial components that distinguished qualitative research from quantitative research. These are some examples: Qualitative research methodology provides a more in-depth understanding of the social world. It employs small samples of between 5 and 20 units; it employs data gathering methods like as interviews, focus groups, and participant observation; and it allows for the exploration of novel themes and concepts. The qualitative research method describes and discusses differences, interactions, individual experiences, and group norms. Its strength is that it is based on respondents' thoughts, feelings, and experiences in their native environment. According to Collis/Hassey (2003), the advantage of qualitative research is that it provides a comprehensive description and analysis of a research topic without limiting the scope of the investigation or the character of participants' responses.

However, the disadvantage of qualitative research methodology is that, while it is rich, it is time consuming and less generalisable. In addition, the researcher is subjectively involved in the topic matter (Miles/Haberman, 1994: 40). Furthermore, Bell (2005) contends that because qualitative research uses small samples, the results cannot be regarded as reflecting the larger community and so cannot be generalised. These flaws in qualitative research, on which this study was founded, were avoided by ensuring that the researcher was not emotionally attached to the research and was as objective as possible in data collection and processing. Fieldwork was carried out between December 2019 and August 2021. As a result, six in-depth interviews were conducted in Chivi district ward 7 with three male and three female respondents. The irrigation project benefited all of the respondents. The researcher also wished to interview the district administrator and the ward Councillor about this program, however they were both unable owing to network issues and other commitments.

The following is how the chapter is organised: First, a backdrop to the project's introduction and the reasons for the program's introduction. Second, an examination of the project's impact on people's livelihoods. Third, the project's contribution to peacebuilding and sustainable development. Finally, the chapter concludes with conclusions and recommendations.

14.6 Analysis: Enhanced Livelihoods in Chivi Ward 7

According to the researcher's findings, the implementation of the Zimari irrigation project in Chivi District, Ward 7, resulted in improved livelihoods, which was consistent with Sustainable Goals 1 and 2. It is therefore no accident that several international civil society organisations collaborated with rural communities to introduce water and sanitation programs aimed at reducing and eventually eliminating poverty and hunger. The building of a solar-powered borehole in Chivi district ward 7 provided a much-needed boost to the drought-stricken community. Because of the borehole drilled by the Food and Agriculture Organisation in collaboration with Heifer, Care International, Red Cross, and local residents, the community of Ward 7 in Chivi district is now growing a variety of crops throughout the year. Since the project's inception in 2019, poverty levels in the ward have been dramatically reduced, as households now have access to not just adequate but also nutritional food. The project also resulted in the achievement of sustainable development goal number 2: Zero Hunger which aims to end hunger, achieve food security, improve nutrition, and promote sustainable agriculture (Goal 2: Zero Hunger—United Nations Sustainable Development). The researcher discovered that community members' food security and nutrition had substantially improved since the project's inception by producing a variety of crops that enhanced and balanced their diet. While in the past, even if community members wanted to improve their food security and nutrition needs, it was not possible because they did not have access to adequate rainfall on which they relied, but with the drilling of a solar powered borehole, water became available for the community members who were eager to grow a variety of crops such as vegetables, maize, beans, and carrots, among others, which ultimately resulted in them managing to achieve food security. It must be underlined in this article that ensuring food security is a critical component of peacebuilding. When people lack food security, they are more prone to be violent. There were many violent disputes prior to the implementation of the irrigation project because community members were hungry and lacked access to food. Because of poverty, some community members ended up committing crimes such as stealing and house-breaking. When people are food insecure, their communities become divided and social cohesion lacks. This irrigation project improved social cohesiveness because community members worked together to drill, manage, and maintain the borehole, which provided a source of livelihood for them. The researcher recognised in this article an intriguing mix of improving lives, achieving sustainable development goals, and fostering a calm environment.

The irrigation project not only helped to achieve Sustainable Goals 1 and 2, but also Goals 3 and 6 even if they are not the focus of this article. Goal 3 seeks good health and wellbeing, ensuring healthy lifestyles and promoting well-being for all ages. Families involved in the irrigation project have been ensured of excellent health by growing a range of crops all year round, as they do not have to go without food as they did in the past due to irregular rainfall, which resulted in crop failures and famine. The sixth sustainable goal ensures the provision of safe drinking water and sanitation. Prior to the implementation of the project, people relied on water taken from wells drilled in rivers, which was unsafe to drink. The water in the rivers was tainted with human waste because the people did not have bathrooms and relied on the bush. This resulted in a high number of cases of water-borne infections in the district. The implementation of the irrigation project coincided with the Red Cross' sanitation project, which urged community members to construct pit latrine toilets in order to reduce cases of water-borne disease, which were on the rise in the area and throughout the country. As a way of encouraging community members to build toilets, the Red Cross provided technical assistance to villages who desired to install these toilets. The researcher interviewed locals and discovered that the prevalence of water-borne infections had decreased in the ward due to the successful implementation of both the irrigation and sanitation projects by international civil society organisations in collaboration with local communities. The interviews with community members also revealed that this partnership had helped the ward to achieve sustainable development goals 1, 2, 3, and 6.

14.7 The Convergence of Sustainable Development and Peacebuilding in Chivi Ward 7

Peace and sustainable development are inextricably linked. The World Commission on Environment and Development (1987) defined sustainable development as development that meets the requirements of the current generation without jeopardising future generations' ability to meet their own needs. The goal of sustainable development is to strike a balance between economic development and environmental stability by ensuring that economic progress does not harm the environment. It was no surprise that peace was included as an integral component of the package when the United Nations Summit on Sustainable Development convened in 2015. Member states, particularly those from the developing world, recognised the close relationship between peace and development. Peacebuilding is defined as the removal of all potential causes of violence. Peacebuilding entails acts aimed at conflict prevention, management, resolution, and transformation, which must be performed thoroughly in order to resolve societal tensions (Galtung, 1996). Poverty and food insecurity are two issues that have been linked to violence around the world. The convergence of sustainable and peacebuilding was made possible by the reduction and maybe elimination of poverty and hunger in Chivi district ward 7 through civil society-local community collaboration. This strong relationship between peace and development culminated in SDG 16, which aims to promote peaceful and inclusive societies. Violence and insecurity impair development and vice versa, according to research from around the world (World Development Report, 2011). The researcher questioned a ward 7 community member who stated that the irrigation plan was a shot in the arm for the community because of the numerous benefits it provided. The member mentioned,

> We are very grateful to FAO and Heifer for the irrigation project that they introduced to us. We are a community which traditionally struggled to make ends meet due to erratic rainfall in the ward. But with the introduction of the irrigation, we produce food all year round and poverty has deserted us.[1]

According to another community member, the ward is no longer impoverished as a result of the availability of water. The community member had this to say about the irrigation scheme,

> Since the introduction of this irrigation scheme, our health has changed tremendously. We grow a variety of nutritious crops in the garden such as carrots, vegetables, maize, tomatoes and onions. Ever since we started this project, I have never gone to sleep on an empty stomach. There is always something to eat in my house.[2]

According to various other interviews with community members, the initiative assisted community members in reducing poverty and improving their health status, as alluded to by the previous respondent who stated that they planted a range of foods to feed themselves.

---

[1] Interview with a community member, 10 Aug 2022.

[2] Interview with a community member, 19 Aug 2022.

Another interviewee stated that the rate of crime in the ward had greatly decreased, which he attributed to the implementation of the irrigation project. He said that before to the implementation of the system, the ward had a high number of goat thefts, but since the implementation of the scheme, the number of goat thefts in the area has decreased.

Without a doubt, the project has tackled the origins or underlying causes of violence in society. The community has been empowered as a result of an excellent collaboration between the community and international civil society organisations. One respondent underlined that the community felt empowered by international civil society organisations, which donated financial resources to the irrigation scheme's implementation. The initiative is community-based and owned, in addition to financial and technical contributions from international civil society organisations. The irrigation scheme's daily programs are developed by community members with no direct involvement from civic society. Community members made decisions about which crops to cultivate and other aspects of the irrigation scheme, as well as the upkeep of the equipment at the irrigation site. For example, each family is responsible for guarding the solar facility and water taps during the day, and this responsibility is performed once a month. At night, the community hired security guards to guard the facility. The security guards were paid by the community members. Each member of the community contributed one dollar (USD) per month towards the payment the security personnel. The importance of communal ownership in this plan cannot be overstated.[3] Social cohesion is a fundamental feature of long-term peacebuilding. The level of connectivity and solidarity among groups in society is referred to as social cohesiveness (Easterly et al., 2006). The concept of social cohesiveness is founded on two dimensions: first, a community's sense of belonging, and second, the interaction between community members themselves. Following the implementation of the irrigation plan, the level of social cohesion in Ward 7 grew dramatically. This project was successful in bringing together 67 families from seven different villages. The anticipation with such a large number of families is that there will be little cohesion, however the way this plan is run demonstrates that there is significant cohesion among the many groups that are part of the scheme.

The level of cooperation among community members on the irrigation issue demonstrates the presence of social cohesion in the ward. Members always encourage one another to participate in irrigation-related activities, and in the event of a breakdown of the irrigation's water taps, community members always work together to ensure that they are fixed. One interviewee underlined that they also consider environmental considerations when growing crops in the irrigation. He stated that they always consult with Agritex officers from the Ministry of Agriculture in order to execute proper farming methods that conserve the environment for future generations to use. He also claims that community members are following the traditional farming methods promoted and recommended by the Ministry of Agriculture, demonstrating a high level of social cohesiveness. The Zimbabwean President, speaking at the launch of the African Women Leaders Network, Zimbabwe Chapter, emphasised

[3] Interview with a community member, 27 Sept 2022.

the importance of social cohesion in achieving peace and sustainable development, noting that peace and social cohesion are prerequisites for achieving the sustainable development goals (United Nations Economic Commission for Africa, 2020). This notion has now been adopted by the residents of Chivi Ward 7.

14.8 Challenges

The ward 7 community members who are working with international civil society organisations on the irrigation project revealed that while the irrigation project was a shot in the arm and had managed to eradicate poverty and hunger among them, they were concerned that they did not have markets to sell their crops that they produced due to lack of access to markets and distance to markets. Members of the community stated that they were no longer producing solely for their own consumption but were also producing in excess in order to sell the excess products. The irrigation is located around 100 km from Masvingo's major town, making it difficult for members to transport their harvest, particularly vegetables, to Masvingo. They cited high transportation expenses as a barrier. Because they could not sell their goods, they had resorted to storing it for future use by employing conventional methods of storage. This was viewed as a challenge because they needed to sell their excess produce in order to make money to purchase some essential necessities that they could not produce with their irrigation. They aappealed to civil society organisations with whom they were partnering to assist them in finding markets for their produce. Besides facing difficulties in getting markets for their irrigation products, they also had domesticated animals such as cattle, donkeys, and goats breaking into the irrigation and destroying their crops, particularly at night. Members interviewed by the researcher complained about agricultural devastation caused by unsafe fence. They utilised barbed wire to enclose the irrigated area because they could not afford a more secure mesh fence. As a result, members' crops have been destroyed by domestic animals. Despite these obstacles, they demonstrated that the irrigation made their lives better than before.

14.9 Conclusion

The installation of a solar-powered borehole and irrigation in Chivi district's Ward 7 as part of a civil society-community partnership came as a relief to community members who had been subjected to periodic erratic rainfall patterns for a long time, failing to meet their sustainable development goals. Erratic rainfall patterns in Ward 7 exposed community members to poverty and water conflicts, which periodically erupted when members attempted to get water for domestic use, despite the fact that the water drawn from the river was contaminated by human waste and health risk. The installation of a solar-powered borehole provided large-scale access to water and horticulture, which considerably contributes to the achievement of Sustainable Development Goals 1 and 2. The ward's reduction and eradication of hunger, as well as access to safe drinking water, serves as a springboard for long-term peace. Crop production all year promotes food security while also reducing poverty and theft, making them peacebuilding measures. The example of Chivi district ward 7 indicates that community-civil society collaboration for sustainable development and peace is achievable. If it is applied in other wards in the district that also face water issues, it has the potential to significantly reduce poverty and community conflict, resulting in a developed and peaceful society. However, at this critical juncture, what is required in Ward 7 is for the communities to hunt for markets where they may sell their produce, something they have so far failed to accomplish. They must perform market research for things that will help them become successful farmers.

References

Bell, J. (2005). *Doing your Research Project*. Buckingham: Open University Press.

Chambers, R., & Conway, G. (1992). Sustainable Rural Livelihoods: Practical Concepts for the 21st Century. *IDS Discussion Paper 296*. Brighton: IDS.

Chipato, F., Ncube, C., & Dorman, S. R. The Politics of Civil Society in Zimbabwe. In M. Tendi, J. McGregor, & J. Alexander (Eds.), *The Oxford Handbook of Zimbabwean Politics* (pp. 0). Oxford University Press. https://doi.org/10.1093/oxfordhb/9780198805472.013.8.

Collis, J., & Hussey, R. (2003). *Business Research: A Practical Guide for Undergraduate and Postgraduate Students*. Hampshire: Palgrave Macmillan.

Creswell, J. W. (1998). *Qualitative Inquiry and Research Design: Choosing among Five Traditions*. London: Sage.

de Coning, C. (2008). Understanding peacebuilding: Consolidating the peace process. Conflict Trends, 4. https://doi.org/10.10520/EJC16026. Accessed 14 June 2022.

Easterly, W., Ritzen, J., & Woolcock, M. (2006). Social cohesion, institutions, and growth. *Economics and Politics, 18*(2), 103–120.

Galtung, J. (1996). Peace by peaceful means: Peace and conflict, development and civilisation. London. Sage Publications.

Ghali, B. B. (1992). An agenda for peace: Preventive diplomacy, peacemaking and peacekeeping. United Nations. https://unrol.org/files/A_47_277.pdf. Accessed 14 June 2022.

Ghani, E., & Lyer, L. (2010). Conflict and development–lessons from South Asia. *Economic Premise, The World Bank*, 31 (Sep.), 2–3.

Huaxianews. (2020). Zimbabwe making progress towards attainment of vision 2030: President. *Xinhuanet*, Dec 16. http://www.xinhuanet.com/english/2020-12/16/c_139594452.htm. Accessed 24 October 2021.

Miles, M. B., & Huberman, A. M. (1994). *Qualitative data analysis: An expanded sourcebook*. CA: sage.

Moyo, T. (2023). Zimbabwean government passes law designed to throttle independent civil society. *Daily Maverick*, January 23. https://www.dailymaverick.co.za/article/2023-01-23-zimbabwean-government-passes-law-designed-to-throttle-independent-civil-society/. Accessed 10 June 2023.

Mvumi. B., Donaldson. T., & Mhunduru. J. (1998). A report on baseline data available for Chivi District, Masvingo Province. Crop post harvest programme Zimbabwe. Final Draft Report. Cranfield University, *Funded by the UK Department for International Development under the*

Renewable Natural Resources Research Strategy 1995- 2005. https://assets.publishing.service.gov.uk/media/57a08d9ced915d3cfd001b16/R6674g.pdf. Accessed 8 June 2023.

OECD. (2014). Fragile states: Domestic revenue mobilization in fragile states. https://www.oecd.org/dac/conflict-fragility-resilience/docs/FSR-2014.pdf. Accessed 8 June 2023.

OECD. (2015). States of fragility 2015: Meeting post- 2015 ambitions- highlights, Mar. https://gsdrc.org/document-library/states-of-fragility-2015-meeting-post-2015-ambitions/. Accessed 8 June 2023.

Office of the President and Cabinet (Zimbabwe). (2019). Vision 2030 key to end poverty, says President. OPC. http://www.theopc.gov.zw/index.php/546-vision-2030-key-to-end-poverty-says-president. Accessed 24 October 2021.

Snape, D., & Spencer, L. (2003). Chapter 1. The foundations of qualitative research. In Richie, J. & Lewis, J. (2003). *Qualitative research practice*. London: Sage.

Sustainable Livelihoods Support Office (DFID). (1999). Sustainable livelihoods and poverty elimination. http://www.livelihoods.org/info/docs/dec99bbfg.htm/. Accessed 3 September 2021.

UN. (2023). Sustainable Development Goals. Goal 2: Zero Hunger. https://www.un.org/sustainabledevelopment/hunger/. Accessed 8 December 2022.

UN. (n.d.). Sustainable Development Goals. Goal 1: End poverty in all its forms everywhere. https://www.un.org/sustainabledevelopment/poverty. Accessed 28 October 2020.

UNDP. (1999). SLA in urban areas. http://www.undp.org/sl/Documents/General%20info/. Accessed 3 September 2021.

United Nations Development Program (UNDP, 2014). "Delivering the Post-2015 Development Agenda: Opportunities at the National and Local Levels." http://www.undp.org/content/dam/undp/library/MDG/. Accessed 29 May 2023.

United Nations Economic Commission for Africa. (2020). Address by the President of the Republic of Zimbabwe and Commander- in Chief of the Defence Forces, his Excellency, DR, E.D. Mnangagwa, at the Launch of the African Women Leaders Network (AWLN) Zimbabwe Chapter. *UNECA*. Feb 25. https://www.uneca.org/sites/default/files/TCND/ARFSD2020/statement-by-president-of-zimbabwe.pdf. Accessed 24 October 2021.

VanDyck, C.K. (2017). Concept and definition of civil society sustainability. Washington DC: *Centre for Strategic and International Studies*. https://csis-prod.s3.amazonaws.com/s3fspublic/publication/170630_VanDyck_CivilSocietySustainability_Web.pdf?QfxMIeHr9U8aeV1kOjFo.FBTsLG76HPX. Accessed 23 August 2021.

World Economic Forum. (2013). The future role of civil society. *World Economic Forum in collaboration with KPMG International*. http://www3.weforum.org/docs/WEF_FutureRoleCivilSociety_Report_2013.pdf. Accessed 23 August 2021.

World Commission on Environment and Development. (1987). *Our Common Future*. Oxford: Oxford University Press.

World Development Report. (2011). Conflict, security and development. World Bank. World development report 2011 : conflict, security, and development - overview (worldbank.org). Accessed 8 December 2022.

Chapter 15
Peace, Security and Sustainable Development Goals in Zimbabwe

Torque Mude

Abstract Peace and security issues are at the core of Sustainable Development Goals (SDGs) that the world seeks to achieve by the year 2030. Broadly conceived, SDGs also known as Global Goals are a global roadmap to end poverty, protect the planet and ensure that all people enjoy peace and prosperity. While there is considerable peace in Zimbabwe, the trajectory of human security in the country leaves a lot to be desired. With recurring violent protests, political violence, inter-party and intra-party latent conflicts, human rights abuses and most of all the military rule that emerged with the military ouster of then-President Robert Mugabe, sustaining peace, which is panacea to the achievement of SDGs in Zimbabwe, hangs in the balance and could remain a phantom on the horizons if 'guns are not silenced' and human security is not prioritised. Against this background, the objective of this chapter is to examine the options available in the pursuit of sustainable peace and human security for the attainment of SDGs in Zimbabwe. The conclusion drawn from this chapter is that both sustainable peace and the security are critical for the attainment of Global Goals in Zimbabwe. Hence, democratisation, reconciliation, transitional justice, the eradication of poverty, protection of the planetary biosphere upon which human survival depends on and sustaining peace are sine qua non to the attainment of SDGs in Zimbabwe. Data for this paper were obtained from secondary data sources including books, newspapers, journal articles and published and unpublished dissertations and theses. Secondary data was then categorised into themes and concepts from which inferences transformed into arguments advanced into this chapter were drawn.

Keywords Sustainable peace · Human security · Sustainable development goals · Zimbabwe.

Torque Mude (PhD) serves as a Postdoctoral Research Fellow at the Centre for Africa-China Studies of the University of Johannesburg and Senior Lecturer in the Department of Public Management and Political Studies at the University of Namibia. He holds a Ph.D. in International Politics and is in the final year of the Bachelor of Laws Degree (UNISA). His research engages African politics, international relations, governance, peacebuilding, international law and development. Email: tmude@unam.na; mudetorque@gmail.com. ORCID: https://orcid.org/0000-0002-9885-6637

J. C. K. Kiyala and N. Chivasa (eds.), *Climate Change and Socio-political Violence in Sub-Saharan Africa in the Anthropocene*, The Anthropocene: Politik—Economics—Society—Science 37,
https://doi.org/10.1007/978-3-031-48375-2_15

15.1 Introduction

Just like with other members of international community, the attainment of SDGs by the year 2030 is on Zimbabwe's priority schedule. At the core of these Global Goals lie peace and security issues which constitute SDG 16 calling for peaceful and inclusive societies premised on the respect for human rights, rule of law and good governance. From ending poverty to strengthening partnerships for global development, SDGs are geared towards achieving development, peace and security. To this end, putting an end to poverty and hunger, promotion of human rights, rule of law and good governance, reduction of income inequality, promotion of economic security, gender equality attainment of quality education and health security among others are critical in the attainment of Global Goals in Zimbabwe. While the foregoing peace and security issues in Zimbabwe leave a lot to be desired, there are promising prospects of attaining them in the realisation of SDGs. Be that as it may, it suffices to indicate that a lot needs to be done to achieve the desirable measures of peace and security that conform to the aspirations of SDGs.

Even though it appears more of an uphill task to achieve enviable levels of sustainable peace and human security that tally with SDG aspirations in Zimbabwe by 2030, it takes political will and commitment to make such headway. While efforts are being made, for instance, to achieve food security through command agricultural programmes, other sectors that are of primary importance in the attainment of SDGs are being neglected. For instance, due to militarisation of the state a lot needs to be done to restore good governance and human rights promotion which are largely undermined by the politicisation and militarisation of domestic governance institutions in the country. More so, the rule of law which is one of the pillars of good governance and human rights protection is heavily compromised. Apart from challenges mentioned above, there is ample room for improvement towards Zimbabwe's attainment of SDGs by 2030. At this juncture, it suffices to conceptualise peace, human security and SDGs. This chapter is organised into four sections. The first section introduces the chapter. In the second section, key concepts are conceptualised and this is followed by a section that discusses findings and lastly the conclusion.

15.2 Conceptual Framework

15.2.1 Unpacking the Concept Peace

Like many other theoretical concepts, the term peace cannot be defined without attracting objections and counter-objections due to different interpretations as well as theoretical bases. As such, the term has often been dismissed as utopian and that it lacks analytical rigor (Royce, 2004). However, there is consensus among scholars and peace practitioners on what peace denotes; tranquillity, harmony, justice and other desirable attributes of society (Taheri and Dehghan, 2014; Webster's Third

New International Dictionary, 1993; Galtung, 1990; Aaron, 1966; Quinchy, 1964). A ground-breaking conceptualisation of peace was proffered by Johan Galtung (1985) who stresses the importance of distinguishing between positive and negative peace in defining peace. For him, negative peace is "the absence of violence, absence of war" and positive peace is "the integration of human society" characterised by harmony, justice and other desirable ways of life (Ibid). For the purpose of this chapter, Galtung's definition of peace will be used to provide theoretical reference of the term and in so far as the term conjures harmonious relations.

15.2.2 The Human-Centred Security Paradigm

Like peace, the term security is the central focus in post-Cold War international politics; yet it is an essentially contested concept. The concept security is also deemed ambiguous (Wolfers, 1952). Classical definitions of security focus on the freedom from war (Bellany, 1981: 102), successful pursuit of national interests (Thunberg, 1982: 50), safeguarding identity and values (Ayoob, 1984) and military capability (Luciani, 1989: 151). Drawing from traditional security paradigms such as realism and pluralism, classical security was understood in terms of the welfare and survival of the state pursued largely through military strength. Drawing from the ideas propagated by Lord Castlereagh and Woodrow Wilson on just equilibrium and self-determination, the end of the Cold War ushered a paradigm shift from the centrality of the state to the accommodation of individuals as an object of security (Rothschild, 1995: 54). As a relatively new security paradigm, human security has become the dominant tool of analysis of the security and sovereignty of the individual, albeit from a liberalist perspective.

Departing from the rigidity of the traditional security paradigm while conceived as complimentary to the traditional understanding of security, human security is understood as a preventative rather than reactionary security paradigm premised on the safety from chronic threats to human welfare such as violence, repression, hunger and diseases among others (Jones, 2014; Gomez and Gasper, 2013; ul Haq, 1995; United Nations Development Programme [UNDP], 1994). However, this has not escaped criticism from political realists who view the human security discourse as a direct insult and affront to the traditional state security paradigm whose narrow focus debunks the totality of the focus of security (Booth, 2007). For example, Buzan (2004) criticises human security as a reductionist and idealistic notion that has little value in the security lexicons.

Even though the traditional conceptualisation of security which puts the state at the apex has maintained its analytical rigor, the emergence of the human security paradigm has altered and broadened the focus of security. It now encompasses threats to the welfare of the individual that transcend military force including availability of food, unemployment, poverty, natural disasters, epidemics, social and political injustice and related socio-economic threats. From a maximalist approach, Buzan

(1991) conceptualises security as the pursuit of freedom and survival through maintenance of independent identity and integrity against hostile forces by both states and individuals. Buzan argues that security is primarily concerned with the fate of human safety and secondarily about the personal welfare of the individual which cannot be divorced from the sovereign territorial state the individual resides in. In his analysis, he does not narrow down the focus of security to either the state or the individual as he recognises the importance of security at both the individual and state levels. This conceptualisation of security provides constructive service in highlighting the holistic approach to security by stressing the importance of both the state and the individual as referent objects of security. However, countries like Zimbabwe view security as a derivative of power.

It is not mere logomachy and sentimental that the understanding of security in the Zimbabwean context is envisaged more from the perspective of traditional understanding of security as the survival of the state from external and internal political forces. Ample evidence of this reality can be found in examining the manner in which the Zimbabwean government prioritises social and economic concerns affecting the people. For instance, the tendency to use of force to suppress protests is sufficient evidence of the politicisation and militarisation of security in Zimbabwe. Instead of redressing the socio-economic concerns of the population, the Zimbabwean government has a tendency of politicising the contentious political activities by purporting that they are part of the Western sponsored regime change agenda in Zimbabwe and resorted to the denial policy response of deploying the police and army to suppress the protests.

With such kind of political orientation, the attainment of peace and security in line with SDGs will remain a phantom on the horizon. Nevertheless, it only takes policy alignment to the tenets of the SDGs to attain the Global Goals and policy alignment takes political will and commitment. Against this background, the main thesis of this chapter is that the attainment of SDGs in Zimbabwe is depended on the achievement of acceptable levels of peace and security in the country. This is attributable to the strong correlation between SDGs and peace and security. The following section conceptualises SDGs and highlights the relationship between Global Goals and peace and security.

15.2.3 A Conceptual Overview of Sustainable Development Goals

SDGs emerged in September 2015 as a new universal agenda building on the Millennium Development Goals (MDGs) with the aspiration to build from what the MDGs failed to achieve by the year 2015. SDGs are conceived as a people-centred set of global and transformative goals and targets designed to free humanity from the shackles of poverty and other undesirable ways of life such as epidemics, pandemics, conflicts, gender inequality, climate change and injustice among others (United

Nations News, 2020). They consist of seventeen Goals that are set to be achieved within fifteen years from the year 2015. These seventeen Goals are no poverty, zero hunger, good health and wellbeing, quality education, gender equality, clean water and sanitation, affordable and clean energy, decent work and economic growth, industry, innovation and infrastructure, reduced inequalities, sustainable communities, responsible consumption and production, climate action, life below water, life on land, peace, justice and strong institutions and partnership for goals (United Nations News, 2020).

Key areas of significant importance for humanity and the planet identified by SDGs evolve around elimination of poverty and hunger, socio-economic security, peace, justice and global partnership for development. These are directly related to sustainable peace and human security. Hence, the first point of departure in attaining SDGs is working towards achieving peace and human security. It also suffices to indicate that peace and security issues and the new set of universal goals are inextricably intertwined. Be that as it may, it is imperative for Zimbabwe to first make efforts to achieve peace and security in order to meet the standards required to meet the SDGs by 2030.

15.3 The Trajectory of Peace and Security in Zimbabwe

To comprehend the state of peace and security in Zimbabwe, it is imperative to explore the political and socio-economic situation in Zimbabwe. This goes a long way in exploring the prospects for achieving sustainable peace and human security which, as maintained in this chapter, are sine qua non to the achievement of SDGs in Zimbabwe. To begin with, the political situation in Zimbabwe is not favourable for the realisation of sustainable peace and security. Power struggles between and within the dominant political parties, Movement for Democratic Change (MDC), Citizens Coalition for Change (CCC) and Zimbabwe African National Union Patriotic Front (ZANU PF); have resulted in political uncertainty, political bickering and a fragile political atmosphere which is disingenuous to positive peace. More so, politically motivated murders have featured prominently in the Zimbabwean political system.

On 15 August 2011, Retired Army General Solomon Mujuru died under mysterious circumstances when his house was gutted by fire at his Alamein Farm in Beatrice approximately 40 km southwest of Zimbabwe's capital Harare (Daily Mail, 2011). Due to power struggles with ZANU PF, Mujuru was allegedly assassinated because he was viewed as a power broker between the ruling party and the MDC; a situation that was likely to intensify turmoil in the ruling party. Furthermore, on 19 June 2013, former Mines Minister Edward Chindori Chininga died in a mysterious car accident along the Raffingora-Guruve road in the Mashonaland Central Province (Nehanda Radio News, 19 June 2013). Although there is no evidence to suggest that he was assassinated, sources privy to government secrets such as Baba Jukwa, a cyber whistleblower, alleged that the former minister was assassinated because he was in the habit of leaking state secrets. Recently, ZANU PF provincial secretary for production and labour Nathaniel Mhiripiri was found dead at his farm in Nyazura in Manicaland Province on 14 Mar 2017 with a bullet in his head (News Day, 2017).

If these deaths are assassinations as purported, it means they fall under the category of torture and extermination which are egregious human rights abuses. Human rights abuses undermine human security which is an important aspect of security. Apart from assassinations, the Zimbabwean government has a bad human rights record. Leaders and members of opposition parties, leaders of civil society organisations and general citizens have also been exposed to intimidation, torture and arbitrary arrest. The arbitrary arrest of leaders of social movements including Evan Mawarire of #ThisFlag, Sten Zvorwadza of National Vendors Union of Zimbabwe, Promise Mkwananzi of #Tajamuka and political activists such as Linda Musarira speak volumes to the state of human rights abuses in Zimbabwe. Most recently, journalists, opposition political activists, anti-corruption activists and civil society leaders have been the targets for abduction, torture, unilateral arrests and unfair detention. Examples include the arrests, abduction and torture of Citizens Coalition Change Chairman Job Sikhala and other opposition politicians including Jacob Ngarivhume, Joana Mamombe, Cecilia Chimbiri and Netsai Marova and the arrest and unfair detention of Hopewell Chin'ono for exposing corruption in the government. If the situation remains unchanged, the likelihood of the attainment of SDGs in Zimbabwe will remain a phantom on the horizon. Hence, sustaining peace and promoting human security through reforming and strengthening institutions of governance is panacea to the country's pursuit of Sustainable Development Goals.

Moreover, the state of the economic situation in Zimbabwe undermines the economic security of the people. Cash shortages, poverty, unemployment, closure of companies and corruption have impacted negatively on the welfare of Zimbabweans. From the observations made by this author, cash shortages and ever-rising food inflation have deteriorated the standards of living of Zimbabweans. Poverty has also rendered Zimbabweans economically disadvantaged to the extent of migrating to other countries such as South Africa, Botswana, Namibia, United Kingdom and Australia among others for better employment opportunities and economic security. While economic problems are not peculiar to Zimbabwe alone, the situation in the Southern African countries are more of a result of policy failure and institutional inadequacies breeding corruption, inefficiency, mismanagement and poor economic governance. Having said this, it suffices to indicate that institutional reforms are overdue in the Zimbabwean political economic system to harness sustainable peace and human security in the pursuit of SDGs.

In conjunction with the above, the question that arises pertains to the manner in which institutional reforms could be executed and the form these reforms could take. Undoubtedly, this is a question of the urgent need for sustainable macroeconomic policies, transitional justice, peacebuilding and reconciliation in Zimbabwe. This against the background of the record of economic mismanagement, electoral fraud and violence which have been and are still a worrisome phenomenon in the Zimbabwean political context especially during election times. During election times, the ruling ZANU PF has a tendency of resorting to coercive tactics to influence election outcomes. Intimidation, torture and enforced disappearance tactics are often employed to force the electorate to vote for the ruling party. Since the turn of the new millennium, presidential elections in Zimbabwe have been marred by pre and post-election violence as well as electoral irregularities including vote rigging. Such were the cases with the 2002, 2008, 2013 and 2018 presidential elections. Efforts at reforming the institutions and mechanisms of vote rigging and political violence have not materialised largely due to the politicisation of the transitional justice, peacebuilding and relationship rebuilding processes in the country. With this said, it suffices to explore the options for ameliorating threats to peace and security in the pursuit of universal goals in Zimbabwe. To this end, the options in the next section are recommended as fundamental in sustaining peace in Zimbabwe.

15.4 Options for Sustaining Peace and Security for Realising Global Goals in Zimbabwe

It can be argued that sustaining peace through the elimination of threats to peace and security is panacea to the attainment of SDGs in Zimbabwe. In other words, by ensuring the prevalence of harmony, justice and development, the subsequent result will be the transformation of human lives. It suffices to highlight that the country failed to meet the MGDs in 2015 due to the failure to address critical peace and security threats including poverty, political instability, violence, unemployment and human rights abuses among other threats. With seven years to the deadline of meeting the recent universal goals (2030), the country needs to focus on addressing peace and security issues to successfully make headway in attaining SDGs.

Emancipating Zimbabweans from the tyranny of poverty should be Zimbabwe's first priority in eliminating socio-economic insecurity. Poverty is a bothersome socio-economic condition bedevilling the Zimbabwean populace to alarming levels. Based on the poverty datum line (PDL) that summarises the standard cost of living for a person who is not poor, the PDL in April 2023 when it was last updated stood at USD550 for an average of five people and USD 102 for one person (Zimbabwe National Statistics Agency, 2023). This means that an average household of five people require the USD550 to purchase food and non-food items to be rendered not poor as of April 2023. Given the salary scale for civil servants in Zimbabwe which is equivalent at an approximate average of USD50 per month (ZWL402 000 as of June 2023) due to continuous depreciation of the local currency, majority of households in the Southern African country are living below the poverty datum line. Against this background, a lot needs to be done by the Zimbabwean government to improve the standard of living for the people. The first point of departure is to raise the salary scale of the people with a view to reach the poverty datum line or to make people live above the poverty threshold. Failure to redress the poverty situation in the country will culminate in failure to meet the first Sustainable Development Goal 1: No Poverty.

More so, ensuring sustainable food security in Zimbabwe will go a long way in meeting the second universal Goal 2: Zero Hunger. The state of food security in Zimbabwe is questionable given the levels of poverty in the country and the negative implications it has for the availability of food to the general populace. The situation was also worsened by the Covid-19 pandemic whose response measures such as lockdown have eroded people's sources of livelihoods. According to United Nations News (2023); World Food Programme (2020), hyperinflation, climate change, economic recession and the coronavirus pandemic have appeared as the main drivers of acute hunger in Zimbabwe. With the income levels of Zimbabwean workers measured against the Food Poverty Line (FPL) which quantifies the minimum consumption expenditure needed to obtain a food basket equivalent to 2100 calories to sustain a household (Zimbabwe National Statistics Agency, 2023) and compounded by effects of the Covid-19 crisis, acute hunger is alarming in the country.

The estimate for FPL in April 2023 for five people stood at USD153 which makes it USD31 per person. Since majority of the Zimbabwean population live in the rural areas, very few can afford to spend USD153 on food. In case of those who live in urban areas, majority are unemployed and rely on informal economic activities as sources of income. Given this food insecurity state of affairs, the Zimbabwean government needs to urgently redress the food security issue should it meet SDGs by 2030. The question that arises pertains to the methodology of doing this.

Perhaps, the mobilisation of available domestic resources for sustainable social development and human security is one of the ways through which food security could be improved in Zimbabwe. Comprehensive policy shift towards prioritising the economic welfare of the people is arguably one of the options to sustain human security in that regard. As a country endowed with abundant natural resources including gold, platinum, coal, lithium, uranium, oil, diamonds and chrome among others, harnessing this advantage to accumulate financial resources to stockpile food resources for the country's sixteen million people suffices to end hunger in Zimbabwe. In addition, good and competent governance of natural resources is needed to combat systemic corruption, mismanagement and looting of natural resources by the political elite in Zimbabwe. This is against the trajectory of state capture, corruption and democratic deficit in the Southern African country (Saunders, 2019; Moyo, 2014).

The availability of food in Zimbabwe has also been scuttled by poor policies compounded by poor agricultural output since the introduction of the land reform policy in year 2000. The land reform minimised productivity in the agricultural sector due to failure by the beneficiaries of the policy to produce sufficient food. As a corrective measure, the government introduced the Command Agriculture Policy operational since the 2016 agricultural season. The policy materialised in the production of grain and improvement of food security in the 2016/2017 season. The subsequent bumper harvest from the season in question sustained the country's food security until 2018. As the 2017/2018 as well as the 2018/2019 seasons were characterised by poor rains attributable to the dynamics of climate change, the food security situation in the country deteriorated again. While the success of the Command Agriculture Policy that the government introduced in the 2016 agricultural season has been substantial in improving the availability of food in the country, rampant corruption characterised the distribution of the Command Agriculture agricultural inputs (Nkomo, 2020). While the continuation of such efforts will obviously culminate in the achievement of sustainable food security to meet the Zero Hunger goal and Good Health and wellbeing, there is urgent need to stem corruption in the country through reforming and strengthening institutional capacity to deal with graft.

On the political front, respect for human rights, rule of law and the prevalence of positive peace are panacea to the realisation of Goal 16: Peace, Justice and Strong Institutions. This author has observed with concern the extent to which the Zimbabwean government has been notoriously involved in perpetrating human rights abuses by sponsoring the arbitrary arrest, harassment, intimidation and torture of political opponents, members and supporters of opposition parties, social movements and civic groups. To succeed in executing such injustices, the Zimbabwean government has installed security and judicial institutions that operate on partisan basis and are loyal to the ruling party. With that said, to realise the sixteenth SDG goal, a lot of effort needs to be channelled towards the elimination of such injustices through the establishment democratic and non-partisan political and judicial institutions. Furthermore, prioritising transitional justice, reconciliation and peacebuilding would deal with historical and contemporary challenges threatening sustainable peace and human security in Zimbabwe.

Since the eruption of the 1981–1984 state-sanctioned Entumbane and Gukurahundi violence that culminated in the death of more than 20,000 people in Matabeleland and Midlands Provinces in Zimbabwe, the trajectory of sustaining peace through the pursuit of truth, justice and reconciliation has remained inadequate. In the aftermath of the Entumbane and Gukurahundi violence, the Dumbutshena and Chihambakwe Commissions of Inquiry were established to investigate the violence but the government is yet to publish the reports until now (2023) largely because the perpetrators of the violence are still in power (Chengeta, 2018). Such hiding of truth and denying justice has threatened peace, justice and reconciliation in Zimbabwe which has culminated in the existence of latent conflicts and hostility between the Zimbabwean government and the people of Midlands and Matabeleland.

Furthermore, despite the establishment of the Organ for National Healing, Reconciliation and Integration (ONHRI) in 2009, National Peace and Reconciliation Commission (NPRC) in 2013 and the Commission of Inquiry into the 1 August 2018 Post-Election in 2018 (Monthlante Commission) for political transition, truth and justice, and reconciliation deficit persist in Zimbabwe. Arguably, this can be attributed to the politicisation of transitional justice processes and subsequent democratic and reconciliation deficit in Zimbabwe. According to Zvobgo (2019), autocratic governments co-opt and manipulate transitional justice institutions to entrench political power and safeguard their rule resulting in what Loyle and Davenport (2016) call transitional injustice. With the above said, it suffices to stress that reforms ought to be urgently implemented to pursue genuine transitional justice. At this juncture, it also suffices to indicate that transitional justice efforts in the country should follow unpretentious political transitions.

Since 1980, political transitions in Zimbabwe have been marred by the co-optation and manipulation of political processes characterising the transitions. The post-Gukurahundi transition that saw the establishment of the Unity Accords to end the hostility between Zimbabwe African National Union (ZANU) and Zimbabwe African People's Union (ZAPU) that had culminated in the Gukurahundi violence did not result in lasting unity. The Unity Accord necessitated the establishment of ZANU Patriotic Front (ZANU PF). While the Unity Accord ushered peace, it resulted in the co-optation of the peace agreement by the former party to weaken the latter. Similarly, the reconciliation and healing process (ONHRI and NPRC) that was introduced in the wake of transition from ZANU PF rule to the 2009 to 2013 Government of National Unity (GNU) were manipulated to buttress ZANU PF's rule. This was also the case with the 2018 Monthlante Commission whose findings ignored the fact that the army used live ammunition on unarmed civilians. Instead of silencing guns, the quasi-judicial institution appeared to have been designed to buttress political power and survival of the ruling government. Having said that, it warrants mention that institutional reforms in the context of peacebuilding is an option required to sustain peace and human security in the pursuit of global goals in Zimbabwe. Institutions such as judiciary, army, police, anti-corruption institutions and truth commissions ought to be reformed and strengthened to sustain peace and human security in harnessing universal goals in Zimbabwe.

15.5 Conclusion

In the final analysis, the thesis of this chapter is that the attainment of sustainable peace and human security is panacea to the realisation by the 2030 of Global Goals. This is attributable to linkages between peace, human security and SDGs. As a result, efforts to achieve sustainable peace as well as economic, food, social and political security subsequently culminate to the realisation of universal goals. This is against the background that the state of peace and security in Zimbabwe leaves a lot to be desired. Suffices to also indicate that while the challenges still exist, there is still ample room for confronting them. Therefore, it is imperative to indicate that redressing these challenges will go a long way in achieving universal goals by the year 2030. Policies and institutions that promote justice, peace, equality, availability of food and economic growth should be at the core of the strategies designed to attain SDGs in Zimbabwe. Institutions such as police and army as well as those checkmating and balancing the operations coercive mechanisms of the state including courts should be reformed. Hence, effort to eliminate threats to sustainable peace and human security would culminate in the attainment of the universal goals by 2030.

References

Aaron, R. (1966). *Peace and war*. New York: Doubleday.

Ayoob, M. (1984). Security in the Third World: The worm about to turn? *International Affairs*, 60(1), 41–51.

Bellany, I. (1981). Towards a theory of international security. *Political Studies*, 29(1), 100–105.

Booth, K. (2007). *Theory of world security*. Cambridge: Cambridge University Press.

Buzan, B. (1991). *People, states and fear*. Hemel Hempstead: Wheatsheaf.

Buzan, B. (2004). A reductionist, idealistic notion that adds little analytical value. *Security Dialogue*, 35(3), 369–370.

Daily Mail.co.uk. (2011). Mugabe loyalist and ex-army chief dies in farmhouse blaze. Daily Mail, 16 Aug. 2011. http://www.dailymail.co.uk/news/article-2026642/Mugabe-loyalist-Mujuru-dies-farmhouse-blaze.html. Accessed 08 April 2017.

Galtung, J. (1985). Twenty five years of peace research: Ten challenges and responses. *Journal of Peace Research*, 22, 141–158.

Galtung, J. (1990). Cultural violence. *Journal of Peace Research*, 27(3), 291–305.

Gomez, O.A., & Gasper, D. (2013). Human security: A thematic guidance note for regional and national human development report teams. United Nations Development Programme Human Development Report office.

Hartland-Thunberg, P. (1982). National economic security: interdependence and vulnerability. *National economic security*, 47–61.

Jones, L. (2014). A critical evaluation of the concept of human security. E-International Relations.

Loyle, C.E., & Davenport, C. (2016). Transitional injustice: Subverting justice in transition and post-conflict societies. *Journal of Human Rights*, 15(1), 126–149.

Luciani, G. (1989). The economic content of security. *Journal of Public Policy*, 8(2), 151–173.

Moyo, S. (2014). Corruption in Zimbabwe: An examination of the roles of the state and civil society in combating corruption. PhD Thesis, University of Central Lancashire.

Nehanda Radio.com. (2013). MP Chindori-Chininga dies in a car crash. Nehanda Radio, 19 June 2013. http://nehandaradio.com/2013/06/19/mp-chindori-chininga-dies-in-car-crash. Accessed 6 April 2017.

News Day.co.zw. (2017). War veteran Mhiripiri found murdered at his farm. News Day, 16 Mar 2017. https://www.newsday.co.zw/2017/03/16/war-veteran-mhiripiri-found-murdered-farm. 9 April 2017.

Nkomo, C. (2020). Zimbabwe: Zimbabwe Anti-Corruption Commission weeds out Command Agriculture looters. All Africa News, Jan 18, 2020. http://www.allafrica.com/zim/2020-Jan-18. Accessed 5 August 2023.

Rothschild, E. (1995). What is security? *Daedalus*, 124(3), 53–98.

Royce, A. (2004). A definition of peace. *Peace and Conflict, 10*(2), 101–116.

Saunders, R. (2019). The politics of resources bargaining, social relations and institutional development in Zimbabwe since independence. United Nations Research Institute for Social Development (UNRISD) Working Paper No. 2019-1. https://www.econstor.eu/bitstream/10419/207017/1/1671843533.pdf. Accessed 8 June 2023.

Solomon, D., & Zvobgo, K. (2019). Co-opting truth: Explaining quasi-judicial institutions in autocratic regimes. *International Studies Quarterly*, 64(4), 785–797.

Taheri, M. A., & Dehghan, M. (2014). Definition of peace and its different types as approached by Halqeh mysticism. *Procedia-Social and Behavioral Sciences, 114*, 56–61.

United Nations Department of Economic and Social Affairs. (2023). Transforming our world: The 2030 Agenda for Sustainable Development. https://sdgs.un.org/2030agenda. Accessed 9 April 2023.

United Nations Development Programme. (2023). Sustainable Development Goals. https://www.undp.org/sustainable-development-goals. Accessed 6 April 2023.

Ul Haq, M. (1995). *Reflections on human development*. New York: Oxford University Press.

United Nations News. (2020). Ravages of acute hunger will likely hit six in ten in Zimbabwe: WFP. http://www.un.org/news/zimbabwe/7899877543457899. Accessed 5 August 2023.
United Nations Development Programme Report. (1994). *New Dimensions of Human Security*. New York: United Nations.
Wolfers, A. (1952). National security as an ambiguous symbol. *Political Science Quarterly*, 67(4), 481–502.
World Food Programme. (2020). Many Zimbabwean families are suffering the ravages of acute hunger. http://www.twitter.com/@WFP/Jul30_2020. Accessed 5 August 2023.
Zimbabwe National Statistics Agency. (2023). Poverty Analysis: Poverty Datum Lines – April 2023. http://www.zimstat.co.zw/sites/default/files/img/publications/Prices/Poverty_04_2016.pdf. Accessed 5 August 2023.

Chapter 16
The Nexus Approach to Climate Action, Planetary Integrity and Sustainable Development in Africa

Alain Jean-Paul Charles Tschudin

Abstract Within the New Way of Working, the model adopted by the United Nations as part of its systemic reform, the nexus approach of integrating interlinking work streams across discrete pillars has been encouraged. Promoting this, in 2020, the Office of the Special Adviser on Africa commissioned a study on accelerated policy implementation of the Sustainable Development Goals (SDGs) across Africa via a nexus approach between peace, development, human rights and humanitarian action. To these pillars was added governance, regarded as key to the nexus and transversal across pillars. The current chapter delimits analysis to focus on Climate Action (SDG13) and related 'planetary integrity' SDGs, Life below Water (SDG14) and Life on Land (SDG15), which, in the light of climate concerns, have been located front and centre of the sustainability debate. There is widespread recognition that climate change leads to human displacement, migration and related human insecurity and conflict risks. Given this scenario, the chapter calls for greater attention to SDG13 relative to the other SDGs, located within a bigger-picture nexus approach. This may enhance critical awareness that can inform policy development and implementation, to mitigate against the negative consequences of failing to consider climate action relative to conflict and natural disasters.

Keywords Sustainable development · Climate action · Conflict · Human security · Nexus approach · Planetary integrity · SDGs

---

Prof. Alain Tschudin is Professor of Peace Studies and Director of the International Centre of Nonviolence based at the Durban University of Technology. He holds Ph.D.s in Psychology and in Ethics. Email: AlainT@dut.ac.za

---

J. C. K. Kiyala and N. Chivasa (eds.), *Climate Change and Socio-political Violence in Sub-Saharan Africa in the Anthropocene*, The Anthropocene: Politik—Economics—Society—Science 37,
https://doi.org/10.1007/978-3-031-48375-2_16

16.1 Climate Change, the Anthropocene and the Sustainability of Life on Earth

From scepticism surrounding climate change to the fatalistic recognition that some significant damage and environmental degradation is irreversible, there is stupefying scientific evidence in favour of climate change, adaptation and the need for sustainability (see Carson, 1962; Thuiller, 2007; Berrang-Ford et al., 2011, Habibullah et al., 2022). The geological period known as the Holocene has lasted for the last 10,000 years and its stability has enabled the flourishing of humans on the earth (Rockström et al., 2009). However, since the first Industrial Revolution, we have entered into the Anthropocene, in which, as Westley et al. (2011) note, human activity has become majorly implicated in determining the future of our planet. Moreover, they suggest that humankind faces a troublesome cocktail of rebound effects from key planetary processes including global nitrogen, sulphur and carbon cycles, diminishing resources (water, crops, energy). This, added to increasingly complex social-ecological systems comprising trade, financial markets and resource extraction, combines to form what they refer to as "threshold behaviour" (2011: 762).

Rockström et al. (2009) note humanity's increasing reliance on fossil fuels and industrialised agriculture. They raise the existential problem of what happens beyond the threshold, once the earth has been tipped out of its stable equilibrium or beyond what they refer to as "planetary boundaries", which constitute the "safe operating space for humanity". The catastrophic consequences of continued failure to act have led to widespread fears of planetary demise and activism led by animated youth such as Greta Thunberg and with increasing social movements such as Extinction Rebellion.

So pronounced is this threat to life, that António Guterres (2022), Secretary-General of the United Nations, speaking on 7 Nov at the 27th United Nations Climate Conference, at Sharm-el-Sheik, commented as follows:

> We are on a highway to climate hell with our foot still on the accelerator…[climate change]…is the defining issue of our age. It is the central challenge of our century. It is unacceptable, outrageous and self-defeating to put it on the back burner. Indeed, many of today's conflicts are linked with growing climate chaos…Human activity is the cause of the climate problem. So human action must be the solution…The science is clear: any hope of limiting temperature rise to 1.5 degrees means achieving global net zero emissions by 2050…[we] are getting dangerously close to the point of no return…That is why at the beginning of COP27, I am calling for a historic Pact between developed and emerging economies – a Climate Solidarity Pact…This is our only hope of meeting our climate goals. Humanity has a choice: cooperate or perish. It is either a Climate Solidarity Pact – or a Collective Suicide Pact.

Much as the following chapter is primarily about Climate Action (SDG13), it will set out from the strong scientific evidence base that climate change and extreme meteorological variation has been induced largely by recent human action, as recognised by the Secretary-General above (2022). Aside from direct actions such as instant nuclear annihilation, one might argue that global warming is today perhaps the most significant threat to the sustainability of life on this planet.

Extreme weather conditions and climactic shifts have led to significant human movement and displacement, in some cases from ancestral lands that predate human recall. This has been noted in several parts of the world, where the drying up of lakes, rivers, water courses and wells has forced nomadic pastoralists to seek out alternative sources for survival. In some instances, these movements have translated into intra-state and international, transboundary conflicts marked by violence and human rights violations (UNECA, 2018). One consequence of human migration, framed by competition over scarce transboundary resources and largely exacerbated by climate change, is therefore the possibility of conflict. As the United Nations Economic Commission for Africa (UNECA) report on New Fringe Pastoralism observes: "Climate change does not directly result in conflict; it can be a catalyst for conflict because it contributes to the shrinking of the environmental resources (water and grazing land for livestock and agricultural production) used for the production and reproduction of livelihood. It affects the adaptive strategies pursued by pastoralists. Thus, competition over scarce resources can escalate into conflict." (2017b: 18)

Indeed, the threat of climate change to peace, security and development is real and present. As a case in point, so concerned is UNECA about the deleterious consequences of conflict for Africa's development, that its reports go so far as to promote models of sustainable development around whether or not conflict is present or has affected member states in question (see UNECA, 2015, 2016, 2017a, b, 2018, 2019). Likewise, in recognising the scourge of conflict, the African Union (2016) championed the cause known as "The Silencing of the Guns", which was intended to culminate in the cessation of all violent conflicts in Africa by 2020. Unfortunately, conflict continues to plague the world, and Africa is not immune from that, especially in the aftermath of developmental backsliding triggered by the global COVID-19 pandemic. Accordingly, the AU has modified its thinking on this topic with its sharing of Vision 2030, by which date it aspires to have addressed conflict on the continent (African Union, 2022).

Cognisant of the imminent danger to life that is presented by climate change, the potential for conflict and the related challenges of escalation and violence, we can drill down to the importance of climate action to sustainable development. Much debate and political negotiation surrounding competing ideologies was involved in the conceptualisation, refinement and finalisation of the SDGs and their targets (Paul, 2008; Ziai, 2016). This in turn required major social, political and economic effort on the part of countries at different stages of development (see Osborn et al., 2015). More recently, however, Adelman (2018) notes how the SDGs serve to legitimisze hegemonic capital, perpetuated under the guise of development, rather than prioritising environmental sustainability.

A recent 'sobering' review of SDGs using an integrated modelling framework (Soergel et al., 2021) concluded that current strategies are inadequate to reach the targets by 2030. They note that, "Given the narrow 2030 time horizon of the SDGs and the risk of losing achievability of the Paris climate targets, it is important to assess how synergies between climate action and other SDGs can be fully exploited through rapid and coordinated action, and which interventions are required to do so." (p. 656). Helpfully, they arrange their SDG indicator outcomes into five clusters, comprising planetary integrity (SDGs 13, 14 & 15); provision of material needs (SDGs 2, 6, 7 & 12); people (SDGs 1, 3, 4, & 5); prosperity (SDGs 8, 9, 10 & 11); and political institutions, peace, partnership (SDGs 16 & 17).

Notably, planetary integrity comes first as "mission critical" and then the other clusters follow; respectively focused on needs provision and resource management, human development, living conditions and governance. It is the notion of 'governing' (p. 659) referred to in the final cluster that merits our attention. For governance, as was recently recognised, while not previously explicitly referred to in the "nexus approach" to peace, security and development, effectively cross-cuts all work streams and themes, and is heavily implicated in the success or failure of sustainable development initiatives (Tschudin, 2022).

The long-standing call for governance reform in global institutions and multilaterals has been intensified since the World Humanitarian Summit (WHS) held in Istanbul, 2016, when the erstwhile UN Secretary-General, Ban Ki-Moon called for "one humanity, with a shared responsibility…to give people a chance at life in dignity" in order to "work better and smarter to prevent, reduce, and meet humanitarian needs." (UN Foundation, 2016). There, in recognition of the vast predominance of "complex emergencies", defined as multifaceted humanitarian crises requiring multisectoral responses (see UNOCHA, 2016), a commitment was made to better integrate humanitarian and developmental support to provide enhanced responses to crises and conflicts, while building social and economic resilience (UN and World Bank, 2018: 254).

As summarised by the UN Foundation (2016), the ten key takeaways from the WHS comprised a people-centred approach; education provision, including for children in crises; a Grand Bargain for greater work efficiency; leaving no-one behind; a truly inclusive approach; improved collaboration; localisation; private sector partnerships; innovation and, finally, risk reduction and increased resilience. In effect, the WHS triggered what has become known as the "New Way of Working" (NWoW), inspiring current SG António Guterres to pursue extensive reforms at the UN.

Giving flight to his predecessor's Agenda for Humanity, a roadmap to reduce global suffering, risk and vulnerability, (see United Nations General Assembly, 2016), the NWoW draws upon its five key responsibilities, including: "political leadership to prevent and end conflicts; uphold the norms that safeguard humanity; leave no one behind; change people's lives and invest in humanity." In striving to make the NWoW workable, further momentum has been provided by the so-called "Grand Bargain" between the UN, the World Bank, donors, aid organisations and NGOs to pool together in building resilience, coordination and cohesive support.

In short, as has been recognised elsewhere, the NWoW makes it easier to align development and humanitarian outcomes, in order to enable local and national systems strengthening, to overcome traditional divisions between humanitarians and developmentalists both in anticipation of, and response to, crises (Tschudin, 2022).

It is within the context of the NWoW that the so-called "nexus" approach has gathered momentum over the last few years, although it has been difficult to define and delimit. The origin of the term stems from the Latin word 'nectere', "to tie or bind together", with nexus referring to "a connected group or series" or a "central or focal point". According to the United Nations Evaluation Group (2018), this approach arose with respect to the initial 'dual' nexus between humanitarian efforts and development. Thereafter, with Secretary-General Guterres emphasising peace, this morphed into a 'triple' nexus, known as the peace-development-humanitarian nexus (CIC, 2019), while others such as Badejo (2020) have referred to a 'quadruple' nexus to include human rights. The nexus study therefore included this latter human rights pillar in a 'quadruple' nexus, demonstrating that it was more significantly implicated than the pillar of humanitarian action (Tschudin, 2022). However, as noted in that study, both of the afore-mentioned pillars were dwarfed by governance, which, if considered as a pillar, would lead to a 'quintuple' nexus. Such is the ongoing evolution of the nexus approach.

For our current purposes, given the severity of the threat presented by climate change to life on earth, whether below water or on land, it seems only appropriate that these so-called 'planetary integrity' variables form the focus of this analysis. Hence, the emphasis lies intentionally on studying climate relative to conflict, peace and development through the nexus approach in Africa, to promote better targeted, ameliorative action that ensures we do not transgress our life or death planetary boundaries.

16.2 Methodology

Authors such as Howe (2019) have argued that the use-value of the triple nexus to meaningfully enable the SDGs has been impeded by a lack of definitional clarity and practical implementation. Mindful of this, and for a comprehensive review, a major study was conducted for the United Nations Office of the Special Adviser on Africa, in which an operationalisation of the nexus was developed around various interlocking pillars, while introducing governance to its core (Tschudin, 2022; see Fig. 16.1). This chapter draws significantly from that research.

Within the UN OSAA study, conducted between October 2020 and March 2022 (15 months), the conventional peace-development-humanitarian-human rights pillars were supplemented by governance and these pillars were held relative to each other and to the SDGs, as well as to the AU's "Africa we want". This approach to the study was undertaken because it was hypothesised that an inter-linked way of working would be pivotal to the greater practical and policy effectiveness required to promote the developmental agenda of both the United Nations and of the African Union's Agenda 2063 (United Nations Economic Commission for Africa, 2018-01).

Accordingly, in the OSAA study, an effort was made to scan and review the literature extensively, to statistically analyse UN data on progressing the SDGs across Africa, in terms of process and means of implementation indicators, to survey African participants from government ministries, NGOs and the educational sector in Africa and to model the nexus approach to accelerating the SDGs using Complex Adaptive Systems (CAS) theory. The study findings supported the value of the New Way of Working, with specific reference to the nexus and the transversal role of governance,

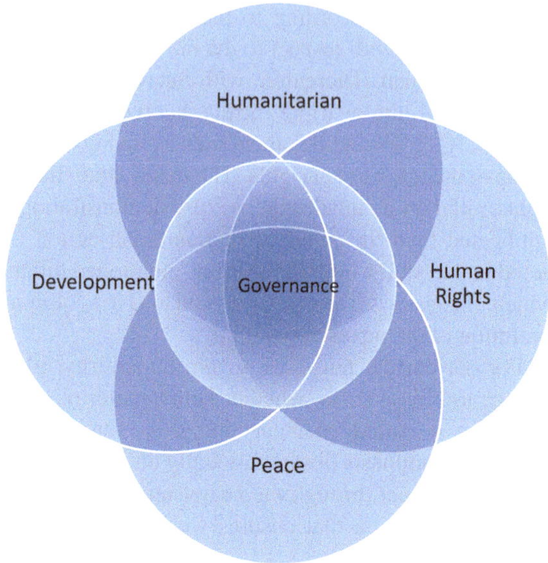

Fig. 16.1 Operational conceptualisation of the Nexus. *Source* Tschudin (2022)

while revealing a gross lack of awareness of the nexus and an urgent need for capacity strengthening and the promotion of practical activities to advance an interlinked way of working.

Distinctively, the current chapter isolates SDG13, namely Climate Action, relative to the nexus pillars and to the other SDGs in order to gain some sense of its perceived relevance among African participants on the recent nexus study. This can be justified given the significance of the climate agenda to the future of the planet and any sustainable development aspirations. Furthermore, in keeping with Soergel et al. (2021), who include Life below Water (SDG14) and Life on Land (SDG15) with SDG13 to form the "planetary integrity" cluster of SDGs, our focus on SDG13 will be supplemented with the inclusion of SDGs 14 and 15. Without the appropriate attention given to these three Goals in combination, the continuation of life on earth will become exceptionally challenging, with competition and conflicts likely to escalate.

The methodological approach comprised the following four steps:

(1) Analysis commenced with a desk review, whereafter the performances of nation-states and Africa as a region on SDGs 13, 14 and 15 were extracted from the UN OSAA study, which used publicly available data from the UN Department of Economic and Social Affairs (DESA) to track progress on the Goals.[1] For each of the SDGs, a process indicator and an means of implementation (MoI) indicator on outcomes was selected, based on maximising coverage for the continent. As Bartram et al. (2018) note, MoI indicators arose after process

[1] See: https://unstats.un.org/sdgs/indicators/database/

indicators in the unfolding of the SDGs and their process of evaluation, and give an idea of whether the outcome is being attained.

(2) Thereafter, relevant analysis of the nexus survey conducted for UN OSAA with African participants conducted in English and French from December 2020 to February 2021 was undertaken and findings for the 'planetary integrity' variables highlighted. Purposive sampling was used insofar as two-thirds of participants were official representatives of African member states and one-third civil society/NGOs drawn from a variety of sectors. Participants were recruited via UN-OSAA and a network of UN agencies such as UNECA, UN Global Compact, UN-Habitat, UNOHRLLS and related contacts specifically in relation to the nexus approach to sustainable development. From 182 views on Survey Monkey, there were 51 responses, 29 in English and 22 in French (the two languages requested by OSAA), with an acceptable response rate of 28% (the norm for surveys is 20–25%). Participants were from 33 African countries, giving the study a majority of coverage of African states (52%). Translations and back translations confirmed the reliability of the survey instruments with the rigor of the study confirmed in the validation workshops. Following this, the findings of perceived performance as recorded by participants modelled relative to actual performance on the SDGs (from the DESA database) were studied, to evaluate positive versus negative deviation in terms of perceived versus actual progress.

(3) Next, the results of the Complex Adaptive Systems (CAS) modelling from the original nexus study were analysed with specific reference to Climate Action, Life below Water and Life on Land and the pillars implicated in their realisation. As Nilsson (2016) recognises, the SDGs are an example of a Complex Adaptive System, since CAS theorising holds that a system is greater than the sum of its parts, thanks to the interaction of parts which leads to emergent properties (see Rogers et al., 2013). Therefore, systems thinking is required to study the SDGs since it considers their relational constitution and complex causality. Related analysis around this allows for the design of a causal network that maps interconnections, directionality and weighting, and the identification of elements that have the largest and smallest effect sizes (see Preiser et al., 2018; Niemeijer/de Groot, 2008; Weitz et al., 2018). The current analysis seeks to interpret the positioning of SDGs 13, 14 and 15 in the CAS generated for Africa in the nexus study, whereby African participants ranked the SDGs in order of priority, while perceived success on performance was ascertained using frequency tables.

(4) Finally, the contributions of participants at two sub-regional capacity strengthening and study validation workshops held in June 2021 were analysed to extract any inputs relevant to Climate Action and the other Planetary Integrity SDGs. These workshops comprised one dedicated to the southern and eastern sub-regions of Africa, and the other to central, north and west African sub-regions. Participants were drawn from member states, civil society and various inter-governmental organisations and were encouraged to comment on the OSAA nexus study. Moderated break-out sessions and plenaries were included to

maximise active participation. Key points were included in the final study recommendations.

16.3 Findings

An analysis of the above findings (Tschudin, 2022) has been delimited to focus on Climate Action (SDG13), with complementary analysis of "planetary integrity" cluster Goals, Life below Water (SDG14) and Life on Land (SDG15), presented by relevant section below. Such an analysis may be useful for understanding climate related issues and the policy innovation, development and implementation needed to promote harmony and prevent conflict in Africa.

16.3.1 UN DESA Indicator Analysis

16.3.1.1 Performance on Climate Action

Of all Process indicators, Climate Action (SDG13) was the best indicator for Gabon only, and did not feature as the worst indicator for any country. This is noteworthy since Gabon, which recently hosted the 2022 Africa Climate Week, is a rare example, globally, of a carbon-negative economy intent on protecting its biodiversity (see World Bank, 2022a).

Meanwhile, in terms of MoI indicators, Climate Action featured as best for three countries, namely Burkina Faso, Malawi and Tanzania, while it fared worst for Burundi and Ethiopia. This is important from an implementation perspective since the World Bank Country Climate and Development Report (CCDR) for Malawi points to the significance of taking climate action and the relevant financing seriously; in the absence of such investment, climate change could contract GDP in that country by 3–9% by 2030, 6–20% by 2040 and 8–16% by 2050 (see World Bank, 2022b).

Likewise, for Burkina Faso, according to the UNDP (2023), the Climate Promise undertaken for that country is bearing fruit, with quantifiable emissions reductions of 29.42% and adaptation actions capable of 30.76% by 2030, arising from inclusive participation including women and youth with a commitment to monitoring and evaluation. Finally, Tanzania, in keeping with international commitments to the UN Framework and Paris Agreement, has progressed from its first National Climate Change Strategy in 2012 to its more recent National Climate Change Response Strategy of 2021 (See URT, 2021).

It comes as little surprise that countries post-conflict or in conflict such as Burundi and Ethiopia have failed to focus on Climate Action. Violent conflict and war have a distinctly negative and destructive impact on the environment.

If we compare across Process and MoI indicators for Climate Action (SDG13), although there is more extensive top-end success for the latter across the continent, unfortunately there is also more extensive bottom level performance for these.

The analysis then advanced to studying a correlational matrix for progress on the SDGs. Climate Action (SDG13) is not implicated in any significant correlational relationship with other SDGs for performance on the SDG Process Indicators (see Fig. 16.2).

Climate action is, however, correlated moderately with Partnerships for the Goals (SDG17) on the MoI indicators with an r-value of 0.42 and $p < 0.01$, confirming the value of a collaborative approach in terms of advancing the climate agenda (See Fig. 16.3). Indeed, already over a decade ago, authors such as Pinkse/Kolk (2012) recognised the value of multi-stakeholder partnerships to advance a climate change and sustainable development nexus, with a focus on governance.

A 2019 SDG Center report notes that, in general, Africa outperforms other global regions in CO_2 emissions reductions, remarking that only one of the world's 20 worst performers is African, namely South Africa, with an output of 467.7 metric tons of CO_2 annually.

SDGs	1	2	3	4	5	6	7	8	9	10	11	12	13	14	15	16	17
1	1																
2	-0.10526	1															
3	0.296447	0.101655	1														
4	0.341379	-0.11763	0.265799	1													
5	0.054813	0.189071	0.028635	-0.14437	1												
6	0.966475	-0.02285	0.308768	0.297144	0.013495	1											
7	0.348193	-0.02122	0.220317	0.105793	0.032965	0.359716	1										
8	0.082415	0.046048	0.249461	0.025695	0.14035	0.088679	0.249422	1									
9	0.106157	0.245726	0.101701	-0.16151	0.018646	0.151408	-0.00634	-0.0392	1								
10	-0.18104	0.435748	0.139622	0.058118	0.348315	-0.08454	0.050067	0.000594	-0.00869	1							
11	0.275477	0.021802	0.414603	0.310731	-0.07924	0.261931	0.001759	0.003527	-0.0544	0.084859	1						
12	-0.31644	0.097803	-0.24546	-0.5651	0.006562	-0.30341	-0.40399	-0.09185	-0.12029	0.016877	-0.25359	1					
13	-0.08492	-0.12716	-0.01598	-0.10655	-0.04173	-0.02359	-0.09358	-0.02784	0.201333	0.074776	-0.06913	0.158422	1				
14	0.091804	-0.10103	-0.03883	-0.11764	0.042096	0.032352	0.096188	-0.00051	0.107835	-0.12943	0.058946	-0.00166	0.02636	1			
15	0.030245	-0.01503	-0.02912	-0.14577	0.137186	0.057423	0.03118	-0.22203	0.133084	-0.0022	0.033329	-0.02615	0.097444	0.11052	1		
16	0.367292	0.085782	-0.0197	0.204529	-0.09298	0.340941	0.00832	-0.21564	0.011731	-0.19839	0.279312	-0.42599	-0.23269	0.285712	0.031893	1	
17	-0.1061	0.034099	0.106658	-0.27833	0.408449	-0.11905	-0.05875	0.267237	0.040536	-0.02804	-0.06564	-0.14018	-0.06911	0.00747	0.002504	-0.07788	1

Fig. 16.2 Correlational matrix with significance values for progress on the SDG Process indicators. *Source* Tschudin (2022)

SDGs	1	2	3	4	5	6	7	8	9	10	11	12	13	14	15	16	17
1	1																
2	-0.11535	1															
3	0.139975	-0.01304	1														
4	-0.07864	-0.03781	-0.01484	1													
5	N/A	N/A	N/A	N/A	1												
6	0.044857	-0.1664	-0.20316	-0.08351	N/A	1											
7	-0.04538	-0.03329	-0.04835	0.309492	N/A	-0.07252	1										
8	-0.02174	-0.01142	-0.00429	0.033116	N/A	0.166414	-0.06485	1									
9	0.353574	-0.02131	-0.00589	-0.02669	N/A	-0.14875	-0.0148	-0.02057	1								
10	0.112833	0.215226	-0.04363	0.056994	N/A	0.160491	-0.10102	-0.09481	0.020551	1							
11	N/A	N/A	N/A	N/A	N/A	N/A	N/A	N/A	N/A	N/A	1						
12	-0.04544	-0.03333	-0.04839	0.309445	N/A	-0.0726	0.999998	-0.06491	-0.01485	-0.10099	N/A	1					
13	-0.24563	0.114535	-0.10267	0.115862	N/A	0.070804	0.15898	0.024917	-0.02974	-0.01876	N/A	0.158391	1				
14	N/A	N/A	N/A	N/A	N/A	N/A	N/A	N/A	N/A	N/A	N/A	N/A	N/A	1			
15	-0.07819	0.47468	-0.02859	-0.05908	N/A	-0.12731	-0.07368	-0.04185	-0.03057	0.24776	N/A	-0.07376	-0.1082	N/A	1		
16	N/A	N/A	N/A	N/A	N/A	N/A	N/A	N/A	N/A	N/A	N/A	N/A	N/A	N/A	N/A	1	
17	0.184751	-0.05084	-0.05649	-0.0529	N/A	0.178082	-0.20978	-0.06274	-0.03239	0.212517	N/A	-0.20991	0.415653	N/A	-0.04139	N/A	1

Fig. 16.3 Correlational matrix with significance values for progress on the SDG MoI indicators (Tschudin, 2022).

16.3.1.2 Performance on Life Below Water and Life on Land

In terms of the other related "planetary integrity" Goals, Life below Water and Life on Land (SDGs 14 and 15), for the Process indicators, Life below Water (SDG14) was the best indicator for Algeria, Cameroon, Comoros and Tanzania, while it was worst for Angola, Djibouti, Eritrea, Gambia, Guinea, Guinea-Bissau, Libya, Mauritania, Sao Tome and Principe, Senegal, Sierra Leone, South Africa and Tunisia, while Life on Land (SDG15) did not feature amongst best or worst performers. Clearly, some of the top process performers have an extensive water mass or coastal profile, while some of the worst performers are landlocked and struggle with desertification. While South Africa might appear as a surprise on the list of worst performers, marine pollution, overfishing and marine and coastal conservation are problematic in that country, where up to 250,000 tons of litter are dumped into the oceans annually, as reported by the United Nations Environment Programme (2020). Evaluatively, in terms of process indicators, Life below Water was more than 3 times likely to score negatively than positively for goal progression on the continent.

By comparison, for Means of Implementation indicators, Life Below Water (SDG14) did not feature as best or worst for any country, while Life on Land (SDG15) was best for Angola, Comoros, Egypt, Eswatini, Gambia, Guinea, Togo, Tunisia and Uganda, and worst for Algeria, Cabo Verde, CAR, Congo, Kenya, Libya, Mali, Mozambique, Namibia, Rwanda, Senegal, Seychelles and Somalia (see Appendix 1, Tschudin, 2022). Once again, a spread of difference is evident in the top performers while several of the worst performers are in a conflict/post-conflict state. For MoI indicators, Life on Land was almost 1.5 times more likely to be a negative than a positive bookend for progress in Africa. Otherwise interpreted, Life on Land is half as likely as Life below Water to feature as an extreme negative on Process indicators. Progress on this Goal is in keeping with what the Brookings Institute (2019) reported, namely that Africa was only on target to meet 3 goals: this goal, SDG15 (Life on Land), SDG5 (Gender Equality) and SDG13 (Climate Action).

Regarding the other "planetary integrity" variables and the correlational analysis, for Life below Water (SDG14), there is a weak positive correlation with Peace, Justice and Strong Institutions (SDG16) on the Process indicator (r-value of 0.29 and $p < 0.05$) which confirms the link between governance and water concerns, while nothing of significance is recorded on the Outcome/MoI indicators. We have already discussed above, the linkage between governance and marine pollution in the case of South Africa and how urgent government action is required to ameliorate the situation in order to not jeopardise the attainment of the goal.

Finally, Life on Land (SDG15) does not reveal any significant correlations for process indicators, however, on the Outcome/MoI indicator, it shares a moderate but statistically highly significant correlation with SDG2, Zero Hunger (r-value of 0.47 and $p < 0.001$). This makes sense, given the relationship between responsible preservation of life on land for food resources and the reduction of hunger. Indeed, Zhang et al. (2021) have written on the necessity of bringing together Goals 2 and 15, although they note how the challenge of preserving life on land can potentially clash with the notion of food production that is obviously needed to reduce and eliminate hunger. Although this can be mitigated by carefully designed cropland allocation strategies, the issue of food security does not only arise due to production concerns, but is affected by a variety of factors in the food value chain, such as post-harvest loss, food waste, etc.

Concerning Life below Water, the SDG Center (2019) suggested that were was a paucity of data to assess progress, noting that most African maritime states do not have protected marine areas. This can be compared with Life on Land, where the continent registers a good performance with protected land to safeguard biodiversity, lower only than Europe and North America. As the report notes, with the implementation and enforcement of "concerted and focused policy interventions", it should make the Goal by 2030.

16.3.2 Nexus Survey and SDG performance

Participants were asked which pillars were implicated in the realisation of each Goal, and the responses for SDG13, along with SDGs 14 and 15 merit discussion. For Climate Action, Development was strongly implicated with Governance close behind, with a sizeable gap to Human Rights, followed by Peace, and another gap to the Humanitarian pillar. Life Below Water showed a varying pattern, with Development way ahead of Governance, then a notable fall to Peace, then Human Rights with Humanitarian at its lowest level across SDGs (together with SDG9). Life on Land repeats the same pattern as SDG13, with Development followed by Governance, albeit with a reduced gap than SDG14, followed by similar scores to SDG13 for Human Rights, then Peace, and finally Humanitarian midway between its position for SDGs 13 and 14. We return to this finding, with weightings under Complex Adaptive Systems modelling.

When ranked in order of priority for achievement, SDG13 was one of 7 Goals that was not ranked as a top priority (the others included SDGs 5, 6, 8, 10, 14 & 15). It was one of 3 SDGs to have second priority as its highest score, together with SDGs 7 and 15, lying ahead only of SDG14 as a third priority, and SDGs 5, 10 & 12 as fourth priorities. This does not reflect well for the significance attached to Climate Action, or Life on Land and below Water.

We then probed for capacity and asked how well the participants felt their governments were performing on the SDGs in the real world. For Climate Action (SDG13) the results were spread from very poor to good. The biggest category was 50/50, followed by poor, then good and very poor. There were no scores in the very good category for this SDG. The spread, which is largely negative to 50/50, indicates that urgent work needs to be done to bootstrap the progress on SDG13. Similar patterns are noted for Life below Water (SDG14) and Life on Land (SDG15), which flag to the reader that significant work will be required to reach these goals (see Fig. 16.4).

When participants were asked to rank the biggest perceived risks to achieving the Africa we Want and the SDGs respectively (see Figures 16.4 & 16.5 below), the results were noteworthy. In both cases, climate risks were not once ranked as the

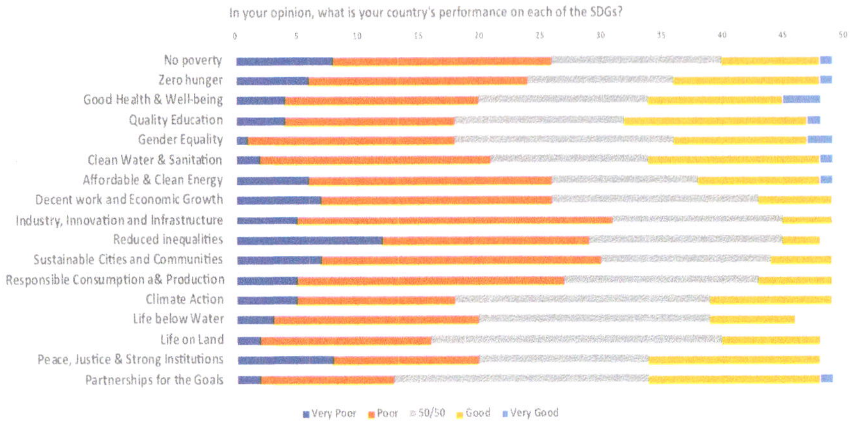

Fig. 16.4 Perceived country performance on the SDGs. *Source* Tschudin (2022)

greatest risk, but rather came in as the second to lowest risk of all (10/11), located only above the escalation of Africa's disease burden. This demonstrates how little of a risk climate change is regarded to be by African participants.

Fascinatingly, this can be contrasted with other visible risks, such as conflict, political challenges and external interference which are regarded as the three major spoilers of the AU's vision, while the results also reveal just how conflict, social and economic inequalities, followed by political challenges are the three greatest threats to advancing the SDGs (see Fig. 16.6).

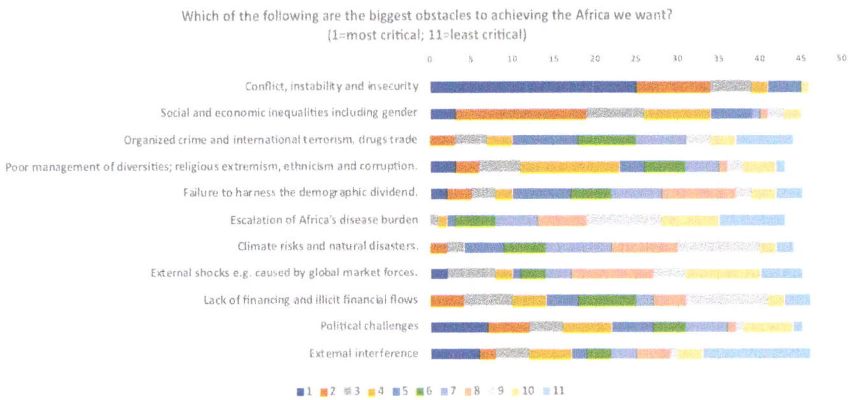

Fig. 16.5 Biggest perceived risks to achieving the Africa we Want. *Source* Tschudin (2022)

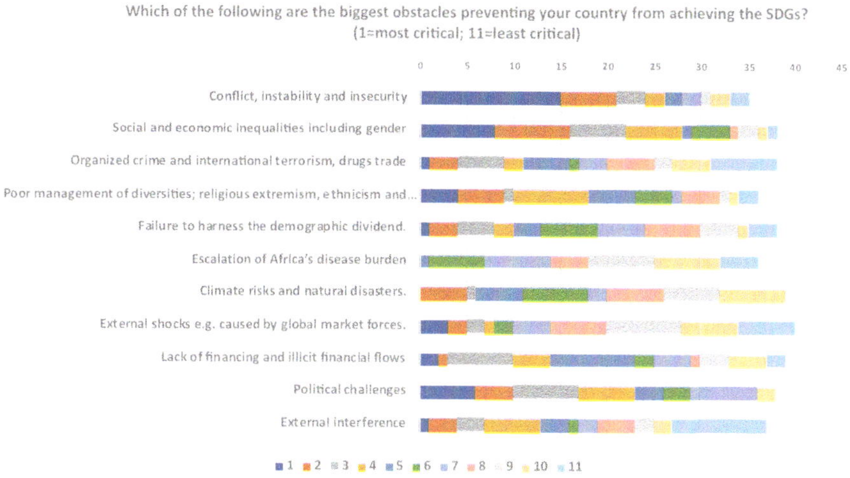

Fig. 16.6 Biggest perceived risks to achieving the SDGs. *Source* Tschudin (2022)

16.3.3 Modelling Actual and Perceived Performance on the SDGs

Generally speaking, participants have a lower opinion regarding advances made than is the case regarding improvements in reality. One of the planetary integrity indicators, Life below Water (SDG14) is only one of three exceptions where opinions are more positive than the situation in reality (the others being SDGs 2 and 10, Zero Hunger and Reduced Inequalities). Moreover, it is concerning that Life below Water (SDG14) and Life on Land (SDG15) represent two of the six worst performers, all below par on actual improvement, indicating that much more targeted policy intervention is required in this domain.

If we re-interpret the data into bands of improvement, or what we have termed "corrected improvement", scores are as follows: very poor (0–19%); poor (20–39%); below average (40–49%) average (50%), above average (51–59%) good 60–79%; very good 80% +). Climate Action (SDG13) registered a slightly "above average" score of 53.7% on corrected improvement (6th highest of the goals), which is higher than the opinion registered regarding progress on this (at 30.7%, 'poor').

Interestingly, of all the "planetary integrity" variables, opinions on progress were the lowest for Climate Action (SDG13), followed by Life on Land (SDG15) (30.8%) and Life below Water (SDG14) (34.8%). While Life below Water is most positively perceived of the three, albeit poorly, is it the worst performer at 0.1% or "very poor", while Life on Land (SDG15) scored "below average" at 47.8% for corrected improvement.

16.3.4 Complex Adaptive Systems Modelling

An outstanding finding, when participants were asked to prioritise or rank order the SDGs, was that Climate Action (SDG13) ranked only as the 12th most important Goal, whereas Life on Land and Life below Water ranked as 15th and 17th least important goals, respectively. From this placement we can infer that the "planetary integrity" goals are not highly regarded when prioritising sustainable development in Africa. This is especially disturbing given the importance of this cluster and the magnitude of the impending existential crisis (see Soergel et al., 2021), and merits further and urgent consideration from a policy development perspective (see Table 16.1).

Concerning the different pillars of the nexus perceived to be involved with Climate Action (SDG13), Development was regarded as most highly implicated, followed closely by Governance (0.9 weighting), with a sizeable gap to Peace and Human Rights (tied 3rd at 0.5 weighting), with the Humanitarian pillar (0.3 weighting) significantly lowest of all (see Table 16.1).

For Life below Water (SDG14), the pattern repeats, with Development leading, followed by Governance (albeit with a reduced perceived influence for this SDG at 0.7 weighting), and once again a large gap to Peace and Human Rights, tied with less impact (at 0.3 weighting), with Humanitarian featuring minimally (at 0.1 weighting).

Finally, for Life on Land (SDG15), Development again leads, with Governance prominent (at 0.8 weighting), bucking the trend for the above findings, with Peace (0.6) ahead of Human Rights (0.5) and the Humanitarian pillar again at the rear, albeit with more significance at an 0.3 weighting.

16.3.5 Sub-regional Workshops

It is fascinating to note that, within the sub-regional capacity-strengthening and validation workshops hosted by OSAA during the nexus study, participants identified climate change as the first challenge to implementing a nexus approach to sustainable development on the African continent. This was in contrast to the survey participants, for whom it was not raised as a concern.

Documentation specifically flagged in this regard comprised the United Nations Office for Disaster Risk Reduction Report (2021) with respect to drought and its management, as well as the UN-OHRLLS Preliminary Report on Lessons learned from implementing the Istanbul Programme of Action for the LDCs (2011–2020), which highlights the impact of climate on these countries, as well as LLDCs and SIDs.

The latter report by OHRLLS, has two priority areas that raise climate as a red flag. The first, Priority 2.2, Agriculture, Food Security and Rural Development, suggests that Least Developed Countries are highly dependent on "climate sensitive sectors", where agriculture is both significant as a source of employment and of revenue.

Table 16.1 Perceived SDG importance ranking with nexus pillars implicated spirce

SDG importance ranking: overall	SDG number and description	Peace rank (weight)	Humanitarian rank (weight)	Development rank (weight)	Human rights rank (weight)	Governance rank (weight)
1	1. No Poverty	4 (0,5)	4 (0,5)	1 (1,0)	3 (0,6)	2 (0,7)
4	2. Zero Hunger	3 (0,6)	2 (0,7)	1 (1,0)	3 (0,6)	3 (0,6)
3	3. Good Health and well-being	3 (0,7)	4 (0,6)	1 (1,0)	4 (0,6)	2 (0,8)
2	4. Quality Education	2 (0,6)	3 (0,3)	1 (1,0)	1 (1,0)	1 (1,0)
9	5. Gender Equality	4 (0,5)	5 (0,3)	3 (0,6)	1 (1,0)	2 (0,7)
5	6. Clean Water and Sanitation	5 (0,4)	4 (0,5)	1 (1,0)	2 (0,7)	3 (0,6)
8	7. Affordable and Clean Energy	3 (0,4)	4 (0,2)	1 (1,0)	3 (0,4)	2 (0,7)
6	8. Decent Work and Economic Growth	4 (0,5)	5 (0,2)	1 (1,0)	3 (0,7)	2 (0,8)
11	9. Industry, innovation and infrastructure	3 (0,4)	5 (0,1)	1 (1,0)	4 (0,3)	2 (0,8)
7	10. Reduced Inequalities	3 (0,7)	4 (0,5)	1 (1,0)	2 (0,9)	1 (1,0)
13	11. Sustainable Cities and Communities	4 (0,4)	5 (0,2)	1 (1,0)	3 (0,5)	2 (0,9)
14	12. Responsible Production and Consumption	4 (0,2)	3 (0,3)	1 (1,0)	3 (0,3)	2 (0,8)
12	13. Climate Action	3 (0,5)	4 (0,3)	1 (1,0)	3 (0,5)	2 (0,9)
17	14. Life below Water	3 (0,3)	4 (0,1)	1 (1,0)	3 (0,3)	2 (0,7)
15	15. Life on Land	3 (0,6)	5 (0,3)	1 (1,0)	4 (0,5)	2 (0,8)

(continued)

Table 16.1 (continued)

SDG importance ranking: overall	SDG number and description	Peace rank (weight)	Humanitarian rank (weight)	Development rank (weight)	Human rights rank (weight)	Governance rank (weight)
10	16. Peace, Justice and Strong Institutions	1 (1,0)	4 (0,6)	3 (0,8)	2 (0,9)	1 (1,0)
16	17. Partnerships for the Goals	3 (0,7)	4 (0,6)	2 (0,9)	3 (0,7)	1 (1,0)

Source Tschudin (2022)

Vulnerability and resilience to climate change become examples of major challenges that can only be addressed by promoting "climate smart" agricultural initiatives, encouraging agribusiness amongst women and youth, improving food security and poverty alleviation through input subsidies, irrigation, land reform and rehabilitation (UN-OHRLLS, 2021: 29–36).

The second, Priority 2.6, concerns multiple crises and other emerging challenges, where communities need strengthened resilience against climate change with concerted investment in substantial disaster mitigation infrastructure, including early warning systems. Health crises such as Ebola, and global pandemics such as COVID-19 are two examples of notable threats to development. This is the case, especially given the predominance of the informal sector, lack of ICT connectivity, reliance on social protection and fiscal reserves, inadequate health care, the interruption of human capital development and economic growth due to the closure of educational facilities in LDCs (UN-OHRLLS, 2021: 60–65).

Participants also requested positive use of the nexus approach in practice, and it is not coincidental that the three relevant examples chosen were drawn from the PDU of the Sahel Alliance, in an area that has been severely impacted upon by climate change. Two of the examples directly mitigate against climate challenges as a driver of conflict while encouraging development and livelihood security. The first comprises water supply, sanitation and pastoral hydraulic facilities for Kanem (above Lake Chad, bordering Niger) and the second, access to drinking water and sanitation in El Chargui and El Gharbi, Mauritania. The third in Tillabéry, Niger, with cross-border implications for Burkina Faso and Mali, focuses on strengthening trust and collaboration through dialogue and inclusive local participation (see Sahel Alliance, n.d.: 1, 2, 3).

Finally, in relation to the request for examples of the nexus approach from non-conflict contexts, two of the examples chosen related to climate change and resource scarcity and management. The first was for Botswana, where climate change has been identified as a major risk to natural capital and tourism revenue, a key sector for the country's economy. Hambira and Saarinen (2015) note that urgent policy intervention is required, suggesting that a lack of information and uncertainty surrounding the phenomenon are hindering progress. The second example stems from Tunisia in the MENA region, with the GIZ (2019) assessment on the nexus synergies surrounding water, energy and food sectors. This report highlighted successful collaboration between government ministries, agencies, unions and partners, such as the League of Arab States (LSA) which led to significant risk reduction related to resource scarcity, while promoting strategic synergy that resulted in policy development and improved delivery.

16.4 Discussion

At the outset of our discussion, it would be prudent to acknowledge the caution voiced by Soergel et al. concerning planetary integrity, "Averting dangerous climate change and preserving the integrity of the biosphere provide the essential foundation for long-term human well-being on planet Earth." (2021: 657). In keeping with this concern, our specific focus in this chapter has been to study the cluster of these "planetary integrity" SDGs, comprising Climate Action, Life below Water and Life on Land, with respect to performance on the sustainability agenda in Africa.

With the inception of the SDGs in 2015, Lu et al. (2015) observed that they are more onerous on the scientific community than their forerunners, the Millennium Development Goals (MDGs), since they require "coordinated global monitoring and modelling of many factors—social, economic and environmental", using the example that while climate-change mitigation happens locally, the reach is global (p. 432). They propose five priorities to assist: devising metrics, establishing monitoring mechanisms, evaluating progress, enhancing infrastructure, and standardising and verifying data. Furthermore, they suggest that "Scientists need to support the SDGs. We must help to integrate monitoring and evaluation mechanisms into policy-making at all levels and ensure that information about our planet is easily available to all" (p. 433).

In as many words, this scientific approach to the SDGs is compatible with the reform-minded NWoW in the international system, which prioritises efficiency, coordination, integration, communication with an eye to enhancing government performance on policy development, implementation and service delivery. As the joint UN and World Bank publication Pathways for Peace recognises, "The New Way of Working advocates for pooled and combined data, analysis, and information; better coordination of planning and programming processes; effective leadership for collective outcomes; and financing modalities to support closer collaboration across humanitarian, development, and peacebuilding operations." (2018: 279).

Roughly halfway into the window of the 2030 Goals, and our research confirms the call made by Pathways for Peace (2018), that cross-sectoral alignment and collaboration are vital for sustainable development to succeed. Notably, while the research highlights the importance of the nexus approach, somewhat ironically, the significance attached to the approach by survey participants was minimal, reinforcing the call for capacity strengthening for actors and stakeholders (as echoed in the sub-regional workshops) with corresponding policy innovation and development.

Likewise, regional bodies such as the African Union have recognised that without integration and partnerships, the sustainability of any growth and developmental efforts will be stymied (see AU and UN, 2018). In this regard, consider the impressive governance and policy innovations required to facilitate Africa's ambitious intra-continental free trade agreement (AU, 2018a), such as intra-African migration (AU, 2018b), the free movement of persons (AU and IOM, 2018) and joint COVID-19 vaccine lobbying by the AU, African Centres for Disease Control and COVAX (Africa Renewal, 2021).

Moving from the general growth and development debate to the specifics surrounding sustainable development in Africa, recall the 2019 Brookings Institute referred to above, namely that the continent is only on target to meet 3 SDGs (Gender Equality, Climate Action and Life on Land). These last two goals represent the primary target and one of two secondary targets of our current analysis. Moreover, and somewhat disturbingly, according to the SDG Center (2019), upon which research the Brookings report was drawn, African countries are unlikely to reach the other 10 Goals for Which data exists (there was no data for SDGs 10-12, 14).

Our own analysis revealed that climate-related SDGs were not highly regarded by survey participants, who were asked to share their opinion on the use of an interlinked approach to accelerate the implementation of policy to achieve the SDGs in Africa. Unlike visible drivers of conflict, such as violent insurrection or war, the threat posed by climate change is that of a largely silent adversary. Whereas conflict, instability and insecurity rule the roost in being identified by participants as the biggest challenges to achieving the SDGs and "The Africa We Want", followed by social and economic inequality and political challenges, climate risks are perceived by them to present only a negligible threat factor to the above agendas.

This finding may relate to a lack of climate awareness (see Simpson et al., 2021) and reinforces the need for more adequate and appropriate information-flows and communications regarding the existential threat that the climate crisis poses to life on the planet. Notably, based on their Sixth Assessment Report of 2021, the Intergovernmental Panel on Climate Change has helpfully suggested that policymakers shift their modelling from "climate risk" to "climate resilience" development, which regards climate, ecosystems and human society as coupled systems (IPCC, 2022: 6).

While it is easier to relate climate change to conflict than to prove a directly causal link from the former to the latter, Homer-Dixon has written extensively on resource scarcity and adverse human relations (e.g. 1991, 2001). Early on, in 1994, he presented case studies in support of the hypothesis that "environmental scarcity causes violent conflict", noting that such conflict would be "persistent, diffuse, and sub-national" and likely to worsen (p. 39). He also noted that ecological degradation was only one of three causes of environmental scarcity, the other two being population growth and unequal resource distribution. Taken together, these diminish the capacity of affected societies to adapt (1994: 40). Homer-Dixon's concern was two predicted consequences: fragmentation leading to monumental outward migration or authoritarian regimes launching attacks to create diversions from their own internal situations. Both disruptive outcomes, he argued, should become key foci for international security. Close on 30 years later, and both state fragmentation and authoritarian aggression remain problematic, as has been witnessed in Africa and abroad (see UNECA reports cited above).

Writing more recently, Stuart (2019: 175) astutely states, "While the growing impact of climate change does not automatically lead to more fragility and conflict, it does raise threats to security because it has an influence on social, economic and political structures". Helpfully, he distinguishes between national security and human security approaches to climate-induced migration. Herein, he cautions us to be wary of automatically translating environmental degradation into a security threat that justifies a hegemonic (Northern) securitisation response towards global inequality. Rather, he proposes that a risk-based approach proves to bear more fruit (2019: 177).

Accordingly, Stuart (2019) draws on seven climate-related risks identified in the G7 report entitled, "A New Climate for Peace" (see Rüttinger et al., 2015) comprising: local resource competition, livelihood security and migration, extreme weather events and disasters, volatile food prices and provision, trans-boundary water management, sea-level rise and coastal degradation and finally, unintended effects of climate policies. Using these risks, he isolates four pathways which impact on insecure communities and which can lead to violent extremisms. It is helpful to share such knowledge with officials concerned at local, national and trans-border levels, in order to advance our understanding of the complexity around human security and climate.

For example, Stuart suggests the following:

1. increasing resource stress results in increasing fragility which can multiply armed groups;
2. environmental degradation leads to livelihood destruction and food insecurity, which results in recruitment to armed groups for survival;
3. climate change can point to increased environmentally-induced migration which can lead to recruitment and financing;
4. finally, climate change promotes increased resource scarcity and fragility which can culminate in the weaponisation of natural resources.

He then applies this matrix to his socio-economic and political contextual analysis of Boko Haram in the Lake Chad Basin and Al-Shabaab in the Horn of Africa, to show how climate change can proliferate extremist movements in Africa, concluding that climate security considerations need to feature more prominently in conflict prevention and management strategies (Stuart, 2019).

Vellturo (2020), in a penetrating analysis, reinforces this notion of complex causality, cautioning strongly against simplistic attributions regarding certain migrant populations, notably the Fulani, as regards conflict and violent extremisms. She calls on scholars to regard the political and economic drivers of violence, rather than the more prevalent ethnic labelling and the drawing of 'false equivalencies' that pay scant regard to history, complexity and context.

Indeed, the findings of the current study draw on participant perceptions relative to actual country data on the SDGs, representing a concerted attempt to map our knowledge more specifically and intentionally in terms of historical context and geopolitical spread.

It is interesting to note that Osborn et al. (2015), for example, in their ranking of the SDGs in so-called "developed countries", placed SDG13 first, followed by SDGs 7 and 12 (Energy and Consumption), then SDG14 (Life below Water), with SDG15, Life on Land tied sixth. If we compare this with the regional findings for Africa as discussed above, we observe a very different pattern; the first places are taken by respectively by SDG1 (No Poverty), then SDG3 (Quality Education) and next, SDG4 (Good Health and Wellbeing). Note that for Africa, Climate Action only appears as the 12th most regarded goal, and of the bottom 3 performers, third worst is SDG15 (Life on Land) and worst, SDG14 (Life below Water), punctuated only by the second worst performer, SDG17 Partnerships for the Goals.

The tragically low position of the three "planetary integrity" variables across the spread of the perceived prioritisation of the SDGs in Africa is concerning. This angst is reinforced by the related study findings which show that both for the achievement of the SDGs and of the AU's Agenda 2063, climate concerns are ranked second-lowest for Africa, only above the continent's disease burden. It might be tempting to become downcast by these empirical findings, which imply that concerns for planetary boundaries are generally thought to receive scant regard on the continent.

In mitigation, note how the sub-regional workshop participants identified climate change as their first challenge to implementing a new way of working and to the success of the nexus approach across Africa. This is a positive takeaway, since our findings effectively reinforce the primary problem of awareness and in particular, the perception-reality gap regarding the SDGs. As Paul (2008: 579) puts it, "The biggest challenge of sustainable development remains the global consciousness from households to boardrooms regarding the importance of tackling the challenges of the Industrial Revolution: a limitless human and environmental exploitation."

It is precisely through increasing consciousness and sharing informed, factual knowledge around the climate crisis that the status quo can be changed. Let us attend, for a moment, to the study participants' perceptions of low progress for country performance on the SDGs, and specifically on SDGs 13, 14 and 15, perhaps informed by the usual tropes and negative stereotypes surrounding development in Africa. Factually, from UN data, we have seen that Climate Action was a top performer across SDG indicators for less than a handful of African countries and a bottom performer for two countries. By comparison, Life below Water registered a much larger bottom-end than top-end performance for process indicators on the continent, compared with Means of Implementation indicators for Life on Land, which was only half as extreme.

These indicator results were mirrored in the positive and negative deviance findings for survey participants. Notably, participants perceived performance on Life below Water (SDG14) to be higher than it actually was (at 'poor' versus the reality of "very poor"), while the other planetary integrity Goals were less highly rated, although they performed better, e.g. Climate Action (SDG13) was perceived worst of the three as 'poor', but in actual fact performed best of them at "above average", while Life on Land (SDG15) was regarded as 'poor' but performed at "below average".

Further to the above, participants revealed a generally negative to average perception of their country capacities to achieve the SDGs in general, regarding them as lacking sufficient capacity to shift the dial on the sustainability agenda. This same pattern applies to the planetary integrity variables, as indicated by the pronounced skew of their answers from very poor to 50–50. This finding, in combination with the low ranking of SDGs 13–15 in the prioritisation of Goals, does not bode well for accelerated implementation and confirms the need for capacity strengthening.

Such initiatives are being undertaken more generally by IDEP, in partnership with OSAA. A strong recommendation from the current, focused study is that interventions specifically pertaining to climate action and planetary integrity be promoted by multi-laterals, inter-governmental organisations, RECs, national and local governments, with support from NGOs, civil society and the private sector. This is precisely where the importance of SDG17 comes to the fore. We should not neglect to note that Partnerships for the Goals is significantly correlated with Climate Action ($p < 0.01$) and moreover, that in the UN OSAA study, SDG17 came out as the best performer on improvement across the Goals in Africa. While participants perceived it as scoring 'poorly' (34.9%), in reality it scored in the "very good" zone (at 86.7%), the only SDG to reach this top zone of improvement. Partnerships are clearly instrumental in advancing positive climate action and such initiatives could be taken up by the UN Global Compact and related entities.

In terms of promoting other aligned SDGs, Peace, Justice and Strong Institutions (SDG16), should be given greater airtime in the climate space. We noted how there was a weak albeit significant correlation ($p < 0.05$) between this Goal and Life below Water (SDG14), which strongly suggests that if greater attention is dedicated to involving institutions in planetary stewardship, there will be positive gains of ethical advancement, and both for justice and peace. Finally, we noted how, amongst the indicators under analysis, the most significant correlation ($p < 0.001$) was between Life on Land (SDG15) and Zero Hunger (SDG2). As suggested earlier, more mindful attention to detail concerning food provision and policy planning around crop planting and food security is totally compatible with preserving terrestrial life. It does suggest that more cross-sectoral collaboration is required and here, national coordination mechanisms and planning ministries need greater synchronicity with their counterparts in agriculture, infrastructure, finance and economic development, together with dedicated organisations such as the UNDP, FAO and WFP.

16.5 Conclusion

Aside from the patently obvious fact that Sustainable Development requires Development (these being Sustainable *Development* Goals), the elephant in the room with respect to the 2030 sustainability agenda is Governance. While Soergel et al.

(2021: 663) hinted at the importance of "integrating governance factors" (quality of political institutions and peace), they admitted to not yet treating them 'endogenously' and suggested that, "As an important next step, this would shed further light on their importance as preconditions (or also bottlenecks) for sustainable development." Meanwhile, Habibullah et al. (2022) in their recent study on climate change and biodiversity loss note that while temperature, precipitation and more frequent natural disasters, along with economic development, result in greater biodiversity loss, "good governance", promoted in conjunction with nature conservation and emissions controls can mitigate this.

Globally, the IPCC report (2022) confirms that evidence with high to very high confidence levels exists for the negative impact of climate change to terrestrial, freshwater and oceans ecosystems structure, species range shifts and phenology. For human societies, particularly significant negative impacts arise on health and well-being, as well as on cities, settlements and infrastructure (with medium confidence levels for negative impacts on water scarcity and food production).

In keeping with such insights, the current chapter, building on the UN OSAA study that located governance at the core of the nexus once again confirms the integral significance of governance to sustainability, which 82% of participants felt was relevant and, predominantly, extremely relevant to the success of the nexus approach (Tschudin, 2022). Zooming in on Climate Action and the urgency of the existential threat facing life on earth if planetary boundaries are transgressed, the chapter has shown how greater attention needs to be given to awareness of and performance on SDGs 13, 14 and 15 in Africa, as a matter of immediate priority.

This is clearly the case from an interlinked nexus perspective, because for Climate Action, as well as for the overall "planetary integrity" SDGs, Complex Adaptive Systems modelling reveals that the perceived work pillars of importance first involve Development, then Governance, and next Peace and Human Rights (with Peace slightly ahead of Human Rights due to the positioning of Life on Land), whereas Humanitarian action is hardly involved. This differs slightly from the picture for all the SDGs taken together in the Africa regional study, where Governance was predominant, followed by Peace, next Development, then Human Rights and finally, the Humanitarian pillar.

Such subtle variations are important and studying the pattern for the planetary integrity cluster Goals compared to that of the overall SDG study for Africa is helpful. This applies insofar as it can sharpen focus and direct policy development, implementation and intervention strategies for these specific SDGs, according to the relevance of the pillars implicated in their achievement. Attention to detail and context specificity are heavily connected with ensuring human security, promoting societal harmony, and minimising conflict and violence.

The five key nuanced takeaways from the chapter may be summed as follows. First, we must confront the negative perceptions that we Africans have of our own progress on the SDGs in general and specifically on Climate Action. While it is true that Life below Water and Life on Land both merit significant attention to realise improvement, the reality is not as bleak as the perception thereof. We should celebrate gains made, along with the many positive innovations arising on our continent and debunk negative stereotypes.

Next, capacity, which has been identified as a major impediment, can be built and strengthened by information sharing and awareness-raising initiatives through educational and continuing professional development initiatives. These initiatives ought to be directed at local and national public servants, along with relevant civil society and private sector stakeholders. They need to be accompanied by communication strategies for inclusive sharing so that broad-based democratic participation is enabled.

Third, in the drive for climate solidarity, the related bootstrapping of SDGs 16 and 17, namely Peace, Justice and Strong Institutions, together with Partnerships for the Goals, cannot be overemphasised. These institutional and collective action-oriented Goals should be regarded as complementary to the realisation of critical Goals such as SDGs 1 and 2, No Poverty and Zero Hunger, and not as detached from them.

Fourth, the unique pillars or work streams involved all have varying roles with differential weightings to play. Specialists in peace and security, development, governance, human rights and humanitarian work need to escape silos, echo chambers and duplication in order to collaborate with one another to compound gains from their respective strengths and comparative advantages, instead of competing in turf wars. Cross-sectoral expertise and transdisciplinary efforts, for example from climate science, the social sciences and ICT, need to be embraced and aligned.

Finally, the success of the above initiatives is both largely enabled by and dependent upon an intentionally inter-connected, systematic and structured, yet agile, adaptive way of working, which builds synergy to promote efficiency, collaboration and coordination. While it is true that the nexus approach relies on financing, infrastructure, human, material and other resources, in the longer term the benefit outweighs the cost.

Such points will hopefully serve as food for urgent, open-minded thought, perhaps, in the face of self-interested short-termism and planetary extinction. Meanwhile, the critical window of opportunity for concerted climate action and mitigation continues to close rapidly upon us.

References

Adelman, S. (2018). The sustainable development goals, Anthropocentrism and Neoliberalism. In D. French & L. Kotzé (Eds.), *Sustainable Development Goals: law, theory and implementation* (pp. 1–23). Cheltenham: Edward Elgar.

Africa Renewal. (2021). COVID-19 vaccines: AU, Africa CDC and COVAX call on donors and manufacturers to accelerate access. Joint statement on actions needed from 1 Jan 2022 on dose donations to African countries. New York: United Nations.

African Union. (2016). African Union Master Roadmap of Practical Steps to Silence the Guns in Africa by Year 2020 (Lusaka Master Roadmap 2016).

African Union. (2018a). Agreement establishing the African Continental Free Trade Area. Kigali, Rwanda: 21 Mar 2018. https://au.int/sites/default/files/treaties/36437-treaty-consolidated_text_on_cfta_-_en.pdf. Accessed 15 February 2023.

African Union. (2018b). Migration Policy Framework for Africa and Plan of Action (2018–2030). African Union Commission, Addis Ababa, May 2018.

African Union. (2022). Decision of the 14th Extraordinary Session of the Assembly of the Union on Silencing the Guns in Africa. https://www.peaceau.org/uploads/ext-assembly-au-dec-1-and-decl-1-xiv-e.pdf. Accessed 15 February 2023.

African Union and International Organization for Migration (IOM). (2018). Study on the benefits and challenges of free movement of persons in Africa. Geneva: IOM. https://au.int/en/documents/20200204/african-union-master-roadmap-practical-steps-silence-guns-africa-year-2020-lusaka. Accessed 15 February 2023.

Badejo, B. (2020). Study on the interlinkages between the development, peace and security, human rights and humanitarian pillars in West and Central Africa. Draft inception report, UNECA, 29 Sep 2020.

Berrang-Ford, L., Ford, J.D., & Paterson, J. (2011). Are we adapting to climate change? *Global Environmental Change*, 21(1), 25–33. https://doi.org/10.1016/j.gloenvcha.2010.09.012

Bartram, J., Brocklehurst, C., Bradley, D., Muller, M., & Evans, B. (2018). Policy review of the means of implementation targets and indicators for the sustainable development goal for water and sanitation. *NPJ Clean Water, 1*(1), 3. https://doi.org/10.1038/s41545-018-0003-0

Carson, R. (1962). *Silent Spring*. Boston: Houghton Mifflin.

Center on International Cooperation. (2019). *The Triple Nexus in Practice: Toward a New Way of Working in Protracted and Repeated Crises*. New York: Center on International Cooperation.

GIZ (2019). Nexus Assessment for Tunisia Synergies of the water, energy and food sectors Executive Summary. Deutsche Gesellschaft für Internationale Zusammenarbeit (GIZ), Tunis, Sep 2019. https://uploads.water-energy-food.org/legacy/nexus_summary_web_version.pdf. Accessed 15 February 2023.

Guterres, A. (2022, Nov 7) "Secretary-General's remarks to High-Level opening of COP27", 27th Conference of the Parties to the United Nations Framework Convention on Climate Change (COP27), Sharm el-Sheikh, Egypt. https://www.un.org/sg/en/content/sg/speeches/2022-11-07/secretary-generals-remarks-high-level-opening-of-cop27. Accessed 15 February 2023.

Habibullah, M. S., Din, B. H., Tan, S.-H., & Zahid, H. (2022). Impact of climate change on biodiversity loss: global evidence. *Environmental Science and Pollution Research, 29*(1), 1073–1086. https://doi.org/10.1007/s11356-021-15702-8

Hambira, W. & Saarinen, J. (2015). Policy-makers' perceptions of the tourism–climate change nexus: Policy needs and constraints in Botswana. *Development Southern Africa,* 32. https://doi.org/10.1080/0376835X.2015.1010716

Homer-Dixon, T.F. (1991). On the Threshold: Environmental Changes As Causes of Acute Conflict. *International Security*, 16(2), 76–116.

Homer-Dixon, T.F. (1994). Environmental Scarcities and Violent Conflict: Evidence from Cases. *International Security*, 19(1), 5–40.

Homer-Dixon, T.F. (2001). Environment, Scarcity and Violence. Princeton: Princeton University Press.

Howe, P. (2019). The triple nexus: A potential approach to supporting the achievement of the Sustainable Development Goals? World Development 124:1–13. https://doi.org/10.1177/0007650311427426. hal-00707337

IPCC. (2022). Summary for Policymakers. In H.O. Pörtner, D.C. Roberts & E.S. Poloczanska (Eds.), *Climate Change 2022: Impacts, Adaptation and Vulnerability. Contribution of Working Group II to the Sixth Assessment Report of the Intergovernmental Panel on Climate Change* (pp. 3–33). Cambridge: Cambridge University Press. https://doi.org/10.1017/9781009325844.001

Lu, Y., Nakicenovic, N., Visbeck, M., & Stevance, A.-S. (2015). Policy: Five priorities for the UN sustainable development goals. *Nature, 520*(7548), 432–433. https://doi.org/10.1038/520432a

Niemeijer. D., & de Groot, R. (2008). A conceptual framework for selecting environmental indicator sets, *Ecological Indicators*, 8, 14–25.

Nilsson, M. (2016). Understanding and mapping important interactions among SDGs. Background paper for Expert meeting in preparation for HLPF 2017: Readying institutions and policies for integrated approaches to implementation of the 2030 Agenda. Vienna, Dec 2016.

Osborn, D., Cutter, A., & Ullah, F. (2015). Universal Sustainable Development Goals: understanding the transformational challenge for developed countries. Report of a study by Stakeholder Forum, May 2015.

Paul, B.D. (2008). A history of the concept of sustainable development: literature review. The Annals of the University of Oradea. *Economic Sciences Series*, *17*(2), 576–580.

Pinkse, J., & Kolk, A. (2012). Addressing the climate change sustainable development nexus: the role of multi-stakeholder partnerships. *Business and Society*, 51(1), 176–210.

Preise, R., Biggs, R., De Vos, A., & Folke, C. (2018). Social-ecological systems as complex adaptive systems: organizing principles for advancing research methods and approaches. *Ecology and Science,* 23, 46. 10.5751

Rockström, J., Steffen, W., Noone, K., Persson, Å., Chapin, F. S., Lambin, E. F., Lenton, T. M., Scheffer, M., Folke, C., & Schellnhuber, H. J. (2009). A safe operating space for humanity. *nature, 461*(7263), 472–475.

Rogers, K. H., Luton, R., Biggs, H., Biggs, R., Blignaut, S., Choles, A. G., Palmer, C. G., & Tangwe, P. (2013). Fostering complexity thinking in action research for change in social–ecological systems. *Ecology and Society, 18*(2). https://doi.org/10.5751/ES-05330-180231

Rüttinger, L., Smith, D., Stang, G., Tänzler, D., & Vivekananda, J. (2015). *A New Climate for Peace: Taking Action on Climate and Fragility Risks.* Adelphi / International Alert / Woodrow Wilson International Center for Scholars / European Union Institute for Security Studies. https://www.newclimateforpeace.org. Accessed 15 February 2023.

Sahel Alliance (N.d. 3) "Common Ground": Strengthening mutual trust and collaboration between communities, local authorities and defence and security forces in the Tillabéry Region (Niger). https://www.alliance-sahel.org/en/projets-pdu/common-ground-strengthening-mutual-trust-and-collaboration-between-communities-local-authorities-and-defence-and-security-forces-in-the-tillabery-region-niger/ Accessed 28 January 2023.

Sahel Alliance. (N.d. 1). Kanem water supply, sanitation and pastoral hydraulic facilities project. https://www.alliance-sahel.org/en/projets-pdu/projet-damenagement-des-ouvrages-dadduction-en-eau-potable-dassainissement-et-douvrages-hydrauliques-pastoraux-dans-la-region-du-kanem/ Accessed 28 January 2023.

Sahel Alliance. (N.d. 2). Project "Access to drinking water and sanitation in the 2 Hodhs regions of Mauritania (G5-2H). https://www.alliance-sahel.org/en/projets-pdu/project-access-to-drinking-water-and-sanitation-in-the-2-hodhs-regions-of-maurtania-g5-2h/ Accessed 28 January 2023.

SDG Center. (2019). Africa 2030: Sustainable Development Goals Three-Year Reality Check. The Sustainable Development Goals Center for Africa. https://sdgcafrica.org/wp-content/uploads/2019/06/AFRICA-2030-SDGs-THREE-YEAR-REALITY-CHECK-REPORT.pdf. Accessed 15 February 2023.

Simpson, N. P., Andrews, T. M., Krönke, M., Lennard, C., Odoulami, R. C., Ouweneel, B., Steynor A., & Trisos, C. H. (2021). Climate change literacy in Africa. *Nature Climate Change, 11*(11), 937–944. https://doi.org/10.1038/s41558-021-01171-x

Soergel, B., Kriegler, E., Weindl, I., Rauner, S., Dirnaichner, A., Ruhe, C., Hofmann, M., Bauer, N., Bertram, C., & Bodirsky, B. L. (2021). A sustainable development pathway for climate action within the UN 2030 Agenda. *Nature Climate Change, 11*(8), 656–664. https://doi.org/10.1038/s41558-021-01098-3

Stuart, J.W. (2019) Climate change and violent extremism in Africa: a contested link. In A. Tschudin, A. Moffat, C. Buchanan-Clarke, S. Russell & L. Coutts (Eds.), *Extremisms in Africa*, Vol 2, (pp. 175–200. Johannesburg: Tracey McDonald Publishers.

Thuiller, W. (2007). Climate change and the ecologist. *Nature,* 448, 550–552. https://doi.org/10.1038/448550a

Tschudin, A. (2022). Strengthening the Capacity of African Countries to Design and Implement Policies that Promote the Nexus between Peace, Humanitarian Work, Development and Human Rights for an Accelerated Implementation of the SDGs. New York: United Nations Office of the Special Adviser on Africa.

United Nations and World Bank. (2018). Pathways for Peace: Inclusive Approaches to Preventing Violent Conflict. Washington, DC. https://doi.org/10.1596/978-1-4648-1162-3. License: Creative Commons Attribution CC BY 3.0 IGO.

United Nations Economic Commission for Africa. (2015). *Conflicts in the Democratic Republic of Congo: Causes, impact and implications for the Great Lakes region.* Addis Ababa: ECA.

United Nations Economic Commission for Africa. (2016). *Human and Economic Cost of Conflict in the Horn of Africa: Implications for a Transformative and Inclusive Post-Conflict Development.* Addis Ababa: ECA.

United Nations Economic Commission for Africa. (2017a). *Conflict in the Sahel region and the developmental consequences.* Addis Ababa: ECA.

United Nations Economic Commission for Africa. (2017b). *New Fringe Pastoralism: Conflict and Insecurity and Development in the Horn of Africa and the Sahel.* Addis Ababa: ECA.

United Nations Economic Commission for Africa. (2018). *Transboundary natural resource disputes in Africa: Policies, institutions and management experiences.* Addis Ababa: ECA.

United Nations Economic Commission for Africa (2018-01). AU–UN framework for the implementation of agenda 2063 and the 2030 agenda for sustainable development. Addis Ababa: UN. ECA. https://hdl.handle.net/10855/24248.

United Nations Economic Commission for Africa. (2019). *Sahel 2043: Towards a resilient, inclusive and prosperous Sahel region.* Addis Ababa: ECA.

United Nations Environment Programme. (2020). South Africa aims to stop marine litter at its source. 4 Sep. https://www.unep.org/news-and-stories/story/south-africa-aims-stop-marine-litter-its-source. Accessed 15 Feb 2023.

United Nations Evaluation Group. (2018). The Humanitarian-Development Nexus – What do evaluations have to say? Mapping and synthesis of evaluations. Working paper. Humanitarian Evaluation Interest Group, New York, Feb 2018.

United Nations Foundation. (2016). 10 Key Takeaways from the World Humanitarian Summit. https://unfoundation.org/blog/post/10-key-takeaways-from-the-world-humanitarian-summit/ Accessed 2 January 2023.

United Nations General Assembly. (2016). One humanity: shared responsibility. Report of the Secretary-General for the World Humanitarian Summit. A/70/709. 2 Feb 2016. United Nations, New York.

United Nations Office for Disaster Risk Reduction. (2021). GAR Special Report on Drought 2021. https://www.undrr.org/publication/gar-special-report-drought-2021. Accessed 15 February 2023.

United Nations Office for the Coordination of Humanitarian Affairs. (2016). *World Humanitarian Data and Trends 2016.* New York: UNOCHA.

United Nations Office of the High Representative for the Least Developed Countries, Landlocked Developing Countries and Small Island Developing States. (2021). Lessons learned: implementing the Istanbul Programme of Action for Least Developed Countries for the Decade 2011-2020. Preliminary Report, OHRLLS. https://www.un.org/ohrlls/sites/www.un.org.ohrlls/files/preliminary_report_lessons_learned_ipoa_implementation.pdf/ Accessed 15 February 2023.

United Republic of Tanzania (URT). (2021). National Climate Change Response Strategy (2021–2026). Vice-President's Office, Division of Environment, Government Printer, Dodoma.

Vellturo, M. (2020) Fulani and Jihad: the argument against simplistic narratives in West Africa. In A. Tschudin, A. Moffat, C. Buchanan-Clarke, S. Russell & L. Coutts, L. (Eds.), *Extremisms in Africa,* Vol 3 (pp 262–283). Johannesburg: Good Governance Africa.

Weitz, N., Carlsen, H., Nilsson, M., & Skanberg, K. (2018). Towards systemic and contextual priority setting for implementing the 2030 Agenda. *Sustain Science,* 13, 531–548.

Westley, F., Olsson, P., Folke, C., Homer-Dixon, T., Vredenburg, H., Loorbach, D., Thompson, J., Nilsson, M., Lambin, E., Sendzimir, J., Banerjee, B., Galaz, V., & van der Leeuw, S. (2011) Tipping toward sustainability: emerging pathways of transformation. *Ambio, 40*(7), 762–780. https://doi.org/10.1007/s13280-011-0186-9

World Bank. (2022a). Every day champions in Gabon are helping turn the tide in the fight against climate change. 31 Aug. https://www.worldbank.org/en/news/feature/2022/08/31/everyday-champions-in-gabon-are-helping-turn-the-tide-in-the-fight-against-climate-change. Accessed 15 February 2023.

World Bank. (2022b). Climate-informed economic development key to Malawi's future growth and resilience. 27 Oct. https://www.worldbank.org/en/news/press-release/2022/10/27/climate-informed-economic-development-key-to-malawi-s-future-growth-and-resilience. Accessed 15 February 2023.

Zhang, Y., Runting, R., Webb, E., Edwards, D. & Carrasco, R. (2021). Coordinated intensification to reconcile the 'zero hunger' and 'life on land' Sustainable Development Goals. *Journal of Environmental Management*, 284. 112032. https://doi.org/10.1016/j.jenvman.2021.112032

Ziai, A. (2016). Development discourse and global history: from colonialism to the sustainable development goals. London: Routledge.

Chapter 17
Social Climate, Social Capital, and Social Goods in Sub-Saharan African Countries: Implications for Sustainable Development

Aliye F. Mataraci and Ozge Buyukdagli

Abstract This work analyses the findings of the 2021 World Happiness Report (WHR) and World Social Capital Monitor (WSCM) for Sub-Saharan African countries to provide a better understanding of the social climate in the region. The WHR measures peoples' subjective happiness in countries around the globe, focusing on seven key factors: happiness, gross domestic product, social support, healthy life expectancy, freedom to make life choices, perception of corruption, and generosity. The WSCM is a worldwide open access Social Capital Survey for citizens to better understand the social environment in their countries. Developed by the Basel Institute of Commons and Economics (http://commons.ch/) for the United Nations (UN) Sustainable Development partnership, the WSCM serves as a tool within the UN framework to promote Sustainable Development Goals. This open access online survey (http://commons.ch/assessing-social-capital/), conducted in 50 languages and 300 cities as of 2021, assesses the social capital of countries worldwide. Timelines and comments, for eight social goods (climate, trust, measures, taxes, invest, helpfulness, friendliness, hospitality) provided by the survey, have been published annually since 2017 for each country affiliated with the WSCM. Both the WHR and WSCM reports aim to measure the social climate by considering a particular set of factors. In this study, two data sets containing the values for these factors are used to analyse the social climate in Sub-Saharan African countries. A range of techniques, including data visualisation, correlation coefficients, and clustering, are employed to identify the characteristics of the countries in the region and the interrelationships between them based on these factors.

Prof Aliye F. Mataraci (PhD), Adjunct Professor, Heads the Social Sciences Department and Coordinator of Social and Political Sciences Program, On behalf of the International University of Sarajevo. Email: amataraci@ius.edu.ba; Orcid ID: http://orcid.org/0000-0001-8289-8763

Prof Özge Büyükdağlı (PhD), Assist. Prof. Dr, International University of Sarajevo, Computer Sciences and Engineering Program, Faculty Member. Email: obuyukdagli@ius.edu.ba; Orcid ID: http://orcid.org/0000-0001-5758-4607

J. C. K. Kiyala and N. Chivasa (eds.), *Climate Change and Socio-political Violence in Sub-Saharan Africa in the Anthropocene*, The Anthropocene: Politik—Economics—Society—Science 37,
https://doi.org/10.1007/978-3-031-48375-2_17

Keywords Social goods · Social climate · Sustainable development · World happiness report · World social capital monitor · Sub-Saharan Africa · Data analysis

17.1 Introduction

This work aims to better understand the social climate in Sub-Saharan African countries based on the results of the World Happiness Report (WHR) and the World Social Capital Monitor (WSCM). World happiness is one of the main objectives of sustainable development which is closely tied to social capital and social climate. Jan-Emmanuel De Neve and Jeffrey D. Sachs, exploring the empirical links between the Sustainable Development Goals (SDG) and wellbeing, found a strong positive correlation: 'Unpacking the SDGs by looking at how each SDG relates to wellbeing shows, in most cases, a strong positive correlation.' (2020). Thus, countries with higher SDG Index scores tend to experience greater subjective wellbeing, from which it is possible to conclude that sustainable development makes populations happier. This study examines the validity of this suggested positive correlation for Sub-Saharan African countries.

Using the WHR and WSMC data for the Sub-Saharan African countries in 2021, this work identifies the similarities, common characteristics, and distinctions among these countries to gain a better understanding of the social climate in the region. Data mining techniques are applied to analyse the social environment, as they are among the most powerful tools for examining diverse types of data. They facilitate the analysis of the characteristics, similarities, and differences between varying entities or factors, allowing for the detection of patterns hidden withing large quantities of data collected for different purposes on different topics over the years.

Among the various data mining techniques, clustering is one of the most useful and practical methods to find data points with similar characteristics and patterns, providing opportunities to group the data based on multiple factors (Abdullah et al., 2022; Aggarwal/Sharma, 2019; Islam et al., 2021; Kapourani/Sanguinetti, 2019). This study employs clustering as a data mining technique to analyse the data, accompanied by a variety of data visualisation methods to better understand and represent the data. Section two presents the factors considered in the WHR 2021 data, their descriptive statistics, and the correlation analyses conducted. Section three discusses the clustering algorithm applied to the WHR 2021 data and the clustering results. Section four elaborates on social climate, social capital and social goods in Sub-Saharan Africa based on the results of the WSCM 2021, in addition to exploring the relationship between SDGs and social climate. All these discussions are wrapped up in concluding remarks.

17.2 Happiness in Sub-Saharan African Countries

The WHR aims to measure the subjective perception of happiness among people in countries around the globe. The WHR and the growing international interest in happiness can be attributed to Resolution 65/309, "Happiness: Towards a holistic approach to development," which was sponsored by Bhutan and adopted by the General Assembly of the United Nations on 19 July 2011. This resolution invites national governments to 'give more importance to happiness and wellbeing in determining how to achieve and measure social and economic development.' Subsequently, 20 March was proclaimed the International Day of Happiness, observed annually since its establishment on 28 June 2012 (Helliwell et al., 2021).

The WHR has been published every year since its inaugural edition on 2 April 2012, celebrating its tenth anniversary in 2022. The Gallup World Poll has consistently been its primary source, in addition to the World Risk Poll and the ICL-YouGov Behaviour Tracker (Helliwell et al., 2021). It considers six main factors that contribute to happiness: gross domestic product, social support, healthy life expectancy, freedom to make life choices, perception of corruption, and generosity. In addition, the report presents survey results for happiness scores and investigates the relationship between these six factors and the happiness scores of each country.

Since 2012, a total of 166 countries have been included in WHR reports, 44 of which are Sub-Saharan African countries. Every year, the WHR incorporates data from different surveys and databases and provides insightful commentary and analysis for all countries. There are several studies in the literature focusing on specific regions, years, and/or applying new data analysis algorithms to extract meaningful observations and conclusions based on WHR results (Zhu, 2022; Khder/Sayfi, 2022; Ulkhaq, 2020).

The WHR 2022, written by a group of independent experts, was published in March 2022. However, the dataset has not yet been made public as of the writing of this chapter. Therefore, this study relies on the dataset from the WHR 2021. The WHR 2021 presents results for seven factors: 1. Happiness score (HS), 2. Gross domestic product (GDP), 3. Social support (SS), 4. Health life expectancy (HLE), 5. Freedom to make life choices (FMLC), 6. Perception of corruption (PC). 7. Generosity (G) (Helliwell et al., 2021). Each of these factors will be explored in the following section. The 36 Sub-Saharan African countries covered by the WHR 2021 are listed in Table 17.1:

Table 17.1 Sub-Saharan African Countries data in WHR 2021 report

Benin, Botswana, Burkina Faso, Burundi, Cameroon, Chad, Comoros, Congo (Brazzaville), Eswatini, Ethiopia, Gabon, Gambia, Ghana, Guinea, Ivory Coast, Kenya, Lesotho, Liberia, Madagascar, Malawi, Mali, Mauritania, Mauritius, Mozambique, Namibia, Niger, Nigeria, Rwanda, Senegal, Sierra Leone, South Africa, Tanzania, Togo, Uganda, Zambia, Zimbabwe

Source Authors

17.2.1 *Introduction to the Factors and Descriptive Statistics Used for The WHR 2021 Data*

Table 17.2 presents a summary of the data used in this study along with its descriptive statistics, while and the complete dataset is given in the Appendix. The average, standard deviation, minimum and maximum values for each of the seven factors is presented for 36 Sub-Saharan African countries included in WHR 2021. World averages for each factor are also included as a reference to better understand the position of Sub-Saharan African countries. It is apparent from the table below that the averages for all seven factors in Sub-Saharan African countries are below the world averages. However, it is important to note that there is at least one country within the Sub-Saharan African region that exceeds the world average for each factor. This observation is promising as it indicates that there are some countries in the region that are excelling in specific areas. In the following sub-sections, detailed explanations, data visualisations, and some observations obtained from the WHR dataset will be presented.

Table 17.2 Summary of the WHR 2021 data

Factors	Average (Sub-Saharan African)	Average (World)	Standard deviation (Sub-Saharan African)	Minimum (Sub-Saharan African)	Maximum (Sub-Saharan African)
Happiness score (HS)	4.49	5.53	0.66	3.15	6.05
Gross domestic product (GDP)	8.08	9.43	0.80	6.64	10.01
Social support (SS)	0.70	0.86	0.10	0.49	0.91
Health life expectancy (HLE)	55.89	64.99	3.84	48.48	66.70
Freedom to make life choices (FMLC)	0.72	0.79	0.09	0.55	0.90
Generosity (G)	0.01	– 0.02	0.13	– 0.25	0.42
Perception of corruption (PC)	0.77	0.73	0.13	0.17	0.92

Source Authors

17.2.2 Happiness Score (HS)

Happiness score, also known as subjective wellbeing, is based on individuals' subjective responses to the following question: "Please imagine a ladder, with steps numbered from 0 at the bottom to 10 at the top. The top of the ladder represents the best possible life for you and the bottom of the ladder represents the worst possible life for you. On which step of the ladder would you say you personally feel you stand at this time?" (Helliwell et al., 2021).

In the WHR 2021, the happiness score value for each country reflects the country's national average response. The global average HS score is 5.53, while for the Sub-Saharan African countries, this score is 4.49. The HS for each Sub-Saharan African country covered by WHR 2021 is presented with a bubble plot in Fig. 17.1. Each bubble represents a country, and the size of each bubble corresponds to the size of their population. The vertical axis shows the HS, and the straight bar represents the world average for the HS factor. Among the Sub-Saharan African countries, Mauritius is the only one with HS (6.05) above the world average.

Figure 17.1 offers a comparison of the HS for all countries covered in the WHR 2021. To better represent this extensive data, countries are grouped according to their geographical regions and the bubble sizes reflect the sizes of the countries' total population. The vertical axis again represents the HS levels. Notably, South Asian countries, encompassing a population of over 1.8 billion, have the lowest average HS with a value of 4.44. They are followed by the Sub-Saharan African countries, with an HS of 4.49 HS and a population of 930 million people. Figure 17.2 allows us to observe a potential negative correlation between population size and happiness levels: People living in the most densely populated areas appear to be the least happy, according to the answers provided by the participants of the WHR 2021.

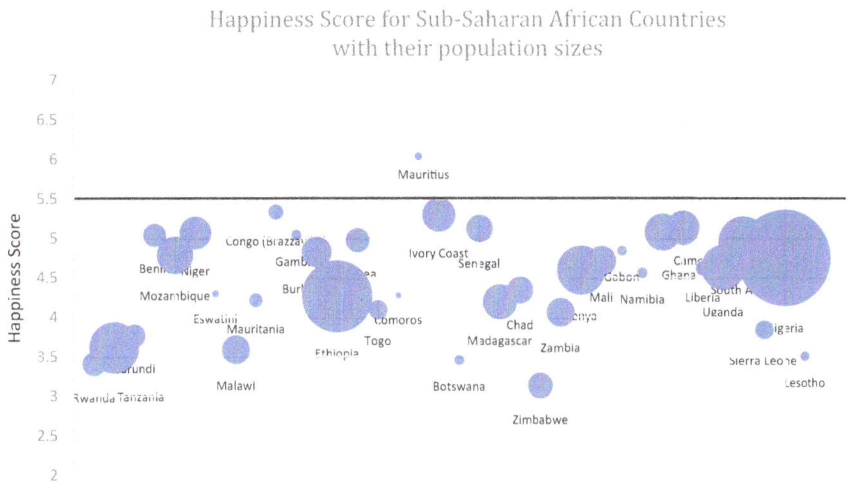

Fig. 17.1 Happiness Score for Sub-Saharan African Countries. *Source* Authors

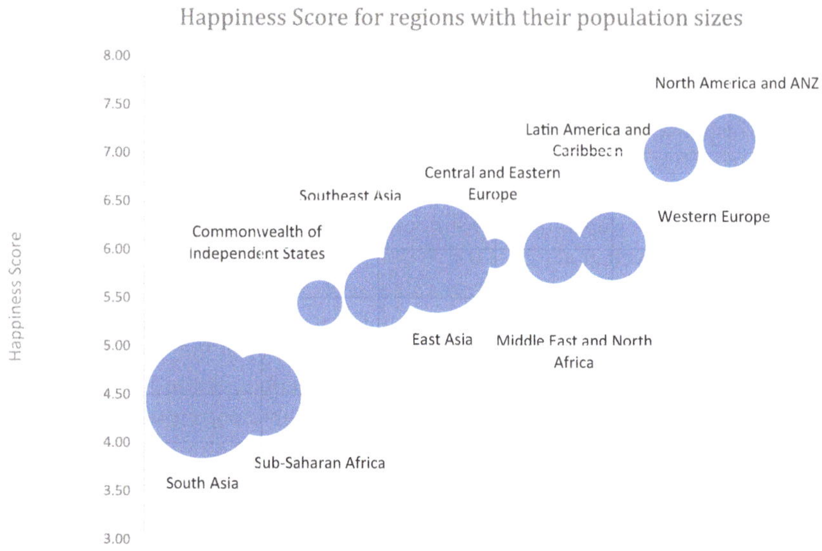

Fig. 17.2 Happiness Score for regions with their population sizes. *Source* The Authors

17.2.3 Gross Domestic Product (GDP)

The GDP per capita data that is presented in the WHR 2021 is collected from other international reports such as the OECD Economic Outlook and the World Bank's Global Economic Prospects. The GDP for each country was forecasted based on the time series data from 2019 to 2020 from the OECD Economic Outlook No. 108 (OECD Economic Outlook: Statistics and Projections (database), 2020). In cases where data was not available in the reported statistics for a specific country, it was estimated using the World Bank's Global Economic Prospects. According to the WHR 2021, only three Sub-Saharan African countries reported a GDP higher than the world average (9.43): Gabon (9.60), Botswana (9.78), and Mauritius (10.01). In Fig. 17.3, the GDP values for all countries included in the WHR 2021 are presented. The countries are grouped according to their regions, each of which is designated by a different color. The vertical axis shows the GDP values, and the world average is marked by a solid line on the plot. The Sub-Saharan African region has the largest number of countries with GDP values below world average out of all the other regions.

17.2.4 Social Support (SS)

Social Support, as measured by the WHR, aims to measure the individuals' perceptions of having someone they can rely on in times of trouble. Participants are asked a

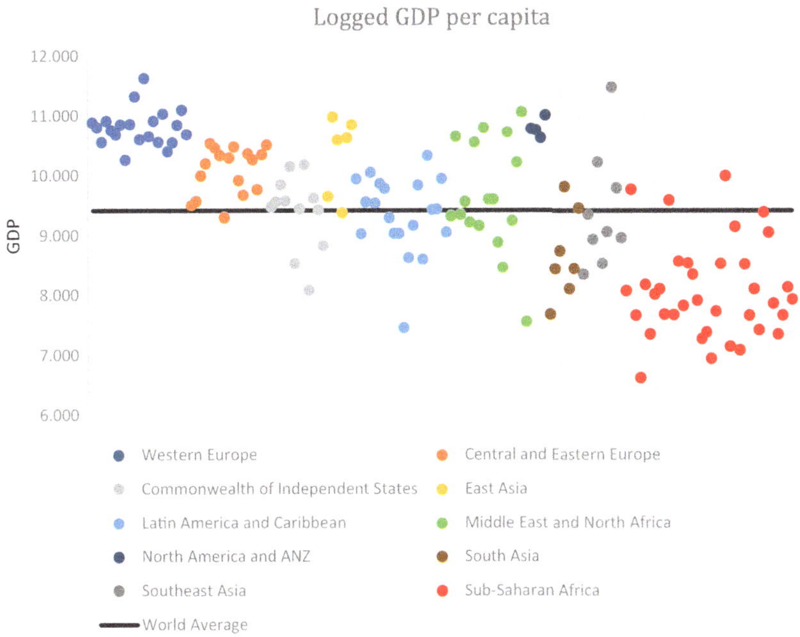

Fig. 17.3 GDP of all countries. *Source* Authors

yes or no question, with response values of 1 and 0, respectively. The question posed reads:

"If you were in trouble, do you have relatives or friends you can count on to help you whenever you need them, or not?"

The world average for SS reported by the WHR 2021 is 0.81. However, this value is 0.70 for Sub-Saharan African countries. Namibia (0.82), Mauritius (0.91), and South Africa (0.86) are the only countries that scored above the world average. Among the 149 countries that participated in the 2021 survey, Mauritius ranked 38th in terms of the highest SS value. After Afghanistan (0.46), the countries with the least values for this factor are three Sub-Saharan African countries: Benin (0.49), Burundi (0.49), and Malawi (0.53).

In Fig. 17.4, a scatter plot is used to visualise the relationship between GDP and SS for Sub-Saharan African countries. Out of all the factors examined in this study, GDP exhibits the strongest relationship with SS, as was shown in Sect. 2.9 as well. The plot reveals that as GDP increases, the SS value also increases. A trendline has also been added to the plot to highlight the direction of the relationship between these two factors.

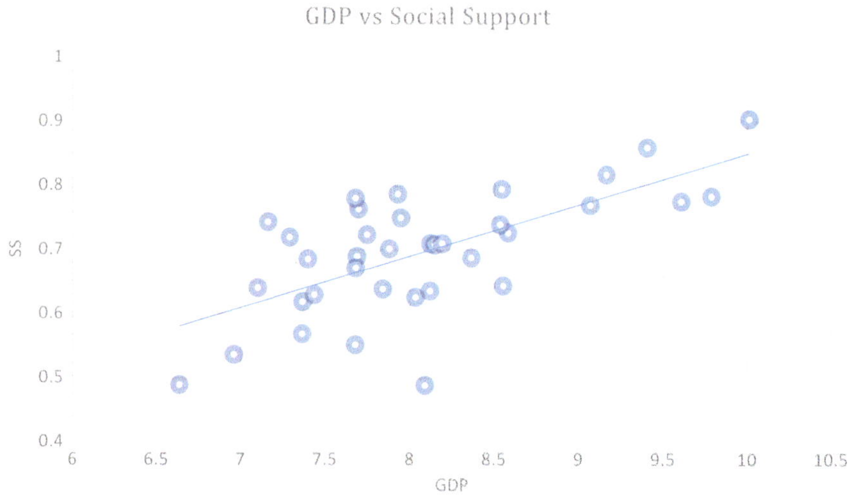

Fig. 17.4 Comparison between GDP and SS. *Source* Authors

17.2.5 Healthy Life Expectancy (HLE)

Healthy life expectancy at birth data is derived from the World Health Organization (WHO) report. According to the WHR 2021 results, all Sub-Saharan African countries other than Mauritius are below the world average in terms of HLE.

Figure 17.5 presents the HLE data of all countries through a filled world map, which uses a color scale to represent HLE values. The lowest value (lightest blue) of 48.48 corresponds to Chad, while the highest value (darkest blue) of 76.95 corresponds to Singapore. All other countries are represented with varying shades of blue based on their rankings in terms of HLE values. As the HLE value increases the shade of the blue darkens. Countries shown in grey do not have data for this factor in the WHR 2021. The figure demonstrates that Africa as a continent contains predominately light blue regions, indicating low values of HLE, with the lightest areas concentrated in the Sub-Saharan region.

Chad, Lesotho, and Nigeria are identified as the 3 countries with the lowest HLE values in the world. Among all other Sub-Saharan African countries considered in this study, Chad has the highest maternal mortality rate, with 1140 deaths per 100,000 live births, as reported by the WHO (WHO et al., 2019). One factor contributing to this high maternal mortality rate can be attributed to the low percentage of births attended by skilled health personnel, as indicated by UNICEF (UNICEF, 2020). Chad again has the lowest percentage for this indicator, with the value of 20%, whereas the average percentage for Sub-Saharan African countries is 70.6%. Lesotho, in comparison to Chad, has better values for these two indicators. It reports a maternal mortality rate of 544 and a percentage of births attended by skilled health personnel of 43.4%, which still falls below the regional averages. A noteworthy indicator value

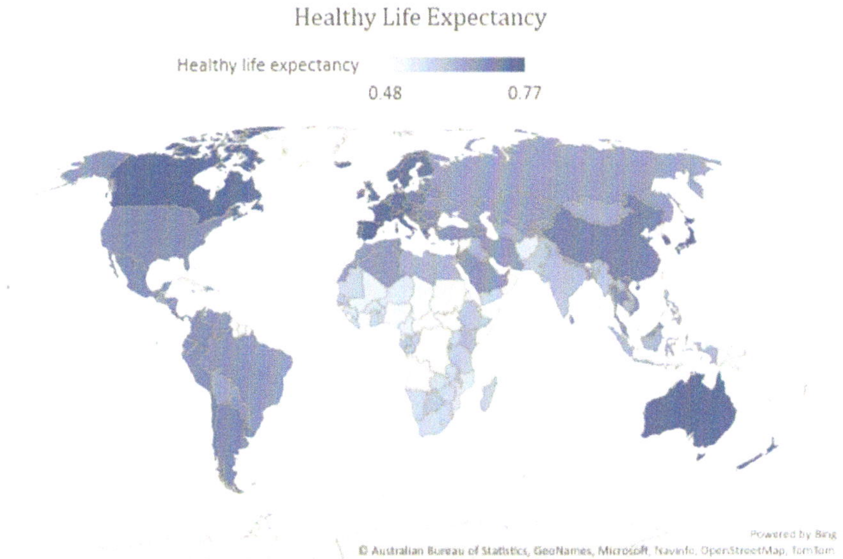

Fig. 17.5: World Map of HLE values for 149 countries from WHR 2021. *Source* Authors

for Lesotho, as reported by UNAIDS, is the number of new HIV infections. Among Sub-Saharan African countries, Lesotho has the highest value for this indicator at 6.43 per 1000 uninfected members of the population (UNAIDS, 2020).

17.2.6 Freedom to Make Life Choices (FMLC)

This factor aims to measure the individuals' perceptions regarding their freedom to make their own life choices. The question provides two response options, 'satisfied' and 'dissatisfied,' which correspond to the values of 1 and 0, respectively. Participants were asked, "Are you satisfied or dissatisfied with your freedom to choose what you do with your life?" The world average for FMLC is reported as 0.79 in the WHR 2021, with the average for the Sub-Saharan African countries covered in the study reported as 0.72. Several countries in the region have FMLC scores above the world average, namely Rwanda (0.90), Mozambique (0.88), Mauritius (0.87), Tanzania (0.83), Botswana (0.82), Ghana (0.81), and Niger (0.81). Although Sub-Saharan African countries mainly rank lower in the FMLC index, Rwanda is exceptional with a ranking of 28th among the 149 countries covered by the WHR 2021.

Figure 17.6 presents the regional average FMLC values for all the countries in the WHR 2021. Following the Middle East and the North Africa, the Sub-Saharan African region has the second lowest average value for this factor.

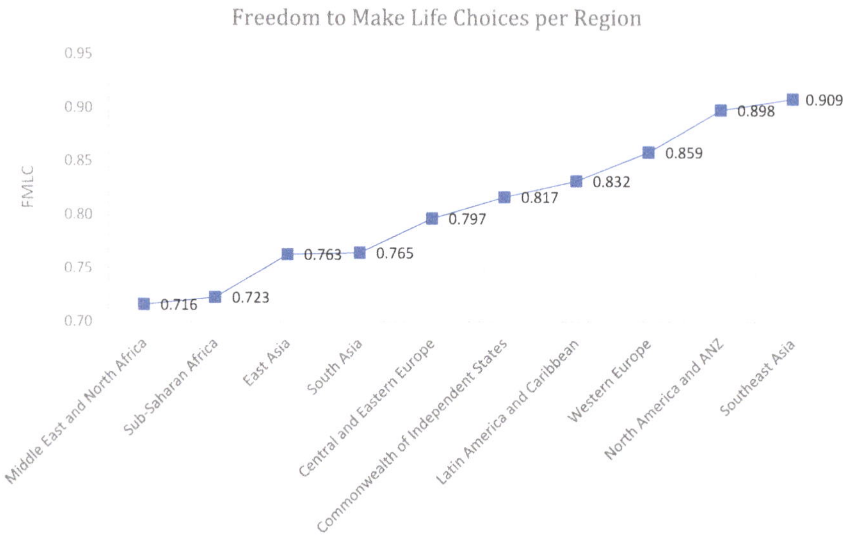

Fig. 17.6 Freedom to make life choices averages per regions. *Source* Authors

17.2.7 *Perception of Corruption (PC)*

Perception of Corruption for each country is determined by calculating the average of participant responses for two different questions that evaluate the perception of corruption at governmental and business levels: "Is corruption widespread throughout the government or not?" and "Is corruption widespread within businesses or not?" Both are yes or no questions, with corresponding response values 1 and 0, respectively. The world PC average is reported as 0.73 in the WHR 2021. Among Sub-Saharan African countries, seven scored below this average: Rwanda (0.17), Tanzania (0.58), Burundi (0.61), Benin (0.66), Mozambique (0.68), Niger (0.69), and Eswatini (0.71). A significant gap can be observed between Rwanda and other Sub-Saharan African Countries in terms of PC. This low score places Rwanda in the second lowest rank of the 149 countries covered by the WHR 2021. Figure 17.7 illustrates the PC values for all Sub-Saharan countries, with the world average highlighted in red. The global rankings of the countries are indicated on the right side of each bar.

17.2.8 *Generosity (G)*

Generosity as a factor aims to measure the donation habits of individuals. The values obtained for this factor are adjusted for income disparities using GDP, in order to provide a fair representation of this a habit regardless of the income levels. To assess this factor, survey participants are asked the following yes or no question, with 'no'

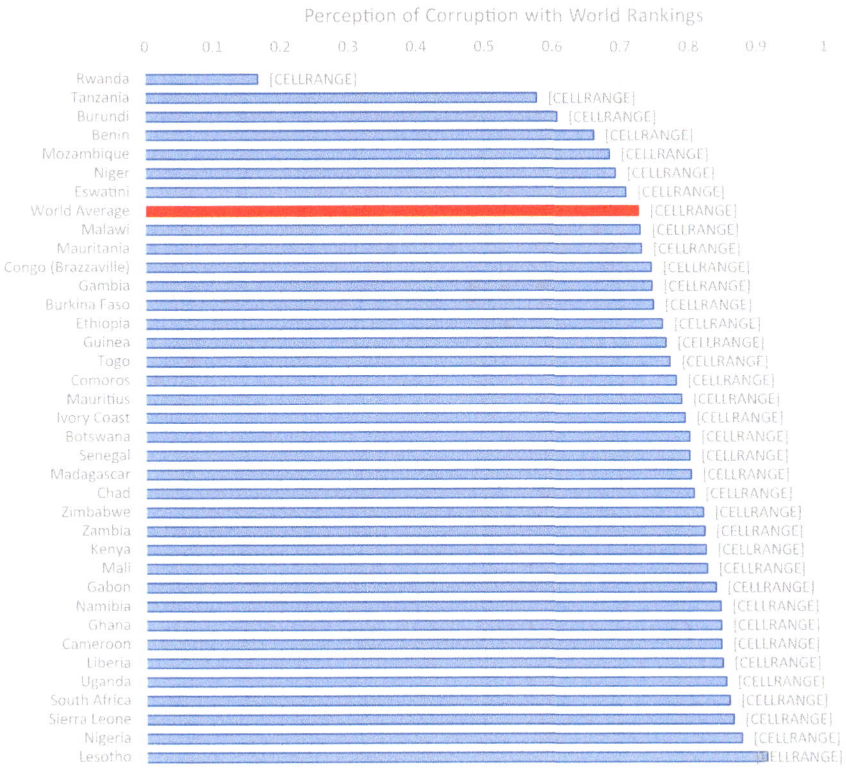

Fig. 17.7 Perception of Corruption with World Rankings. *Source* Authors

evaluated as 0 and answer 'yes' as 1: "Have you donated money to a charity in the past month?"

A scatter plot is presented in Fig. 17.8 to illustrate the relationship between GDP and G. There are many factors that can affect a country's level of generosity, such as cultural and religious beliefs, government policies, and individual attitudes towards giving. It is also expected that a country's GDP may influence its capacity for generosity in terms of providing the resources necessary for aid and charitable giving. In the figure, GDP is represented on the horizontal axis, while the vertical axis represents the G values. The world average G value is – 0.02. Almost 60% of Sub-Saharan Africa countries are above this average, as depicted by the blue dots shown in Fig. 17.8. The countries marked by red triangles are those with generosity values below the world average. It is interesting to observe that as GDP increases, G value decreases. The countries with the highest GDP in the region have the lowest generosity values. This inverse/negative correlation is also discussed with supporting numerical results in the subsequent sub-section.

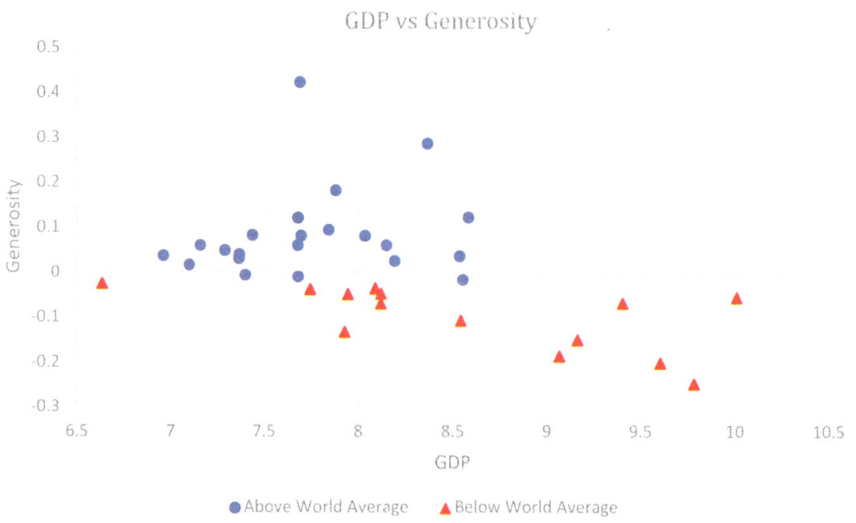

Fig. 17.8 GDP versus Generosity. *Source* Authors

17.2.9 The Correlation Between Factors for Sub-Saharan African Countries

The analysis of correlations between data series is valuable as it demonstrates any sort of relationship that might exist between data sets. This statistic offers a variety of analytical insights: it can help us understand how one factor influences another, make predictions about the future, and develop new or alternative strategies to impact the targeted factor. The correlation coefficient quantifies the strength and direction of a linear relationship between two variables or factors on a scale from − 1 to 1. A coefficient of − 1 indicates a perfect inverse (negative) correlation, + 1 indicates a perfect direct (positive) correlation, and values between − 1 and 1 represent varying degrees of correlation. A coefficient close to 0 signifies little correlation between the variables, while values close to − 1 or 1 indicate stronger correlation. The independence of variables is suggested by a correlation coefficient of 0.

A heat map displaying the calculated correlation coefficient values for the seven factors introduced in the previous sub-sections is presented in Fig. 17.9. The objective is to illustrate the relationships between these factors if any exist. The correlation coefficients are calculated based on the WHR 2021 data for 36 Sub-Saharan African countries.

The highest correlation observed is between GDP and SS, with a value of 0.67, indicating a strong positive or direct correlation. The second highest correlation is between PC and SS, with a moderate correlation value of 0.47. In this case, the correlation coefficient indicates a positive relationship, suggesting that people tend to rely on one another more in societies where trust in the government or business sector is low. The third significant relationship is between GDP and G, which is a negative/

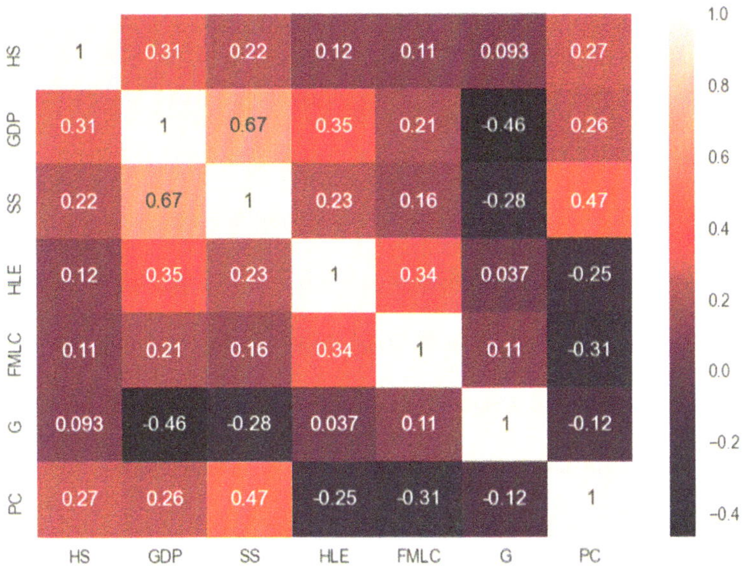

Fig. 17.9 Heat map for the correlation values for the seven factors. *Source* Authors

inverse correlation with a coefficient of − 0.46. This negative/inverse correlation implies that as a country's GDP decreases, people are more willing to help and donate to those in need. This finding may be attributed to a greater awareness among people in poorer countries of the struggles and hardships faced by others, leading them to be more inclined to help whenever they are able.

17.2.10 The Relationship Between the Happiness Score and SDG Index for Sub-Saharan African Countries

The SDG Index (SDGI) was created to evaluate countries' progress in achieving the SDGs. It involves selecting several indicators to track progress towards each goal, which are assigned scores on a scale from 0 (worst) to 100 (target). The overall SDGI score for a country is calculated by averaging its scores across all SDGs, with equal weight given to each goal.

De Neve and Sachs (2020) investigate the relationship between SDGs and subjective wellbeing, namely HS. They conclude that there is evidence of an increasing marginal impact of subjective wellbeing to sustainable development. In this section, a similar analysis is conducted to examine the relationship between Happiness and SDGs for Sub-Saharan African countries.

In Fig. 17.10, a scatter plot depicts SDGI Scores and Happiness Scores for 2021. Sub-Saharan African countries are represented in blue to distinguish them from all other countries with SDGI and HS data available for 2021, which are marked in orange. The figure clearly shows that countries with higher SDGI tend to have higher HS. The orange trendline also supports this observation as it has a noticeably increasing trend. However, upon closer examination of the Sub-Saharan African countries, the same pattern is less pronounced than in other countries. The blue trendline, derived solely from Sub-Saharan African countries' data points, has a slope which is much less steep than the general trendline. This observation suggests that Sub-Saharan African countries may have a distinct relationship between SDGs and HS as compared to other countries.

To further investigate this relationship, the correlations between SDGI and HS were calculated for both Sub-Saharan African countries and countries worldwide. A robust correlation of 0.723 was observed between SGDI and HS for all countries, whereas the same value is only 0.096 when only Sub-Saharan African countries are considered. This is a noteworthy observation given the general perception that SDGI and HS have a strong positive correlation in general (De Neve/Sachs, 2020).

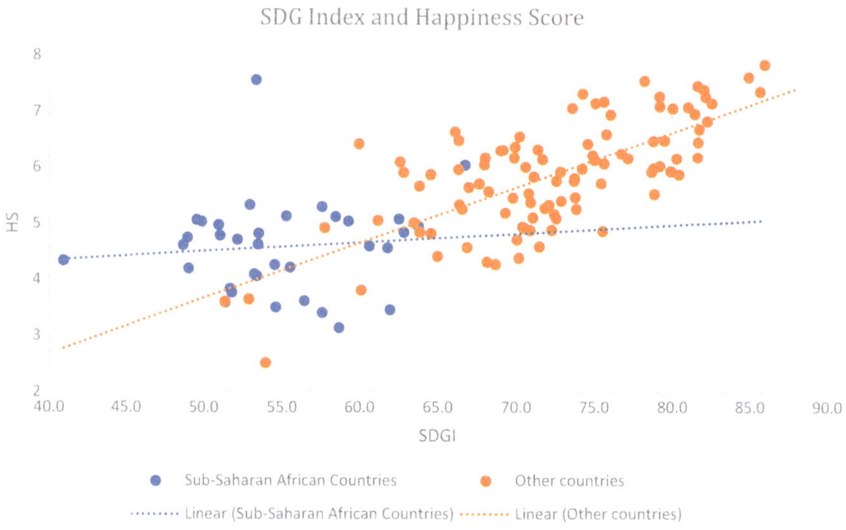

Fig. 17.10 SDG Index and HS. *Source* Authors

17.3 Cluster Analyses of Sub-Saharan African Countries with Similar Tendencies in Social Goods

17.3.1 Clustering

Data sets consist of data points which may have multiple dimensions. In our case, the data points correspond to countries, and the dimensions correspond to the factors measured by the WHR 2021 for each country included (the data set can be found in the Appendix). The goal of clustering algorithms is to divide the dataset into clusters, where data points within a cluster share similar values. This amounts to grouping the countries according to their similar values and characteristics concerning the factors. The more dimensions (factors) there are for each data point (country), the more challenging it is for researchers to detect similarities. Clustering algorithms thus help us identify similarities and group data points with similar factor values for further analysis.

There are various clustering algorithms in literature, with k-means being one of the most commonly used due to its simplicity and effectiveness. The idea behind the basic k-means algorithm is straightforward: There is a pre-determined parameter, k, which corresponds to the number of clusters. In our case, it is the number of groups containing countries with similar characteristics. The algorithm determines the optimal positions of the k cluster centers and identifies the data points that are closest to that center. Every data point in the dataset has a distance to the closest cluster center. The objective of the algorithm is to minimise the sum of these distances.

Figure 17.11 presented below provides a simple example to better understand the basic principle of k-means algorithm. The visualisation in Fig. 17.11a represents the sample data set. Given the k-means algorithm with the parameter k set as '2' (for this example), it finds the best two positions for cluster centers and locates two distinct groups by calculating the distances of each data point to the cluster centers. In Fig. 17.11b, cluster centers are marked by a plus sign within a circle and cluster sets are represented by the circles themselves.

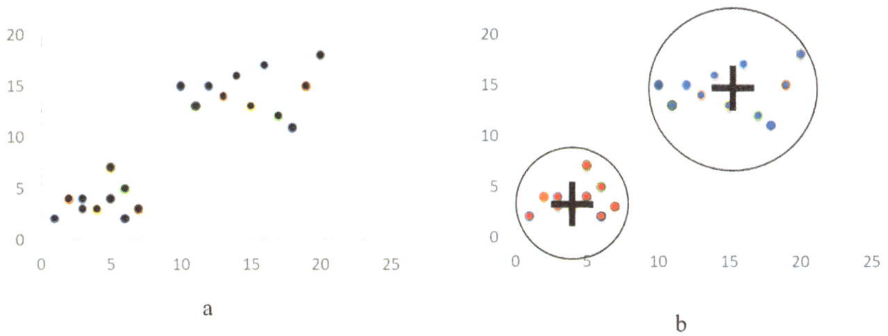

Fig. 17.11 Sample data set and clusters. *Source* Authors

The initial parameter can be determined by the analyst using the domain knowledge or by using preliminary research about the topic. For example, for in a study which aims to group countries according to their GDPs and other economic parameters, the number of groups can be set as as '3' to represent three categories such as "high-income countries", "middle-income countries", and "low-income countries" based on prior expectations. However, if there is no intuitive answer to the question of how many groups the data should be divided into, there are also optimisation algorithms that can help the analyst in finding this parameter by analysing the distribution of the data points. For this study, the Elbow Method is used to determine the optimal number of clusters, denoted as 'k,' as the parameter for the k-means clustering algorithm. The Elbow Method entails running the k-means algorithm for increasing values of k, from 1 to a pre-defined upper limit. The total distance and fit time for each k value are recorded. The point at which the distance and the fit time are the best provides the optimal k value and thus the optimal number of clusters for that dataset (Bholowalia/Kumar, 2014).

17.3.2 Dataset, Data Preparation, and Assumptions for Sub-Saharan African Countries

This study uses the seven factors reported in the WHR 2021 to cluster the 36 Sub-Saharan African countries. Each of the factors has their own scale, ranging from yes/no responses with assigned values between 1 and 0, to wider response scales between 0 and 10. In implementing the clustering algorithm, all the factors should have the same scale in order to avoid analytical bias and numerical misrepresentation. For this reason, normalisation is applied to the factor values. This simple calculation assigns all the values to the interval [0,1] where 0 represents the lower boundary and 1 represents the upper boundary for each factor within the selected country set. In our case, this is all 36 Sub-Saharan African countries listed in Table 17.3.

After normalisation, the data is prepared for implementation into the clustering algorithm. The main objective is to group the Sub-Saharan African countries based on all 7 factors: HS, GDP, HLE, FMLC, SS, PC, and GNR. Countries grouped together can be understood to have similar values for these factors, indicating potentially comparable social climates within their societies. Once the groups are established, it is possible to further explore the reasons behind these similarities, such as examining potential geographical impacts, or identifying other commonalities among the countries. Essentially, the goal is to determine what factors contribute to their similar social environment. As discussed in Sect. 3.1, the number of groups that are to be formed should be determined beforehand. For this study, no predefined number of groups is assumed, but rather the Elbow Method is applied to ascertain the optimal number of groups suggested by the data.

Table 17.3 k-means clustering for Sub-Saharan African countries under the 7 factors

Cluster 1	Cluster 2	Cluster 3		Cluster 4
Tanzania	Chad	Congo (Brazzaville)	Burkina Faso	Mauritius
Malawi	Eswatini	Ivory Coast	Mozambique	South Africa
Rwanda	Comoros	Cameroon	Nigeria	Gabon
	Mauritania	Senegal	Mali	Namibia
	Madagascar	Ghana	Uganda	Botswana
	Togo	Niger	Liberia	
	Sierra Leone	Gambia	Kenya	
	Burundi	Benin	Ethiopia	
	Lesotho	Guinea	Zambia	
	Zimbabwe			
Total: 3 countries	Total: 10 countries	Total: 18 countries		Total: 5 countries

Source Authors

The resulting graph, depicted in Fig. 17.12, indicates the optimal number of groups is 4. Thus, the countries under investigation can be divided into four groups, characterised by similar values for the selected factors. These four groups would effectively represent the countries, capturing their similar characteristics.

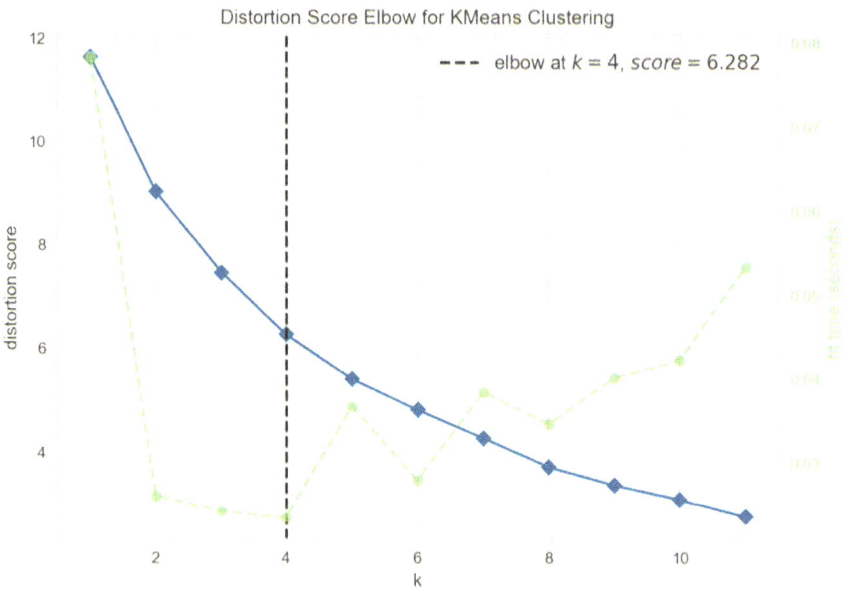

Fig. 17.12 Elbow method results, optimal k being equal to 4. *Source* Authors

17.3.3 Clustering Results

After determining the parameter for the number of groups/clusters to be formed, the k-means clustering algorithm is applied to the data. Both the Elbow Method and k-means clustering algorithm are implemented using the Python programming language, employing the yellowbrick.cluster and sklearn.cluster libraries. The resulting clusters are presented in Table 17.3. It is worth noting that there are no other factors considered in the grouping these countries, other than the normalised numerical values of the seven factors previously outlined, each of which carries equal weight, with no factor prioritised over any other.

The geographical distribution of the clusters is presented in Fig. 17.13, wherein each cluster of countries is distinguished by a specific color. It can be observed from the map that Clusters 1 and 4 primarily consist of neighboring countries. Further comments and discussions regarding each cluster are given in the following subsections.

The average values of the factor values for each cluster are given in Table 17.4. In the last row, the overall average for each factor across all Sub-Saharan African countries is also provided to allow for a clear comparison between the clusters and overall average.

To further investigate and better visualise the characteristics of these clusters, a radar chart was created, incorporating each cluster and displaying their average factor values, as shown in Fig. 17.3. The seven factors are placed around the radar chart, and the average values for each factor within the clustered countries are presented. It should be noted that the values in the axes are normalised values for these factors, ranging from 0 to1. Every cluster is represented by a distinct color. It is evident from the radar chart that there is no simple rule linking happiness or a better social climate to specific factors. For example, countries with high GDP may have low scores for G, and high scores for FLMC may not reflect a similar level of happiness (Fig. 17.14).

17.3.4 Common Characteristics of Clusters of Sub-Saharan African Countries

Cluster 1

Cluster 1 is the smallest cluster, consisting of three neighboring countries: Tanzania, Malawi, and Rwanda. These countries have the lowest average values for GDP, SS, and PC among all Sub-Saharan African countries. They have the lowest scores for the economy-related factors but the highest averages for FMLC and G. Nevertheless, these three neighboring countries have the lowest HS averages of all the clusters. This leads to the important observation that the group of countries with the lowest economic indicators is also the most unhappy.

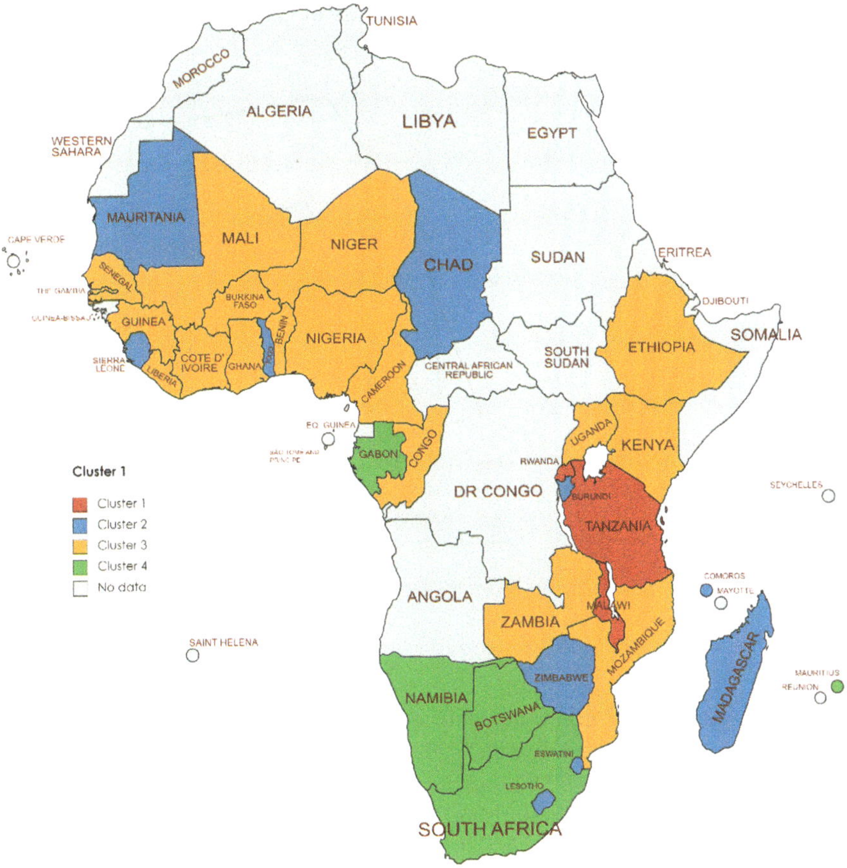

Fig. 17.13 Map view of clusters (created by using mapchart.net). *Source* Authors

Table 17.4 Average factor values for clusters

	HS	GDP	SS	HLE	FMLC	G	PC
Cluster 1	3.55	7.50	0.60	59.12	0.84	0.09	0.49
Cluster 2	3.98	7.77	0.67	53.80	0.62	– 0.03	0.78
Cluster 3	4.86	7.92	0.69	55.39	0.74	0.07	0.79
Cluster 4	4.78	9.59	0.83	59.93	0.78	– 0.14	0.83
All Sub-Saharan African	4.49	8.08	0.7	55.89	0.72	0.01	0.77

Source Authors

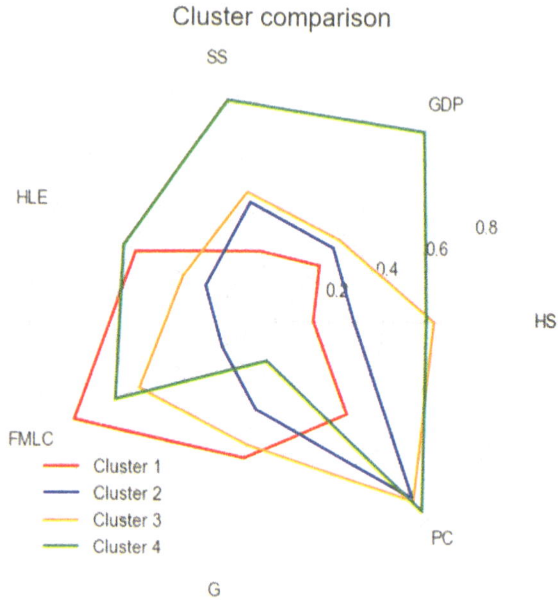

Fig. 17.14 Radar chart for clusters and their factor averages. *Source* Authors

Cluster 2

Cluster 2 comprises countries with highly agriculture-dependent economies. Among them, Chad and Sierra Leone have the highest percentage contribution of agriculture, forestry, and fishing to their GDPs, based on World Bank data for 2021(World Bank, 2021). According to the WHR 2021, the residents of these countries have the lowest perception of FMLC among the 36 Sub-Saharan countries, with a value of 0.62. This factor is the most distinctive characteristic of this cluster, as it has the lowest value among all the factors evaluated for Cluster 2. Additionally, there is a significant gap between Cluster 2 and other clusters in terms of the average FMLC scores, which can be seen in Table 17.4. This cluster also has the lowest average HLE score among all clusters and includes two countries with the lowest HLE score in the world: Chad and Lesotho.

Cluster 3

Cluster 3 is the largest cluster, encompassing 18 countries with a combined population of over 674 million people. Most of the factor values in this cluster fall withing moderate levels in comparison to other Sub-Saharan African countries. Cluster 3 has the highest happiness score out of all clusters and exemplifies the fact that there is no simple rule for happiness. The happiest group does not necessarily have the highest healthy life expectancy or GDP.

Cluster 4

The countries in Cluster 4 are the only "upper middle-income countries" among the 36 Sub-Saharan African countries examined, according to the data from the World Bank and OECD National Accounts. All other countries in Sub-Saharan Africa considered in this study are classified as lower-middle income or low-income countries. As expected, there is a substantial gap in GDP values between this cluster and others. Higher income levels also correspond to higher social support values, according to the results of the analysis of WHR 2021 data. Interestingly, Cluster 4 exhibits the lowest score for the G factor among these countries, while the PC factor is the highest in the governmental and business sectors. This group has the highest average unemployment rate among all clusters, with a value of 22% (International Labour Organization, 2022). Nonetheless, it ranks as the second happiest cluster after Cluster 3 and has the highest HLE score.

17.4 Social Climate, Social Capital and Social Goods in Sub-Saharan Africa

Social climate is typically defined as the perceptions of a social environment shared by a group of people. It is a dynamic and distinct phenomenon that can vary both across and within groups. It is rooted in perceptions, i.e., how people view certain social processes and interactions in their environment. Social climate research has evolved from the early experiments conducted by White/Lippitt (1960), which compared democratic, autocratic, and laissez-faire leadership in small groups of children. Since then, the concept and measurement of social climate have been applied across diverse disciplines, both within and beyond the field of psychology (Bennett, 2010).

Social capital, as defined in the Encyclopedia Britannica by Poteyeva (2018), is a concept "that involves the potential of individuals to secure benefits and invent solutions to problems through membership in social networks. Social capital revolves around three dimensions: interconnected networks of relationships between individuals and groups (social ties or social participation), levels of trust that characterise these ties, and resources or benefits that are both gained and transferred by virtue of social ties and social participation." A social good, also referred to as the common good, is typically understood as an action that provides some sort of benefit for the general public, such as clean air, clean water, healthcare, and literacy (Social Change Central, n.d.). The notion of social good can be traced back historically to Ancient Greek philosophy and implies a positive impact on individuals or society as a whole. It also underlies charitable and philanthropic endeavours which aim to enhance the wellbeing of individuals and communities (Kenton, 2022).

The World Social Capital Monitor (WSCM) is an open access survey assessing the social capital of countries around the globe in order to better understand their social climate. It is developed by the Basel Institute of Commons and Economics (BICE) for the UN Sustainable Development Partnership. The WSCM has been published annually since 2017 and became a UN partnership project in 2016. The monitor, which operates on a voluntary basis without a budget, is aligned with the SDGs, specifically SDGs 8 (Decent Work and Economic Growth), 10 (Reduced Inequalities), 11 (Sustainable Cities and Communities), and 16 (Peace, Justice and Strong Institutions). Nevertheless, serves as a tool for promoting all 17 SDGs, as they all depend on social goods. Assessing and enhancing social goods is one of the few approaches to achieving all 17 SDGs collectively (Dill, n.d.).

The WSCM emphasises the measurement of social goods, considering them major non-material assets that lead to the provision of new instruments for achieving the 17 SDGs, including addressing poverty and crises: 'Almost all public goods, such as security, environmental protection, education and health can be built entirely on social goods and therefore require not only funding. If a community is able to resolve conflicts peacefully through compromise, it does not need weapons and armies. When neighbors help each other in natural disasters and economic crises, they can cope with their consequences much more easily. In many rural areas of Africa, this so-called social capital exists. When most people do not have monetary income and assets – and this is the reality across Africa-barter and gifts play the main role.' (Basel Institute of Commons and Economics, 2021).

The WSCM 2021 covers 300 cities, including results from 177 cities in Africa. It features 400 comments from the respondents and includes case studies for towns in Afghanistan, Albania, Bangladesh, Bosnia and Herzegovina, Brazil, Cape Verde, Chile, Ethiopia, Germany, Montenegro, Nepal, North Macedonia, Pakistan, and Serbia. (Mataraci, 2022) The WSCM survey, which is available in 50 languages, provides timelines and comments for eight social goods for each participating country: social climate, helpfulness, friendliness, hospitality, interpersonal trust, austerity measures, willingness to pay taxes, local investment. Participants in the survey evaluate these social goods in their countries on a scale from 1 to 10, with 1 representing the lowest and 10 representing the highest score (Dill, 2021). An advantage of the WSCM is its ability to generate localised results, ranging from small settlements to large cities, making it a relatively inclusive and representative tool. The WSCM 2021 includes results for 125 cities and towns and 35 administrative units, such as municipalities and communities, from 32 Sub-Saharan African countries, which are listed in Table 17.5. Comoros, Eswatini, Gambia, Ivory Coast are the Sub-Saharan African countries covered by the WHR 2021, but not the WSCM 2021.

Table 17.5 Sub-Saharan African Countries included in WSCM 2021 report

Benin, Botswana, Burkina Faso, Burundi, Cameroon, Chad, Congo (Brazzaville), Ethiopia, Gabon, Ghana, Guinea, Kenya, Lesotho, Liberia, Madagascar, Malawi, Mali, Mauritania, Mauritius, Mozambique, Namibia, Niger, Nigeria, Rwanda, Senegal, Sierra Leone, South Africa, Tanzania, Togo, Uganda, Zambia, Zimbabwe

Source Authors

Table 17.6 Summary of the WHR 2021 data

Social goods	Average (Sub-Saharan African Countries)	Minimum (Sub-Saharan African Countries)	Maximum (Sub-Saharan African Countries)
Social climate	5.89	2.25	8.23
Interpersonal trust	5.61	4.00	9.00
Austerity measures	6.00	4.20	10.00
Willingness to pay taxes	5.78	3.55	8.30
Local investment	5.99	3.00	7.67
Helpfulness	6.49	3.25	8.47
Friendliness	7.36	3.75	9.70
Hospitality	7.69	6.30	9.20

Source Authors

17.4.1 Social Goods and Social Capital in Sub-Saharan Africa

Table 17.6 summarises the average, minimum, and maximum scores for all social goods in Sub-Saharan African countries participating in the WSCM 2021. The overall average scores for all social goods range from 5.61 to 7.69, indicating a relatively high level of social capital and positive social climate in the region. Among the eight social goods assessed, hospitality has the highest average score, followed by friendliness, helpfulness, and austerity measures, all of which fall within the range of 6 to 7.69. These scores are followed by local investment, social climate, willingness to pay taxes, and interpersonal trust, ranging between 5.99 and 5.61 in descending order. These scores reinforce the existence of social capital in Sub-Saharan Africa and bode well for the WSCM's central claim that social goods provide the foundation for the development of nearly all public goods. In this regard, higher scores for social goods indicate the presence of social capital, which can facilitate the process of sustainable development by encouraging stronger investment towards sustainable development goals.

17.4.2 Social Climate in Clusters of Sub-Saharan African Countries

When comparing the social climate across the four clusters of Sub-Saharan African Countries, Cluster 1 appears to have the best social climate. Recalling the characteristics of Cluster 1, it is the smallest of the clusters, consisting of Tanzania, Malawi, and Rwanda, which have the lowest average values for GDP, SS, and PC values among all of the Sub-Saharan African countries. Additionally, they have the lowest scores for the economy-related factors but the highest averages for FMLC and G. Despite having the highest average social climate, this cluster records the lowest levels of happiness. These findings suggest a potential negative correlation between GDP and social climate: as GDP decreases, the social climate tends to improve, which is in line with one of the WSCM's assumption that social capital functions more effectively in poverty-stricken and crisis-ridden areas.

Cluster 4, which consists of "upper middle-income countries" in Sub-Saharan Africa and ranks as the second happiest cluster, emerges as the cluster with the second-best social climate, as demonstrated in Fig. 17.15. This finding supports the expectation that higher income correlates with higher social support, as revealed by the WHR 2021 results. However, it is worth noting that the G factor is the lowest in this cluster, while the PC and HLE factors are the highest.

Cluster 3, which exhibits the highest happiness score among all clusters, ranks third in terms of social climate. As a reminder, this cluster is the largest, comprising 18 countries with a total population of over 674 million people. The majority of the factor values in this cluster are moderate in comparison to the other clusters. The proximity of the average scores for happiness (4.86) and social climate (5.84) may suggest a potential positive correlation between social climate and happiness, supporting the expected relationship between these two factors.

Cluster 2, which consists of countries with highly agriculture-dependent economies in Sub-Saharan Africa and records the lowest scores for FMLC and HLE, ranks last when in terms of social climate. Nevertheless, even the lowest average score (5.60) for social climate among the Sub-Saharan African countries participating in the WSCM 2021 falls within the range of the average scores (between 5 and 6) for the majority. These findings underscore the potential for improvement in social

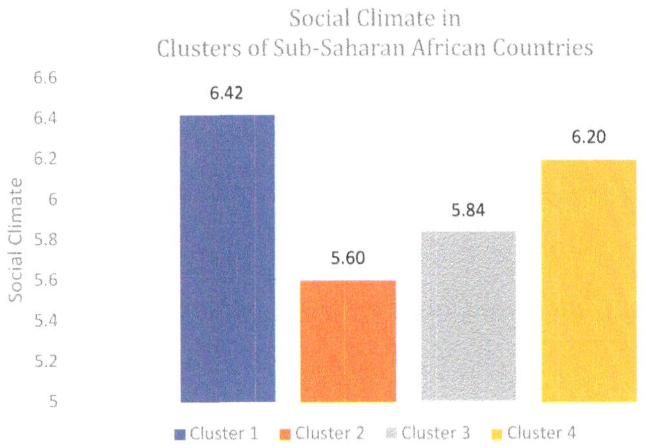

Fig. 17.15 Social Climate in Clusters of Sub-Saharan African Countries. *Source* Authors

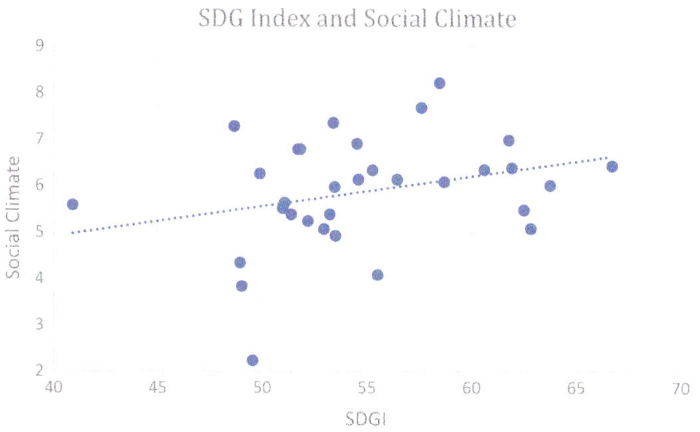

Fig. 17.16 SDG Index and Social Climate. *Source* Authors

goods and, consequently, the social climate to enhance sustainable development in the region.

17.4.3 SDG Index Score and Social Climate

Figure 17.16 presents a scatter plot illustrating the SDG Index and Social Climate in Sub-Saharan African Countries. The horizontal axes represent the SDGI values, while the vertical axes represent the Social Climate values. A slight trend can be observed, indicating that as SDGI increases, there is a tendency for Social Climate values to increase as well. The trendline in this plot is steeper than the trendline shown for Happiness Score in Sect. 2.10. As expected, the correlation coefficient between Social Climate and SDGI is higher than the coefficient for HS and SDGI, which is 0.296. However, this correlation is not significant enough to furnish conclusive evidence of any relationship between these two factors. Chad has the lowest SDGI value of 40.9, and yet its Social Climate value is 5.6, which is close to the average Social Climate value for Sub-Saharan African countries.

17.5 Conclusion

This study aimed to analyse the social climate in Sub-Saharan Africa by utilising data from the WHR 2021 and WSCM 2021. The WHR 2021 considers seven primary factors, namely happiness, gross domestic product, social support, healthy life expectancy, freedom to make life choices, perception of corruption, and generosity. When comparing the average scores of Sub-Saharan African countries to the global average for these factors, it is evident that the region's score is generally lower. However, it is essential to highlight that there are certain countries in the region that surpass the global average for each of these factors. This observation is significant in that it demonstrates progress in specific areas and offers hope for the region's development, albeit at varying rates and in different aspects.

Several noteworthy correlations between these factors are also evident in Sub-Saharan African countries. The highest positive correlations are observed between SS and two other factors: GDP and PC. This indicates that as GDP increases, so does SS, and as PC decreases, SS tends to increase in these communities. Additionally, a negative correlation is detected between GDP and G, suggesting that, as GDP increases, generosity levels decrease. These correlations provide valuable insights into the relationships between different factors analysed in Sub-Saharan African countries and can be useful for developing policies to improve the region's social climate.

Through clustering analysis, Sub-Saharan African countries are grouped into four clusters. The examination of the clusters reveals that there is no simple rule for determining happiness: countries with higher GDP have lower levels of generosity, and populations with a greater sense of freedom to make life choices may have low happiness scores.

The analysis of settlements in the 32 Sub-Saharan African countries that participated in the WSCM 2021 indicates that the overall average scores for the eight social goods reflect a relatively high social capital and positive social climate in the region. Among the eight social goods, hospitality has the highest average score, followed by friendliness, helpfulness, austerity measures, local investment, social climate, and willingness to pay taxes, with interpersonal trust having the lowest average score. These scores affirm the existence of social capital in Sub-Saharan Africa and provide hope for the WSCM's main assertion that most public goods can be built on social goods. Furthermore, even the slight tendency for Social Climate values to increase with SDGI, as shown in Sect. 4.3, is a source of optimism for sustainable development in the region despite existing challenges.

Appendix

The WHR 2021 related data set used in this study:

Country name	HS	GDP	SS	HLE	FMLC	G	PC
Benin	5.045	8.087	0.489	54.713	0.757	– 0.034	0.661
Botswana	3.467	9.782	0.784	59.269	0.824	– 0.246	0.801
Burkina Faso	4.834	7.678	0.672	54.151	0.695	– 0.009	0.748
Burundi	3.775	6.635	0.49	53.4	0.626	– 0.024	0.607
Cameroon	5.142	8.189	0.71	53.515	0.731	0.026	0.848
Chad	4.355	7.364	0.619	48.478	0.579	0.041	0.807
Comoros	4.289	8.031	0.626	57.349	0.548	0.082	0.781
Congo (Brazzaville)	5.342	8.117	0.636	58.221	0.695	– 0.068	0.745
Ethiopia	4.275	7.694	0.764	59	0.752	0.082	0.761
Gabon	4.852	9.603	0.776	59.962	0.731	– 0.2	0.84
Gambia	5.051	7.686	0.69	55.16	0.697	0.424	0.746
Ghana	5.088	8.58	0.727	57.586	0.807	0.123	0.848
Guinea	4.984	7.838	0.639	55.008	0.697	0.095	0.766
Ivory Coast	5.306	8.551	0.644	50.114	0.741	– 0.016	0.794
Kenya	4.607	8.361	0.688	60.704	0.779	0.287	0.825
Lesotho	3.512	7.926	0.787	48.7	0.715	– 0.131	0.915
Liberia	4.625	7.288	0.72	56.498	0.735	0.05	0.85
Madagascar	4.208	7.396	0.686	59.305	0.552	– 0.005	0.803
Malawi	3.6	6.958	0.537	57.948	0.78	0.038	0.729
Mali	4.723	7.744	0.724	51.969	0.697	– 0.036	0.827
Mauritania	4.227	8.542	0.795	57.161	0.561	– 0.106	0.731
Mauritius	6.049	10.008	0.905	66.701	0.867	– 0.054	0.789
Mozambique	4.794	7.158	0.744	54.706	0.882	0.061	0.684
Namibia	4.574	9.161	0.818	56.799	0.719	– 0.149	0.847
Niger	5.074	7.098	0.641	53.78	0.806	0.018	0.693
Nigeria	4.759	8.533	0.74	50.102	0.737	0.037	0.878
Rwanda	3.415	7.676	0.552	61.4	0.897	0.061	0.167
Senegal	5.132	8.118	0.71	59.802	0.695	– 0.046	0.801
Sierra Leone	3.849	7.434	0.63	51.651	0.717	0.084	0.866
South Africa	4.956	9.403	0.86	56.904	0.749	– 0.067	0.86
Eswatini	4.308	9.065	0.77	50.833	0.647	– 0.185	0.708
Tanzania	3.623	7.876	0.702	57.999	0.833	0.183	0.577
Togo	4.107	7.362	0.569	54.914	0.619	0.032	0.772
Uganda	4.636	7.677	0.781	56.101	0.709	0.122	0.855

(continued)

(continued)

Country name	HS	GDP	SS	HLE	FMLC	G	PC
Zambia	4.073	8.145	0.708	55.809	0.782	0.061	0.823
Zimbabwe	3.145	7.943	0.75	56.201	0.677	− 0.047	0.821

Source The authors

The WSCM 2021 data set used in this study:

Country name	CL	TR	ME	TA	LO	HE	FR	HO
Benin	6.28	5	4.2	4.03	5.03	5.93	6.18	6.55
Botswana	6.4	6.3	6.8	6.3	5.6	7	7.4	7.4
Burkina Faso	4.94	4.02	6.42	5.82	6.14	6.98	7.84	7.44
Burundi	6.8	6.33	5.6	7.27	6.33	7.07	7.07	8.13
Cameroon	6.35	5.57	6.13	5.97	6.78	7.03	7.47	7.62
Chad	5.6	5.6	5.4	5.4	5.3	5.7	6.3	7.3
Congo (Brazzaville)	5.09	4.53	6.67	5.57	6.23	5.74	6.57	7.91
Ethiopia	6.92	6.62	7.35	7.85	7.67	8.47	8.92	8.83
Gabon	5.1	4.1	5.3	4.6	4.1	5.7	5.6	6.3
Ghana	5.5	5.07	5.15	5.68	6.41	6.08	7.38	6.43
Guinea	5.53	6	6.87	6.53	6.03	7.2	7.13	8.33
Kenya	6.36	5.51	5.85	5.61	7	7.21	7.94	8.14
Lesotho	6.15	6.4	6.65	5.15	5.25	5.65	8.15	8.4
Liberia	7.3	5	6.8	7	7.3	5.8	8.3	9
Madagascar	3.85	4	4.45	3.55	3	4.65	6.2	6.95
Malawi	5.4	5.53	7.23	7.03	7.05	7.53	8.3	8.23
Mali	5.26	5.54	4.88	4.78	5.48	6.34	7.42	7.78
Mauritius	6.45	5.45	4.38	5	6.25	6.8	6.95	6.58
Mauritania	4.1	5	4.9	4.7	4.9	7	7.9	8.1
Mozambique	5.65	4.85	5.5	4.65	7	6.5	6.5	6.5
Namibia	7	9	10	7	6.5	7.5	7.5	7.5
Niger	2.25	6.5	6	5.75	5.5	3.25	3.75	6.75
Nigeria	4.36	4.14	5.5	5.29	4.93	5.57	7	6.71
Rwanda	7.7	7	8	8.3	6.75	7.55	7.75	8
Senegal	8.23	7.83	7.23	7.57	7.17	7.93	8.27	9.17
Sierra Leone	6.8	5.3	5.5	6	5.8	7.5	9.7	9.2
South Africa	6.03	4.71	5.76	5.59	6.3	6.14	6.83	6.89
Tanzania	6.15	6.67	6.4	6.15	6.55	6.83	8.06	8.43
Togo	5.4	5.1	4.2	4.5	5.2	4.8	6.2	6.8
Uganda	5.99	5.11	5.33	5.64	6.06	5.84	7.89	8.1

(continued)

(continued)

Country name	CL	TR	ME	TA	LO	HE	FR	HO
Zambia	7.38	6.35	5.9	5.58	6.48	7.18	9.08	8.45
Zimbabwe	6.1	5.33	5.63	5.2	5.47	7.3	7.93	8.23

Source The authors

References

Aggarwal, D. & D. Sharma. (2019). Application of clustering for student result analysis. *International Journal of Recent Technology and Engineering (IJRTE), 7*(6C).

A Zhu. (2022). An Analytical View on World Happiness with Unsupervised Machine Learning. *European Journal of Humanities and Social Sciences, 2(3)*, 1–8. https://doi.org/10.24018/ejsocial.2022.2.3.216

Abdullah, D., Susilo, S., Ahmar, A.S., Rusli, R., & Hidayat, R. (2022). The application of K-means clustering for province clustering in Indonesia of the risk of the COVID-19 pandemic based on COVID-19 data. *Quality and Quantity, 56*(3), 1283–1291. https://doi.org/10.1007/S11135-021-01176-W

Basel Institute of Commons and Economics (BICE). (2021). *The World Social Capital Monitor, 300 cities and their local social goods to achieve the SDGs, Progress Report of the World Social Capital Monitor to the UN SDG Partnerships*. https://sustainabledevelopment.un.org/content/documents/commitments/3914_11706_commitment_300_Cities_and_their_social_goods_to_achieve_the_SDGs.pdf\. Accessed 6 August 2023.

Bennett, J.B. (2010). Social Climate Research. In *The Corsini Encyclopedia of Psychology* (pp. 1–4). Wiley. https://doi.org/10.1002/9780470479216.corpsy0885

Bholowalia, P., & Kumar, A. (2014). EBK-means: A clustering technique based on elbow method and k-means in WSN. *International Journal of Computer Applications, 105*(9), 975–8887.

De Neve, J.-E., & Sachs, J. D. (2020). The SDGs and human well-being: a global analysis of synergies, trade-offs, and regional differences. *Scientific Reports, 10*(1), 15113. https://doi.org/10.1038/s41598-020-71916-9

Dill, A. (n.d.). *Measuring Social Capital for SDG – methodology and practice of the World Social Capital Monitor*. Retrieved April 6, 2023. http://commons.ch/wp-content/uploads/World_Social_Capital_Monitor_Methodology.pdf. Accessed 6 August 2023.

Dill, A. (2021). *Understanding the methodology of the World Social Capital Monitor*. http://commons.ch/wp-content/uploads/Understanding_the_Methodology_of_the_World_Social_Capital_Monitor.pdf. Accessed 6 August 2023.

Helliwell, J.F., Layard, R., Sachs, J., & De Neve, J.-E. (2021). *World Happiness Report 2021*.

International Labour Organization. (2022). *ILO Modelled Estimates and Projections database (ILOEST)*.

Islam, M.K., Ali, M.S., Miah, M.S., Rahman, M.M., Alam, M. S., & Hossain, M.A. (2021). Brain tumor detection in MR image using superpixels, principal component analysis and template based K-means clustering algorithm. *Machine Learning with Applications, 5*. https://www.sciencedirect.com/science/article/pii/S2666827021000220. Accessed 6 August 2023.

Kapourani, C.A., & Sanguinetti, G. (2019). Melissa: Bayesian clustering and imputation of single-cell methylomes. *Genome Biology, 20*(1). https://doi.org/10.1186/S13059-019-1665-8

Kenton, W. (2022). *Social Good: Definition, Benefits, Examples*. https://www.investopedia.com/terms/s/social_good.asp. Accessed 6 August 2023.

Khder, M., & Sayfi, M. (2022). Analysis of World Happiness Report Dataset Using Machine Learning Approaches. *Appl, SW Fujo – Int. J. Advance Soft Compu., 14*(1). https://doi.org/10.15849/IJASCA.220328.02

Mataraci, A. F. (2022). Social Climate in the Western Balkans (2018-2021). *İmgelem, 6*(10), 69–88. https://doi.org/10.53791/imgelem.1140838

OECD Economic Outlook: Statistics and Projections. (database). (2020). *OECD Economic Outlook No 108 (Edition 2020/2).*

Poteyeva, M. (2018). *Social Capital.* https://www.britannica.com/topic/social-capital. Accessed 6 August 2023.

Social Change Central. (n.d.). *What is Social Good?* Retrieved April 6, 2023. https://www.socialchangecentral.com/what-is-social-good/. Accessed 6 August 2023.

Ulkhaq, M.M. (2020). Clustering Countries According to The World Happiness Report. *Statistica & Applicazioni, 18*(2).

UNAIDS. (2020). UNAIDS Data 2020 Data. https://www.unaids.org/sites/default/files/media_asset/2020_aids-data-book_en.pdf. Accessed 6 August 2023.

UNICEF. (2020). *State of the World's Children, Childinfo, and Demographic and Health Surveys.* https://data.worldbank.org/indicator/SH.STA.BRTC.ZS\. Accessed 6 August 2023.

WHO, UNICEF, UNFPA, & World Bank Group, and the U. N. P. Division. (2019). *Trends in Maternal Mortality: 2000 to 2017.* https://data.worldbank.org/indicator/SH.STA.MMRT. Accessed 6 August 2023.

World Bank. (2021). Agriculture, forestry, and fishing, value added (% of GDP). *World Bank National Accounts Data, and OECD National Accounts Data Files.* https://data.worldbank.org/indicator/NV.AGR.TOTL.ZS. Accessed 6 August 2023.

Chapter 18
Occupational Health and Safety of Artisanal Miners in Kolwezi, Democratic Republic of Congo: A Problem for Goals 3 and 8 of Sustainable Development

Lufuluabo E. Maweja, Thokozani Mbonane, Shalin Bidassey-Manilal, Charlotte Mokoatle, Claris Siyamayambo, Bheki Magunga, and Martha Chadyiwa

Abstract In recent years artisanal mining has become more prevalent compared to several other mining activities in the City of Kolwezi, in the Democratic Republic of Congo (DRC). A cross-sectional study was conducted among 546 workers, selected using simple random sampling from five different artisanal mines—namely Kasulo, Mutoshi, Kamilombe, Uck Drain, and Tilwezembe *et Intermediaires*. A closed-ended

Dr Lufuluabo E. Maweja is a medical doctor from the University of Lubumbashi (2008) and hold a Master's degree in Public Health (Occupational and Environmental health) from the University of Johannesburg (2022). Email: ebenmaweja78@yahoo.com

Dr Thokozani Mbonane (PhD) is currently appointed as a Head of Department and Senior Lecturer in the Department of Health, University of Johannesburg. Email: tmbonane@uj.ac.za; ORCID ID: http://orcid.org/0000-0001-5802-4265

Shalin Bidassey-Manilal is a lecturer in the Environmental Health Department (Faculty of Health Sciences, University of Johannesburg) and is currently completing her Doctoral Degree in Environmental Health.; Email: shalinb@uj.ac.za; ORCID ID: http://orcid.org/0000-0002-3007-3142

Charlotte Mokoatle holds a Master's degree in Environmental Health and currently in the final year of her doctoral studies in Environmental Health; Email: charlottem@uj.ac.za; ORCID ID: http://orcid.org/0000-0002-6884-9225

Claris Siyamayambo is a mathematical statistician and a researcher (PhD in Statistics (2021), M.A. in Statistics (2012), PGDE (2004), BSc General in Mathematics and Statistics, (2001) trained at the Northwest University, South Africa; Email: clariss@uj.ac.za

Bheki Magunga is a lecturer in the Department of Environmental Health, Some of the Module includes sustainable development and ecology as well as research project. He is in the final stages of his PhD in Environmental Health which he is pursuing with the Central University of Technology (CUT). Email: bhekim@uj.ac.za

Martha Chadyiwa is a registered EHP with the Health Professions Council of South Africa. She is a holder of a Bachelor of Science Degree in Environmental Health. She is finalising a PhD in Public Health at Witwatersrand University focusing on Occupational Health and Safety; Email: mchadyiwa@uj.ac.za; ORCID ID: https://orcid.org/0000-0001-6263-4402

J. C. K. Kiyala and N. Chivasa (eds.), *Climate Change and Socio-political Violence in Sub-Saharan Africa in the Anthropocene*, The Anthropocene: Politik—Economics—Society—Science 37,
https://doi.org/10.1007/978-3-031-48375-2_18

questionnaire was administered to collect the data. The data were analysed using both the Statistical Package for Social Sciences (SPSS) version 25, and the Centre for Disease Control (CDC) program EPINFO 7.2. The results revealed a significant relationship between socio-demographic characteristics and possessing professional mining cards, and that an artisanal miner who possesses a professional mining card has a better understanding of the Occupational Health and Safety (OHS) and the DRC safety and mining regulations. Knowledge of the artisanal mining bank can play a role in promoting safer and more sustainable practices, which can contribute to achieving SDG 3 and SDG 8.

Keywords Artisanal miners · Occupational health and safety · Democratic Republic of Congo · Mining regulations · Artisanal mining card · Sustainable development goals

18.1 Introduction

In the late 1990s and early 2000s public mining enterprises in the Democratic Republic of Congo (DRC) such as MIBA (Société Minière de Bakwanga), and GECAMINES (La Générale des Carrières et des Mines) collapsed. This was due to a combination of factors, such as mismanagement, corruption, civil unrest, and economic instability. MIBA experienced a decline in production in the late 1990s, while the civil war that began in 1996 significantly affected GECAMINES. Both companies were privatised in the early 2000s, as part of a broader programme of economic reform in the DRC (Putzel et al., 2008). A large proportion of the mining activities has been taken over by actors in the informal sector, who exploit the abandoned concessions, or those that formerly belonged to public enterprises (Kamundala, 2017). As of 2002, much artisanal mining activity has been happening, due to an increased demand for minerals, and high rates of unemployment. This vertiginous progression of artisanal mining activities is not without unfortunate consequences. For instance, several young people who engage in artisanal mining are exposed to all kinds of occupational hazards, leading to the daily loss of human lives, due to non-compliance with safety regulations (Kouame, 2015).

In 2006, 13 accidents were reported at the artisanal exploitation site of Ruashi Mining, which resulted in 35 deaths and 16 injuries, indicating an annual victim rate of 0.4 to 0.5% (Tsurukawa et al., 2011). In June 2019, 40 artisanal miners died in Kamoto Copper Company (KCC) in Kolwezi, and in October 2019, 21 artisanal miners died when the gold mine in Maniema Province collapsed (Cuvelier, 2020). Occupational health and safety in the mining industry has improved significantly over the past years (due to various factors, including advances in technology, changes in industry practices, and increased regulatory oversight). Occupational health and safety in mining has also improved because the Governments and industry bodies have established more stringent safety regulations and standards, which reduced accidents and injuries. New technologies such as automation, remote sensing, and drones, have been adopted, which have enhanced safety, as workers are now less likely to perform hazardous tasks. Over the years, many mines have invested in training and education programmes to ensure that workers are aware of, and can mitigate safety risks. This has improved the safety culture, which prioritises the health and safety of workers above everything else, as the trend has been towards continuous improvement, with growing recognition of the importance of maintaining safe working conditions for all workers in the mining sector. However, more measures must still be implemented to ensure a safe working environment for all the workers (Burgess/Connell, 2006). Therefore, this study aimed to explore the knowledge, attitudes, and behaviours of artisanal miners in Kolwezi toward OHS and mining regulations to bridge their knowledge gaps.

The working environment in artisanal mining is unsafe, because of non-compliance to safety standards – for instance, miners are not given personal protection equipment (PPE) to protect them against the effects of radiation, injuries, and other accidents that may occur (Burgess/Connell, 2006). Article 14 of the International Labour Organization (ILO) Convention No.176 states that under national laws and regulations, "workers should perform their specific duties following their training guidelines which include compliance with the prescribed safety and health measures in consideration of how possible carelessness can put not only their lives but also that of others in danger" (Organisation International Labour, 2018). Therefore, occupational health and safety laws and regulations are crucial for protecting employees from professional risks, which have negative effects environmentally, economically and socially.

The DRC has developed regulations adapted to their context to ensure occupational safety, based on Title 7 of the Labour Code, Article 207 of the Mining Code, and Article 492 of the Mining Regulations (Michel-Kabamba et al., 2021). The Congolese Labour Code requires that an occupational health and safety system be implemented and promoted. It is in this perspective that the Ministry of Labour took various actions to set up a promotional framework in matters of health and safety, specifically establishing a better regulatory framework that aims to improve health and safety at work (Michel-Kabamba et al., 2021). In addition to implementing these fundamental tools governing the mining sector in the DRC, operators involved in formalising and improving artisanal mining in the DRC produced several official texts such as: orders, instructions, circulars, and deeds of engagement. According to the Organisation International Labour (2018) examples of such tools include:

1. The Extractive Industries Transparency Initiative (EITI), which is a global standard for transparency and accountability in the extractive sector. EITI produces official texts such as the EITI Standard and the EITI Implementation Guide, which provide instructions and guidance to countries and companies on disclosing and reporting their extractive activities.
2. The Kimberly Process Certification Scheme, is an international mechanism that aims to prevent the trade of conflict diamonds: This scheme produces official texts such as the Kimberley Process Certification Scheme Manual, which provides

guidance on implementing the certification scheme and conducting diamond exports.

3. The International Council on Mining and Metals (ICMM) is a global industry association that provides guidance and support to mining companies on sustainable mining practices. The ICMM produces official texts such as the ICMM Mining Principles and the ICMM Performance Expectations, which provide guidance on improving environmental and social performance in the mining sector.
4. The World Bank is a major funder of mining projects around the world, and produces official texts such as the World Bank Environmental, Health and Safety Guidelines for Mining, which provide guidance on managing environmental and social risks associated with mining activities.

The rule of law requires all stakeholders involved to observe and apply all the provisions of the law. It is a fundamental principle of governance that ensures that everyone is subject to the law, and that no one is above it (Tegnan et al., 2021). It requires that the laws be clear, consistent, and predictable, and that they are enforced impartially and without bias. This means that all stakeholders involved in the mining sector; including governments, mining companies, and communities, ought to observe and apply all the provisions of the law: including the laws related to environmental protection, labour rights, and human rights. The rule of law is essential to ensure that the mining sector operates in a transparent and accountable manner, and that the rights of all stakeholders are respected and protected. However, in practice, regulations are not complied with, as several studies have highlighted. For instance, it was found that many artisanal miners do not have artisanal miners' professional cards (Debrah et al., 2014), and that most of the accidents and injuries happening in artisanal mines are due to the non-use of PPE (Aram et al., 2021). For instance, of the 58 artisanal mining sites visited in Haut-Katanga and Lualaba provinces, only two had PPE (De Haan/Mcquilken, 2023). The general working conditions in artisanal mining are extremely bad and, often, dangerous. For example, diggers, including children, work barefooted, without PPE, in unventilated wells (Ahmed, 2010). In general, artisanal mining operations lack safety equipment, and they do not comply to international safety standard (Tsurukawa et al., 2011).

All UN member states accepted and adopted in 2015 the agenda for sustainable development, and identified 17 goals and 169 targets. All the developed and developing countries were encouraged to collaborate, and form a partnership to ensure that these goals are achieved. The main objective of these goals is to eradicate poverty, enhance education and health, scale down inequality, and escalate economic growth. These sustainable development goals will be reviewed (whether they have been achieved or not) in 2030 (Shulla et al., 2021). For the purposes of this Chapter, we focused on the following goals:

1. Goal 3 (indicator 3.9) states that by 2030 signatories should have substantially reduced the number of deaths and illnesses due to hazardous chemicals and air, water, and soil pollution and contamination. Miners who are not trained, and do

not comply with the applicable acts and regulations are susceptible to deaths and illnesses from the hazardous illegal mining practice (Landrigan et al., 2022).

2. Miners are also exposed to toxic chemicals such as mercury and engage in unsafe work practices. Thus, poor occupational health and safety practices in the mining sector can lead to a range of health risks, musculoskeletal injuries, and exposure to hazardous chemicals (Wireko-Gyebi et al., 2022). Artisanal mining also poses a health risk to the surrounding communities. In addition, these unfavourable working and living conditions enable respiratory conditions, HIV, and several types of zoonotic diseases (Allan-Blitz et al., 2022).
3. SDG 8 promotes sustained, inclusive, and sustainable economic growth, full and productive employment, and decent work for all. Despite contributing tremendously to the country's economic growth, artisanal miners are underpaid, work in extremely hazardous conditions, and have no form of protection – whether social or occupational (World Bank, 2017). They are exposed to substandard occupational health and safety conditions, and have no access to proper healthcare facilities. This makes it difficult for them to enforce their labour rights or social protection—hence injuries and accidents are rarely reported. The lack of proper laws in place makes artisanal mining a hub for illegal activities, such as the promotion of child labour (Singo, et al., 2022a, b).

Non-compliance with regulations leads to an increasing number of accidents in artisanal mining. As such, there is a great need to compile data that can help us elucidate on the reasons regulations are not complied to. Therefore, the objectives of our review were:

1. to analyse the socio-demographic characteristics of artisanal miners, their gender, age, marital status, the highest level of education, and professional experience in relation to the possession of professional cards.
2. To evaluate the knowledge of artisanal miners of the Occupational Health and Safety regulations.
3. To appraise the knowledge of artisanal miners of the DRC mining regulations.
4. To analyse the attitudes of artisanal miners towards the DRC mining regulations; and
5. to assess the association between compliance to safety regulations and possessing the professional mining cards.

We are of the view that strategies that would help resolve occupational health and safety issues in the mining industry include promoting SDG 3 and SDG 8 in the mining sector. It is important, in this regard, to implement policies and regulations that would prioritise worker safety, provide proper training and education to miners, and promote the use of appropriate safety equipment and technology. Additionally, it is crucial to encourage partnerships between mining companies, government, and local communities to foster responsible mining practices that prioritise safety, health, and sustainability. Community engagement and education are also essential for raising awareness and promoting safety in the mining sector. Finally, effective enforcement mechanisms and accountability frameworks should be implemented to ensure that all stakeholders comply with safety regulations, and that those guilty of non-compliance are held responsible for their actions. Implementing these strategies will create a safer and more sustainable mining industry that enhances the lives and livelihoods of all stakeholders involved.

Few other strategies can be adapted from primary health care to encourage and improve compliance to existing OHS legislations in the informal mining sector. However, these strategies require that the informal mining sector and government institutions, especially the health and mining sector adopt collaborative approach to enforce compliance. First, primary prevention activities such as occupational health education and awareness, the introduction of adaptative occupational management systems, occupational health and hygiene, as well as medical screenings should be promoted. Second, curative services (secondary prevention) to artisanal miners in local facilities must be provided, because such services are non-existent within the informal mining sector. Some services can be tailored, such as the diagnosis of lead poisoning among artisanal miners, which this population is susceptible to. Last, tertiary preventions services such as rehabilitation, which cater for artisanal miners who have been disabled due to artisanal mining activities for them to be integrated back into their communities. Furthermore, a One Health approach would ensure that these strategies are adapted and implemented as the approach promotes collaborative effort from different stakeholders.

Knowledge of workplace hazards and how to control them.

Some of the major knowledge banks, attitudes and behaviours of artisanal miners, which demonstrate an understanding of Occupational Health and Safety are as follows:

- an understanding of the relevant laws and regulations related to OHS;
- an awareness of the importance of PPE;
- knowledge of emergency procedures in the event of an accident or injury;
- a proactive attitude in identifying and reporting hazards and potential risks;
- an understanding of the importance of workplace ergonomics,
- a willingness to participate in OHS training and other educational programmes;
- consistently adhering to the safety protocols and procedures;
- focusing on continuous improvement and an conducting ongoing evaluation of OHS policies and practices; and
- promoting the culture of safety in the workplace.

To improve occupational health and safety, and provide insights into the DRC regulations, some measures should be taken, such as the continuous awareness and training on OHS and DRC mining regulations. A link exists between artisanal mining knowledge, attitudes, and behaviours and the achievement of SDGs 3 and 8. Artisanal mining can significantly impact the health and wellbeing of miners and the surrounding communities alike. In addition, improving working conditions and accelerating economic opportunities in the artisanal mining sector can contribute to achieving the goals of SDG8. The attitudes and behaviours of both miners and other stakeholders also play a role in determining the impacts of artisanal mining on health, wellbeing, and economic growth.

This consideration is essential for meeting Goal 3 of the UN 2030 Sustainable Development Goals (SDGs) "to ensure healthy lives and the promotion of wellbeing for all ages." Ensuring that all miners have professional mining cards enhances SDG 8, which promotes sustained, inclusive, and sustainable economic growth, full and productive employment, and decent work for all.

18.2 Rationale

The rationale for investigating the occupational health and safety of artisanal miners in Kolwezi, Democratic Republic of Congo, is based on the fact that there are significant health and safety risks confronting artisanal miners in the region. Artisanal mining is often associated with unsafe working conditions, environmental degradation, and human rights violations. Inadequate safety measures, exposure to toxic substances, and a lack of proper equipment and protective gear are some of the major occupational health and safety challenges confronting artisanal miners in Kolwezi. These issues not only threaten the lives and health of the miners, but also affect their families and local communities. Additionally, the lack of safety measures undermines the social and economic development of the region and hinders the goals of sustainable development. Thus, this study investigated the occupational health and safety of artisanal miners in Kolwezi, Democratic Republic of Congo to identify the challenges and opportunities for improving safety and health in the mining industry, and to promote sustainable and responsible mining practices that align with Goals 3 and 8 of the Sustainable Development Goals.

The inquiry was also conducted to highlight the current and perceived occupational health and safety issues among artisanal miners in Kolwezi City. The findings of the inquiry can be used as a baseline to address the challenge of non-compliance among the local population and other similar areas elsewhere. The findings of the presented reports are based on the knowledge, attitudes, and behaviours of artisanal miners in relation to occupational health and safety, and the reasons for the non-compliance to OHS standards of artisanal miners in the area under study.

18.3 Conceptual Framework

This study employed two concepts that guided the entire research process—namely occupational health and Safety, and sustainable development.

18.3.1 Occupational Health and Safety

Occupational Health and Safety (OHS) is a multidisciplinary field that concerns itself with the prevention of work-related injuries, illnesses, and deaths (Hermanus, 2007; Tetzlaff et al., 2021). A significant body of research has identified and assessed different types of hazards in the workplace – including chemical, physical, and biological hazards (Yi et al., 2021). In this regard, the findings of this research can inform risk assessment and hazard control measures. The concept "safety culture" refers to the values, attitudes, and behaviours that influence safety performance in the workplace (Tetzlaff et al., 2021). Studies have shown that promoting a positive safety culture can lead to improved safety outcomes and reduced injury rates. Literature on OHS has highlighted the need to protect workers from hazards in the workplace. This includes using PPE, being trained on safety issues, and ergonomic interventions (Kiran, 2021). The literature in this regard also considers the legal and policy frameworks that govern workplace safety. This includes regulations, standards, and enforcement mechanisms that aim to promote safer workplaces and hold employers accountable for ensuring safe working conditions. Research into emerging OHS issues, such as the impact of new technologies on workplace safety, the mental health and wellbeing of workers, and the effects of climate change on occupational health and safety is ongoing (Šolc et al., 2022). Thus, this research is essential in guiding the relevant stakeholders in developing effective strategies to prevent work-related injuries, illnesses and deaths, and to ensure safe and healthy working conditions for all.

18.3.2 Sustainable Development: Goals 3 and 8

Briant Carant (2017) has analysed the evolution of the Millennium Development Goals (MDGs) into the Sustainable Development Goals (SDGs) and has argued that the latter represent a more inclusive, participatory and ambitious agenda for sustainable development. The scholar highlights, in this regard, the role of civil society in shaping the SDGs, and calls for all stakeholders to be more engaged and accountable. Budhathoki et al. (2017) explored the potential of health literacy to address SDG 3, which aims to ensure the healthy living and wellbeing of all the people. In this study the researchers aimed to review the existing evidence of health literacy interventions in Nepal and argue that such interventions can contribute to achieving SDG 3 targets by increasing people's knowledge, skills and confidence to manage their health. Akolgo (2018) examined the lessons that can be learned from the narrative of the MDGs in the context of poverty eradication in sub-Saharan Africa. The scholar argues that the MDGs' narrow focus on a few priority areas, and the lack of contextualisation, contributed to limited progress in poverty reduction. The scholar suggests further that the SDGs, with their more comprehensive and integrated approach, can address some of the shortcomings of the MDGs. To this end,

Guégan et al. (2018) discuss the need for more integrative thinking to achieve SDG 3, emphasising the interdependence of human, animal, and environmental health. The scholars argue that an interdisciplinary and One.

Health approach is essential to address the complex challenges related to the global health and wellbeing of the people.

Williams and Hunt (2017) examined the challenges of ensuring accountability and human rights in the implementation of SDG 3. The scholars argue that the availability of reliable and disaggregated data on health outcomes and inequalities poses a significant obstacle to monitoring and evaluating progress towards SDG 3. To this end, the researchers call for greater attention to human rights principles, and involving the marginalised groups in the implementation of SDG 3.

Linking occupational health and safety with Goals 3 and 8 of sustainable development involves recognising the critical role of health and safety in achieving sustainable development. Sustainable Development Goal 3 talks to ensuring healthy lives and promoting the wellbeing of all, while Goal 8 aims to promote sustained, inclusive, and sustainable economic growth, full and productive employment, and decent work for all (Brymer/Lacaze, 2021). Occupational health and safety is an essential component to achieving both goals, as a safe and healthy work environment is crucial for ensuring the wellbeing of the workers, enhancing productivity, and supporting sustainable economic development (Fonseca/Carvalho, 2019). Therefore, linking occupational health and safety with Goals 3 and 8 of sustainable development involves promoting responsible and sustainable mining practices that prioritise the safety and health of the workers.

18.4 Methodology

18.4.1 Study Design

This study adopted a descriptive quantitative approach, and employed a cross-sectional research design. Cross-sectional studies generally seek to ascertain the prevalence of a problem, phenomenon, and attitude by recording the data without manipulating the study environment, and generally describe the characteristics of a community (Brink, 2006). The data were compiled between 2 and 4 December 2020 in Kolwezi City, Lualaba Province, in the Democratic Republic of Congo. The main objective of the survey was to determine the knowledge of occupational health and safety that the Congolese artisanal miners workers display when they perform their daily activities on the various mining sites, and their compliance with mining safety regulations.

18.4.2 Study Site

Kolwezi is a mining town in Lualaba Province, DRC, situated on the Manika Plateau in southern Congo at the geographic coordinates 26° East longitude, and 1° South latitude. Several people have qualified the town of Kolwezi as the "economic lung" of the DRC, because it has rich mineral deposits of copper and cobalt, and remains the first industrial and mining centre of the DRC.

Kolwezi was established in 1937, as an average city in colonial times, and extends to 102 ha. It is well planned and equipped, and compose of European managers' districts and popular African quarters (it was one of the first camps to provide housing for workers). An impressive demographic and spatial boom saw Kolwezi evolving into an important city that it is today, where the old quarters are caught in the tide of self-construction by individual owners. This has invaded the mining concessions. A growing part of the inhabited space has escaped the boundaries of GECAMINES, which formerly owned the city, and which has plans to expand to the city to exploit the newly identified deposits (Bruneau/Mansila, n.d.).

Demographically, Kolwezi is inhabited by a population that displays a continuous and almost uninterrupted growth rate. Industrial and artisanal mining activities impact the growth of the population, as different mining companies offer job prospects that attract the intellectual cream from different cities. Unemployed young people from different regions also come to artisanally exploit minerals; sometimes their various movements escaping the control of the authorities. The whole city of Kolwezi has 572, 942 inhabitants. Different foreign nationals are identifiable in the city; among them Chinese, Belgians, French, Lebanese, Indians, Americans, Australians, Greeks, Indonesians, British, Zimbabweans, and South Africans.

18.4.3 Population

A study population is a collection of objects, events, or individuals with certain common characteristics that are of interest to the researcher. Arias-Gómez et al. (2016) define a study population as the theoretically specified aggregation of the elements of the study. The study population consisted of artisanal workers selected from the five artisanal mining organisations in Kolwezi.

18.4.4 Sampling Techniques

The actual population size of artisanal miners operating in the Lualaba Province is unknown. However, according to the latest estimation of artisanal miners' population size by the "Société Civile de Kolwezi" (Kolwezi Civil Society) the number of artisanal miner workers stood at 140,000 in 2019 (Hu et al., 2020). The study sample was large enough to ensure the calculation of a 95% confidence interval for the various statistics to be calculated, with a margin of error being 5% and 80% respectively to achieve statistical testing power. The minimal sample size, which was chosen, using EPI Info version 7.2.3.0, needed to be 383 individuals. However, for this study, the sample size was increased to 546 artisanal miners.

18.4.4.1 Sample Size

The researchers employed simple random sampling to select 546 individuals from the larger pool of artisanal miner workers in Kolwezi. The data were collected through face-to-face interviews, covering the following aspects: (1) socio-demographic characteristics, (2) knowledge of occupational health and safety, (3) knowledge of the DRC mining safety regulations, (4) attitude towards the DRC mining safety regulations, and (5) behaviour towards the DRC mining safety regulations.

18.4.4.2 Inclusion Criteria

All artisanal mining citizens and residents of Kolwezi, with or without professional mining cards, who had worked in the mining sector for at least one year, were included in the study. In addition, age groups from 15 years were also included in the study – even though the Mining Code does not allow persons under the age of 18 to work in this sector. Including persons under the age of 18 in this study was justified by the fact that artisanal miners under 18 in the DRC are found in many artisanal mining activities. The United Nations Children's Fund (UNICEF) estimated in 2014 that around 40,000 young boys and girls worked across the mines in southern DRC, and that many of them were involved in cobalt mining (Murray/Newby, 2012). If the study had excluded persons under 18, it would have missed a very important health and safety issue in this area.

18.4.4.3 Exclusion Criteria

Non-citizen, visitors, and miners who had worked for less than a year in the mining sector were excluded from the study. In addition, individuals who refused or failed to give their consent to participate in the study, workers who were not soberminded, as well as workers who were aggressive and non-cooperative were excluded.

18.4.5 Data Collection

The data were collected using a closed-ended questionnaire, validated by public health experts, which covered the following aspects: (1) socio-demographic characteristics, (2) knowledge of occupational health and safety regulations, (3) knowledge of the DRC safety regulations, (4) attitudes towards the DRC mining regulations, and (5) the attitudes of artisanal miners towards the DRC mining regulations. The questionnaire was translated into French and administrated by a team of four (4) trained investigators, who were undergraduate students at the time. The questionnaire was administered to the randomly selected artisanal miners in Kolwezi City.

18.4.6 Data Analysis

Data were cleaned, coded, and processed using the IBM SPSS Statistics Version 25 to generate descriptive statistics such as the *Mean, Median, SD, Percentages,* and *Frequency* tables. Frequency distribution was used to compare the relationship between the different variables. The strength of association was calculated, using odds ratios (OR) and Epi Info 7, with probability of statistical significance set at a p-value of < 0.05, with a confidence interval (CI) of 95%. The results were presented in frequency tables and bar graphs. Regression analysis was performed to determine if possessing miner professional cards influenced the knowledge, attitudes, and the behaviours of artisanal miners towards occupational health and safety guidelines, and the DRC mining regulations.

18.4.7 Ethical Standards Observed

Permission to conduct the study: The proposal was submitted to the University of Johannesburg, Ethics Committee, and the Higher Degrees Committee for review and approval. Ethical clearance was granted by the University of Johannesburg—while permission to conduct the study was requested from, and granted by the D.R. Congo Provincial Ministry of Mines.

Informed consent: Participants were given an information letter, which explained the purpose of the study, the significance and the objectives thereof, for them to read and acknowledge by signing the letter. Mine managers assisted participants under the age of 18, after parents had consented to their participation in the study. Miners were also given consent forms, which explained the research procedures in a language appropriate for the individual's level of maturity. We explained to participants in the forms that they were not forced to participate in the study, and that they could withdraw from the study at any time without any consequences. However, once the data had been submitted and captured, participants would not be able to withdraw from participation, because at that stage participants would be anonymous, as they would have been given study identity numbers.

The principle of respect for autonomy: This principle considers participants' self-determination, and forms the basis for informed consent, and respect for confidentiality in healthcare practice (Dhai/McQuoid-Mason, 2011). Researchers in this study upheld the right of participants to self-determination by obtaining their informed consent. In this way, participants had the right to decide whether or not to participate in the study.

Confidentiality: The researchers ensured confidentiality by protecting participants' identity, privacy, self-worth, and dignity. Thus, participants in this study had the right to remain anonymous. Their responses thereof, would not be disclosed to other participants. Further, the study participants were not asked to provide their names, and each were identified by a study identity number.

Right to privacy: The data collected were kept confidential, and the anonymity of each responder was ensured. In this regard, the researchers would ensure that all the completed questionnaires were kept in a locked cabinet for five years, before they would be destroyed. The researchers also made it clear to participants that they would derive no direct benefits from their participation, and that they would not be exposed to any physical or psychological harm.

18.5 Results

This chapter reports on the findings of a study conducted between 2 and 4 December 2020 in Kolwezi City, Lualaba Province, in the Democratic Republic of Congo. Table 18.1 (in the appendices section) shows the age distribution of participants. The mean age for a male artisanal worker was 30.77 years (SD (8.37)), while the average age for a female artisanal worker was 29.52 years (SD (6.46)). A T-test for the mean difference was performed ($t = 1.52$), and no significant difference (5% level of significance and p-value $= 0.13$) was found in terms of the mean age between male and female miner workers (All the tables have been moved to the last section of this chapter under Appendices section).

Table 18.1 Age descriptive statistics

	Overall mean	Male mean	Female mean	t-test for equality of means		
N	546	467	79	t	df	p-value
Mean	30.59	30.77	29.52	1.524	126.988	0.130
Std. deviation	8.129	8.372	6.455			

Source The Authors

Tables 18.2 and 18.3 (in the Appendices section) summarise the socio-demographic characteristics stratified by possession of professional mining cards. The significant findings were in relation to the workers' age groups (30 to 34, 35 to 39, and 40 years and above), which were more likely to have professional cards than the 25 to 29 age group, with OR = 4.43 (CI: 2.13 – 9.22), OR = 3.10 (CI:1.40 – 6.88), and OR = 9.12 (CI:4.03 – 20.67), respectively. The divorced, widowed, or separated workers were significantly more likely to have professional cards than the married: OR = 8.82 (CI: 1.93 – 40.36).

Tables 18.4, 18.5 and 18.6 (in the Appendices section) present Knowledge of occupational health and safety and the DRC mining regulations, stratified by possession of professional mining cards. Workers with no knowledge or irrelevant knowledge

Table 18.2 Frequency distribution for socio-demographic characteristics

Characteristics	Total		Professional card yes		Professional card no	
	n	%	n	%	n	%
Total	546	100	162	29.7	384	70.3
Gender						
Male	467	85.5	144	30.8	323	69.2
Female	79	14.5	18	22.8	61	77.2
Age group						
15–19	62	11.4	1	1.6	61	98.4
20–24	77	14.1	8	10.4	69	89.6
25–29	73	13.4	10	13.7	63	86.3
30–34	172	31.5	71	41.3	101	58.7
35–39	91	16.7	30	33.0	61	67.0
40 – >	71	13.0	42	59.2	29	40.8
Educational level						
No formal education	34	6.2	9	26.5	25	73.5
Primary	140	25.6	25	17.9	115	82.1
Secondary	335	61.4	112	33.4	223	66.6
College	37	6.8	16	43.2	21	56.8
Marital status						
Single	160	29.3	10	6.3	150	93.8
Married	367	67.2	141	38.4	226	61.6
Divorced/Widowed/Separated	19	3.5	11	57.9	8	42.1
Professional experience						
1–5 years	248	45.4	14	5.6	234	94.4
6–10 years	147	26.9	46	31.3	101	68.7
> 10 years	151	27.7	102	67.5	49	32.5

Source The Authors

Table 18.3 Crude odds ratio and adjusted odds ratio for socio-demographic characteristics of research participants

Characteristics	Crude odd ratio	95% confidence intervals	Adjusted* odds ratio	95% confidence intervals
Gender				
Male	1.51	0.86–2.65	1.14	0.57–2.29
Female	Reference	Reference	Reference	Reference
Age group				
15–19	**0.10**	**0.01–0.83**	**0.10**	**0.01–0.88**
20–24	0.73	0.27–1.97	0.48	0.16–1.45
25–29	Reference	Reference	Reference	Reference
30–34	**4.43**	**2.13–9.22**	1.27	0.52–3.11
35–39	**3.10**	**1.40–6.88**	0.76	0.28–2.03
40 and higher	**9.12**	**4.03–20.67**	1.65	0.59–4.65
Educational level				
No formal education	0.47	0.17–1.29	0.48	0.12–1.88
Primary	**0.29**	**0.13–0.62**	**0.21**	**0.08–0.55**
Secondary	0.66	0.33–1.31	**0.39**	**0.16–0.94**
College	Reference	Reference	Reference	Reference
Marital status				
Single	**0.11**	**0.06–0.21**	0.64	0.27–1.52
Married	Reference	Reference	Reference	Reference
Divorced/ Widowed/ Separated	**8.82**	**1.93–40.36**	1.12	0.40–3.16
Professional experience				
1–5 years	**0.03**	**0.02–0.05**	**0.04**	**0.02–0.09**
6–10 years	**0.22**	**0.13–0.36**	**0.28**	**0.16–0.49**
> 10 years	Reference	Reference	Reference	Reference

*Adjusted for age, gender, education level, marital status, professional experience
Source The Authors

on non-certified sites were significantly more likely to have a professional card: OR = 1.91 (CI: 1.04 – 3.51), AOR = 4.38 (CI: 1.84 – 10.39), and OR = 8.91 (CI: 2.69 – 29.48), respectively. Compared to miners with the knowledge that untrained workers should not practice, artisanal miners who believe or did not know that untrained workers are not allowed to practice were significantly more likely to have professional cards—AOR= 8.9 (CI:2.69 – 29.48), and AOR =1.91 (CI:1.04 – 3.51), respectively.

Table 18.4 Distribution and crude odds ratios for participants for knowledge of occupational health safety stratified by possession of professional mining card

Characteristics	Total		Professional card yes		Professional card no		Crude odd ratio	95% confidence intervals
	n	%	n	%	n	%		
Total	546	10.0	162	29.7	384	70.3		
Have you ever heard of occupational health safety?								
Yes	319	58.4	126	77.8	193	50.2	Reference	Reference
No	227	41.6	36	22.2	191	49.8	**0.29**	**0.19–0.44**
OHS is								
An area of medicine aimed at preventing and combating work-related diseases,	236	43.2	97	41.1	139	58.9	Reference	Reference
Workers' union	1	0.2	0	0.0	1	100.0	–	–
A health insurance	19	3.5	4	21.1	15	78.9	0.38	0.12–1.19
I don't know	43	7.9	17	39.5	26	60.5	0.94	0.48–1.82
Other	20	3.7	8	40.0	12	60.0	0.96	0.38–2.43
I have never heard	227	41. 6	36	15.9	191	84.1	**0.27**	**0.17–0.42**
PPE is								
Equipment worn to minimize exposure to work hazards	241	44.1	105	43.6	136	56.4	Reference	Reference
Equipment for minerals detection	4	0.7	0	0.0	4	100.0	–	–
Equipment for mining extraction	4	0.7	0	0.0	4	100.0	–	–
I don't know	56	10.3	19	33.9	37	66.1	**0.67**	**0.36–1.22**
Other	14	2.6	2	14.3	12	85.7	**0.22**	**0.07–0.97**
I have never heard	227	41.6	36	15.9	191	84.1	**0.24**	**0.16–0.38**

Source The Authors

Table 18.5 Frequency distribution for knowledge of the DRC regulation

Characteristics	Total		Professional card yes		Professional card no	
Categories	n	%	n	%	n	%
Total	546	100	162	29.7	384	70.3
Are you aware of DRC mining safety regulations?						
Yes	225	41.2	75	33.3	150	66.7
No	321	58.8	87	27.1	234	72.9
How did you learn about mining safety regulations?						
TV/Radio	29	5.3	10	34.5	19	65.5
Internet	11	2.0	2	18.2	9	81.8
Colleagues miners	16	2.9	1	6.3	15	93.8
At the mines site	141	25.8	59	41.8	82	58.2
Other	28	5.1	3	10.7	25	89.3
I have never heard about it	321	58.8	87	27.1	234	72.9
Is PPE wearing compulsory?						
Yes	160	29.3	62	38.8	98	61.3
No	50	9.2	6	12.0	44	88.0
Don't know	15	2.7	7	46.7	8	53.3
Never heard about it	321	58.8	87	27.1	234	72.9
Are explosives and chemicals allowed?						
Yes	8	1.5	4	50.0	4	50.0
No	215	39.4	71	33.0	144	67.0
Don't know	2	0.4	0	0.0	2	100.0
I have never heard about it	321	58.8	87	27.1	234	72.9
Are tunnels longer than 30 meters allowed?						
Yes	90	16.5	35	38.9	55	61.1
No	105	19.2	29	27.6	76	72.4
I don't know	30	5.5	11	36.7	19	63.3
I have never heard about it	321	58.8	87	27.1	234	72.9
Is artisanal mining allowed on non-certified sites?						
Yes	16	2.9	12	75.0	4	25.0
No	135	24.7	34	25.2	101	74.8
Don't know	74	13.6	29	39.2	45	60.8
I have never heard about it	321	58.8	87	27.1	234	72.9

(continued)

Table 18.5 (continued)

Characteristics	Total		Professional card yes		Professional card no	
Categories	n	%	n	%	n	%
Are untrained workers allowed on mining sites?						
Yes	71	13.0	24	33.8	47	66.2
No	87	15.9	28	32.2	59	67.8
Don't know	67	12.3	23	34.3	44	65.8
I have never heard about it	321	58.8	87	27.1	234	72.9
Reason for not knowing the mining safety regulations						
It is for industrial mining	67	12.3	17	25.4	50	74.6
No popularization	103	18.9	34	33.0	69	67.0
No interest	72	13.2	15	20.8	57	79.2
No reason	71	13.0	21	29.6	50	70.4
Other	8	1.5	0	0.0	8	100.0
I know about the regulations	225	41.2	75	33.3	150	66.7

Source The Authors

Table 18.6 Crude odds ratios and adjusted odds ratios for knowledge of the DRC mining safety regulations

Characteristics	Crude odd ratio	95% confidence intervals	Adjusted odds ratio	95% confidence intervals
Are you aware of DRC mining safety regulations?				
Yes	Reference	Reference	Reference	Reference
No	0.72	0.50–1.03	0.74	0.51–1.07
How did you learn about mining safety regulations?				
TV/Radio	0.73	0.32–1.69	0.13	0.02–1.10
Internet	0.31	0.06–1.48	0.30	0.02–3.80
Colleagues miners	**0.09**	**0.01–0.55**	**0.09**	**0.01–0.72**
At the mine site	Reference	Reference	Reference	Reference
Other	**0.17**	**0.05–0.58**	0.56	0.05–5.84
I have never heard about it	**0.52**	**0.34–0.78**	0.18	0.02–1.38
Is PPE wearing compulsory?				
Yes	Reference	Reference	Reference	Reference
No	**0.22**	**0.09–0.54**	**0.19**	**0.07–0.51**
Don't know	1.38	0.48–4.00	1.79	0.58–5.52

(continued)

Table 18.6 (continued)

Characteristics	Crude odd ratio	95% confidence intervals	Adjusted odds ratio	95% confidence intervals
I have never heard about it	**0.59**	**0.390.88**	0.14	0.03–0.69
Are explosives and chemicals allowed?				
Yes	1.67	0.91–3.05	1.64	0.37–7.27
No	Reference	Reference	Reference	Reference
Don't know	1.52	0.64–3.58	–	–
I have never heard about it	0.94	0.57–1.54	0.79	0.03–21.66
Are tunnels longer than 30 meters allowed?				
Yes	1.67	0.91–3.05	1.63	0.84–3.16
No	Reference	Reference	Reference	Reference
I don't know	1.52	0.64–3.58	1.16	0.43–3.16
I have never heard about it	0.94	0.57–1.54	1.02	0.05–19.70
Is artisanal mining allowed on non-certified sites?				
Yes	**8.91**	**2.69–29.48**	1.63	0.83–3.16
No	Reference	Reference	Reference	Reference
Don't know	**1.91**	**1.04–3.51**	**4.38**	**1.84–10.39**
I have never heard about it	1.10	0.70–1.75	0.91	0.57–1.43
Are untrained workers allowed on mining sites?				
Yes	1.08	0.55–2.10	**8.91**	**2.69–29.48**
No	Reference	Reference	Reference	Reference
Don't know	1.10	0.56–2.17	**1.91**	**1.04–3.51**
I have never heard about it	0.78	0.47–1.31	0.75	0.04–15.63
Reason for not knowing the mining safety regulations				
It is for industrial mining	Reference	Reference	Reference	Reference
No popularization	1.45	0.73–2.88	1.39	0.69–2.79
No interest	0.77	0.35–1.71	0.74	0.33–1.65
No reason	1.24	0.58–2.62	1.31	0.61–2.83
I know about the regulation	1.47	0.79–2.72	–	–

Source The Authors

Figure 18.1. accounts for the various reasons artisanal miners are not aware of DRC mining regulations.

About 42% of artisanal workers professed not to be aware of the DRC mining safety regulations. Meanwhile, 18.9% of the miners thought that the regulations were not popularised, while 12.3% of them stated that the regulations were only applicable to industrial mining. About 13% of the miners had no interest. or did not see the need to familiarise themselves with the regulations (Fig. 18.2).

Tables 18.7, 18.8 and 18.9 (in the Appendices section) account for the attitudes and behaviours of artisanal miners toward OHS and the DRC mining regulations.

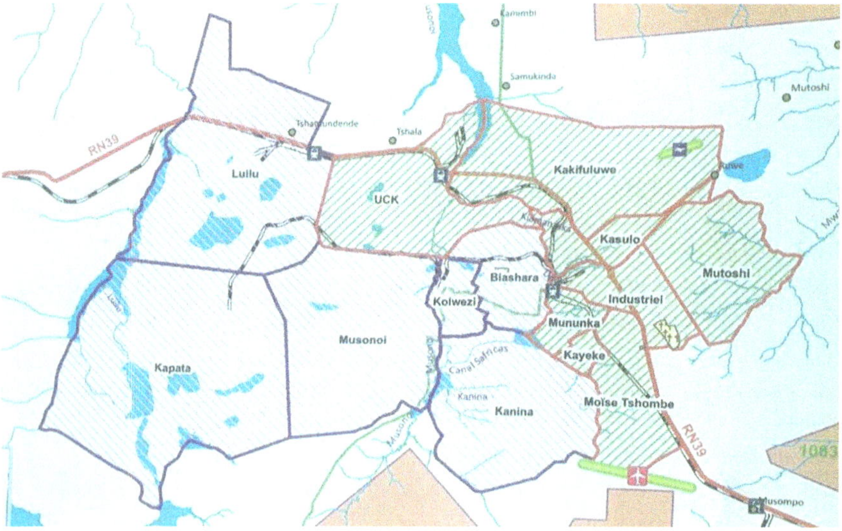

Fig. 18.1 Site map. *Source* Gregory Mthembu-Salter (2009)

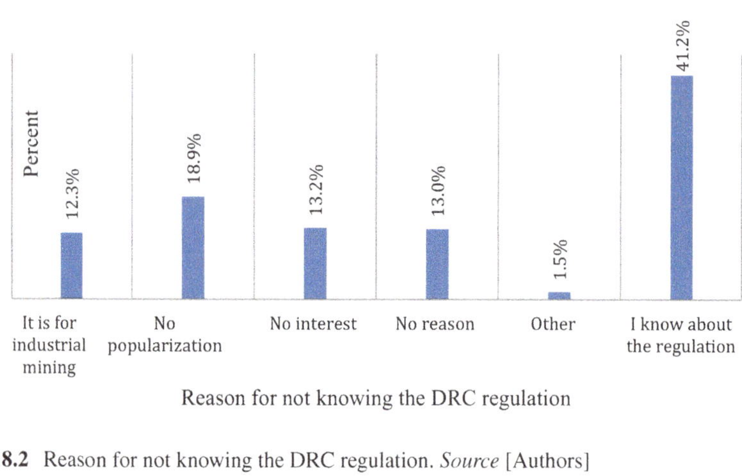

Fig. 18.2 Reason for not knowing the DRC regulation. *Source* [Authors]

Table 18.7 Frequency distribution for attitude toward DRC mining safety regulations

Characteristics	Total		Professional card yes		Professional card no	
	n	%	n	%	n	%
	546	100	162	29.7	384	70.3
Is PPE protective from accidents?						
Never	153	28.0	37	24.2	116	75.8
Rarely	76	13.9	19	25.0	57	75.0
Sometimes	147	26.9	41	27.9	106	72.1
Very often	98	17.9	31	31.6	67	68.4
Always	72	13.2	34	47.2	38	52.8
Solution for applying the DRC mining safety regulations						
By creating awareness	425	77.8	135	31.8	290	68.2
Supervision	25	4.6	10	40.0	15	60.0
Sanction	28	5.1	5	17.9	23	82.1
Nothing can be done	64	11.7	11	17.2	53	82.8
Others	4	0.7	1	25.0	3	75.0

Source The Authors

Table 18.8 Crude odds ratios and adjusted odds ratios for attitudes towards DRC mining safety regulations

Characteristics	Crude odd ratio	95% confidence intervals	Adjusted odds ratio	95% confidence intervals
Categories				
Total				
Is PPE protective from accidents?				
Never	**0.36**	**0.20–0.65**	0.61	0.29–1.30
Rarely	**0.37**	**0.19–0.75**	0.61	0.26–1.42
Sometimes	**0.43**	**0.24–0.78**	**0.44**	**0.21–0.94**
Very often	**0.52**	**0.28–0.97**	**0.44**	**0.20–0.98**
Always	Reference	Reference	Reference	Reference
Solution for applying the DRC mining safety regulations				
By creating awareness	Reference	Reference	Reference	Reference
Supervision	1.43	0.63–3.27	1.26	0.53–3.03
Sanction	0.47	0.17–1.26	0.49	0.16–1.46
Nothing can be done	**0.45**	**0.23–0.88**	**0.33**	**0.15–0.73**
Others	0.72	0.07–6.95	0.84	0.8–8.52

Source The Authors

Table 18.9 Crude odds ratios and adjusted odds ratios for behaviour and compliance towards DRC mining safety regulations

Characteristics	Crude odd ratio	95% confidence intervals	Adjusted odds ratio	95% confidence intervals
Conducting safety checks before work?				
Never	0.64	0.37–1.11	0.55	0.27–1.14
Rarely	2.05	0.13–32.94	3.40	0.16–74.12
Sometimes	**0.17**	**0.04–0.73**	0.23	0.05–1.11
Very often	0.51	0.20–1.28	0.64	0.21–1.92
Always	Reference	Reference	Reference	Reference
Wearing PPE during work				
Never	0.61	0.32–1.15	0.73	0.33–1.59
Rarely	0.17	0.02–1.40	0.35	0.04–3.31
Sometimes	**3.14**	**1.02–9.65**	2.71	0.72–10.12
Very often	**5.69**	**2.38–13.61**	**7.36**	**2.72–19.88**
Always	Reference	Reference	Reference	Reference
Applying the DRC mining safety regulations				
Never	**0.29**	**0.16–0.53**	**0.20**	**0.09–0.48**
Rarely	**0.43**	**0.20–0.94**	**0.16**	**0.06–0.47**
Sometimes	0.53	0.25–1.11	**0.22**	**0.08–0.59**
Very often	1.77	0.15–20.82	2.67	0.18–39.55
Always	Reference	Reference	Reference	Reference

Source The Authors

Overall, 153 (28.0%) of the 546 participants were of the view that PPE does not, in any way, offer protection against accidents, while 147 (26.9%) believed that PPE sometimes offers protection against accidents. Only 72 (13.2%) were of the view that PPE offers protection during accidents, while 424 (77.6%) of participants were of the view that raising an awareness of the use of PPE in mining operations would solve the problem of artisanal miners who do not comply with the DRC mining safety regulations. Workers who "sometimes", or "very often" wear the PPE were significantly more likely to have professional cards than workers who were 'always' wearing PPE with OR = 3.14 (CI: 1.02 – 9.65). OR = 5.69 (CI: 2.38 – 13.61), and AOR =7.36 (CI: 2.72 – 19.88) respectively.

When asked what they believe may help curb non-compliance with mining safety regulations, artisanal miners stated many reasons. Figure 18.3 outlines the main factors that may provide solutions to the challenge of non-compliance.

Most of the artisanal workers (77.8%) believed that the challenge of non-compliance could be addressed if an awareness is raised among artisanal mining workers' community of the mining safety regulations, while 11.7% of them believed that nothing could be done to address the problem.

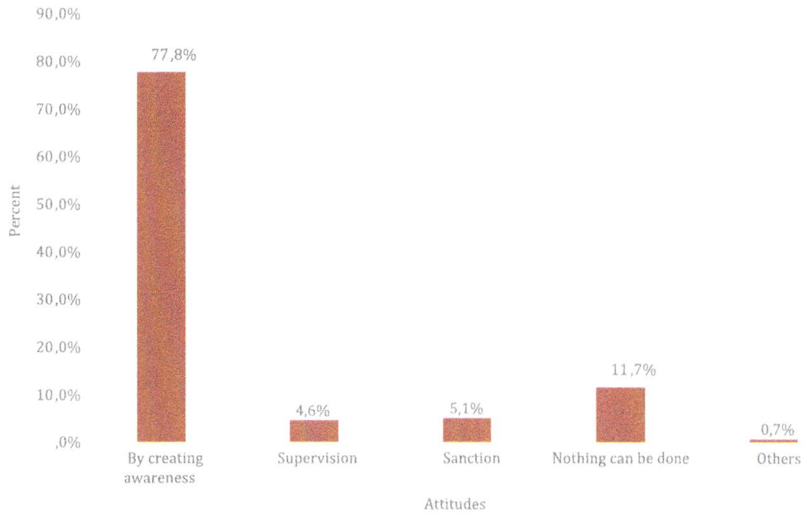

Fig. 18.3 Solutions for applying the DRC mining safety regulations. *Source* The Authors

18.6 Discussion

This study was carried out in the five artisanal mines in Kolwezi City, in the Democratic Republic of Congo, and aimed to assess the knowledge, attitudes, and behaviours of artisanal miners towards OHS and the DRC mining safety regulations. The study compared miners with professional miners' cards to those without professional mining cards, using SPSS version 25, and the CDC program EPINFO 7.2.

The study found that age group, educational level, marital status, and work experience influences the decision to own or not own a professional card. Moreover, the older the artisanal miner (40 years and older), the higher their educational level (up to college level), and that the more professional experience they possess, the more they are likely to comply with this regulation. The study also found that not having a professional miner's card impacts negatively on the knowledge, attitudes, and behaviours of artisanal miners towards OHS, and the DRC mining safety regulations.

18.6.1 Socio-demographic Characteristics

In assessing the relationship between the socio-demographic characteristics of participants and possession of the professional miners' cards, we found that age group, educational level, marital status, and professional experience play a major role. For the age group variable, the crude odds for an artisanal mining worker between 15 and 19 years to have a professional miner's card were approximately 10% less than that of a worker aged between 25 and 29 years. This may be indicative of a non-negligible lower level of compliance to the mining safety regulations. However, if the age category 30 to 34 is considered; the odds for workers who have professional mining cards are 4.43 times higher than the odds for the workers between the ages of 25 and 29. Similar observations were made with other age categories, seeming to indicate that the older the artisanal miners, the more likely they are to comply with safety regulations (Ajith/Ghosh, 2019).

Young men are also, more likely to be pressured by their institutions to produce more ore than their older colleagues. In this regard, the findings of this study show that participants who completed primary school are 0.29 less likely to have professional miners' cards, compared to those with a College education – and that artisanal miners who have completed secondary school were 0.39 times less likely than participants with a College education to have professional miners' card. It can be argued, then, that being knowledgeable on health and safety issues is influenced by educational level. The findings of another study found a significantly high correlation between occupational safety knowledge and the level of education – that is the higher the level of education, the higher the chances of being knowledgeable (Williams/Hunt, 2017). Respondents who were highly knowledgeable had attained better education. This is in agreement with the rational model of health promotion, which states that overall, people with lower literacy levels are not in good health (Gibney et al., 2020).

The study findings also reveal that the divorced, widowed, or those who are separated were 8.8 times significantly more likely to comply with the safety regulations, evidenced by the fact that they possessed professional miners' cards, as opposed to married participants. This finding contradicts the findings of a study carried out in Nigeria, where married artisanal miners were more likely to comply with safety regulations (Afolayan et al., 2021). Workers tend to be more careful in what they do when they have more social responsibilities (de Haan/Geenen, 2016). In a typically African context, marriage can give one a sense of responsibility and accomplishment. It is expected that married workers would be more conscious about health and safety, and that they would tend to comply with safety regulations because of the responsibilities they assume because of their status—unlike unmarried persons, whose primary responsibility might just be to cater for themselves. In the DRC, artisanal mining is a daily and survival activity with little income, not exceeding US$ 10 for most of them (Sovacool, 2021). Our assumption is that married artisanal miners simply do not want to be identified as such to avoid paying the multiple taxes that the state demand—hence they do not have miners' cards. Thus, we conclude that there is a relationship between socio-demographic characteristics and possession of professional mining cards.

18.6.2 Knowledge of Occupational Health Safety

One of the most common problems in the artisanal mining sector is non-compliance with occupational health and safety regulations. Many small-scale and artisanal mining operations do not have in place safety standards, and enforcement of mine safety requirements are not aware of the risks inherent in mining, and have no access to better equipment (Ruppert, 2004).

This study found that the odds of an artisanal miner without a professional card, who have never heard of occupational safety and health are 29% as opposed to the odds for an artisanal miner with a professional card. It appears that artisanal miners with professional cards are more knowledgeable about occupational health and safety regulations than miners without professional cards. Similarly, the odds for artisanal mining workers without professional cards who have no idea what PPE is, or who have never heard of it, is 22%—while the odds for artisanal workers with the relevant knowledge about PPE and possess professional miners' cards is 24%. Once more, it appears that having a professional miner's card makes a significant difference when it comes to occupational health and safety standards.

18.6.3 Knowledge of the DRC Mining Safety Regulations

Most artisanal miners lack information and knowledge on much of the DRC legislation. During the baseline survey, we asked the formal and informal artisanal miners if they have some knowledge of the mining safety regulations. We found that 58.8% were not aware of them. This was evident when reference was made to the requirements of Articles 111 and 112 of the mining code, which stipulate that *after reading the Mining Code eligible persons agree to comply with regulations on environmental protection, hygiene and safety in the artisanal mining areas* (Tsurukawa et al., 2011), before being issued the artisanal miner's card, to which the artisanal miners were not conversant, in relation to the DRC mining safety regulations.

When assessing the relationship between the variables, we found no difference in knowledge between workers with and without professional miners' cards in respect of an awareness of mining safety regulations, their opinions on the use of explosives, or the maximum length of tunnels – yet Annex V of the Mining Code stipulates that they should be familiar with all these provisions (Tsurukawa et al., 2011). The Congolese mining safety regulations are very complex, and contain several appendices, which have not been popularised, and are, therefore, not comprehensible. Moreover, until today, no clarity has been provided in relation to the other appendices relating to occupational health and safety. This complexity of the regulations hinders mining workers' acquisition of the knowledge. In fact, these regulations have more details relating to industrial mining, and very few provisions on the health and safety of the artisanal mine operator. Indeed, they focus more on-site preparation operations, the use of wood and water supply needs for the operation, and the conservation of vegetation cut during mining operations (Tsurukawa et al., 2011). This situation refers to the statement of Salihu, which stipulates that occupational safety regulations must be developed in such a way that everyone will understand them. When the requirements of a regulation are very complex and bulky to be understood, the target group is likely to resist compliance, due to their misinterpretation of the law (Tziaferi et al., 2011). Furthermore, the findings of this study suggest that the best place for artisanal miners to learn about mining safety regulations is at the mining site, as opposed to learning from miners' colleagues or other sources.

Contrary to expectations, mining workers who understand that untrained workers are not allowed on mining sites, and those who do not know if untrained workers are allowed or not, were 8.91 and 1.91 times more likely to have professional miners' card respectively. This finding has shown that artisanal miners who read the "code of good conduct in artisanal miners" before being registered, do not clearly understand the provisions. A study conducted by Tziaferi et al., 2011) among Australian mineworkers corroborates the findings of this study, as they also found that many coal mineworkers have a knowledge of mining regulations, but upon further investigation, he found that they were not able to articulate the regulations stipulated, and how they should be implemented (Tziaferi et al., 2011).

18.6.4 Attitudes Toward DRC Mining Safety Regulations

Attitude plays an important role in influencing the thinking and the general practice of a person (Hou, 2014). Two aspects were considered in assessing the attitudes of artisanal mining workers in respect of the DRC mining safety regulations—namely wearing PPE and solutions to enhance compliance to regulations. Relative to workers who believe that PPE offers guaranteed protection in the case of an accident, the odds for workers without professional miners' cards, and those who believed that they can benefit marginally or not at all from wearing PPE are substantially lower (between 36% and 52%) than the odds for artisanal workers with professional miners' cards, who believe that wearing PPE is beneficial to them.

This statistically significant association between attitude and the protection offered by PPE is evidenced by the fact that the workers with no professional miners' cards did not perceive themselves as being vulnerable to mining hazards. The main driver for these artisanal miners is going where they can find the richest ores and get the greatest profits without worrying about their health and safety, or the unstable ground, caused by previous mining activities in the area, which render additional operating activity dangerous. The findings of a study conducted in Dubai have shown that workers who perceived that they were exposed to hazards had significantly higher self-claimed use of PPE (Koja-Odongo/Mostert, 2013). In this study, mining workers' perceptions that they are vulnerable to hazards serve as a key motivator for them to wear PPE.

The findings of our study also point out that many (77.8%) artisanal miners believed that awareness can mitigate non-compliance to the DRC mining safety regulations. To address compliance with mining regulation, mining workers who were of the view that nothing can be done to improve the situation were 45% less likely to have professional miners' cards, as opposed to the mining workers who believed that raising awareness can be a solution to the problem. The attitude towards action is motivated mainly by the existing knowledge. According to WHO, the rational model of health promotion provides that increasing a person's knowledge breeds a positive attitude and results in a change in behaviour (Hou, 2014). In general, poor knowledge negatively affects attitudes (Han et al., 2021).

18.6.5 Behaviour and Compliance of Artisanal Miners Towards the DRC Mining Safety Regulations

Good behaviour and compliance for a number of subjects can be determined by being well informed in relation to good attitudes towards these particular issues (Zin/Ismail, 2012). The findings of this study show that in relation to the behaviour of artisanal mining workers and their compliance to the DRC mining safety regulations, the odds for workers who believed that conducting regular safety checks is useful are 0.27 times as high as the odds for the other workers who did not believe that these regular checks are always useful. In relation to believing that workers must always wear PPE at work, the odds that workers sometimes, or very often wear PPE, are 3.14 and 5.69 times higher respectively when artisanal workers have professional miners' cards, than when they do not have cards. While providing personnel with PPE is a fundamental obligation in a mining company, in the artisanal sector everyone is responsible for their life, therefore, acquiring PPE is at their expense; and yet, this is of paramount importance in the safety of personnel working in the mines.

The findings of this study show that OHS specialists never supervised artisanal work, as stated in the 87.4% of the responses. This could negatively impact the behaviour of workers, as the authorities are not assuming responsibility for their safety. Global Witness (2009) mentions that the absence of official control and supervision in artisanal mines, the non-intervention of local agents in the mines in respect of helping the diggers and advising them on risks, or measures to prevent accidents, the government's refusal to take responsibility for the safety of the diggers, and their failure to enforce the laws and regulations can explain the high accident rate in artisanal mines (Driscoll et al., 2020). Although a statistical association was not found, it is important to note that the descriptive statistics indicate a very high frequency (85.7%) of artisanal miners who have never been supervised by OHS inspectors, contrary to the prescriptions of the labour code. This situation could be due to the fact that there is no clear regulation governing artisanal mining, as well as the authorities' negligence of the artisanal mining industry.

18.6.6 Implications of the Knowledge, Behaviours, and Attitudes of Artisanal Miners on the Achievement of SDG Goals

The Health Belief Model (HBM) predicts that people's perceptions of a threat of disease or illness determine their likelihood of adopting a healthy behaviour. This is a theoretical model that provides that people make the decision to adopt a particular health behaviour when they anticipate a negative impact of their behaviour—hence, they tend to take measures aimed at self-protection (Kambris et al., 2019). Before commencing work, artisanal miners should, in accordance with the DRC mining regulations, receive training in the *Service d'Assistance et d'Encadrement de l'Exploitation Minière Artisanale et à Petite Échelle SAE*MAPE in respect of the Mining Code, the Code of Conduct, environmental responsibility, techniques, safety, and hygiene. The training should be based on an evaluation test conducted by the SAEMAPE, who should forward the results to the Head of the Provincial Division of Mines to issue professional cards. Only artisanal miners who pass the test are given cards to show that they received training in OHS, mining regulations, and have been made aware of the danger that they will be exposed to. Nonetheless, the artisanal miners with no professional cards in this study did not perceive themselves as vulnerable to accidents; hence they displayed negative attitudes to compliance to OHS guidelines and the DRC mining safety regulations.

Poor communication and popularisation of the policies did not motivate artisanal miners to act (to take preventive measures) to protect themselves from hazards, and to comply with the regulations. Knowledge of the modalities for undertaking an activity is very important because it is the guiding principle and the power that enable one to work effectively and defend their position. Popularising training is crucial as artisanal miners can avoid accidents that may happen during their mining activities. It also removes the barriers that prevent miners to undertake proper training. Lack of legislative compliance, improper establishment of policies, and the availability of training programmes may contribute to the high rates of accidents, which could, otherwise, be avoided. Various factors, such as the knowledge of safety regulations, may influence workers' decision to use or not use PPE.

The attitudes and behaviours of artisanal miners towards occupational health and safety in Kolwezi can have a significant impact on the achievement of SDG 3: "Ensure healthy lives and promote wellbeing for all at all ages." Poor occupational health and safety practices in the mining sector can lead to a range of health risks that includes respiratory problems, musculoskeletal injuries, and exposure to hazardous chemicals. Artisanal miners in Kolwezi are less likely to be exposed to these health risks, and more likely to maintain their physical and mental wellbeing if they display positive attitudes and behaviours towards occupational health and safety. This can contribute directly to the achievement of SDG 3 as it will ensure the promotion of healthy lives and reduce the incidence of health problems in this population. Conversely, if artisanal miners in Kolwezi display negative attitudes and behaviour towards occupational health and safety, they are more likely to engage in practices that put their health at risk. This can negatively impact the achievement of SDG 3 as there will be increased incidences of health problems and reduced overall wellbeing in this population. Therefore, promoting positive attitudes and behaviours towards occupational health and safety among artisanal miners in Kolwezi can serve as a key step to achieving SDG 3.

The attitudes and behaviours of artisanal miners on occupational health and safety in Kolwezi can also have an impact on the achievement of SDG 8: "Decent Work and Economic Growth."Occupational health and safety is an important aspect to decent work, as workers would be able to carry out their tasks in a safe and healthy environment. Artisanal miners in Kolwezi are less likely to be exposed to work-related injuries and illnesses if they display positive attitudes and behaviours towards occupational health and safety. This can, in turn, enhance their ability to work and contribute to the local economy, and in the process, contribute to the achievement of SDG 8 in the sense that it will promote sustained, inclusive, and sustainable economic growth. Conversely, artisanal miners in Kolwezi are more likely to be vulnerable to work-related injuries and illnesses, which can negatively impact their ability to work and contribute to the local economy, if they display negative attitudes and behaviours towards occupational health and safety. This, in turn, can negatively impact the achievement of SDG 8 in the sense that their attitudes and behaviours will negatively impact economic growth and the availability of decent work opportunities in the region. Therefore, promoting positive attitudes and behaviours towards occupational health and safety among artisanal miners in Kolwezi will not only contribute to their health and wellbeing, but contribute to the achievement of SDG 8, as well as accelerating economic outcomes in the region.

The findings of this study can serve as a basis for further research on artisanal mining in the DRC or similar settings in other countries, particularly on the impact of being registered as a miner, and how this contributes to compliance with regulations. Thus, the findings of this study will, hopefully, help identify the gaps in the knowledge of OHS and mining regulations, as observed in Kolwezi, due to the fact that some artisanal miners are not aware of the regulations.

18.7 Limitations of the Study

This study aimed to assess the knowledge, attitudes, and behaviours of artisanal miners in Kolwezi City, the Democratic Republic of Congo, towards occupational health and safety. The study focused on artisanal workers in one city—therefore, the findings cannot be generalised to all artisanal mining in the DRC. As this was a quantitative study, participants could not elaborate on their responses, as the close-ended questions did not allow for descriptive answers, which may, sometimes, be very useful. Another limitation is that the study did not explore the reasons some artisanal miners did not possess professional miners' cards.

18.8 Conclusion

It is widely acknowledged that accidents in artisanal mines are under-reported, or not reported at all, because illegal operators do not want to draw attention to themselves, and the fact that there is often, no form of compensation or social security for the injured. A combination of factors such as (a) not being knowledgeable on OHS, or the non-enforcement of safety rules, (b) a lack of awareness of OHS regulations, (c) illiteracy, (d) the influence of age, (e) lack of training, and (f) inadequate equipment suggest the likelihood of more accidents in many artisanal mining operations than in larger, more formal and public mines. The hazardous nature of the mining environment calls for workers to be extremely vigilant and to comply with safety standards. The government and all stakeholders should, in this regard, put more effort to improve the work environment and upgrade mining safety regulations to minimise the occurrence of accidents in artisanal mining, which could be avoided.

This study has identified a disconnect between the established mining safety regulations and their implementation; and that these regulations are not adequately communicated to stakeholders, including government agents. This issue must be adequately addressed to ensure better health and safety among artisanal mining workers. Mining authorities should promote education and awareness on the use of personal protective equipment (Aram et al., 2021).

18.9 Recommendations

The following are recommendations, based on the findings of the study:

18.9.1 The DRC Authorities and Stakeholders

Immediate steps should be taken to ensure more effective control of the artisanal mining sector, improve the working conditions of artisanal miners, and facilitate the implementation of the existing legislation, including the Mining Code and Labour Code. The Ministry of Mines, in coordination with the Ministry of Labour, should assume full responsibility for policies governing this very critical area. The police should cooperate in this regard, to enforce regulations in the mines. A basic training programme should be in place that caters for all artisanal miners, which should provide as much information as possible on health and safety issues. The work programme, led by the SAEMAPE, should prioritise this. The artisanal miner registration process ought to be simplified, while multiple illegal taxes should be abolished, psychological support and counselling offered to artisanal workers at the mine precinct to ensure that mineworkers are at optimum mental health to perform their duties. Regular enforcement activities by the inspectorate such as inspections, compliance audits, and the use of policy instruments such as sanctions and prohibitions to influence and ensure compliance with safety regulations should be highly encouraged and deployed. Greater attempts should be made in collaboration with education and training to promote occupational health and safety issues. This should be integrated and be made compulsory in the post-primary curriculum.

18.9.2 To the Artisanal Miners

Highly informative courses on occupational health and safety, or the related sessions should be organised for artisanal miners. In addition, artisanal miners ought to wear PPE in mining sites. There should be a bunch of easy mechanisms to help and encourage artisanal miners to report any practice that could put their health and safety at risk. Finding strategies to improve working conditions and create a conducive environment and lobby for employers to improve safety and health at the workplace should be encouraged.

Ethical Issues Permission to conduct this study was obtained from the University of Johannesburg, Ethics Committee (REC-616-2020), the Higher Degrees Committee (MPH HDC-01-46-2020), and the DRC authorisation (CAB.MIN/MINES/LBA/227/CNK/2020). The participants voluntarily signed a consent form before participating in the study.

Acknowledgments The authors wish to thank Mr. J. M. Tshizainga Sanama, the provincial ministry of Mines, Lualaba, and the research assistants for working tirelessly on this project.

Funding This study was not commissioned, and authors funded it themselves.

References

Afolayan, D.O., Onwualu, A.P., Eggleston, C.M., Adetunji, A.R., Tao, M. & Amankwah, R.K. (2021). Safe mining assessment of artisanal barite mining activities in Nigeria. *Mining*, *1*(2), 224–240. https://doi.org/10.3390/mining1020015

Ahmed, I. (2010). Effects of motivational factors on employees job satisfaction: A case study of University of the Punjab, Pakistan. *International Journal of Business and Management*, *5*(3), 70–80. https://doi.org/10.5539/ijbm.v5n3p70

Ajith, M.M. & Ghosh, A.K. (2019). Comparison of parameters for likelihood and severities of injuries in artisanal and small-scale mining (ASM). *Safety Science*, *118*, 212–220. https://doi.org/10.1016/j.ssci.2019.04.010

Akolgo, I.A. (2018). Agenda 2030 in sub-Saharan Africa: What the Millennium Development Goals' narrative teaches about poverty eradication. African Review of Economics and Finance. *African Review of Economics and Finance*, 10(*1*), *3–22*.

Allan-Blitz, L.-T., Goldfine, C. & Erickson, T.B. (2022). Environmental and health risks posed to children by artisanal gold mining: A systematic review. *SAGE Open Medicine*, *10*, 1–10 205031212210769. https://doi.org/10.1177/20503121221076934

Aram, S. A., Saalidong, B. M., Appiah, A. & Utip, I. B. (2021). Occupational health and safety in mining: Predictive probabilities of Personal Protective Equipment (PPE) use among artisanal goldminers in Ghana. *PLoS ONE, 16*(9 Sep), 1–14. https://doi.org/10.1371/journal.pone.0257772

Arias-Gómez, J., Villasís-Keever, M.Á. & Miranda-Novales, M.G. (2016). The research protocol III. Study population. *Revista Alergia Mexico, 63*(2), 33–47. https://doi.org/10.29262/ram.v63i2.181

Briant Carant, J. (2017). Unheard voices: A critical discourse analysis of the Millennium Development Goals' evolution into the Sustainable Development Goals. *Third World Quarterly, 38*(1), 16–41. https://doi.org/10.1080/01436597.2016.1166944

Brink, H. (2006). *Fundamentals of social research methodology for health care professionals.* Cape Town : Juta & Company Ltd.

Bruneau, J. C. & Mansila Fu-Kiau. (n.d.). KOLWEZI: l'espace habitée et ses problèmes dans le premier centre minier du zaïre. https://horizon.documentation.ird.fr/exl-doc/pleins_textes/pleins_textes_4/sci_hum/23506.pdf. Accessed on 8 May 2021.

Brymer, E. & Lacaze, A.-M. (2021). An overview of SDG 3. In B. McCullough, T. Kellison & N. Melton (Eds.), *The Routledge Handbook of Sport and Sustainable Development* (1), 13. Routledge. https://doi.org/10.4324/9781003023968

Budhathoki, S. S., Pokharel, P. K., Good, S., Limbu, S., Bhattachan, M. & Osborne, R. H. (2017). The potential of health literacy to address the health related UN sustainable development goal 3 (SDG3) in Nepal: A rapid review. *BMC Health Services Research, 17*(1), 237. https://doi.org/10.1186/s12913-017-2183-6

Burgess, J. & Connell, J. (2006). Temporary work and human resources management: Issues, challenges and responses. *Personnel Review, 35*(2), 129–140. https://doi.org/10.1108/00483480610645786

Cuvelier, J. (2020). Making sense of repressive actions against artisanal miners in southeastern Democratic Republic of Congo. *Congo Research Briefs,* 6.

De Haan Dales, K. & Mcquilken, J.J. (2023). A Mapping Artisanal and Small-Scale Mining to the de Haan, J. & Geenen, S. (2016). Mining cooperatives in Eastern DRC: The interplay between historical power relations and formal institutions. *Extractive Industries and Society, 3*(3), 823–831. https://doi.org/10.1016/j.exis.2016.05.003

Debrah, A. A., Watson, I. & Quansah, D. P. O. (2014). Comparison between artisanal and small-scale mining in Ghana and South Africa. The regional frameworks: Yaoundé Vision and the MMSD in the AMV. *The Journal of The Southern African Institute of Mining and Metallurgy, 114*(NOV), 913–921. http://www.scielo.org.za/scielo.php?script=sci_arttext&pid=S22256253201400l100009

Dhai, A., & McQuoid-Mason, D. (2011). Bioethics, human rights and health law: principles and practice. Juta and Company Ltd.

Driscoll, T., Rushton, L., Hutchings, S. J., Straif, K., Steenland, K., Abate, D., Abbafati, C., Acharya, D., Adebayo, O. M., Afshari, M., Akinyemiju, T., Alahdab, F., Anjomshoa, M., Antonio, C. A. T., Aremu, O., Ataro, Z., AyalaQuintanilla, B. P., MattarBanoub, J. A., Barker-Collo, S. L., … Lim, S. S. (2020). Global and regional burden of disease and injury in 2016 arising from occupational exposures: A systematic analysis for the Global Burden of Disease Study 2016. *Occupational and Environmental Medicine, 77*(3), 133–141. https://doi.org/10.1136/oemed-2019-106008

Fonseca, L. & Carvalho, F. (2019). The reporting of SDGs by Quality, Environmental, and Occupational Health and Safety-Certified Organizations. *Sustainability, 11*(20), 5797. https://doi.org/10.3390/su11205797

Gibney, S., Bruton, L., Ryan, C., Doyle, G. & Rowlands, G. (2020). Increasing health literacy may reduce health inequalities: Evidence from a National Population Survey in Ireland. *International Journal of Environmental Research and Public Health, 17*(16), 5891. https://doi.org/10.3390/ijerph17165891

Global Witness. (2009). Face à un fusil, que peut-on faire? La guerre et la militarisation du secteur minier dans l'est du Congo [Report], 27.

Gregory Mthembu-Salter. (2009). Occasional Papers no 35. *Natural Resource Governanc9e* 1(185)1–14. https://issafrica.s3.amazonaws.com/site/uploads/Paper185.pdf. Accessed 8 August 2023.

Guégan, J.F., Suzán, G., Kati-Coulibaly, S., Bonpamgue, D. N. & Moatti, J.P. (2018). Sustainable Development Goal #3, "health and well-being", and the need for more integrative thinking. *Veterinaria Mexico, 5*(2)1–18. https://doi.org/10.21753/vmoa.5.2.44

Han, B., Son, S. & Kim, S. (2021). Measuring safety climate in the construction industry: A systematic literature review. *Sustainability, 13*(19), 10603. https://doi.org/10.3390/su131910603

Health: A Global Access Science Source, 21(1), 1–13. https://doi.org/10.1186/s12940-022-00877-5

Hermanus, M. (2007). Occupational health and safety in mining-status, new developments and concerns: Annual number and rate of fatalities in mining. *The Journal of the Southern African Institute of Mining and Metallurgy, 107*, 531–538.

Hou, S.-I. (2014). Health education: Theoretical concepts, effective strategies and core competencies. *Health Promotion Practice, 15*(5), 619–621. https://doi.org/10.1177/1524839914538045

Hu, L., Zhang, M., Yang, Z., Fan, Y., Li, J., Wang, H. & Lubale, C. (2020). Estimating dewatering in an underground mine by using a 3D finite element model. *PLoS ONE, 15*(10 Oct), 1–22. https://doi.org/10.1371/journal.pone.0239682

Kambris, M.E.K., Khan, S. & Al Falasi, S.N. (2019). Perceptions of health and safety among workers in the automotive repair industry in Dubai (United Arab Emirates): A Cross-sectional exploratory study. *Journal of Ecophysiology and Occupational Health, 19*(3 & 4), 126. https://doi.org/10.18311/jeoh/2019/23357

Kamundala, G. (2017). *Exploitation minière Industrielle et artisanale au Sud Kivu: Possibilités d'une cohabitation pacifique? Etude de base sur la situation socio-économique des ménages vivant dans et autour des sites miniers du Sud-Kivu View project Projet Or: formalisation, 1,* 1–22. https://ipisresearch.be/wp-content/uploads/2012/12/20130314_GKamundala_SKivu.pdf. Accessed 6 August 2023.

Kiran, S. (2021). Occupational health could be the new normal challenge in the trade and health cycle: Keywords analysis between 1990 and 2020. *Safety and Health at Work, 12*(2), 272–276. https://doi.org/10.1016/j.shaw.2020.11.003

Koja-Odongo, R. & Mostert, R. (2013). Information seeking behaviour: A conceptual framework. *South African Journal of Libraries and Information Science, 72*(3), 145–158. https://doi.org/10.7553/72-3-1112

Kouame, K. J. A., Feng, Y., Jiang, F. & Zhu, S. (2015). Evasion of children in Ivory Coast artisanal mining activities. *Journal of Sustainable Development, 8*(9), 24. https://doi.org/10.5539/jsd.v8n9p24

Landrigan, P., Bose-O'Reilly, S., Elbel, J., Nordberg, G., Lucchini, R., Bartrem, C., Grandjean, P., Mergler, D., Moyo, D., & Nemery, B. (2022). Reducing disease and death from Artisanal and Small-Scale Mining (ASM)-the urgent need for responsible mining in the context of growing global demand for minerals and metals for climate change mitigation. *Environmental health, 21*(1), 78.

Michel-Kabamba, N., Ngatu, N. R., Leon-Kabamba, N., Katumbo-Mukemo, A., Mukuku, O., Ngoy-iMukonkole, J., Ngoie-Mwamba, G., Kilolo-Ngoie, E., Bwana-Kangulu, I., Kafusthi-Mukemo, D., Banza-Ndala, D. B., Kabila-Mutombo, D., Balela-Kabasu, M. C., Kanyiki-Katala, M., Syed-Mahfuz, A. H., Murakami, A., Kanda, K., Mashima, Y., Oscar-Luboya, N., & Hirao, T. (2021). Occupational COVID-19 prevention among Congolese healthcare workers: Knowledge, practices, PPE compliance, and safety imperatives. *Tropical Medicine and Infectious Disease, 6*(1), 1–11. https://doi.org/10.3390/tropicalmed6010006

Murray, C. & Newby, H. (2012). Data resource profile: United Nations Children's Fund (UNICEF). *International Journal of Epidemiology, 41*(6), 1595–1601. https://doi.org/10.1093/ije/dys185

Organisation International Labour. (2018). *Key principles underlying sanctioning procedures applicable to occupational safety and health violations* 4 (26) 227–249. https://www.ilo.org/wcmsp5/groups/public/---ed_dialogue/---lab_admin/documents/publication/wcms_797461.pdf. https://doi.org/10.1007/978-981-105206-4_26. Accessed 6 August 2023.

Putzel, J., Lindemann, S. & Schouten, C. (2008). *Drivers of change in the Democratic Republic of Congo: The Rise and Decline of The State and Challenges for Reconstruction Crisis States Working Papers Series No.2 Drivers of Change. 1797*(26), 1–15.

Ruppert, D. (2004). The elements of statistical learning: Data mining, inference, and prediction. *Journal of the American Statistical Association, 99*(466), 567. https://doi.org/10.1198/jasa.2004.s339

Shulla, K., Voigt, B.-F., Cibian, S., Scandone, G., Martinez, E., Nelkovski, F. & Salehi, P. (2021). Effects of COVID-19 on the Sustainable Development Goals (SDGs). *Discover Sustainability, 2*(1), 15. https://doi.org/10.1007/s43621-021-00026-x

Singo, J., Isunju, J. B., Moyo, D., Bose-O'Reilly, S., Steckling-Muschack, N. & Mamuse, A. (2022a). Accidents, injuries, and safety among artisanal and small-scale gold miners in Zimbabwe. *International Journal of Environmental Research and Public Health, 19*(14), 2–21. https://doi.org/10.3390/ijerph19148663

Singo, J., Isunju, J. B., Moyo, D., Steckling-Muschack, N., Bose-O'Reilly, S. & Mamuse, A. (2022b). Hazards and Control Measures among Artisanal and Small-Scale Gold Miners in Zimbabwe. *Annals of Global Health, 88*(1), 21. https://doi.org/10.5334/aogh.3621

Šolc, M., Blaško, P., Girmanová, L. & Kliment, J. (2022). The development trend of the occupational health and safety in the context of ISO 45001:2018. *Standards, 2*(3), 294–305. https://doi.org/10.3390/standards2030021

Sovacool, B. K. (2021). When subterranean slavery supports sustainability transitions? Power, patriarchy, and child labor in artisanal Congolese cobalt mining. *The Extractive Industries and Society, 8*(1), 271–293. https://doi.org/10.1016/j.exis.2020.11.018

Sustainable Development Goals HIGHLIGHTS. In *Newark DE: University of Delaware (Minerals, Materials and Society program in partnership with Pact).* Newark DE: University of Delaware (Minerals, Materials and Society program in partnership with Pact).

Tegnan, H., Karjoko, L., Barkhuizen, J. & Bajrektarevic, A. H. (2021). Mining corruption and environmental degradation in Indonesia: Critical legal issues. *BESTUUR, 9*(2), 90. https://doi.org/10.20961/bestuur.v9i2.55219

Tetzlaff, E.J., Goggins, K.A., Pegoraro, A.L., Dorman, S.C., Pakalnis, V. & Eger, T.R. (2021). Safety culture: A retrospective analysis of Occupational Health and Safety Mining Reports. *Safety and Health at Work, 12*(2), 201–208. https://doi.org/10.1016/j.shaw.2020.12.001

Tsurukawa, N., Prakash, S. & Manhart, A. (2011). *Social impacts of artisanal cobalt mining in Katanga, Democratic Republic of Congo Social impacts of artisanal cobalt mining in Katanga, Democratic Republic of Congo III*: Öko-Institut e.V, Freiburg.

Tziaferi, S. G., Sourtzi, P., Kalokairinou, A., Sgourou, E., Koumoulas, E. & Velonakis, E. (2011). Risk assessment of physical hazards in Greek hospitals combining staff's perception, experts' evaluation and objective measurements. *Safety and Health at Work, 2*(3), 260–272. https://doi.org/10.5491/SHAW.2011.2.3.260

Williams, C., & Hunt, P. (2017). Neglecting human rights: Accountability, data and Sustainable Development Goal 3. *The International Journal of Human Rights, 21*(8), 1114–1143. https://doi.org/10.1080/13642987.2017.1348706

Wireko-Gyebi, R.S., Arhin, A.A., Braimah, I., King, R.S. & Lykke, A.M. (2022). Working in a risky environment: Coping and risk handling strategies among small-scale miners in Ghana. *Safety and Health at Work, 13*(2), 163–169. https://doi.org/10.1016/j.shaw.2022.02.004

World Bank. (2017). Decent work and economic growth: Promote sustained, inclusive, and sustainable economic growth, full and productive employment, and decent work for all. In *Atlas of Sustainable Development Goals 2017: From World Development Indicators* (pp. 44–49). The World Bank. https://doi.org/10.1596/978-1-4648-1080-0_ch8

Yi, G., Jeong, J., Bae, Y., Shin, J., Ma, H., Lee, N., Park, S.-H., & Park, D. (2021). Workers' exposure to Indium compounds at the electronics industry in Republic of Korea. *Safety and Health at Work*, *12*(2), 238–243. https://doi.org/10.1016/j.shaw.2020.11.002

Zin, S. M. & Ismail, F. (2012). Employers' behavioural safety compliance factors toward occupational, safety and health improvement in the construction industry. *Procedia - Social and Behavioral Sciences*, *36*(June 2011), 742–751. https://doi.org/10.1016/j.sbspro.2012.03.081

Part IV
Justice and Peacebuilding

Chapter 19
Russia-Ukraine War: Impacts on Energy Market and Food Security in Africa

Sunday Israel Oyebamiji, Muesiri Obero Ashe, and K. Jean Chrysostome Kiyala

Abstract The supremacy of Russia in Eastern Europe, including its hegemony in the Warsaw Pact until the collapse of socialism in the 1990s created a national pride that the country is determined to sustain at any cost. The conflict between this pride and NATO's expansionary agenda has been a key factor in the ongoing war in Ukraine. Apart from the loss of lives, the war is affecting a host of strategic issues, including food security, energy supply and sustainable development, both regionally and globally. A scholarly study of its impact can, therefore, serve as one means of reimagining new geopolitics and geostrategy for the realization of security, economic and political goals. This chapter is a study along this line. It challenges the original notion of democratic peace in trying to define the true impact of the war as it identifies some subjectivity in the original notion. As a way forward, it recommends the creation of a platform for a global consciousness to seek and foster regional and international stability based on the UN alarm 'War No More', to privilege peaceful coexistence, enhance sustainable development, fight climate change, eradicate hunger and poverty, and make energy available, particularly in Africa. Secondly, the chapter recommends that the UN can improve on its Charter for the equality of all nations to prevent inter-nations wars that often end in human tragedies and ecological catastrophes.

Dr Sunday Israel Oyebamiji is an expert in history and strategic studies, international relations and gender XE "Gender studies. Specifically, on women migration, climate, and environmental studies. University of KwaZulu-Natal, South Africa XE "South Africa" Email: Oyebamijis@ukzn.ac.za, ORCID: https://orcid.org/0000-0001-5949-6563

Dr Ashe, Muesiri Obero, PhD. University of KwaZulu-Natal, South Africa. Emails: muesiri@yahoo.com, Ashem@ukzn.ac.za, ORCID: https://orcid.org/0000-0001-9314-2101

Dr Jean Chrysostome K. Kiyala holds a PhD in Public Administration: Peace Studies from DUT; he is a Senior Lecturer at the International Centre of Nonviolence (DUT), Associate Professor and visiting lecturer at the Université de Bandundu and the Université Evangélique en Afrique. (Democratic Republic of Congo); Emails: JeanK@dut.ac.za; kljeanchrysostome@gmail.com; ORCID: https://orcid.org/0000-0002-9587-0230

J. C. K. Kiyala and N. Chivasa (eds.), *Climate Change and Socio-political Violence in Sub-Saharan Africa in the Anthropocene*, The Anthropocene: Politik—Economics—Society—Science 37,
https://doi.org/10.1007/978-3-031-48375-2_19

Keywords Energy market · Food security · Democratic peace · NATO's expansionary agenda · Russia's war in Ukraine · UN Security XE "Security Council

19.1 Introduction

The impact of the Russian-Ukraine war has taken different dimensions. Yet an empirical understanding of a number of themes in this regard is still vague because most of the issues in such themes are yet to receive scholarly analysis. An aspect of this is the effect of the conflict on the energy and food security sectors in Africa. Some studies have shown that most sectors of the food industry in Africa depend on global food supply chains, while international agencies are now involved in projects designed to overcome food insecurity in many countries of the continent (Ashe, 2019; Burney, 2018). At the same time, existing data have revealed the depth of the adverse effect of the war on the same sector in the European Union (EU) (Korosteleva, 2022). In this context, it is empirical to postulate that the energy and food sectors in Africa cannot be completely shielded from the crises emanating from the ongoing War. This chapter is meant to address the reality of the situation. For instance, Europe is said to be on the lookout for alternative sources of energy supply as a result of the war (Daggash, 2023). This alone is bound to affect the energy sector in Africa, either positively or negatively. Therefore, the focus of the chapter is to examine the impact of the war on the energy sector and food security in Africa, partly within the framework and intricacies surrounding its global impact.

The analysis commenced with a brief discussion of the historical trends in the international system that led to the rivalry and the subsequent outbreak of war. This is followed by some explanation of different stages in the rivalry as well as the role of the North Atlantic Treaty Organisation (NATO), partly as an intermediary and partly as a rival to Russia and its allies. This provides a background for the conceptual framework required to examine the impact of the war. The key analysis is centred on the exploration of its impact on energy supply and food security in Africa as an aspect of the threat to global security and within the platform of how the rethinking of regional geopolitics and geostrategies in Europe is likely to affect energy and food security in Africa. These, therefore, define the overall analytical focus in the search for an objective conclusion, as each has played a valuable role in sustaining the trend that leads to the conclusion.

19.2 Prelude to the Rivalry

The Russian-Ukraine war has its underlying origins in the global demonstration of military superiority between two rival military camps, which was believed to have ended with the collapse of the Soviet economy in the 1990s. The first of these rival camps is NATO, with headquarters in Brussels and Britain and the USA as the key pivoting powers. The second is the defunct Warsaw Military Organisation (Warsaw Pact), with headquarters in Warsaw and the Union of Soviet Socialist Republics (USSR), now Russia, as the key sponsor. It all started with the emergence of the USSR as a world power after the Bolshevik Revolution of 1917, resulting in the rapid spread of the socialist economic system, which became a serious threat to Western capitalist powers by the end of World War II in 1949. The capitalist powers conceived socialist ideology as a threat to the existence of capitalism and formed NATO in 1949 to counter the influence of the USSR and its allies. The USSR responded by forming the Warsaw Pact in 1955. NATO is presently marking out fresh strategies to increase its member states, but the Warsaw Pact was dissolved in July 1991 following the collapse of the Soviet economy (Shvangiradze, 2022). The major impact of the operation of the two rival camps is the creation of an atmosphere prone to tension in the international system.

The fact is that the rise of tension in Eastern Europe since the collapse of the Soviet economy, culminating in the ongoing Russian-Ukraine war, is at least an indirect outcome of the rapid, aggressive expansion of NATO since 1997. It is significant to note that NATO has incorporated fourteen more countries into its fold since 1997, increasing its member states to thirty-one, and that all of the fourteen new states except North Macedonia were formerly under the umbrella of the Warsaw Pact. More significant is that even Poland, the former headquarters of the Warsaw Pact, is among these fourteen states, as shown in the map below (HHS, 2022; Ishika, 2022).

Whatever the viewpoints of scholars and analysts on international relations, one thing stands out clearly in this context: NATO and the Western powers have actually exploited the collapse of the Russian economy to expand the horizons of their influence to the fullest. Therefore, this has constituted a serious threat to Russia and increased the rate of tension in Russia's relationship with the neighbouring states formerly under its influence. Therefore, under perpetual threat, Russia was compelled to annex Crimea in 2014, apparently in a desperate bid to, at least, do something to redeem its dwindling image, which is now being exploited by its rivals to erode every remnant of its influence and take all that belongs to it. From the map below, it would appear that Russia is pleading with NATO and its Western allies to leave only these two states (i.e., Ukraine and Belarus) for it having taken all others (Shvangiradze, 2022; HHS, 2022). Therefore, Russia's reaction to Ukraine's engagement with NATO is natural in this context.

There is a need to point out that humanity has come a long way with new ideas in liberalism and inter-state relations, from the sub-regional and continental levels to the global level. As such, modern international relations should be able to operate above the apparent crude pattern of pre-World War II, where rivals are seen as enemies instead of opponents. The rival you perceive as an enemy will surely reciprocate by taking you as an enemy. Therefore, the pre-World War II culture of the winner takes all, and the loser loses all should by now be replaced with a structure where rivals in the international system see themselves as opponents rather than enemies, and even when there is a winner, the loser is not humiliated. Therefore, even though the Cold War ended in the 1990s, the current Russia-Ukraine War is largely an outcome of the entanglements created in international politics by the instruments employed in the pursuit of Cold War politics (Fig. 19.1).

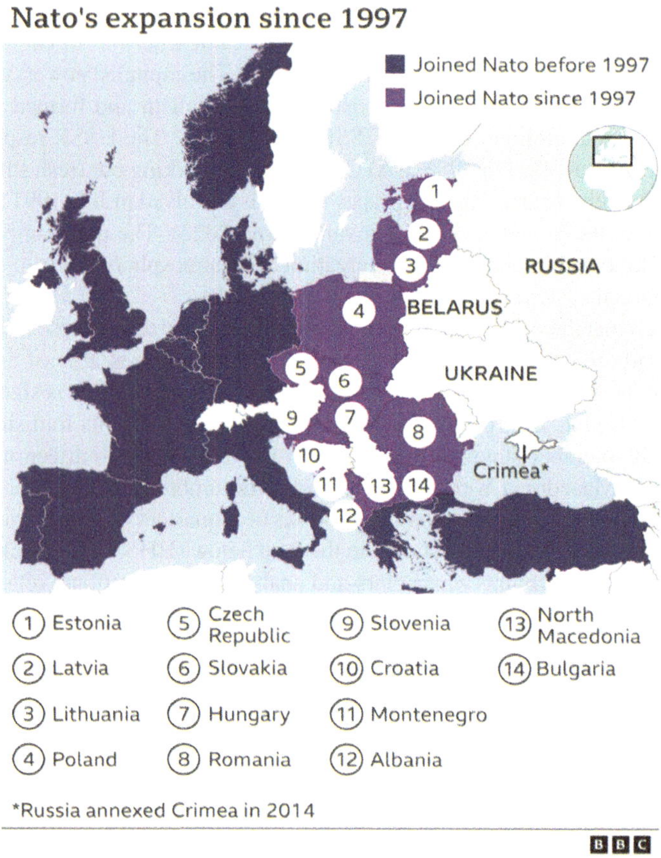

Fig. 19.1 NATO's expansion cartography. *Source* Analysis of the Russian-Ukraine conflict, Compiled by U.S. Department of Health and Human Services, Mar 2022. https://www.hhs.gov >files

19.3 The Rivalry

Overall, Russia is very uncomfortable with the presence of NATO and the USA along its border. At the same time, one of the only two remaining states along its border, Ukraine, was just warming up to come under the umbrella of NATO. When the situation became apparent, Russia, in response, began to search for excuses. One of the pretexts was that Ukraine had lost its sovereignty to NATO's expansionist policies, which created a platform exploited by the USA to engage Ukraine as a "battering ram" against Russia and destabilise the peace of the sub-region. In the context of national sovereignty and accepted principles in international politics, the veracity of both claims is highly questionable. President Vladimir Putin added further that Russia was responding to alleviate the plight of separatist groups in Ukraine and other parts of the sub-region, which have appealed to it for assistance against the militarisation and nazification of the area. This is, of course, another controversial claim. He indeed makes it clear that NATO's operations backed up by US military power in the area are structured to antagonise Russia and its supporters like "the Russian-Belarusian Union State, the Collective Security Treaty Organization and the Commonwealth of Independent States (CIS)" (Leigh, 2022: 5).

One of the key Russian political figures, Vyacheslav Volodin, holding the office of speaker in the legislature, even declared categorically that Ukraine had been annexed by NATO and "has become a colony of the US" (Leigh, 2022: 5). It has been observed that these reactions demonstrate that Russia is determined to curb the rapid advancement of NATO into its area of influence at almost any cost (Leigh, 2022). This and related developments also indicate that, apart from the territories already under Russia's control, it will not accept a cease-fire and peace negotiation until there is a structure in Ukraine, such as a change of government, which enables it to have control over the affairs of that state (Leigh, 2022).

These are reiterated by NATO and its Western powers, while Russia recites the rapid advancement of NATO since 1997, encroaching into territories which were formerly under Russia's influence. NATO and its allies labelled Russia's military action against Ukraine as aggressive, portraying it as "the biggest security threat in a generation" (CSR, 2022: 3). But Russia is of the view that it is only striving to retain a measure of its influence amidst the rapidly unilateral expansion of NATO's horizon now being exploited by the Western powers, particularly the USA, to threaten Russia's own existence. Sanctions against Russia have been imposed individually by different countries, not under the umbrella of NATO. But Russia believes in the existence of an underground interrelationship through NATO that creates a common platform for those imposing sanctions on its economy. It is significant to note that the expansion of NATO since 1997 and the role of the USA and Britain in the entire scene have created in Russia the impression that NATO is set to erode every remnant of its influence, even in Eastern Europe, and to create allies along its borders that could be used to threaten and intimidate it, while the USA and its Western allies dictate the pace of development in global affairs (CSR, 2022).

For instance, Russia's annexation of Crimea and the subsequent military campaign in Ukraine commenced with the emergence of pro-Western leadership in the Ukrainian government. Military analysts are of the view that the ultimate initial plan of the Russian army was to capture the capital city of Ukraine, Kyiv, within the first two months (Barros et al., 2022). But this plan failed, and as the war dragged on, new tactics and weapons were introduced. Among these was the use of Iranian drones by the Russian forces. Meanwhile, some member states of the EU and the USA were compelled to provide Ukraine with some assistance. This includes both food and arms, as well as ammunition. The humanitarian support from some Western countries, the USA and some member states of the EU has been of significant help to both the government and the masses in Ukraine. Politically, the EU has shown significant commitment in terms of support for the Ukrainian government through advice and sensitisation on the right policy focus. The EU commitment to military aid has served as one means of reducing the plight of Ukraine, but there are a number of EU member states that have not been fully committed to this and can hardly be blamed as they also have their own domestic problems in one way or another. The same is true in the aspect of financial aid to Ukraine. However, these are issues that may require critical assessment. This is because justice and equity in international relations accommodate various indices whose analytic relationships are often intricate and complex when placed on an empirical platform for the purpose of objectivity. This support for Ukraine by the Western powers including the USA and other NATO member states, is also employed by Russia to buttress its claim of the creation of satellite states along its border by NATO to erode its power and global influence.

19.4 Conceptual Framework

Studies on the history of conflicts reveal that conflict is probably as old as the existence of mankind and, that considering the level of incompatible interests and approaches to issues surrounding the sharing of resources, it is unrealistic to think of a society that is completely free from conflict. This is more so in a world where resources are scarce in relation to the needs of a growing population. This alone has led to the development of different and sometimes conflicting approaches to wealth creation, which is also reflected in the methods and systems of distribution and management of resources. For instance, until recently, NATO conceived the capitalist system and ideology as the best way of creating wealth and managing resources, while its rival Warsaw Pact, sponsored by the USSR, now Russia, subscribed to the socialist economy and ideology for the same purpose. This was a key source of conflict, as both often clashed in different regions as they tried to export their ideologies to the rest of the earth. Such conflicts were recorded in the histories of East and West Germany, South and North Korea, and, to some extent, Africa (Sklar, 2008). Even when an issue like national pride plays a role in warfare, resources are still the primary consideration in assessing its impact. This is, of course, also well reflected in the present Russian-Ukraine war, which is why its impact on resources concerning ecology, energy, food security, and sustainable development deserves significant attention.

19.4.1 Democratic Peace

The democratic peace theory (DPT) claims that democracy promotes the emergence of peaceful societies. This is based on the premise that a democratic state cannot declare war without the approval of its legislature and its citizens. The concept of peace may attract some controversy when analysed from the opposing platforms of democracy and dictatorship. It is, therefore, not surprising that the theory has attracted some significant criticism (Longley, 2022). One problem is the Western perception of DPT on the platform of equating democracy with capitalism and dictatorship with socialism. Nevertheless, the DPT is apparently gaining wide acceptance in most regions of the world, including Europe and North America. Parmar (2013: 251) contends that DPT is easy to promote in the USA because, apart from its "great symbolic resonance" and reaffirmation of "positive ideas about American national identity… it also serves as an excellent rationalising device for establishment forces that wish to promote the consolidation of American power using the cover of promoting democracy and eliminating brutal dictatorships." This clearly implies that the national interests and pride of the pivoting powers are, in one way or another, integrated into the democratic package structured for a globalised peace system. This introduces some complexities about the implementation of democracy as a political system. It also tends to imply that what is obtained at any point in history and in a particular region is only a brand.

The traditional concept of democracy emphasised freedom of speech and expression as its most essential characteristics. This is expected to ensure majority support and is well incorporated into the capitalist democracy adopted by NATO, which is being exported to different regions of the world. But the socialist bloc identified a gap, which it exploited to dominate many parts of the world until the collapse of the Soviet economy in the 1990s. It is simply the fact that historical experiences since World War II demonstrates that for the masses, which constitute over 80% of the population in most regions of the world, particularly the less developed countries, access to the basic necessities of life constitutes an essential feature of democracy. In any case, most of these people are hardly literate and enlightened enough to imbibe the ideas of freedom of speech and expression, more so when they completely lack the economic power to sustain them when legally implicated. The practical aspect of it is that survival comes first. Such individuals do not have access to the media or other means that enable their voices to be heard. This is what the Soviet Union exploited to expand the horizons of socialism like wildfire, as the socialist ideology created a structure that accommodates this feature of democracy. It is significant to note that the decline of socialism was not in any way connected to the perceived dictatorship advocated by its rival camp but rather to the collapse of the Soviet economy. This implies that whatever the perception of the economic and political system of the socialist bloc, it has its own brand of democracy, which, though criticised by its rival camp, appeals to the masses of many regions of the world in the presence of a sponsor and pivoting power to co-ordinate, mobilise, and provide the necessary economic aid to member states, just like the operating structure in its rival capitalist camp.

Therefore, the DPT is bound to attract criticism until these conceptual contradictions are well resolved. This implies that the theory should operate beyond the platform of equating democracy with capitalism. These are issues to be considered right now in resolving the current warfare because, on the surface, it is a conflict between Russia and Ukraine, yet its origin has much to do with the rivalry between the two camps. In other words, the ongoing democratic peace sponsored by the USA and other powers of NATO must identify and fill this practical gap that sustained the socialist bloc. In practice, access to the basic necessities of life now stands as the first principle of democracy to stabilise the masses and sustain majority support because it is the primary means of sustaining life. This is, in fact, why resources come first in assessing the impact of war, and it is in light of this that the present study examines the impact of the war on resources, including markets, energy and food security in Africa, from the perspective of the democratic peace theory.

19.4.2 Challenges of the UN Charter and the Security Council

The United Nations Security Council faces several challenges in handling the demands of the belligerents, particularly Russia, and in managing the complications emanating from the war (Hovell, 2022). In the first place, the resolution of the UN Security Council was antagonised by Russia, using its position as one of the five states with veto power in the Security Council. Thus, the Council had to hold another meeting in respect of Russia's reaction to its decision, which declared the attempt to annex Ukraine or parts of its territory as a breach of the UN Charter and "a threat to international peace and security" (UN News, 2022: 1). But Russia rejects the majority view about its occupation of Ukraine, claiming to have held a referendum in the four territories of Ukraine under its control to determine their stand on self-determination. They include Donetsk, Kherson, Luhansk and Zaporizhzhya. In its response, the UN Security Council "described the so- called referendums held by Russia in the four regions of Ukraine, which Moscow now regards as sovereign territory…as illegal and an attempt to modify Ukraine's internationally recognised borders" (UN News, 2022: 3). An earlier draft resolution rejecting the unlawful annexation of the aforementioned four territories in Ukraine enjoyed the support of precisely 10 out of 15 members of the Security Council. Four members, Brazil, China, Gabon and India, preferred to remain neutral, while Russia voted against it. This was followed by a UN General Assembly meeting in April 2022 to deliberate on Russia's position on the issue, with particular reference to the veto power advantage and its general implications for addressing the present situation. As stated earlier, Russia's attempt to annex Ukraine is largely a response to the rapidly expanding horizons of NATO since 1991, following the collapse of the Soviet economy, while the pivoting powers of NATO and the capitalist bloc also enjoyed this veto power that was being exploited by Russia to protect its interests.

Therefore, from an objective point of view, while Russia is generally aggressive, there are lapses arising from the interests of individual states and regions. The historical trend indicates that there are underlying factors in the origin of the ongoing conflicts, which on their own constitute a challenge to the UN Security Council as some of these give credence to Russia's claim and, therefore, introduce more intricacies into the entire scene. Overall, the real danger of the crisis is the likelihood of compounding the management of global peace and security. This is because NATO is fortifying its camp, just as Russia and its allies are apparently not ready to grant any concessions. It would appear that the UN has done its best within the strength and limitations of its present charter. But if its best is considered insufficient for the ongoing scenario, then it is expected to consider the restructuring of its charter, including identification of any lapses with the veto power clause, to align it with the changing trend in global security outlook.

19.5 Impacts of war

The impact of the war is now a key issue attracting the attention of the UN, regional organisations and international agencies. In a globalised system with clear evidence of interrelationships between the forces at play, a discussion of its impact on any aspect of the African economy and social system requires some understanding of the trend in Europe, the host of the conflict. As explained below, in trying to define this, the geostrategy and geopolitics of the EU is among the primary focus areas. This itself requires some understanding of the extent to which Ukraine is able to align its policies with the diplomatic trend of the EU vis-à-vis the threat the war poses to the global multilateral system.

Since the military campaign of February 2022, the existing figures from the war front indicate that "an estimated 4900 civilians have died in Ukraine, some 6000 have been injured, and more than 5.6 million have left for countries in Europe, creating the region's biggest refugee crisis in a generation" (Yuhas, 2022: 4). The figures provided by a physicist of Ukrainian origin, George Gamota, currently based in the United States also shed more light on this. According to the source, "another 6.3 million people are internally displaced, and among those affected are the country's researchers" (Yuhas, 2022: 4). About one-quarter of these were compelled to flee the territory (Yuhas, 2022; Oyebamiji et al., 2022). The chart below provides details of the number of lives lost and those injured in Ukraine since 24th February 2022 when the war started, until Jan 2023 (Fig. 19.2).

But the Russian side has also lost many lives. Official data from Russia stated that Russian troops had lost 1353 of their members as of 08 August 2022. But the USA

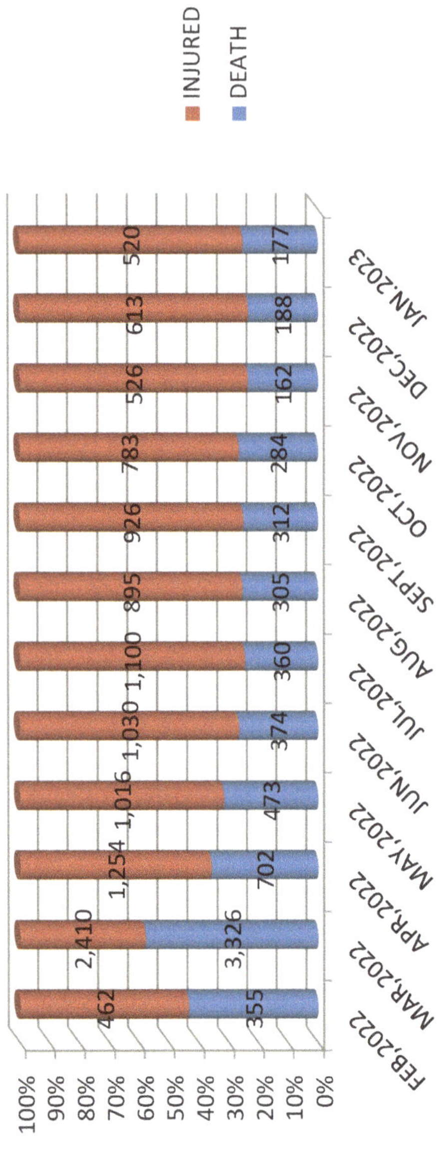

Fig. 19.2 Injuries and death toll statistics of Russia-Ukraine war. *Source* Drawn from figures compiled by Statista Research Department and United Nations High Commissioner for Human Rights (OHCHR), Feb. 6, 2023

official estimate placed this at 5000. The source states that as of 08 August 2022, Russia had lost a total of 25,000 lives, including troops at the war front (Yuhas, 2022). The war is said to have done much harm to Ukraine. For instance, it has halted the evolution of modern research structures in the country largely through the demolition of various structures and facilities already put in place for this purpose and partly through the exodus of experts in different careers. Ukrainian institutions of learning are also affected. Among these tertiary institutions are universities and specialised institutes like the Kharkiv Institute of Physics and Technology. One of its research units was also devastated by the war as the Russian forces introduced incessant bombing of different places, much of which was said to be indiscriminate (Yuhas, 2022; Gaind et al., 2022).

19.5.1 Geopolitics and Geostrategy of the European Union

The war has had a significant impact on the geopolitics and geostrategy of the EU, as well as the policy focus and strategic approach adopted by some other regions in the world, with some level of interrelationship among the forces at play. Therefore, an understanding of the European scene in this context provides a valuable means for a precise analysis of the impact of the war on Africa. This is because the regional policy focus and strategy of the EU in trying to cope with the prosecution, politics and crises of the war have been largely instrumental in the new development in the energy sector that has had an adverse effect on Africa, but with some elements that may be exploited by African countries to enhance their economies.

At the continental level, it would appear that the impact of the war has, in certain aspects, taken a number of member states of the EU by surprise, as it is argued that "the Ukraine war... is an exogenous shock, which, like the pandemic, accentuates divergences among EU countries" (Barros et al., 2022: 6). This is related to the issue of sanctions against Russia vis-à-vis Russia's economic policies and strategic plans to counter the sanctions. For instance, the energy sectors of many member states of the EU were in a precarious condition following the unanimous decision to look for alternative sources of energy supply and, thus, do away with Russia as a source of energy supply for Europe. New resolutions of member states of the Organisation of Petroleum Exporting Countries (OPEC) to keep production low as a means of controlling the global price of crude oil have, in one way or another, compounded the EU's effort of realising its objective. It is observed that inflation was a major problem in the economies of EU's member states before the war. This has been a key impediment to the success of economic policies and strategic efforts put in place by different members of the EU to revive their economies following the gradual demise of COVID-19. Trying to combine this with sanctions against Russia is already having adverse effects on different sectors of their economies. Moreover, global trade had not been operating smoothly for some years before the war, as it had been interrupted by different unforeseen occurrences. The COVID- 19 pandemic is among the most recent ones. But even before the pandemic, commercial activities at global and even regional levels had experienced occasional disruptions for one reason or another (Barros et al., 2022).

Amidst the insufficient energy supply to meet the demand of citizens at home, Europe is receiving refugees from the war, which further compounds the scarcity of essential products, including energy, at home. Public authorities support hosting refugees from Ukraine, which has played a noticeable role in creating an enabling environment to protect the social and legal rights of the refugees. As a result of the positive response of constituted authorities, civil societies have taken it upon themselves to protect refugees, while issues of hospitality are receiving due attention from a host of bodies from the national to the grassroots level. The funding has enjoyed the support of the EU, and it is mobilising funds whose utilisation can filter from the national level down to the local level. This is done in relation to the management of refugees "admitted under the EU's temporary protection directive" (Gabriel, 2022: 3). This is only a short-term device to reduce tension created by the emerging need to define a structure for the reception and protection of refugees. As stated above, it is apparently too early to examine the impact of this because the available data indicate that refugees will continue to arrive. This invariably implies additional demand on basic public utilities, ranging from health services and shelter to transport and education.

It has been observed that, in the long run, the arrival of refugees and its attendant challenges in relation to the inadequacy of resources would test the potency of cooperation and unity among members of the EU. At the moment, some states have received more than others, and this lack of uniformity in the distribution of refugees has not attracted any significant negative reaction from those that host most of them, and whatever controversy arises from this is said to be negligible compared to the experience in previous situations like the 2015 Syrian crisis and its refugee scene. However, it is argued that as the number increases, increasing prices of goods, and shortages of essential consumables, including food items, coupled with election issues and the burden on basic social amenities, can result in a change of attitude in some of the states.

The issues discussed above are among those to consider in determining the extent of the role the war is likely to play in shaping the future of the European social system, politics and economy, as well as the nature of the approach to international affairs, both at the regional and inter-state levels. It has been observed that the overall focus, aspirations and expectations of the EU as a regional body from all member states are "continued unity, open-handedness, willingness to accept sacrifices and defence of fundamental values" (Leigh, 2022: 5). But will these be fully realised? For instance, Germany has decided to do away with the Russian energy supply, notably hydrocarbons, in addition to a total restructuring of its energy sector with the hope of replacing nuclear energy with another alternative. This has implications for the entire region, as the country may also have to compete for available energy supplies from other regions in Europe. Therefore, apart from prioritising energy security, the EU will need to look once more at the drawing board to develop more practical strategies and mechanisms for satisfying energy demand in Europe without partial dependence on Russia.

It is probably too early to predict the future outcome of the war in relation to the development of new trends in the energy sector, given Russia's war policy and its probable post-war energy policy and Ukraine's war effort vis-à-vis its interest in the EU and NATO. This is because the actors are social beings (i.e., human beings) whose actions are difficult to predict. Any expected future outcome is based on existing facts and figures derived from the actions of the belligerents. But in an international system where states are only propelled by permanent interests, even at the expense of allies, these actions are subject to change as new interests and intricacies develop. In trying to assess the likely outcome of the war, a number of questions emerge:

1. What is Russia's strategy and projection for the future?
2. Is Russia likely to have its way at the end of the day, or is there likely to be a compromise?
3. What are the plans and strategies of the EU and NATO for the future of the war, and what are their expectations?
4. What do NATO and the EU have at stake in the conflict? Are they merely interested in global peace, or is the war another revival of the traditional global rivalry between NATO and the Eastern bloc previously under the Warsaw Pact?

As the answers to these questions unfold, one will have a better understanding of the role Africa can play in the energy sector and, therefore, the depth of anticipated opportunity open to the continent to exploit in the sector. As already explained above, Russia's foreign policy and participation in international affairs are anchored on the principle of classifying all the states along its immediate borders as territories under her influence and are therefore very uncomfortable with the closeness of any, including Ukraine, to the West. It is significant to point out that the scene is compounded by the aspirations of some of these countries to join NATO or the European Union. This is largely due to the rapid expansion of NATO since 1997, bringing the majority of member states of the former Warsaw Pact under its umbrella and even including Poland, which once occupied a strategic position in the military intelligence of the Warsaw Pact (HHS, 2022). Some Russian political figures view Ukrainian sovereignty as illegitimate or as a relatively recent invention. President Putin said that the principle of self-determination has to be adhered to by all stakeholders and justified his claim by citing territories that demand separation from Ukraine. He contends that these separatist territories seek the assistance of Russia, adding that "his goal was to protect people subjected to bullying and genocide" and that the overall aim is "demilitarisation and de-Nazification of Ukraine" (HHS, 2022: 3). NATO rejects Russia's claim in totality, describing the war on Ukraine as "the biggest security threat in a generation" (Barros et al., 2022: 6).

Moreover, analysts are of the view that international collaborations in science ultimately tend to follow geopolitical alignments and are, therefore, among the indices for consideration. So a long-lasting Western diplomatic split with Russia is likely to be reflected in the pattern of regional and global cooperation in education, research and general scholarship. This also has its own dimension in the energy sector. As Europe works out new modalities for its research and education at both the national and regional levels, Russia will also seek new partners in the global arena. It has been observed that Russia is likely to support the creation of new platforms that can enhance its collaboration with a number of Asian states, particularly China and India. It is, however, argued that this may not take the expected dimension, as China is not likely to give it significant support until it is sure of how much benefit it can derive from it. Recent studies on the geopolitics of global science are looking at the entire scene from different perspectives, though with some element of controversy. In one of the studies, researchers from the Harvard Kennedy School in Cambridge, Massachusetts, and some other institutions are of the view "that China's leadership would gain more from maximising global scientific collaboration than it would from risking damage to its Western partnerships by entering a bipartite research engagement with Russia, a country with an ailing position in international science" (Gaind/Abbott, 2022: 442).

The war may add some element to the unity of Europe in the future, as it is now claimed that "Russia's invasion of Ukraine brought an uncommon degree of unity to the often-fractious EU, as its 27 countries provided Kyiv with military support, strengthened their own defences, imposed sanctions on Russia, welcomed several million Ukrainian refugees, and took the first step towards possible EU membership for Ukraine" (Gabriel, 2022: 4). The president of the European Commission, Ursula von der Leyen, called the EU's restrictions on Russia "the largest sanctions package in our Union's history" (Leigh, 2022: 4). European countries that were previously reluctant to approve such sanctions against Russia cooperated with ease. Among them were Germany and Italy. They created a structure supported by the USA and the United Kingdom for the coordination and implementation of the sanctions which actually had adverse effects on the exchange rate of the Russian currency.

According to Josep Borrell Fontelles, the 2022 Ukraine war saw "the belated birth of a geopolitical EU," bringing a "leap forward on EU security and defence" (Leigh, 2022: 4). The future impact of all these developments on the politics of the energy sector can only be determined with time.

19.5.2 The Russian-Ukraine War and the Energy Sector in Africa

The Russian-Ukraine war has had major effects on the energy sector in Africa. On the negative side, the war has led to an increase in the price of crude oil globally, soaring above US$100 per barrel for the first time since 2014 (Hatab, 2022a). Though African countries do not import oil from Russia, business ventures operating on the continent have to bear the high cost of energy for the production of goods and services. Similarly, household consumption is adversely affected as the masses also have to buy at the same price. In West Africa, 40% of electricity for industrial and domestic use is derived from diesel-powered generators (Auth/Moss, 2023). 50% of the electricity consumed in Senegal is generated from petroleum. These have serious adverse effects on businesses and the general cost of living at the grassroots level, as their impact often has a multiplier effect. An increase in the price of gas has a direct effect on the cost of electricity generation in countries like Côte d'Ivoire, Ghana and Nigeria, which is passed on to consumers. In the case of Nigeria, business ventures are still finding it difficult to devise means of staying afloat as the country has recorded an increase of over 200% in the price of diesel, recently (SRD, 2023; Addeh, 2022). In addition, Russia is a major investor in African countries, particularly in the development of infrastructure (Daggash, 2023). With the numerous sanctions and counter sanctions involved in the war, projects associated with such investments could be abandoned or their completion delayed, which may have direct or indirect effects on the energy sector (Daggash, 2023).

Beyond this, however, the impact of the war on the energy sector globally may be strategically exploited by African countries to enhance the energy industry on the continent. Among the key issues here is that African oil-producing countries can exploit the situation to increase national revenue by devising means to increase the production of crude oil (Auth/Moss, 2023). But from the existing data the African member states of the Organization of Oil Producing and Exporting Countries (OPEC), including Nigeria, lack the infrastructural and other requirements to increase production, except probably in the long run (Daggash, 2023; Addeh, 2022). The war has altered the geopolitics and geostrategy of the European Union (EU) and other regions of the world that have socio-economic elements that can aid the expansion of energy production by African states and thus enhance their economies. The major issue, as discussed above, is the unanimous decision by EU member states to boycott Russia's energy products.

In light of the scarcity of energy products that accompanies this development, each member state of the EU is adopting different devices to protect the well-being of its populace. This approach has not been uniform, as each country has certain peculiarities in the crises experienced in the energy sector, and therefore, the methods employed and policies adopted have been structured in relation to the nature of the challenges encountered in each country, taking into consideration the peculiar intricacies involved in the effort to diversify the sources of energy supply (Korosteleva, 2022). The EU probably underestimated the gravity of the challenges likely to be created in the energy sector in relation to the adoption of sanctions against Russia as one means of making the latter accept negotiation. For instance, the protest in France over what was considered a little increase in the price of energy was said to have taken the leadership of the EU and other constituted authorities in charge, including the government of France, by surprise. Presently, the effort of many member states of the EU to source funds has compelled them to accept the alternative of securing debts. This arises from the fact that the lapses in the energy sector tend to compound the challenges emanating from the implementation of other economic policies like those meant to address disruption in commerce (Bearak et al., 2022).

It can, therefore, be said that the impact of the war on the economies of EU member states could have been relatively minimal in the absence of the crisis emanating from the energy sector. The available facts and figures demonstrate that it is extremely difficult to find a replacement for the huge quantity of gas imported into Europe from Russia annually. The annual figure for gas inflow into Europe from this source amounts to precisely 160 billion cubic metres. On the financial aspect of the transaction, Russian exporters receive a minimum of 800 million euros from Europe daily, and financial analysts and statisticians are of the view that with this structure, the EU is already financing the war for Russia, which invariably dwarfs the effort of the EU with regard to its support for Ukraine's defence, whether in terms of materials, finance, or the supply of arms and ammunition (Korosteleva, 2022).

Germany started a gas production project some years before the war, and this could have been of some help at this point in time. But it is yet to be completed, even though the project is only meant to produce liquefied natural gas (LNG), which is by far more expensive compared to pipeline gas. LNG is not yet readily available for consumers. Interconnector network devices between EU member states for gas transportation are still far from being completed. As stated above, all countries have been on the lookout for alternative sources of energy supply. The USA has also stepped into the situation with President Joe Biden promising to increase the supply of LNG to Europe. But it was found that extra costs would also be incurred to get more of the product across to Europe, and the producers were not likely to respond to the demand until an increase in price to cover the additional costs is effected (Korosteleva, 2022).

In the search for future source of energy, Africa major producers of gas are among the key alternatives the EU has in mind. These include Algeria, Angola and Nigeria (Ndigwe, 2023; Bearak et al., 2022). At the same time, the European Union's regional leaders have pledged to give financial support for the exploitation of recent gas discoveries in some African countries, namely Equatorial Guinea, Mozambique and Senegal (Abiodun, 2023). However, the EU has the intention of investing in the development of renewable energy with a proposed investment package of €210 billion. But it has been observed that "it will take years for the EU to realise these targets, so it still needs fossil fuels in the interim" (Daggash, 2023: 4). This implies that there are gaps that African countries can fill to expand energy production and invariably create more sources of income that can enhance national development.

Basically, African countries involved in the gas production industry, particularly Algeria, Libya and Nigeria, can help increase gas supply to Europe. It is estimated that these countries have a capacity of up to 65 billion cubic metres (bcm) of gas for export in the absence of certain limiting factors arising from previous export engagements, in addition to demand for domestic consumption (Ndigwe, 2023). Their effective response would, therefore, be of some help to the EU's search for new sources of energy to make up for the 155bcm of gas shortage arising from the sanctions and counter sanctions between the EU and Russia in the prosecution of the war vis-à-vis the EU's adjustment of its geopolitics and geostrategy to accommodate a vision of energy sufficiency even in the future without any reliance on Russia (Daggash, 2023; Bearak et al., 2022). However, lack of investment in the industry coupled with poor or absent maintenance is among the challenges or limiting factors against the expansion of production in Algeria and Nigeria. Yet previous trade agreements have committed Nigeria to supply contracts with a number of countries in Europe and may, therefore, only be able to give out a fraction of its total product. In addition, expansion of production is a long-term process, as "longer-term investments such as the Trans-Saharan and Nigeria-Morocco pipelines would take decades to build" (Daggash, 2023: 6).

Therefore, Africa cannot derive any significant benefit from this apparent vacuum created by the war without expanding production through investment. This involves creating the right atmosphere to attract investors, such as good infrastructure and good regulation policies in the industry. Overall, the expected benefits for Africa are wrought with great challenges. African countries thinking of exploiting this opportunity will need to be very strategic in different ways to derive the best from it. First and foremost, they must understand and play along with the EU's geopolitics and geostrategy underpinning the entire package. It is the EU's ongoing policy and strategy to create alternative sources of energy supply. This can change in the future in response to new developments in the energy sector and new trends in global politics.

For instance, the EU was the key exponent of the policy of global decarbonisation until the war (Auth/Moss, 2023). As a result of this, African countries with poor energy supplies trying to develop fossil fuel projects like gas were denied access to the global development fund by Western powers (Auth/Moss, 2023). Thus, before the Russia-Ukraine war, African leaders expressed their viewpoints about these policies and other devices by Western powers to deny them access to such funds, which they considered frustrating and hypocritical. This was among the issues discussed by the president of Uganda in one of his public speeches on the theme. The president of Malawi and vice president of Nigeria equally expressed the same viewpoint (Gbadamosi, 2022). Some Western countries, including the USA, claimed to operate a policy that marked some cases as exceptions to this restriction on the release of energy development funds. But it was observed that this was not reflected in practice with regard to support for African countries (Auth/Moss, 2023). But the policy of global decarbonisation is now being jettisoned to address the shortage of energy supply in Europe.

In light of Western geopolitics and geostrategy, African countries need to have robust policies and strategies in place, particularly those that can sustain production at home and protect investors' interests against the adverse effect of changes in global politics since it is now well established that all actors in the international system only have permanent interests and hardly any permanent friends. An essential aspect of such policies is the creation of a platform or structure that can provide and sustain a good market for the product at home. The war has also created a platform for African countries to participate in the production of low-carbon energy devices through involvement in the technology and mining of critical minerals (Daggash, 2023). Russia is among the key producers of these minerals, and the ongoing development that requires the participation of African countries arises from the current policy of the EU member states to boycott Russia's products. In particular, Central and Southern Africa have mineral resources that can be utilised to be part of the anticipated low-carbon technology culture. Manganese is found in commercial quantities in the Democratic Republic of the Congo (DRC), South Africa and Zimbabwe. Cobalt, lithium and platinum are also among the key mineral resources of these countries by global rating. African countries can also upgrade their overall industrial plans, source more funding to enable them to operate above extractive models, such as mining, and engage in refining their minerals. This will also facilitate the creation of a structure that can support the development and sustenance of a local market for the products.

19.5.3 Food Security in Africa

Food supply and security problems have been compounded by several factors emanating from the war. The obvious one is the collapse of the Ukrainian food and related sectors creating hunger in the country in addition to halting the exportation of goods, including food, to neighbouring states. The second is the need to curtail Russia's aggression through sanctions, which invariably involve the reduction of commercial transactions with her to the barest minimum. As a result, EU member states have placed embargoes on the importation of energy and food products from Russia. Looking at the probable future trend of the conflict, the secretary-general of the UNO, António Guterres, attempted some analysis of the war in April 2022 based on the number of deaths, population displacement and the depth of harm to public utilities. He argued that "the war could throw one-fifth of humanity (1.7 billion people) into poverty, destitution and hunger on a scale not seen in decades" (Grajewski, 2022: 6). The war has disrupted the food industry in Ukraine and this is also endangering food sufficiency in Europe as there is obstruction of the exportation of food and fuel from Ukraine and Russia. This situation is also compounded by the ban placed on exports by some other nations, which could have partially made up for the gap. Ukraine received global attention in financial and material aid, and the intricacies surrounding this, in addition to the increasing fear of food insecurity, are now among key indices in global developmental funding (Grajewski, 2022). But there are speculations that the food scarcity and insecurity created by the war are likely to create an atmosphere for exploring new insights in research in the food and energy sectors along lines that were initially neglected, including the production of fertilisers (Gaind et al., 2022). Nevertheless, whatever benefit is expected to accrue from this, the adverse effects far outweigh the gains.

The food industry in Africa depends on global food supply chains (FSCs) to sustain the sufficiency of certain food items (Hatab, 2022a), as most countries on the continent are still constantly looking at the drawing board in search of devices to sustain food security. In some countries, international agencies have been involved to support the funding of the food sector (Ashe, 2019). This largely explains why any major conflict like the present Russian-Ukrainian war often has adverse effects on food security in Africa. It has been observed that "many nations in East, West, Middle and Southern Africa import a substantial amount of wheat, maize, sunflower, barley, fertiliser and vegetable oils from Russia and Ukraine" (Nweke et al., 2022: 71). The existing records show that about 50% of these imports go to Algeria, Egypt, Kenya, Nigeria, Sudan, Tanzania, and South Africa. Therefore, these countries receive most of their imports from Russia and Ukraine.

Russia and Ukraine are also the source of over 40% of certain grains, particularly wheat, exported to Cameroon, Sudan, Tanzania, and Uganda (Nweke et al., 2022). At the same time, "in 2019 and 2020, the dependency of Egypt, Sudan, Kongo, Senegal and Tanzania on Russian and Ukrainian wheat imports stood at 82%, 75%, 69%, 66% and 64%, respectively" (Hatab, 2022b: 40). It is, therefore, not surprising that within the first four months of the war, its adverse effects on FSCs have already been felt in some African sub-regions (Hatab, 2022b). As a result of this, African countries that depend partially or fully on the supply of fertilisers from Russia, combined with inconsistencies in the supply of energy as the energy market had links with products from Russia, were threatened in one way or another. Overall, there has been a general disruption in commercial and related links with Russia and Ukraine, including the shipping sector. Yet sanctions imposed on Russia and her counter-reactions have their own external dimensions and, therefore, "negative third-party effects" (Hatab, 2022b: 3).

Affected African countries have not found it easy to manage the situation because the war started when Africa was just recovering from the economic, including food, crisis of the COVID-19 pandemic. Furthermore, in East Africa, the latter stage of the pandemic was accompanied by a locust plague on farmlands that further endangered the structures put in place to sustain food security (Salih, 2020). It is significant to point out that data compiled by the Food and Agricultural Organisation (FAO) revealed that the war had compounded food insecurity in different regions of the world. The figures show that this was partly due to the fragile structure left behind by the COVID-19 pandemic, whose food crisis left 130 million people undernourished globally, of whom 40 million were from Africa (Hatab, 2022b).

Food insecurity and hunger will likely be increase globally and particularly in Africa as the Kremlin, on Monday, July 17, withdrew from the grain/food deal brokered by Türkiye and signed in July 2022. As a results, many African countries, already dealing with high inflation, will have to continue struggling to cereals to feed their respective populations.

19.6 Conclusion and Recommendation

History has revealed that conflict can hardly be totally avoided in human society because of incompatible interests and approaches to issues surrounding the sharing of resources, and the preservation of individual and national pride among different states in various regional and global affairs. The war has elements of national and ideological pride in its origins. NATO's and the EU's expansion since 1997 is threatening Russia and its allies. The future of the war for all stakeholders can hardly be precisely defined at the moment, nor is its end time easy to predict. The trend so far tends to indicate that the war is likely to throw many people into poverty and divert Russia's attention in terms of alignment and focus on socio-economic issues in East and Sub-Saharan Africa.

Narrowing the impact down to Africa, the soaring price of energy globally is contributing to the increasing cost of production and the invariably high cost of living in African countries. In addition, the war has had an adverse effect on the food sector in Africa, as the continent depends on global FSCs with most of its countries even obtaining fertilisers from Russia as well as grains from Russia or Ukraine. On the positive side, the EU's energy policy vis-à-vis sharp adjustments in its regional geopolitics and geostrategy to cope with the exigencies of the war has created some intricacies in global energy supply and an apparent vacuum that Africa can exploit to enhance its energy production and improve national income.

For African countries, the present study recommends the looking at the drawing board for new regional operative food and energy policies that can ensure self-sufficiency during major crises or conflicts in other regions of the world. As already explained above, the changes in the energy sector that tend to favour Africa should be exploited for improvement and innovation in the sector as one means of ensuring self-sufficiency and creating economies that can absorb shocks emanating from crises in other regions of the world in the future.

The study, however, admits that the best approach is the adoption of the right policies and strategies at the global level, considering the intricate interrelationships in commerce and economic activities that demonstrate that no region or individual can successfully exist or operate in isolation from the global system. Therefore, based on the trauma, economic crises and total number of lives lost so far, the present study recommends the need for world powers to give peace and security of life priority over military and ideological superiority often exploited to support selfish geostrategic and geopolitical ambitions, including expansionist and conservatist politics meant to sustain national or regional pride.

For instance, how can regional or national pride compensate society for the numerous lives wasted within a few months most sophisticated and technological warfare? From a study of major trends in the history of these events, what can we actually say is the tangible benefit of the defence of military and ideological superiority to the average man and the society at large? Is military supremacy still a justification for warfare while military stockpiles of several decades have been depleted in the first year of war? The idea of waging war was ludicrous and adventurist because the date one decides to invade can be known, but the death tolls of civilians and belligerents are not calculable before wars. The immensity of the Russian war in Ukraine is already the source of intergenerational hostilities and enmity between the two sister nations. That is irreversible. The ecological, human and financial costs of war are too enormous to justify. The United Nations once defined 'security' as security for people and not for land space. One cannot reconcile the ongoing waste of lives with this definition.

Russia's failure to achieve its apparent master plan of capturing Kyiv, the capital city of Ukraine, within a few days or weeks, is one lesson that emerged from the war, which should serve as a guide in the future when the use of arms is considered mandatory by a major power. In this case, Russia was unable to do what it successfully did in Prague in 1968 and Kabul in 1979 and what the USA did in Afghanistan in 2001, Iraq in 2003 and Libya in 2011. It is argued that the Russian side was grossly over-optimistic in planning and that this primarily accounts for the failure as "the initial attacks displayed hallmarks of over-confidence and wishful thinking that the Ukrainians would not offer serious resistance" (Grossman, 2022: 35).

Yet, the lesson thereof is that the world is changing even more rapidly beyond the expectation of humanity, and every scene and terrain in global politics and diplomacy must be taken as a new ball game and strictly on its merit, while every nation must be careful not to rely on past glory. There is a need for the world powers to transcend the horizons of existing ideological principles, precepts and established political theories based on historical analogies. In other words, the fact that they worked then does not necessarily mean they are good enough for the present situation. Military and ideological superiority rooted in regional and national pride may be good enough for World War I and World War II and to justify the loss of lives in the rivalry between NATO and the defunct Warsaw Pact up to 1991 and the post-Cold War rivalry between NATO and Russia up to the end of the last century. But now humanity is questioning the security thereof, even at the regional level considering the sporadic enormous waste of lives in the name of regional security. Therefore, the tenacious resistance of Ukraine constitutes one means of questioning the validity of this security model.

References

Abiodun, S. (2023). ECOWAS, EU Move to boost West African energy sector, *The PUNCH*, June 1. www.punching.com. Accessed 20 June 2023.

Addeh, E. (2022). Inflation: diesel prices rise by 215%, gas 70.6%, petrol 17% in one year, THISDAY, Nov. 29. www.thisdaylive.com. Accessed 20 June 2023.

Ashe, M.O. (2019). International Agencies and the Quest for Food Security in Nigeria, 1970-2015, *Ubuntu: Journal of Conflict and Social Transformation*, Special Issue, March, *8*(1), 251–274.

Auth, K. & Moss, T. (2023). How Russia's Invasion of Ukraine Will Impact Africa's Energy Transition, *Africa in Transition* https://www.cfr.org/blog/how-russias-invasion-ukraine-will-impact-africas-energy-transition, Accessed 14 June, 2023. Accessed 16 June 2023.

Barros, G., Hird, K., Bailey, R. & Kagan, F.W. (2022). Russian Offensive Campaign Assessment, Institute for the Study of War & AEI's Critical Threats Project 2022, Oct 26. https://understandingwar.org/backgrounder/russian-offensive-campaign-assessment-october-26. Accessed 8 July 2023.

Bearak, M., Eddy M., & Searcey, D. (2022). Power Balance Shifts as Europe, Facing a Gas Crisis, Turns to Africa for Help, *The New York Times*, Oct. 27. www.nytimes.com. Accessed 18 June 2023.

Burney, S. (2018). UNDP Highlights Food Security in First Africa Human Development Report, https://www.sustainweb.org/pdf/wto.pdf. Accessed 6 October 2022.

CSR (Congressional Research Service) (2022). Russia's War in Ukraine: Military and Intelligence Aspects, CRS Updated September 14, 2022. https://crsreports.congress.gov. Accessed April 4 2023.

Daggash, H.A. (2023). Analysis: Impact of Russia's Invasion of Ukraine on Africa's Energy Transition, *Centre for Journalism Innovation and Development (CJID)*. https://thecjid.org/analysis-impact-of-russias-invasion-of-ukraine-on-africas-energy-transition/. Accessed 15 June 2023.

Gabriel, S. (2022). Putin's war immediately united the EU and NATO. https://www.project-syndicate.org/onpoint. Accessed 17 March 2023.

Gaind, N., Abbott, A., Gibney, E., Witz, A., Tollefson, J., Irwin, A., & Van Nordon, R. (2022). How the war in Ukraine is changing global science. https://www.nature.com/articles/. Accessed 22 March 2023.

Gbadamosi, N. (2022). Africans Decry Europe Energy Hypocrisy, *Foreign Policy's African Brief*, July 20. www.foreignpolicy.com. Accessed 20 June 2023.

Grajewski, M. (2022). The latest on Russia's war on Ukraine, What Think Tanks are Thinking, Strategy and Innovation Unit, European Parliamentary Research Service, May 21. https://www.europarl.europa.eu/RegData/etudes/BRIE/2023/747910/EPRS_BRI(2023)747910_EN.pdf. Accessed 20 July 2023.

Grossman, D. (2022). Russia's War in Ukraine: What are the Emerging Military lessons. https://www.rand.org/about/people/g/grossman. Accessed 12 March 2023.

Hatab, A.A. (2022a). Africa's Food Security under the Shadow of the Russia-Ukraine Conflict, *Strategic Review for Southern Africa, 44*(1), 37-46 Nordic Africa Institute, Uppsala, Sweden, & Department of Economics, Swedish University of Agricultural Sciences, Uppsala, Sweden. https://doi.org/10.35293/srsa.v44i1.4083. Accessed 20 June 2023.

Hatab, A. A. (2022b). Russia's invasion of Ukraine jeopardizes food security in Africa. Shocks to global food supply chains could lead to social and political unrest, *NAI Policy Notes,* The Nordic Africa Institute, July.

HHS (Health and Human Services) (2022). Analysis of the Russian-Ukraine Conflict, Compiled by U.S. Department of Health and Human Services, Mar 2022. https://www.hhs.gov>files. Accessed 8 April 2023.

Hovell, D. (2022). Council at War: Russian-Ukraine and the UN Security Council, Blog of the European Journal of International Law, Feb. 25. https://www.ejiltalk.org/council-at-war-russia-ukraine-and-the-un-security-council/. Accessed 8 August 2023.

Korosteleva, J. (2022). The Implication of Russia's Invasion of Ukraine for the EU Energy Market and Businesses. British Journal of Management, Sept. 7, 33(4):1678-1682.

Leigh, M. (2022). The EU's response to the war in Ukraine. Michael Leighhttps://www.gisreportsonline.com/e/leigh-micheal/. Accessed 18 March 2023.

Longley, R. (2022). What is Democratic Peace Theory? Definition and Examples, January. https://www.thoughtco.com>demo. Accessed 15 February 2023.

Masters, J. (2022). Ukraine: Conflict at the Crossroads of Europe and Russia. https://www.cfr.org/bio/jonathan-masters. Accessed 18 June 2023.

Ndigwe, C. (2023). Africa's Gas Supply to hit 585 Billion Cubic Meters by 2050, *BUSINESS DAY*, Feb. 7. https://www.businessday.ng. Accessed 19 June 2023.

Nweke, CC, Obi, CB and Igweike, OJ (2022). Russia-Ukraine War, its Impacts on Food Security in Africa, *Global Journal of Social Sciences Studies*, *8*(2):70–81.

Oyebamiji, S. I., Oyebade, O. O., Cohen, J. H., & Abraham, D. O. (2022). Echoes of Colour Discrimination in Refugee Protection Regime: The Experience of Africans Fleeing the Russia-Ukrainian War. *Migration Letters*, *19*(5), 695–707.

Parmar, I. (2013). The 'knowledge politics' of democratic peace Theory, *International Politics,* 50, 231–256. https://doi.org/10.1057/ip.2013.4

Salih, A.A., Baraibar, M., Mwangi, K.K., and Artan, G. (2020). Climate Change and Locust Outbreak in East Africa. *Nature Climate Change 10*(7), 584–585.

Shvangiradze, T. (2022). NATO vs. Warsaw Pact: How 2 Powers Opposed Each Other, *THE COLLECTOR*, May 7. https://www.thecollector.com>nato. Accessed 11 March 2023.

Sklar, R.L. (2008). Beyond Capitalism and Socialism in Africa, *The Journal of Modern African Studies*, *26*(1), 1–2.

SRD (2023). Monthly percentage change in diesel prices in Nigeria, 2019-2023, published by Statista Research Department (SRD), April 18. https://www.statista.com>statistics. Accessed 20 June 2023.

UN News. (2022). Russia vetoes Security Council resolution condemning attempted annexation of Ukraine regions, Oct. https://news.un.org/en/story/2022/09/1129102. Accessed 10 April 2023.

Yadav, I. (2022). Russia-Ukraine Crisis and Role of NATO: 7 Pionts, *Hindustan Times,* Feb. 25. https://www.hindustantimes.com. Accessed 12 March 2023.

Yuhas, A. (2022). Thousands of Civilian Deaths and 6.6 Million Refugees: Calculating the Costs of War, Aug. 24. http://www.nytimes.com/by/alan-yuhas. Accessed 11 February 2023.

Chapter 20
Child Soldiers, Justice and Peacebuilding: A Dialectical Analysis of the Conventional Criminal Justice System and African Indigenous Jurisprudence

Jean Chrysostome K. Kiyala

Abstract Justice processes involving children accused of war-related crimes and other violations of international humanitarian and human rights laws, as a result of their participation in hostilities, often end in contested outcomes because of the child soldiers' paradoxical portrayals as victims, perpetrators and heroes. Similarly, the lack of justice endangers security and peace processes in war-affected communities. While domestic and international legislation set to adjudicate accusations levelled against children associated with hostilities grapples with striking a balance between the violations of the child's rights (unlawful recruitments), serving the interests of justice for victims of child soldiers' violent acts, and child soldiers' criminal responsibility, some countries in Sub-Saharan Africa have made headway in holding child soldiers accountable, attaining justice for victims via traditional and restorative approaches and paving the way for community post-conflict sustainable peace with child soldiers. The refusal of the ICC to consider the Ugandan traditional mechanisms of accountability in the sentencing of Dominic Ongwen, the abducted child soldier of the Lord's Resistance Army who perpetrated war crimes and crimes against humanity, has raised serious concerns about the future of traditional or indigenous jurisprudence in handling justice for former child soldiers. This chapter engages a dialectic argument drawn from literary research to argue for a justice model that takes into account cultural context, traditional values and the precedent of African jurisprudence, a compounded interplay of restorative retributivism and restorative sanctions against the logic of the mainstream conventional criminal justice system.

Dr Jean Chrysostome K. Kiyala holds a PhD in Public Administration: Peace Studies from DUT; he is a Senior Lecturer at the International Centre of Nonviolence (DUT), Associate Professor and visiting lecturer at the Université de Bandundu and the Université Evangélique en Afrique (Democratic Republic of Congo); Emails: JeanK@dut.ac.za; kljeanchrysostome@gmail.com; ORCID: https://orcid.org/0000-0002-9587-0230

J. C. K. Kiyala and N. Chivasa (eds.), *Climate Change and Socio-political Violence in Sub-Saharan Africa in the Anthropocene*, The Anthropocene: Politik—Economics—Society—Science 37,
https://doi.org/10.1007/978-3-031-48375-2_20

Keywords Armed conflicts · Child soldiers · Community-based jurisprudence · Indigenous justice · Peacebuilding · Restorative justice · Sub-Saharan Africa

20.1 Introduction

> A theory, however elegant and economical, must be rejected or revised if it is untrue; likewise laws and institutions, no matter how efficient and well-arranged, must be reformed or abolished if they are unjust.
>
> (John Rawls, 2004, A Theory of Justice, p. 3).

The changing nature of contemporary warfare has maintained and increased child vulnerability over the years, going back to World War II and in ongoing conflicts in different parts of the world. The child soldiering phenomenon has gained prominent attention in psychology, law, and social and political sciences, and research is ongoing about the involvement of persons below the age of 18 in militarised conflicts. The complexity of the problem regarding the use of child soldiers stems from the paradoxical portrayals of children associated with armed forces and groups when they are perceived as victims, perpetrators, or heroes (Drumbl, 2016; Kiyala, 2018a; Wessells, 2006). Recruitments of persons under the age of eighteen have been observed in countries such as Bangladesh, Germany, India, the United States, the United Kingdom, Canada, Australia, and China; in the African region, in the Democratic Republic of Congo (DRC), Angola, Mozambique, Liberia, Sierra Leone, Uganda, and Rwanda; and in the Asian Region, particularly, but not limited to, Timor-Leste, the Philippines, Cambodia, Nepal, and Papua New Guinea (Drumbl, 2012b). Recently, children have been recruited and trained in terrorist operations by ISIS in Syria and Iraq (Almohammad, 2018; Nyamutata, 2020). Commonly known as 'children associated with armed groups and armed forces' (CAAFAG), the involvement of children with ISIS has led to a new description of such category of children as 'children associated with terrorist groups' (CATGs) (Jørgensen, 2019).

Child soldiering is an ongoing phenomenon. ReliefWeb (2022) published a report which covers the Secretary General's evidence for child enrolments in armed conflicts in Myanmar [S/2020/1243], Syria [S/2021/398], Afghanistan [S/2021/662], Yemen [S/2021/761], and the Central African Republic [S/2021/882].

Amidst unceasing recruitment and use of children as soldiers in Afghanistan, the Central African Republic, Colombia, the DRC, Iraq, Mali, Myanmar, Somalia, South Sudan, Sudan, Syria and Yemen, many recommendations are being made that call on these countries to abide by the legal normative frameworks pertaining to the protection of children, such as international humanitarian law (IHL) and human rights law (IHRL), to stop hostilities and the use of child soldiers, prioritise the reintegration of these children into society by upholding international juvenile justice standards, to ensure that counterterrorism operations and mechanisms do not impede humanitarian relief, to treat child soldiers primarily as victims, halt their unlawful detention, promote and protect child rights and well-being, and empower them to be active in peacebuilding efforts (ReliefWeb, 2022).

The concept of recruitment of child soldiers is linked to the involuntary enrolment of children into "armed forces or armed groups, or the threat of it, which has been used as one way of petrifying and extorting civilians" (International Committee for the Red Cross [ICRC], 2012). To a certain extent, the recruitment of these children has had a multi-variant purpose, whereby the recruiters shift the recruitment motivations into various activities, including acting as "spies, messengers, domestic servants, sexual slaves, etc., rather than using them as directly involved in the fighting" (ICRC, 2012). Unfortunately, all the measures suggested by the UN Security Council are proving ineffective in halting the use of child soldiers, nor are they deterring either the children or the adults who enlist them. Thus, it is essential to move from the conventional legalistic approach of dealing with the child soldiering phenomenon and to resort to cultural and locally-based approaches that are anchored in the contexts where children are recruited and used as soldiers. This requires firstly determining the effects of globalisation on the problem (Kiyala, 2018b), and establishing how local cultures could be used tactically to promote peace between child soldiers and their wounded communities (Among, 2013; Honwana, 1997; Kiyala, 2016).

CAAGs or CATGs are confronted with a myriad of challenges, ranging from their reintegration into society (Kiyala, 2015; Schotsmans, 2012), to recovery from posttraumatic stress disorders (PSTDs) (Betancourt et al., 2013a, b; Derluyn et al., 2004; Forstmeier et al., 2009), access to psychosocial well-being support (Kiyala, 2020; Wessells, 2012), establishing their criminal responsibility for alleged violations of international human rights and humanitarian law (Kiyala, 2018a; Veale, 2006), and dealing with epistemic injustice to which they are exposed, especially under Article 1(F)(a) of the 1951 Convention Related to the Status of Refugees (Kiyala, 2021). Furthermore, holding CAAFAGs accountable under the conventional criminal justice system and international criminal law for crimes and human rights violations they allegedly carried out during their participation in armed conflicts poses a critical conundrum for the administration of justice (Drumbl, 2012b; Holá/Bouwknegt, 2019). The corollaries of children's involvement in armed conflict cannot be exhausted; however, it is important to note the complicated process of reconciliation with victims (Baines, 2007; Bayer et al., 2007), the rehabilitation path for those children (Bainomugisha, 2011; Coppens et al., 2012; Kiyala, 2020), and providing mental health support (Betancourt/Borisova et al., 2013a; Kohrt et al., 2010).

The question which remains essential in the prosecution of children recruited to fight in armed conflicts and who later become adult perpetrators of war crimes is to determine at which stage their victimhood ends and their self-regulatory capacity begins when under the subjugation of a higher military commander, such as Joseph Kony of the LRA; this dilemma is epitomised by the indictment and conviction of Dominique Ongwen on changes of war crimes and crimes against humanity (Ramos, 2019). The intricacies of balancing the criminal responsibility of child soldiers with their status as victims make it more difficult to adjudicate their charges under international human rights and humanitarian law. Such a predicament is supported by the controversial legal arguments regarding the determination of *actus reus* and *mens rea* in the circumstances under which child recruitment is conducted, their state of mind under military superiors and the psychological frameworks of military training, and how that affects children's engagement at the front line and when carrying out other activities associated with child soldiering under the Paris Principles, Article 2(1), which includes a range of activities in which children take part which allow them to be considered as CAAFAGs, which include fighting, cooking for troops, carrying military hardware and other materials, being messengers, spying and being used for sexual purposes. Article 2(1) does not only refer to a child who is taking or has taken a direct part in hostilities (UNICEF, 2007).

What is the level of freedom and the willingness of abducted children to perform all these soldiering activities? Besides the whole problematic criminalisation of child soldiers' offensive actions in war time, several communities where this category of youths have carried out heinous human atrocities prefer traditional approaches of accountability which are proving effective in fostering reconciliation and building sustainable peace with former child soldiers. Such has been the results of experiments conducted in the DRC, using the *Baraza* institution (Kiyala, 2016, 2018a, 2018b, 2021); in Sierra Leone, the *fambul tok* has been utilised to foster reconciliation between former child soldiers and their communities (Graybill, 2010; Hoffman, 2008); and local communities, traditional, religious, and political leaders in Uganda have been advocating *mato oput* as the instrument of pursuing accountability of the LRA's former child soldiers (Among, 2013; Bangura, 2008; Institute for Global Leadership, 2009).

Against this backdrop, this chapter explores several experiments in child soldiering justice in Sub-Saharan Africa. It establishes the extent to which such local and community-based justice mechanisms are imbued with contemporary restorative justice tenets to build peaceful and harmonious conditions for child soldiers' social reintegration and return to society. The multi-layered factors that wreak havoc on the well-being of countless numbers of children involved in armed conflicts and terrorist organisations call for adequate strategies to mitigate the detrimental effects on the lives and emotional and human development of CAAFAGs and CATGs. Taking this investigation further, an overview of contemporaneous peacebuilding and justice administration rooted in African humanism (*ubuntu*) and communitarianism (Dolamo, 2013; Gade, 2013) is dialectically analysed by opposing the conventional retributive criminal justice system and the contemporary restorative justice accountability model. The dialectic reasoning that leads the thesis statement of this chapter is drawn from a pluralist methodological approach and multi-methodology, blending various qualitative methods and conceptual frameworks supporting the overarching argument for non-adjudicatory procedures as the pathway to justice for child soldiers and post-conflict societal peacebuilding. A combination of qualitative and literary research methods is employed here, and thematic, content and discourse analytical methods within the constructivist worldview are applied to literary and empirical sources.

This chapter argues for the pursuit of indigenous or traditional and restorative justice approaches as pathways to justice when children used as soldiers are accused of perpetrating war-related crimes; it is based on the epistemological assumptions that credit traditional restorative justice and conventional restorative justice practices with the potential to overcome the legal constraints faced by the prosecutorial procedures aimed at holding these children accountable. This thesis statement is evidenced by several studies that showcase the viability and effectiveness of restorative justice in facilitating justice for child soldiers and their victims, as in the case of *Baraza or barza* in the Eastern DRC (Clark, 2008a; Kiyala, 2016, 2018a) and community-based restorative justice to attain reconciliation between former child soldiers and their communities, as reported in post-civil war Sierra Leone (Baines, 2007; Graybill, 2010; Hoffman, 2008). Similar experiments have been tested and proved promising and satisfactory in northern Uganda among the Acholi local communities under the *mato oput* "drinking the bitter root" rituals (Afako, 2006; Bangura, 2008; Institute for Global Leadership, 2009). In a similar vein, the traditional cleansing rituals known as *ku pahla or ukupahla* ("communicating with the ancestors") (AfroSavvy, 2020), was used to substitute the criminal justice approach in the reintegration of former soldiers in the aftermath of the Mozambican civil wars (Bangura, 2008; Honwana, 1997). The relevance of traditional approaches to justice cannot be better advocated if the inadequacy of court rooms to dispense justice is sustained by the impasse in and the unconvincing end of adjudicatory procedures for former child soldiers such as Dominic Ongwen; He was sentenced by the Trial Chamber IX on 6 May 2021 to 25 years of imprisonment (ICC, 2021). Subsequently, the defence appealed that conviction on 21 Jul and 26 Aug 2021 (ICC, 2021). We will return to Ongwen's case in the discussion centred on the enigmatic prosecutions of former child soldiers.

The chapter consists of six main sections. The introduction lays out the background to the challenges faced by child soldiers; the second section discusses the concept of child soldiers and underlines a few key issues pertaining to the participation of children in armed conflicts; the third section looks into the theory of justice and the justification and the means for holding former child soldiers accountable. The fourth section explores the conundrum faced by the mainstream criminal justice system in holding accountable two prominent former child soldiers who committed human atrocities after the age of 18, notably Dominic Ongwen and Thomas Kwoyelo. The fifth section of this chapter examines the application of restorative justice and the legal ground for its application to processes of holding former child soldiers accountable. The final section provides some concluding remarks to this essay.

20.2 Child Soldiers: Conceptual Analysis and Legal Implications

The definition of child soldiers in this chapter is based on the Paris Principles, Article 2(1), which states that:

> A child associated with an armed force or armed group refers to any person below 18 years of age who is or who has been recruited or used by an armed force or armed group in any capacity, including but not limited to children, boys and girls, used as fighters, cooks, porters,

messengers, spies or for sexual purposes. It does not only refer to a child who is taking or has taken a direct part in hostilities (UNICEF, 2007).

Article 2 of the African Charter of the Rights of the Child (ACRC) defines a child as "every human being below the age of 18 years" (UNICEF, 1999). According to the Convention on the Rights of the Child (CRC), Article 1, "[...] a child means every human being below the age of eighteen years unless under the law applicable to the child, majority is attained earlier" (United Nations, 1989). Furthermore, Article 1 of the Optional Protocol to the Convention on the Rights of the Child on the involvement of children in armed conflicts (OPAC) interdicts the recruitment of persons below the age of eighteen. This provision follows the recommendation of the twenty-sixth International Conference of the Red Cross and Red Crescent in December 1995 that advises parties in conflict to take every possible step to guarantee that children below the age of 18 years old do not participate in hostilities and supports the June 1999 International Labour Organization Convention No. 182 on the Prohibition and Immediate Action for the Elimination of the Worst Forms of Child Labour. All these instruments maintain 18 as the minimum age for recruitment and participation in armed forces and prohibit forced or compulsory enlistment of those under 18 in hostilities (ILO, 2002). Article 1 of OPAC enjoins State Parties to use necessary mechanisms to prevent youngsters who have not attained the age of 18 from participating directly in hostilities (United Nations, 2000).

The parameters of this definition are problematic and reflect what hermeneutic injustice against child soldiers is (Kiyala, 2021): both child soldiers and society lose sight of the real world of child soldiering, which is often wrongly presented by the media (they are depicted as heroes, victims or perpetrators). Child soldiers are coerced by armed commanders to murder and obediently execute the orders they receive (Wessells, 2006). As a result, the involvement of children in armed groups has dire consequences, causing them to suffer severe trauma that motivates their psychic behaviours to resort to violence (Beresford, 2005). Circumstances in which violence is committed by child soldiers found in armed conflict areas are identified as being high levels of consumption of drugs or alcohol (Drumbl, 2012a). It is believed that not all child soldiers perpetrate acts of violence leading to killing or torture. Only a few cases of holding child soldiers accountable have been witnessed, instead of a continuous legal opinion considering some child soldiers as victims of war while they were part of armed groups.

Such interpretation and misrepresentation of the child's life, either in armed conflicts or terrorist organisations, leaves such a child vulnerable to various treatments amounting to epistemic injustice (Kiyala, 2021). This deepens their trauma and worsens their vulnerability. Are these youngsters (children and teens) genuinely qualified for soldiering and military responsibilities? This question begs for understanding, which is grounded in a recruitment age based on international standards.

20.2.1 Age Matters

The "child age" is a determinant factor, either in the enrolment of young people to fight in militarised conflicts, or in establishing their criminal responsibility before and after they have reached the threshold of "18-year-old" according to the international standards relating to the age of responsibility in involvement in labour and military service (Table 20.1). Furthermore, the "child age" has become a controversial and politicised question in humanitarian law and the administration of justice (Drumbl, 2016; Kiyala, 2018a; Rosen, 2007, 2009).

The age provision for the involvement of youths in labour as provided by International Labour Organization (ILO) (1996–2017) disqualifies recruitment of persons below the age of 18 in soldiering activities or military operations because of the risks posed by such engagement. This is well elaborated in the ILO under the Minimum Age Convention, 1973 (No. 138), and the Convention concerning Minimum Age for Admission to Employment (entry into force: 19 June 1976), Article 3, paragraphs 1 and 3. For instance, this instrument states the following, in 3 (3):

> Notwithstanding the provisions of paragraph 1 of this Article, national laws or regulations or the competent authority may, after consultation with the organisations of employers and workers concerned, where such exist, authorise employment or work as from the age of 16 years on condition that the health, safety and morals of the young persons concerned

Table 20.1 Minimum age of child involvement in labour

Standards	Geneva Convention, Protocol I; Article 77. Protection of children; # 2 Protocol II Article 4: Fundamental guarantees, # 3, d	African Charter of the Rights and Welfare of the Child (ACRWC) Article 2	The Rome Statute of the International Criminal Court; Article 8, # 2 (b) (xxvi)	The Paris Principles; # 1, 14	ILO Article 3 1. C138: Minimum Age Convention, 1973 (No. 138
Age	15	18	15	18	18
Service	Parties in conflicts	All actions concerning the child	Conscription in national armed forces or using them to participate actively in hostilities	Advocating for States to raise the minimum age of recruitment or use to 18 in all circumstances	The minimum age for admission to any type of employment or work; no risks to health, safety or morals of young
Nuances	Priority to the oldest, between 15 and 18	Not directly related to military operations			16 years on condition that the conditions of 18 are preserved

Source Author

> are fully protected and that the young persons have received adequate specific instruction or vocational training in the relevant branch of activity (International Labour Organization (ILO), 1996–2017).

In the case of child soldiers, there is neither vocational nor professional training before sending them to the frontline or to camps to execute soldiering activities as defined by the Paris Principles Article 2(1), to which we alluded earlier. Table 20.1 captures the major legal provisions related to the age of involvement in labour force whether in civilian or military engagement.

Scholars such as Rosen (2007) dispute the globalisation of *childhood* that is vulgarised by human rights advocates and humanitarian organisations in their efforts to foster the regulation about the use and end of child soldiers. He notes: "Treaty-making efforts reflect an emerging 'politics of age' that shapes the concept of 'childhood' in international law, and in which different international, regional, and local actors make use of age categories to advance particular political and ideological positions" (p. 296). Rosen (2007) points out that the concept of child soldiers forged by international treaties and laws did not emerge from the cultural, political and social contexts where young soldiers are found in the battlefield—thus, making child soldiers as a contemporary humanitarian crisis—but he argues that the problem of child soldiers derives "from an emerging transnational 'politics of age' that shapes the concept of 'childhood' in international law" (Rosen, 2007: 296). In his view, rather, various ideologies and political positions have fostered the notion of a "universal childhood" propagated by international law sustained by the United Nations, local and regional actors who describe child soldiers based on the "the politics of age", thus making child soldiers a politically and socially-construed concept (Rosen, 2007). He observes: "The representations of childhood by all these groups are political constructs used to support legal and political agendas, and they discount the more varied and complex local understandings of children and childhood found in anthropological research" (Rosen, 2007: 296).

In a similar vein, Honwana (2006) describes the concept of childhood as a social and cultural construct. She further expounds how persons below the age of 14 take earlier responsibilities (gardening, fishing, hunting, etc.) in Mozambican and Angolan traditional societies (Honwana, 2006). In the same line of argument, Van Rooyen et al. (2006) found that teenagers grow to adulthood via rites of passage in African societies such as the Southern Ndebele of South Africa. They write: "Both boys and girls, ranging from twelve to twenty years of age, undergo initiation. Initiation is also offered to adolescents who are about to marry and have not yet attended any initiation school" (p. 26). In the initiation process, the boys or young men are set aside and made to wear "hornlike structures made out of grass and reed ('*umshojo*'), which are placed on their foreheads for a period of a month, prior to initiation" (Van Rooyen et al., 2006: 27). At this important age, the initiates are entrusted with economic and social advantages in the community (Van Rooyen et al., 2006).

It is an established fact that the rite of passage that incorporates initiation to adulthood takes place at different ages. Circumcision, as an activity of the initiation rite is undertaken on youths aged between 8 and 20 among the Babukusu ethnic group of western Kenya (Peltzer et al., 2008). While several traditional societies interdict the involvement of children in warring activities, such as the Zulu nation of South Africa (Honwana, 2006), and the Rega and Shi societies, of the Eastern DRC (Kiyala, 2016), the Mai-Mai utilise children and medicine men and they play an active role in fighting (Jourdan, 2011; Kiyala, 2018a). A current appraisal of the chief grounds of violence between individuals in sub-Saharan Africa (Hove/Harris, 2019) encompasses various aspects pertinent to South Africa. A twofold opinion suggests that the male-controlled underpinning of the social order and its related understanding of maleness first relates to "attitudes and behaviours to be a man" suggesting the fundamental meaning embedded in youngsters' socialisation of manhood in Africa is connected to the justification and encouragement of utilising violence and threats as justifying manhood (Dianzenza/Harris, 2022). Secondly, and particularly in the South African context, many South Africans are disturbed by suffering interactive and organisational violence originating from the demised apartheid regime, showing itself in anger and a quick resort to violence (Dianzenza/Harris, 2022). These perspectives can critically challenge international humanitarian law and treaties that portray children used in armed conflict as primarily victims and perpetrators of war-related crimes. The contribution of international legal standards pertaining to the treatment of children accused of human atrocities in war time is to nuance the circumstances in the determination of guilt and application of remedial actions (justice) for victims and child soldiers themselves. For instance, in the Beijing Rules, Rule 2(a) states: "a juvenile is a child or young person who, under the respective legal systems, may be dealt with for an offence in a manner which is different from an adult" (United Nations, 1985). Furthermore, Rule 2(2) stipulates:

> It should be noted that age limits will depend on, and are explicitly made dependent on, each respective legal system, thus fully respecting the economic, social, political, cultural and legal systems of Member States. This makes for a wide variety of ages coming under the definition of 'juvenile', ranging from 7 years to 18 years or above. Such a variety seems inevitable in view of the different national legal systems and does not diminish the impact of these Standard Minimum Rules (United Nations, 1985)

Rule 2 should be interpreted in conjunction with rules 3 and 4; they affirm that minors in the age category of 7–18 years who are found guilty should be treated in agreement with "equitable and humane justice for children in conflict with the law" (United Nations, 1985); and in the case of Rules 4 (1) "in those legal systems recognising the concept of the age of criminal responsibility for juveniles, the beginning of that age shall not be fixed at too low an age level, bearing in mind the facts of emotional, mental and intellectual maturity" (United Nations, 1985). It is evident that Rules 2, 3, and 4 suggest that the notion of childhood should be weighed with child agency and criminal responsibility in the context of domestic and international legislation (Kiyala, 2018a). Thus, "[f]orced and voluntary recruitment neither alter the reality of childhood nor negate the role and responsibility of society vis-à-vis this category of people" (Kiyala, 2018a: 10).

Knowledge of war is brought to children by the society where this is experienced. As time goes by, pictures of war and heroism begin to attract children; the imposing look of police officers and soldiers, flying fighter-jets, and the display of armoured tanks, sink into children's minds and become part of their imagination. The media present heroes of war, extolling fierce combatants and their exploits, and all these portrayals and the exposure to scenes of war, including toys, such as guns, tanks, and armoured vehicles that are given to children and adolescents, help them develop inclinations for military engagement. Thus, children become familiar with the concept of war (Vriens et al., 1999).

According to Wessells (2006: 21), "conflict often erupts along ethnic or religious fault lines in divided societies that extend privileges and status to save groups while denying them to others". In this line of thinking, the desire to fight comes from feelings of humiliation, exclusion, oppression, exploitation and segregation, and in the sense that fighting becomes a "justified and legitimate" means of retaliation, of pushing towards transformation, political, social, and economic reforms, and independence or liberation. This situation endangers the lives of children who likely become involved in armed conflict under the influence of parents, families, and teachers, who encourage them to pursue the struggle (Wessells, 2006: 21). Wessells asserts that "this poisonous mixture of fear and hate heighten perceptions of evil and wrongdoing by the other and strengthens children's sense of victimhood, motivating attacks and encouraging on-going cycles of violence" (Wessells, 2006: 22).

When children participate in armed conflicts, they are identified with soldiers, troops, and terrorists, and called "child soldiers" or "children associated with armed groups and armed forces" (Betancourt et al., 2012; Kiyala, 2018a) and since the rise of terrorist movements, some children are also designated as "children associated with terrorist groups" (Jørgensen, 2019). The various depictions of children involved in militarised conflict raises enormous questions around their criminal responsibility and their victimhood, created by the unfortunate circumstances of their enrolment, either by abduction—a recruitment approach mainly used by war lords such as the LRA in Northern Uganda (Pham et al., 2008; ReliefWeb, 2007)—or conscription ("compulsory enrolment and induction into the military service"). It is reported that "[c]hildren and youth – some as young as 7 and 8 years old – have been forced to mutilate and kill civilians, including members of their own families and communities" by the LRA, an experience described as traumatic (ReliefWeb, 2007: 1).

In the final analysis, the universalisation of the concept of childhood as starting as childbirth and ending at 18, in accord with the Convention on the Rights of the Child (CRC) (Article 1) is problematic; a better conceptualisation of childhood should overcome the antithetical frames of humanitarian law and anthropology, and should rather be developed within a cultural and social context (Rosen, 2007, 2009). However, because of the globalisation of the child soldiering phenomenon and the legal principle of complementarity between the domestic and the international, nations should deal with the definition of childhood as suggested by Article 1 of the CRC, which sets 18 as the threshold, while giving the responsibility to different nations or contexts to define a child as "every human being below the age of eighteen years unless under the law applicable to the child, majority is attained earlier" (United Nations, 1989).

Beyond the legal humanitarian doctrine and anthropology, psychology should be evoked to establish the impact of armed conflicts on persons below the age of 18 to support the universalisation of the concept of childhood and its contextualisation—a context-based definition. This openness creates a further conundrum and challenge to the administration of justice, the pursuit of the child's best interests and protection, which will then be decided upon by the multiplicity of contexts. Thus, the socio-cultural heterogeneity of ages of involvement in warriorhood in Africa and around the globe, where childhood and military engagement are neither exclusive nor contradictory of each other; and despite the fact that, in given circumstances, soldiering may be a sound opportunity or a needful and positive choice for many children, their vulnerability and victimhood are not underestimated (Rosen, 2007). Thus, there is a need to deconstruct the idea of the "child soldier" as a non-specific prototype of humanitarian discourse on child soldiering—to concentrate concretely on the conflicts and the environment in which children undergo severe brutality and trauma (Rosen, 2007). In addition, rethinking the child soldiers concept on the basis of the diverse circumstances, cultures and historical contexts in which childhood and warring life are not separated, is necessary to reframe the paradoxical portrayals of child soldiers, which present this category of youths as both victims and heroes, and redefine the parameters of their criminal responsibility under international humanitarian and human rights law. But the nature of their recruitment, 'voluntary' or "coerced/abduction" is one of the key issues affecting justice processes involving this category of youths.

20.2.2 Voluntary and Non-voluntary Participation in Hostilities

The concept of voluntary and non-voluntary enlistment (or enrolment) and conscription is a significant aspect of the debates on child soldiering practice (Kiyala, 2018a; Wessells, 2006). Voluntary enlistment reflects the common understanding that children approach armed commanders, for various reasons, and ask to be enlisted to fight, whereas coercive enlistment suggests the presence of agents to use power or authority to get children to fight. The use of power is often comprehended along the line of force exerted on children via abduction, threats, or intimidation, while authority often alludes to the influence of community leaders or community members or parents who encourage a child to take arms and fight, as it happens in the context of community auto-defence militia in the Eastern DRC (Kiyala, 2018a, b).

Article 3 (3) of OPAC does not proscribe the recruitment of persons below the age of 18: "States Parties shall take all feasible measures to ensure that members of their armed forces who have not attained the age of 18 years do not take a direct part in hostilities" (United Nations, 2000). However, in paragraphs (a), (b), (c) and (d) of the same article, OPAC provides the grounds on which such recruitment should take place; these are stated here:

> (a) Such recruitment is genuinely voluntary; (b) Such recruitment is carried out with the informed consent of the person's parents or legal guardians; (c) Such persons are fully

informed of the duties involved in such military service; (d) Such persons provide reliable proof of age prior to acceptance into national military service (United Nations, 2000).

The law leaves some loopholes regarding age, because the Rome Statute of the International Criminal Court criminalises the recruitment of persons below the age of 18 (Table 20.1), which is also prohibited by Protocol I and II of the Geneva Convention (Table 20.1), and is considered among "war crimes" and grave breaches of the Geneva Conventions of 12 Aug1949, namely, either the conscription or enlistment of children under the age of 15 years or their use in active hostilities of both international and non-international armed conflicts (International Criminal Court, 2002).[1] Prohibiting the recruitment of persons who have not attained the age of 18 years to take a direct part in hostilities is problematic, because it may suggest that such persons could be used in other military activities covered by the definition of CAAFAGs under Article 2(1) of the Paris Principles (UNICEF, 2007).

Voluntary participation is contested because of the sum of factors that dictate children's decisions. The fact that a child's free will and informed decision are blurred by the contingency of factors on which many decisions are finally taken, indicates that voluntary enlistment should be admitted with a certain level of scepticism. This does not preclude the firm conviction displayed by some children, who dream of a career in military service or soldiering. Because of this uncertainty, child protection advocates "argue for a straight standard, preventing both compulsory and voluntary recruitment under eighteen" (Coomaraswamy, 2010, p. 540).

Social and economic elements, the multifaceted nature of contemporary warfare—such as inter-community militarised hostilities, proxy-warfare, the rise of ideologically-oriented community defence armed groups and militia, emerging under the alibi of protecting a community under threat by a rival community, propagandist advertising of terrorism and radicalisation, the erosion of traditional moral values—are all to be blamed for ongoing recruitment and use of child soldiers. Whether or not there is the presence of pull and push factors (Wessells, 2006), there are no grounds to consider children as soldiers, in the strict sense of soldiering as a profession, as Honwana (2006) argues.

The arguments around voluntary enlistment of persons below the age of 18 in armed groups have been considerably discussed by scholars. Such forced conscription and abductions are reported on Ugandan child soldiers' recruitment (Veale/Stavrou, 2007; Vindevogel et al., 2011); it is a common practice in the DRC (Kiyala, 2018a); and it was also utilised in Sierra Leone (Aning/McIntyre, 2005), and in other parts of the world (Wessells, 2006).

It follows that the concept of voluntary enlistment poses a serious problem when attributing it to the child soldiering decision-making process. In most cases, push and pull factors inform children's motivations to enrol with fighting factions. Children would not become soldiers if there were no war, despite living in abject poverty. War is the immediate trigger to children becoming soldiers though poverty remain one of the key factors from children living in conflict zones as Brett and Specht (2004, p. 14) aver: "young people who are not living in impoverished circumstances are

[1] Article 8(2)(b) ((xxvi) of the Rome Statute of International Criminal Court.

unlikely to join the armed forces or armed groups, although there will always be exception". This observation indicates that war which finds children in their homes is the main manipulator factor in their minds, which, associated with other realities, pushes or attracts them to join fighting forces, in the context child self-enlistment (we intentionally refrain from using the disputed concept of voluntary enlistment). Voluntary enlistment is problematic in the absence of studies that can establish the freedom of a child's decision-making processes to assume their responsibility fully and reasonably, and their capacity to defend knowingly the steps leading to a final decision to seek enlistment by either regular or non-government armed groups or forces. Children who are left unprotected in a conflict environment, without work and education, are more likely to join armed groups.

Regarding the nature of enlistment—by abduction or voluntary decision, and the legal implications and determination of how they are or should be treated before the international law, it has been established that the participation of persons below the age of 18 in armed conflicts imperils their mental and physical integrity (Amnesty International 2008). The nature of recruitment is immaterial in establishing the negative impact of soldiering on persons below the age of 18 and their long-standing association with armed groups and forces beyond the age of 18, and the extent of post-soldiering effects are invariably perceptible in those children when they exit hostilities. Some suffer significant post-traumatic stress disorder; others emerge strong, to the extent of commanding their fellow soldiers, and even becoming perpetrators of war crimes and crimes against humanity (Baines, 2009; ICC, 2021; Nortje, 2017). All child soldiers, regardless of the nature of their recruitment, suffer indiscriminately the same fate with regard to war-related traumatic experiences, and criminal responsibility for acts of violations of international human rights and humanitarian law when they carry out such transgressions at an age that does not insulate them from criminal responsibility under the international law. That was the argument used by the ICC to indict and convict Dominic Ongwen (International Criminal Court, 2021b). I will come back to Ongwen vs. the Prosecutor of the ICC.

20.2.3 Pull and Push Factors Motivating Child Soldiers to Join Armed Groups

Research points out that there are myriad factors that motivate children to join different factions which give them certain privileges they lack in their respective social environments. Push factors such as socio-biological, political and educational restrictions are identified as being at the core of young people's voluntary recruitment, while the dearth of family social support and the available military infrastructure in the areas where children live are believed to play a crucial role in the motivation for children to be recruited (Kiyala, 2019). Additional factors that push children to join armed groups include the disappearance of protective social structures that used to care and shelter children from adults' abuses. The system value that prevailed in

Sub-Saharan African countries in wartime have disintegrated as a result of protracted and prolonged conflicts, and even the changing nature of warfare—war is brought to civilians and communities are entangled in hostilities.

Researchers in the domain of child soldiering also demonstrate a developing interest in the supply and demand characteristics of engaging children in the soldiering market. Factors described include poverty, a dearth of education and a lack of political prospects, and are believed to play a role in supplying child soldiers (Dupuy, 2010) Voluntary recruitment that was conducted by the Revolutionary Armed Forces of Colombia (FARC) in Colombia exposed the reality that most children that joined were motivated by issues of 'scarcity, joblessness, retribution, evading fierceness from the competing troops, and the attraction of the armed lifetime' as key drive reasons (Sanin, 2010). On the one hand, the contribution of the aspect of voluntary joining of armed groups by children has been partly justified by the prospects of finding work in the environment where they dwell, and the uncertainty and insufficiency of rudimentary provisions like food (Wessells, 2006). On the other hand, the demand side has much emphasised the cost-effectiveness of using child soldiers (Lasley/Thyne, 2015). The advantage of the cost-effectiveness of using child soldiers instead of adults to those who recruit them is their willingness to be retained: they are risk-takers, but also tend to remain for longer periods without deserting (Gates, 2017).

To a certain extent, other child soldier recruiters find it convenient to remain attached to children as elements to use on the ground because of the incentives they generate during actions known as "spoils of war". It is a common practice that rebel groups become used to enriching themselves from the looting of resources in the areas they control. Unfortunately, leaders of the rebel groups get an advantage from child soldiers as they limit the plunders of war to those who are eligible, rewarding the others disproportionately (Beber/Blattman, 2013). The mobility of child soldiers is higher and they are more easily recruited while they live in refugee camps and schools compared to adult rebels (Gates, 2017). Research reports have, however, suggested that there has been a significant decay in recent years, though this practice of child recruitment has not stopped in its entirety, especially on the African continent. Worldwide participation of children in some conflicts was noticed, while their utilisation in front-line fighting in armed organisations as well in groups perpetrating terrorism has been widely witnessed.

20.2.4 Exit from Hostilities

Most children exit hostilities either by self-demobilisation (Juma, 2012; Kiyala, 2015) or through Disarmament, Demobilisation and Reintegration (Kiyala, 2018a), and such children were mostly abducted and underwent appalling hardships.

Faced with the tough realities of the battlefield, many children self-demobilise, especially in the context where there is no formal programme of reintegration into society, which often takes place as step to Disarmament, Demobilisation and Reintegration (DDR) (Kiyala, 2015). However, reintegration is oftentimes executed in the period that follows the conflict, and the employment of an agreement leading to peace between the parties. Reintegration is presented as a procedure that consists of children going through an evolution to rejoin "civil society and enter meaningful roles and identities as civilians who are accepted by their families and communities in a context of local and national reconciliation" (UNICEF, 2007). Sadly, the transition from soldiering to civilian life is frequently met with enormous challenges, ranging from a lack of adequate accountability mechanisms (Kiyala, 2015) to unattended issues of psychosocial well-being which impede a meaningful resettlement in society (Kiyala, 2020; Wessells/Monteiro, 2001), and dealing with mental health problems (Betancourt et al., 2012; Coppens et al., 2012). These three problems are only the few of the myriad residual post-war consequences for children and adolescents.

In the aftermath of armed conflicts, society grapples with these issues. A study conducted in Nepal on previous child soldiers' participation in rebel groups, researching their psychosocial wellbeing, concluded two important aspects to be considered: "inter-subjective determinants of well-being" that are part of the close social environment in developing familiarity (family and community) and "structural influences on well-being" related to their survival after they come back to civilian life (Medeiros et al., 2020). The outcomes generated in this study identified a trend of the dialectic connection between the "social, political, economic and emotional aspects of resilience" vis-à-vis the violent settings of young children.

20.2.5 The Mainstream Criminal Justice System versus African Indigenous Jurisprudence

Like any system, both the mainstream criminal justice system and indigenous justice systems have their inherent limitations; these two systems are presented in dichotomic terms, which are conveyed by the narrative that one is a worse alternative of justice, whereas the other is more humanising and the best option. A dialectical analysis of both reveal they share some means of rebuking errors, antisocial behaviour and crimes by means of punishment. However, the ends pursued by conventional criminal justice diverge from the African worldview on justice that we discuss in this section.

The main points of controversy are the following: firstly, the use of the ICC to prosecute international crimes which lie beyond the purview of several indigenous jurisprudence and tribal courts due to the propensity of violence that the ICC has jurisdiction over. Secondly, the modern conception of child soldiers did not feature in the system of traditional justice in pre-and-post colonial Africa until the emergence and increasing scale of the child soldiering phenomenon. Thirdly, the ICC lacks the jurisdiction to prosecute persons under eighteen at the time of the commission of the crime, according to Article 26, pertaining to the exclusion of jurisdiction over persons under eighteen: "The Court shall have no jurisdiction over any person who was under the age of 18 at the time of the alleged commission of a crime" (International Criminal Court, 2002). While the ICC sets out in Article 1 the overall jurisdiction of the Court, the Rome Statute of the ICC does not cover the questions of former child soldiers who become commanders and turn perpetrators of the crimes that require a legal action by the ICC. This point begs the following question: Do former child soldiers whose rights were violated under the international law by being recruited unlawfully and who grow into adults and commit atrocities fall under the category outlined by Article 1 of the Rome Statute of the ICC? It states:

> An International Criminal Court ("the Court") is hereby established. It shall be a permanent institution and shall have the power to exercise its jurisdiction over persons for the most serious crimes of international concern, as referred to in this Statute, and shall be complementary to national criminal jurisdictions. The jurisdiction and functioning of the Court shall be governed by the provisions of this Statute (International Criminal Court, 2002).

The lack of clarity on that question leaves former child soldier such as Dominic Ongwen and Thomas Kwoyelo at the mercy of the court's interpretations. The sentencing of such victim-perpetrators is problematic and raises both moral and ethical questions. The nature of these shortcomings experienced by the ICC can be dealt with by the statement of Rawls (2004), who advocates the reform or abolition of laws and institutions, regardless of how "efficient and well-arranged" they are, because of the risks they run of institutionalising injustice (p. 3).

The prosecutions of criminal activities in the Eastern DR Congo have been labelled as lacking cultural resonance among rural communities; therefore, there is more interest in using traditional justice mechanisms as well as finding solutions at customary level. Reports suggest that most of the population in remote settings are not able and are reluctant to access the country's criminal justice mechanism because of the inaccessibility of the areas, the intricacies of the process and its unaffordability (Aroussi, 2016). The common practice of justice in the Eastern DR Congo tends to involve reparative and restorative aspects believed to be at the hearts of those who suffered fatalities.

The practical means of identifying the limitations in the judicial system in the Eastern DR Congo is to consider the practices of the state justice system. Among the difficulties or limitations in accessing the justice system are aspects identified by Aroussi (2016) that include:

- The high cost of the justice services and the remote location of the communities living in rural areas.
- Payment of high fees to access the services for each activity and for the issuance of an authorisation of arrest for justice to be granted.
- The scarcity of indictment of higher army officials of the Armed Forces of the DRC (FARDC) whose impunity actions remain high.
- People who are convicted are oftentimes seen escaping from places of incarceration due to crumbling custodial circumstances, lack of safety and fraud in penitentiaries countrywide and;

- Recurrent non-payment of court-awarded compensation to the victim due to the release of the criminal, the non-appearance of public compensation resources and the difficulty and price of the process.

Still, in DR Congo, one of the most important limitations of the customary law remains the challenge faced by the country in putting in place more "customary laws and practice more traditional and cultural practices" due to the dearth of subsidies (Mudyara, 2022), but also the insufficiency of courts, especially in the rural areas. On the international front, child soldiers' prosecution has been one of the main concerns of society, in terms of understanding ways of socially reintegrating them, but also of them being accepted by communities, whether they played a key role in the violence or not. Looking at the global level of criminal liability, their prosecution remains obstructed by different laws from an international perspective. As a matter of fact, under Article 26 of the ICC, the Court of law does not give guarantee the jurisdiction for processes of prosecuting people below 18 years of age, though they may be found in the situation they were involved in committing the crime (International Criminal Court, 2002). As a result, the shortcomings of the judiciary system raise a couple of thoughts:

- Child soldiers who are involved in armed group activities are considered adult criminals. Though they might benefit from a certain degree of clemency, they remained accused perpetrators. The "best interest of those children" (Amnesty International, 2000) should be emphasised, and apply to avoid the kind of severe punishment reserved for adult soldiers who commit war-related atrocities.
- Child soldiers are indicted for past crimes committed after reaching the threshold of criminal responsibility age—at 18, according to the international law. Adult commanders who recruit and force them to violate international human rights and humanitarian law should shoulder the greatest responsibilities for their alleged crimes, because they accomplish instructions they receive from senior commanders, at times under death threats (Popovski/Arts, 2006).

The limitations of the local and international criminal justice system toward child soldiers in DR Congo, Sierra Leone and Uganda are generally explained by the understanding of war-related crimes pursuant to the Rome Statute of the ICC, which has changed the overall judicial configuration and landscape of the administration of justice. The majority of crimes defined as war crimes, crimes of genocide and crimes against humanity have never been handled by traditional justice mechanisms. For instance, the crimes for which Dominic Ongwen was prosecuted and sentenced are beyond the capacity and power of a *mato oput* process (Manoba/Sehmi, 2021). However other LRA rebels have undergone that process (Negara, 2012). Because of the lack of historical precedent on *mato oput* dealing with crimes of international amplitude, "the application of traditional justice mechanisms must be considered based on the value they may add during certain parts of the proceedings" (Manoba/Sehmi, 2021). Furthermore, such adaptation must be context specific, and demands the full incorporation of the views of victims of mass atrocities in their choice to engage (or not) with these traditional processes; such traditional justice procedures must guarantee the complete participation of women and social groups traditionally marginalised, and curb the perpetuation of culturally-entrenched marginalisation (Manoba/Sehmi, 2021). Similarly, instead of holding narrow conceptions of crime, alternatives should be considered in extremely complex cases such as that of Mr Ongwen and Mr Kwoyelo. Macdonald and Porter (2016, p. 718) and Porter (2016, p. 718) contend: "The case of Kwoyelo shows the importance of looking at war crimes and available options for redress in the political, socio-economic and institutional context in which they occur, rather than considering them as divorced from the rest of lived realities". This argument underlines the necessity to consider a holistic model of accountability for former child soldiers, which is adapted to socio-cultural realities, and framed within human rights principles, especially with respect to war survivors' rights, those of both victims and perpetrators.

20.2.6 Underlying Justification of Child Soldiering Accountability

Justice has been described as an inviolable and fundamental characteristic? of the nature of each person, that even society as a whole cannot countermand (Rawls, 2004). Society should uphold this inviolability in assuring that justice is maintained in any situation, condition, or circumstance. It becomes evident that former child soldiers possess an inviolability founded on justice and protected by justice, to the same degree as victims who complain that their rights and dignity were violated by soldiers who might have been younger (or older) than 18. While in war time, the scale of atrocities and soldiers' movements do not guarantee the identification of perpetrators, in some instances, perpetrators become known to victims when guns fall silent and war survivors, both victims and perpetrators become aware of who did what to whom. What to do with persons whose rights and dignity have been transgressed? What should be the fate of aggressors who have survived the brutality of war? How should we treat persons who are defined as children under the international humanitarian law when they allegedly violated international human rights and humanitarian law, and those who were aware at the time of the commission of international crimes?

Firstly, it is a legal requirement of international law that children accused of violating human rights law and humanitarian law who have exited armed groups and forces be called to account, according to Article 8(9) of the Paris Principles pursuant to the treatment of children within justice mechanisms. The Rome Statute of the International Criminal Court excludes in its jurisdiction lawsuits over any person who was under 18 at the time when he/she allegedly committed a crime, and proscribes criminal prosecution by international court or tribunal while pursuing prosecution and punishment solely for armed forces and group leaders.[2] It also holds that children accused of crimes under international or national law, supposedly perpetrated while

[2] Art. 8(6); 8(7) UNICEF, The Parish Principles.

associated with armed forces and groups, be treated in accordance with international standards for juvenile justice.[3]

Children who are accused of breaking the law have the right to legal help and fair treatment in a justice system that respects their rights. Governments are required to set a minimum age below which children cannot be held criminally responsible and to provide minimum guarantees for the fairness and quick resolution of judicial or alternative proceedings (Clark, 2008b).

Article 8(9) of the Paris Principles advocates respect in the use of all relevant international laws and standards, with consideration to the defendants' status as children. It states: "Alternatives to judicial proceedings should be sought for children at the national level";[4] and also makes the following provision: "If national judicial proceedings take place, children are entitled to the highest standards of safeguards available according to international law and standards and every effort should be made to seek alternatives to placing the child in institutions."[5]

To conclude this legal justification for adequate justice mechanisms for juvenile soldiers, the law requires: "Children associated with armed forces or armed groups who return to communities without undergoing any judicial or other proceedings should be closely monitored to ensure that they are not treated as scapegoats or subjected to any processes or mechanisms that contravene their rights".[6]

Secondly, the process of child soldiers accounting for their actions provides the opportunity to dissociate child victims from perpetrators, thus avoiding the stereotype of *demobi* for all former child soldiers, regardless of whether they have actively participated in hostilities or not; it thus facilitates their social reintegration. Thirdly, it discourages the phenomenon of voluntary re-enrolment, defections and switching between different armed groups; and finally, war-affected communities are reluctant to accept child soldiers back into the community without them going through a proper process of accountability for their alleged acts of violence when soldiering. In the same context, self-demobilised child soldiers fear for their lives and are afraid of reprisals and unlawful arrests, due to the fact that there is no juvenile justice structure to protect them and to ensure that they are heard fairly regarding alleged atrocities that they might have been responsible for (Kiyala, 2015). Demobilised young soldiers in this inquiry generally did not agree that they should be held accountable in a court of law for their actions performed under the orders of their superiors and armed group commanders. Any attempt to pursue a criminal justice approach would force them back into armed groups and militias that have become safe havens for them. In addition, affected communities suggested that a fair structure of accountability would protect returnee child soldiers and facilitate their reintegration into society, a process that is impeded by the lack of fair justice procedures (Kiyala, 2015).

---

[3] Art. 8(6)(7)(8)(9) UNICEF, The Paris Principles.

[4] Art. 8(9)(9)(0) UNICEF, The Paris Principles.

[5] Art. 8(9(1) UNICEF, The Paris Principles.

[6] Art. 8(11) UNICEF, The Paris Principles.

20.2.7 Child Soldiers in International Law

It is believed that although there are specific advantages in human rights and traditions linked to humanitarian activities concerning global and non-global conflicts, some discrepancies exist. These inconsistencies are entrenched in the justification of the unpreparedness for the development of and rise in children joining armed groups leading to flaws, gaps in the law, and a dearth of intelligibility affecting the whole issue of child soldiers (Fox, 2005).

One of the concerns in prosecuting children that participate in armed conflict (where some are victims of forcible recruitment or join voluntarily by playing a certain role in the activities) is whether there are solid grounds to find them guilty. The criminalisation of child soldiers by ways of customary national or international law has always generated critical debates. Considering the criminality of child soldiers and various laws (local or international) that should regulate the responsibility of those that join armed groups voluntarily or by coercion, several processes are considered. This makes it difficult to find these child soldiers guilty, whether they were actively involved in atrocities or not. It is believed that the process of accountability for child soldiers encompasses an approach connected to criminal justice and the international criminal law for those who are found infringing the law. Prosecuting acts of violence and other atrocities taking place during a war, in which child soldiers take part as victims and offenders, meet a lot of resistance due to the child soldiers' two-fold identity but also because of their recruiters not considering their age. Research in international law suggests that instead of considering such children as 'perpetrators', they should be rather placed under the column of 'victims'.

Article 40 of the Convention on the Rights of the Child (CRC) stipulates the cognisance of the rights of each child who is accused of any infringement (in this case child soldiers who have participated in the armed groups) to be fairly treated. The CRC addresses this by highlighting the key considerations of abiding by human rights and ultimate liberties, including the age and the willingness of those child soldiers for full reintegration into society (United Nations, 1989). A better emphasis on criminal responsibility related to persons who are underage could be approached at two different levels: crimes committed as a perpetrator, and crimes that consider those involved as victims. Before this discussion, let us consider the opinions raised in this domain by academics which consider the illegitimate recruitment of those whose age does not permit them to be recruited lawfully. Conventional and customary international humanitarian law and international human rights law forbid the unlawful recruitment or utilisation of someone who is under the age of fifteen in armed groups (Drumbl, 2012b).

On one hand, many organisations in civil society, as well as other UN-related structures, have supported increasing the age to eighteen, while Rosen (2009) suggested a worldwide ban on recruiting children under eighteen. In addition, Wessells (2006) reported on the strides made by the Coalition to Stop the Use of Child Soldiers in leading a significant campaign to establish eighteen as the acceptable age for recruitment in military service. Therefore, it is significant that difficulties encountered in prosecuting children involved in armed groups meet challenges, as laws advocated by various organisations make it difficult to establish the whole process of whether or not they are guilty.

Child soldiers' responsibility in committing an offence against local communities needs to be first decided at the national level before any international opinion becomes involved. Yet, the lack of such accountability bodies or their weakness in holding these behaviours unlawful should not exercise pressure on national structures to leave the prosecution to international structures, and the non-execution of the law to prosecute such children should not be encouraged and taken over by regional or international organisations. Unfortunately, in post-conflict situations, several domestic criminal courts rely heavily on the provisions of the ICC, as is the case with the Ugandan ICD (Seelinger, 2017), the Special Court for Sierra Leone, which was the first international tribunal to try and convict persons for the use of child soldiers (Special Court for Sierra Leone, 2000, 2009), and the International Criminal Tribunal for Rwanda (ICTR), which was the prime international court set to prosecute high-ranking individuals involved in massive human rights violations in Africa, particularly the prosecution of those allegedly responsible for the 1994 Rwandan Genocide.

To discourage the illegal recruitment of child soldiers and their use in the activities that are planned by their recruiters, Drumbl (2012b) presents the legal reference to restricting those who practice this business by considering:

- International Human Rights Law and Humanitarian Law determine the legal age limit for recruiting active persons to be part of armed groups. This also includes aspects related to "hard law", and "soft law, policy and best practices".
- International Criminal Law prohibits the recruitment of children under the age of fifteen in international or non-international conflicts.
- Specific cases in this situation include: the Special Court for Sierra Leone and the International Criminal Court (with the possibility of deciding on a criminal sanction).
- International criminal law that is a more practicable approach is considered in order to dissuade the continuity of armed groups in destabilising their respective nations for the benefit of those that fund their activities.

These laws can only work if they are sufficiently practised in eliminating or deterring all possible networks providing funding, expertise and weaponry at the regional, national and international levels. The fuelling of armed conflicts or wars in so-called unstable states will only work if accountability is considered in the sources of finance, training and skills transfer, and provision of fighting materials.

20.2.8 Criminal Responsibility from a Victim Perspective

Global and UN standards state that victimhood can only be established by the process of courts, and these standards are linked to ideals such as the UN Declaration of

Basic Principles for Victims of Crime and Abuse of Power ("the Victims' Declaration") and the UN Basic Principles and Guidelines on the Right to Remedy and Reparation for Victims of Gross Transgressions of International Humanitarian Law and International Human Rights Law (Amnesty International, 2008). Determining responsibility in matters concerning crimes committed by child soldiers, or because of their involvement, is always constrained by legal procedures. Amnesty International (2008) infers, based on international law, that any prosecution of a victim, whether an individual or a group of individuals that suffered a variety of physical or mental discrimination of any kind, is to be presented before the jurisdiction of a court as a consequence of crimes. Referring to the definition of a child soldier as someone who is below the age of eighteen who has been recruited into or utilised by an armed group (Harvey, 2003), irrespective of the manner in which they were engaged in an armed group, he/she is identified as a victim (ibid). it is important to highlight that child soldiers' involvement in an armed group has dire consequences in terms of their bodily and emotional wellbeing.

The typology of victims has been addressed differently, depending on the researcher. On the one hand, there are those commonly known as "direct victims" (people who have undergone direct suffering and finally died as a result of the attacks, corporal and emotional mistreatment, etc. originating from "indirect victims" (people who are hurt due to indirect cruelty committed by those they are close to). On the other hand, Holá and Bouwknegt (2019) reveal a three-layered victimhood that considers child soldiers as "legal victims" (requires that their age, recruitment and use be legally proven), "participating/benefiting victims" (this category of victims are privileged to have their parents take part in various proceedings), and "symbolic victims" (identified as real victims irrespective of their status of victim-perpetrator).

20.3 The Conundrum of the Conventional Justice System

The development of strategies/policies to deal with child soldiers joining armed groups or when they decide to leave for reintegration or to benefit from any other existing programmes goes into various procedures. In the case of dealing with those that have played a key role in any form of violence culminating in the loss of life, many challenges to prosecuting them exist in the justice system. This is because there is a myriad of theories, practices and laws that are propagated, while controversial interpretations of these laws make things complicated.

Challenges faced in dealing with child soldiers in terms of their protection or prosecution lie in the cohort of lawful considerations that work for the betterment of the situation. While protection or prosecution might be granted to victims or perpetrators who played any type of role in armed conflicts, some child soldiers evade the law being enforced. Taking into consideration this argument, various legislations challenge whether the prosecution of child soldiers should take place or not.

The Western logic of retribution is propagated by a legalistic mindset on the assumption that harsh punishment inflicted on wrongdoers will yield deterrence; this punitive judicial rhetoric is antithetical to the African worldview and conception of justice. Henberg (1990) notes: "Despite our moral misgivings, retributive canons of justice—the return of evil to evildoers—remain entrenched in law, literature, and popular moral precept" (p. n.p.). The author contends that "the persistence and pervasiveness" of retributive practice is comprehended from a perspective of "evolutionary naturalism", and its origin is traced in human biology and psychology; he further explains the historical use of retribution in divergent cultural expressions, notably as "law codes, scriptures, drama, poetry, philosophy, and novels" (Henberg, 1990, p. n.p.). He contends: "understanding retributive thought as the quest for solace in the face of suffering helps to explain its variable nature. Since there is no single defensible moral criterion for identifying exact retaliation, culture is more important than nature in selecting among retributive practices" (Henberg, 1990, p. n.p.). His argument is supported by the thought that certain forms of retribution are culturally tolerated, while others are condemned; thus, in lieu of the erroneous tendency to consider legal punishment as morally justified, Henberg sustains that "legal punishment should be thought of as morally permitted" (Henberg, 1990). This reasoning is acceptable in that punishment is accepted across cultures, religions, myths and traditions. However, because each culture defines the nature of retribution that is accepted in a given society, the legal universalism pursued by the ICC and the Dutch Legal Framework is wrong and untrue, and to adopt Rawls (2004)'s view, "[a] theory, however elegant and economical, must be rejected or revised if it is untrue; likewise laws and institutions, no matter how efficient and well-arranged, must be reformed or abolished if they are unjust" (p. 3).

A study conducted by Carlsmith (2006, p. 437) to examine the underlying motives for people to desire punishment found "converging evidence that people punish primarily on the basis of retribution" (p. 437) rather than seeking deterrence or incapacitation. Society often pursues punishment without this being driven by the desire to educate wrongdoers to refrain from recidivism. Thus, Carlsmith (2006) found that utilitarian responses should be considered as punitive sanctions, as he writes: "Retribution focuses on the completed act, and does not concern itself with the likely consequences of the punishment. By contrast, utility expressly and exclusively focuses on the future benefits that can result from punishment" (p. 449).

The psychology of retribution requires examining the relationship between the perpetrator and the punisher, and what is useful and susceptible of influencing the objectives of punishment (Carlsmith, 2006). For this reason, the process of holding former child soldiers accountable should be cognisant of the context and the environment where offences were committed, and the cultural understanding and purpose of sanctions, which should be defined and implemented by the given society. Thus, retribution for former child soldiers should be context-based and enriched by the social, cultural and mythical etiquettes, traditions and religious beliefs of a given society. African traditional societies, such as the Acholi, inflicted severe personal punishment on individuals or groups of persons found to have breached societal norms (Manoba/Sehmi, 2021). In the eastern DRC, the *baraza* applied educational punishment to wrong-doers and children who misbehaved (Kiyala, 2016).

The complexity of holding former child soldiers accountable under international law is the understanding of child soldiering practice in a contemporary world and the established of the ICC, which appears at odds with the approaches adopted by several local communities which hold to longstanding traditions of restorative justice, which are not synonymous with non-punitive sanction at all. There has been no such total exclusion of admonition and punishment in indigenous mechanisms of dispute resolution and holding wrongdoers accountable. Discussing this, Aja (1997, p. 366) asserts: "every crime or offense attracts commensurate punishment aimed at reinstating the community's ontological equilibrium. This, in the main, constituted justice in the indigenous Igbo culture". It is argued that the African legal mindset has consciousness of fairness and justice, and this is explained by Agbakoba and Nwauche (2006, p. 73) in the following terms: "Central to our understanding of the law is an appreciation of the socio-political structure of which the law is part". It is within this context that responsibility and punishment are framed and understood (Agbakoba/Nwauche, 2006). In Uganda, Manoba and Sehmi (2021, p. n.p.) note: "While not documented, the values, infractions (crimes) and punishments, respectively, were often passed down orally from generation to generation through the elders of the community". Sadly, in post-colonial Africa, where Uganda is no exception, communities, families and victims have been disempowered, and the state has taken control of conflict management, thus causing the erosion of some aspects of traditional justice mechanisms, and also taking away the role of adjudication and dealing with conflicts formerly held by community elders and traditional justice leaders (Manoba/Sehmi, 2021).

Several scholars embrace this view, and we question the "legal universalism" of the criminal justice system and its relevance in exporting cases of community-rights violations to an international criminal court which may lack the necessary socio-political, ethical and religious foundation to lead fair and adequate hearings of witnesses and be unable to apply sanctions, if necessary, that are accepted in the environment where the offences have been committed. Rethinking the ICC's approaches becomes necessary, as we will develop in the cases of Dominic Ongwen and Thomas Kwoyelo in the subsequent sections.

Religious beliefs are essential, especially in Sub-Saharan Africa, where the nexus of relation between the living, the dead and the land are so critical for peace. While Western psychology focuses on modern psychotherapy for traumatic post-war experiences, with punitive justice, Sub-Saharan African communities rely on restorative approaches which are deemed necessary to transform wounded relationships and heal the land infected by crimes. To this end, ex-combatants in Mozambique had to be cleansed through *Ka Pahla/ku pahla* rituals which aimed to ease their reintegration into society after their contact with dead bodies an blood and their committing atrocities (Honwana, 1997). In Northern Uganda, among the Acholi people, cleansing and reparation rituals take place under *mato oput* (Baguma, 2012; Bangura, 2008). *Mato oput* is understood as "a process and ceremony involving two clans bringing together the victim and the perpetrator to establish, through mediation, the 'truth' and pay compensation" (Manoba/Sehmi, 2021, p. n.p.). One distinctive dimension of the *mato oput* is the freedom recognised in perpetrators' initiative to accept the role they played in the crime and participation in the process is fully voluntary or not done at all (Baines, 2005). It requires setting a fixed a date for the *Lapid Kwo* (negotiator of the compensation) to hand over the compensation and an agreement with an *ajwaka* (traditional healer or spirit medium) who is then called on to invoke the dead whose spirit will determine that the qualified person in the offended or affected clan will receive the compensation (Manoba/Sehmi, 2021).

In the Eastern DRC, the *baraza* has been utilised as a restorative platform to hold wrongdoers accountable and repair harm in families and in the community (Kiyala, 2016); *baraza* is referred to as locally-based jurisprudence to handle atrocities committed by former child solders (Kiyala, 2021) and seeks to bring non-punitive justice to rural communities (Mudyara, 2022).

African cosmology and beliefs foster wholeness and a holistic approach to dealing with antisocial behaviours. The cleansing dimension of justice is absent in the conventional Western criminal justice pursued by the ICC and similar criminal courts established on the continent to prosecute persons accused of war atrocities, which under the Rome Statute of the ICC could amount to war crimes, crimes of genocide and crimes against humanity. It follows that the ICC cannot be the sole court with universal jurisdiction to pursue legal actions against accused war criminals.

The trial of Dominic Ongwen has raised many questions regarding the ability of the ICC to prosecute war crimes committed in African armed conflicts and civil wars and its relevance to rural sub-Saharan African communities such as the Eastern DRC, Northern Uganda, etc. For instance, Benyera (2021, p. 1) avers: "In order for justice to be done, the International Criminal Court's (ICC) geography and logic must be shifted epistemologically from the Global North". He advocates the use of local justice structures such as *mato oput* which resonate with the survivors of the LRA rebellion's atrocities, both victims and perpetrators, and contends that if *mato oput* justice was instituted first, the 'victims turned perpetrators paradox' would have been addressed via cleansing rituals, granting pardon, and reintegration into the community (Benyera, 2021). Such a traditional community justice would address the psychosocial needs of former child soldiers, foster reconciliation, and ease social reintegration, but it should embrace the rights of children and safeguard young returnee soldiers; it would prevent rejection and marginalisation (Kiyala, 2021; Veale/Stavrou, 2007). A complementarity is a justice scheme advocated by Greenawalt (2009), who calls for an integration approach that merges the principles of the ICC and local initiatives. However, Sehmi (2021) adopts a broader interpretation of the real causes and effects of conflicts, which rest in socio-economic structures which should be incorporated in the courts' pronouncements as justice for victims; the Office of the Prosecutor must recognise the spectrum of socio-economic violence in the charges brought forward against the accused, and finally, the ICC Trust Fund for Victims should be accelerated to provide immediate assistance to victims and reparation must be ordered to pull down "pervasive systems of structural inequalities" (Sehmi, 2021, p. 1).

While proponents of traditional justice strive to either support fully the viability of traditional justice for post-conflict accountability, or call for an integrated model that combines the ICC's tenets and the indigenous jurisprudence, others have misgivings about the potential of locally-based transitional justice mechanisms to ensure neutrality of procedures. To illustrate this point, for example, Branch (2014) suggests that "ethno-justice"—the type of transitional justice experimented with in Northern Uganda and which has become prominent in the post-conflict era in Africa—should be probed as far as its "political and practical consequences" are concerned (p. 608). He is apprehensive of the fact that ethno-justice in the Acholi region "can end up extending forms of unaccountable, patriarchal power within Acholi society, funded and supported by the Ugandan state and international donors" (p. 608). Besides, ethno-justice being a beneficiary of the Ugandan Government could make the State escape accountability for its violence during the war (p. 608). Notwithstanding this concern, accountability processes such as *mato oput,* which incorporate the principles of rights of the victims and reparation for harm, and address the psycho-social wellbeing of war survivors (former child soldiers and victims of their atrocities), are likely to restore the relationship within the community affected by war, as has happened in Northern Uganda, where former LRA soldiers underwent *mato oput* justice (Negara, 2012).

20.3.1 The Dilemma of Conventional Criminal Justice: Dominic Ongwen and Thomas Kweyolo

Taking former child soldiers through a process of accountability is necessary from a humanitarian and legal perspective. On the humanitarian level, this benefits children who are unjustly treated as delinquents and criminals and as a result, suffer marginalisation and exclusion and eventually return to armed groups (Kiyala, 2015). From a legal perspective, child soldiers' accountability can be advocated for on the ground of the circumscribed actor model.

According to Drumbl (2012b, p. 17), "a circumscribed actor can do, has the ability not to do, and the ability to do otherwise than what he/she actually has done". This model corresponds to the condition of many child soldiers. It disillusions public opinion, according to which all acts carried out by child soldiers are remotely commanded and child soldiers do not always have choices to do otherwise. Some of their accounts reveal that they also take initiatives in their own capacities to carry out violent actions, though most of them are ordered by their superiors (Kiyala, 2015).

The use of restorative justice via peacemaking circles and *baraza* is aligned to the above-mentioned legal propositions, which consider adjudicative or non-adjudicative measures as a vital step towards family reunification and social reintegration of former child soldiers (Kiyala, 2018a). International law, in this regard, sets accountability of former child soldiers as a preventive measure against child soldiering and a protective disposition of demobilised young soldiers.

Demobilised child soldiers see incarceration, imprisonment, and death-threats as too severe and unacceptable. To understand their thinking, it should be noted that these are not just ordinary children. They have soldiered, even commanded their peers on the battleground. Their soldiering mentality is not yet managed, to allow for integrating them into society as civilians. This transition is not easy, but it is possible, by allowing children to account for their past antisocial conduct. The most appropriate platform for them should be restorative justice, since the Convention of the Rights of the Child outlaws incarceration and punitive sanctions against demobilised child soldiers for crimes they committed while in armed groups (Drumbl, 2012b). Also, the CRC, Article (40) (1) stipulates:

> States Parties recognize the right of every child alleged as, accused of, or recognized as having infringed the penal law to be treated in a manner consistent with the promotion of the child's sense of dignity and worth, which reinforces the child's respect for the human rights and fundamental freedoms of others and which takes into account the child's age and the desirability of promoting the child's reintegration and the child's assuming a constructive role in society (United Nations, 1989).

While there exist some cases of infringement of this provision, such as in the DRC (Braithwaite, 2015), the main concern remains the administration of justice for former child soldiers who turned perpetrators as adults. Two cases in point are that of Dominic Ongwen (Baines, 2009; Burke, 2016; International Criminal Court, 2021a), who was sentenced to 25 years imprisonment by the ICC for war crimes and crimes against humanity, and Thomas Kwoyelo, who was kidnapped at the age of thirteen and forced into rebellion to serve as a child soldier by the LRA. Kwoyelo is standing trial at the International Crimes Division (ICD) of the High Court of Uganda, and is represented by James D. Pirtle, a naval aviator and aspiring attorney (Pirtle, 2018). Like his fellow Dominic Ongwen, Kwoyelo rose to the rank of commander, and is accused of war crimes and crime against humanity. Kwoyelo's defence team won his case on appeal on equal protection grounds (though the government has refused to release him); he is still in custody in Uganda.

20.3.2 Dominic Ongwen

The case of Dominic Ongwen stands as a poignant illustration of the war-post effects on former child soldiers; the fact of being abducted or being pushed to fight by personal circumstances does not change the reality of child soldiering, especially its effects on the future of those persons who were abducted and compelled to fight. Ongwen became the Brigadier General of the LRA; he participated in devising and implementing plans to assault civilian populations and was charged with the same crimes he himself was a victim of (Pangalangan, 2017). Before the International Criminal Court (2016) Dominic Ongwen was accused, "pursuant to articles 25(3) (a) (direct perpetration, indirect perpetration and indirect co-perpetration), 25(3) (b) (ordering), 25(3) (d) (i) and (ii) and 28(a) (command responsibility) of the Rome Statute", of allegedly committing war crimes (torture, sexual slavery, attacks against the civilian population, cruel treatment, rape, murder and attempted murder, looting, the conscription and use of persons below the age of 15 to participate actively in hostilities, etc.) and crimes against humanity, namely various human atrocities, not limited to enslavement, coerced marriage, inhumane acts, persecution, and other crimes (International Criminal Court, 2016). Dominic Ongwen, the alleged Brigade Commander of the Sinia Brigade of the LRA, was charged with 70 counts of crimes against humanity and war crimes which he was accused of having committed after 1 Jul 2002 in Northern Uganda (International Criminal Court, 2021b). Ongwen's trial began on 6 Dec 2016, and on 4 Feb 2021 Trial Chamber IX found Dominic Ongwen guilty on a total of 61 counts, which encompassed crimes against humanity and war crimes that he perpetrated in Northern Uganda between 1 Jul 2002 and 31 Dec 2005 (International Criminal Court, 2021b).

The former child soldier's fate was decided without taking into consideration the brutal approach used by the LRA to indoctrinate recruited children to execute commanders' orders and live up to this, even after crossing the "magic" 18 years of age, which humanitarian law doctrine calls the end of childhood. For instance, it is reported that Ongwen's defence counsel, Krispus Ayena Odongo, described Ongwen's decision as "a devil's choice" between "compliance or certain death", and Ongwen survived the harsh reality of military life in the LRA by "playing smart" as a rebel fighter (Ross, 2016). So, it was the name he earned for himself, several promotions he managed to gain through adaptation to life in the bush, and that notoriety, according to Ross (2016), which precipitated his indictment by the ICC. Borrowing from the political scientist Adam Branch in his book about the conflict, *Displacing Human Rights*, (Ross, 2016) cites: "Justice in the wake of mass violence means reform towards a new social and political order, and it is this understanding that the ICC disciplines out of existence by equating justice with criminal justice" (p. n.p.). Unfortunately, the ICC's doctrine of justice is criminal justice, which ignores social and political transformation which is instrumental in preventing mass violence. Scott's view of justice embraces the tenets of transitional justice framed by Macdonald and Porter (2016), which is interpersonally oriented and seeks to guarantee societal harmony by promoting restorative processes and reconciliation. This perspective is entrenched in the Acholiland, where most atrocities were carried out by the LRA in Northern Uganda, and where Dominic Ongwen comes from.

The sentencing of Ongwen, which is on appeal (ICC, 2021), is a strong indication of the shifted responsibility which an abducted child soldier takes on, which he would not have if he was not conscripted to fight with the LRA. In a similar vein, the complex and paradoxical portrayals of child soldier in the Lord's Resistance Army's (LRA) as victim, perpetrator, or both, as a result of child abduction such as in the case of Dominic Ongwen, who turned warlord, poses serious moral, ethical and legal challenges to the fairness of the criminal justice system. Society has a moral and legal obligation to protect this category of persons and this suggests a rethinking of the international criminal justice system in light of Dominic Ongwen's conviction (Benyera, 2021). This scholar argues: "In order for justice to be done, the International Criminal Court's (ICC) geography and logic must be shifted epistemologically from the Global North to the sites where these atrocities were committed" (Benyera, 2021: 1). He also points out that the Roman-Dutch law logic which extensively apprises the ICC's jurisprudence provides little existential reasoning based on the principles of validity to the thousands of the victims of the LRA's operations spanning something missing – a range of years? (Benyera, 2021). Accountability for war survivors, which includes both victims and perpetrators, should be context-based and cognisant of the needs, created by war, of war-affected persons and communities, including the perpetrators, via locally-based systems such as *mato oput,* as noted here:

> The logic of the international criminal justice systems in general and that of the ICC need to shift away from western-centric legalism towards the inclusivity of other forms of historical accountability. The logic of historical accountability is that the applicable mechanisms must resonate with the victims whose humanity was harmed. This implies that the needs of the victims inform[s] both the logic and the mechanisms used (Benyera, 2021: 14).

Several scholars have advocated a balance between the victimhood of child soldiers and their responsibility for violations of international human rights law as adults, as Nortje (2017: 16) observes:

> This cross-over from being a victim to becoming a perpetrator of crimes under international law under these unique circumstances should not be brushed aside by the ICC, thereby ensuring that Ongwen's rights as both a victim and perpetrator of crimes under international law can be dealt with legitimately.

Using similar reasoning, Pangalangan (2017) argues: "The Court in Ongwen faces the difficult task of determining where the realm of the victim ends and that of the perpetrator begins" (p. 634); thus, he infers: "Court must confront not only the hefty burden of adjudication but the sensitivities of ensuring that the long-term objective of protecting the child soldier is neither abandoned nor shifted" (p. 635).

This argument corroborates the views held by traditional leaders, political leaders and religious leaders among the Acholi people of Northern Uganda (Bangura, 2008; Benyera, 2021; Ramos, 2019; Simons, 2016). Unfortunately, the request to adopt Ugandan traditional justice approaches in sentencing the LRA commander Dominic Ongwen at the International Criminal Court was declined; that plea was brought to the attention of the ICC by the legal representation for victims, which noted the potential of these traditional mechanisms to avoid the marginalisation and discrimination of perpetrators (Manoba/Sehmi, 2021).

20.3.3 Thomas Kwoyelo

Thomas Kwoyelo was 13 when he was abducted by the LRA rebels who trained him to fight within their ranks (Pirtle, 2018), and he rose to become a colonel in the rebellion led by Joseph Kony. Thomas Kwoyelo, alias Latoni, is facing 93 counts of war crimes and crimes against humanity which he allegedly committed between Jan 1995 and Dec 2005 as a colonel in Northern Uganda; the long-awaited trial commenced on 24 Sep 2018 and it is reported that the first two prosecution witnesses testified from Mar 12 to 14, 2019, and the court started hearing testimonies from five more prosecution witnesses from 1 to 18 Jul 2019? (Ogora, 2020). Thomas Kwoyelo was enlisted into the LRA rebellion and underwent a complete military training under the eyes of Joseph Kony; later, Kwoyelo became a war criminal charged on 53 counts before the Ugandan war crimes courts (ICD) in Gulu in Aug2010; in addition, his charges included the violation of the Fourth Geneva Convention, pursuant to Art.147 of Uganda's 1964 Geneva Conventions Act. Colonel Thomas Kwoyelo, who is now 52, has been in custody for 13 years, a very long pre-trial detention, after he unsuccessfully asked to be transferred to the International Criminal Court (Matsiko, 2022).

The court has been having difficulties when it comes to victim participation in Kwoyelo's case. In the ICD, victims' lawyers present their views and concerns to the court, and should suspects be convicted, there may also be an opportunity for reparations. Government officials, however, say resources to allow victims to participate fully in the proceedings are insufficient (Nyeko, 2018–2022). The Constitutional Court, in its ruling No. 36 of 2011, concluded that the defendant, namely Thomas Kwoyelo, was entitled to amnesty because he had renounced his rebel activities (International Committee of the Red Cross, 2015). However, that judicial pronouncement was not executed, thus creating scepticism about the fairness of the accused's prosecution.

The test of the word missing of the International Crimes Division of the Court was to demonstrate whether criminal courts in Uganda could utilise customary international law as grounds to conduct the prosecution of war crimes, crimes against humanity and genocide (Seelinger, 2017). One more concern was the inability of the ICD to directly implement customary international law, and this could divest the Ugandan courts of the power to pursue legal actions against some categories of crimes or to prosecute them to the maximum extent (Seelinger, 2017). Furthermore, Thomas Kwoyelo's alleged crimes took place between 1992 and 2005, in the middle of the passage of the Geneva Conventions Act (1964) and the International Crimes Act (2010) (Seelinger, 2017). It was a period during which Uganda's written law did not cover war crimes and crimes against humanity carried out in the context of non-international armed conflict; thus, the customary international law becomes an instrumental source of law for the ICD to lead action against (Seelinger, 2017). It follows that the ICD is entirely dependent on the customary international framework to prosecute crimes of violations of international human rights and humanitarian law which fall under the jurisdiction of the ICC; thus, the ICD is left with limited options for prosecuting non-international crimes that occurred between 1964 and 2010. So, "direct application of customary international law is one way to enable the domestic prosecution of atrocities committed in Northern Uganda during the 1980s and 1990s as war crimes or crimes against humanity" (Seelinger, 2017: 39).

The question which has been asked is the following: Why were all the senior LRA commanders granted amnesty while this has been denied to Kwoyelo, who was a child at the time he was abducted by Joseph Kony's LRA, and was conscripted against his will into the rebel group?[7] (Matsiko, 2022). Furthermore, there have been interrogations as to why Kwoyelo would be the only senior LRA commander put on trial while others have been given amnesty. There is no other LRA commander who has been tried by the ICD for war crimes and crimes against humanity since

[7] Question raised by advocate Moses Sserwanga, Mr Kwoyelo's Counsel.

the inception of the domestic amnesty law in 2000; and an estimated 12,900 former LRA combatants have benefited from this (Matsiko, 2022). This situation of injustice against Mr Kwoyelo ensuing from the conventional justice parameters is equally raised by Macdonald and Porter (2016), who point out that other officers, even more senior to Kwoyelo, had been granted amnesty under Uganda's 2000 Amnesty Act, including the former LRA spokesperson Sam Kolo, who surrendered in 2005. Kwoyelo submitted that he was "a victim not only of unequal treatment, but also of gross government negligence" (Macdonald/Porter, 2016: 701). Kwoyelo's deposition refers to the point we raised earlier, regarding the failure of the State to protect children such as him; he stated: The same state, "which should have protected me from abduction as a child, is the same state that is prosecuting me when I have since renounced armed rebellion"[8] (Macdonald/Porter, 2016: 701).

As a backdrop to this analysis, it is important to interpret and understand the context of the Acholiland, where egregious atrocities were committed by the LRA, the consciousness around the response to crimes and the impact of justice and its outcomes (punishment, redress or restoration) on the future of societal harmony (Macdonald/Porter, 2016). To curb the uneven and unequal administration of justice, and the discriminatory and preferential treatment of some alleged war criminals, several scholars and law practitioners continue to advocate the use of restorative justice and indigenous approaches of accountability for violations of international human rights and humanitarian law (Kiyala, 2022) or even what Branch (2014) calls 'ethnojustice' in the context of Uganda. The aims of innovative jurisprudence in handling former child soldiers' accountability is reflected in the following argument by Macdonald and Porter (2016: 706):

> Thus characters such as Kwoyelo no longer have an individual identity; rather, they take on a broader social and collective identity – that of the innocent child, abducted against his or her will, deprived of personal agency, forced to commit terrible crimes, and whose fate is now subject to the vicissitudes of a hostile central state.

In Gulu, where Kwoyelo comes from, people expressed mixed feelings and expressed their difficulties in reconciling compensation, punishment, forgiveness and incarceration, because of the paradoxical identity of this former child soldier, a rebellion commander, not standing accused of heinous crimes he committed (Macdonald/Porter, 2016). To overcome this conundrum, some Ugandans suggested that victims of his atrocities should be compensated and Kwoyelo could then be taken though the *mato oput* community justice (Macdonald/Porter, 2016). This view is closer to the suggestion made by his counsel, advocating *mato oput* justice instead of the international criminal justice system (Pirtle, 2014). The social harmony evoked by Macdonald/Porter (2016) resonates with 'ethnojustice' and is also described as traditional justice which "claims to promote social reconstruction, peace and justice after episodes of war by rebuilding traditional order" (Branch, 2014, p. 608). The prospects of a transitional justice that can foster peace and social harmony should be considered as the best alternative in seeking redress for past injustices with large-scale

---

[8] Affidavit in support of the Constitutional Reference No. 36 of 2011, Thomas Kwoyelo, Respondent (Macdonald/Porter (2016, p. 706).

violence involving young people who grew up in armed groups and are now accused of international crimes.

The truth about Acholiland, with regard to the Kwoyelo's trial, comes down to two things, according to Macdonald and Porter (2016: 171): "[...] first, on the moral jurisdiction and probity of the institution responsible for forming judgment and overseeing punishment; and second, on the impact these processes will have on social harmony in post-conflict life". The pursuit of justice should take into consideration the social environment where victims and offenders will continue to live; such awareness should determine the way justice procedures should be conducted, namely, addressing the needs of victims, facilitating the reintegration of offenders and creating the conditions for peace in the aftermath of brutal and violent conflicts.

20.4 Child Soldiering and Justice

Justice is a complex and multifaceted concept. Its complexity emerges from its nature, its objectives and the means to achieve its goals, and finally who decides on the approach to render justice—society and the justice system, victims of injustices and atrocities, or perpetrators? Who is the beneficiary of justice? Who should define the parameters of a satisfactory justice procedure? These questions are crucial to pave the way for an adequate administration of justice, particularly in the context of persons recruited to fight when they are below 18 years of age.

The existence of community rests upon the benefit of justice, because the understanding of truth remains a system of thoughts that leads society. Rawls (2004) postulates that no matter the sophistication and efficiency of a theory, it needs to be forbidden or reviewed if it is incorrect, similarly considering also that "laws and institutions" should be renewed or eradicated, even if they are efficient and well-established, if they are not just for the people. Justice is also presented as an operational establishment of a society that considers the principal societal institutes to dispense the essential civil liberties as well as responsibilities to regulate the division of benefits from communal collaboration (Rawls, 2004). This makes justice the prerequisite for managing the affairs of a nation by using the principles of morality as the key to encouraging the distribution of fair and equitable justice for all. It is the responsibility of state institutions to privilege equality in matters concerning "social, economic, institutional, and political rights" (page reference?) to avoid the dire consequences of the failure to implement such rights. The question that has ever preoccupied the legal mind, child protective advocates, policymakers and sociologists is what model of justice for young people who exit fighting forces they had joined when they were below the age of 18? This is what this chapter will discuss in sections that follow.

20.4.1 Justice Theory

Attention has been paid to conceptualising the concept of justice connected with the issue of child soldiers concerning how this social group is treated by recruiters, the conditions of their recruitment and also their rights to withdraw from such activities and be protected by society. Oftentimes, a serious look is given to dealing with child soldiers' circumstances when recruited (willingly or by coercion) but also to societal services that implement activities related to restoration, social justice, and transitional justice, especially specific models matching African contexts of justice. This is part of the restorative justice foundation that has evolved from the 1970s, with the mission of exploring the results of arbitrations between those who committed crimes and their victims. Johnstone (2013) considers restorative justice as a type of "social movement" pursuing the drive of transforming the modern social world issues related to lawbreaking and disruptive behaviour.

The review of justice mechanisms in this essay is conducted under the assumptions espousing the argument put forward by Rawls (2004: 3): "A theory, however elegant and economical, must be rejected or revised if it is untrue; likewise, laws and institutions, no matter how efficient and well-arranged, must be reformed or abolished if they are unjust". Rawls starts with the principle that "[e]ach person possesses an inviolability founded on justice that even the welfare of society as a whole cannot override" (p. 3); thus, justice does not permit that the "loss of freedom for some is made right by a greater good shared by others. It does not allow that the sacrifices imposed on a few are outweighed by the larger sum of advantages enjoyed by many" (p. 4). He further asserts a just society is "a society in which (1) everyone accepts and knows that the others accept the same principles of justice, and (2) the basic social institutions generally satisfy and are generally known to satisfy these principles" (p. 4). It follows that the imperative of justice in society serves to regulate interpersonal conduct and the relationship between individuals and society in such a way that fairness remains the goal that justice pursues. Rawls frames his thoughts about the social corporations, the distribution of interests and the pursuit of shared good as social justice:

> In this case while men may put forth excessive demands on one another, they nevertheless acknowledge a common point of view from which their claims may be adjudicated. If men's inclination to self-interest makes their vigilance against one another necessary, their public sense of justice makes their secure association together possible (Rawls, 2004: 4–5).

When pursuing justice against a category of people in society, such as children accused of war-related crimes, various opinions emerge; some advocate locally-based approaches of dispute resolution which encompass cleansing rituals, healing and reconciliation processes; others demand the establishment of the criminal justice system apparatus to adjudicate the cases of human rights and humanitarian law violations.

There is, however, a myriad of ways restorative justice has been presented by different academics and its practice has been widely used worldwide. From the viewpoint of most experts in the domain, restorative justice allows a diversity of people to use a dissimilar approach to answer some concerns related to harm committed in any disputable matter. Oftentimes, it encompasses key principles like "voluntariness, safety, inclusion, respect, dignity, responsibility, accountability, truth-telling and honesty – and the prudential values are informed in any given case by the master value of individual choice" (Kirkwood, 2022: 1). These values sustain change of identity in restorative justice processes and forge a better understanding of the problem, persons involved, and the outcome of justice emerge from dialogue and the meaning given to an event by stakeholders.

In general, justice should not be pursued for other ends than those that concerned with the survival of humanity. This reasoning echoes the justice conception of Rawls (2004, p. 5); he asserts: "Among individuals with disparate aims and purposes a shared conception of justice establishes the bonds of civic friendship; the general desire for justice limits the pursuit of other ends". He similarly observes that a person conceives of justice as what constitutes the "fundamental charter of a well-ordered human association" (p. 5). Unfortunately, established societies are rarely well-ordered in this sense, because what is just and unjust is often disputable; people disagree about which values should determine and explain the basic terms of their association (p. 5). This proposition entails that a just society is so defined when it does not support arbitrary distinctions between persons in conferring basic rights and duties and when a proper balance between competing claims to the advantages of social life is defined by rules, despite this controversy around a conception of justice, and "there are other fundamental social problems, in particular those of coordination, efficiency, and stability" (Rawls, 2004: 5).

Stability is a key element of a viable human community that should not be disregarded when institutions strive to define and pursue justice. In the context of former child soldiers, their rights should not be overridden by those of the supposed victims, and the end of justice should be fairness and the best interest of society. For this reason, children accused of violating the rights of others should be heard, and their own victimhood should be balanced with accusations of violations of others' rights. The next section explores the contentious issue of accountability of former child soldiers.

20.4.2 Restorative Justice and Restoring Peace

Available academic writing in the field of restorative justice is wide in range and varies depending on the context. The easiest way that this has been approached is from a summary of the dominant beliefs, suggesting that it encompasses meeting between affected parties and authors of harm, making reparations and moving towards transformation of relationships (Johnstone/Van Ness, 2011). Restorative justice has gained momentum in the last few decades in the sense that it offers alternatives to the existing legal criminal justice system.

From the African perspective, observations have been made that conclude that the justice processes at local levels reflect methods of restorative justice. However, a legal opinion suggests adopting the concept of African restorative justice to combine the existing global way of acting with the inclusion of features that make it more African-related. An important viewpoint proposes considering both local justice systems mostly considered as tradition-related and the international justice system which could bring together these practices to help communities caught in various disputes (Murambadoro, 2015), and encourage the significance of marrying both the local and international justice systems to help surpass the apparent logical thinking that suggests the rejection of local justice, privileging the acceptance of the modern system of justice. Restorative justice has always been used in different programmes such as "victim-offender mediation, community and family group conferencing, sentencing circles, and peacemaking circles combined with reparation probation and community boards panels" (United Nations, 2006: 11). The process of justice for former child soldiers should consider reconciliation, reintegration and building lasting peace in the community where such young soldiers have carried out antisocial actions, including war-related crimes.

Considering the development of the African justice perspective that privileges a community harmony-related style of life and the interrelation of all people concerned, the section following will address some case studies of African restorative justice involving child soldiers. Three countries are concerned: the DRC, Sierra Leone and Uganda. Among the concerns that restorative justice gives much attention to is the attainment of a well-balanced decision that motivates the acceptance of offenders by communities that underwent violence perpetrated by children or any member of the community. In a way, restorative justice takes on retributive justice to bring together those who wronged the community to assume responsibility in the community. There is no meaningful re-settlement of former child soldiers in society without supportive peacebuilding structures that can ensure an ongoing process of relationship-mending in the aftermath of the inhumanities that former young soldiers are often accused of.

Kiefer et al. (2020) divulge the process that encompasses the apology of those who were involved in wrongdoing and finally have the courage to correct the damage done in the community. The offender offers, therefore, to contribute to the promotion of forgiveness by the community (not a goal followed by restorative justice), resolve damages, as well as develop a positive attitude to the justice system. These acts are susceptible of constructing durable peace and creating an environment that is secure, not hostile, to demobilised young soldiers.

The child soldiers' restorative justice case in question takes sources from the practices that offend local and international law concerning the age of recruitment, but also to a large extent the way these children are recruited (voluntarily or by coercion). It is believed that there are serious concerns on moral and legal grounds for persons under the age of 18 being part of recruitment in armed conflict zones. The controversial practice goes in contravention of existing policies, legislature, fairness and the processes that enter into recruiting these children into militarised spaces. The dire consequences of recruiting children by any means (whether by their own will or by coercion) demonstrate the deficiency of the existing strategies and legal protective instruments linked to Children Associated with Armed Forces and Armed Groups (CAAFAG) or child soldiers.

The consequences of child soldiers' recruitment always culminate in prejudicial handling, biases, lack of advantages and socio-economic imbalances. To the above aspects demystifying the malicious disproportional treatment of these child soldiers should be added a variety of predicaments. These should be drawn from the Internationalised Criminal Domestic Justice (ICDJ) formal systems, including the confrontation of matters related to the impartiality of child soldiers' answerability procedures while facing justice, when they are considered as either (infrequently) defendants or sometimes as victims or eyewitnesses (Holá/Bouwknegt, 2019). The alternative of such a restorative justice system based on international norms remains profitable to parties concerned in the areas affected by armed conflicts. However, from a justice system developing procedures where offenders are brought to circles of community justice, oppressors are likely to feel more relieved than in other adjudication procedures, because they are given the opportunity to explain what happened to the victims and why they committed such violence against civilian populations (Kiyala 2018a, 2021). In the case of DR Congo, helping the local justice system to respond to epistemological social injustices toward child soldiers via Local Baraza and Provincial Baraza or intercommunity Baraza (Kiyala, 2016) remains a more practical and appropriate way to deal with local problem-solving for those children or adults who played a key role in the violence, and must be given preference. The functioning of Baraza remains one of the structures of the *Tribunaux Coutumiers* (Customary Tribunals) with constituents playing the role of de facto jurisdiction, dissimilar from the *Tribunaux de Droit Ecrit* (De Jure Tribunals). Part of De Jure's jurisdiction is the Appeal Courts, High Courts and *Tribunaux de Paix (Tripaix)* or Peace Courts, and Magistrate's Courts, while Customary tribunals function at the district level and the Chieftainship/Collectivity level, but also function as supplementary courts of law (Rubbers/Gallez, 2015).

Restorative justice may be a new concept in North Kivu, but its practice through Victim-Offender Mediation has been carried out in both Baraza and transit and orientation centres, under the patronage of Caritas Goma. When children appear at a Baraza, they remain innocent and the responsibility for the armed violence they are accused of is imputed to their parents, who failed to prevent their enrolment in fighting forces, and to armed group commanders, who recruited and used them as soldiers.

Karp (2004: 8) asserts that "accountability is demonstrated through expression of remorse and commitment and through the completion of tasks negotiated as part of the sanctioning process". However, genuine apology, remorse and acts of generosity may prove more effective than simply prosecuting and jailing perpetrators (who sometimes show no remorse for their odious acts), as making proportionate reparation for injuries and human rights violations is complex, as observed by Villa-Vicencio (2008: 12): "The death of a child, the rape of a spouse, the destruction of a family home, a business, a community is irreparable".

Restorative justice emphasises a variety of aspects that are related to re-integrating, restoring, and re-educating the people who have wronged society in different ways. Practical considerations of restorative justice are believed to deal with key aspects that are meant to repair "moral harm" (Ward/Durrant, 2021). Restorative justice, therefore, is differentiated from many other legal practices that come from various responses to criminal injury as part of conventional criminal justice (Daly, 2016). To a certain extent, Brookes/McDonough (2006) believe that harm must be considered as different from a conflict in which someone has been aggrieved. Interestingly, a distinguishing feature of restorative justice gives importance to orienting the type of ethical damage, and handles answerability and obligation, while its final intention is to address the harm.

The focus of the above discussion expresses the fact that crime is a desecration of persons and relationships, which at a later stage requires giving attention to principled morals of obligation and responsibility to prudently seek reparation for the damage caused. Practically, Walker (2006) sums up moral repair by considering the following aspects:

- Making the person who committed the offence take responsibility for his acts.
- Admitting and finding ways to resolve the victim's harm.
- Declaring the violation of the norms by the oppressor to allow the offence committed to be looked at by the assurance of the community.
- Re-establishing confidence among the victims by following existing norms and the practices that express them.
- Inspiring optimism about the standards and the persons accountable for their support to be well-intentioned of trust.
- Re-building acceptable ethical associations with previous enemies who wronged the community, which furnish a prescriptive agenda that helps participants engage directly with the type of harm caused.

Such practical ways to bring together the aggressor and the victim within the restoration of justice require the consideration of frank and openminded communication; interaction that is susceptible of providing a space to understand the implications and ramifications of harm and the arising moral obligation to repair the inflicted or heal sustained wounds; and the responsibility to be undertaken by the aggressor to refrain from harming again. Adopting and voluntarily adhering to this path of responsibility, of restoring communication, of showing openness to taking responsibility for one's antisocial conduct, can forge the path to forgiveness, reconciliation and peacebuilding.

Oftentimes, restorative justice is equated with non-punitive justice. This is a misconstrued conceptualisation. Proponents of this theory of justice argue that restorative justice can use punitive sanctions, but that it intends to yield restorative outcomes; it is not aimed at humiliating the offender, but rather at helping them reform their conduct (Dignan, 2011; Walgrave, 2011a, b). It is within these teleological parameters that Walgrave (2011b) writes about "restorative retributivism" which is distinct from "punitive retributivism", because restorative retributivism ensued from restorative processes and outcomes seek to confront an offender with the responsibility he/she holds for the harm caused, whereas "punitive retributivism" entails that "the offender is confronted with his responsibility by others, and must submit to the consequences imposed by the criminal justice system" (Walgrave, 2011a: 14). An integrated approach of restorative justice supposes including some principles connected to the criminal justice system (Walgrave, 2011a). It follows that the principles of criminal justice are susceptible of leveraging a consistent model of rights and liberties that can shield restorative justice from power abuses (Walgrave, 2011a).

The ultimate aim of restorative justice is to amend attitudes and behaviour, and heal broken relations, in a spirit of shalom, which Zehr (2002: 19) describes as "all-rightness with each other, with the creator and with the environment". Restorative justice's finality contrasts with the criminal justice system, which seeks to hold offenders accountable by means of the punishment they deserve (Zehr, 2002: 6). Restorative justice is grounded in the pursuit of harmonious interpersonal relations after a conflict and the remedying of the distortion of peace induced by a crime which harms peace. Restoring justice entails restoring peace between parties affected by a conflict or crime. Restorative justice adopts a more creative technique when dealing with offences, offenders, victims and society; it highlights the adverse effects of punishment that do not mean true accountability (Zehr, 2002). So, the mainstream criminal justice system takes crime as a breach of law, a transgression against the state and a violation of law and the constitution, while restorative justice considers crime as violations against individuals that occasion liabilities and obligations (Zehr/Mika, 2003).

Restorative justice looks beyond crime to see stakeholders and the obligations to repair harm and ensure everything returns to "all-rightness"; it conceives crime as a contravention of people's relations, which generates obligations for reuniting victims, offenders and community members and healing affected parties and mending broken relations—that is not the goal of the mainstream criminal justice system (Zehr, 2002; Zehr/Mika, 2003). In a similar vein, Kiefer et al. (2020: 105) aver: "Restorative justice provides an alternative to retributive justice by seeking to restore criminal offenders to be responsible members of the community". The communitarian dimension of restorative justice and its goal are similar to the African worldview of justice. Conflicts are dealt with in cultural context and the future of the victim-offender relation is always at stake. Repairing harm and making a commitment to refrain from re-offending constitute the basis of stability and lay the foundation for harmonious relationships between offenders and aggressors. Such attitudes and engagement are deemed indispensable for durable peace.

20.4.3 Restorative Justice in International Law

Emerging as a theoretic method, restorative justice is the system that deals with criminal affairs with consideration to attaining reparation of harm to the wronged person. Youths who fight in armed groups, militias and regular forces may come from known or identified families, and therefore need to be protected by different structures of the society, starting with the family (as the initial institution that is meant to take care of and provide for children's wellbeing). There is, however, an inability to understand the future risks that children undergo in becoming part of the active social groups involved in activities of armed groups. This has become a common reality in some well-known nations where children become the source of recruitment, which has become unavoidable, as circumstances that contribute to such decisions are diversified. Whether or not the responsibility of caring for and providing for children remains a parental duty, many educational or behavioural traits acquired by children are received in the society/community. Therefore, when such children become child soldiers (or street children, as is common), the failure to provide education and the means to bring up our children remains a societal responsibility.

The Western logic of justice contrasts with African perspectives, especially in the case of children used as soldiers accused of human atrocities in war time (Kiyala, 2022). African societies have a longstanding social ethos anchored in the inalienable relation between a person and the community, which is known as African communitarianism (Christians, 2004; Eze, 2008) and is sustained by *ubuntu* ethics which deny retaliation, retribution, and punishment; they rather promote reconciliation, healing, social cohesions and restorative justice (Christians, 2004; Gade, 2013; Schoeman, 2016), and in *ubuntu* ethics, justice entails fairness (Letseka, 2014) and promotes human rights (Metz, 2011). Among the alternatives to punitive justice (criminal accountability), international law allows for restorative approaches for handling children accused of violating international humanitarian and human rights law. This is stated in the Paris Principles, Article 3(6), which states:

> Children who are accused of crimes under international law allegedly committed while they were associated with armed forces or armed groups should be considered primarily as victims of offences against international law, not only as perpetrators. They must be treated in accordance with international law in a framework of restorative justice and social rehabilitation, consistent with international law which offers children special protection through numerous agreements and principles.[9]

Furthermore, Article 3(7) also establishes: 'Wherever possible, alternatives to judicial proceedings must be sought, in line with the Convention on the Rights of the Child and other international standards for juvenile justice'.[10] The Convention of the Rights of the Child (CRC) states:

> States Parties shall take all appropriate measures to promote physical and psychological recovery and social reintegration of a child victim of: any form of neglect, exploitation, or abuse; torture or any other form of cruel, inhuman or degrading treatment or punishment; or armed conflicts. Such recovery and reintegration shall take place in an environment which fosters the health, self-respect and dignity of the child.[11]

Unfortunately, in the DRC, child soldiers who exit hostilities without proper processes for accountability and reintegration into society face a myriad of challenges (Kiyala, 2015). Many child soldiers in North Kivu Province are self-demobilised in an environment marked by a paralysed justice system and the absence of juvenile justice (Kiyala, 2015). What affects them most is the lack of peacebuilding mechanisms during their transition from soldiers to civilian life. The real problem is how

---

[9] Art. 3(6) UNICEF, The Paris Principles.

[10] Art. 3(7), *Ibid.*

[11] Art. 39 UN Convention on the Rights of the Child, GA Res. 44/25, 20 November 1989.

to reintegrate these former child soldiers. Should they be held accountable for their actions as child soldiers as a condition of being accepted back by their communities and, if so, what mechanisms should be employed? Should they be forgiven? Will the accountability and acceptance mechanisms convince them not to re-join an armed group in the future? It should be noted that the process has implications, not only for former child soldiers themselves, but hopefully will also influence other children who might become child soldiers. That is, it has a preventative potential.

Former child soldiers were concerned about the harsh way the local criminal justice system deals with them. This is known among child soldiers as *kufungwa* (a Kiswahili word used for incarceration), and they detest these structures, preferring local jurisprudence (Kiyala 2016, 2021). But experiments conducted in 2021 to facilitate reconciliation between former child soldiers and their communities yielded promising results and turn ex-young combatants to community peacebuilders (Kiyala 2021).

20.4.4 Community-Based Justice and Holding Child Soldiers Accountable

There have been numerous traditional methods utilised on the African continent to deal with aspects related to resolving conflict, mostly using customary courts of law, boards of elders, family talks, and courts at the levels of the community (Kiyala, 2016, 2021). The majority of these methods of encouraging justice, forgiveness, reconciliation, etc. using community-based initiatives began to increase because of the conflicts and violence on the continent. Civil armed conflicts and wars with national backgrounds but motivated by regional or transnational interests have contributed to the problem, leaving many people displaced, dead or affected by violence. These armed conflicts or wars in some parts of the continent have laid a path to implementing various alternatives to law addressing social rights desecrations or culpability and the enormously insensitive handling of people in society.

Experts and academics involved in transitional justice and other related forms of dealing with the business of conflicts after the war and the peace mechanisms have contributed to reducing tensions in many countries. Unfortunately, these processes are challenged by various difficulties at the national and international levels in reconciling and rendering justice to parties concerned in the conflict. These mechanisms help to repair communal relations and institutions in many parts of countries working to avert the reappearance of violence. An all-inclusive observation in most cases of violence and the solutions provided by different bodies suggest the existence of various incongruities between the way critics are presented by academics or field NGOs and the certainty of what stands as the truth about the affected communities in their knowledge of transitional justice (McEvoy/McConnachie, 2012).

In cases where violence was perpetrated in areas where armed conflicts erupted, local initiatives to reconcile, forgive and accept those who were involved in the fighting have adopted successful key mechanisms. The limitations observed in restorative justice and the practices of traditional justice show the gaps between the local justice system and the international justice system. The slow pace and limitations of the international courts to produce speedy trials prompted local initiatives imbued with African traditional restorative principles to satisfactorily conduct the hearings of persons accused of war crimes, crimes of genocide and crimes against humanity (Kiyala, 2022). This is simply because there are active traditional mechanisms for resolving such problems with care and success within community social groups, in the DRC (Kiyala, 2016, 2021, 2022), in Uganda (Baguma, 2012; Bangura, 2008; Institute for Global Leadership, 2009), in Sierra Leone (Graybill, 2010, 2017; Hoffman, 2008) and in Mozambique (Honwana, 1997). The experiments with traditional restorative justice anchored in African cultural values have proved to be effective, especially in the context of formal and informal transitional justice, and they should be considered as a better alternative of justice for former soldiers in the environment of civil wars and other armed conflicts (Kiyala, 2022). Macdonald and Porter (2016: 718) point out that "procedural pluralism" has influenced a popular and a "'holistic' transitional justice across the continent, which incorporates formal 'retributive' criminal processes and informal 'restorative' reconciliatory processes".

The reason for communities leaning towards mending relations and settling conflicts via alternative approaches such as traditional and community restorative justice mechanisms is the longstanding belief and moral obligation that the person has towards the community and vice-versa. The implications of this interdependency are the ethical importance accorded to *palaver* (Bujo, 1998; Ngoyi, 2006; Scheid, 2011) and African humanism, also known as *ubuntu* philosophy (Nabudere, 2005; Udo, 2020). In Sub-Saharan Africa, where interrelations are conceived only within a community (Wiredu, 2008), dealing with the aftermath of violence, conflicts and large-scale human atrocities is not done without figuring out what social harmony and stability will be in the future. Such consciousness defines the process of resolving a conflict and addressing justice where there have been injustices and violations of human rights. Restorative justice imbued with the spirit of *ubuntu* has prevailed in most of the countries that have suffered massive human atrocities, such as Mozambique, Uganda, Sierra Leone and the DRC.

The Baraza mechanism of accountability and conflict settlement and prevention, as well as restorative justice peacemaking circles, have been singled out as viable mechanisms to pursue justice for violations of international law by former child soldiers in the DRC, because these two approaches of conflict settlement are imbued with restorative principles (Kiyala, 2016, 2018a, 2021). The proposed double-track mechanism of child soldiering accountability has been experimented with in the context of African post-conflict societies when such approaches have yielded healing in Mozambique (Honwana, 1997; Wamba, 2004), reconciliation in Sierra Leone (Graybill, 2010; Hoffman, 2008), peace and social reintegration of former child soldiers of the LRA in Northern Uganda (Baguma, 2008, 2012; Institute for Global Leadership, 2009), and reconciliation and social reintegration of child soldiers in the East of the DRC (Kiyala, 2018a, 2021).

20.5 Conclusion

The debate about child soldiers' involvement in armed conflicts in the context of Sub-Saharan Africa and a few other well-known countries that have experienced socio-political and economic challenges has drawn much attention, both in the political and academic fields. While circumstances leading to children joining armed conflicts remain controversial and varied, national and international judicial structures have been battling to render justice in order to maintain the rule of law when it is jeopardised by disastrous events that culminate in loss of life, political instability and the destruction of infrastructure, education systems, health facilities, the judiciary, etc. The context of child soldiers' involvement in fighting forces, whether in a regular state army or an armed group or militia, claims equal share of national resources and opportunities when holding them accountable.

Despite the existence of domestic judiciaries in countries affected by war, the choice of an adjudication system or legal procedures to address the adversities of war, under the international law, lies with the international community, which dictates the rules and venues for procedures. The responsibility of the ICC to administer justice to all citizens and discourage mass atrocities by upholding the principles of democracy, has come under scrutiny, as the ICC gets entangled in procedures that are disputed by legal experts, human rights activists, policymakers and the communities where violations of international human rights and humanitarian law have occurred.

The national and international conventions and protocols developed as watchdog frameworks for the cause of law-abiding practices are commonly unconsidered, and the rights of victims-turned-perpetrators are jeopardised by the notion of legal universalism conveyed by the mainstream criminal justice system represented by the ICC. Therefore, the incongruities in applying the law that governs child soldiers' recruitment and the prosecutions of persons who are no longer children for infringing international law as adults need global attention, particularly in Sub-Saharan Africa, where such cases have come to light as the adjudicating of Mr Ongwen and Mr Kwoyelo is being contested. As a result, the unlawful recruitment of child soldiers and the violations due to young persons' rights being ignored by the activities of the court (particularly the detention of Mr Kwoyelo for thirteen years in pretrial by the ICD, and the sentencing of Mr Ongwen to 25 years imprisonment) put the ICC in the spotlight and voices are raised in urging a rethink of the ICC's fairness on both legal and ethical grounds.

Looking at the various experiments in child soldiering justice from a Sub-Saharan African perspective, the multi-layered nature of factors connected to child soldiers' participation in armed and terrorist groups calls for better handling procedures in their prosecution. However, their liability as child soldiers in participating in armed groups remains challenging as they are both oppressors and victims according to their age, and their prosecution in the local and community-based systems faces lawful constraints. To a certain extent, the indictment and conviction of Ongwen, the unjust privation of Mr Kwoyelo's liberty in pretrial and the denial of his amnesty rights under the Ugandan Amnesty Law typify the insufficiencies and the retribution logic of the mainstream criminal justice system that needs to be reformed, because of it becoming unjust.

A consideration to utilising alternatives to criminal justice mechanisms, as a way to bring restorative justice in armed conflict situations where communities have been wounded, is advocated in this essay. Bringing victim communities together with oppressors (child soldiers) to restore justice has been successfully experienced; this was demonstrated by the cases of the Local Baraza and Provincial Baraza in the DR Congo, the community-based experiment to reach reconciliation between former child soldiers and communities in post-war Sierra Leone, the use of *ethnojustice* in the Acholi local communities under *mato oput* in Uganda, and the implementation of the traditional cleansing rituals known as *ku pahla* or *ukupahla* (communicating with the ancestors), which was used to substitute the criminal justice approach in the reintegration of former soldiers in the aftermath of the Mozambican civil wars; all stand as positive means of restorative justice and restoring peace in these African countries.

Finally, justice frameworks constitute the pathways for dealing with the accountability process of accused child soldiers; thus, restorative justice stands as the future for child soldiering justice, to overcome the limitations of the conventional criminal justice system in holding child perpetrators accountable. This argument is founded on the experiments exploring customary justice of four African countries, namely the DRC, Liberia, Sierra Leone and Uganda. It is therefore necessary to frame justice for former child soldiers within the context of future relations in a social environment where victims and perpetrators must start a new journey of transformed relationships and identities, from victims and perpetrators to partners in nation-building and societal peacebuilding; they should be conscious of the need to build social harmony and maintain peace and stability. The state should address socio-economic inequalities and develop essential programmes of reintegration, which consider the psychosocial wellbeing of survivors of war and militarised conflicts. Necessary measures must be taken to contextualise the principles of international law and adapt it to the socio-cultural reality to preserve fairness and maintain social stability in post-war societies. Such a hybrid strategy of justice is conducive to peace-making and peacebuilding, which are the prerequisite for security and development.

References

Afako, B. (2006). Traditional drink unites Ugandans BBC Focus on Africa. Friday, 29 Sep 2006, 07:50 GMT 08:50 UK. http://news.bbc.co.uk/2/hi/africa/5382816.stm. Accesse 6 August 2023.

AfroSavvy. (2020). UkuPhahla – Communicating with your Ancestors. https://www.afrosavvy.com/post/ukuphahla. Accessed 6 August 2023.

Agbakoba, J., & Nwauche, E. (2006). African conceptions of justice, responsibility and punishment. *Cambrian L. Rev.*, 37, 73.

Aja, E. (1997). Crime and punishment: an indigenous African experience. *The Journal of Value Inquiry*, *31*(3), 353–368.

Almohammad, A. (2018). ISIS child soldiers in Syria: The structural and predatory recruitment, enlistment, pre-training indoctrination, training, and deployment. *The International Centre for Counter-Terrorism–The Hague, 8*(4), 1-29. https://doi.org/10.19165/2018.1.14

Amnesty International. (2000). Child soldiers: criminals or victims? https://www.amnesty.org/en/documents/ior50/002/2000/en/. Accessed 8 August 2023.

Amnesty International. (2008). Victims' rights: Fact Sheet Four 2007-2008. https://www.amnesty.org/en/documents/ior50/002/2000/en/. Accessed 8 August 2023.

Among, H. (2013). The application of traditional justice mechanisms to the atrocities committed by child soldiers in Uganda: A practical restorative justice approach. *African Human Rights Law Journal, 13*(2), 441–463.

Aning, K., & McIntyre, A. (2005). From youth rebellion to child abduction: The anatomy of recruitment in Sierra Leone. In A. MacIntyre (Ed.), *Invisible stakeholders: Children and war in Africa* (pp. 67–86). Pretoria: Institute for Security Studies.

Aroussi, S. (2016). Women, Peace and Security and the Democratic Republic of the Congo: Time to Rethink Sexual Violence as Gender-Based Harm? *Politics and Gender, 10*(3), 1–28.

Baguma, C. (2012). When the traditional justice system is the best suited approach to conflict management: the Acholi Mato Oput, Joseph Kony, and the Lord's Resistance Army (LRA) in Uganda. *Journal of Global Initiatives: Policy, Pedagogy, Perspective, 7*(1), 31–43.

Baines, E. (2005). Roco Wat I Acoli. Restoring Relationships in Acholi-land: Traditional Approaches to Justice and Reintegration. Liu Institute for Global Issues and Gulu District NGO Forum. https://sppga.ubc.ca/wp-content/uploads/sites/5/2016/03/15Sept2005_Roco_Wat_I_Acoli.pdf. Accesed 8 August 2023.

Baines, E.K. (2007). The haunting of Alice: local approaches to justice and reconciliation in Northern Uganda. *International Journal of Transitional Justice, 1*(1), 91–114.

Baines, E.K. (2009). Complex political perpetrators: reflections on Dominic Ongwen. *The Journal of Modern African Studies, 47*(2), 163–191.

Bainomugisha, A. (2011). Child Soldiers in Northern Uganda: An Analysis of the Challenges and Opportunities for Reintegration and Rehabilitation University of Bradford]. Bradford. https://bradscholars.brad.ac.uk/handle/10454/5284.

Bangura, A.K. (2008). The Politics of the struggle to resolve the conflict In Uganda: Westerners pushing their legal approach versus Ugandans insisting on their Mato Oput. *Journal of Pan African Studies*, 2(5), 142–178.

Bayer, C.P., Klasen, F., & Adam, H. (2007). Association of trauma and PTSD symptoms with openness to reconciliation and feelings of revenge among former Ugandan and Congolese child soldiers. *Jama, 298*(5), 555–559.

Beber, B., & Blattman, C. (2013). The logic of child soldiering and coercion. *International Organization, 67*(1), 65–104.

Benyera, E. (2021). Child victim, loyal war spirit medium or war criminal: shifting the geography and logic of historical accountability in Dominic Ongwen's ICC trial. *African Identities*, 1–16. https://doi.org/10.1080/14725843.2021.2005537

Beresford, S. (2005). Child Witnesses and the International Criminal Justice System: Does the International Criminal Court Protect the Most Vulnerable? *Journal of International Criminal Justice, 3*(3), 721–748.

Betancourt, T. S., Borisova, I., Williams, T. P., Meyers-Ohki, S. E., Rubin-Smith, J. E., Annan, J., & Kohrt, B.A. (2013a). Research Review: Psychosocial adjustment and mental health in former child soldiers–a systematic review of the literature and recommendations for future research. *Journal of Child Psychology and Psychiatry, 54*(1), 17–36.

Betancourt, T.S., Newnham, E.A., McBain, R., & Brennan, R.T. (2013). Post-traumatic stress symptoms among former child soldiers in Sierra Leone: follow-up study. *The British Journal of Psychiatry, 203*(3), 196–202.

Betancourt, T., Zaeh, S., Ettien, A., & Khan, L. (2012). Psychosocial adjustment and mental health services in post-conflict Sierra Leone: experiences of CAAFAG and war-affected youth, families and service providers. In I. Derluyn, C. Mels, S. Parmentier & W. Vandenhole (Eds.), *New Series on Transitional Justice.* Antwerp: Intersentia. Vol. 11 (pp. 403–436). Cambridge: Intersentia.

Braithwaite, J. (2015). Paralegals changing lenses. *Restorative Justice, 3*(3), 311–324. https://doi.org/10.1080/20504721.2015.1109332

Branch, A. (2014). The violence of peace: ethnojustice in northern Uganda. *Development and Change, 45*(3), 608–630. https://doi.org/10.1111/dech.12094

Brett, R., & Specht, I. (2004). *Young soldiers: Why they choose to fight.* London: Lynne Rienner.

Brookes, D., & McDonough, I. (2006). The differences between mediation and restorative justice/practice. https://www.researchgate.net/publication/335704100_The_Differences_between_Mediation_and_Restorative_JusticePractice. Accessed 19 February 2024.

Bujo, B. (1998). *The ethical dimension of community: The African model and the dialogue between North and South.* Limuru: Paulines Publications Africa.

Burke, J. (2016). Child soldier to war criminal: the trial of Dominic Ongwen. Ugandan Ongwen was abducted by the Lord's Resistance Army at the age of nine. Twenty-five years later, a trial is to establish whether the victim turned perpetrator. The Guardian. Sunday 27 Mar 2016 14.49 BST. https://www.theguardian.com/law/2016/mar/27/dominic-ongwen-war-crimes-trial-former-child-soldier-icc-uganda. Accessed 8 August 2023.

Carlsmith, K.M. (2006). The roles of retribution and utility in determining punishment. *Journal of Experimental Social Psychology, 42*(4), 437–451. https://doi.org/10.1016/j.jesp.2005.06.007

Christians, C. G. (2004). Ubuntu and communitarianism in media ethics. *Ecquid Novi: African Journalism Studies, 25*(2), 235–256.

Clark, P. (2008a). Ethnicity, leadership and conflict mediation in Eastern Democratic Republic of Congo: the case of the Barza inter-communautaire. *Journal of Eastern African Studies*, 2(1), 1–17.

Clark, P. (2008b). Law, Politics and Pragmatism: The ICC and Case Selection in the Democratic Republic of Congo and Uganda. *Courting Conflict*, 37–44.

Coomaraswamy, R. (2010). The optional protocol to the convention on the rights of the child on the involvement of children in armed conflict–towards universal ratification. *The International Journal of Children's Rights, 18*(4), 535–549.

Coppens, K., Vindevogel, S., Derluyn, I., Loots, G., & Broekaert, E. (2012). Psychosocial care in rehabilitation centres for former child soldiers in northern Uganda. In I. Derluyn, C. Mels, S. Parmentier & W. Vandenhole (Eds.), *Re-member: Rehabilitation, reintegration and reconciliation of war-affected children.* Vol. 11 (pp. 329–362). Cambridge: Intersentia.

Daly, K. (2016). What is restorative justice? Fresh answers to a vexed question. Victims & Offenders, *11*(1), 9–29.

Derluyn, I., Broekaert, E., Schuyten, G., & De Temmerman, E. (2004). Post-traumatic stress in former Ugandan child soldiers. *The Lancet, 363*(9412), 861–863.

Dianzenza, D., & Harris, G. (2022). Building the Capacities of Young South African Men in Responsible, Loving and Nonviolent Fathering. *African Journal of Inter/Multidisciplinary Studies*, 4(1), 53–63.

Dignan, J. (2011). Juvenile justice, criminal courts and restorative justice. In G. Johnstone & D. W. V. Ness (Eds.), *Handbook of restorative justice* (pp. 269–291). London: Routledge.

Dolamo, R. (2013). Botho/Ubuntu: the heart of African ethics. *Scriptura: International Journal of Bible, Religion and Theology in Southern Africa, 112*(1), 1–10.

Drumbl, M. (2016). The Kapo on Film: Tragic Perpetrators and Imperfect Victims. *Griffith Journal of Law & Human Dignity, 6*(1), 229–271.

Drumbl, M. A. (2012a). Child soldiers and clicktivism: Justice, myths, and prevention. *Journal of Human Rights Practice, 4*(3), 481–485.

Drumbl, M.A. (2012b). *Reimagining child soldiers in international law and policy.* Oxford: Oxford University Press.

Dupuy, K. (2010). They Fight Like Soldiers, They Die Like Children: The Global Quest to Eradicate the Use of Child Soldiers. *International Journal, 66*(1), 228–230.

Eze, M. O. (2008). What is African communitarianism? Against consensus as a regulative ideal. *South African Journal of Philosophy, 27*(4), 386–399.
Forstmeier, S., Kuwert, P., Spitzer, C., Freyberger, H. J., & Maercker, A. (2009). Posttraumatic growth, social acknowledgment as survivors, and sense of coherence in former German child soldiers of World War II. *The American Journal of Geriatric Psychiatry*, 17(12), 1030–1039.
Fox, M.-J. (2005). Child soldiers and international law: Patchwork gains and conceptual debates. Human Rights Review, *7*(1), 27–48.
Gade, C.B. (2013). Ubuntu and Restorative Justice: addressing the strife and divisions of the past in post-apartheid South Africa [online] (PhD). Aarhus University, Department of Culture and Society, Denmark.
Gates, S. (2017). Membership matters: Coerced recruits and rebel allegiance. *Journal of Peace Research,* 54(5), 674–686.
Graybill, L. (2010). Traditional practices and reconciliation in Sierra Leone: the effectiveness of Fambul Tok. *Conflict Trends*, *1*(3), 41–47.
Graybill, L.S. (2017). *Religion, tradition, and restorative justice in Sierra Leone.* Indiana: University of Notre Dame Press.
Greenawalt, A.K. (2009). Complementarity in crisis: Uganda, alternative justice, and the International Criminal Court. *Va. J. Int'l L., 50*, 107.
Harvey, R. (2003). Children and Armed Conflict: A guide to international humanitarian and human rights law. The International Bureau for Children's Rights. https://resourcecentre.savethechildren.net/document/children-and-armed-conflict-guide-international-humanitarian-and-human-rights-law/. Accessed 8 August 2023.
Henberg, M. (1990). Retribution: Evil for evil in ethics, law, and literature. Temple University Press. https://philpapers.org/rec/HENREF. Accessed 8 August 2023.
Hoffman, E. (2008). Reconciliation in Sierra Leone: local processes yield global lessons. *Fletcher Forum of World Affairs*, *32*, 129–141.
Holá, B., & Bouwknegt, T.B. (2019). Chapter 16: Child soldiers in international courtrooms: unqualified perpetrators, erratic witnesses and irreparable victims? In M.A. Drumbl & J.C. Barrett (Eds.), *Research Handbook on Child Soldiers* (pp. 350–373). Edward Elgar Publishing. https://doi.org/10.4337/9781788114486.00027
Honwana, A. (1997). Healing for peace: traditional healers and post-war reconstruction in Southern Mozambique. *Peace and Conflict: Journal of Peace Psychology*, 3(3), 293–305.
Honwana, A. (2006). Child soldiers in Africa. Philadelphia: University of Pennsylvania Press.
Hove, M., & Harris, G. (2019). *Infrastructures for Peace in Sub-Saharan Africa.* Cham: Springer.
ICC. (2021). Ongwen case: hearing on the Defence appeals against verdict and sentence – Practical Information. https://www.icc-cpi.int/news/ongwen-case-hearing-defence-appeals-against-verdict-and-sentence-practical-information. Accessed 8 August 2023.
ILO. (2002). Eliminating the worst forms of child labour: A practical guide to ILO Convention No. 182. International Labour Organization and Inter-Parliamentary Union 2002. http://www.ipu.org/PDF/publications/childlabour_en.pdf
Institute for Global Leadership. (2009). Community perspectives on the Mato Oput process: a research study by the Mato Oput project [online] The Institute for Global Leadership, TUFTS University and Institute of Peace and Strategic Studies Gulu.
International Committee for the Red Cross (ICRC). (2012). Child soldiers and other children associated with armed forces and armed groups. file:///C:/Users/PC2/Documents/CHILDSO%209.pdf
International Committee of the Red Cross (ICRC). (2015). Thomas Kwoyelo alias Latoni v. Uganda, Supreme Court, 8 Apr 2015: Consitutional Appeal No. 01 of 2012 – Thomas Kwoyelo alias Latoni v. Uganda, Supreme Court. https://ihl-databases.icrc.org/en/national-practice/thomas-kwoyelo-alias-latoni-v-uganda-constitutional-court-22-september-2011. Accessed 19 February 2024
International Criminal Court. (2002). Text of the Rome Statute circulated as document A/CONF.183/9 of 17 Jul 1998 and corrected by procès-verbaux of 10 Nov. 1998, 12 Jul 1999, 30 Nov 1999, 8 May 2000, 17 Jan 2001 and 16 Jan 2002. The Statute entered into force on 1 Jul 2002.

International Criminal Court. https://www.icc-cpi.int/sites/default/files/RS-Eng.pdf. Accessed 19 February 2024.
International Criminal Court. (2016). Situation in Uganda: The Prosecutor v. Dominic Ongwen – ICC-02/04-01/15. Geneva. https://www.icc-cpi.int/sites/default/files/CourtRecords/CR2016_05118.PDF. Accessed 19 February 2023.
International Criminal Court. (2021a). Dominic Ongwen sentenced to 25 years of imprisonment. ICC-CPI-20210506-PR1590. https://www.icc-cpi.int/Pages/item.aspx?name=pr1590. Accessed 8 August 2023.
International Criminal Court. (2021b). Ongwen Case: The Prosecutor v. Dominic Ongwen (ICC-02/04-01/15_Trial). https://www.icc-cpi.int/uganda/ongwen. Accessed 8 August 2023.
International Labour Organization (ILO). (1996-2017). C138 – Minimum Age Convention, 1973 (No. 138): Convention concerning Minimum Age for Admission to Employment (Entry into force: 19 Jun 1976). https://www.ilo.org/dyn/normlex/en/f?p=NORMLEXPUB:12100:0::NO::P12100_ilo_code:C138. Accessed 8 August 2023.
Johnstone, G. (2013). *Restorative justice: ideas, values, debates*. London: Routledge.
Johnstone, G., & Van Ness, D. W. (2011). *Handbook of restorative justice*. London: Routledge.
Jørgensen, N. H. (2019). Children associated with terrorist groups in the context of the legal framework for child soldiers. *Questions of International Law*, 60, 5–23.
Jourdan, L. (2011). Mayi-Mayi: young rebels in Kivu, DRC. *Africa Development*, *36*(3–4), 89–112.
Juma, M. (2012). Community-based approaches to the reintegration of self-demobilised child soldiers: the case of the Democratic Republic of Congo. In I. Derluyn, C. Mels, S. Parmentier & W. Vandenhole (Eds.), *Re-member: Rehabilitation, reintegration and reconciliation of war-affected children* (pp. 377–401). Cambridge: Intersentia.
Karp, D.R. (2004). Introducing restorative justice to the campus community. In D.R. Karp & A. Thom (Eds.), *Restorative justice on the college campus: Promoting student growth and responsibility, and reawakening the spirit of campus community* (pp. 5–15). Springfield: Charles C. Thomas.
Kiefer, R.P., Worthington Jr, E.L., Wenzel, M., Woodyatt, L., & Berry, J.W. (2020). Apology and restitution in a role-play restorative justice experiment: Multiple perspectives, multiple measures. *Journal of Psychology and Theology*, *48*(2), 105–117. https://doi.org/10.1177/0091647120911114
Kirkwood, S. (2022). A practice framework for restorative justice. *Aggression and Violent Behavior*, *63*, 101688. https://doi.org/10.1016/j.avb.2021.101688
Kiyala, J.C.K. (2015). Challenges of reintegrating self-demobilised child soldiers in North Kivu Province: prospects for accountability and reconciliation via restorative justice peacemaking circles. *Human Rights Review*, 16(2), 99–122. https://doi.org/10.1007/s12142-015-0361-7
Kiyala, J.C. (2016). Utilising a traditional approach to restorative justice in the reintegration of former child soldiers in the North Kivu Province, Democratic Republic of Congo. *Africa Insight*, 46(3), 33–50.
Kiyala, J.C.K. (2018a). *Child Soldiers and Restorative Justice: Participatory Action Research in the Eastern Democratic Republic of Congo*. Springer International Publishing. https://doi.org/10.1007/978-3-319-90071-1
Kiyala, J.C.K. (2018b). The impact of globalisation on child soldiering: challenges and opportunities in the context of the Democratic Republic of Congo. *International Social Science Journal*, *66*(221–222), 271–284. https://doi.org/10.1111/issj.12132
Kiyala, J.C.K. (2019). Combining restorative justice and social justice: Prospects of child soldiering transitional justice. *African Security Review*, 27(3–4), 193–211. https://doi.org/10.1080/10246029.2018.1559868
Kiyala, J.C.K. (2020). Dynamics of Child soldiers' Psychosocial Rehabilitation and Well-being: Perspectives from Bioecological Systems Theory in the Democratic Republic of Congo. *Child Psychiatry & Human Development*, *52*(3), 376–388. https://doi.org/10.1007/s10578-020-01022-4

Kiyala, J.C.K. (2021). Responding to Epistemic Injustice against Child Soldiers. *The International Journal of Children's Rights*, 29(3), 662–700. https://doi.org/10.1163/15718182-29030006

Kiyala, J.C.K. (2022). Indigenous Approaches of peacebuilding and Accountability of Former Child Soldiers: African Case Studies. In J.C.K. Kiyala & G.T. Harris (Eds.), *Civil Society and Peacebuilding in Sub-Saharan Africa in the Anthropocene: An overview* (pp. 423–464). Springer Nature Switzerland. https://doi.org/10.1007/978-3-030-95179-5_18

Kohrt, B.A., Jordans, M.J., Tol, W.A., Perera, E., Karki, R., Koirala, S., & Upadhaya, N. (2010). Social ecology of child soldiers: child, family, and community determinants of mental health, psychosocial well-being, and reintegration in Nepal. *Transcultural Psychiatry*, *47*(5), 727–753.

Lasley, T., & Thyne, C. (2015). Secession, legitimacy and the use of child soldiers. *Conflict Management and Peace Science*, *32*(3), 289–308.

Letseka, M. (2014). Ubuntu and justice as fairness. *Mediterranean Journal of Social Sciences*, 5(9), 544.

Macdonald, A., & Porter, H. (2016). The trial of Thomas Kwoyelo: opportunity or spectre? Reflections from the ground on the first LRA prosecution. *Africa*, 86(4), 698–722. https://doi.org/10.1017/S000197201600053X

Manoba, J.A., & Sehmi, A. (2021). What the Dominic Ongwen trial could mean for traditional justice mechanisms in the ICC. Africa at LSE. http://eprints.lse.ac.uk/111627/1/africaatlse_2021_07_26_dominic_ongwen_trial_traditional_justice.pdf. Accessed 8 August 2023.

Matsiko, G. (2022). Uganda: Kwoyelo, 13 years in custody without trial https://www.justiceinfo.net/en/89875-uganda-kwoyelo-13-years-custody-without-trial.html Accessed 8 August 2023.

McEvoy, K., & McConnachie, K. (2012). Victimology in transitional justice: Victimhood, innocence and hierarchy. *European Journal of Criminology*, *9*(5), 527–538.

Medeiros, E., Shrestha, P.N., Gaire, H., & Orr, D.M. (2020). Life after armed group involvement in Nepal: A clinical ethnography of psychological well-being of former "child soldiers" over time. *Transcultural Psychiatry*, 57(1), 183–196. https://doi.org/10.1177/1363461519850338

Metz, T. (2011). Ubuntu as a moral theory and human rights in South Africa. *African Human Rights Law Journal*, *11*(2), 532–559.

Mudyara, K. (2022). 5 Decolonization and Transitional Justice. Policing and the Courts Through the Looking Glass: The Road to Transitional Justice and Decolonization in Zimbabwe. In M. Asadullah, C. Cortez, G. Holding, H. Said, J. Smith, K. Schick, K. Mudyara, M. Korchak, N. Kimber, N. Shawush & S. Dawndyck (Eds.), *Decolonization and Justice: An Introductory Overview* (pp. 45–55). online: University of Regina OER Publishing Program. https://opentextbooks.uregina.ca/decolonizingjustice/chapter/decolonization-and-transitional-justice/. 20 June 2023.

Murambadoro, R. (2015). 'We cannot reconcile until the past has been acknowledged': Perspectives on Gukurahundi from Matabeleland, Zimbabwe. *African Journal on Conflict Resolution*, *15*(1), 33–57.

Nabudere, D. W. (2005). Ubuntu philosophy: memory and reconciliation. https://repositories.lib.utexas.edu/bitstream/handle/2152/4521/3621.pdf. Accesed 20 June 2020.

Negara, K. (2012). Bitter Root: two former Lord's Resistance Army commanders seek tribal justice in order to be granted atonement for their crimes [online video]. Al Jazeera 28 Jun 2012 06h36 GMT. https://www.youtube.com/watch?v=oPYXdTL-j_I. Accessed 20 June 2023.

Ngoyi, A.T. (2006). M. Tshiamalenga Ntumba: a philosopher attentive to the problem of African theology. In B. Bujo & J.I. Muya (Eds.), *African Theology in the 21st Century. the contribution of the pioneers* (Vol. 2, pp. 161–181). Nairobi: Paulines Publications Africa.

Nortje, W. (2017). Victim or villain: exploring the possible bases of a defence in the Ongwen case at the International Criminal Court. *International Criminal Law Review*, *17*(1), 186–207. https://doi.org/10.1163/15718123-01701002

Nyamutata, C. (2020). Young terrorists or child soldiers? ISIS children, International Law and Victimhood. *Journal of Conflict and Security Law*, *25*(2), 237–261. https://doi.org/10.1093/jcsl/krz034

Nyeko, O. (2018–2022). A Test Case for Justice in Uganda: Government Should Signal More Commitment to Uganda's International Crimes Division. https://www.hrw.org/news/2018/11/15/test-case-justice-uganda. Accessed 7 August 2023.

Ogora, L.O. (2020). Four Witnesses Testify in Thomas Kwoyelo Trial; Widow Gives Testimony About Husband's Death. https://www.ijmonitor.org/2020/03/four-witnesses-testify-in-thomas-kwoyelo-trial-widow-gives-testimony-about-husbands-death/. Accessed 7 August 2023.

Pangalangan, R.L.A. (2017). Dominic Ongwen and the Rotten Social Background Defense: The Criminal Culpability of Child Soldiers Turned War Criminals. *Am. U. Int'l L. Rev., 33*, 605. https://digitalcommons.wcl.american.edu/auilr/vol33/iss3/6. Accessed 7 August 2023.

Peltzer, K., Nqeketo, A., Petros, G., & Kanta, X. (2008). Traditional circumcision during manhood initiation rituals in the Eastern Cape, South Africa: a pre-post intervention evaluation. *BMC Public Health, 8*(1), 1–6. https://doi.org/10.1186/1471-2458-8-64

Pham, P.N., Vinck, P., & Stover, E. (2008). The Lord's Resistance Army and forced conscription in northern Uganda. *Hum. Rts. Q., 30*, 404.

Pirtle, J. (2014). Traditional Justice as an Alternative for Child Soldiers. https://cedar.wwu.edu/fairhaven_wif/2013-2014/2013-2014/4/. Accessed 7 August 2023.

Pirtle, J. (2018). Everyone Will Hate You For This: How I Came to Represent Former Ugandan Child Soldier Thomas Kwoyelo in His War Crimes Trial (Kindle Edition ed.). Oregon: Hellgate Press.

Popovski, V., & Arts, K. (2006). *International criminal accountability and children's rights*, (Vol. 4). The Hague: The Hague Academic Press.

Pranis, K. (2013). Peacemaking circles. In G. Johnstone (Ed.), *A restorative justice reader* (2nd ed., pp. 117–121). London: Routledge.

Ramos, C. (2019). Dominic Ongwen on trial: problematizing definitional boundaries and exploring the possibilities of socialization. In M.A. Drumbl & J.C. Barret (Eds.), *Research Handbook on Child Soldiers* (pp. 374–389). Cheltenham: Edward Elgar Publishing. https://doi.org/10.4337/9781788114486

Rawls, J. (2004). A theory of justice. Revised Edition. In Ethics (pp. 229–234). Cambridge: Belknap Press of Harvard University Press.

ReliefWeb. (2007). Abducted: The Lord's Resistance Army and forced conscription in Northern Uganda. https://reliefweb.int/report/uganda/abducted-lords-resistance-army-and-forced-conscription-northern-uganda. Accessed 7 August 2023.

ReliefWeb. (2022). Children and Armed Conflict. (3 Jan 2022). https://reliefweb.int/report/afghanistan/children-and-armed-conflict-monthly-update-Jan-2022. Accessed 7 August 2023.

Rosen, D. M. (2007). Child soldiers, international humanitarian law, and the globalization of childhood. *American Anthropologist*, 109(2), 296–306. https://doi.org/10.1525/AA.2007.109.2.296

Rosen, D. M. (2009). Who Is a Child-The Legal Conundrum of Child Soldiers. *Conn. J. Int'l L., 25*, 81.

Ross, S. (2016). Dominic Ongwen and the Search for Justice. https://www.academia.edu/29221028/Dominic_Ongwen_and_the_Search_for_Justice. Accessed 7 August 2023.

Rubbers, B., & Gallez, E. (2015). Réformer la «justice de proximité» en RD Congo. Une comparaison entre tribunaux coutumiers et tribunaux de paix à Lubumbashi. *Critique Internationale, 66*, 145–164.

Sanin, F.G. (2010). Organizing minors: The case of Colombia. In S. Gates & S. Reich (Eds.), *Child soldiers in the age of fractured states* (pp. 121–142). Pittsburgh: University of Pittsburg Press.

Scheid, A.F. (2011). Under the Palaver Tree: Community Ethics for Truth-Telling and Reconciliation. *Journal of the Society of Christian Ethics, 31*(1), 17–36.

Schoeman, M. (2016). The African concept of Ubuntu and restorative justice. In T. Gavrielides & V. Artinopoulou (Eds.), *Reconstructing restorative justice philosophy* (1st ed., pp. 311–330). Lindon: Routledge. https://doi.org/10.4324/9781315604053

Schotsmans, M. (2012). No return home: the (non-) reintegration of youth ex-combatants in Sierra Leone as a challenge to the contextualisation of DDR and transitional justice. In I. Derluyn,

C. Mels, S. Parmentier & W. Vandenhole (Eds.), *Re-Member: Rehabilitation, reintegration and reconciliation of war-affected children* (pp. 215–241). Cambridge: Intersentia.

Seelinger, K.T. (2017). Uganda's Case of Thomas Kwoyelo: Customary International Law on Trial. *Calif. L. Rev. Online, 8*, 19. https://doi.org/10.15779/Z38HT2GB28

Sehmi, A. (2021). Emphasising socio-economic narratives of truth, justice and reparations in The Prosecutor v. Dominic Ongwen. *The International Journal of Human Rights*, 1–24. https://doi.org/10.1080/13642987.2021.2007373

Simons, M. (2016). Dominic Ongwen, Ugandan Rebel Leader, Denies Charges at ICC Trial. Available via The New York Times. https://www.nytimes.com/2016/12/06/world/africa/dominic-ongwen-uganda-rebel-hague-icc.html. Accessed 7 August 2023.

Special Court for Sierra Leone. (2000). Statute of the Special Court for Sierra Leone. http://www.rscsl.org/Documents/scsl-statute.pdf. Accessed 7 August 2023.

Special Court for Sierra Leone. (2009). Trial Chamber II. Case No.: SCSL-04-16-T http://www.rscsl.org/Documents/Decisions/AFRC/613/SCSL-04-16-T-613s.pdf. Accessed 7 August 2023.

Udo, E.J. (2020). The Dialogic Dimension of Ubuntu for Global Peacebuilding. In E. Essien (Ed.), Handbook of Research on the Impact of Culture in Conflict Prevention and Peacebuilding (pp. 302–322). IGI Global. https://doi.org/10.4018/978-1-7998-2574-6.ch018

UNICEF. (1999). The African Charter on the Rights and Welfare of the Child. Document CAB/LEG/24.9/49 of 1990 promulgated in 1990 by the OAU, Document AFR 59/01/97 of Sep 1997. https://www.corecommitments.unicef.org/kp/the-african-charter-on-the-rights-and-welfare-of-the-child. Accessed 20 February 2024.

UNICEF. (2007). The Paris Principles, Principles and Guidelines on Children Associated with Armed Forces or Armed Groups, Feb 2007. http://www.ohchr.org/EN/ProfessionalInterest/Pages/CRC.aspx. Accessed 7 August 2023.

United Nations. (1985). Standard Minimum Rules for the Administration of Juvenile Justice ("The Beijing Rules"). Adopted by General Assembly resolution 40/33 of 29 Nov 1985. http://www.ohchr.org/Documents/ProfessionalInterest/beijingrules.pdf. Accessed 7 August 2023.

United Nations. (1989). Convention on the Rights of the Child. Entry into force: 2 Sep 1990. UN Treaty Series, 20 Nov 1989. Vol. 1577, p. 3. http://www.refworld.org/docid/3ae6b38f0.html. Accessed 7 August 2023.

United Nations. (2000). Optional Protocol to the Convention on the Rights of the Child on the Involvement of Children in Armed Conflicts. UN General Assembly Resolution. Document A/RES/54/263. Adopted and opened for signature, ratification and accession on 25 May 2000 entered into force on 12 Feb 2002 http://www.refworld.org/pdfid/47fdfb180.pdf. Accessed 7 August 2023.

United Nations. (2006). *A Handbook of restorative justice programmes. Criminal justice handbook series.* Vienna: UN Office on Drugs and Crime (UNODC).

Van Rooyen, L., Potgieter, F., & Mtezuka, L. (2006). Initiation School Amongst the Southern Ndebele People of South Africa: Depreciating Tradition or Appreciating Treasure? *International Journal of Adolescence and Youth*, 13(1–2), 13–41. https://doi.org/10.1080/02673843.2006.9747964

Veale, A. (2006). The criminal responsibility of former child soldiers: contributions from psychology. In V. Popovski & K. Arts (Eds.), *International criminal accountability and the rights of the child* (pp. 97–108). The Hague: Hague Academic Press.

Veale, A., & Stavrou, A. (2007). Former Lord's Resistance Army child soldier abductees: Explorations of identity in reintegration and reconciliation. *Peace and Conflict: Journal of Peace Psychology*, 13(3), 273.

Villa-Vicencio, C. (2008). Transitional justice, restoration and prosecution. In D. Sullivan & L. Tifft (Eds.), *Handbook of restorative justice* (pp. 387–400). London: Routledge.

Vindevogel, S., Coppens, K., Derluyn, I., De Schryver, M., Loots, G., & Broekaert, E. (2011). Forced conscription of children during armed conflict: Experiences of former child soldiers in northern Uganda. *Child Abuse and Neglect, 35*(7), 551–562.

Vriens, L., Raviv, A., Oppenheimer, L., & Bar-Tal, D. (1999). Children, war, and peace: a review of fifty years of research from the perspective of a balanced concept of peace education. How children understand war and peace: A call for international peace education, 27–58.

Walgrave, L. (2011a). Integrating criminal justice and restorative justice. In G. Johnstone & D. W. V. Ness (Eds.), *Handbook of restorative justice* (pp. 559–579). London: Routledge.

Walgrave, L. (2011b). Towards restoration as the mainstream in youth justice. In E. Elliott & R. M. Gordon (Eds.), *New directions in restorative justice* (pp. 3–25). London: Routledge.

Walker, M.U. (2006). *Moral repair: Reconstructing moral relations after wrongdoing.* Cambridge: Cambridge University Press.

Wamba, L. (2004). Complementary Strengths: Western Psychology & Traditional Healing. Rebuilding hope for child soldiers in post-war Mozambique. Minneapolis: The Center for Victims of Torture.

Ward, T., & Durrant, R. (2021). Practice frameworks in correctional psychology: Translating causal theories and normative assumptions into practice. *Aggression and Violent Behavior*, 58, 101612.

Wessells, M. (2006). *Child soldiers: from violence to protection.* Cambridge: Harvard University Press.

Wessells, M. (2012). Psychological well-being and the integration of war-affected children: toward a community resilience approach. In I. Derluyn, C. Mels, S. Parmentier & W. Vandenhole (Eds.), *Re-Member: rehabilitation, reintegration and reconciliation of war-affected children* (11th ed., pp. 57–75). Cambridge: Intersentia.

Wessells, M., & Monteiro, C. (2001). Psychosocial interventions and post-war reconstruction in Angola: Interweaving Western and traditional approaches. In D.J. Christie, R.V. Wagner & D. D.N. Winter (Eds.), *Peace, conflict, and violence: Peace psychology for the 21st century* (pp. 262–275). Prentice Hall/Pearson Education. https://psycnet.apa.org/record/2002-17740-022. Accessed 8 August 2023.

Wiredu, K. (2008). Social philosophy in postcolonial Africa: Some preliminaries concerning communalism and communitarianism. *South African Journal of Philosophy*, 27(4), 332–339.

Zehr, H. (2002). *The little book of restorative justice.* PA: Good Books, Intercourse.

Zehr, H., & Mika, H. (2003). Fundamental concepts of restorative justice. In E. McLaughlin, R. Fergusson, G. Hugues & L. Westmarland (Eds.), *Restorative justice: critical issues* (pp. 40–43). London: Sage Publications.

Chapter 21
"Honour Culture" and Rape Denialism in the Southern Sudanese Culture: A Prospective Response from Ubuntu and Restorative Justice

Josephine Kimanu Mauwa and Jean Chrysostome K. Kiyala

Abstract Denialism over rape cases deprives survivors of their human rights and dignity. The South Sudanese's "honour culture" is aimed at protecting the reputation of local communities affected by rape incidence, and hiding the evils that may tarnish the good image of the South Sudanese society. At the same time, the lack of acknowledgement of the suffering caused by such violence against women throws rape survivors into a condition of perpetual victimisation and denies them justice. In this chapter, an examination is conducted on the conventional notion of honour culture linked to rape. The focus is on the attitudes of survivors and local individuals regarding the root causes of honour culture in relation to rape cases. It is noted that the silencing of victims and their lack of access to public justice exacerbate this issue. The public adjudication of sexual violence cases would garner significant attention, which goes against the principles of honour culture that the South Sudanese Society seeks to protect. These perspectives emerged from in-depth interviews and focus-group discussions with 36 participants, including political leaders, traditional leaders, women leaders, among whom were rape survivors and international NGO leaders. Thematic and content analytical methods were employed to examine and interpret the data set. It emerged from this study that the predicament of criminal justice that perpetuates rape injustices via honour culture preservation can be overcome by using the *ubuntu* restorative justice (URJ) practice which promotes dialogue, acknowledgement of guilt, empathy, repair of harm and human rights violation in

Dr Josephine Kimanu Mauwa holds a PhD in Management Sciences, in Public Administration: Peace Studies from the International Centre of Nonviolence, Faculty of Management Sciences, Durban University of Technology, South Africa. She is a lecturer in the Faculty of Social Sciences, Peace and conflict Transformation Department, Université Evangelique en Afrique, Democratic Republic of Congo; Email: mauwakimanu@yahoo.fr

Dr Jean Chrysostome K. Kiyala holds a PhD in Public Administration: Peace Studies from DUT; he is a Senior Lecturer at the International Centre of Nonviolence (DUT), Associate Professor and visiting lecturer at the Université de Bandundu and the Université Evangélique en Afrique (Democratic Republic of Congo); Emails: JeanK@dut.ac.za; kljeanchrysostome@gmail.com; ORCID: https://orcid.org/0000-0002-9587-0230

J. C. K. Kiyala and N. Chivasa (eds.), *Climate Change and Socio-political Violence in Sub-Saharan Africa in the Anthropocene*, The Anthropocene: Politik—Economics—Society—Science 37,
https://doi.org/10.1007/978-3-031-48375-2_21

view of restoring justice and human dignity. URJ prevents retaliation, encourages forgiveness and empowerment of victims – giving them a voice and addressing their violated rights and facilitating restoration after their ordeals in wartime.

Keywords Honour culture · Rape denialism · Rape survivors · Restorative justice · South Sudan · *Ubuntu* restorative justice

21.1 Introduction

Since the eruption of the third civil war in South Sudan in 2013, sexual violence and rape have become endemic, reaching the level of what is referred to as "genocidal rape" (Pinaud, 2020). In 2018, the United Nations Mission in South Sudan gave an estimate of 175 women and girls who were victims of rape and other sexual attacks and physical assaults in Northern Unity state of South Soudan in Bentiu (Nasiwa, 2021). This report was supported by Doctors Without Borders in Bentiu, alleging that over 125 cases of sexual assaults suffered by women were rape-related, within a period of two weeks (Nasiwa, 2021). A similar report estimated that 65% of women and girls were subjected to physical or sexual violence (Nasiwa, 2021). In 2021, health authorities in South Sudan reported that they attended to approximately 330 cases of rape, and other types of physical violence, and other issues related to gender-based violence (GBV) (Takpiny, 2021). Reliefweb reported on the role that radio programmes broadcasting can play in fighting and violence, for example, in Northern-Bahr-el-Ghazal in South Sudan (Kele, 2020). For instance, Rebeca Abour, former Minister of Child Welfare and Social Development observed: "The role of media is great when it comes to fighting these alarmingly common issues by raising awareness about them. Messages about ending violence against females have been translated into local languages spoken here to maximise their reach and impact", cited in Kele (2020). While such information diffusion process is useful in efforts to curb all sorts of violence, it also provides needed data to further analyse rape incidence in South Soudan. Furthermore, we will explore the influence of radio broadcasting in reevaluating the concept of honour culture and potentially reshaping its purpose towards promoting a culture of respect and dignity for every individual underscored by the philosophy of *ubuntu*, which is rooted in the African social ethics, and sustained by URJ (Oko Elechi et al., 2010). *Ubuntu* conveys the notion of African humanism (Gaylard, 2004).

Furthermore, a survey conducted by Ellsberg et al. (2020), based on interviews with 2244 women aged between 15 and 64, revealed an important result: 50% (Protection of Civilian sites (PoCs) camps in Juba) to 65% (PoCs in Juba and Rumbek) of the total number of rapes occurred as a result of physical or sexual violence either by a partner or non-partner during their lifetimes (Ellsberg et al., 2020). In addition, about 35% of respondents suffered rape, attempted rape or other forms of sexual aggression by a non-partner during their lifetime (Ellsberg et al., 2020). Finally, the same investigation shows that women with lifetime-partners experienced a "lifetime prevalence of physical and/or sexual partner violence [ranging] between 54% in the Juba PoCs camps and 73% in Rumbek. Restrictive marital practices and gender norms, and experiences of conflict were major drivers of both partner and non-partner violence" (Ellsberg et al., 2020:1).

It follows that women and girls in South Sudan are among the most affected persons by rape and sexual violence in the world (Ellsberg et al., 2020). Consequently, rape survivors experience the harshest reality of disintegration of their relationships with the communities to which they belong and seek to integrate. This is due to the imposition of a dishonoured identity caused by negative perceptions of rape by the community against rape survivors. They also bear stigmas (Takpiny, 2021). Unfortunately, the process of rejoining their own communities after being exposed to rape and other sexual aggression is often met with hostility by society, mainly because of cultural factors related to honour culture which opposes "shame culture". This reality becomes the focus of confrontation between political and traditional authorities who hold different views on how to manage the challenges faced by rape survivors. Collins Dictionary online (2007) defines 'honour' as "doing what you believe to be right and being confident that you have done what is right". The tensions we referred to above are explained by the fact that most of societies (whether in Africa or outside the continent) perceive rape against women as the destruction of honour of victims; they are further alleged to have brought shame to their respective families or communities – this view is widely spread and culturally justified by the South Sudanese society (Watts/Zimmerman, 2002).

Even though South Sudanese males and females are influenced by Arabic countries' culture, their belief in honour culture related to women is more protective and different from the rest of the Arabic world where, in gender issues, honour culture lead up to killing and that is called "honour killing" (Gibbs et al., 2019). In our view, this is extreme violence against women.

Rather than looking at the extreme violence, the focus of this chapter is on honour culture related to rape survivors, or rape management as an issue of contention between political leaders and traditional authorities. The latter claim to be the custodians of a good cultural practice that protects women who have suffered rape to continue to be perceived as dignified persons despite their traumatic experiences of rape. Traditional leaders preserve this culture because they believe in its ability to help affected rape survivors to reestablish their position in the community by remaining anonymous.

Since the eruption of war in South Sudan, the notion of honour culture related to women protection has been fragilised, caught up into political and traditional protection interests making women, especially women rape survivors, vulnerable. The Sudanese civil war that started in 1955 has affected peace and social harmony (1955 to 1972; 1982–2005) up to the secession of Sudan in 2011. In 2013, the newly independent country went through a drastic increase in violence towards civilians compounded with a higher level of abuse against women. For instance, 24 500 South Sudanese women and girls were reportedly at risk of sexual violence (Sudan Human Security Baseline Assessment, 2019). According to the Global Women's Institute (2017), in some areas up to 70% of women have been victims of sexual and/or physical intimate partner violence; and one in three women have experienced some form of sexual abuse (including rape and transactional sex). This statistic shows how the situation of violence against women has surpassed the critical levels.

The failure of political authorities has led to a drastic impunity of rape perpetrators. As asserted by Lowenstein (2010) "very weak rule of law institution as and insufficient attention by (government of South Sudan) authorities to rule of law issues have given rise to an environment of impunity". Thus, the institutional support to government officials, military forces and male criminals have increased violence committed against women without any fear of punishment (Human Rights Watch, 2009). Not only has the issue of rape and sexual violence been denied by authorities (The Nuhanovic Foundation, 2016), but this violation also rarely comes to the attention of the authorities (Lowenstein, 2010).

Against this background, the underlying factors that weaken gender relations based on the rape crisis have to be elucidated, especially the relationship between rape survivors and their aggressors; that needs to be taken into consideration how to help survivors. Numerous studies have investigated the factors contributing to the violent attitudes and behaviors of individuals who commit rape against women. However, there is a limited amount of research exploring the factors that contribute to the vulnerability of rape survivors' relationships and their challenges in reintegrating into the community, particularly in the context of honor culture. The political settlement and customary institutions in South Sudan are concerned with preserving their societal reputation. Unfortunately, doing so at the expenses of rape survivors does indeed dishonour victims whose rights are violated, especially when they are silenced to preserve the dignity of the South Sudanese culture.

Honour culture emerged as the theory that could explain the attitudes and motivations behind preventing publicity of rape incidence. As a response to these negative impacts of rape culture, we explored the philosophy of *ubuntu*–African humanisms that seeks to promote the dignity of the human person, restorative justice for antisocial acts, forgiveness and reconciliation between parties affected by an offence.

The overall objective of this study was to explore how the desire to preserve the dignity and reputation of the South Sudanese traditional society 'unconsciously' infringes the rights of rape survivors, and the extent to which URJ could help overcome injustices suffered by rape survivors in South Sudan. Concealing rape incidence to preserve honour and the reputation of society was investigated, and the following specific aims were set for the study:

- To explore the issue of rape by analysing its nature, functions, root causes and the extent to which its survivors relate to their aggressors.
- To examine and comprehend various factors that weaken the management of rape incidence in the community; and the extent to which those factors exacerbate victims' vulnerability in socialisation.
- To investigate the impacts of honour culture in reporting and dealing with rape.
- To establish how and to which degree *ubuntu* philosophy and restorative justice principles can inform rape management in the South Sudanese society.

This study is based on the assumption that adopting a restorative approach to address honour culture would promote openness in addressing rape cases. By creating platforms where rape survivors can be heard and their stories shared, the study aims to replace the horrors of honour culture with a focus on upholding human dignity. This approach aligns with the principles of *ubuntu* social humanism, which are rooted in African ethos and emphasise the value of human life and interconnectedness.

This study is very important, because it is concerned with a country where it is reported that some 65% of women and girls have experienced sexual or other physical violence, according to the United Nations Children's Agency's 2019 press release, in which the UN reported that an estimate 88% increase was seen in conflict-related sexual violence for the first quarter of 2021 (Mednick, 2021). The conviction of soldiers who raped a woman and her daughter-in-law in Yei gives Impression that honour culture is becoming irrelevant in face of atrocious acts of sexual violence. As Michael Machar Malual, head of civilian-military relations for the army in Central Equatoria state expressed his views over the sentencing of rapist soldiers in Yei: "We apologise, we won't let it happen again, and we'll arrest people who do it" (Mednick, 2021). Several women hoped that verdict would stimulate more survivors to come out and speak publicly about their sexual ordeals in a country where sexual assault is described as a scourge (Mednick, 2021).

The main components of this chapters are an introduction followed by a discussion on the rape incidence (causes, consequences and denialism). Then, the theoretical framework that guides the research is explained before research design and methods section. Research findings are presented followed by a discussion that precedes this study's conclusions and recommendations.

21.2 Rape Incidence

The ongoing conflict in South Sudan has exacerbated poverty, and instability throughout large parts of the country. This has affected the educational system but also the system making the rule of law difficult to apply. Despite the fact that South Sudan government is bound by the Geneva Convention of 1949 and their additional protocols and acceded to several International Human Rights treaties such as the African Charter on Human and People's Rights and the Convention on the Elimination of All Forms of Discrimination against Women and its Optional Protocol that states: "crimes against humanity and war crimes are not criminalised as such in South Sudanese law, meaning that sexual violence cannot be prosecuted in South Sudan under these crimes" (Soccol, 2018: 3).

Several studies of GBV have identified rape as "a weapon of war" (Card, 1996; Maedl, 2011). The UN Secretary-General report A/SS/163-S/2000/712 on the Promotion and Protection of the rights of children states that rape has become a "weapon of warfare" as bullets and machetes (UN, 2020). In the absence of quantitative and qualitative studies that provide a definition of rape as a "weapon of war", the UN describes this phenomenon as a "tactic of war to humiliate, dominate, instil fear in, disperse and/or forcibly relocate civilian members of a community or ethnic group." (UN Doc. S/2000/712, A/55/163 cited in Maedi, 2011: 130). The use of rape as "a weapon of war" conveys the idea of how it is widespread in the Democratic Republic of the Congo (DRC) and the number of persons who perpetrate this war crime and crime against humanity. Children and adults are all victims of this odious wartime practice.

International NGOs that work on the ground and are in direct contact with rape survivors allege that the South Sudanese government has developed a range of practices to protect individual and the country's reputation by avoiding publicity around rape incidence. As a result of the protective measures to safeguard culture and reputation, injustice has become the byproduct of such societal stance.

Victims are silenced by frightening them to provide proof, evidence of their perpetrators, yet most of rape perpetrators in wartime belong to armies, militias and security services which are powerful tools to discourage victims to report their ordeals. The culture of impunity allows perpetrators to keep on committing offences without fear of being held accountable which implicitly promotes a culture of rape. As asserted by Gravelin et al. (2019: 15),

> rape culture is maintained by the norm of silencing victims of rape, [...] Failure to report rape not only protects perpetrators from punishment but also communicates a tolerance for sexual assault that delegitimises victims' experiences and perpetuates victim blaming.

Muting rape victims in South Sudan affects the management of this offence. In our society, silence continues to symbolize a lack of power (Ahrens, 2006; George et al., 1992; Koss, 1985). Rape serves as a means to further reinforce the powerlessness of women, who are often denied agency and authority (Brownmiller, 1975; MacKinnon, 1987).

The lack of support from governmental institutions of justice to get through restorative justice (United Nations Office on Drugs and Crime, 2006) and processes or compensation for rape survivors lead affected communities to resort to international NGO structures and local community as structures for assistance. Restorative justice gives a voice to both victims and offenders and generates restorative sanctions in a manner that the conventional and adversarial justice processes do not (McGlynn et al., 2012). This is done by empowering survivors where the processes of accountability for rape offences are diverted from governmental judicial institutions to NGOs.

21.2.1 Causes

The impact of the war in Southern Sudan on socio-traditional norms related to rape within the community is significant. Wartime rape has weakened and undermined efforts to combat rape practices, compounded by the lack of enforcement of laws. This is primarily due to the fact that the perpetrators of rape often belong to various factions, including government forces, militias, and rebel groups (Soccol, 2018). Their dual role as both offenders and enforcers of the law creates complexity in ensuring justice. In this context of wartime rape, the suffering inflicted upon women and girls within the community has unfortunately become a distressing norm (Isikozlu/Millard, 2010).

There are instances where government actors protect the country's reputation through rape denialism. The reluctance to establish a specific court in the country to investigate the issue could be due to the fact some members of the regular military services would be responsible in a certain number of cases (Soccol, 2018). Though, prosecution of those responsible for human rights violations and "conflict entrepreneurs" can eventually yield stability and accountability (Snyder/Vinjamuri, 2003), the justice system has been slow in dealing with rape crime. Sexual violence has been ongoing, partly because the Government of South Sudan has made little progress in implementing essential aspects of the 2018's Revitalised Peace Agreement which laid the foundation for dealing with persistent conflict and insecurity, and that has become the basis for the recurrence of conflict-related sexual violence (United Nations, 2022). Women and girls persistently suffer the scourge of sexual violence that has become endemic in South Sudan (United Nations, 2022). This situation needs redress.

21.2.2 Consequences

African pastoralist communities have always been characterised by violence and hostility, which are often employed to defend or expand territorial holdings or to replenish depleted populations (Scoones, 2020). South Sudanese social norms, applicable to both pastoralist tribes and peasant pastoralists, encompass a culture of aggression where revenge against perpetrators of rape, even up to the point of murder, is deemed acceptable. While pastoralist tribes in South Sudan have certain regulations pertaining to honor culture, including extreme honour culture, the fundamental importance placed on honour drives their response. Consequently, resorting to extreme violence becomes the final solution to reclaim their tarnished honour and restore their sense of dignity. Herders are committed to preserving their value and worth (Cohen et al., 1996), the concept of preserving honour through acts of violence, specifically in cases of sexual violence and rape against women within the community, is distinct from the practice of "honour killing" or killing to preserve the honour of the culture.

In the honour killing philosophy like in Middle Eastern culture, a very sensitive society, dominated by the culture of honour (Pitt-Rivers, 1965), female victims of sexual code deviation whether victims or perpetrators are exposed to killing and are killed (Gibbs et al., 2019; Lamis Abu et al., 2007; Navratan, 2014; Tahira, 2006). In South Sudanese tradition, despite being rooted in a herding philosophy and influenced by Arabic and Middle Eastern cultures where honour killing is more prevalent, there is a notable distinction. In cases of sexual misconduct or deviation, rather than targeting females, the focus shifts to the exposure and punishment of male perpetrators, often resulting in their own killings. Here, honour culture is understood through the lens of the herding economy, where the protection of one's property, worth and wealth becomes a symbol of honour (Cohen et al., 1996). Hence, any action undertaken by male members of the community to protect women victims is perceived as act of bravery, and that honour men in the community. Thus, culture honour conveys both a positive and negative connotation. Thus, it is essential to examine this concept as a symbol of protection within South Sudanese communities, as the safety of women and girls largely depends on how society interprets honour, the consequences of rape. When a rape occurs, affected families or communities may seek to eliminate the offender as a means of protecting women from retaliation. In such cases, women are shielded from reprisals by the offender's family, often leading to unreported incidents of rape. However, this protection of women's honour is limited to preserving the reputation of the family or community.

Nevertheless, when rape occurs within the family or in the context of intimate partner violence (IPV), the acceptance and condoning of violence against women by both women (82%) and men (81%) is prevalent. This form of violence, including rape, physical abuse and psychological harm, is seen as a means of maintaining family unity (CARE, n.d.). Women are expected to tolerate such violence. This represents the third stage in the evolution of honor culture (Mosquera, 2013; Shackelford, 2005), where sexual violence and rape are largely disregarded and women lack protection.

21.2.3 Denialism

The existence of rape cannot be denied simply because it is denied by individuals. However, it is often not acknowledged due to the negative consequences it has on the victim's sense of self-worth and the tarnishing of family honour, as mentioned earlier. In reality, rape victims experience a range of reactions, with common responses including victimisation, fear, anxiety, loss of control, flashbacks, difficulty concentrating, guilt, depression, loss of interest in sexual activity and disrupted relationships. The withdrawal behaviour resulting from sadness and depression leads to a breakdown in relationships, as the victim feels embarrassed and ashamed (Loyola University Maryland, 2021). This disruption extends beyond personal relationships, affecting families, communities and society as a whole, causing the victim to feel embarrassed. Both the victim and the perpetrator experience embarrassment, albeit for different reasons. The victim's disruption of relationships leads to a sense of inferiority, shaped not only by her own perception but also by society's negative perception of rape and the stigma attached to it. On the other hand, the perpetrator's withdrawal behaviour stems from fear of facing extreme violence as a consequence of their actions and the subsequent punishment.

Although the reasons for disrupted relationships may differ, the consequences of their actions on tiers such as family members remain the same: a diminution of their family reputation and the tarnishing of their honour. Due to these repercussions of rape, as well as the imperative of upholding their honour and that of their families, individuals in the community are reluctant to acknowledge this offence. Thus, the denial of rape in South Sudan community has multiple facets. The denial of rape could, to some extent, indicate non-acceptance within the community or a reluctance to acknowledge it as such. Instead, a euphemistic approach may be employed to describe the action, aiming to downplay the severity of the issue. Consequently, rape is often substituted with alternative terminology, and is called *zina* (Batha, 2015). Fricke and Amir (2007) assert that rape has the same meaning as one of the offences of the *zina*, namely, adultery or sexual immorality in which most of the cases end by confession of the perpetrators to protect victim from being punished or stoned. As *zina*, the fault is attributed more to victim rather to the perpetrator and leads to specific punishment. In this case, any rape in the marriage is denied but is called *zina* (Ministry of Legal Affairs and Constitutional Development, 2008).

Denial can be individual, personal, psychological, private, shared, social, collective and organised, cultural, historical or official (Cohen, 2002). Despite the existence of tangible evidence, such as the widespread occurrence of rape, authorities are aware of what is happening but deny their responsibility. This denial can be seen as a way for these actors to avoid confronting difficult truths and shift their focus away from an unpleasant and painful reality (Ashkenas, 2013) giving the impression of encouraging the perpetration of rape to continue. With government denial of rape cases in the country, some of the NGOs interviewed accused the government of tacitly perpetuating the culture of rape.

21.3 Theoretical Framework

Employing grounded theory mode of inquiry for the case of rape incidence in South Soudan, we conducted collected abundant empirical data that we analysed to generate the report we organised in themes after the process of coding data as part of thematic analytical research approach. Recognising the a-theoretical dimension of grounded theory research, we also tried to understand rape narratives and complexities by analysing the concept of culture and honour. Later in the inquiry we imagine how *ubuntu* philosophy could be used to avert the adversarial nature of the prosecutorial justice system that partly justified the tendence to conceal rape cases. Therefore, we were able to develop the concept of "honour culture" that we discussed earlier as the theoretical framework that emerged from this inquiry. To make this theory relevant and divert it from the original meaning it was given by participants, we related it to ubuntu to develop a "honour culture" framework that is imbued with the spirit of *ubuntu* to make culture a protective instrument of the most vulnerable social group in wartime, notably women, girls and children.

Two theories inform this inquiry. Firstly, "honour culture" that emanated for the inquiry and *ubuntu* restorative justice (URJ). A combined application of both theories would avert the negative implication of "honour culture" that tended to conceal rape to preserve the reputation and honour of the South Sudanese communities. We used the past here because of rape denialism is slowly disappearing since rape cases are being prosecuted in South Sundan.

21.3.1 Honour and Culture: Conceptual Analysis

Theoretically, the explanation of honour culture is well addressed under socio-psychological and anthropological approaches. According to the evolution of the notion of honour culture, the debate is based on three major explanations (Mohamed, 2018): herding economy protection; a culture of demonstrating dominance; and maintaining one's honour and culture honour to be protected (Cohen et al., 1996), any threat to these foundations of honour leads to violence. Considering South Sudanese people's pastoral culture, it aligns with the herding explanation of honour culture, where the value and importance placed on honour are particularly elevated (Legassicke, 2013). However, their honour eager is very particular when it comes to the issue of women.

Honour is a multifaced and multidimension term which can be understood through different studies, generally in psychology, anthropology or sociology. In social psychology, the term honour is defined as a specific determinant of emotion and behaviour containing different reactions in different cultures (Mosquera et al., 2002b). It involves the reputation of a person: "one's worth in one's own eyes and in the eyes of others". As virtue, the term honour itself does not express any wrong meaning or connotation; it is related to integrity, good moral character and altruism; and as culture, it is based on reputation, self-worth is a male characteristic as it is extremely important for them to maintain their reputation for being competent providers and strong protectors (Cohen et al., 1996). Placed in honour culture, a man's feeling of being a "real man" links himself and his family to a successful reputed defence (Lindsey/Ryan, 2011: 1612). In this sense, difference can be made in an honour culture regarding gender. Men in the honour culture strive to build and defend reputations for strength, bravery and an intolerance for disrespect (Cohen/Nisbett, 1994), yet women strive to build and protect their reputation for loyalty and sexual purity (Vandello/Cohen, 2003). In the cultural code, women's sexual honor is considered an indicator of male relatives' honour, whereas men's honour is determined by the opinion of their parents, close relatives, friends, and peers. The values and priorities assigned to these two forms of gender honour differ significantly.

Masculine honour follows a certain trend of gender role. In this sense, men are honoured by being protector and provider, hard-working, having sexual experience, defending oneself against insults; it refers to values such as respect, bravery, fearlessness, charisma, and pride (Guerra et al., 2012). The feminine honour code emphasises values such as chastity and conformity, sexual purity, modesty, decorum in dress and discretion in social relations (Christianson et al., 2020; Sen, 2005). Even though feminine honour is also related to female gender role, it is more strictly defined and emphasises female sexuality but tends to relax restrictions on male sexuality (except in religious societies) (Thomas, 1959). Women ought to maintain their reputation by behaving with sexual modesty and chastity even though a dishonourable behaviour can disgrace both male and female and tarnish both families' names (Christianson et al., 2020; Cihangir, 2012; Mosquera et al., 2002). In this regard, many theorists relate the notion of honour to shame as equally important. Wilkan cited in Gill (2014) asserts that individuals in communities that value honour are not only motivated by a desire to obtain and maintain honour, but equally concerned with avoiding shame.

Atari (2018) asserts that honour is self-worth based on an individual's (usually a man's) reputation and his own assessment of what others think. However, reputation ought to be considered not only at the individual level, but in both individual and social settings, the latter including group reputation (family values, harmony, respect). According to Shackelford (2005), the culture of honour revolves around the protection of one's reputation, which is regarded as the central aspect. This type of culture places significant importance on a man's readiness and capability to respond to threats or insults against his reputation or that of his family, often through violent or lethal means. It emphasises on "upholding and defending reputation and person of oneself and one's family" (Nisbett/Cohen, 1996).

This chapter emphasises the theoretical aspect of honour culture and provides some explanations on how it destabilises survivors' relationships with their respective community members. There are significant contrasting views perceptions by men and women on honour culture and its fundamental values. On the structural level, the contrast between traditional leaders and political leaders, the issue of shame is put in forefront to decide on any attitude and management to protect survivors. In other words, political structures protect men's honour by shaming rape survivors. Yet, traditional leaders act on the principle of protecting rape survivors to avoid their shame. Therefore, there is a clash between collective protection of honour and individual and relational protection of honour.

21.3.2 Ubuntu and Restorative Justice

The concept of *ubuntu*, deeply embedded in African communal life, emphasises the fundamental importance of recognising and respecting the dignity of human rights among individuals. *Ubuntu* serves as a guiding principle for promoting mutual respect, preventing violence, addressing harm, fostering reconciliation and facilitating healing within human interactions. Originating in the Zululand in South Africa, *ubuntu* denotes care and forgiveness that are critical to the maintenance of sociability and societal harmony; *ubuntu* underscores African humanism that is embedded in both the dialogic structure of community communication and it lies at the heart of the ontological being of an African – the indivisibility of the inner-person and its actions that are enmeshed together to make the human person an ethical being (Kiyala, 2022a). *Ubuntu*, when interpreted as a justice framework within the African context, seeks to attain the objective of "justice-making". This entails not only the restoration of victims and facilitating the social reintegration of offenders but also prioritises the restoration of relationships and the reestablishment of social harmony that may have been disrupted by conflict. This aspect of African justice, as highlighted by Oko Elechi et al. (2010) underscores the significance of repairing relationships and promoting overall social harmony: "[...], *ubuntu* better captures the underlying African worldview that expresses Africa's egalitarian, humanistic, interconnectedness, communitarian and participatory democratic values" (p. 73).

Ubuntu has been viewed as African jurisprudence underpinning the justification of diversion from the criminal justice system where restorative justice is advocated. This consideration is explained by the fact that restorative justice approaches embody the values held by *ubuntu*, notably, "consensus, agreement, and reconciliation" (Anderson, 2003: 24). The ethical dimension of *ubuntu* – setting this philosophy as a moral obligation in interpersonal relationships that hinges on African communitarianism and a humanistic worldview – creates more opportunities for mending broken relations. From an African cosmological vantage, *ubuntu*'s extensive and diverse attributes cover the realm of theology because the relation between person and community is intrinsically beyond the realm of material occurrences; the two are bonded by an extraterritorial relation that is guided and sustained by a spiritual force. Archbishop Desmond Tutu articulated this reality in terms of existing bond between community and personhood and the triune God, the Godhead of three persons, the Father, the Son and the Holy Spirit, and human beings exist as persons in community because they are created in the image of the triune God (Battle, 2009). This scholar notes: "This is the context we are called to understand ubuntu – an African way of redefining relationships: what is broken is made whole, what is cast down is raised up"; it resonates with the Christian way (Battle, 2009: 13). For that reason, Archbishop Desmond Tutu, contended that that forgiveness is better than retribution, both from a spiritual and political perspective (Battle, 2000).

In the aftermath of human atrocities experienced in Africa as a result of the erosion of traditional ethical principles, *ubuntu* is the pillar of regeneration of a more humane society. It addressed equity (Bennett, 2011), human rights issues (Metz, 2011), hospitality (Gathogo, 2008), and peace-building (Arthur et al., 2015). Faced with genocide, horrendous human rights violations and recurring armed conflicts, xenophobic violence, social and economic inequalities, African communities need to reinvigorate *ubuntu* transformative power to re-create societal harmony. The social upheaval on the African continent and the increase in crimes and destruction of human by war, racism, direct violence, and human atrocities worldwide is characteristic of a world where the culture of life has eroded, and the way to restore the dignity of the human person is by upholding the *ubuntu* principles.

While the criminal conventional justice system has presented a serious predicament to forgiveness, reconciliation, truth-telling, acknowledgement and reparation of harm, restoration of victims and perpetrators, URJ appears as a realistic approach to overcome the adversarial system of justice that maintain hostility between victims and offenders and deepen enmity. Our proposition to consider *ubuntu* as peace-building and transitional justice deemed able to transform rape survivors and perpetrator's identity is sustained by post-war community-based restorative justice initiatives in Sub-Saharan Africa that have yielded promising results. Those include the *fambul tok* (family talk in *krio* local language of Sierra Leone) that has facilitated reconciliation and social reintegration of persons accused of committing war atrocities in Sierra Leone (Graybill, 2004; 2010; Hoffman, 2008). Similarly, *mato oput* rituals (consumption of bitter root) by former combatants of the Lord's Resistance

Army (LRA) in Northern Uganda facilitate the dialogue between former soldiers and war survivors in the Acholi nation (Baguma, 2012; Bangura, 2008; Institute for Global Leadership, 2009). A case worth being mentioned here is that of the Baraza dialoguing institution that facilitated the reinsertion of former child soldiers in the DRC (Kiyala, 2018, 2021). These African experiments of justice are all imbued with restorative justice which resonate with African *ubuntu* humanistic philosophy. *Ubuntu* is antithetical to revenge and victimisation. This humanistic philosophy has the potential to preserve peace and minimise threats to security (Akinola/Uzodike, 2018).

While acknowledging the complexities and intricacies of pursuing justice for egregious violations of human rights and wartime atrocities, it is nonetheless necessary to examine the prospects of justice that is attainable in such in circumstances. The common ways of dealing with crime is the use of force or violence means to hoped to achieve deterrence. Walgrave (2011b: 4) argues: "when voluntary processes cannot be achieved or judged to be sufficient, pressure or force must be considered". However, in this convoluted debate stemming from rape incidence, when coercion is employed on offenders to comply with accountability measures and eventual sanctions imposed on them, as one would think, we predicate that the argument laid by Walgrave (2011b) on "punitive retributivism" and "restorative Retributivism" is more balanced. Punitive retributivism, as a restorative approach challenges the offender to take responsibility for his/her wrongdoing, whereas in "punitive retributivism", "the offender is confronted with his responsibility by others, and must submit to the consequences imposed by the criminal justice system" (Walgrave 2011b: 14). The offender is subject to "voluntary deliberation" or 'coercion' in the process leading to restorative retributivism or restorative retributivism in order to generate restorative outcomes, fair results, and prevent adverse effects that may taint the procedure and cause it to be deemed unfair and thus, engender hostile attitudes instead of restoration on the side of the offender. We assert that restoration of the aggressor/perpetrator should remain the ultimate aim of holding affenders accountable for their antisocial actions.

Walgrave (2011a) further maintains:

> if the power balance is unequal – the offender is not cooperative, the victim is too revengeful, the community is too exclusionary and/or the facilitator is too interventionist – restorative justice processes may yield very unjust outcomes (p. 461).

The process of holding a perpetrator accountable using restorative justice should be cognisant of the dispositions and attitudes shown by both the victim and the offender to ensure that power balance is observed to reach restorative outcomes. The criminal justice system is not concerned with such procedural predicament. Zehr (2002: 6) affirms: "The criminal justice system is concerned about holding offenders accountable, but that means making sure offenders get the punishment they deserve". This approach runs the risk of failing to lead aggressors/perpetrators to become aware of the impacts of their actions; simultaneously miss the opportunity to equip offenders with enough psychological and ethical resources, spaces and change to breed and express empathy to victims (Kiyala 2018). Subsequently, offenders may get disheartened, feel alienated by the accountability process and society, and finally expostulate the entire restoration exercise (Kiyala 2018). This occurs particularly in criminal litigation and legal proceedings that lend offenders in prison. The response of restorative justice is to deal with such unfortunate circumstances is adopting a creative approach to responding to criminal offenses, other antisocial acts and their authors. It contrasts the adversarial effects of punishment used as means of justice and outcome of accountability while in fact failing to transform the offender and repair broken relationships. Zehr (2002) objects to the criminal justice approach which he describes as lacking a true accountability.

In a pathetic act of grace, Brandt Jean showed love and forgiveness to Amber Guyger, the police officer who killed his—an event that sparked "A Debate Over Forgiving" as he said:

> "I don't know if this is possible, but can I give her a hug, please?" Brandt Jean asked the judge, after telling Guyger that his main desire wasn't for her to go to jail but to "give your life to Christ." (Chappell/Gonzales, 2019: n.p.).

These words convey the essence of *ubuntu* that Archbishop Tutu used to underscore principal fundamentals of the African philosophy and social ethos of conflict transformation and peacebuilding during the works of the Truth and Reconciliation Commission (TRC) that sough the truth, justice and accountability, repair, forgiveness, reconciliation, social cohesion and nation-rebuilding in the aftermath of henous crimes committed by and during the Apartheid rule in South Africa. The archbishop wrote:

> Anger, resentment, lust for revenge, even success through aggressive competitiveness, are corrosive of this good. To forgive is not just to be altruistic. It is the best from of self-interest. What dehumanises you, inexorably dehumanises me. Forgiveness gives people resilience, enabling them to survive and emerge still human despite all efforts to dehumanise them (Tutu 1999: 35).

In addition to the approaches taken by Brandt Jean and Archbishop Tutu, Emanual Levinas' conceptualisation of humanism overlaps *ubuntu* philosophy also known as African humanism. It opposes retaliation and deployment of force and violence as an answer to violence or a means to resolve conflicts. His experience of the Jewish Holocaust that affected him so profoundly and the loss of his family in the Nazi concentration camps moulded his ethics and philosophy of 'alternity'. He responded to the losses of millions and millions of Jews by offering forgiveness; Cohen (2006: vii) reports on that in the following statment:

> [...] Levinas returns love for hate, the "wisdom of love", the "humanism of the other," against the "hatred of the other"—a hard love, yes, a "difficult freedom," no doubt, but the subtleties no less than the brutalities of contemporary anti-humanism have never more sophisticated.

It follows that retaliation is antithetical to humanism and the very essence of *ubuntu* philosophy and ethics, because African humanism is grounded in the relational communication, interactive and dialogic characteristic of social engagement, even when people have experienced inhumanities that are inimical to the core values held by *buntu*.

Against the backdrop we conducted this research under the assumptions that ubuntu and restorative justice have the potential to transform the relationships between rape survivors and their perpetrators, and their communities. We will discuss the extent to which *ubuntu* and restorative justice can be applied to minimise the deviations of honour culture and restore the dignity of rape victims.

The following section deals with the methodological approach adopted in this study to explore the effects of honour culture on the perpetuation of rape denialism and lack of measures to hold perpetrators accountable.

21.4 Design and Methods

This chapter reports the findings of an inquiry that was designed as a case study and literary research based on the South Sudanese traditional adherence to honour culture. The qualitative approach was used to collect data. The use of qualitative methods was informed by the research objectives to provide in-depth knowledge and understanding of lived experiences by women who have experienced gruesome sexual violence, especially rape. The combination of existing literature on this issue and empirical research was deliberately chosen to prove a comprehensive awareness of how this injustice is lived by survivors and how the South Sudanese government officials and traditional leaders perceive and justify silence over this matter that is disguised as honour culture.

21.4.1 Design

While a case study can use mixed methods (Guetterman/Fetters, 2018; Kaplan/Duchon, 1988), this specific case study design was purely based on qualitative methods of data collection and analysis via in-depth interviews, focus groups and participatory observation in view of furnishing insightful descriptions of the problem under investigation. We utilised qualitative grounded theory in the specific case of South Sudan. Grounded theory is here understood as qualitative research design from which the inquirer generates a general explanation (a theory) of processes, actions and interactions that are shaped and framed by the views of a large number of participants (Creswell, 2007; Strauss/Corbin, 1998). This procedure entails that a researcher starts a study with no a priori theoretical orientation but develops a theoretical framework that describes the patterns of attitudes, views and behaviours explaining the existence of a certain worldview.

This qualitative research incorporated interpretive paradigm following simultaneous data collection and thematic analysis at data; this approach helps to develop theories about social processes that are definitely grounded in real-life experiences of people (Charmaz, 2006; Glaser/Strauss, 1967; McClement/Harlos, 2008; Strauss/

Corbin, 1990 cited in Rieger, 2019: 1). According to Creswell et al. (2007), a qualitative case study design proceeds with in-depth, descriptive questions aimed at developing a comprehensive understanding of the way in which different cases provide awareness of a single case. South Sudan is the context in which rape is dealt with.

While there exist various aspect of grounded theory, namely "classic Glaserian grounded theory, Straussian grounded theory, and constructivist grounded theory" (Rieger, 2019: 1), we relied on the constructivists grounded theory coupled with thematic analytical methods to deepen the understanding of "honour culture" concept. Later, we explored the relation between the existing notion of "honour" and "culture" to derive from it the meaning of "honour culture" from the South Sudanese local community's perspectives. Then, we found some linkages between this concept and the way rape incidence was perceived and dealt with in the South Sudanese society, particularly in the research sites.

The use of literary research tools and techniques was essential in understanding rape incidence in South Sudan, exploring *ubuntu* philosophy and its relation to restorative justice, exploring legal instruments pertaining to human rights, particularly the rights of victims of crimes and abuses of power. The source of this information comprised online sources, and printed materials from the library and scholarly web engines such as Google scholar and ResearchGate. These techniques and tools are explained and endorsed by Towheed (2010).

21.4.2 Methods

During the research process, the nature of rape, the root causes of rape and the response of institutions (governmental and non-state organisations) for rape management were explored by means of qualitative methods for empirical data collection, analysis and interpretation. Participatory observation took place between 07 and 30 Dec 2022, while interviews and focus group discussions were conducted by the first author from 12 to 23 Dec 2022. Several items were considered on the list for understanding, namely feelings of persons affected by rape incidence; perceptions of traditional leaders and civil authorities about rape survivors; reasons for silencing rape victims; protective structures in society to avoid publicising rape incidence; cultural justification of muting rape victims; voices of rape survivors; voice of the community; voice of traditional and tribal leaders; voice of government officers; voice of NGOs; and the justice system's attitudes and inaction regarding rape incidence. All these elements were examined by observation, interviews and focus groups discussions.

The investigations during this stage provided an overview of the reasons for rape and showcased the degree to which interpersonal relationships have been destroyed at various levels of society.

To unpack the information meaningfully, the investigation comprised three groups: the main survivors of rape, political and institutional leaders such as international non-governmental organisations, traditional leaders and political authorities.

Table 21.1 Participant demographics

Research instruments	Participants	Occupation	Gender
Interviews	2	Anonymous Humanitarian organisation Mercy Corps	1 Male 1 Female
	2	Political leaders Ambassador and ministry of gender	Males
	5 from 11	Rape survivors	Females
Focus groups	15	Community leaders	Female
	6	Traditional court Community leaders	Male
	6	Radio broadcasting programmed/debate	Male

Source The Author

We conducted fieldwork with those three groups to get the overview of the issue, its management strategies, the level of rape survivors' vulnerability to handle their relationship, Various factors that emerged from our investigation are presented here, including an integration of research-based findings and the literature on the matter.

21.4.2.1 Research Sites and Target Population

The research was conducted by the authors of this chapter within the following selected site of Upper Nile region in Malakal town, the eastern part of South Sudan and in Juba, the capital city. It includes three main groups of population: rape survivors, males and females, community leaders, traditional leaders and institutional leaders, both humanitarian and political leaders. Based on a nonprobability sampling technique, we collected data from four political institutional leaders, 21 community leaders (traditional and women leaders) and five rape survivors (Table 21.1).

There were two translators who assisted during the local language interviews that were conducted in Arabic, Dinka, Nuer and Shilluk as languages used by participants/informants in this study. The sample was set for 50 participants; however, only 36 responded positively and agreed to participate because they were available and accessible to the researchers. Five out of 11 participants we identified as rape survivors came forward and shared their experiences on the matter. Others were reluctant to acknowledge being rape survivors for private and undisclosed reasons.

21.4.2.2 Sampling Techniques

We used nonprobability sampling techniques (Etikan et al., 2016), based on a purposive and convenience sampling (Teddlie/Yu, 2007), and snowballing (Davies, 2007). These techniques helped to select and meet key respondents, informants (rapists, victims and traditional leaders). The collection of information was limited to participants who were accessible and available since only a few were known in their communities. By the consent of local authorities, we have been allowed to meet and identify rape survivors. But because of pressure put on rape survivors by their aggressors, some rape survivors have denied being victims, and only victims of the war in Malakal shared their ordeals. Participants were identified through snowballing and those who were available were encouraged to participate by members of a local NGO who assisted with meeting different informants. All were recruited by convenience or directed to progressively by the first participants. They also played an important role in introducing us to rape survivors.

21.4.2.3 Data Collection

Two sets of qualitative data were collected, namely empirical (interviews and focus group discussions) and radio broadcasting data generated from a debate on the South Sudan Tribune in which different stakeholders participated. These included pro-government speakers and listeners who tended to contradict the description of sexual violence suffered by female persons in various communities.

We used unstructured interviews, involving a little standardisation, and semi-structured interviews using an interview guide. These were completed with follow-up questions when it was necessary. In order to ensure that interviews were of acceptable quality, we carefully chose research sites. The language barrier was overcome, because we employed two local persons who assisted with the translation from local dialect into English, throughout the process of data gathering, because local languages used in interviews were Arabic, Dinka, Nuer and Shilluk.

21.4.3 Data Analysis

Constructivism paradigm or a worldview which is interpretive was employed to grasp the meaning of information provided by research participants (Adom et al., 2016). Empirical data from interviews, focus group discussions, and transcribed radio broadcasting debated were coded–labelled and organised to identify different emerging themes and the relationships between them. Coding was necessary and useful to find common themes and concepts that made possible a thematic analysis of data sets. Thematic analysis extracts the codes and develops the themes of the outputs. This method is used in social sciences research (Saldaña, 2021) and discourse analysis (Fairclough, 2001). The adoption of qualitative content analysis entailed interpreting the content of textual data "through the systematic classification process of coding and identifying themes or patterns" here reflects the definition provided by Hsieh and Shannon (2005: 1278); thus, researchers' subjective minds were essential in interpreting the meaning of narrated and observed experience, while remaining faithful and as objective as possible to ensure the report represent the meaning given by participants.

Content analysis allows for interpretation of the context connected with the use of the word or phrase in informant speeches, and the task of the researcher is to explore the use of words or discover the range of meanings attributed to those in normal use (Hsieh/Shannon, 2005). Our understanding of sexual aggression, rape and cases of denial to protect the honour culture and the reputation of perpetrators was drawn from the local perceptions of violent sexual behaviours which often go unaccounted for.

21.4.4 Research Ethics

Research into a sensitive issue such as gender-based violence requires discipline, empathy, patience, objectivity, fairness in interaction with different stakeholders, honesty, but moreover, abiding by high ethical standards. In this inquiry, we observed underlying principles and ethical guidelines that are required in social sciences and ethical scholars to prevent abuses and help investigators in their responsibility to remain ethical scientists, as outlined and explained by Bless, Higson-Smith and Sithole (2013: 29) who further explain: "research ethics places an emphasis on the humane and sensitive treatment of research participants who may be placed at varying degrees of risks by research procedures".

Key ethical standards that underscored our responsibilities as investigators focused on the principle of non-maleficence by ensuring no participant was harmed or hurt as a result of participating in the inquiry; beneficence by promoting justice for participants and making their voice heard; autonomy by ensuring all participants were free to provide information they were free to release and they were also free to end their participation at any time of the inquiry if they decided to do so. Justice was upheld by fostering equal treatment of participants and avoiding discrimination. Fidelity and trustworthiness by keeping the promises and engagement as agreed between researchers and participants; respect for participants' rights and dignity and avoiding infringing their rights, These standards were inspired by the work of Bless et al. (2013).

To adhere to established ethical norms (Bless et al., 2013), we followed a procedure of obtaining verbal informed consent from participants who voluntarily chose to take part in the study. No participants were compelled or coerced into participating against their will. To maintain confidentiality, all sensitive information obtained was safeguarded and kept strictly confidential, with no disclosure to any third party. To ensure participant anonymity, we assigned codes to represent each individual, thus dissociating their identity from their information. Anonymity was further protected by removing any links between participants and their data to prevent identification. Additionally, appropriate referral measures were in place, as a trained counselor from the Mercy Corps Organisation accompanied our research team and provided assistance in addressing any potential negative effects that may have arisen during our interactions. Lastly, it is important to note that all participants involved in the study were over the age of 18.

21.5 Results

There were two sets of qualitative data, namely empirical data collected via interviews and focus group discussion, and radio broadcasting debates related to the ill-treatment of rape survivors that is embedded in structures such as the justice system, security services and the state. We start by presenting empirical data, which is followed by data that originated from radio broadcasting that took place on 19 December 2020 at Juba. The main themes that emerged from the analysis of data include causes and various perceptions of rape, namely rape as an offensive and embarrassing act, as a weapon of war, rape denialism, rape management, mitigating actions and resilience.

21.5.1 Perceptions and Causes of Rape

The analysis of the interview data suggest that rape is a form of taboo in the South Sudan communities, a source of offence that could result in the killing of the perpetrator(s). It was found that raping a girl is akin to insulting the family of the victim, which will seek retribution or compensation from the perpetuator.

Participants expressed their different perspectives on rape. The following statement originated from a focus group with traditional leaders, TL_1:

> In our tradition as Dinka, if a person rapes a girl, they will not let him go. People will fight, until they make sure that the offender is killed. Or fight with the offender's parents because rape does not exist in our tradition. If the girl reports to the mother that she was raped, the mother report to the family, her brothers or family. The big people, elder or family will then go to fight for the victim's family, they consider it as if the offender wanted to play with the victim's family, kind of undermining them. That is why they fight, or they will ask the person to marry the girl as way of solving the problem[1].

The statement made by TL_1 shows clearly that rape is not acceptable in the Sudanese society. The community several times resorted to "mob justice" to settle scores with people accused of rape.

The following statement derived from the focus group with a group of women leaders:

> An offender if is caught, brothers of the victim could kill him unless the offender have a hidden place somewhere to save his life, otherwise he will be killed. Such person does not have right to live in the community because he will be killed. For the perpetrator, he cannot let himself be seen in the community because he is almost the suspicious person. The perpetrator looks like not well brought up. This is the overall point, the cultural and traditional perception whether for Dinka, Nuer, shinu or whatever culture here in South Sudan, this how our perception is built up[2].

Sharing similar sentiments, and from a cultural perspective, another of the participants interviewed saw rape as a form of embarrassment. It was extracted from

[1] Traditional community leaders, focus group, Malakal town, 15 December 2020.

[2] Women leaders, focus group, Malakal town, 22 December 2020.

the interview that rape undervalue the victim(s). Consequently, the perpetuator was considered persona non-grata in the communities and had to become a fugitive to survive, or be killed if found. Both the women leader and the community leader expressed their dissatisfaction with the issues of rape in the South Sudan community. They both reinforced their claim that rape is alien to the Sudanese culture and therefore unacceptable as it diminishes the reputation of the victims and the perpetrator's families.

Focus group discussion in which men and women was recorded as follows:

> The issue of rape is embarrassing in our culture. It is even making you feel lower, having no value, you are not important in the community, you are considered like a weak person. In ancient time, rape was not at all! Rape does not exist in our tradition and culture because people were afraid of the consequences like tarnish, besmirch, sully the name of the father of the family. The reputation of the family will go damaged. Even when someone from the offender family wants to marry another girl, the girl's family will object because it presumes that all the members of aggressor are violent and "bad people", because of the rape incidence. People were therefore afraid of sexual misconduct.[3]

Other observations were made by informants, and they were also recorded and are also thematically presented.

Drawing from the above narrative, one could deduce that rape is a serious offence for which punishment in the South Sudan culture is death. As such, rape is a forbidden and unforgivable sin in the culture of South Sudanese. Besides, the South Sudan law prohibits rape and offenders are liable to a 10-year jail term if found guilty.

Member 1_MCO gave his opinion as stated below:

> I think is not allowed, culturally we forbid it. Culturally issue of rape is not accepted though they know that it forbidden. If someone rapes in the town, because we have legal services, they will take the case to legal services as we have law in place in South Sudan. If someone rapes, he will be punished, I think for ten years according to law in SSD.[4]

The traditional prohibition of rape has been strengthened by South Sudanese legislation with the aim of preventing and minimising its repercussions on both perpetrators, victims, and the wider community. It is well-known that this crime extends its destructive impact beyond the immediate parties involved. The South Sudanese government has established a maximum sentence of 14 years for rape offences (Ministry of Legal Affairs and Constitutional Development, 2008). However, ensuring the implementation and effectiveness of this law remains a challenging issue, as discussed below.

21.5.2 Theme 1: Weapon of War

Previous studies have shown that rape increases during conflict and is commonly utilised as a weapon of war (Danjibo, 2019; International Criminal Court, 2011;

[3] Men and women community leaders, focus group, Malakal town, 19 December 2020.

[4] International Mercy Corps worker, Interview, Malakal town, 13 December 2020.

Isikozlu/Millard, 2010) Participants attributed the recurrence of rape cases to the war in the country. It was extracted from the data that the military used rape as a weapon of war.

It emerged from another focus-group discussion conducted with community leaders, Traditional Leaders (TL)_2:

> In our tradition, the cause of rape is war. From the time of war, we have been witnessing the recurrence of rape in our community. Military or armies when they get to a village where there are not many people, they do rape women and sometime took them to their camps. Moreover, those armies know that no one will charge them especially as it is during war; they take advantage of the crisis.[5]

Member 1_MCO reported:

> During the war, people are using it like weapon of war. If they get a woman, they rape her as they were targeting women as well. Sometimes, they were killing victims after raping them. But some of them were just raped and released.[6]

From the above narrative, it is apparent that the military fighting in the conflicts takes advantage of a failing system to perpetuate their acts. This assertion is strengthened by the revelation made by International Mercy Corp team participants interviewed on the connection between lawlessness and rape cases in South Sudan.

Member 1_MCO expressed his views in the following terms:

> During the time of war, there is no punishment, there is no legal services standing apart from people who are wearing guns, they are the ones who taking the law. As we can see, when there is war there is no punishment and people who have guns, have power. If you have power, you can do anything you need to do, and they are using it like part of weapon of war. Because if you rape a person and is in a certain tribe that you are fighting against, that mean you will see their weaknesses.[7]

Apart from the rape perpetrated by military men during conflicts, it was found that women living in the United Nations Protection of Civilian (PoC) sites across the country have become victims of rapes in the war-torn country. This rape incidence is said to be perpetuated by civilians in these sites and are mostly opportunistic attacks on women. The international NGO leader interviewed illuminated this. Member 1_UNFPA said:

> Civil raping women is happening in town or in refugee camp. The reason of being raped in the camp is just that men target whether a girl or woman when they get opportunity. It happens most of the time among young people. Some did it while they are drunk. Nowadays, rape has diminished its scale compared to the time of war. Nevertheless, GBV is there, we cannot deny that fact.[8]

---

[5] Traditional community leaders, focus group, Malakal town, 28 December 2020.

[6] International Mercy Corp workers, Focus group and training, Malakal town, 27 December 2020.

[7] International Mercy Corp workers, Focus group and training, Malakal town, 27 December 2020.

[8] Interview with a member of the United Nations Population Fund (UNFPA), Malakal town, 14 December 2020.

The above view supports the assessment of UN and other NGO's reports according to which there is widespread of sexual violence among women residing in the PoCs sites, especially in Malakal. However, due to the shortcomings of South Sudanese legislation, the prosecution of sexual violence acts under domestic law fails to fully encompass the extent and seriousness of the violations committed during the conflict (Soccol, 2018). Furthermore, the issue is complicated by the interchangeable use of the concepts of rape and *Zina*, as discussed earlier. While international NGOs cannot deny the ongoing existence of rape, whether in PoC sites or urban areas, political authorities face difficulties in acknowledging this reality.

21.5.3 Theme 2: Rape Denialism

South Sudan has experienced its fair share of conflicts and wars. Following the establishment of the new country, there were concerns among commentators that violence, whether driven by ethnicity or militia groups, could potentially undermine the stability of the nation. There are several reasons that can be identified to explain the political denial of rape against women, with a significant focus on protecting the country's image, reputation, and honor.

One prominent political authority working with Awut Deng[9] denied the existence of rape as stated below:

> Rape does not exist in South Sudan! From our culture, up to the government or political authority, no one will tell you that issue of rape exists in South Sudan, and you have to consider it like that. There are no rape victims in South Sudan. Because if they tell you to bring or to show us one of them, you won't be able to give us proof. Reports that come from international NGO are just kind of machination trying to dirty our country at international level. Yet that issue does not exist.[10]

As a new nation, South Sudan political authorities are much more concerned about protecting their reputation. Acknowledging the existence of rape against women means exposing the failure of the judiciary system.

21.5.4 Theme 3: Rape Management

Participants expressed their different perspectives on rape – a complex offence that calls for a variety of responses aimed at protecting the reputation of the South Sudanese Society and persons affected as well. The following statement originated from a focus group with traditional leaders, TL_1:

---

[9] Awut Deng Acuil is a politician lady, Minister of foreign affairs of South Sudan from 2019 to 2020.

[10] Interview with prominent figure of South Sudan politicians, Juba, 12 December 2020.

> In our tradition as Dinka[11], if a person rapes a girl, they will not leave him go. People will fight, until they make sure that the offender is killed. Or fight with the offender's parents because rape does not exist in our tradition. If the girl reports to the mother that she was raped, the mother reports to the family, her brothers or family. The big people, elder or family will then go to fight because for the victim's family, they consider it as if the offender wanted to play with the victim's family, kind of undermining them. That is why they fight, or they will ask the person to marry the girl as way of solving the problem.[12]

It appears that the Dinka tribes protected women by putting in place stringent sanctions to deter the perpetration and perpetuation of this odious act. We have observed that rape is a consequence of the erosion of the indigenous Dinka's value systems, which has been further intensified by years of civil wars, resulting in increased vulnerability among women and girls. Consequently, in order to protect the dignity of the violated individual, the Dinka society established the obligation that the perpetrator of the rape should marry the survivor. This practice is seen as a way to prevent dishonor and stigma that could otherwise harm both the survivor's life and reputation.

Part of the inquiry of this article was to find mechanisms that potentially foster reconciliation and social reintegration of rape survivors, victims of rape crime.

According to Member 1_MCO,

> The government is frightening victims letting them deny the issue of rape. The fact that the government keep on denying the existence of rape in South Sudan, the government is preventing victims to testify how victims they are as proof of rape existence. The transitional justice does not exist (not having power) because of the government position against victims and services, departments and NGO working on the issue of GBV and related cases.[13]

Another notable leader asserted that managing rape issues within the community in the past, included very harsh punishment such as breaking the legs of the perpetrator or killing him.

TL_3 observed:

> In our culture, it is known that if a man impregnated a girl, his legs had to be broken, or if he raped, he would be killed, because the offender has destroyed the girl's life (means, community, traditional death penalty against offender is equal to the destruction of victim life).[14]

In the South Sudanese culture, the girl child is considered a property of the father. As such, violating her without the consent of the father is viewed as trespassing into another man's territory without authorisation.

Marrying the victim to the perpetrator was another measure of addressing and managing rape issues in the community. It is seen as the best measure given the fact that raping a woman dehumanised and destroyed her worth, particularly if she was a mature girl. This violation of the intimacy and privacy of a women was considered as

[11] The Dinka are the largest ethnic tribe in South Sudan; they are known for their traditional practices of agriculture and livestock raising.

[12] Interview with a member of the United Nations Population Fund (UNFPA prefer to hide the identity), Malakal town, 18 December 2020.

[13] International Mercy Corp workers, Focus group and training, Malakal town, 27 December 2020.

[14] Traditional community leaders, Focus group, Malakal town, 27 December 2020.

a person seek to own another human being by means of rape. That justifies why such perpetrators were obligated to many their victims, because the marriage opportunity of rape survivors was compromised.

The same traditional leader, TL_4 argued:

> If it happened that a mature girl has been raped, there is a decision to be taken that, she must be given to the perpetrator. He must marry her because he has destroyed her. When someone has been raped, she has been destroyed because culturally the offender is now owing her dowry[15]. If she has been raped, then she becomes useless. As we know who has raped her, then we say, you have to marry her.[16]

Requiring a rape survivor to marry her aggressor is a strategy that could increase rape incidence, and perpetrators of such violent acts would expect to be give their victims as partners.

21.5.5 Mitigating Actions and Community's Response to Rape

Data collected from members of NGOs working on research sites further revealed how rape was perceived and how the community responded when it occurred. TL_ 5 commented:

> 90 percent are reporting because we are maintaining our policy like confidentiality, non-discrimination, do not harm, which will make and encouraged them to report. The respect of that policy makes them report to get good services, and most of women who do not get treatment most of the time attempt to commit suicide, find themselves useless, some of them are afraid of losing their husband, family, etc.[17]

Based on interviews and information gathered from online sources such as the South Sudan Tribune, it is evident that political authorities at higher levels of the government not only deny the existence of rape but also employ strategic tactics to discourage survivors from reporting their cases. This approach aims to eliminate evidence and prevent the truth from coming to light, as we will explore in more detail later on. Member 1_UNPFA remarked:

> Politician have put their path away to see what is happening, what is taking place in the time of the war. Unfortunately, the government fined themselves as the ones that who are practicing, encouraging rape. That is why if there are some reports of women who have been raped, most of the time they (government) are denying the fact. These days those things are not happening (international NGO leaders). As the government is denying rape facts, in this way, even perpetrators could not be prosecuted to wipe, clean the issue of rape.[18]

---

[15] An amount of property or money brought by a bride to her husband on their marriage.

[16] Traditional community leaders, Focus group, Malakal town, 27 December 2020.

[17] Traditional community leaders, Focus group, Malakal town, 27 December 2020.

[18] Interview with a member of the United Nations Population Fund (UNFPA), Malakal town, 14 December 2020.

It was queried that if there were no rape cases, how come NGOs and hospitals in our "one stop centres" kept on getting and assisting women victims of rape. Elaborating further, the international NGO leaders exposed the extent to which the government has gone to silence the reporting of rape in the country.

> The government has even organised a committee in charge of denying the existence of mass rape against women in South Sudan. The minister of gender advised the UN representative to talk such issue with her before taking the issue to the government or before doing any report. Because there is information that should be shared and the ones that must not be shared about victims of rape in South Sudan.[19]

The above attempt to cover the incidence of rape cases resonates with Madut Jok's (2013) commentary that many women who may have the courage to report such violations of their basic rights are often taunted, laughed at and sometimes dismissed by the police. Madut Jok (2013) attributes this to what he termed cultural biases, which makes it difficult for evidence to be established against the perpetuator, should the victims decide to legally pursue the case. Another motive for denying or hiding the incidence of rape by the political class was to protect themselves from prosecution in the future.[20]

Member 2_UNPFA asserted:

> They need to hide the issue of rape and denying it as their names won't be looking good. Most of the GBV are crimes and those crimes will be punished, politicians will be punished in the future for what the crimes have done in the community. They just want to save the good face of the government.[21]

Women leaders argued that South Sudan justice system is distorted and against women:

> The traditional way was safeguarding women because of the "community death penalty" mechanism of dealing with perpetrator. The perpetrator as well couldn't be freed because he could be killed in the community because he has destroyed the victim life. Nowadays, the perpetrator is not killed but arrest that's why as community, we as women, we are lacking what to do? We are perplexed. That's our big issue now. For us women feel vulnerable because of the judiciary procedure of managing rape issue (Women leader).[22]

The lack of political and legal initiatives to address rape in South Sudan has amplified the vulnerability of women in their pursuit of justice. However, the growing awareness of the widespread occurrence of rape in the country is a positive development that offers hope for transforming the lives of countless individuals who have endured the horrifying experiences of this violence. Furthermore, the ramifications of gender-based violence (GBV) are profound, as some individuals may

[19] International NGO (UNPFA), (prefer to hide the identity), Interview, Malakal town, 22 December 2020.

[20] Madut Jok, South Sudan Tribune, observers' side, Juba, 19 December 2020.

[21] International NGO (UNPFA), (prefer to hide the identity), Interview, Malakal town, 22 December 2020.

[22] Community Women leaders, Focus group, Malakal town, 27 December 2020.

carry the psychological and emotional traumas throughout their lives. One of such consequences of this aggression against women is children born out of rape.

The next section continues to provide insights that come from different sources, namely radio broadcasting discussions. News aired on experiences of people touched by rape has the potential to halt this brutal behaviour (Kele, 2020). For this reason, broadcasts that conveyed news to the public was a useful source of data.

21.6 Results from Radio Broadcasting Programme

In a comment posted on the South Sudan Tribune editorial platform by internet users, a discussion took place regarding the government's denial of the reported 125 cases of rape (Sudan Tribune, 2018). This debate, which occurred one year prior to our data collection for this current article, involved both government supporters and civil observers, who engaged in a lively exchange of contrasting perspectives on the factual truth surrounding the allegations made by international NGOs regarding rape against women. The dialogue presented below highlights the differing viewpoints and reactions between the government and NGOs. Observers claim that the South Sudanese government is complicit in serious rape violence against women, a claim that is refuted by supporters of the regime.

Speaker 1 commented: "Juba government has never accepted such crimes (of rape) report since then, South Sudan government is not faithful, murdering, robbing, looting are the most common practices.[23] Speaker 2 affirmed:

> We are all looking for facts to back the sexual attacks. MSF[24] said the rapes occurred between Nov 19, 2018, to Nov 29, 2019 during food distribution .... We know now there was no food distribution .... between Nov 19th and 29th. ... I said that last time that, the report was a make up by Doctors Without Borders. It was impossible such a number to be raped within 10 days. This Agency is not telling the truth because their agenda is not treating people, but they are making experiment on human beings on fake thing against others human in the world. They are damaging image of South Sudanese people and the country at large.[25]

Speaker 3 uttered the following:

> Listen man, Drs never lie on something like that. Unless they are untrained professionals. I know this because it's the fact. Gangs of Jeng area truly did this. I am very sure one day one time every guilty. God is watching.[26]

Speaker 4 also stated:

> The report was created purposely to tarnish the image of the government since peace has been signed. No place on earth where such huge rape occurred within less than one week.

---

[23] Midit Milot, South Sudan Tribune, observer's side, Juba, 19 December 2020.

[24] Médecins sans Frontières (MSF) mentioned by Speaker during the radio broadcasting programme.

[25] Mayendit, South Sudan Tribune, government's side, Juba, 19 December 2020.

[26] Madut Jok, South Sudan Tribune, observers' side, Juba, 19 December 2020.

Even women and girls in Bentiu laughed at the report and some called it insult to them. Some mindless pseudo-intellectuals seem to believe the rape drama by MSF cone men and women and their surrogates in between. … It is a coordinated strategy by foreign agents in South Sudan to come up daily with damning stories to shame the government and people of South Sudan and perhaps soon to lobby for UN trusteeship again. They will never leave South Sudan alone. False rape 125 cases…, CTSAMN Monitors robbed and stripped naked in Luri, etc. and many more such fake reports to come soon.[27]

Speaker 2, on the government's side mentioned:

It is just the beginning, and more things would follow if Juba's government were doing nothing about those Agencies operating in South Sudan. …, because it is bad to hear people reporting lies, exaggeration and propaganda just to tear image of other people or country.[28]

Speaker 5:

... I'm not surprised the fake investigation was also led by a tribal woman Awut Deng[29]. The women raped were Nuer and the investigator was Tribal Dinka who is under the regime when the scale of rape amounted to the description of rape as weapon of war on the Nuer Land.[30]

The Dinka tribe was accused of aiding the government and carrying out war crimes without facing any legal repercussions in the nation. Due to this mentality and the lack of punishment, rapes against women of other tribes, such as Nuer women, have escalated.

Speaker 1 made an observation from a religious point of view; he said: "God is watching, those who are denying the fact will face it in turn, Awut Deng cannot tell the truth since the victims are different from her tribesmen".[31]

As part of a new cabinet named in April 2016, As a result, accusations of rape against women were mostly rejected since they seem to deter the government, country reputation. Speaker 6 noted: Whoever expected anything than this finding from a team led by Awut Deng Acuil should be leaving (living) in Mars …! The bond between Kiir and Awud is still intact!"[32]

From this broadcast debates and confrontations, the issue at stake we identified is the denial of rape. We will explore further as we discuss the findings and see the implication of the denialist attitude and its implication for the honour of the South Sudanese Society and the dignity of women and girls.

---

[27] Ayul, South Sudan Tribune, observers' side, Juba, 19 December 2020.

[28] Eye see, South Sudan Tribune, observers' side, Juba, 19 December 2020.

[29] Eastern, South Sudan Tribune, observers' side, Juba, 19 December 2020.

[30] Madut Jok, South Sudan Tribune, observers' side, Juba, 19 December 2020.

[31] Mayendit, South Sudan Tribune, government's side, Juba, 19 December 2020.

[32] Mayendit, South Sudan Tribune, government's side, Juba, 19 December 2020.

21.7 Discussion

The key findings of this investigation revolve around two main themes: honour culture, which we have identified as a theory used to justify the non-publicity of rape incidents, and rape denialism. We delved into the underlying causes and consequences of rape, attributing them to the erosion of cultural values and the societal fear of publicising such cases. Regrettably, while the concept of *ubuntu*, originating from South Africa, did not emerge directly from the data, we have introduced it into the discussion to underscore its relevance in addressing the negative implications of criminal court proceedings when dealing with rape cases. We established the relations between URJ theory and legal instruments pertaining to the 1985 UN Declaration on the Rights of Victims of Crime and Abuses of Power (United Nations, 2005) and its draft convention (Groenhuijsen, 2008; South Africa Department of Justice and Constitutional Development, 2006). We found that perpetuated rape denialism serves the purpose of preventing the outside world from undermining the local domestic culture of the South Sudanese people. This societal worldview constitutes a hindrance to accountability procedures for survivors and their restoration. That theory conveys the meanings, perceptions, and descriptions of a phenomenon according to various informants within a particular context. Such theory represents the view held by a larger population over a certain social practice. Grounded theory from data and categories emerging from actions and interactions of the population involved in a study, analysed and interpreted by the researcher. Rieger (2019: 8) notes: "Constructivist grounded theory researchers acknowledge the subjectivity of data and data analysis, a stance that aligns CGT with constructivism".

21.7.1 Rape Denialism

Madut Jok (2013) noted that the doomsday predictions outraged many of the South Sudan leaders who simply dismissed them as nothing more than the wishes of the countries' enemies. The said reason may be attributed to why the government continued to deny the incidence of rape and GBV in the country.

It is reasonable to argue that both victims and perpetrators of rape are both victims of the social and political collapse of the South Sudan society which remains up to present a big challenge and was not covered because of the difficulties related to meeting persons with persons with sexual or rape offence histories. Agreeing with Madut Jok (2013), the only way to avert a disaster rooted in history is by reckoning with the past, designing a reconciliation process, training the security forces in ethics that men and women in uniform are supposed to abide by, and undertaking a concerted effort to return to an open democratic society. In consonance with Madut Jok (2013), it is sufficient to assume that leadership that is based on suspicion, fear and gender discrimination is leadership looking into an abyss.

In exploring suggestions from the participants on ways to manage the rampant incidence of rape cases among women in South Sudan, it was found that the traditional way of addressing rape issues, which is "death penalty for the perpetrator", safeguards the women. In contrast, women feel vulnerable while seeking justice and recompense via the judiciary procedures involved in managing rape issues.

At the community level, traditional approaches based on extreme physical violence, justified by the violation of property, were employed as a deterrent to prevent rape against women. These approaches involved the participation of traditional leaders who facilitated restorative justice, and their role was generally well-regarded in the community (Haki, 2011), despite the use of extreme physical violence.

Supporting the findings of Ellsberg et al. (2020), it was observed that in some cases, the devaluation of a girl through rape might lead to families of the victim and perpetrator agreeing to a marriage. However, it is important to note that society does not endorse rape as a strategy to force the victim and perpetrator into a marital union. Rape cannot be morally accepted as a means to obtain a female person's quick agreement to a marriage.

Two factors motivate rape survivors to feel that their relationships are being re-established: the restoration of honour through extreme violence against their offenders and the policies of international NGOs in addressing the treatment of rape survivors. The notion of a survivor whose dignity has been destroyed serves as the main motivation in traditional courts, customary law, and the community to take drastic and severe actions in order to restore the survivors' honour and provide them with a sense of restorative justice. Honour restoration through traditional restorative justice has huge impact on survivors as it can prevent them from suicide for instance. In fact, South Sudan society is among the rare communities that apply death against the perpetrators of sexual violence. The application of this practice is made possible in the case of child sex offences (Mancini/Mears, 2010) which contrasts with South Sudanese society where the death penalty based on sex offence is imposed for any form of sexual offence against any victim. When rape involves an adult, there could be a way to negotiation for marriage as a solution to the problem.

The "marry her sex offender" approach to problem-solving may be encouraging the rape of young and adult women since the rapist may use it to get a wife of his choosing against the wishes of the rape victim. Knowing in advance alternatives of the management, some sex offenders take advantage of the managerial system to achieve their own goals and needs. On the other hand, survivors could also condone it to protect their reputation, save face, and maintain good relationships in the community, despite the fact that they are not doing it for themselves, but for others to have a positive perception and to reestablish their relationships that were disrupted by the issue of rape. Given that survivors feel powerless, they prefer, albeit reluctantly, to adopt and comply with the accepted social norms that protect them from further increasing their socio-relational vulnerability.

21.7.2 Preservation of Honour Culture and Justice for Rape Survivors

Rape has disrupted war survivors' relationships in the community where people have sufferent the brutality of rape. Looking ahead, the pathways to eradicate injustices suffered by rape survivors in the name of culture of honour, giving survivors the option of marriage would minimise shame and the harsh punishment inflicted on perpetrators. Such measures should be applied only in circumstances where the survivor is an adult and willing to enter into a conjugal relationship without coercion. However, cleansing rituals, healing, and reconciliation should be taken into consideration to prevent the surviving rape partner from experiencing further abuses and psychological illness.

Finally, though URJ was not tested to back the argument on the viability and effectiveness of this justice paradigm to overcome inimical justice system, based on past studies of post-conflicts peace-building mechanisms (Kiyala, 2016; 2021; 2022b; Honwana, 1997; Graybill 2010), we maintain that further studies could prove very significant to support the use of *ubuntu* justice in the context of South Sudan. Parther "marrying her perpetrator" is not helpful to the survivor herself to live freely even though the aim is to save her reputation and restore her relationships. This seems to save perpetrator's life instead of saving that of the survivor. The international NGOs such as International Mercy Corp, UNFPA and World Vision focus more on trying to bring comfort to survivors covering their stigma and shame, while enhancing their reputation through the relief that they get from the international NGOs. These mechanisms try to address the lack of appropriate management of the problem by the justice institutions.

Judicial mechanisms to restore rape survivors' reputation and honour have contrasted with those applied by customs and international NGOs. To garner support from governmental judiciary institutions, the culture of honour has been prioritised to uphold the nation's reputation locally and internationally, especially in addressing grave violations and crimes against girls and women during times of war. They have been using the South Sudan Tribune to control information, as in any crisis, authorities have a responsibility to understand the type of crisis and determine a strategic response that will maximise their reputational protection (Coombs, 2007). Governmental bodies have been employing information manipulation theory (Levine, 2014) to control information, where the use of "lies, exaggerations, half-truths, secrets", and diversionary responses is more prevalent in the conversation. This is applicable whether the information comes from international NGOs, customary practices, the community, or even from rape survivors themselves. Through these tactics, the controllers shape the communication and thoughts of the recipients at a cognitive level (Druckman, 2001).

However, there is a noticeable improvement of rape survivor's treatment since the indictment and sentencing of a sudanese soldier for rape by a military court in South Sudan has sentenced 10 soldiers to jail terms ranging from seven years to life for a rampage in which a journalist was killed and foreign aid workers raped—these odious crimes were committed during an assault in the Terrain Hotel in the capital Juba in 2016 (BBC News, 2018). The same court instructed South Sudan's government to pay each rape survivor $4000 (£3000) in compensation (BBC News, 2018). The same report notes that this was the first time that justice and reparation were sought and obtained for a rape victim (BBC News, 2028). Such judicial procedures should be pursued along restorative justice—processes imbued with the spirit if *ubuntu*, to enhance healing, reconciliation and societal harmony.

21.7.3 Honour Culture and Ubuntu Restorative Justice

The main contribution of this chapter clearly appears as the development of "honour culture theory" which we explored extensively and demonstrated how it has become judicially and ethically a cultural deficit in the management of wartime and post-armed conflicts atrocities in South Sudan.

The main contribution of this inquiry was the discussion of the concept of honour culture and its underlying roots found in the cultural aspirations the people of South Sudan to protect their reputation and their cultural etiquette. Unfortunately, that comes at a cost, namely the violation of human and victims' rights. Rape survivors have been forced to conceal the truth about the damning atrocities they endure in wartime. The lack of access to the conventional justice system has been singled out as what paradoxically sustains and justifies honour culture theory. It is believed that exposure of sexual violence against women will attract national and international publicity, and that would be detrimental to the credibility and reputation of affected societies. Thus, honour culture has become a hindrance to the implementation of the UN Declaration of Basic Principles of Justice for Victims of Crime and Abuse of Power (United Nations, 1985).

The contribution of Groenhuijsen (2008) which is deemed crucial in this chapter is the provision of the Basic Principles of Justice for Victims of Crime and Abuse of Power, particularly Principle 7 which alludes to restorative justice mechanism. It states: "Informal mechanisms for the resolution of disputes, including mediation, arbitration and customary justice or indigenous practices, should be utilised where appropriate to facilitate conciliation and redress for victims" (United Nations, 1985). In the same convention draft, non-adjudicative mechanisms are advocated in ensuring that all victims benefit from comprehensive judicial cover. Restorative justice is explicitly mentioned in Principle 9 (1):

> States Parties shall endeavour, where appropriate, to establish or enhance systems of restorative justice, that seek to represent victims' interests as a priority. States shall emphasise the need for acceptance by the offender of his or her responsibility for the offence and the acknowledgement of the adverse consequences of the offence for the victim (Groenhuijsen, 2008; South Africa Department of Justice and Constitutional Development, 2006).

In addition, the convention provides for assistance and compensation of those victims (United Nations, 1985). As per the draft convention, the URJ paradigm could be a model of non-punitive response to victims of crime; it has the potential to avert the negative consequences of the criminal justice system that deepens conflicts and interpersonal animosity, inter-family and intercommunity enmity, because victims and perpetrators do not go to court and come out as friends or partners. Since the rights of victims cannot be respected without knowing who they are and what happened to them, it becomes imperative to use non-adjudicative mechanisms, even those traditional approaches of conflicts resolution rooted in local cultures that are imbued with the principles of *ubuntu*.

Ubuntu is anchored in African philosophy and social ethos where, for instance, pursuing justice is intended to restore victims and facilitate the reintegration of the offender back into society or their respective community, and seeking the restoration of interpersonal relationships and societal peace that have been compromised by conflicts (Oko Elechi et al., 2010). The African approach to justice originates from the collaborative and participatory nature of engagement in which all stakeholders have equal opportunities to access the conflict resolution process and participate in it as equal partners (Oko Elechi et al., 2010). Those procedures are based on the communitarian principles that emphasise the importance of individual wellbeing that sustains the vitality of the community (Oko Elechi et al., 2010). That is the basis of human rights and dignity (Metz, 2011; 2012).

21.8 Conclusion and Recommendations

This chapters explored the scourge of rape, its diverse ramifications and varying treatments received by survivors, which, in most cases, are appalling. To analyse the data collected, we used the framework of honour culture theory, which allowed us to gain insights about the socio-psychological and anthropological understanding honour's functions at individual, collective, group and national levels. Additionally, we incorporated a socio-political and social ethical perspective based on *Ubuntu* philosophy and restorative justice to analyse data and explore eventual alternatives to the criminal justice to reduce honour-induced violence.

The nature, functions, root causes, the complexities of rape survivors-perpetrators were examined to determine the key factors that weaken the management of rape incidents in the community. Furthermore, the vulnerability of victims in socialising and the impacts of honour culture in reporting and dealing with rape were examined. The rise and persistence of rape in the South Sudanese community under study was alarming and calls for immediate actions which include:

1. Enhancing the implementing of the 2018 Revitalised Peace Agreement which laid the basis for addressing insecurity and conflicts that perpetuate sexual atrocities against girls and women.
2. The concept of honour culture should be understood as respect of human rights and dignity; it should focus on protection of women, addressing the plight of

rape victims, addressing their immediate and long-term needs; and providing psycho-social support to affected women and children.
3. Radio broadcasting programmes that foster awareness about honour culture that rather emphasises restoration of human dignity and justice for the violence they suffer in wartime.
4. *Ubuntu* approach of justice that aims to restore and heal broken relationships should be tried to mend victim-aggressor relationships where rape cases are reported. Such trials may yield positive results since community-based restorative justice in post-conflict context in Sub-Saharan Africa have sowm promising outcomes in the DRC, Uganda, Mozambique and Sierra Leone.
5. Women and girls should be given safe spaces within their respective communities to speak about their sexual ordeals within protective and supportive structures to encourage healing of memory.
6. Rape survivors should be provided with needed legal, psychological, medical, social and economic support that can enhance their wellbeing and accelerate their recovery.
7. As argued by Groenhuijsen (2008), there is need for a more effective statutory convention such as the drafted "UN Convention on Justice and Support for Victims of Crime and Abuse of Power" which could ensure the rights of rape survivors are respected and justice is served such victims and other for egregious crimes and power abuse.
8. An integrated approach of justice that combines retributive sanctions and restorative sanctions should envisaged as an adapted model that bridge the local wisdom of justice anchored in the philosophy of *ubuntu* with the conventioanal model of justice. But the ultimate end of such strategy should be restoration rather than humiliation and punishment of rape perpetrators, based of Walgrave's (2011a, b) theorisation of punitive retributivism and restorative restributivism. Restorative retributivism should prevail; it can be more successful that punitive retributivis, particularly if the process is conducned and imbued with *ubuntu* spirit.

Rethinking honour culture is possible by diverting criminal cases involving rape perpetrators to restorative justice approaches since these are imbued with the spirit of *ubuntu*. Rape incidents must not be concealed, but rather denounced and dealt with via societal structures that spare victims from shame and restore their human dignity. There is no single strategy to deal with this crime that dehumanises many people in war-torn spaces. Rape victims must not be muted. The real meaning of honour culture should focus on the protection of women and girls' rights and dignity.

Finally, linking the South Sudan's experiences of war atrocities to other works such as Kiyala (2016, 2022a, b) on the usefulness of indigenous approaches of justice to forge the path of reconciliation via the Baraza traditional conflict resolution and peace-building forum, healing and reconciliation in the aftermath of large-scale violence, we suggest that further research be conducted in the context of South Sudan. Such studies should aim to test, quantitatively, the impacts of restorative justice and locally based traditional models of justice to ensure there is peace and reconciliation in society, and justice for rape survivors and rehabilitation of perpetrators. These are crucial steps to be taken in order to foster societal cohesion and harmony, particularly between war-affected persons and former soldiers. We suggest conducting studies aimed to establish how and to which degree *ubuntu* philosophy and restorative justice principles could inform an effective and humanistic management of rape incidence in South Sudan and beyond.

Acknowledgements To Open Society Foundations of London for their financial support during the tough time of COVID-19; to Dr. Jimmy Juma Lwamba for his supportive academic orientation to get this grant; and to Dr. Sylvia Blanche Kaye for her inputs to this work; and to the International Centre of Nonviolence (Durban University of Technology).

Funding This work was supported by the Open Society Foundations (OSF) under Grant IN2019-53941.

References

Adom, D., Yeboah, A., & Ankrah, A.K. (2016). Constructivism philosophical paradigm: Implication for research, teaching and learning. *Glob. J. Arts. Hum. Soc. Sci., 4*, 1–9.

Ahrens, C.E. (2006). Being Silenced: The Impact of Negative Social Reactions on the Disclosure of Rape. *Springer, 38*, 263–274.

Akinola, A.O., & Uzodike, U.O. (2018). Ubuntu and the quest for conflict resolution in Africa. *Journal of Black Studies, 49*(2), 91–113. https://doi.org/10.1177/0021934717736186

Anderson, A. (2003). *Restorative justice, the African philosophy of Ubuntu and the diversion of criminal prosecution.* Paper presented at the 17th international conference of the International Society for the Reform of Criminal Law, The Hague, Netherlands. https://www.readkong.com/page/restorative-justice-the-african-philosophy-of-ubuntu-and-4649952?p=1. Accessed 28 February 2019.

Anderson, R. (2007). *Thematic content analysis (TCA): Descriptive Presentation of Qualitative Data.* http://rosemarieanderson.com/wp-content/uploads/2014/08/ThematicContentAnalysis.pdf. Accessed 8 August 2023.

Arthur, D.D., Issifu, A.K., & Marfo, S. (2015). An analysis of the influence of Ubuntu principle on the South Africa peace building process.

Ashkenas, R. (2013). The Dangers of Denial. *Harvard Business Publishing.*

Atari, M. (2020). Culture of honor. In V. Zeigler-Hill & T.K. Shackelford (Eds.), *Encyclopedia of personality and individual differences* (pp. 977–980). Cham: Springer. https://doi.org/10.1007/978-3-319-24612-3_650

Baguma, C. (2012). When the traditional justice system is the best suited approach to conflict management: the Acholi Mato Oput, Joseph Kony, and the Lord's Resistance Army (LRA) in Uganda'. *Journal of Global Initiatives: Policy, Pedagogy, Perspective, 7*(1), 31–43.

Bangura, A.K. (2008). The Politics of the struggle to resolve the conflict in Uganda: Westerners pushing their legal approach versus Ugandans insisting on their Mato Oput. *Journal of Pan African Studies, 2*(5), 142–178.

Batha, E. (2015). Sudan changes law that left rape victims punished for adultery: TRFN. https://www.reuters.com/article/ozatp-us-sudan-rape-law-idAFKBN0NF15J20150424. Accessed 12 March 2023.

Battle, M. (2000). A theology of community: The ubuntu theology of Desmond Tutu. *Interpretation, 54*(2), 173–182. https://doi.org/10.1177/002096430005400206

Battle, M. (2009). *Ubuntu: I in you and you in me*: Church Publishing, Inc.

BBC News. (2018). South Sudan soldiers jailed for rape and murder. https://www.bbc.com/news/world-africa-45432454 Accessed 23 May 2023.

Bennett, T.W. (2011). Ubuntu: an African equity. *Potchefstroom Electronic Law Journal/ Potchefstroomse Elektroniese Regsblad, 14*(4).

Bless, C., Higson-Smith, C., & Sithole, S.L. (2013). *Fundamentals of social research methods: an African perspective* (5th ed.). Cape Town: Juta & Company.

Brownmiller, S. (1975). *Against our will: Men, women, and rape.* New-York: Fawcett Columbine.

Card, C. (1996). Rape as a weapon of war. *Hypatia, 11*, 5–18.

CARE. (n.d). *South Sudan, Gender in brief*. https://www.care.org/wp-content/uploads/2020/05/Gender20in20Brief20South20Sudan20.pdf 12 March 2023.

Chappell, B., & Gonzales, R. (2019). *Brandt Jean's Act Of Grace Toward His Brother's Killer Sparks A Debate Over Forgiving.* NPR Newsletter, October 3, 20194:21 PM ET. https://www.npr.org/2019/10/03/766866875/brandt-jeans-act-of-grace-toward-his-brother-s-killer-sparks-a-debate-over-forgi. Accessed 9 August 2023.

Christianson, M., Teiler, Å., & Eriksson, C. (2020). A woman's honor tumbles down on all of us in the family, but a man's honor is only his": young women's experiences of patriarchal chastity norms. *International Journal of Qualitative Studies on Health and Well-being, 16*, 1–14. https://doi.org/10.1080/17482631.2020.1862480

Cihangir, S. (2012). Gender specific honor codes and cultural change. *Group Processes & Intergroup Relations, 16*(3), 1–15. https://doi.org/10.1177/1368430212463453

Cohen, D., & Nisbett, R.E. (1994). Self-protection and the culture of honor: Explaining southern violence. *Personality and Social Psychology Bulletin, 20*, 551–567.

Cohen, D., Nisbett, R. E., Bowdle, B. F., & Schwarz, N. (1996). Insult, aggression, and the southern culture of honor: An "experimental ethnography.". *Journal of personality and social psychology, 70*(5), 945–960.

Cohen, S. (2002). States of Denial: Knowing about Atrocities and Suffering. *Criminal Justice, 2*(3), 345–355.

Collins, B. (2007). English dictionary-Collins dictionaries. California: Collins dictionaries publishing. https://www.collinsdictionary.com/dictionary/english/honour Accessed 23 Jan 2023

Coombs, W.T. (2007). Protecting Organization Reputations During a Crisis: The Development and Application of Situational Crisis Communication Theory. *Corporate Reputation Review 10*, 163–176. https://doi.org/10.1057/palgrave.crr.1550049

Creswell, J.W., Hanson, W.E., Clark Plano, V.L., & Morales, A. (2007). Qualitative research designs: Selection and implementation. *The counseling psychologist, 35*(2), 236–264. https://doi.org/10.1177/0011000006287390

Danjibo, N. (2019). Rape as a weapon of war against women and girls. *African Journals Online, Gender and Behaviour, 17*(2).

Davies, M. B. (2007). *Doing a successful research project: using qualitative or quantitative methods.* New York: Pagrave MacMillan.

Druckman, J. N. (2001). 'The implications of framing effects for citizen competence', Political Behavior. *Political behavior, 23*(3), 225–256.

Edgerton, R. (Ed.) (1971). *The individual in cultural adaptation: a study of four East African peoples.* Berkeley: University of California Press.

Ellsberg, M., Ovince, J., Murphy, M., Blackwell, A., Reddy, D., Stennes, J.,... Contreras, M. (2020). No safe place: Prevalence and correlates of violence against conflict-affected women and girls in South Sudan. *PLoS one, 15*(10), 1–24. https://doi.org/10.1371/journal.pone.0237965

Etikan, I., Musa, S.A., & Alkassim, R.S. (2016). Comparison of convenience sampling and purposive sampling. *Am. J. Theor. App. Stat.*, 5(1), 1–4. https://doi.org/10.11648/j.ajtas.20160501.11

Fairclough, N. (2001). Critical discourse analysis as a method in social scientific research. *Methods of critical discourse analysis, 5*(11), 121–138.

Fricke, A. L., & Amir, A. (2007). Law without justice. https://www.refworld.org/pdfid/47a6eb870.pdf. Accessed 2 July 2021.

Gathogo, J. (2008). African philosophy as expressed in the concepts of hospitality and ubuntu. *Journal of theology for Southern Africa, 130*, 39.

Gaylard, R. (2004). Welcome to the world of our humanity: (African) humanism, Ubuntu and black South African writing. *Journal of Literary Studies, 20*(3_4), 265–282. https://doi.org/10.1080/02564710408530357

George, L., Winfeld, I., & Blazer, D. (1992). Sociocultural factors in sexual assault: Comparison of two representative samples of women. *Journal of Social Issues, 48*, 105–125.

Gibbs, A., Said, N., Corboz, J., & Jewkes, R. (2019). Factors associated with 'honour killing' in Afghanistan and the occupied Palestinian Territories: Two cross-sectional studies. *Plos one, 14*(8), e0219125.

Gill, A. K. (2014). Introduction:'Honour'and 'honour'-based violence: Challenging common assumptions. *'Honour'Killing and Violence: Theory, Policy and Practice*, 1-23. https://doi.org/10.1057/9781137289568_1

Global Women's Institute. (2017). *What works to prevent violence against women and girls research programme*. London: department for International Development.

Gravelin, C.R., Biernat, M., & Bucher, C.E. (2019). Blaming the victim of acquaintance rape: Individual, situational, sociocultural factors. *Front Psychol.*, 9, 2422.

Graybill, L.S. (2004). Pardon, punishment, and amnesia: three African post-conflict methods. *Third World Quarterly, 25*(6), 1117–1130.

Graybill, L.S. (2010). Traditional practices and reconciliation in Sierra Leone: the effectiveness of Fambul Tok. *Conflict Trends, 1*(3), 41–47.

Groenhuijsen, M. (2008). The draft UN convention on justice and support for victims of crime, with special reference to its provisions on restorative justice. *Int'l Annals Criminology, 46*, 121–136.

Guerra, V.M., Roger, G.-S., & Milica, V. (2012). The importance of honor concerns across eight countries. *Group Processes & Intergroup Relations, 16*(3), 298–318. https://doi.org/10.1177/1368430212463451

Guetterman, T.C., & Fetters, M.D. (2018). Two methodological approaches to the integration of mixed methods and case study designs: A systematic review. *American Behavioral Scientist, 62*(7), 900–918. https://doi.org/10.1177/0002764218772641

Haki. (2011). Combatting Gender-Based Violence in the customary courts of South Sudan. http://www.hakinetwork.org/blog/2014/9/3/combatting-violence-against-women-in-the-customary-courts-of-south-sudan Accessed 12 March 2023.

Hoffman, E. (2008). Reconciliation in Sierra Leone: local processes yield global lessons. *Fletcher Forum of World Affairs, 32*, 129–141.

Honwana, A. (1997). Healing for peace: traditional healers and post-war reconstruction in Southern Mozambique. *Peace and Conflict: Journal of Peace Psychology, 3*(3), 293–305. https://doi.org/10.1207/s15327949pac0303_6

Hsieh, H.-F., & Shannon, S.E. (2005). Three approaches to qualitative content analysis. *Qualitative health research, 15*(9), 1277–1288.

Human Rights Watch (2009). There is no protection: Insecurity and human rights in Southern Sudan. https://www.hrw.org/report/2009/02/12/there-no-protection/insecurity-and-human-rights-southern-sudan Accessed 12 March 2023.

Institute for Global Leadership. (2009). *Community perspectives on the Mato Oput process: a research study by the Mato Oput project* The Institute for Global Leadership, TUFTS University and Institute of Peace and Strategic Studies Gulu.

Isikozlu, E., & Millard, A.S. (2010). Wartime Rape: Identifying Knowledge Gaps their Implications. *JSTOR, 28*(1), 35–41. https://www.jstor.org/stable/24233006 Accessed 28 October 2022.

Kaplan, B., & Duchon, D. (1988). Combining qualitative and quantitative methods in information systems research: a case study. *MIS quarterly*, 571–586.

Kele, E. (2020). Radio programming can play crucial role in ending gender-based violence in Northern-Bahr-el-Ghazal. *reliefweb, 11 Dec 2020.* https://reliefweb.int/report/south-sudan/radio-programming-can-play-crucial-role-ending-gender-based-violence-northern. Accessed 1 March 2023.

Kiyala, J.C.K. (2016). Utilising a traditional approach to restorative justice in the reintegration of former child soldiers in the North Kivu Province, Democratic Republic of Congo. *Africa Insight, 46*(3), 33–50.

Kiyala, J.C.K. (2018). *Child Soldiers and Restorative Justice: Participatory Action Research in the Eastern Democratic Republic of Congo.* Cham ZG: Springer International Publishing. https://doi.org/10.1007/978-3-319-90071-1

Kiyala, J.C.K. (2021). Responding to Epistemic Injustice against Child Soldiers. *The International Journal of Children's Rights, 29*(3), 662–700. https://doi.org/10.1163/15718182-29030006

Kiyala, J.C.K. (2022a). Religion, Tradition, and Restorative Justice in Sierra Leone: by Lyn S. Graybill, Notre Dame, Indiana, University of Notre Dame Press, 2017, viii+ 324 pp. In: Taylor & Francis.

Kiyala, J.C.K. (2022b). Underlying moral justification of Indaba and Baraza dialogic institutions in African social ethics and philosophy. In L.C. Rodrigues & J.O. Chimakonam (Eds.), *African Ethics: A Guide to Key Ideas* (pp. 159–184). London: Bloomsbury Publishing Plc.

Koss, M. (1985). The hidden rape victim: Personality, attitudinal, and situational characteristics. *Psychology of Women Quarterly, 9*, 193–212.

Lamis Abu, N., Fatmeh, A.-T., Oraib, A.-M., Suad, I., Kifah, R., & Amal, A. A. (2007). Crimes of women's killing in Palestine in the period 2004–2006: Palestine Non-Government Organisation Against Domestic Violence Against Women, Beit Jala. Fatmeh A-T, Oraib A-M, Suad I, Kifah R, Amal AA (2007). Crimes of women's killing in Palestine in the period 2004–2006: Palestine Non-Government Organisation Against Domestic Violence Against Women, Beit Jala.

Legassicke, M. (2013). *Cyclical Violence in Jonglei State: The Deadly Shift in the Practice of Cattle Raiding,*. (Doctorate). University of Waterloo, Ontario.

Cohen, R.A. (2006). Introduction: Humanism and anti-humanism–Levinas, Cassirer, and Heidegger. In E. Levinas (Ed.), *Humanism of the Other* (pp. vii-). Chicago: University of Illinois.

Levine, T.R. (2014). Truth-Default Theory (TDT): A Theory of Human Deception and Deception Detection. *Journal of Language and Social Psychology, 33*(4), 378–392. https://doi.org/10.1177/0261927X14535916

Lindsey, L.O., & Ryan, P.B. (2011). Culture of honor and violence against the self. *Journal of personality and social psychology, 37*(12), 1611–1623.

Lowenstein, A.K. (2010). Gender-based violence in Southern Sudan: Justice for women long overdue. *Enough Project.* https://enoughproject.org (12 Mar 2023).

Loyola University MaryLand. (2021). Common reaction to sexual assault. https://www.loyola.edu/department/counseling-center/services/students/concerns/sexual-assault/reactions. Accessed 20 June 2023.

MacKinnon, C. (1987). *Feminism unmodified: Discourses on life and law. Cambridge, MA: Harvard University Press.* MA: Harvard University Press.

Madut Jok, J. (2013). State, Law, and Insecurity in South Sudan. *the fletcher forum of world affairs, 37*(2), 69–80.

Maedl A (2011). Rape as weapon of war in the Eastern DRC? The victims' perspective. Human Rights Quartely, *33,* 128.

Mancini, C., & Mears, D. (2010). To execute or not to execute? Examining public support for capital punishment of sex offenders. *Journal of Criminal Justice 38*(5), 959-968.

McGlynn C, Westmorland N, Godden, N (2012). 'I just wanted him to hear me': Sexual violence and the possibilities of restorative justice. *J. Law Soc., 39*: 213–240.

Mednick, S. (2021). Rare conviction of South Sudan soldiers for rape raises hope. *AP News, Jan 15, 2021.* https://apnews.com/article/africa-south-sudan-crime-d1349141044a4fae7257588cb8ad4d05 Accessed 1 March 2023.

Metz, T. (2011). Ubuntu as a moral theory and human rights in South Africa. *African Human Rights Law Journal, 11*(2), 532–559.

Metz, T. (2012). African conceptions of human dignity: Vitality and community as the ground of human rights. *Human Rights Review, 13*(1), 19–37. https://doi.org/10.1007/s12142-011-0200-4

Ministry of Legal affaires and constitutional development. (2008). *Acts supplement No 1, The penal code act, 2008*. Juba: South Sudan Government.
Mohamed, A. (2018). Encyclopedia of personality and individual differences. In: Spring International Publishing.
Moritz, M. (2008). Acritical examination of honor cultures and herding societies in Africa. *African Studies Review, 51*(2), 99–117. https://doi.org/10.1353/arw.0.0052
Mosquera, P.M.R. (2013). In the name of honor: On virtue, reputation and violence. *Group Processes & Intergroup Relations, 16*(3), 271–278. https://doi.org/10.1177/1368430212472590
Nasiwa, J. (2021). South Sudan: No Justice for Women, No Peace in South Sudan. https://allafrica.com/stories/202103190887.html Accessed 22 February 2022.
Navratan, F. (2014). *Honour killing in Canadian context*. UK-London: Lambert Academic Publishing.
Nisbett, R.E., & Cohen, D. (1996). *Culture of honor: The psychology of violence in the South*. Westview: Boulder, CO.
Oko Elechi, O., Morris, S.V., & Schauer, E. J. (2010). Restoring justice (ubuntu): an African perspective. *International Criminal Justice Review, 20*(1), 73–85. https://doi.org/10.1177/1057567710361719
Osch, V., Breugelmans, S.M., Zeelenberg, M., & Bölük, P. (2013). A different kind of honor culture: Family honor and aggression in Turks. *Group Processes & Intergroup Relations, 16*(3), 334–344.
Pinaud, C. (2020). Genocidal Rape in South Sudan: Organization, Function, and Effects. *Human Rights Quarterly, 42*(3), 667–694.
Pitt-Rivers, J. (1965). *Honra e posição social [Honor and social status]*. Lisbonia: Fundação Clouse Gulbenkian.
Redress and KCHRED. (2008). Reforming Sudan's legislation on rape and sexual violence. https://redress.org/wp-content/uploads/2018/01/Sep-08-Reforming-Sudans-Legislation-Position-Paper.pdf (12 Mar 2023).
Rieger, K. L. (2019). Discriminating among grounded theory approaches. *Nursing inquiry, 26*(1), 1–12. https://doi.org/10.1111/nin.12261
Saldaña, J. (2021). *The coding manual for qualitative researchers*. Los Angeles: sage.
Scoones, I. (2020). Pastoralists and peasants: perspectives on agrarian change. Taylor and Francis Group, London.
Sen, P. (2005). *Crimes of honour, value and meaning*: Zed Books Ltd.
Shackelford, T.K. (2005a). An evolutionary psychological perspective on cultures of honor. *Evolutionary psychology*, 381–391.
Snyder, J., & Vinjamuri, L. (2003). Trails and errors: Principles and fragmentism in strategies of international justice. *International security* 2, 8.
Soccol, A. (2018). Sexual and Gender-based Violence in South Sudan: a Step Towards Justice. *Groningen Journal of International Law*.
South Africa Department of Justice and Constitutional Development. (2006). DRAFT UN Convention on Justice and Support for Victims of Crime and Abuse of Power. https://www.justice.gov.za/vc/docs/international/2006_Draft%20UN%20Convention%20Victims.pdf.
Spoo, S., Kaylor, L.E., & Schaaf, S. (2018). Spoo S, Kaylor LE, Schaaf S, et al. Victims' Attitudes Toward Sex Offenders and Sex Offender Legislation. *International Journal of Offender Therapy and Comparative Criminology, 6*(11), 3385–3407. https://doi.org/10.1177/0306624X17740537
Strauss, A., & Corbin, J. (1998). *Basics of qualitative research techniques*. Thousand Oaks: Sage Publications, Inc.
Sudan Human Security Baseline Assessment (2019). *Women's Security and the law in South Sudan*: https://land.igad.int/index.php/documents-1/countries/south-sudan/gender-5/999-women-s-security-and-the-law-in-south-sudan/file. 12 March 2023.
Sudan Tribune. (2018). Reports about sexual attacks in Bentiu are false: South Sudanese government. https://sudantribune.com/article64940/ Accessed 17 May 2022.

Tahira, S.K. (2006). *Beyond Honour, A Historical materialist Explanation of honour Related Violence* New-York: Oxford University Press.

Takpiny, B. (2021). South Sudan reports over 300 cases of gender-based violence in 2021. https://www.aa.com.tr/en/africa/south-sudan-reports-over-300-cases-of-gender-based-violence-in-2021/2483745 Accessed 24 February 2022.

Teddlie, C., & Yu, F. (2007). Mixed Methods Sampling: A Typology With Examples. *Journal of Mixed Methods Research, SAGE Publications, 1*(1), 77–100.

The Nuhanovic Foundation. (2016). Accountability for sexual violence committed by armed men in South Sudan. https://www.legalactionworldwide.org/wp-content/uploads/Legal-Action-Worldwide-Report-on-Accountability-for-Sexual-Violence-Committed-by-Armed-Men-in-South-Sudan.pdf Accessed 12 March 2023.

Thomas, K. (1959). The double standard. *Journal of the History of Ideas, 20*(2), 195–216.

Towheed, S. (2010). Tools and techniques for literary research: Using online and printed sources. In D.d.S. Correa & W.R. Owens (Eds.), *The Handbook to Literary Research* (2nd ed., pp. 9-35). London: Routledge.

Tutu, D.M. (1999). *No future without forgiveness*. London: Rider.

United Nations. (1985). Declaration of basic principles of justice for victims of crime and abuse of power. https://www.unodc.org/pdf/compendium/compendium_2006_part_03_02.pdf Accessed 12 November 2017.

United Nations. (2005). Basic Principles and Guidelines on the Right to a Remedy and Reparation for Victims of Gross Violations of International Human Rights Law and Serious Violations of International Humanitarian Law. Adopted and proclaimed by General Assembly resolution 60/147 of 16 Dec 2005. http://www.ohchr.org/EN/ProfessionalInterest/Pages/RemedyAndReparation.aspx Accessed 6 September 2014.

United Nations. (2020). Children and armed conflict. Report of the Secretary-General on the Promotion and protection of the rights of children Security Council. Fifty-fifth year General Assembly, Fifty-fifth session, Item 112 of the provisional agenda.

United Nations. (2022). Conflict-related sexual violence against women and girls in South Sudan: Conference room paper of the Commission on Human Rights in South Sudan. A/HRC/49/CRP.4. https://www.ohchr.org/sites/default/files/2022-03/A_HRC_49_CRP_4.pdf Accessed 2 March 2023.

United Nations Office on Drugs and Crime. (2006). Handbook on restorative justice programmes. United Nations, Vienna.

Vaismoradi, M., Turunen, H., & Bondas, T. (2013). Content analysis and thematic analysis: Implications for conducting a qualitative descriptive study. *Nursing & health sciences, 15*(3), 398–405.

Vandello, J. A., & Cohen, D. (2003). Male honor and female fidelity: Implicit cultural scripts that perpetuate domestic violence. *Journal of Personality and Social Psychology, 84*, 997–1010.

Walgrave, L. (2011a). Integrating criminal justice and restorative justice. In G. Johnstone & D.W. van Ness (Eds.) *Handbook of restorative justice* (pp 559–579). London: Routledge.

Walgrave, L. (2011b) Towards restoration as the mainstream in youth justice. In E. Elliott & R.M. Gordon (Eds.) *New directions in restorative justice* (pp. 3-25). London: Routledge.

Watts, C., & Zimmerman, C. (2002). Violence against women: global scope and magnitude. *the Lancet, 35*(9), 1232–1237.

Zehr. H. (2002). *The little book of restorative justice*. Pennsylvania: Good Books.

Chapter 22
Effects of State Capturing of Truth and Reconciliation Commissions in Zimbabwe on Sustainable Development

Torque Mude

Abstract The objective of this chapter is to explore ways through which the trajectory of the state capture of truth and reconciliation commissions in Zimbabwe could be addressed in the pursuit of sustainable peace. Since the attainment of independence in 1980, the country has experienced successive episodes of political violence and human rights abuses. From the Gukurahundi massacres in Matabeleland and Midlands Provinces in the 1980s to the 2008 and 2018 post-electoral violence, the state has been responsible for unleashing violence on political opponents. In all the cases, state-led efforts that have been purported to redress the past have largely been politicised to uncover the truth, compel collective forgetting and deny atrocities. The Dumbutshena and Chihambakwe commissions established to uncover the truth and the Organisation for National Healing, Reconciliation and Integration (ONHRI), National Peace and Reconciliation Commission (NPRC) established for reconciliation following the 2008 post-election violence and the Commission of Inquiry into the 1 August Post-election Violence have all been unable to induce sustainable peace arguably due to their politicisation. Consequently, latent conflicts, tribal divisions, impunity, political instability, violent political contestations, disunity, recurrence of political violence and wanton human rights abuses have continued to threaten sustainable peace in Zimbabwe. Against this background, it is the aim of this chapter to examine historical and recent patterns of transitional justice and reconciliation in Zimbabwe to not only identify complexities, contexts and contingencies thereof, but to rely on historical accounts and recent expressions as well as perspectives in mapping endogenous strategies for addressing the state capture of transitional justice and reconciliation processes towards sustaining peace. Data for this paper

Torque Mude (PhD) serves as a Postdoctoral Research Fellow at the Centre for Africa-China Studies of the University of Johannesburg and Senior Lecturer in the Department of Public Management and Political Studies at the University of Namibia. He holds a Ph.D. in International Politics and is in the final year of the Bachelor of Laws Degree (UNISA). His research engages African politics, international relations, governance, peacebuilding, international law and development. Email(s): tmude@unam.na; mudetorque@gmail.com. https://orcid.org/0000-0002-9885-6637

J. C. K. Kiyala and N. Chivasa (eds.), *Climate Change and Socio-political Violence in Sub-Saharan Africa in the Anthropocene*, The Anthropocene: Politik—Economics—Society—Science 37,
https://doi.org/10.1007/978-3-031-48375-2_22

were obtained from secondary sources including books, journal article and newspaper articles as well as through primary research including the observation of the researcher.

Keywords Transitional justice · Truth and reconciliation commissions · Sustaining peace · Zimbabwe

22.1 Introduction

Zimbabwe's short post-independence history has been characterised by worrisome episodes of political violence, atrocities, and human rights violations. Although Zimbabwe has instituted five (5) truth and reconciliation commissions since 1980 to facilitate transitions from episodes of violence and human rights abuses, sustainable peace has continued to traverse a difficult terrain due to the politicisation of these transitional justice and reconciliation efforts. Efforts at redress, because they were state-led, resulted in politicised transitional justice and reconciliation efforts and concomitant negative peace in the country. In this context, two important questions shape the political and moral agenda for transitional justice and reconciliation towards sustaining peace in Zimbabwe where accountability mechanisms are largely state-controlled. First, how can the politicisation of quasi-judicial institutions be redressed in the pursuit of transitional justice and reconciliation towards sustainable peace? And second, how can justice, national healing and viable relationship-rebuilding be consolidated from a past of violence and human rights abuse to sustain peace in Zimbabwe?

To answer these questions, it can be argued that unless and until comprehensive reform measures are espoused to hybridise transitional justice with a view to countering the politicisation of truth and reconciliation commissions, sustaining peace in Zimbabwe will remain a pipedream. Hybridising transitional justice denotes combining, in a transformational manner, government-led and non-governmental efforts at pursuing transition from widespread violence and human rights abuses. Hence, the unique contribution of this chapter is its focus on the historical patterns of transitional justice and reconciliation in Zimbabwe to not only identify complexities, patterns, contexts and contingencies thereof, but to rely on historical accounts and recent expressions as well as perspectives in mapping endogenous strategies for sustaining peace in the country.

With the above said, this chapter shifts the gaze of analysis from mainstream studies on the questions of transitional justice and reconciliation in Zimbabwe that are more evaluative, exploratory and descriptive, rather than prescriptive. Moreover, the study also tilts the angle of analysis from evaluating governmental and CSO independent transitional justice and reconciliation posturing to the hybridisation impact of their triangulated efforts. The three central questions this chapter to address are: To what extent have Zimbabwe's five truth and reconciliation commissions formed to investigate political violence and human rights violations been politicised and with what implications? How can future truth and reconciliation interventions be reconfigured to best suit the needs and conditions of sustainable peace in Zimbabwe? How can transitional justice and reconciliation mechanisms be strengthened to counter their politicisation in the quest for sustainable peace? These questions will broaden conversations on transitional justice and reconciliation in Zimbabwe. This is because they transcend questions on the government's reluctance to uncover the truth and be accountable by focusing on the relatively under-covered theme of what ought to be done to sustain peace in the country's highly shrunk transitional justice and reconciliation processes.

22.2 Historicising the Politicisation of Truth and Reconciliation in Zimbabwe

Since the Zimbabwe's attainment of independence in 1980, a total of five (5) truth and reconciliation commissions have been established in pursuit of transitional justice and reconciliation following successive interregnums of excessive state-sanctioned violence and the ensuing gross human rights abuses in the country. These are the Dumbutshena Commission of Inquiry into the Entumbane Uprising (1981); Chihambakwe Commission of Inquiry into Gukurahundi Massacres (1983); Organ for National Healing, Reconciliation and Integration (ONHRI) (2009); National Peace and Reconciliation Commission (NPRC) (2013) and Commission of Inquiry into the 1 August 2018 Post-election Violence (2018). Nevertheless, transitional injustice in the form of atrocity denial and collective processes of forgetting as well as reconciliation deficit still haunts Zimbabwe. The implications have been latent conflicts, tribal divisions, impunity, political instability, political intolerance, disunity, recurrence of political violence and wanton human rights abuses which are continuing to threaten sustainable peace.

Arguably, this is attributed to the politicisation of truth and reconciliation commissions and subsequent co-optation of transitional justice illustrated by the reluctance to publish inquiry reports and implement inquiry findings, as well as co-opting truth. Consequently, the implications of this state of affairs has been transitional injustice and negative peace. From the Gukurahundi massacres of the 1980s to the subsequent incidents of political violence and human rights abuses in 2005 and 2008, 2018, the government-led efforts at transitional justice and reconciliation through truth and reconciliation commissions have appeared disingenuous to sustainable peace. The efforts have taken the forms of self-investigations and victor's commissions whose interests are to manipulate transitions for the purpose of entrenching power and survival at the expense of uncovering truths, achieving justice and reconciling relations.

The above confirms Solomon and Zvobgo (2019)'s observation that truth and reconciliation commissions are co-opted by autocratic governments that perceive transitional justice as threats to their rule and are interested in buttressing their power, securing their survival and longevity of their rule while weakening their rivals through politicising transitions from atrocities. Since Zimbabwe's state-led accountability mechanisms are co-opted, CSOs, NGOS and community leaders have been leading efforts to pursue justice and reconcile broken relationships through community engagement, advocacy and litigation. To maintain the cycle of uncovering truths and atrocity denial, these non-state efforts have also been frustrated and deliberately disabled by the government through exclusionary actions and rhetoric, making transitional justice and reconciliation unattainable.

Against this backdrop, the chapter aims to generate relevant knowledge on viable strategies for sustaining peace in Zimbabwe through pursuit of genuine transition from violence and human rights abuses. Zimbabwean narratives and conversations on transitions from violence and atrocious politics have largely been monologues on challenges to transitional justice and reconciliation as well as how CSO's community engagement has closed the peacebuilding gap. This has largely eclipsed the extent to which government-CSO transitional justice synergies could be harnessed as an antithesis to primary reliance on government-controlled mechanisms whose politicisation has proved disingenuous to sustainable peace for four decades. Therefore, this chapter endeavours to suggest strategies for depoliticising transitional justice and reconciliation processes to foster sustainable peace in Zimbabwe.

22.3 Theoretical Framework

The theoretical foundations of this chapter are premised on the complexity theory of sustaining peace which makes reference to the nexus between complex systems, resilience and sustainability. The theory illuminates on how socio-political systems collapse, how they regain order and what can be done to strengthen their resilience so that they may be able to mitigate relapse into violence or cope better with its effects (de Coning, 2016). The theory is used to explain that for peace to be sustained in Zimbabwe, resilient transitional justice and reconciliation institutions and processes should be re-organised, strengthened, demonstrate holism, non-linearity and reconfigured through hybridisation.

The chapter conceptualises sustaining peace from the perspective of the UN Report of Advisory Group of Experts (2015) as efforts to reclaim peace through reinforcing the structures, institutions and attitudes that underpin it. Sustainable peace denotes a situation characterised by the absence of physical violence, the elimination of unacceptable socio-political forms of discrimination and propensity to enhance the constructive transformation of conflicts (Reychler/Paffenholz, 2001). This chapter is premised on the view that key to sustaining peace in Zimbabwe are transitional justice and reconciliation. Transitional justice encompasses judicial and non-judicial mechanisms and approaches employed to address human rights violations and the wrongdoings of repressive regimes (Bell, 2015; Cobban, 2007; Teitel, 2002). Reconciliation is conceptualised as a peacebuilding process that involves the mutual acknowledgment of the past accompanied by changing destructive attitudes and behavior into constructive relationships towards sustainable peace, truth and justice (Rosoux, 2008; Brouneus, 2003; Lederach, 1997).

22.4 Suggested Reforms for Truth and Reconciliation in Zimbabwe

Existing knowledge on transitional justice and reconciliation in Zimbabwe has concentrated more on the impediments to the success of peacebuilding processes without delving into the extent to which transitional justice and reconciliation processes in question have been politicised. Above all, existing literature on transitional justice and reconciliation in Zimbabwe fall short of illuminating on response strategies to counter the politicisation of quasi-judicial institutions towards sustaining peace. Hence, by relying on historical patterns and the contemporary complexities, contexts and contingencies to generate strategies for confronting the politicisation of transitional justice and reconciliation to sustaining peace in the country, the chapter constitutes an epistemic break by generating the urgently needed academic relevant knowledge on how to counter the politicisation of truth and reconciliation commissions in the quest for sustainable peace in Zimbabwe.

Scholars have concluded that the Zimbabwean government has since the 1980s been unwilling to commit to addressing the violent past and more recent atrocities for fear of being implicated and losing political power (Chengeta, 2018; Murambadoro, 2015; Hapanyengwi-Chemhuru, 2013; Mashingaidze, 2010). A shocking development in line with these observations pertains to the NPRC's claim that the Dumbutshena and Chihambakwe reports of the 1980s Matabeleland massacres are missing (Mthiyane, 2018). This can be said to be deliberate moves to co-opt transitional justice processes by covering atrocities which Loyle and Davenport (2016) call transitional injustice emanating from subverting justice in the pursuit of political survival.

It can be argued that the first area to redress in addressing the state capture of truth and reconciliation processes in Zimbabwe is governance. This is attributable to the challenges in respect of democratic deficit in Zimbabwe. Given the current dynamics of the state capture of the judicial and legislative branches of government as well as quasi-judicial institutions such as the National Peace and Reconciliation Commission, the pursuit of truth and reconciliation depends on strengthening democratic governance which has remained precarious since 1980 and became more unpalatable since 2017 coup in Zimbabwe. What presently exists in the Zimbabwean context is dominant political party democracy. This phenomenon has infested Zimbabwean politics since 1980. According to Choudhry (2010), one of the pathologies of a dominant political party democracy is the capturing of independent institutions meant to check the exercise of political power by the dominant party, enmeshing them in networks of patronage. One unprecedented ramification of this phenomenon in Zimbabwe is that the rule of law, separation of powers and constitutionalism as well as separation of state and party have remained a phantom on the horizons. As a result, organs of state and reconciliation institutions largely operate as agency of state and government whose mandates and outcomes are directly influenced by the state. Therefore, democratising the country's political system can arguably go a long way as a foundation in setting the stage for the establishment of institutions, processes, frameworks and mechanisms for genuine truth and reconciliation processes in Zimbabwe.

Having said the above, suffices to reflect on the urgent need to introduce governance reforms in Zimbabwe. The need for political and legal reforms in Zimbabwe is a discourse that has dominated the country's political landscape since the era of the Government of National Unity (GNU) from 2009 to 2013. In essence, the GNU period was meant to foster the much needed reforms in the country. The establishment of the Organ for National Healing, Reconciliation and Integration in 2009 was one of the positive steps taken during the GNU tenure. While it was a positive step in the reconciliation direction, Organ for National Healing, Reconciliation and Integration was not accorded an independent mandate to facilitate reconciliation efforts. It remained dependent on the state for funding, power and mandate. Due to the inherent shortcomings of the ONHRI, the GNU embarked on a reform agenda that culminated in the launch of National Peace and Reconciliation Commission in 2013 to succeed the former. The reforms, however, have fallen short of sustainability because the later operates in the same manner as the former in relation to dependent on the state. This being said, the underlying roots cause of this predicament can be said to be the legal and political trajectory of Zimbabwean politics that is characterised by too much concentration of power, and subsequently reconciliation policy discretion, in the government. A complete overhaul of this system through promulgation of corrective legal frameworks suffices to make reconciliation a domain which is not influenced by and dependent on the government.

Against the above background, legal and policy frameworks aimed at depoliticising truth and reconciliation processes ought to be adopted urgently in Zimbabwe. The question that arises at this juncture relates to the ways in and extent to which this could be achieved. In terms of legal frameworks, the adoption of the new constitution in 2013 can be said to have laid a foundation for political and legal reforms necessary for democratic governance. However, many of the clauses of the new constitution are yet to be aligned with legislation and other laws relating to peacebuilding issues. For instance, new electoral laws outlined in the new constitution are yet to be aligned with the Electoral Act. Furthermore, the legislation empowering the National Peace and Reconciliation Commission is yet to be aligned with the Constitution due to lack of political will on the part of the government. This is largely because the Zimbabwean government appears not willing to have truth about past atrocities publicised just as it is averse to pursue genuine relationship building for fear of prosecution, reprisal and losing power. Hence, policies and laws assuring and reassuring freedom from persecution could go a long a long way in convincing perpetrators of past who are still in government to promulgate rules and policies designed to achieve genuine reconciliation. The first port of call in this regard is the parliament. The parliament should execute its law-making mandate free from interference by the executive; doing so would set the stage for the adoption of mechanisms to prevent the state capture of reconciliation policies, laws and institutions.

From a radical point of view, it can also be argued that democratic and peaceful regime change is an option that can be considered in setting agenda for addressing the state capture of truth and reconciliation institutions in Zimbabwe. The rationale behind such a radical stance is against the background that the current ruling regime, ZANU PF, which has been in power since 1980 has dismally failed to advance truth and reconciliation following successive periods of atrocities to which its officials has perpetrated. This is largely attributed to this regime's power hunger and authoritarian tendencies. Instead of fostering unpretentious truth and reconciliation, the regime has for the past forty years been consolidating its power base through co-opting justice, denying truth and create further divisions between perpetrators and victims of its atrocities. For instance, reports of the Gukurahundi atrocities have been hidden since 1984. Whilst there has not been closure for the atrocities of the 1980s, the regime sponsored another wave of political violence and human rights abuses in the aftermath of the 2008 elections whose aftermath was also characterised by concealing evidence and denying truth.

In 2018, the same regime presided over the deployment of the military to shoot protesting unarmed civilians in the aftermath of the 2018 Presidential and Parliamentary elections. This culminated in the establishment of the Commission of Inquiry into 1 August 2018 Post-Election Violence whose processes were largely influenced by the government. The appointment of commissioners and funding for the commission were government-led. Furthermore, the personnel appointed had ZANU PF links and the manner in which the inquiry was carried out reflected pure politicisation. Moreover, the report of the commission has not be implemented; illuminating ZANU PF's precedence of co-opting truth and reconciliation commissions which serve as their agents for legitimacy. More worrisome is the regime's capture of all state and governmental institutions including the military which it uses as a coercive tool to intimidate and torture the people which it has come to perceive as political opponents. This entanglement between the state, government and military was attested by the ZANU PF political commissar Patrick Chinamasa's declaration that the bond between the army and ZANU PF is unbreakable (Chinamasa, 2021). All these are indications that the elephant in room is ZANU PF and dictatorial tendencies. Hence, best way to address the predicament under discussion is to effect regime change through democratic and peaceful processes.

More so, relying on public opinion in espousing constructive measures to depoliticise truth and reconciliation processes in Zimbabwe. Conceptualised as the thoughts and views of the public on a particular issue (Pethe, 2018), it can be argued that public opinion is useful in mapping the desires, wants and thoughts of Zimbabweans on what could be done to address the state capture of truth and reconciliation commissions in Zimbabwe. The collective opinion of people on a specific topic of concern is significant in mapping changes along the public's policy preference (Smith, 2008). Relying on public opinion in addressing the politicisation of truth and reconciliation commissions in Zimbabwe is one of the ways to illuminate the trajectory of democracy in the country. It has for long been observed that public opinion serves as the foundations of democracy illustrating the sovereignty of the people which is the very basis of democracy (Pethe, 2018; Verma, 1956; Kelsen, 1955). Moreover, public opinion will go a long way in fostering the socio-political change that is sine qua non to addressing the state capture of truth and reconciliation commissions in Zimbabwe. It is maintained that public opinion is a significant agent of socio-political change without which no law, actions and policy aimed at socio-political change can be successful (Hanitzsch et al., 2016). Hence, the desired aims of addressing the state capture of truth and reconciliation commissions can be secured by obtaining public opinion on how to reform and change this status quo of concern in Zimbabwe.

Furthermore, this chapter illuminates on how harnessing international, continental and regional frameworks on transitional justice and reconciliation could go a long way in depoliticising truth and reconciliation processes in Zimbabwe. Therefore, utilising and adapting to UN Approach to Transitional Justice (2010), African Union (AU) Silencing Guns Roadmap (2016) and AU Transitional Justice Framework (2018) in countering the state capture of quasi-judicial institutions towards sustaining peace. Among other theoretical and practical steps, these frameworks were designed to serve as tools and actionable guides to sustain peace through ending repressive rule, fighting impunity, pursuing accountability and consolidating justice through strengthening national institutions, ensuring accountability and effectively managing political transitions to rid Africa of human rights violations and cycles of violence (AU Transitional Justice Framework, 2018; AU Master Roadmap of Practical Steps to Silence the Guns in Africa by Year, 2016; UN Transitional Justice Approach, 2010).

Moreover, drawing lessons from South Africa, Cambodia and Guatemala among other states that experienced varying degrees of success of truth and reconciliation processes, it is crucial to illuminate on best practices for ex-post policy refinement to sustain peace in Zimbabwe. These lessons will be utilised in line with the international best practices outlined in the United Nations and African Union reconciliation policy frameworks. For instance, there is need for multi-sectoral, transformative and accommodative truth and reconciliation processes in Zimbabwe. Civil society formations, non-government organisations and other relevant stakeholders should be given enough normative and political space to influence reconciliation in Zimbabwe. Hence, the practice of perceiving the contributions of these non-state actors as inimical should be given the contempt is warrants. At this juncture, it also suffices to indicate that this also depends on the degree of democratic governance in Zimbabwe.

22.5 Conclusion

This chapter explored the measures necessary in addressing the state capture of truth and reconciliation commissions towards sustaining peace in Zimbabwe. It is maintained in this chapter that of all the measures mooted to resolve the truth and reconciliation imbroglio in Zimbabwe, the most urgent and critical one is reforming the country's governance architecture whose power configurations are at the apex of militating against genuine truth and reconciliation processes. Hence, it is argued that democratic governance is sine qua non to depoliticising domestic laws, institutions and mechanisms towards the pursuit of sustainable peace through genuine truth and reconciliation processes. In its current state, governance dynamics in Zimbabwe are not conducive due to democratic governance deficit characterised by lack of constitutionalism, rule of law and separation of powers. Therefore, instituting democratic reforms will set the stage and foundation upon which depoliticising truth and reconciliation processes will be hinged.

References

Bell, J. (2015). Understanding transitional justice and its two major dilemmas. *Journal of Interdisciplinary Conflict Science, 1*(2), 1–17.
Brouneus, K. (2003). Reconciliation – Theory and practice for development cooperation. A report from the Swedish International Development Cooperation Agency.
Chinamasa, P. (2021). Cited in Bulawayo24 News. 2021. ZANU PF won't surrender bond with military. Available at: https://bulawayo24.com/index-id-news-sc-national-byo-205441.html.
Choudhry, S. (2010). 'He had a mandate': The South African Constitutional Court and the African National Congress in a dominant party democracy. *Constitutional Court Review, 2*(1), 1–86.
Cobban, H. 2007. *Amnesty after atrocity: Healing nations after genocide and war crimes*. Boulder: Paradigm Publishers.
de Coning, C. (2016). From peacebuilding to sustaining peace: Implications of complexity for resilience and sustainability. *International Policies, Practices and Discourses*, 4(3), 166–181.
Hanitzsch, T., Hanusch, F., & Lauerer, C. (2016). Setting the agenda, influencing public opinion and advocating for social change. *Journalism Studies, 17*(1), 1–20.
Hapanyengwi-Chemhuru, O. (2013). Reconciliation, conciliation, integration and national healing: Possibilities and challenges in Zimbabwe. *African Journal of Conflict Resolution, 13*(1), 79–99.
Honebein, P. C. (1996). Seven goals for the design of constructivist learning environments. In W. G. Brent (Ed.). *1996. Constructivist learning environments: case studies in instructional design. Educational Technology Publications*. New Jersey: Englewood Cliffs.
Kelsen, H. (1955). Foundations of democracy. *Ethics, 66*(1, Part 2), 1–101.
Lederach, J.P. (1997). *Building peace: sustainable reconciliation in divided societies*. Washington, D.C: United States Institute of Peace Press.
Loyle, C.E, & Davenport, C. (2016). Transitional injustice: Subverting justice in transition and post-conflict societies. *Journal of Human Rights, 15*(1), 126–149.
Mashingaidze, T.M. (2010). Zimbabwe illusive national healing and reconciliation processes: From independence to the inclusive government, 1980-2009. *Conflict Trends*, 2010(1), 19–27.
Mthiyane, V. (2018). NPRC: Dumbutshena, Chihambakwe reports missing. Available at: http://www.cite.org.zw/dumbutshena-chihambakwe-reports-missing.
Murambadoro, R. (2015). Perspectives on Gukurahundi from Matabeleland, Zimbabwe. *African Journal of Conflict Resolution, 1*, 1–15.
Pethe, R. (2018). Why should public opinion matter in a country? *International journal of Advance Research and Development, 3*(1), 66–69.
Reychler, L., & Paffenholz, T. (2001). *From conflict to peace building: conceptual framework*. Boulder: Lynne Reinner Publishers.
Rosoux, V. (2008). *Reconciliation as a peacebuilding process: Scope and limits*. London: Sage.
Smith, E.R.A. (2008). What is public opinion? *Journal of Politics and Society, 10*(1), 95–105.
Solomon, D. & Zvobgo, K. (2019). Co-opting truth: Explaining quasi-judicial institutions in autocratic regimes. *International Studies Quarterly, 64*(4), 785–797.

Teitel, R.G. (2002). Transitional justice in a new era. *Fordham International Law Journal*, 26(4), 893–906.

United Nations. (2010). Guidance Note of The Secretary-General United Nations Approach to Transitional Justice. https://www.un.org/ruleoflaw/files/TJ_Guidance_Note_Mar_2010FINAL.pdf. Accessed 20 June 2023.

United Nations. (2015). The challenges of sustaining peace: Report of the Advisory Group of Experts on the Review of the Peacebuilding Architecture. UN Doc. A/69/968-S/2015/490, 30 June 2015.

Verma, V.P. (1956). Public opinion and democracy. *The Indian Journal of Political Science*, *17*(4), 311–321.

Chapter 23
Handling Indiscipline Involving Children at Home and Schools: A Restorative Justice Perspective to Unlearn Violent Behaviours in Zimbabwe

Ntombizakhe Moyo-Nyoni and Dorothy Moyo

Abstract Disciplining children using restorative justice approaches can be one way of contributing to building peaceful societies in Southern Africa. This article explores restorative justice approaches to disciplining children within the endogenous systems of the Ndebele society in Bulawayo, Zimbabwe. The researchers Ntombizakhe and Dorothy argue that using violent means by adults (parents and teachers) at home and school to enforce discipline among children gives rise to outcomes of violence, with children becoming violent in their adulthood and impeding the creation of sustainable peace. Data for this study were collected through focus group discussions. Three focus group discussions were held with ten teachers, ten parents and twelve learners previously trained in conflict resolution by the Alternative to Violence Project, a peacebuilding programme. The study reveals that violence is a learnt behaviour, which can be unlearnt as children and communities learn and adopt non-violent practices from traditional endogenous practices of curbing indiscipline at the family and community level. The article concludes that parents and school authorities should be empowered and encouraged to cultivate restorative justice strategies for redressing indiscipline in children. If homes and schools adopt restorative justice approaches to handling indiscipline, the same constructive ways might be cultivated in communities and at the national level for sustainable, peaceful communities.

Keywords Disciplining children · Corporal punishment · Restorative justice · Endogenous systems · Ndebele society · Zimbabwe

---

Dr Ntombizakhe Moyo-Nyoni (PhD) is a lecture in Peace Studies at the Midlands State University in Gweru, Zimbabwe. She holds a Ph.D. in Management Sciences Specialising in Public Administration (Peacebuilding Programme) from Durban University of Technology, South Africa. Email: zakhemoyo@gmail.com

Dr Dorothy Moyo holds a BSc Sociology degree and an MSc in Peace Leadership and Conflict Resolution and a PhD in Management Sciences specilaising in Public Administration—Peace Studies. Email: dorothyhopemoyo@gmail.com

---

J. C. K. Kiyala and N. Chivasa (eds.), *Climate Change and Socio-political Violence in Sub-Saharan Africa in the Anthropocene*, The Anthropocene: Politik—Economics—Society—Science 37,
https://doi.org/10.1007/978-3-031-48375-2_23

23.1 Introduction

> There can be no keener revelation of a society's soul than the way in which it treats its children
>
> (Nelson Mandela 1995).

Disciplining children using restorative justice approaches can be one way of building more peaceful societies. However, the use of retributive methods such as corporal punishment is widespread, despite the fact that it has been condemned in most parts of the world (Chiramba/Harris, 2020; Dari et al., 2018). 'Difficult', 'disturbed' or 'disruptive' behaviour in communities, homes and schools is commonly dealt with through punitive measures. This threatens the creation of a culture of peace in schools and the wider community. Children suffer exclusion, suspension, expulsion, harsh mandatory punishments, corporal punishment, and surveillance, among other forms, in the name of discipline (Gershoff, 2013; Payne/Welch, 2013). Such punitive measures prevent children from reflecting on needs and obligations created by wrong behaviour, as the focus is shifted to pain brought about by punishment (Suvall, 2009). Corporal punishment harms children and advances adult interests (Umubyeyi/Harris, 2017; Makura et al., 2015). Children learn that violence is a normal way of resolving differences and correcting wrongdoing.

Studies such as Dari et al. (2018) and Umubyeyi and Harris (2017) give overwhelming evidence of the negative effects of corporal punishment, which affects the health and behaviour of children with children internalising violence as a way of life. Grill (2005: 100) posits that "collective childhood experiences of a society is probably the most important factor driving decisions made at political, business and social level". Grill's observation reveals the importance of exploring non-violent restorative approaches to handling discipline, given that childhood experiences affect decisions in adulthood.

This paper explores restorative approaches to disciplining children, which promote positive and peaceful practices from the endogenous traditions of the Ndebele society. The traditional methods of disciplining children used in the Ndebele society have elements of restorative justice, which can be leveraged to reduce punitive methods. Corporal punishment, although used to discipline children, is used reservedly as a last resort.

Restorative justice, from where restorative approaches derive, engages victims, offenders and everyone affected by wrong behaviour in seeking solutions for the problem (Zehr, 2019; Wearmouth et al., 2007; Payne/Welch, 2013). Restorative justice is an old traditional concept embedded in various African cultures, including Zimbabwe (Moyo, 2016), although modern restorative justice practices can be traced back to the New Zealand Maori community (Wearmouth et al., 2007). One of the earliest advocates of restorative justice, Braithwaite (1999), states that children should be disciplined using restorative justice because the approach uses reintegrative shaming, which presupposes that if a child is made aware of the consequences of his/her behaviour, the child will feel ashamed because of his action and avoid getting into a similar predicament. This is likely to restore the child who is given a chance to express shame and remorse (Braithwaite, 1999; Wearmouth et al., 2007).

23.2 The Zimbabwean Context

Internationally and regionally punitive forms of discipline are regarded as violence against children and a violation of their fundamental rights. Regional and international conventions outlaw the outright use of corporal punishment and similar degrading and dehumanising punitive methods. Zimbabwe is a signatory to regional and international child rights conventions, such as the Convention on the Rights of the Child (CRC) and the African Charter on the Rights and Welfare of the Child (ACRWC), which protect children from harsh punitive treatment. Moreover, Zimbabwe's Constitution 2013 (Section 53) outlaws corporal punishment or any cruel, inhumane, or degrading treatment or punishment of children.

The message from the international community and the Government of Zimbabwe is that violence against children in corporal punishment is unacceptable and should not be allowed (Chiramba/Harris, 2020; Mushowe, 2018). However, despite these Conventions, societies in Zimbabwe, as is the case in most parts of the world, believe in punitive forms of enforcing discipline (Dari, 2018; Makura et al., 2015; Makwenya et al., 2012). Those advocating for punitive measures believe these are the best ways to instill acceptable behaviour and morals and deter wrong behaviour (Dari et al., 2018; Umubyeyi/Harris, 2017; Gershoff, 2013). Those against punitive measures believe that corporal punishment is child abuse, which can contribute to poor mental health, aggression, anti-social behaviour and abuse of others in adulthood (Gershoff, 2013).

23.3 Theoretical Framework

The restorative justice theory influenced this study, focusing on identifying restorative justice elements of disciplining children in the Ndebele society. Identifying these positive elements can instil a sense of learning from what is done and how it is done. The idea is similar to the idea of reflexivity explained by Francis (2002), which refers to the ability to both look outwards and inwards to ourselves, as a way of learning from experience, and becoming more aware of our emotions, our perceptions, our assumptions, our values, our skills and knowledge. Restorative justice theory assumes that the involvement of all stakeholders in resolving a conflict is key in transforming factors fueling conflicts (Zehr, 2019). In this case, the assumption is that if parents and teachers can adopt restorative justice approaches in addressing children's indiscipline instead of violence, this can restore the offending child and instil a desire to do right.

There is a wide range of literature on the restorative justice theory, which is said to be founded by Zehr (2002). However, much of the literature is largely in the prison's context (Zehr, 2019; Wearmouth et al., 2007; Zehr, 2002). The works of Wearmouth et al. (2007) and Zehr (2019) indicate that restorative justice works, especially in addressing issues of crime. If restorative justice approaches can transform prisoners usually rejected by societies, can they fail to transform children's misbehaviour at school and home? Restorative justice allows dialogue between the offender, the offended and the community around them (Zehr, 2019). In disciplining children, the adults and children can discuss factors contributing to the undesired behaviour, its effects and the possible solutions to the problem. Such practices can restore and strengthen the relationship between the children and adults, such that the lines of communication between them remain open for engagement.

23.4 Restorative Justice and Potential to Unlearn Violence

Workman (2008) gives an example of the applicability of the restorative justice theory in New Zealand. He notes that domestic conflicts, prison management, prison reintegration and gang conflicts had been managed through the use of restorative justice. Principles of restorative justice have been adopted by schools as well. All these initiatives did not happen instantly but were gradually incorporated into the system as positive outcomes linked to the application of restorative justice theory were realised. Restorative justice theory enables a contextual approach to justice instead of reliance on procedural justice, and the offender can see the impact of behavioural change. Braithwaite (1999) notes that people must understand that restorative justice is a wider strategy for confronting injustice in any area where injustice occurs.

Jenkins (2006) presents an empirical study on the use of restorative justice theory by the people of Gullah Island in North Caroline. Jenkins observed that for the people of Gullah Island, restorative justice became a way of life, where the community worked together (the young and the old) at all levels of life. The community took up the responsibility as parents of educating their children about moral living. Jenkins argues that restorative justice triumphs over violence because of three factors: co-operation versus competition, community building rather than being individualistic, use of spiritual knowledge rather than scientific sources and valuing *Ubuntu* within every individual. Jenkins argues that focusing on community building and keeping people in the same spirit is key to preventing violence. Hence, this study advocates that using restorative justice approaches to discipline can prevent violence because, once children are taught to resolve conflicts through co-operation, they will do so up to adulthood.

As indicated above, it is noted that in most instances, restorative justice has been used in the context of the criminal justice system, which might raise doubts on whether the theory can be applied effectively to issues of disciplining children. This study demonstrates that this is possible.

23.5 Research Design and Methods

A case study research design was used to collect data for this study. A case study was chosen because it offers a variety of research methods to produce a complete portrayal of an identified subject (Davies, 2007). A case study was useful in attaining detailed knowledge and information indicating the beliefs and widespread use of restorative justice approaches to addressing indiscipline in children within the community under study. This research design was chosen because there was a need to explore extensively, traditional non-violent ways of enforcing child discipline in Bulawayo, Zimbabwe. Bulawayo was chosen because it is a city at the heart of the Matabeleland region. Thus participants were conversant with the endogenous traditions of the Ndebele society for instilling discipline in children.

23.6 Methodological Approach

The study used qualitative methods of data collection. Bashir et al. (2008: 43) note that "qualitative research seeks to understand, as completely as possible, the phenomena under study". According to Davis (2007), qualitative research is a process where an in–depth enquiry is made into a situation or people to comprehend it better. Thus, an in-depth inquiry was conducted with the participants to comprehend their perceptions regarding the use of restorative justice approaches in the community when disciplining children. The qualitative data approach was helpful in acquiring data about the feelings, opinions, attitudes and perceptions of the respondents. Golafshani (2003: 600) posits that qualitative research uses a naturalistic approach that seeks to understand phenomena in context–specific settings, where the researcher does not attempt to manipulate the phenomenon of interest but allows the phenomenon to unfold naturally. As is the case with focus group discussions, participants were given a chance to discuss the subject in a natural setting.

23.6.1 Data Collection Approaches

Data were collected through focus group discussions. As observed by Hofstee (2006), focus group discussions are a type of qualitative research methodology where small homogenous groups of people are brought together to discuss specific topics informally under the guidance of a moderator. Three focus group discussions were held with ten parents, ten teachers and twelve secondary school learners trained in Alternatives to Violence Project (AVP) basic workshops. AVP offers workshops to equip participants with creative ways to address conflicts instead of resorting to violence (Aubrey, 2007). The thirty-two participants were randomly drawn from over one hundred people trained in AVP. The AVP workshops exposed participants to non-violence practices, which enabled the participants of this study to have the capacity to identify non-violent practices from endogenous practices of the Ndebele society. Focus group discussions provided interaction between the participants, which was essential in generating data for this study. Participants were stimulated to discuss the topic openly, as multiple lines of communication were created. The moderator asked open-ended questions, allowing participants to express their opinions, reasons for certain behaviours, descriptions of practices, and beliefs.

23.6.2 Sampling Techniques

A non-probability purposive sample was used. This sampling technique was necessitated by collecting data from individuals well-versed in the traditional practices, customs and values of people from Matabeleland who had undergone AVP training in creative conflict resolution. The information of the participants was obtained through attendance registers of the AVP workshops.

23.6.3 Ethical of the Inquiry

All participants were informed about the purpose of the research before the beginning of the study, and all the participants agreed to participate. Every participant was treated with dignity and respect, and all contributions were valued, acknowledged and given attention. The ethical principles that McNamara (1994) highlights, including doing no harm, confidentiality and anonymity, were adhered to. The principle of doing no harm was handled with care, seeing that the selection of participants could have been assumed to fuel tribalism, because Bulawayo is a residence of people from both Matabeleland and Mashonaland, and the study was focusing only on people from Matabeleland. It was emphasised that participation was voluntary, and those uncomfortable could opt out without penalty at any point during the study. Participants were informed that discussions would be recorded using an audio recorder for later transcription and analysis. They gave consent to the recording.

23.6.4 Data Analysis

The data analysis thoroughly described the processes, characteristics, transcription and contents of restorative and non-violent approaches practised within the Ndebele society in disciplining children. Data were separated into manageable themes, patterns, trends and relationships, later formulated into theoretical concepts (Mouton, 2001: 108; Davies, 2007: 135). Focus was put on understanding the primary context of the messages by understanding the participants' values, attitudes and behaviours. Audio-taped data were first transcribed, then merged with the moderator's notes and coded under different themes. The themes were carefully explored to capture the finer details, meanings and insights not captured during the first coding process. Thick descriptive data were captured to give a comprehensive understanding of the context of the study and the findings.

23.7 Findings

23.7.1 Perspectives from Teachers

Participants indicated that punitive methods of discipline were commonly used by teachers and parents when disciplining children. Seventy per cent (70%) of the teachers indicated that corporal punishment was the common method to disciplining children; slapping, beating, spanking, and pinching. Teachers indicated they were aware that corporal punishment was banned in schools but found themselves with no choice but to administer it as learners got out of control and misbehaved. They indicated that they use methods other than corporal punishment, which they considered less severe, such as making learners stand on one leg for an interval, expulsion from the class, and dialogue, among other ways. In the words of one teacher,

> …however, it's not all the time where children are beaten; there is counselling in the form of talking with the learner about the bad behaviour... time out where learners are excluded from certain activities, assigning duties during break or lunchtime or after school. There are also cases of parental involvement where parents are engaged about the behaviour of the misbehaving child. There is also positive discipline where the teacher and the learner sit and set goals/targets and keep checking whether the learner's behaviour is on track with set targets[1].

Most teachers were confident that there are restorative justice methods of disciplining children that can be drawn from the Ndebele traditions, such as counselling. Restorative methods were considered more effective than retributive ones in the Ndebele society. Teachers gave testimonies of how they had used alternatives to violent methods such as dialogue, mediation and other methodologies drawn from the traditional Ndebele societies. Participants referred to the AVP workshops, which reminded them of non-violent methods of handling conflicts, especially the principles of *Ubuntu* which are key principles in the Ndebele tradition. A teacher at a boy's high school narrated how he and other teachers who took part in the alternative to violence project training previously had a strong belief that boys could only comply after corporal punishment. However, after training in AVP, they decided to engage in dialogue with boys involved in substance abuse. The teacher pointed out that after engaging with the boys, they got to the root of the cause of the problematic behaviour.

[1] Mrs Grace Dube, Female, 29 years, FGD with teachers, 31 October 2021.

The boys were counselled on the dangers and consequences of substance abuse, and there was attitudinal and behavioural change as the boys embraced the principles of *Ubuntu*, such as caring, empathy and humanness.

Teachers suggested positive discipline methods, which they divided into two categories: methods to prevent misbehaviour and methods from redressing misbehaviour. Teachers averred that the Ndebele tradition emphasised preventing misbehaviour as evidenced by sayings such as "isigogo sigoqwa sisemanzi" [discipline a child whilst he/she is still young]. The saying places emphasis on guiding children on how they should behave in the early years of life. The teachers indicated that one way of avoiding corporal punishment was to prevent misbehaviour. Preventing misbehaviour involves effective listening, taking the concerns of children seriously and guidance. The idea of listening to children was linked to the Ndebele saying, "okuluma indlebe ngowakho" [meaning that the one who warns you is your friend].

23.7.2 Perspectives from Parents

All parents were of the view that children should be punished for wrongdoing and guided in doing what is right. Thirty-four per cent of the parents indicated that they use corporal punishment all the time, 26 % indicated that they use corporal punishment occasionally, and 40% indicated that they did not use it at all. Children were said to be exposed to corporal punishment as early as two years. In addition to corporal punishment, parents scolded, threatened, and restricted play and other favourite activities or assigned additional household chores as part of discipline. It was noted that some parents believed there is a difference between corporal punishment and spanking, with spanking being less harsh when compared to corporal punishment. This view was held mainly by parents who purported to be against corporal punishment. In the words of one parent,

> We discipline our children by beating them, although I know beating does not correct the behaviours *otshaywayo ucina esetunyile...kasakhali loba etshaywa* (the one that is beaten often gets used to the beating and ends up not even crying, showing that beating loses its purpose). I believe that there is truth in the Ndebele saying that says, isigogo sigoqwa sisemanzi [discipline a child whilst he/she is young…the problem is that most of our children are not taught the right morals. So spanking them once in a while is not violence.[2]

On counselling, parents reported that the Ndebele tradition encouraged parents to refer their children for guidance to aunts, grandparents and pastors. The process was said to be proactive and done before encountering misbehaviour problems. Corporal punishment for disciplining children was discouraged in the Ndebele society, as evidenced by sayings such as "*induku kayakhi umuzi*", meaning beating someone will never bring the desired change hence the need for counselling. However, some parents

[2] Ms Lindani Maphosa, Female, 36 years, FGD with parents. 20 November 2021, Bulawayo.

indicated that the ban on corporal punishment had disempowered them from disciplining children as children threatened to report them to the police. One participant said,

> We no longer know how to deal with our children; we often end up watching them misbehave...we are overwhelmed as we feel the responsibility of guiding our children is taken away from us. We fear we are becoming more Western than African.[3]

Similarly, another parent justifying corporal punishment on religious grounds said,

> What worries me is that once I stop disciplining my child using the stick, I will be failing to follow biblical teachings, which instruct us to train children in the way they should grow so that they won't depart from it. The Bible states that we should use a stick to discipline children.[4]

We also gathered data that help to understand various perspectives from learners regarding violence and nonviolence. Their views are presented in the section that follows.

23.7.3 Perspectives from Learners

Learners indicated that both parents and teachers beat them. One stated that "the closure of schools due to Covid-19 lockdowns removed us from the frying pan and threw us direct to the fire", indicating that they are beaten at school and home.

Learners stated that many sayings from the Ndebele traditions indicate that nonviolent parenting is possible. Some sayings mentioned included: "Induku kayakhi muzi" (a knobkerrie will not build a home), "okoniwa ngomlomo kulungiswa ngomlomo" (when people wrong each other in conversation, they should use dialogue to restore the relationship). Learners noted that their parents knew all these sayings but did not apply them. One learner in emphasising the non-use of restorative methods by parents of said,

> Dialogue, guidance and counselling seem far from our parents, but we should remember that we do not all have modern parents, some of our parents don't know these new methods; they refer back to how they were disciplined.[5]

Similarly, another learner said,

> Parents punish us using painful ways because it is the easy way out, and they may defend themselves by saying they have no choice. But they do have a choice between talking to us or beating us; they beat us because it's easy for them to do so."[6]

---

[3] Mrs Maureen Moyo, Female, 47 years, FGD with parents, 20 November 2021, Bulawayo.

[4] Mrs Nkosinhle Ndiweni, Female, 52 years. FGD with parents, 20 November 2021, Bulawayo.

[5] Nohle Ngulube, Female 16 years, FGD with learners, 28 October 2021. Bulawayo.

[6] LeeRoy Mhlanga, Male 17 years, FGD with learners, 28 October, Bulawayo.

The majority of learners noted that punitive methods of discipline were a violation of their rights. They indicated that corporal punishment and other punitive methods of discipline humiliated them and made them bitter against those administering the punishment, especially when administered in their peers' presence. One learner responded that corporal punishment did not work for those used to pain. He said, "I am a rugby player, and I am used to physical pain. So, a stick once in a while will not make much difference to me".[7]

The findings reveal that all participant learners believed that they would respond better to their parents and teachers if they had a voice to articulate their views on issues without fear or intimidation.

23.8 Discussions of the Findings

23.8.1 Discipline versus Punishment

The findings of this study revealed that the traditional Ndebele approaches to disciplining children were more restorative than retributive. Discipline aims to stop undesired behaviour, teach the desired appropriate behaviour, and motivate healthier decision making in the misbehaving child (Mullet, 2014). Punishment temporarily restrains the child from behaving in a certain way but fails to teach an inner sense of self-discipline (Gershoff, 2013). A punishment which in most cases is punitive, makes the child obey the instructions and rules in the presence of the enforcer, but this does not guarantee that the misbehaviour will not be repeated in the absence of the enforcer (Gershoff, 2013; Amstultz/Mullet, 2015).

Discipline has a place in the development of children, while violent punishment negatively affects the child's development. Article 28(2) of the United Nations Convention on the Rights of the Child affirms the role of school discipline in child education. In that vein, it states that Member States are required to "take all appropriate measures to ensure that school discipline is administered in a manner consistent with the child's dignity and in conformity with the convention" (United Nations, 1989). Therefore, teachers have a role in maintaining discipline at schools, albeit through restorative non-violent methods.

23.8.2 Restorative Alternatives to Discipline

It is important to note that the ban on corporal punishment is insufficient; it is essential to equip teachers and parents with positive alternative methods of disciplining children (Pinheiro, 2006). Legislation without any accompanying change in attitudes and

[7] Themba Ndlovu, Male 16 years, FGD with learners, 28 October, Bulawayo.

mindset is insufficient to result in positive outcomes. Restorative justice approaches to disciplining children can complement legislation in achieving positive goals as they offer a constructive method that focuses on the harm done and provide an opportunity to understand the nature of misbehaviour and make things right. Restorative justice seeks healthy outcomes by giving voice and power to those harmed by the offence, healing and repairing relationships, encouraging accountability, reintegrating the offender and creating caring climates (Mullet, 2014:1, 3).

Restorative approaches to discipline promote accountability and responsibility among learners, thus creating a conducive learning environment (Chiramba/Harris, 2020; Payne/Welch, 2013; Wearmouth et al., 2007). These approaches allow learners to learn from their mistakes and for the creation of new relationships (Chiramba/Harris, 2020). An observation made by Mushowe (2018) is that children need the discipline that teaches them self-control, alternatives for undesirable behaviour, motivation for acting differently, understanding the repercussions of wrongful behaviour and awareness of what they ought to be doing right. This can be achieved through a collective effort from individuals, families, schools, and communities, where a culture of nonviolence is cultivated in all spheres of life.

23.8.3 Factors Inhibiting the Use of Restorative Approaches

Corporal punishment results from factors such as religious beliefs, cultural beliefs, wanting quick-fix solutions and lack of alternatives for disciplining children (Dari et al., 2018; Umubyeyi/Harris, 2017). Chiramba and Harris (2020) observe that whilst corporal punishment in schools is outlawed, the Ministry responsible for education has not suggested alternative methods, let alone train teachers on such alternatives.

Positive parenting is essential for establishing sustainable peace in Africa. As indicated by Grill (2005), decisions made by adults at political, social and economic levels are influenced by childhood experiences of decision making. The implication is that the much-needed sustainable peace can be attained through changed mindsets which will seek to implement non-violent strategies as a way of life.

23.9 Conclusion

This study advocates for restorative justice approaches in disciplining children. It is argued that most Ndebele endogenous systems have non-violent restorative approaches to disciplining children, which families, schools and communities can adopt. It is envisioned that children can replicate such methodologies as they grow, to the extent of influencing decisions for peacebuilding at the national level.

23.10 Recommendations

The study presents the following recommendations:

- Research should be conducted to draw out non-violent lessons from endogenous practices to influence practice.
- Societies should enforce and equip parents with non-violent parenting skills.
- Teacher training colleges should include, in their curriculum, the use of non-punitive methods of disciplining children.
- Children should be empowered with values and norms to reduce incidents exposing them to violent treatment by adults.

References

Amstultz, L.S., & Mullet, J.H. (2015). *The little book of restorative discipline for schools: Teaching responsibility; creating caring climates*. Intercourse: Good Books.

Audrey, R. (2007). *AVP prison strategy*. MN: AVP Inc.

Babbie, E., & Mouton, J. (2001). *The Practice of social research*. Cape Town: Oxford Press.

Braithwaite, J. (1999). *A future where punishment is marginalized: realistic or Utopian*? UCLA

Chiramba, E.N., & Harris, G. (2020). Restorative Discipline Practices: An Action Research Project in Zimbabwean Primary Schools. *Educational Research for Social Change*, *9*(2), 118–122.

Dari, N.K.W., Mawodza, O., Mingo, & Olson, B.D. (2018). Empowerment and Accountability and Multi-layered Contexts of Influences: Parenting and the Avoidance of Harm in Harare, Zimbabwe. *Australian and New Zealand Journal of Family Therapy*, *39*(3), 380–393.

Davis, M.B. (2007). *Doing a successful research project using qualitative or quantitative methods*. England: Palgrave Macmillan.

Gershoff, E.T. (2013). Spanking and child development: We know enough now to stop hitting our children. *Child Development Perspectives, 7*, 133–137. https://doi.org/10.1111/cdep.12038.

Grille, R. (2005). *Parenting for a peaceful world*. Richmond, Australia: The Children's Project.

Jenkins, M. (2006. Gullah Island dispute resolution: An Example of Afrocentric Restorative justice. *Journal of Black Studies*, *37*(2): 299–319.

Makura, E., Chikwiri, E., Musiyiwa, J., Mandewo, J., & Hodzi, C.J. (2015). Discipline, the law and practical alternatives. In *The Zimbabwe Child Monitor 2015*. Harare: Women's University in Africa.

Makwanya, P., Moyo, W., & Nyenya, T. (2012). Perceptions of the stakeholders towards the use of corporal punishment in Zimbabwean schools: A case study of Bulawayo. *International Journal of Asian Social Science*, *2*(8):1231–1239.

Mouton, J. (2001). *How to succeed in your master's and doctoral studies*. A South African guide and resources book. Pretoria: Van Schaik.

Moyo, N. (2016). *Peacebuilding among ex-prisoners and their families: Enhancing the impact of the Second Chance Rehabilitation Centre, Zimbabwe*. PhD thesis, Durban University of Technology.

Mullet, J.H. (2014). Restorative disciplining. From getting even to getting well. *Children and Schools*, 36(3), 157–162.

Mushowe, B. (2018). A ray of hope for the outlawing of corporal punishment in Zimbabwe: A review of recent developments. *University of Zimbabwe Law Journal*, *1*(1).

Payne, A.A., & Welch, A. (2013). Restorative justice in schools: The influence of Race on Restorative discipline. *Youth & Society* XX(X) 1–26. https://doi.org/10.1177/0044118X12473125

Pinheiro, P.S. (2006). *World report on violence against children*. Geneva: United Nations Secretary-General's Study on Violence against Children.
Suvall, C. (2009). Restorative Justice in Schools: Learning from Jena High School. *Harvard Civil Rights-Civil Liberties Law Review*, *44*, 547–569.
The Zimbabwe constitution. (2013). Harare: Government printers.
Umubyeyi, B., & Harris, G. (2017). Promoting non-violent parenting among refugee mothers in Durban. In Kaye S, Harris G (eds.), *Building peace via Action Research: African case studies* (pp. 53–72). Addis Ababa: UN University of Peace.
Wearmouth, J., Mckinney, R., & Glynn, D. (2007). Restorative justice in schools: A New Zealand example. *Educational Research*, *49* (1) 37-49. https://doi.org/10.1080/00131880701200740
Workman, K. (2008). *The future of restorative justice-control, co-option and co-operation*. Address to the 2008 restorative justice Aotearoa conference [unpublished]
Zehr, H., (2019). *Restorative justice beyond crime: A vision to guide and sustain our lives*. UPPADO, 21.

Chapter 24
Schools as Agents of Change: Impacts of Restorative Circles at Mbambangwe School in KwaZulu-Natal, South Africa

Sibusisiwe Mlambo and Jean Chrysostome K. Kiyala

Abstract The exposure of children to all forms of violence has resulted in schools becoming sites of violence. This kind of violence is embedded in the broader South African society. The problem is that South Africa is perceived as a violent country, and schools are subsequently violent places. Since 1994, efforts have been made towards ending violence in schools, but more attention has been given to halting corporal punishment. While punitive sanctions are predominantly the most utilised response to violence in the broader South African society, restorative justice approaches as a means of handling school violence and peace-building has been lacking. Very little has been done vis-à-vis rebuilding and restoring interpersonal relationships that were destroyed many years ago through colonisation and apartheid. Ongoing violence is experienced in the educational environment that some tie to past legacy of institutionalised violence that characterised the Apartheid regime. The target population was Grade 9 learners in 2019 who were going to Grade 10 in 2020. The study aimed to establish the most effective strategy for curbing school violence. Interviews were conducted with teachers, learners, parents, School Governing Body (SGB) members and community stakeholders. The findings revealed that restorative justice and Ubuntu ethics are principles that can be used to address school discipline issues in order to avert learners' violent behaviours. The inquiry further found that non-violent conflict resolution mechanisms have the potential to settle disputes between learners and learners and teachers constructively, and lead to the restoration of relationships between affected parties. Issues around trust between parents and learners, open channels of communication between the parents and their children and encouraging good working relationship between learners, parents and teachers

Dr Primrose Sibusisiwe Mlambo is a researcher, trainer and facilitator with over 25 years' experience in the development sector, peace, violence and HIV and AIDS related programmes. She holds a master's degree in Community Development and PhD in Public Administration: Peace Studies. Email: feladies@gmail.com. ORCID: https://orcid.org/0009-0000-2042-7544

Dr. Jean Chrysostome K. Kiyala holds a PhD in Public Administration: Peace Studies from DUT; he is a Senior Lecturer at the International Centre of Nonviolence (DUT), Associate Professor and visiting lecturer at the Université de Bandundu and the Université Evangélique en Afrique (Democratic Republic of Congo); Emails: JeanK@dut.ac.za; kljeanchrysostome@gmail.com. ORCID: https://orcid.org/0000-0002-9587-0230

J. C. K. Kiyala and N. Chivasa (eds.), *Climate Change and Socio-political Violence in Sub-Saharan Africa in the Anthropocene*, The Anthropocene: Politik—Economics—Society—Science 37,
https://doi.org/10.1007/978-3-031-48375-2_24

are paramount to building peace and transforming the school environment – making it peaceful and conducive to learning.

Keywords Action research · Conflict transformation · Mbambangwe · Peace-building · Restorative justice · School violence · Ubuntu · KwaZulu-Natal

24.1 Introduction

South Africa is one of the most violent countries in the world, particularly for young people growing up in poor urban settings where violence is often a part of their daily lives (Scorgie, 2017). The history of violence has victimised all South Africans, including children, young people, parents, communities and the business community. In 2012, the Department of Basic Education reported that more than 14% of the nearly 14 million learners experienced violence, corporal punishment or verbal abuse at school (Grobler, 2018). The sad reality is that corporal punishment continues to be a legal means of disciplining children in a third of the world countries (Gershoff, 2017). Furthermore, there have been gruesome attacks such as stabbings in and around schools and that have led to pupils' deaths. Inter-learner violence occurs frequently. For instance, in February 2023, a grade 10 learner from Geluksdal Secondary School in Brakpan, Ekurhuleni, in Gauteng Province was stabbed to death; the learner, identified as Tumelo Jwili, 17, was in Sebokeng when he died (South Africa Government Communication and Information System, 2023: n.p.). During February 2023, a female student from Tshwane University of Technology (TUT), known as Ntokozo Xaba (20-year-old) was stabbed to death and was found dead in her residence at TUT (Kgosana, 2023). These are just two cases among many. To mark the beginning of the annual *16 Days of Activism for No Violence against Women and Children Campaign*, van Zyl (2022: n.p.), a South African MP from the Democratic Alliance party stated:

> The crime statistics for the second quarter of 2022/23 revealed the horrifying reality of 83 rapes and 19 murders committed on the premises of educational facilities, including primary, secondary and high schools, day care facilities, special schools and tertiary institutions. There were six rapes at day care centres, five at special needs schools and tertiary institutions and 67 at schools.

In addition, she demonstrated how gender-based violence and femicide (GBVF) has become endemic and pervasive in South Africa as confirmed by the Statistics South Africa (StatsSA). van Zyl (2022) pointed out that in 2022, 90,037 girls aged 10–19 years bore children between March 2021 and April 2022. There were also disturbing cases of 258 assault and grievous bodily harm (GBH); 22 cases of attempted murder that occurred on educational premises, and 411 gang-related incidents notified by the Department of Basic Education in 2022. van Zyl (2022) contended that schools have become battlegrounds.

While punitive sanctions are predominantly the most used approach at schools and the larger society, restorative justice approaches and peace-building tools are lacking. The South African Department of Justice and Constitutional Development (2011: 4) underlines the discrepancies between retributive justice and restorative justice in the following terms:

> In Conventional Criminal and Civil Justice Processes, the victim and offender are positioned as adversaries, and expected to remain passive while all the key decisions are made by professionals (judges, lawyers, social workers, correctional officers etc.). In restorative justice approach, the victim and offender are part of the process/programme and the key decisions that are taken are influenced by them.

Restorative justice (RJ) is an empowerment of the community whereas retributive justice excludes the community and the offender who can only be used as witnesses without having a decision in the sentencing procedures (Van Ness et al., 2003). With RJ, the offender is involved in the sentencing process; and victim and offender are not antagonists but partners in finding solutions and adopting the final agreement to address the hurt. RJ eases the stress of victims and offenders that become entangled in security and the social costs of rigorous criminal laws (Lanni, 2005).

However, there seems to be limited systematic scholarly research on the use of peace-building to combat school violence in South Africa. Most peace-keeping approaches to school violence from disruptive learners rely on peace through strength strategies by which teachers deter hostile learners with the threats of consequences, so that learners comply through fear (Chiramba/Harris, 2020). Some of the challenges that African countries have when it comes to conflict resolution is the adoption of the Western methods that in most cases have proven to be inappropriate, ineffective and costly. John (2018) argued that African traditional mechanisms have been weakened by the Western approach. What we seek to do in this study is to determine how to create an environment for people to be able to engage and feel respected. Since people are not inherently violent, this, therefore, suggests that there is a possibility for the school community, namely, the parents, teachers, learners and the community to unlearn what has been learned in the past and relearn new ways of resolving conflict. Conflict transformation approaches focus on restoring and strengthening relationships constructively within a social context (Hemphill et al., 2018).

Healing the nation requires processes like conflict transformation and restorative approaches – something that rebuilds relationships, restores human dignity, empowers human beings and discourages violent strategies of resolving conflict.

One of the ways in which conflict can be transformed in schools is through schools becoming agents of change. The rationale behind this is that the school is a place where adolescents spend most of their time, interacting, socialising with each other and where most of their behaviour is influenced (Belle, 2017). Schools are the most important environment outside of the learners' families; therefore, they are a primary domain for promotion of healthy relationships (Grobler, 2018). We see schools as an enabling environment that nurtures the holistic development of learners where they feel they belong (Mathikithela, 2020).

24.2 Context of the Study

The study was conducted in Mbambangwe High School (MHS) in the rural community of Bhekulwandle, located in the magisterial district of Umbumbulu falling under the Ethekwini Municipality. The Umbumbulu area has had a very long history of violence, and, during the past few decades, it has been well known for political faction fights. The violence of the 1980s and 1990s in South Africa, and specifically in Umbumbulu, saw local actors being easily moved from their roles as war leaders in local faction fights and freedom fighters participating in political violence to holding leadership positions in the current government (Mathis, 2013: 421). Bhekulwandle is a rural community under the traditional leadership of Inkosi Hlengwa. According to Mathis (2013: 421), the report on the Ethekwini Municipality Framework Planning Report (2010), the area is still underdeveloped, and the issue complicating the many challenges being faced by the residents is the administrative context because the area partly falls under two local municipality wards and partly within the Traditional Tribal Authority Area.

24.3 Peace Versus Violence: Concepts Analysis

There are two perceptions of peace; negative peace as the absence of (intended) physical or psychological violence, and positive peace as a constructive state of being. Positive peace deals "with the underlying, deep-rooted causes and conditions of a conflict which might develop, or has developed, into manifest violence" (Sandole, 2010: 9 cited Bojesen, 2018: 15). Where positive peace prevails, there is human security and the life conditions are such that human beings are allowed to live satisfying lives and are able to reach their full potential. Stefanie (2017) stated that 'positive peace' is a term coined by peace scholar Johan Galtung (1969), who differentiated between negative and positive peace. Negative peace denotes the absence of direct (physical) violence, often found in the form of a post-war ceasefire whereas positive peace describes more than just the absence of violence and instead describes a situation in which peace is also a state of social justice (Stefanie, 2017).

Cremin and Bevington (2017) argued that positive peace is long-lasting and requires the removal of indirect structural and cultural violence; positive peace describes not only the absence of direct violence but the absence of indirect violence and the presence of harmony and social justice. On the other hand, negative peace describes a state or condition where there is an absence of direct violence (Cremin/Bevington, 2017). Bojesen (2018) stated that to get from a violent situation to a state of positive peace, conflict transformation has to take place. Conflict transformation should be peaceful, as violence breeds violence and forcefully suppressing violence will probably not resolve any conflict in the long run. When handled effectively, conflict provides opportunities to strengthen relationships and assist all parties in developing peaceful conflict resolution skills (Hemphill et al., 2018). This is, obviously, unique to each type of conflict, and the more complex and institutionalised a conflict is, the more extensive a model for peace-building becomes.

24.3.1 Consequences of School Violence

Violence in schools is widespread and affects various levels of educational spaces (primary, secondary and higher learning institutions). It is discernible in the use substances, drugs; sexual violence, harassment, abuse, use of guns, weapons and gangs (Le Roux/Mokhele, 2011). It affects South Africans as school violence reflects what occurs in the larger society. To elucidate this point, Le Roux/Mokhele (2011: 327) note: "School-based violence is a reflection of violence in society and the two cannot be separated. Violence in the home makes learners anxious, depressed, insecure and vulnerable". Workplaces are not spared of brutal behaviours and attitudes (Najafi et al., 2018). Violence results in the gradual deterioration of social skills, contributing to depression, isolation, diminished self-worth, humiliation, and anger (UNESCO, 2009). These negative experiences provoke a desire for revenge, perpetuate societal prejudices, and contribute to high school dropout rates. Additionally, violence fosters persistent anxiety and aggressive behaviour, especially when empathy and compassion towards others have diminished. The most destructive and perilous consequences of school violence is early pregnancy, sexual and gender-based violence, exacerbated by increasing infections with sexually transmitted disease, unwanted pregnancy, low self-esteem, and deterioration of performance in educational endeavours (UNESCO, 2009).

From a legal perspective, there have been International Standard-Setting Instruments and Normative Frameworks that aim to minimise if not eradicate violence and its consequences on children and youths. Similarly, UNESCO has suggested that educational institutions refer to international standards, legal instruments and normative frameworks to build structures that will prevents violence against children. These include:

- International Bill of Human Rights
- The Convention on the Rights of the Child (1989)
- UNESCO Conventions and Recommendations in the field of Education
  - UNESCO Convention against Discrimination in Education (1960)
  - UNESCO Recommendation concerning Education for International Understanding, Cooperation and Peace and Education relating to Human Rights and Fundamental Freedoms (1974)

Notwithstanding the role that these legal instruments play to foster peace on educational environment and promote good learning conditions, violence remains a real challenge that many schools must deal with.

24.3.2 The Dakar Framework for Action

The Dakar Framework for Action (DFA) promotes education for all and exhorts governments and educational institutions to provide safer learning spaces by eliminating all sorts of violence, conflicts and exploitations. Article 8(v) of the DFA states: [...] meet the needs of education systems affected by conflict, natural calamities and instability and conduct educational programmes in ways that promote mutual understanding, peace and tolerance, and that help to prevent violence and conflict; (UNESCO, 2000: 9). Article 3(35) acknowledges the risks and threats faced by young people, particularly adolescent girls, which constrain their learning opportunities and challenge education systems (UNESCO, 2000). Moreover, those youths are confronted with "exploitative labour, the lack of employment, conflict and violence, drug abuse, school-age pregnancy and HIV/AIDS" (UNESCO, 2000: 16). However, the DFA (UNESCO, 2006: 16) also encourages that youth-friendly programmes may be made available and accessible to furnish needed "information, skills, counselling and services needed to protect them from these risks". Thus, UNESCO provides education institutions with both strategies and legal frameworks to confront violence. However, the complexity of the problem and the diversity of contexts where violence takes place are evident in the way that each institution attempts to respond to antisocial conduct. Various methods are used for the same purpose. These include campaigns for safety and non-violence on schools and campus colleges.

Schools, colleges and university campuses have the tradition of providing students with opportunities and facilities to explore new thoughts for the betterment of humanity. Technological advancement and innovations explain this aspiration. However, real progress presupposes creating a safe and peaceful environment where students can communicate and interact fearlessly with each other, without intimidation or any form of threat. Unfortunately, schools, colleges and campuses have turned out to be unsafe places marked by all types of violence. Sexual assault has been identified as the second most common form of sexual crimes at colleges. It is often carried out by a person known to the victim and takes place in victim's residence or in off-campus student housing. Another sexual offence regularly notified is sexual harassment, which in most cases goes unreported probably for fear of reprisals, marginalisation and distortion of relationships, and psychological effects like stigma.

In the face of such threats, the predominant trend in handling discipline and fighting violence in educational institutions, has been the use of punitive justice. Several schools are innovatively improving security around campuses, such as surveillance cameras. Sanctions have increased because students' "internal controls" cannot be trusted (Karp, 2004).

All forms of violence infringe the fundamental right to education and unsafe learning environments reduce the quality of education for all learners (Ferrara et al., 2019: 3). General Social Survey indicates that 70% of adults agree that a "good hard spanking" is sometimes necessary (Vaughan-Eden et al., 2018: 43). Lester et al. (2017: 187) observed that the long-term consequences of school violence can be 'very costly to the whole society'. Ferrara (2019: 2) concurs with Lester, Lawrence and Ward in stating that violence in school can have a physical impact and can cause psychological distress, with an invariably negative social and economic impact.

24.3.3 Agents of Change Inside and Outside School

Schools should take more responsibility about prevention of, and intervention in, violence in schools, currently schools have no strategies to deal with violence (Uzunboylu et al., 2017). Given that emotions and relationships strongly influence learning and that these are by-products of how learners are treated at school, as well as at home and in their communities a positive school climate is at the core of a successful educational experience (Darling-Hammond/Cook-Harvey, 2018). It is important that schools provide a positive learning environment that provides a measure of security and support that maximises learners' ability to learn social and emotional skills as well as academic content (Darling-Hammond/Cook-Harvey, 2018). Increased understanding and empathy often lead to increased respect and appreciation for differences and contributions of all classroom members, knowing that they serve as a protective factor for teachers, learners and parents. Teachers further emphasised the indispensability of certain personal qualities that they felt all peace educators require the prerequisites included 'above all, love': for their learners, for humanity, for nurturing the good in others and themselves (Clarke-Habibi, 2018).

24.4 Schools as Agents of Change

The education sector is increasingly recognised as a crucial space for sustainable transformations of post-conflict and post-disaster societies (Lopes Cardozo et al., 2022). Taylor (2020) concurs with Lopes Cardozo et al. (2022) by stating that learners may build on interpersonal prosocial behaviours to promote structural and cultural change and have the potential for constructive impact on the conflict around them. This can be done through the introduction of peace education in classrooms. Peace education is the process of acquiring the values, the knowledge and developing the attitudes, skills and the behaviour to live in harmony with oneself and with the natural environment (Okoro Okoro, 2018; Supriyanto/Wahyudi, 2018). Peace education has a role to play in character development through character education. It boosts confidence in learners and helps them to develop skills and abilities to interact positively with their environment and peers. Rather than being peripheral to education, peace education is central. Lopes Cardozo et al. (2022) believe that peace education is an 'ensemble' influenced by numerous cultural, political and economic dimensions and moments that operate beyond intra-classroom dynamics.

As human beings, we are naturally connected to each other, and this leads to formation of relationships and therefore able to caringly respond to others through paying attention and listening (Grobler, 2018). By introducing peace education, the school community begin to humanise the learning and teaching processes thereby changing teachers and learners' behaviour, attitudes and the way they relate to each other. The objective for having schools as agents of change is because learners are able to engage in dialogue and understand strategies to deal with and even how to resolve conflicts and problems that may occur (Wahyudin, 2018).

24.4.1 Challenges in Peace-Building in Schools

Grobler (2018: 18) argue that what is missing is the 'teachers' voice' and exploring the ethical response of fostering caring relationships between teachers and learners against a context of school violence. The teachers' voice was important because it will inform research processes about the correct strategies and interventions. The buy-in of the teachers as well as their support is very important for such inventiveness. The challenge is that the manner in which the policies are introduced is such that they are imposed on teachers since there are no prior engagements. According to Wessel-Powell et al. (2019), policies are formal, top-down school reform policies legislated, created, implemented and regulated by national, state and local governments.

24.4.2 Teachers' Perspectives on Schools Being Agents for Peace-Building

Teachers have long emphasised the importance of school climate and community building to the social and emotional wellbeing, behaviour and competency of learners (High, 2017). According to teachers, educating for peace necessitates a willingness to learn and to think differently; to become aware of their own personal worldviews and to strive to orient themselves towards inclusive and integrative classrooms (Clarke-Habibi, 2018). Teachers may be expected to play a range of roles as both agents and subjects of peace-building in conflict-affected contexts: from technocrats to reflexive practitioners, as well as agents of transformation, democratisation, healing, peace and/or resistance (Clarke-Habibi, 2018).

24.5 Theoretical Framework

Restorative justice is being used throughout the world in various complex circumstances to bring accountability and address injustices (Karp, 2004; Karp/Dagle, 2013; Karp/Sacks, 2014). Restorative justice has been effective in addressing youth violence, especially among child soldiers and their community (Kiyala, 2018), and seeking redress for different types of injustices that affect children and adolescents who exit armed conflict in the Democratic Republic of Congo (Kiyala, 2021). Restorative justice perceives crime as a breach of people's interpersonal relationships, acknowledging that these violations engender obligations and liabilities. This requires restorative justice to heal and restore harmony by putting "right the wrongs" (Zehr/Mika, 2003: 41, 42), and finally "creating future accountability plans and agreements that will prevent the same thing from happening again" (Noll, 2003: 284).

24.5.1 Understanding Restorative Justice

Restorative justice is a paradigm of justice that was developed in the mid-seventies originating from indigenous, traditional and customary responses to criminal acts, and evolved to a modern practice in the mid-1970s in Canada as a response to victim-offender mediation. It started as a preferred trial for young offenders and was broadened to pre-sentencing where an offender and a victim could formulate a proposed sentence for the judge to deliberate upon (Van Ness et al., 2003). It is also defined as "a way of responding to criminal behaviour by balancing the needs of the community, the victims and the offenders" (United Nations, 2006: 11). Some scholars depict it as community justice, transformational justice, peace-making criminology and relational justice (Bazemore/Walgrave, 1999 cited in Latimer et al., 2005).

According to the South African Department of Justice and Constitutional Development (2023: n.p.):

> Restorative Justice is an approach to justice that aims to involve the parties to a dispute and others affected by the harm (victims, offenders, families concerned and community members) in collectively identifying harms, needs and obligations through accepting responsibilities, making restitution, and taking measures to prevent a recurrence of the incident and promoting reconciliation.

Other concepts used to describe this model of justice include communitarian justice, making amends, peace-making, positive justice, reconciliation, redress, relational justice, reparative justice and restitution (Miers, 2001). This mechanism of justice presents several advantages, such as preventing re-offending; victims' empowerment; offering and making criminal justice agencies accessible to victims; and enhancing community involvement in processes of dispute resolution (Department of Justice and Constitutional Development, 2023).

This model of justice is gaining ground in criminology and being used as a preferred option to conventional justice. Its flexibility makes it a paradigm whose time has come for complex situations such as juvenile justice, justice for inmates (prisons), justice for pupils and students, and justice in the aftermath of bloody conflict. It is being implemented worldwide in Australia, Canada, England and Wales, New Zealand, USA and many other countries (Braithwaite, 2013; Walker, 2012), and it is also applied in the educational context and acknowledged for its transformative dimension (Winn, 2020).

24.5.2 Restorative Justice Programmes

Karp (2004) presents different models of restorative justice, namely victim-offender mediation, conferencing (family group conferencing or family group decision-making, sentencing circles, peace-making circles), and boards (integrity boards, reparation boards or community panels). The conferencing model emphasises dialogue between the offender and hurt parties. After discussion, both parties decide the course of action to be taken by the offender to repair the harm. Circle models aim to achieve peace or healing of affected persons who have a stake in a conflict; justice boards or panels often bring together comprising members from college faculty, staff and students. The board may proceed without disputant parties and make a sentencing statement, while administrative hearings integrate RJ principles in administrative hearings (Karp, 2004; Karp/Sacks, 2014).

These programmes can be adapted to various institutional and organisational contexts, including dispute resolution, discipline issues, need for reconciliation and violence prevention. The versatile nature of RJ makes it an effective mechanism for "reducing recidivism by encouraging change in individual offenders and facilitating their reintegration into the community" (United Nations, 2006: 11). The focus is on restoration, transformation, healing and prevention of further harm (Umbreit et al., 2002). RJ embraces a multidimensional transformation society, targeting the entire legal system, lives in our families, our behaviour at workplace and the way we engage in politics (Braithwaite, 2003: 1). He sets the reduction of injustice as the main target of RJ as stated here: "so conceived, it targets injustice; to see the goal simply as crime reduction impoverishes its mission".

24.5.3 Restorative Justice Sanctions

Restorative justice involves multiple forms of sanctions, namely, apology, restitution, changed behaviour or generosity (Centre for Justice and Reconciliation, 2008). Noll (2003: 284) argues that RJ holds perpetrators of crime and any offender accountable, as he states: "Restorative justice includes recognising the conflict or harm, repairing the damage (physical and relational) as much as possible, and creating future accountability plans and agreements that will prevent the same thing from happening again". However, in this process the focus remains on transformation, healing and prevention of further harm (Umbreit et al., 2002). So, when applying RJ, lawbreakers and rule-breakers do not get away with their antisocial behaviours, but the sentence passed on them is used to repair harm, restore and reintegrate them into the community. For Karp (2004: 8), RJ accountability is "demonstrated through expression of remorse and commitment, and through the completion of tasks negotiated as part of the sanctioning process".

The use of RJ and its adaptability to the education environment are believed to have transformative effects (Bakalar, 2019). The theory of restorative justice has been favoured as an approach that responds to the needs of transformative justice – switching from usual retributive practices of inflicting shame, punishment and sustaining guilt to restorative sanctions that reflect a more subtle interpretation of harm, the needs of persons affected by harm, and those responsible for that harm (Winn, 2020). The conceptualisation of restorative justice as transformative justice had been explained earlier to show how versatile and adaptable this paradigm of justice is.

A study of restorative justice reparation mechanisms shows that some of its modes of reparation and punitive sanctions overlap. Restorative justice uses apology, restitution, enlightened community service, changed behaviour and generosity (Center for Transitional Justice, 2008; Karp/Sacks, 2014). Punitive sanctions include restitution, community/university service, conduct probation and conduct review (Texas A&M University, 2013). Are these concepts applied with the same understanding? To respond to this question, I should say there are differences on both semantic and pragmatic interpretation of these sentencing approaches. Restorative reparations denote rehabilitation of offenders through participation and dialogue in restorative processes. In essence, RJ reminds offenders of their wrongdoings and give them the chance to take their responsibility to repair the harm and show evidence that they are apt to become "good citizens" and will refrain from committing the same reprehensible acts that have caused hurt to others (Karp, 2004: 10).

Restorative justice is the umbrella term under which restorative discipline falls under which there are different practices. Restorative justice is a process whereby all the parties with a stake in a particular offence come together to resolve collectively how to deal with the aftermath of the offence and its implications for the future (Lyubansky/Barter, 2019). It draws on principles from democratic citizenship and peace-building education (Parker/Bickmore, 2020). Obadire and Sinthumule (2021) argue that the lack of discipline among learners in schools has been one of the foremost problems facing schools over the years.

24.5.4 Ubuntu Restorative Justice and Peace-Building

Several studies of RJ have linked this model of justice to the concept of Ubuntu that is also described as "African Humanism" (Gaylard, 2004). The notion of humanism points out human potential particularly in their capacity for reason, and that entails "the idea of a 'core' or shared humanity which is transcultural and transhistorical" (Gaylard, 2004: 265). It is applied in a diversity of life spheres, as a jurisprudence in legal or judicial processes (Kiyala, 2016, 2018, 2021, 2022); it is also associated with religion and theology (Battle, 2000; Metz, 2011); it is used in education and business management (Gaylard, 2004; Lutz 2009; Newenham-Kahindi, 2009; Oviawe, 2016) as a model of conflict resolution (Akinola/Uzodike, 2018); it is perceived as model for human rights and the dignity of the human person (Dolamo, 2013) in philosophy and ethics (Christians, 2004; Louw, 2008). The richness of Ubuntu does not permit limiting its application to the fields mentioned here. It is extended as an approach to humanise the justice system (Anderson, 2003; Rautenbach, 2015). Ubuntu transcends boundaries of ethics, encompassing a boundless social ethos and philosophy that extends to every aspect of life. The belief in Ubuntu is supported by an extensive collection of scholarly publications that convey its comprehensive thematic and conceptual framework.

Ubuntu is intrinsically linked to individual and collective wellbeing and wellness that is generally affected or promoted via interpersonal relationships. Thus, transforming interpersonal relationships and building peace are the essential goals of Ubuntu which ties it to the goals of RJ and peace-building. Ubuntu informs, orients and sustains the sound health of human interrelations that is critical for human development. Education being one of the crucial factors of human development, it is, thus, important to also discuss school violence from an Ubuntu perspective.

Schirch (2004) contends that transforming relationships is an essential principle of all peace-building programmes, and there is a close tie between peace-building and RJ, in principle. She affirms: "peace-building seeks to transform individuals, families, communities, businesses, structures, and governments away from destructive expressions of conflict and toward constructive growth and development" (Schirch, 2004: 45).

Restorative justice in essence peace-building, because it originated from the Judeao-Christian tradition. In Hebrew, peace is Shalom; it is also covered by the Gospel of Jesus such as: "Peace I leave with you; my peace I give you. I do not give to you as the world gives. Do not let your hearts be troubled and do not be afraid" (John 14:27). Drawn from the article "Shalom" by Timothy Keller, "Shalom experienced is multidimensional, complete wellbeing – physical, psychological, social and spiritual; it flows from all of one's relationships being put right – with God, with(in) oneself, and with others" (cited in New International Version, 2023: n.p.). This religious dimension of peace features in the restorative justice literature. For instance, Zehr (2002: 19) refers to "Shalom" that implies "all-rightness" with each other, with the creator and with the environment". In a similar vein, Noll (2003) explains peace-making as deriving from the ancient Hebrew word "Shalom" which gives to peace-making a broad connotation that entails healthy human interrelations, and the essence and goals of that are peace.

Restorative justice aims at peace and transformation of interrelations, in such a way that those who cause harm to become aware of their responsibility and liability to meet human demands to guarantee respect of individual rights, and this create an opportunity for forgiveness, reconciliation and healing (Schirch, 2004; Zehr/Mika, 2003). This is the objective RJ peace-making processes; they seek to create a space for meeting between victim and offender after a crime, by using acknowledgement of the harm inflicted and offering reparation, instead of punishing the offender. Thus, justice requires accountability that brings peace. So, the concepts of transformation and addressing injustices are central in the RJ literature and this is what restorative justice aims at as expressed by Braithwaite (2003: 1):

> It aspires to offer practical guidance on how we can lead the good life as democratic citizens by struggling against injustice. It says we must conduct that struggle while seeking to dissuade hasty resort to punitive rectification or other forms of stigmatising response against.

In essence, restorative justice reminds offenders of their wrongdoings and gives them the chance to take their responsibility to repair the harm and show evidence that they can become "good citizen" and will refrain from repeating the same reprehensible acts that have caused hurt to others (Karp, 2004).

24.5.5 Restorative Justice Circle and Its Application in Education

An RJ circle is a process that creates a respectful space where an offender and victim meet to address the hurt caused by a crime or an offence. It consists of two types: a peace-making circles and a sentencing circle. Peace-making circles are also called healing circles because they provide a platform that lead offenders to share the pains of the victims, and by taking responsibility for past wrongdoing, an offender is led to reform their life and healing is enhanced (Pranis, 1997, 2013). The discussion is conducted by a facilitator who passes around a "talking peace" that restricts the talking only to the person holding the paper in order to maintain a mutual listening atmosphere. The focus is on amending relationships, healing and prevention of further harm (Umbreit et al., 2002). This approach helps limit recurring antisocial behaviour. It also encourages transformation and brings change in individual offenders.

The basic premise of the classroom sharing circle as a proactive restorative practice is that learner behaviour, character, social and emotional competence is improved when they feel connected to an inclusive, supportive and respectful community (High, 2017). Restorative peace circles have value in classrooms for both post-incident peace-making and as a proactive pedagogy for citizenship, academic learning and developing awareness (Parker/Bickmore, 2020). Lyubansky and Barter (2019) stated that as learners become familiar with the new justice system, it will become part of their normal routine, available when needed but otherwise in the background as academics and relationships take their proper position at centre stage.

There is a pedagogical dimension in this model of justice: "restorative processes help educate community members about the need for civic commitment and build student capacity for evaluating the impact of their behaviour on the community" (Karp, 2004: 7). These proceedings lead students to rationalise institutional policies and accept them as law-abiding because they understand the consensus that is reached on behavioural standards and why it is important to address failures to comply with them that result in misconduct (Karp, 2004).

Restorative justice has become a popular practice useful in handling misconduct and tackling violence in schools and institutions of higher education (Karp/Dagle, 2013; Kidde/Alfred, 2011; Wilson, 2006). The core argument for the viability of RJ is the fact that it promotes dialogue and reparation. In many schools, colleges and universities, restorative justice programmes such as restorative circles, victim-offender mediation, sentencing circles are employed to achieve justice and discourage all forms of violence we alluded to earlier (UNESCO, 2006). This could become a common practice for South Africa.

Peace can only be achieved if indirect structural violence and cultural violence are addressed alongside direct violence (Cremin/Bevington, 2017). Advocates of restorative practices also highlight the effectiveness of sharing circles as a pedagogical tool for teaching social and emotional competencies such as active listening, empathy and respect for others (High, 2017).

Peace circles have some constraints. Herrera/McNair (2020: 3) postulated that circles are the primary goal of developing learners' social and emotional skills and competencies; they primarily focus on building and reaffirming relationships. In healing circles, all parties involved are part of the conversation and sit in a form of a circle (this is not easy in a classroom set up), while one person talks at a time while the facilitator leads the discussion. Cremin/Bevington (2017) identified the need for peace-keeping, peace-building and peace-making concepts to be clarified and updated for the twenty-first century since the nature of global conflict and war is evolving. Time is one of the challenges for schools as the core reason for being at school is learning and teaching. Therefore, there is a lack to clarity about how peace-keeping, peace-making and peace-building fit together holistically in schools (Cremin/Bevington, 2017).

This study used peace-making circles which is relevant to both justice processes and amending relationships that have been affected by harm.

24.5.6 Limitations of Restorative Justice

Crime paradoxically affects not only the victim but the perpetrator as well and their environment/community. It is important to restore all parties touched by the conflict and its consequences. Restoration is preferred to punishment because it has a transformative impact on both victim and offender and helps both parties to heal from the hurt. In this process, the offender is held accountable and led to take responsibility for the offence committed and they accept the need for reparation, amending their conduct, refraining from relapsing into the same reprehensible behaviour and being reintegrated into the community. The concept of justice and reintegration are seen here as key elements of the process because crime cause alienation and estrangement that can be broken only when justice is applied, and a community member alienated by the offence committed is restored into community membership. The following are some key studies conducted to probe the effectiveness of RJ compared to non-restorative processes.

When all attempts to lead the offender student or learner to compliance and reform have failed, expulsion of a student/learner from a school or college campus becomes a necessity. However, Karp/Sacks (2014) still maintain that the process should be reimagined. Discussing the "Model code sanctions", these campus pundits maintain that RJ must be pursued. They start by distinguishing "the code mode" from "restorative justice" as articulated here:

> The model code is more a formal process, emphasising authority and carefully managed procedures as a way to guarantee fairness and legitimacy. Restorative practices instead focus on creating social support in order to elicit honest dialogue and personal investment in the process (Karp/Sacks, 2013: 2).

Karp/Sack (2013:2) argue that model code processes vary and encompass a large array of punishment philosophies that encompass:

> Retribution, deterrence, and incapacitation and is often a model of progressive restriction and exclusion—with sanctions that begin with restrictions such as losing access to the gym or participation in a club, to fines, to removal from campus housing or suspension from classes, and finally to expulsion from the institution.

However, they contend that restorative practices should prevail and "instead focus on strategies to repair the harm caused by the offence and tasks that will rebuild trust between the respondent and the campus community". They advocate reframing the problem or questions such as "should the student be suspended because of the gravity of the offence?" – A question that officers ask under the model codex, or "What actions could the student take to reassure us that he or she will be a responsible member of the community?" – A type of question that is posed while pursuing restorative justice practice. Karp/Sacks (2014), thus, suggest that reframing the question may lead the student to reforming his/her conduct. Students' suspension and expulsion must be kept thought they are anti-communitarian devices; their use should be minimised wherever possible (Karp, 2004). Sending a student away from campuses implies transferring the problem to another community or institution without settling it (Karp, 2004). We support explusion measure for the two following reasons:

- Firstly, "colleges are not correctional facilities", and a student who poses a serious threat to the safety of others must be expelled.
- Secondly, a refusal by a student to respond to "judicial proceedings", and failure to duly accomplish the "sanctioning tasks" should make it a necessity that they be removed (Karp, 2004).

Schools, colleges and university campuses should always strive to promote dialogue. Giving warnings, restricting learners or students' privileges, and using restorative sentencing should be prioritised. RJ emphasises inclusion, because an offender needs to come to understand that education space provides those who find their way in life with the opportunity to be active member of the community. RJ plays that educative role by promoting reintegration of offenders after taking them through restorative programmes instead of "social separation" (Karp, 2004: 9). The danger of expulsion is that the evicted and expelled person may become bitter and shame may drive them to drinking bouts and substance abuse (different drugs) as a means of coping with shame, frustrations, loss of perspectives for better future. In this case, they may commit even more crimes as they are sent back into society as wounded beasts.

24.5.7 Stumbling Block to Non-violence and Peace in Education

Karp (2004) discusses the interrelatedness of diverse and multiple dimensions of misconduct by students. These range from inability to cultivate internal control mechanisms to adjust their behaviour; anxiety to make friends and to create a certain identity; disparities between student culture and the general societal and legal norms regarding drugs and alcohol use; reliance to coercive approach to impose compliance with college policies and criminal procedures in dealing with students' misconduct (Karp, 2004). Barton (2023) draws from studies in schools that found that the use of restorative justice and peace-making circles had been instrumental in reducing violence in schools.

In order to make schools and institutions of higher education a free-from-harm and sympathetic environment, it is necessary to take some steps to protect children, strengthen women's safety and wellbeing on campus, develop a wide range of interventions and assistance for victims of sexual assault, dating and domestic violence and stalking, and involving. When harm occurs, restoration justice should be chosen instead of criminal justice or referring offender to the judicial service.

24.5.8 Education for Peace

The interrelatedness of conflict and violence has been established as a fact (Schirch, 2004). Conflict can be dealt with in two ways: "constructively" when people initiate an approach that satisfies the interests of parties involved; and "destructively" when people use violence to tackle conflict in a way that harms relationships which results in frustration when the needs of other parties are denied and overlooked. This becomes a trigger to violence in the attempt to "do justice or undo injustice" (Schirch, 2004: 22). In order to avert violence, people should study "peace-building strategies" and adopt a culture of peace, because "successful peace-building process can harness both local and external resources to prevent and reduce violence, and to transform and help people recover from it" (Schirch, 2004: 23). Peace-building processes include "waging conflict non-violently", "reducing direct violence", transforming relationships", and "building capacity". Schirch (2004: 6) presents these strategies, and asserts that violence can be prevented though education and training and constructing structure that strengthen a culture of peace that will endure.

This principle applies to the educational environment. Kaman/Harris (2000) conducted a study to evaluate to what extent an academic study of peace would influence student's behaviour. They employed experimental control group design with identical groups of which one was exposed to peace studies. The basis for this study was laid by previous studies conducted in the USA (Kaman/Harris, 2000).

- Experimental study using control groups (Harris, 1991, 1994 cited in Kaman/Harris 2000). Participant observation showed that students who study peace changed their attitudes by adopting a "peaceful lifestyle".
- Harris (1999 cited in Kaman/Harris, 2000) argued that peace education that encompasses violence prevention, conflict resolution, non-violence "moves students away" from thoughts and actions that encourage a "warring world" that is hostile to peace (Harris, 1999: 679).

Based on their research, Kaman/Harris (2000) concluded that there is reason for optimism in believing in the possibility of creating a "peaceful and just society". This was evident from the learners' increased ability to critically analyse issues and their willingness to approach them from a holistic perspective. Therefore, the introduction of peace education in schools and higher educational institutions has the potential to transform the educational environment into a peaceful and secure space, promoting active and participatory learning.

24.6 Research Methodological Approach

The overall aim of the study was to empower members of the school community – teachers, learners and parents – with knowledge and skills that could help to build a culture of peace in a specific school. The specific objectives were:

- To explore the nature, extent, causes and effects of violence in two schools in the rural Bhekulwandle community and Lovu Township.
- To identify past and present efforts to resolve violence in the schools and to design and implement a programme in one school aimed at reducing the level of violence.
- To conduct peace-making circles and evaluate their effects on transforming learner's perceptions on violence and empowering them to become peacemakers.

The underlying assumptions of this inquiry were that leaders were responsive to peace education and exposing them to peace-making circles would help them become peace-builders and agents of change.

The study was designed as action research and used mixed methods. It consisted of twelve sections each broken down into sub-sections. It started with the introduction, provided the context of the study and an analysis of the concepts of violence, its various dimensions and peace. The subsequent sections explore the impacts of schools as instruments of change before presenting the theoretical framework that guides this investigation. That is followed by research design section, population sample, methods, results and the discussion of the findings which incorporate, the significance of the findings for peace at schools. Finally, the last section embodies recommendations for ending violence and making the school environment a safer space for learning.

This study is action research (AR) of a descriptive and exploratory nature where the mixed methods approach was applied, that is, a qualitative and quantitative approach. To elucidate the usefulness of AR in this inquiry, we were inspired by Kostaris et al. (2017) who conducted AR in Australia on a flipped classroom model for improving the learner-centred use of face-to-face sessions that were intended to minimise the use of the lecture style by teachers and increase learners' active learning and collaboration. The results revealed that teacher-facilitated face-to-face sessions provided learners with unique learning experiences through direct access to both their classmates (for collaborative activities) as well as to feedback and scaffolding by their teacher (Kostaris et al., 2017). Another example was AR conducted in Tuncel (2017) which showed that in a culturally diverse environment teachers have to be culturally responsive individuals. Tuncel (2017) further stated that the action part of the research allowed teachers to get to know their own culture, learn about the cultural backgrounds of their learners, gain better understanding of social, economic, political context, and improve their skills and willingness to use culturally appropriate management strategies and their commitment to create inclusive classrooms.

In addition, Kiyala (2018) conducted a longitudinal AR in the DRC. This inquiry used restorative justice peace-making circles (RJPCs), as a model of transitional justice for former child soldiers. Restorative justice evaluation was based on its outcomes. The intervention was efficient as observed: (1) the greater majority of children below the age of 18 (97.2%) exposed to RJPCs, who intended to join armed groups before, changed their mind and never joined or re-joined armed groups after seven months; (2) apologies by former child soldiers were accepted and they were forgiven, and (3) support for prosecution of child soldiers dropped after RJPCs (Kiyala, 2018). Finally healing former child soldiers was a critical step towards change of identity, the transition from soldiering to civilian life, necessary for meaningful reintegration into society. That implied addressing these child soldiers' psychosocial wellbeing and creating an environment where peace prevails, and adequate accountability measures are in place and effective. The overall results reveal that RJPC yielded empathy, vicarious justice, forgiveness, reconciliation, and deterrence of child soldiers (Kiyala, 2018).

Considering the effectiveness of RJ in responding to youth violence both in the context armed conflicts and school violence, peace circles were employed as AR intervention. The outcomes were evaluated to establish pre-and-post intervention results to establish whether or not the intervention was efficient, and whether that is owed to peace circles or other factors.

24.7 Methods

Creswell (2014: 32) highlighted that qualitative research is an approach to "exploring and understanding" the meaning that individuals or groups "ascribe to a social or human problem". Qualitative research, thus, places emphasis on exploration and understanding the meaning (Almalki, 2016: 4). Through a qualitative inquiry, the research team was able to explore and understand the definition of violence and the causes and effects of violence in the area of Bhekulwandle and was able to contextualise the experiences of the school community. The qualitative method enabled a contextualised understanding of subjective experiences (Crowe et al., 2015: 1). This was done through in-depth descriptive questions and probing questions put to participants during interviews.

24.7.1 Population Sample

Purposive sampling technique was employed to select participants. Purposive sampling is the "deliberate choice of participants" due to the qualities the participants possess (Etikan, 2017: 2). Participants were purposively selected because of their knowledge and experience of school violence in the targeted school. Population to the study included the principal and the deputy principal, eight teachers, 110 Grade 9 learners, four parents, four SGB members and six community stakeholders. The stakeholders consisted of a local business owner, members of the Community Policing Forum (CPF), a local councillor, traditional leadership, non-government organisations operating in the area and one member of a faith-based organisation.

The most important criterion for selecting participants to the study was their ability to "provide relevant information" (Asiamah, 2017: 9). The school principal, deputy principal and the SGB members were deliberately selected to participate in this study because of their knowledge in the management of the school. Grade 9 learners were selected because the majority of Grade 9 learners are in their second year of schooling in high school. The assumption was that Grade 9 learners would be in Grade 10 and still be in the same schools the following year when the action part of the research was executed and when the evaluation was conducted on the intervention strategy. Grade 10 teachers were purposefully selected because the learners who were in Grade 9 in 2019 were going to Grade 10 in 2020 and would be taught by the current Grade 10 teachers when the programme would be implemented. Therefore, the current Grade 10 teachers were responsible for implementing the intervention on Grade 10 learners.

24.7.2 Qualitative Data

Multiple data collection instruments were used. The use of more than one approach to data collection "validated" the collected data and enabled researchers to get richer, fuller data which helped to confirm the results. Semi-structured interviews, focus-group discussions and observations were used for qualitative data collection. The interview framework was created as a list of questions, which helped the interviewer to direct conversation towards the research topic and to achieve the richest possible data. Focus-group discussions and semi-structured interviews were conducted with the school principal, teachers, parents and stakeholders. Focus-group discussions were used to gain in-depth understanding of school violence and how it affects

Table 24.1 Data collecting instruments and methods

Group	Number	Type	Language
Principals	2	Interview	English
Teachers	8	Interview and focus group discussion	English
Learners	8	Focus group discussion	English and Zulu
Learners	480	Questionnaires	English and Zulu
Parents	4	Interview	English and Zulu
Stakeholders	4	Interview	English and Zulu

Source The Authors

learners, parents and the community. Observations were done inside the classrooms with the teachers and the learners, on the school premises, during interview sessions and focus-group discussions as well as in the area around the school. What we observed was the relationships between the learners, teachers and the parents and the type of visitors that visited the school. Their levels of engagement, how they related to each other, incidences of violence and the nature of their conversations were observed.

The literature review helped the researcher to formulate the questions to be included in the data collection instrument. For validity and reliability, multiple approaches were used to enhance the accuracy of the findings. Research was conducted very closely with the support group that was formed by experts in the field of research and education. The experts included two school principals from other schools with high levels of violence (one having recently acquired her PhD in peacebuilding), former school inspector and one former teacher who is now a university lecturer and a PhD candidate. A pilot study was conducted in another school where we were able to test the validity of the research questions.

Having a panel of experts' support helped to validate the questions. The questionnaires were administered by the research team and dispatched to all the Grade 9 learners. It would not have been feasible for the research team to interview each learner individually. At the same time, it was important to get the learners' input on their understanding on school violence because they are a bigger part of the school community.

Table 24.1 illustrates the various characteristics of participants in this inquiry, the features of the sample size and the instruments utilised to collect data. We used both English and IsiZulu local language of the KwaZulu-Natal to conduct this investigation.

24.7.3 Validity and Reliability

The main reason for use of two research methods was to "retain maximum validity" while adhering to the rules inherent in each paradigm (Morse/Niehaus, 2016: 10).

The use of two methods provided "strong research findings, produced results that are broader and of more significant impact" (Morse/Niehaus, 2016: 10). According to Fusch/Ness (2015: 6) triangulation goes a long way towards enhancing the reliability of results. Triangulation was used to increase credibility and validity and comprehensive set of research findings, especially the triangulation of empirical data and data supplied by literary research and review of the literature. Various findings of this inquiry were confirmed by similar studies in South Africa (Grobler, 2018).

24.7.4 Research Ethical Standards Observed

One of the issues relating to ethics was the participation of the grade nine learners in the study who were younger than 18 years of age. Most grade nine learners in South Africa are between 14 and 16 years of age. It was important for the researcher team, therefore, to seek the parents' consent for the learners' participation in the study for ethical reasons. What is paramount for researchers in the research process is the "protection of the interests" of the research participants (Flick, 2014: 48). The interests of the young learners were protected through requesting the parents' consent for the learners' participation. The interests of all the participants of the study were protected by ensuring privacy during the interviewing sessions. Interviews have to be conducted according to "ethical approval" policies in a private and quiet space (Crowe et al., 2015: 2). The researcher team requested to be allocated in a private room for the interviews such that during the interviewing process, there was privacy and participants were free to voice their views on school violence. The interests of the participants were also protected by not revealing the identity of the participants. The information collected has been kept confidential in a safe and secure place, and it will be kept as such for at least 5 years after conducting the study. For ethical considerations, before the study was conducted, the researcher team administered consent forms to all the participants and ensured that each participant had understood and agreed with the content of the consent forms.

Permission was sought from the school management team. The principles of non-maleficence, beneficence, autonomy, justice, respect for participants' rights and dignity, rights to withdraw, informed consent and voluntary participation, the principle of confidentiality, anonymity, etc. (Bless et al., 2013) were observed. A letter was written to the gatekeeper, the principal who allowed the study to take place in that institution. Since children are considered as a vulnerable group (Bless et al., 2013: 33), the research was sensitive to their needs: the school principal was the guarantor of learners' safety and all data collection took place within school premises.

24.8 Results

The findings of this study are subdivided into two main categories, namely observations of violence-related narratives and the outcomes of learners' participation in peace circles. All interviews and focus group discussions were conducted at Mbambangwe Secondary School at Bhekulwandle. In this section we first present the results and then discuss them under "discussion" sections. This option is meant to hear the voices of participants before we engage their perspectives with the literature, other studies, draw conclusion and make recommendations.

24.8.1 Violence-Related Narratives

Several themes were developed from empirical data collections (focus groups discussions and interviews that provided a deep understanding of various aspects, dimensions and impacts of violence). It was surprising that almost all the participants viewed school violence as being perpetrated by the learners against the teachers. Hence, they defined school violence as follows:

> [School violence is] when the situation is not conducive for teaching and learning due to learners' ill-discipline, when learners become unruly and fight with each other, when learners do not do their homework on time, and when learners do not listen to their teachers and to each other (Interview with a teacher,Bhekulwandle interview, 3 June 2019).

School violence comes in different forms: "it can be a debacle between learners and teachers and between learners themselves" (Teachers' focus group discussion, 29 Apr 2019).

Teachers stated that "school violence can be seen as when learners use weapons, vulgar language and abusive words to other learners. When they do that, those who are offended retaliate" (Teacher interview, 30 Apr 2019). One teacher further defined school violence as follows:

> [School violence is] when the situation is not conducive for teaching and learning because of learners' ill-discipline, when learners don't do as they are told, when learners don't listen to teachers and to one another, and when learners bully other learners. School violence can also be within educators where they don't honour their duties, when teachers are not punctual for classes or coming unprepared and not showing respect to other teachers that affects the primary goal of learning and teaching. When parents don't take active participation in the educational development of their children, instead promote laziness, disrespect, and child delinquency (Teachers' focus group discussion, 5 Jul 2019).

Another definition of school violence was given as the acts of learners fighting with teachers and learners carrying weapons and drugs to school and being beaten by teachers in return. Some teachers stated that "school violence can be between educators when they do not honour their duties, when they are not being punctual for classes and coming unprepared and disrespect among themselves, which affects the primary goal of learning and teaching" (Teacher interview, 2 May 2019). One teacher further stated that "school violence is when parents don't actively participate in the educational development of their children" (Teacher interview, 6 June 2019). Another teacher defined school violence as a "lack of communication or when there is misunderstanding between learners and teachers" (Teacher interview, 22 Apr 2019). She further explained that "school violence can be verbal or physical abuse, child battering and learners' retaliating." In most instances, boys are portrayed as the perpetrators of school violence, as echoed by teachers who stated that sometimes boys feel as though "they have power over girls and have to impose themselves over them, they feel superior to girls" (Teacher interview, 9 May 2019).

According to teachers who participated in focus groups, the most common forms of violence amongst learners were bullying, insults and the use of unacceptable language. If violence involved two learners, it was easier to resolve the violence because the teachers disciplined them, but if the violence involved more than two learners, it became complicated or difficult to resolve because it ultimately turned into a war.

24.8.1.1 Bullying and Rebellious Learners

Out of the 116 learner participants, 28 responded by stating that bullying was the most common form of violence in their school (Learners' survey, 10 June 2019). Teachers echoed the sentiment that learners fought against others in the classroom or sometimes after school outside the premises. Sometimes learners used scissors, mathematical instruments, pens and rulers to injure others in the absence of other weapons. According to the teachers, two years earlier, learners had brought illegal weapons, such as knives, to school (Teachers' focus group discussion, 5 Jul 2019). They would also bring and use drugs inside the school premises, in toilets and in front of teachers and would also stab one another. Learners would come to school whenever they pleased and left at any time of the day. Because the learners were armed, it was difficult for the teachers to reprimand them when misbehaving.

24.8.1.2 Gang Violence

Two groups of rival gangs were identified. According to teachers, the names given to the gangs in their school are "metro-skhotheni and the rich gangs; they fight both inside and outside the school premises" (Teacher interview, 29 Apr 2019). The problem with the gangs was that when they fought outside the school premises, the fight spread to innocent learners, teachers, and gangs inside the school premises, and when the conflict started inside the school premises, it spread to the community. The groups formed themselves into factions, and this led to violence as they fought for dominance inside the school premises.

According to parents, the formation of gangs in communities occurred when there were fights between learners at school. In addition, learners who were siblings, friends or relatives of the affected parties became involved by taking sides. This violence spilled over into the community, the family members of the affected parties became involved, and the violence ultimately turned into gang violence or faction fights, and the entire village or tribal area was affected. The worst aspect of faction fights and gang-perpetuated violence was that they could last for long periods. Sometimes, when school is closed, mostly for school holidays, the gangsters came and waited for the learners at the school gate to attack them when they left the school premises.

24.8.1.3 Gender-Based Violence

The high rate of poverty and unemployment pushed people to act in ways that were unimaginable. "There are instances where learners as young as 14 years old are impregnated by men in their forties, old enough to be their fathers, and the parents and the family of the girl have given the consent because of the promise of lobola payment" (Principal's interview, 12 July 2019). In one such case, the principal had to intervene to protect the female learner. He called the learner to his office to discuss the issue, and the learner's response was "*uthe uzongilobola*, meaning he has made the promise to pay lobola" (Principal's interview, 12 July 2019), with the parents looking forward to the lobola negotiations. In such cases, sometimes the teachers and the school principal had to intervene, putting their own lives at risk.

Some of the causes of school violence involved boys fighting over girls. When a boy professed his love to a girl who was having an affair with another boy, boys were led to fight. Sometimes, it was the girl's brother who fought the proposing boy, mainly because the brother felt that he had the power and responsibility to protect his sister from other boys. According to stakeholders, boys were more violent than girls because in most cases they wanted to show that they were in control. Gender-based violence was also rife in families. One female learner, whose mother had tolerated abuse for many years, stated, "I think what makes my mother to stay for long in this relationship is because she is financially dependent on my father. My father beats us (mother and children) every day whenever he comes back from work" (Female-learner focus group discussion, 4 July 2019).

24.8.1.4 School Vandalism and Burglary

MHS had experienced burglary and vandalism from the surrounding communities. One such incident happened while the intervention programme was under way. When the learners arrived at 6:00 am, the school library was in flames, and the learners quickly put out the fire. There was an overnight security guard who saw the thugs coming in and had to hide because of the fear of being attacked. According to the principal, they expected the community to protect the school since it is a valuable asset, but the very same people who were supposed to protect the school were destroying it. The members of the community sometimes stole the schools' furniture and used it in their homes. In recent years, local community members had started invading the school land by building their houses inside the school premises.

24.8.1.5 Corporal Punishment

Even though corporal punishment has been banned in South African schools and families, unfortunately, some teachers and parents still use it to discipline learners at school and children at home, respectively. The incorrect ways of resolving violence at home were also used at school: 37 out of 116 learners felt that corporal punishment was the most common form of violence in their school (Learners' survey, 10 June 2019). There is a narrative that a Black child cannot be disciplined without the use of corporal punishment, which is the reason why teachers and parents strongly believe that they cannot do without corporal punishment.

According to some teachers, "besides corporal punishment, sometimes the way teachers talk to learners is degrading and full of harassment. Some teachers provoke learners the way they talk to them, and learners respond rudely" (Teacher interview, 8 June 2019). According to some teachers, "the government has a major contribution to school violence and causing misbehaviour in learners. Learners do not want to be disciplined because they know they have rights. The government has scrapped corporal punishment and this cause unruly behaviour in learners" (Teacher interview, 18 June 2019).

24.8.1.6 Physical Fights, Stabbing and Pickpocketing

Out of the 116 learners, 28 learners indicated that physical fights were the most common form of violence in their school (Learners' survey, 10 June 2019). Out of 116, only two learners indicated that stabbing was the most common form of violence in their school.

24.8.1.7 Violent Parents and School Governing Body

Some of the stakeholders felt that the government's strategy for the employment of teachers was causing violence among the school community. Teachers echoed the sentiment that most, if not all, members of the community within the SGB were not qualified to manage the process of employing professionals. They felt that some of the SGB members had ulterior motives when the employment of teachers was concerned. On the other hand, some SGB members stated that there was friction among the school management team fighting for positions. They also believed that it was because of this friction that the community members who were part of the SGB would find a loophole to divide the teaching staff even further.

Teachers shared their frustration when a teacher was gunned down and killed in a nearby school inside the school premises while the researchers were still collecting data. The assumption was that there was a contestation for a senior position in that school. The deceased teacher was killed early in the morning after parking his car in full view of primary school learners and his colleagues. On the day, during interviews, some members of staff had to leave for the memorial service for the deceased teacher.

Another complaint was launched regarding the SGB members and how they caused violence through funds being allocated to the school by the Department of Basic Education. The SGBs sometimes had a vested interest in projects that were introduced at school and would want to allocate funds in ways that would benefit them. It became a problem when the *Delangokubona* (illegal local structures demanding the allocation of government tenders to local people) come to the school and demand that any project happening in the schools must be given to them. Sometimes the community would intentionally destroy something inside the school premises and demand that it be fixed – this had led almost closing down.

24.8.1.8 Political Violence

Stakeholders stated that the hostile political environment in the community sometimes led to school violence as learners brought issues from outside into the school environment. According to one of the stakeholders:

> Violence can be politically motivated, and these can be divided into two. Firstly, there is the pre-democracy political violence and the post-democracy. The post-democracy political violence is more about crime. This is another form of violence that affects both communities more so toward election times (Stakeholder interview, 29 May 2019).

Additional data was collected to uncover the underlying drivers of violence, basically related to society.

24.8.2 Underlying Causes of a Societal Nature

The following sub-sections present the underlying causes of school violence that are significant in South African society, namely, intergenerational transfer of trauma and violence and the media.

A teacher stated that school violence was sometimes caused by the "uniqueness of individuals and different family backgrounds" (Teacher interview, 30 Apr 2019). The teachers further stated that this resonated with learners since learners came from different backgrounds, and violence in their society spilled over to the school. According to one of the parents, the main cause of school violence was that "boys fight for girls, this is more about love affairs and drugs" (Teacher interview, 3 May 2019). Some stakeholders stated that school violence could be triggered by peer pressure, gangsters, a lack of parental guidance, a lack of communication among school community members, and drug and substance abuse. According to community members, the main reasons for school violence was because there were no activities to keep learners busy after school and during weekends in the community; there were no libraries or recreational and sports facilities. Therefore, young people looked for alternatives to keep themselves busy.

Some parents attributed school violence to teachers dating learners, drug abuse, a lack of parental love, advanced technology and media. Other parents attributed school violence to domestic violence and family disputes. One parent observed:

> When we were going through divorce, my child's academic performance dropped drastically. Since then, it has been very hard to discipline my son and to get him to focus. I can see that he is crying out for help through his behaviour. As a female, it is very hard for me to know and understand his needs because he is a male and I am female. I have been called to several meetings at school to discuss his ill-discipline. I really do not know what to do because I have tried everything, even talking to his father to help, but he doesn't care (Parent interview, 6 June 2019).

Some parents stated that "when the government passed the law to scrap corporal punishment, it stripped us, as parents, of the powers we had over our children." Another comment was that "as teachers, we are powerless to discipline learners because some teachers have been killed and stabbed by the learners in other schools" (Teacher interview, 18 June 2019). Learners did as they pleased in front of the teachers, and this caused the teachers' morale to decrease. The causes of school violence can be grouped into three different categories, namely, the underlying causes of a societal nature, underlying causes of a local nature, and facilitating causes. These are presented in Table 24.2.

Poverty was identified as a major reason for school violence. The high rate of unemployment leading to poverty was one of the reasons for school violence. According to teachers, "the high level of poverty in the community forces other learners to steal from their peers or taking other learners' possessions without their permission in order to be at the same level with them" (Teacher interview, 30 Apr 2019).

Violence that occurs as a result of learners' impoverished backgrounds:

> The conditions under which the learners live are unbearable. Some learners live in one-roomed households, and, as teenagers, they begin to feel uncomfortable sharing the same sleeping space with their parents and siblings. Learners, therefore, resort to sleeping over with their friends. This exposes them to sexual and criminal activities since this is an easy way out of poverty. Learners fight for possessions like money, cellphones and branded items like backpacks. Bringing to school branded and valuable items like clothing and backpacks by the learners that can afford such items is the reason for conflict, and the learners who

Table 24.2 Causes of school violence

Underlying causes of a societal nature	Underlying causes of a local nature	Facilitating causes
Poverty	Lack of recreational facilities	Communication gap
Intergenerational transmission of violence	Social ills	Drug and substance abuse
Media	Learners' upbringing	Division among teachers
	Peer pressure	
	Domestic violence	

Source The Authors

cannot afford [them] would steal these items. Learners who are part of the gangs inform fellow gang members who are outsiders what other learners' possessions are, and then the outsiders target the learners and take away those possessions by force (Teacher interview, 5 Jul 2019).

There was evidence of the high rate of poverty as, according to the participants, some of the local people could be seen digging from the dustbins in the neighbouring suburban areas looking for something to eat. The high rate of poverty and unemployment forced residents to consider selling alcohol as one of the means of generating an income. The presence of taverns in the neighbourhood meant that alcohol was easily accessible. Because of the desperation of tavern owners, it did not matter how they made a profit or to whom they sold alcohol, which could even be sold to underage children and learners wearing their full school uniforms.

The teachers observed: "the reason why there is so much violence in communities is because some learners' behaviour is directly linked to their parents' behaviour. Parents have wrong ways of resolving conflict, and learners behave exactly like their parents" (Teacher interview, 29 May 2019). Learners had a mentality of retaliation, and this originated from their homes. When parents were called to school to assist in the learners' behaviour, it was very easy to see why their children behaved the way they did because the parents sometimes reacted violently for no reason. One of the stakeholders stated that "parents had moved from their positions and responsibilities. This makes learners to be angry because there is no one to take care of them" (Stakeholder interview, 2 Jul 2019). One of the parents commented that "we used to believe that a child belongs to the village and [that] parents and adults have to take responsibility of all the village children equally without discrimination" (Parent interview, 22 May 2019).

One of the parents felt that the media contributed to school violence through the programmes that are shown on television. "Programmes like Yizo Yizo and news from other parts of the country and the world make a contribution to the violence we see in communities because young people begin to emulate activities they have seen on television" (Parent interview, 16 May 2019). Another parent argued that "I think the 1976 uprisings that we see through the eyes of Sarafina has a lot to do with the violence in schools" (Parent interview, 12 Jul 2019).

24.8.3 Underlying Causes of a Local Nature

The following sub-sections present the causes of school violence in the context of the local community where MHS is established. These encompass the lack of recreational activities, social ills, learners' upbringing circumstances, peer pressure and domestic violence.

Research findings have revealed that there are various upbringing and parenting styles within the communities of Bhekulwandle. According to teachers, "some learners are heading households, and such learners carry a lot of burden with them that cause them to be full of anger, hatred, and aggressiveness" (Teacher interview, 3 May

2019). Learners who do not have both parents are forced to assume the role of being household heads. They feel burdened because they are robbed of the opportunity of being children like their peers" (Teacher interview, 30 Apr 2019).

According to teachers, the common households in these communities are child-headed households, households with younger parents (parents who had their children when they were teenagers), single-parent households, households with absent parents and grandparent-headed households. "Most children from such households do not get the parental love and care as they are supposed to" (Teacher interview, 12 June 2019). A teacher reported that "we are prepared to give learners the love that they don't receive from parents and families but lack know-how. Sometimes the learners' unruly behaviour makes it difficult for teachers to give learners the love they need" (Teacher interview, 29 Apr 2019).

The problems affecting Bhekulwandle community are interrelated, and the community find themselves trapped in a vicious cycle. According to teachers, because of the high prevalence of teenage pregnancies in both communities, some of the learners' parents are very young. They are young enough to be their children's siblings. The effects of teenage pregnancy can be seen, where some of the learners' parents were in their late twenties and already had teenage children. The principal commented that:

> Even if you ask a learner to bring the parent to school, they just pitch up with young adults in their late twenties. This makes it hard to relate at the same level with parents because they themselves still need guidance as youngsters (Teacher interview, 5 Jul 2019).

This affects the entire school community because the parents are not mature enough to take care of their own children, and this makes it difficult for teachers to enforce discipline in learners or even to reason with the parents.

Parents lacking parenting skills is one of the causes of school violence. According to a teacher:

> Learners from such households, especially boys, feel pressured when they come across their peers because they don't have father figures and, therefore, try to assume the role of being protectors and get used to being in authority and want to exercise that authority even to outsiders (teacher interview, 30 Apr 2019).

One of the female teachers commented that "boys from such households have a tendency of being rude to male teachers mainly because of the anger they are harbouring against their fathers" (Teacher interview, 22 May 2019). Parents stated that some female single parents are "scared of their children because of violent behaviour displayed by their children, and, in such instances, it becomes very hard for the parent to discipline the child" (Parent interview, 14 Jul 2019).

Teachers believed that learners were violent because there was no love in the families because there were no parents. Teachers echoed that the prevalence of HIV/AIDS caused people to die at a young age, leaving their children behind to be taken care of by their grandparents. One teacher stated that "the majority of parents and grandparents are semi-literate. When it comes to assisting learners with homework and other school-related matters, it becomes difficult for those learners who are growing up without parents" (Teacher interview, 25 June 2019).

Another teacher added that "because of the learners' eagerness in wanting to fit [in] and for acceptance in certain groups, they end up misbehaving" (Teacher interview, 11 Jul 2019). Parents stated that "peer pressure causes learners to want to show off to their peers, and, in the process, they lose respect and become unruly, and this leads to conflict. Their environment shapes them to become the people that they are" (Parent interview, 3 Jul 2019).

Learners' exposure to domestic violence caused them to be violent. According to teachers, "there is no peace in families because of domestic violence, and children from such families are made to believe that violence is the only way of resolving conflict because their parents lack problem-solving skills" (Teacher interview, 14 May 2019). Because of the violence learners experience at home, they tended to be violent toward other learners and teachers. According to stakeholders, "some learners beat their parents that makes it easier for them to beat teachers as well" (Stakeholder interview, 2 Jul 2019). A stakeholder further remarked, "Mbambangwe is no exception to other schools; the violence that we see is also common to other schools in the area." He further stated, "sometimes children do these things because they are in adolescent stage."

24.8.4 Enabling Causes

The following sub-sections present the causes that are the enablers of school violence at MHS. These include the communication gap between stakeholders, drug and substance abuse and division among teachers,

On the communication gap between stakeholders, one of the teachers observed that:

> There is a communication gap between stakeholders, that is, the learners, parents, teachers and other community members. [The] communication gap between stakeholders causes the school community to pull in different directions because of lack of understanding what the other party is doing (Teachers' focus group discussion, 5 Jul 2019).

The measures that the schools use for discipline were what the parents and the SGB have agreed upon. It was also reported that some parents did not attend parents' meetings when they were invited, this led to a lack of understanding on what forms of discipline the parents who had attended the parents' meetings had agreed upon. Parents who did not attend parents' meetings went against what had been agreed upon in meetings, and this caused conflict between such parents, learners, the SGB, teachers and the principal. Sometimes the conflict escalated to the community, and the SGB members became targets and their lives were put at risk. According to another teacher:

> Some parents are not cooperative; they always take the side of the learners against the school policies without verifying the information. They don't come to parents' meetings when they are invited and are always against resolutions that are taken in meetings (Teacher interview, 29 Apr 2019).

Parents' meetings created a platform for communication between the school community. Parents don't attend parents' meeting even when they are invited" (Teacher interview, 21 May 2019). This caused a problem between the parents and the school because the decisions taken at parents' meetings were binding on all parents. When those decisions were implemented, the parents who did not come to the meetings became rebellious toward the principal, staff and SGB. As a result of not attending parents' meetings, one of the parents felt that the school starting time (6 A.M.) was too early for learners, even though this was something that had been agreed upon at a parents' meeting with other stakeholders. She further stated that even though this had dramatically improved the matric results, it made learners vulnerable to attacks by thugs in the early hours of the morning.

Regarding drug and substance abuse, sadly, drugs and alcohol have become a way of life for most people in this community. Some use drugs because they want to escape from the reality of their socioeconomic conditions for a short while. Sometimes people become involved with drugs because they view drugs as a quick way of making money. It is unfortunate that when some learners finish matric, there is nothing else to do because families are poverty-stricken, except for the possibility of selling drugs. High rates of illiteracy, poverty and unemployment mean that some people resort to selling drugs and alcohol for survival. Hence, the level of drug and alcohol abuse is very high for some parents. "Alcoholic parents lack discipline and tend to have wrong parenting styles" (Parent interview, 17 May 2019). According to teachers, "such parents, sometimes when they are called to intervene when their children have done wrong, they fight with the teachers" (Teacher interview, 14 June 2019).

One of the parents stated that the learners had easy access to drugs and alcohol, sometimes because of their parents. It was surprising to realise the degree to which drug and substance abuse were prevalent among learners, even among girls. On the other hand, while collecting data, on a number of occasions, policemen would come to arrest learners who had been found carrying drugs at school, and, in one such case, a female learner had been found carrying drugs. According to the teachers, such girls were being socialised by their fathers to become involved in drugs.

According to one teacher, most of the drug users came from informal settlements. One stakeholder commented that "we have taverns around the school premises, and they are easily accessible by learners as well" (Teacher interview, 10 Jul 2019). As a result of easy access to taverns, some learners came to school drunk. The rate of drug and substance abuse was very high. One of the traditional leaders stated that "the government sold us out when they allowed the foreign nationals into our country. These foreigners sell drugs and use our children for drug trafficking" (Traditional Leader interview, 11 June 2019).

Concerning division among teachers, the SGB did not have the power to correct such corruption because the Department of Basic Education was not sufficiently supportive. One of the parents alluded to the following fact:

> Teachers' unions also contribute to lack of governance because they favour the teachers who are affiliated with them. Some union organisations have bigger say in allocation of teacher

positions, and this causes a lot of conflict amongst teachers and SGB (Parent interview, 14 May 2019).

At the end of various narratives collected, participants made recommendations regarding the way school violence could be prevented and ended. These are presented in the next section.

24.8.5 Stakeholders' Contribution to New Interventions

All stakeholders (the councillors, school principals, parents, teachers, business community, traditional leaders, and church leaders) agreed that they were prepared to participate in any activities that would help reduce school violence. Some teachers volunteered to ensure that life coaches were brought to their school to give motivational talks to learners. Some teachers offered to avail themselves to make stakeholders aware that violence is destructive to the image of the school and to the teachers' and learners' morale. They also mentioned that it was important to understand that people need to live in harmony with one another, and were diverse individuals that need to accept others irrespective of that diversity. One of the teachers stated:

> I can avail myself at least two to three times a week to contribute in teaching life skills to learners. Since love is lacking in families, we can contribute by giving learners the love they need. We are so focused on our families, but we also need to be concerned with what is going on in communities (Teacher interview, 29 Apr 2019).

A teacher's response was:

> I am a teacher, but I am also a parent. As a parent, I want to show learners the negative effects of school violence. I can have one-on-one sessions with the problematic learners. With such engagements, I will be able to identify the root cause of the problem and help to resolve it (Teacher interview, 29 Apr 2019).

Another teacher stated that "I can have seminars with learners every Thursday to teach them about violence and how to avoid and overcome it" (Teacher interview, 22 May 2019). One of the parents stated:

> As parents, we should be working hand-in-hand with the teachers, and we should be more interested about our children's performance at school. I feel we are overburdening teachers with our children. We need to build a strong relationship with the school. We also need to get closer to those children who don't [have] parents (Parent interview, 14 May 2019).

Parents' visibility in schools would cause children to be reluctant to act wrongfully. It would make children realise that the parents were also part of the school community. One of the teachers stated that:

> I will always ascertain that when learners are in my hands, they are safe physically and academically. If the community is part of the school, there could be opportunities created for them through outreach programmes like the introduction of Adult Basic Education and Training (ABET) for those who are illiterate and semi-literate (Teacher interview, 5 Jul 2019).

Another teacher stated that "dealing with learners individually helps. I can deal with parents and learners and sometimes teachers" (Teacher interview, 2 May 2019). One of the parents stated that "I can contribute financially, with my time and in whatever way" (Parent interview, 14 May 2019). Some teachers committed themselves as follows: "we can call community meetings and convince people about peace" (Teacher interview, 22 May 2019). Another teachers stated that:

> [They could] take a leading role to motivate other teachers about the importance of teaching and going to classes. We also commit to motivate learners about being punctual, doing their homework with or without teachers, help in arbitration, mediation in disputes that arise in school, training of parents on school code of conduct and South Africans School Act (SASA), [and] networking with other neighbouring schools with good results to promote team teaching within and outside the school (Teacher interview, 30 Apr 2019).

Some parents stated that they would love to be part of any initiative taken to reduce violence at school:

> I am a community leader, former schoolteacher, former school principal, headed the province in the Department of Education, [and] as human beings, we don't stop learning more so as leaders; therefore, I am prepared to learn new ways of reducing school violence. I am also willing to impart the knowledge I have learned to others (Interview with one parent, 3 Jul 2019).

The local councilor alluded to the fact that:

> As political leadership, the first thing to do is service delivery, and whatever issues affect people at grass root level, we have to address it. We need to focus on infrastructural development, and we need a structured relationship with the traditional leadership. An example of this is when we have a football team with the youngsters, the traditional leadership should focus on grooming them to manhood. We have to revive street committees through councilors in order to bring back the culture of playing sports in communities. Small businesses should ensure that they support and assist the formation of these teams (Stakeholder/one councilor, interview, 16 May 2019).

The section that follows presents various views of participants insofar as the consequences of school violence are concerned.

24.9 Effects of School Violence

Almost all teachers felt that there was nothing good about violence. However, some teachers stated that one good aspect about school violence was that it united people with the same ideas, and, because people did not want violence, they united and fought against it, and the police were always visible. Another view was that:

> The positive side of school violence is that when learners are suspended from school, they are required to bring their parents, and we get to know more about the learner, their background and the family background. We also engage with the parents to try and correct the behaviour of a child (Teacher interview, 30 Apr 2019).

What was positive about school violence was that it brought awareness of the type of learners concerned. Even though school violence was perceived as something unfavourable, there were positive results of violence. Firstly:

> [School violence] opens our eyes as the school community that we need to understand each other, our differences, learn to tolerate each other. I see it as something that can bring positive change and causes us to adapt to good living standards. It can also teach individuals to avoid mistakes of the past (Teacher interview, 5 Jul 2019).

Even though becoming acquainted with each other and learning to be tolerant of one another should not be something that happens as an aftermath of violence, when there was violence, more attention was given to the school, and one had an opportunity to learn about conflict resolution skills.

According to teachers, once violence had started, it lasted several months. The negative side of this was that during that period, learners did not come to school, and this affected the learning and teaching schedule. According to one teacher:

> Some learners get demotivated, and when schools re-open, they don't go back to school. Female learners fall pregnant, and some learners end up identifying themselves with the gangs or thugs for protection. Everyone lives in fear because they don't know what is going to happen (Teacher interview, 29 Apr 2019).

The teachers feared coming to school, learners feared that they might be killed, and the parents feared that their children might be killed. According to one of the teachers, "the investor confidence in the school goes down" (Teacher interview, 5 Jul 2019). He further stated that:

> When requesting funds from neighbouring companies for the improvement of the academic performance for learners, the potential funders turned us down because they believe that it is a high risk to be investing in a school with high levels of violence.

The consequences of school violence have been presented quite extensively throughout this report.

24.9.1 How School Violence Affects Teachers

Teachers felt that when school violence disrupted their equilibrium, it was then difficult for them to impart the content to learners. This was due to the fact that they felt unsafe and fearful in engaging with learners. According to another teacher:

> Previously, when there was violence, it would take longer periods, and classes would be suspended, and that would mean that teachers are unable to come to school and do their work. Whenever there is violence, there is no trust between teachers and learners. Teachers are scared to allocate lower marks to learners because of the consequences of being attacked physically by learners because the learners would accuse teachers of giving them failing marks (Teacher interview, 29 Apr 2019).

Teachers further echoed the sentiments that school violence left scars that were very difficult to heal. It delayed any beneficial process, particularly in teaching and learning. Another view of the teachers was reflected in the following statement:

> Whenever there is violence, teachers get traumatised because violence is unexpected. Teachers are demotivated, fearful and their safety is not guaranteed. School violence promotes ill-discipline on teachers and laziness and high rate of absenteeism, take long leave from work because they avoid going to school, they die of stress, and they become enemies of the community (Interview with a teacher, 30 Apr 2019).

Teachers were not proud to be associated with MHS as violence was prevalent and they had a fear of being embarrassed. According to some teachers, "colleagues from other schools were undermining us because of poor academic performance" (Teacher interview, 2 May 2019). One teacher related an incident when she was having a discussion with colleagues from other schools and stated that the pass rate at Mbambangwe would improve in 2019. The response from the other teachers was "you mean Mbambangwe, Mbambangwe" (Teacher interview, 2 May 2019), which indicated that it was impossible for Mbambangwe to improve their results because of the high level of school violence. The discussion became a joke as Mbambangwe teachers were mocked by their colleagues. "No one had confidence in our school, including the Department of Education, parents, and the community" (Teacher interview, 24 May 2019). Teachers further stated that school violence affected the curriculum coverage.

Teachers emphasised that "it is uncomfortable for teachers to be in classrooms with learners when there is school violence. It is not easy for teachers to drive around the community, and this makes it difficult for them to come to school" (Teacher interview, 2 May 2019).

Another comment by a teacher was "we are teachers, but we are also parents, so when learners get hurt or killed, we are heartbroken as well" (Teacher interview, 24 May 2019). Teachers felt that "when there is violence at school, people believe that the school management is failing to govern the school, and this tarnishes the image of the school" (Teacher interview, 30 Apr 2019). One of the parents commented that "we feel sorry for the teachers because there is shame and embarrassment on teachers and management; this takes away their pride and dignity" (SGB interview, 29 Apr 2019). Whenever there was violence at school, teachers were not assertive enough because they were scared.

24.9.2 How School Violence Affects Learners

Teachers stated that "low morale affects learners' academic performance; this also reduces cooperation between teachers and learners" (Teacher interview, 22 May 2019). It was indicated that "learners who are the cause of violence become victorious since their main aim is to disturb learning and teaching" (Teacher interview, 22 May 2019). They further stated that learners who were committed to learning became worried and stressed since they were concerned about their education, and this led to confrontation with other learners. Whenever there was school violence, classes would be suspended and learners would not be able to go to school. According to parents, teachers and community members, the matric pass rate was very low to the extent that there would not be a single learner who would receive a matric pass. The number of learners dropping out of school increased.

Sometimes, learners who became engaged in violent action were killed, and some were arrested and served time in prison, while others stopped attending school. One parent confirmed that learners had died as a result of school violence. One of the teachers stated that "it is very traumatic to witness a learner dying before you," and this was what had happened previously (Teacher interview, 30 Apr 2019). Sometimes, when learners were at home because of school violence, some learners fell pregnant since they spent time being idle, and some did not go back to school because they had fallen pregnant or because they were discouraged. This was a loss to learners because they might have to repeat their grades, and this became more of a financial burden to the parents. Sometimes the school was vandalised when classes were suspended for long periods, and the school was in danger of being completely shut down. Teachers added that "the prospects for the learners' future is not certain. Learners lose purpose and sense of direction, and their future looks bleak" (Teacher interview, 5 Jul 2019). Teachers expressed several views, including what is covered that this statement:

> Because of school violence, learners get wrong role models, and the cycle of poverty is repeated. Sometimes learners have to identify and associate with the thugs for safety and protection. They become school dropouts, the pass rate drops, learners get into drugs, join gangs and become community misfits.

Violence in schools is a significant problem to the extent that some learners did not recover from the effects.

24.9.3 How School Violence Affects the Community

Teachers observed that "whenever there is school violence, there is loss of confidence between the school, teachers and community members. Because of loss of confidence in the school, enrolment drops as some parents move their children to other schools" (Teacher interview, 22 May 2019). One of the stakeholders added that "when learners are taken to other schools, the community cannot use the school as a resource that is of benefit to them, and this becomes an economic burden to the parents because of the travelling costs involved" (Teacher interview, 29 Apr 2019). An SGB member alluded to the fact that "when learners die due to school violence, and when the parents have to bury their children who die as a result of violence, this also becomes an economic burden to the family" (SGB interview, 29 Apr 2019). According to teachers, "school violence doesn't affect local communities only but the whole nation is affected because school violence produces school dropouts to the society who become an economic burden to the state" (Teacher interview, 22 May 2019).

According to teachers, "parents become worried because the main intention is to have proper education for their children, but whenever there is school violence, parents get worried about their children's safety" (Teacher interview, 30 Apr 2019). An SGB member argued that "when school violence is prolonged, outsiders get involved, leading to no-go areas in our community, and hatred is created in learners and community members" (SGB interview, 21 June 2019). Another SGB member at added that "school violence creates violence in the community as parents take the sides of their children" (SGB interview, 29 May 2019). Teachers stated that "family poverty is prolonged because the parents invest in their children by taking them to school; they believe that their children will have a better future than their parents" (Teacher interview, 5 July 2019). A stakeholder member of the local council commented:

> Whenever there is school violence, the community members live in fear of the unknown; it promotes a cycle of poverty, it promotes a community dependence on drugs, it promotes a divided community without guidance in their children because they are afraid of their children's actions who are gang members (stakeholder interview, 9 June 2019).

One parent stated that "some parents intervene through influencing their children to fight more, and they even provide their children with dangerous weapons" (Parent interview, 14 May 2019). A stakeholder opined that "business people are out of business and out of income" (Stakeholder interview, 21 May 2019). Some parents echoed that "sometimes the learners end up losing respect for their parents and adults because of anger and frustration caused by violence. The loss of respect in children causes pain to parents" (Parent interview, 14 May 2019).

24.10 Response to School Violence

The following sub-sections involve the common responses from teachers and learners whenever school violence took place.

24.10.1 Suspension and Expulsion of Learners from School

Most of the school dropouts were learners who had been expelled from school because of unruly behaviour. According to teachers, the school dropouts could be seen standing on street corners and being idle on a daily basis at any time of the day. Zero tolerance, expulsion, and suspension were presented as last resort measures and a way of handling unruly learners who would not reform their behaviours. Teachers confirmed that the learners who are dropouts are a problem for their schools and the community. Because they were out of school, out of employment, not involved in training and idle, they were vulnerable to being used for drug trafficking and criminal activities.

Learners who were expelled from school came to school during break time and stood outside the school premises to sell drugs, rob and attack learners who were still at school. Because the learners were scared of them, they sometimes voluntarily gave away their possessions, such as money and cellphones. While conducting interviews, the researchers learned that some learners were robbed of their possessions on their way to school. The school community identified the thieves as former learners who had been previously expelled.

Sometimes, when learners were asked to bring their parents to school when there was a case, they dragged anyone from the street to come and represent them. The problem with suspension was that learners were deprived of progressing with their schoolwork.

24.10.2 Response of Learners When There is Violence

According to one of the teachers, "the reaction of other learners creates an attitude that conflict can be resolved through violence whereas there are other options" (Teacher interview, 25 June 2019). One teacher noted:

> Some learners sympathise with those who are part of the action and try to intervene by calling the class representatives who would call the teachers. Sometimes, violence is perpetuated through encouragement by the bystanders' noise and other learners taking sides. When learners take sides, it is sometimes according to the areas where they come from relations and friendship (Teacher interview, 2 May 2019).

Additionally, another teacher observed:

> When learners take sides during the fight, this ends up dividing the class, and this becomes a problem when learners are given a task in class. Because of the enmity created among the learners, group work becomes a problem because they don't want to work with those that are perceived to be their enemies (Teacher interview, 30 Apr 2019).

Teachers echoed that "whenever there is violence, some learners would be absent from school because of [the] fear of being attacked, and others would end up dropping out of school" (Teacher interview, 22 May 2019).

We also attempted to find out which characteristics are presented by learners who show some tendencies to commit violent actions. The sections that follows describe them.

24.10.3 Characteristics of Learners Who Are the Most Likely to be Violent

There were conflicting views by the participants on the learners who were most likely to be engaged in violence. One of the stakeholders felt that "violent learners are from all walks of life. They can be from poor background and well-off families. Mostly learners from single-parent household are violent, and most of the single-parent households are headed by females" (Stakeholder interview, 2 Jul 2019). The observation by teachers was that some of these learners came from single-parent households, "others from poor household, others come from child-headed households" (Teacher interview, 29 Apr 2019). There was a consensus among the teachers that most of the perpetrators were boys, and the girls became the victims. Even though the younger boys were also victims, the majority of the victims were female learners with 47 out of 116 learners stating that boys were more violent than girls (Learners' survey, 10 June 2019).

Learners who were likely to engage in violence were those who are not performing well academically and tended to disturb others; in some cases, it was the older boys who were violent toward the younger ones. Most of the learners who were perpetrators were those who were behind in their grades because they were older and more powerful than the learners in their grade, and the younger ones feared them. Some of the boys who were more violent were those without any father figure in their families. According to one teacher:

> Sometimes the violence is directed to teachers, especially female teachers, even though some of the boys direct their anger to male teachers. They do this because they carry a lot of anger against their fathers who are absent in their lives and become angry against any adult male. Sometimes these learners do this because they are used to being over-protective of their families in the absence [of] adult males (Teacher interview, 2 May 2019).

Parents stated that "learners who come from families where parents are alcoholics, where there is violence and with absent parents, especially fathers, are also in the category of violent behaviour" (Parent interview, 3 Jul 2019). Perpetrators of school violence could be males from the community; sometimes it was boys who had been expelled from school. Previously, violence used to be a daily occurrence, sometimes throughout the day, and, in some instances, it would last for months (Parent interview, 3 Jul 2019).

24.10.4 Action Taken in the Past to Reduce School Violence

Various statements were taken that refer to actions taken in the past to minimise school violence. They are described by participants in what is recorded here.

According to one teacher:

> What has been done in the past is to keep learners occupied all the time using class representatives to report any incidents of violence taking place in classes. As teachers, we believe that learners are able to articulate their views better if there are class representatives. This has helped to keep learners busy all the time (Teacher interview, 2 May 2019).

This was the reason why most learners stated that the Learners' Representative Council (LRC) members were most likely to be less violent. When discussing the behaviour of the LRC members with the president and vice president of the LRC, they believed that when they were given a position, they felt the need to be positive role models for other learners. Some of them who had not been on their best behaviour before they attained those positions started changing their behaviour once they were given the positions.

The traditional leaders, community members and parents had been involved in the past to resolve school violence. The then Ethekwini mayor in 2016 was asked to adopt the school into her foundation such that she would be understand the needs of the school. According to the stakeholders:

> This had positive results since the mayor refurbished the school and made other donations, like fencing, to keep criminal elements away from school premises and to make it more attractive to the learners and the community. A security guard was hired to ensure that no criminal elements come inside the school (SGB interview, 21 June 2019).

One of the teachers mentioned that "we motivate learners and show them the negative effects of school violence." According to teachers, learners with serious drug problems would be sent to rehabilitation institutions. Community structures were established that were mainly focused on violence.

The Department of Basic Education was positive in dealing with the matter and gave the school the support they needed. The tribal authority and the South African Police Services (SAPS) played a pivotal role in reducing school violence. The tribal authority provided the school with warriors who helped by reprimanding anyone who was getting out of line through their visibility at the school and the SAPS through implementation of random searches. The presence of surveillance cameras and armed guards were presented as additional violence preventative measures to curb the recurrence of antisocial acts on school premises. Because learners feared the warriors, they started behaving appropriately. In 2016, new management took over. Teachers stated that "presently, the school management is very strict with learner admission; we have to be able to communicate with the parents, and we are trying to broaden the communication scope to the wider community." In addition, one teacher stated:

> Motivational speakers were invited to talk to learners about the risk of not going to school and gangsters. Peace campaigns were organised in the community, popular musicians and sport people were invited, soccer clinics were organised during sports days, sport activities between the school and SAPS were organised, [and] the Ethekwini mayor was invited for motivational talks and donations in order to promote good image for the school. As teachers, we also try to be exemplary to learners for them to see what is expected of them (Teacher interview, 2 May 2019).

One of the traditional leaders commented:

> Learners were sent back home to bring back their parents because we wanted to understand why the learners were engaging in violence. The strategy that we used was to scare learners for them to stop violence. We told learners that if they want war, they must come and face us instead of being violent to other learners and teachers. That is what stopped the violence in this school. The last incident was when a learner was killed when there were interschool cultural activities (Interview with one traditional leader, 27 June 2019).

According to another stakeholders:

> After a learner was killed, the community had to look for solutions to stop revenge from the rival group. In order to stop revenge from the side of the deceased learner, the families of the perpetrators contributed R200 each that went towards the funeral costs (Stakeholder interview, 21 May 2019).

A stakeholder further stated that stakeholders (parents, teachers, traditional leadership, learners, ministers of religion, and ward councilors) were invited to meetings that had not been fruitful until traditional leadership threatened learners that they would beat them if they ever became violent.

> This is what we are used to. We grew up with intimidation, and it is part of who we are, our culture. It works, but it is not sustainable (Stakeholder interview, 21 May 2019).

Following various statements regarding actions taken in the past to respond to violence at the same high school, we also engaged participants to find out what consequences disciplinary actions had on learners. We report their answers in the next section.

24.10.5 Consequences that Disciplinary Action Had on Learners

According to the teachers, there were many positive results from the suspension and transfer of learners. Firstly, the learners did not lose out on their education. When learners were transferred to other schools and suspended for a few days, this brought peace and harmony within the school because the rate of violence decreased. When learners were transferred to other schools, they were removed from their familiar environment, and because they were not familiar with the new environment, it was difficult for them to maintain their violent behaviour. When they were suspended, the parents were called in to discuss the matter with the school principal, and this action sometimes led them to change their behaviour for the better. However, sometimes the expulsion created an impression on other learners that since they had been expelled from school, they were comfortable and could do as they pleased since they were no longer at school. However, it was also a wake-up call for other learners in the sense that they might find themselves in the same position as those who were expelled or suspended if they got themselves into trouble.

The cleanliness of the school and the matric pass rate had improved. According to teachers, parents, the SGB, learners and stakeholders, the introduction of new management had brought about a reduction in the level of violence. They had been able to restore discipline and order inside the school.

The same sentiment was shared by parents and other community stakeholders. Parents further stated that this had impacted the matric results where they had witnessed dramatic improvement. In the previous years, the matric results had been poor to the extent that there was not a single learner obtaining a matric pass. According to some parents and teachers, "since 2018, there has been a huge change in terms of discipline and behaviour of learners" (Teacher interview, 30 April 2019, and Parent interview, 14 May 2019). The campaigns and the involvement of the CPF, community and traditional leadership and Umbumbulu Education Forum had brought some relief to the school. According to the traditional leadership, "ever since the traditional leadership and *izinduna* intervened, there has never been violence; learners respect teachers, and the matric results have improved" (Stakeholder interview, 21 May 2019).

Learners were now respectful of teachers, there was no longer fighting as before, smoking and drug abuse had decreased, learners were always punctual for school (the school started at 6:00 am and ended at 4:00 pm), learners had enough time for homework, and parental involvement had improved (they attended meetings for monitoring learners' schoolwork). However, there were still those learners who went out of their way to conduct illegal activities, such as selling drugs and carrying dangerous weapons to school. Nevertheless, according to the parents, the initiative was a success in the sense that learners were now scared to act wrongfully because they knew that they would be punished.

24.11 Action Research Interventions

From what the participants stated in focus group discussions and interviews, we were able to design a programme relevant to the challenges causing violence. The intervention strategy was informed by what the participants had stated in the collected data. This section explains the step-by-step process that was followed when implementing the intervention. It also explains why the programme was introduced to Grade 9 learners in 2019 and why the implementation process was carried out in the following year in 2020 for Grade 10 learners and 2021 for Grade 11 learners. Teachers' role in the implementation process is also explained.

24.11.1 Implementation of the Intervention

The need for peace education has become even greater in South Africa because of the lack of skills, such as those of conflict management, conflict resolution, negotiation and mediation. South Africa needs restorative practices to bring back people's dignity that has been damaged by the previous regimes. South Africans need to move away from retributive practices because these degrade human dignity and go hand-in-hand with the structural violence that was part of the previous government policy. With the introduction and implementation of the intervention programme, the study intended and hoped to restore learners' dignity by making them realise how important they are. Learners must be affirmed and know that they are valued, appreciated, needed, loved and cared for so that they can also learn to act in the same manner toward others. It was believed that for South Africa to heal, the nation requires processes such as conflict transformation and restorative approaches to rebuild relationships, restore human dignity, empower human beings and discourage violent strategies of resolving conflict.

The older South African generation, including parents and teachers, have been victims of the past regime where they were made to believe that authoritarianism and control were the only alternative of restoring law and order. It is most likely that such beliefs become transgenerational if not challenged. Therefore, it is important for South Africans to explore alternative means of dealing with conflict. Since people are not inherently violent, the suggestion is, therefore, that there is a possibility for parents, teachers, learners, and the community to unlearn what has been learned in the past and relearn new ways of resolving conflict. The programme that was introduced to learners, therefore, discourages violent ways of resolving conflict and instead focuses on rebuilding relationships, restoring human dignity and the empowerment of individuals. The intervention involved two components – peace-making circles and transformative learning based on the Caregivers' Model. Appendix 1 provides a summary of the intervention carried out, including the purposes, the date(s), and the duration of the intervention.

The overall aim of the intervention was healing or peace-making circles where the conventional disciplinary or punitive approaches to reduce school violence had failed. The implementation of the programme was divided into two parts, the morning peace-making circles and the transformative learning during the Life Orientation (LO) period. On Mondays and Fridays, there is a morning assembly where all learners from Grades 8 to 12 come together to pray with their teachers. On Tuesdays, Wednesdays and Thursdays, the class teachers conduct morning prayers in their respective classrooms for 15 minutes. We believed that there is time to pray and that there is time to relate. The research team suggested that morning prayers in the classrooms from Tuesdays to Thursdays be changed to peace-making circles in classrooms to set the tone for the day and to build relationships.

24.11.2 Intervention: Peace-Making Circles

Peace-making circles were conducted to facilitate emotional healing and build healthy relationships amongst learners and between learners and their teachers. Conversations with teachers have helped children to develop new pathways that will lead to new habits (Chiramba/Harris, 2020: 6). In addition to promoting emotional healing, the purpose was to teach learners new ways of dealing with emotional issues. The research team started with the peace-making circles on 18 February 2020. The research team introduced itself to the learners and introduced the peace-making circles and the procedure to be followed in engagements. The circles were conducted during the time (15 minutes) allocated for morning prayers in classrooms. Peace-making circles were conducted on Tuesdays in Grade 10A and on Wednesdays in Grade 10B. The main researcher led the sessions in the presence of the class teachers. The reasoning behind involving the class teachers was for them to gain knowledge on the issues that affect learners and for them to be in a position to lead the sessions without the help of the research team.

In these sessions, the research team emphasised the importance of speaking about how one feels and being able to be in touch with one's innermost feelings and emotions. Learning to identify how one feels and the reasons for feeling the way that one does are important. It teaches one to learn to deal with negative or undesirable feelings correctly and also to learn to let go of the undesirable emotions and feelings. Having negative or undesirable feelings and not knowing how to deal with them could lead to stress, anxiety, frustration and anger, which could lead to aggressiveness and violent behaviour. According to Bartsch (1996: 28):

> Healing from the emotional pain starts with talking because if we don't talk, the negative emotions accumulate causing us not to cope which can lead to spill over. When these emotions spill over, those who are around us get affected by the effects of the spill over.

The other reason for conducting peace-making circles is for children to learn to be a resource for one another, that is, learning that they can rely on each other as classmates and on their teachers for support, and also learning to support each other. In the circles, learners learned about the importance of talking about their feelings in order to avoid stress and traumatic stress that, in most cases, leads to spillovers and affects other people negatively (Bartsch, 1996: 28). The circles cause learners to begin to know that they are valued, and this leads them to have healthy relationships with themselves and with others. If learners do not value and appreciate themselves, it becomes very difficult to value and appreciate others because human beings cannot give what they do not have. Lacking self-regard could lead to dysfunctional human beings. Classroom engagement taught learners to trust others and to deal with their emotions constructively. It also made them aware of what their peers were going through and allowed for opportunities to show empathy and sympathy to others.

The importance of confidentiality was emphasised where learners had to keep whatever was shared confidential in order to protect each of them. It was highlighted that people should always take care of others by keeping their secrets within their sacred space and not share someone else's story without their permission. Hence, one of the ground rules for the circles was confidentiality. In the circles, the participants learned to relate to other people in healthy ways and learned to respect and value others on the basis that they are all human beings, they are worthy of respect, and they are a resource for all. They learned the importance of listening to others and not interrupting other people when talking because they would also want to be listened to. People want others to feel valued, important, and needed in their lives since people are who they are because of others around them or next to them.

The outcomes of the interventions were evaluated after the intervention by collecting data from the same participants to establish where participation in peace-making circles influenced in learners. Several recommendations were made by teachers and learners; they merged with strategies put forward by other stakeholders involved in this inquiry. The section that follows captures emerging approaches that participants found effective to reduce school violence.

24.12 Emerging Strategies to Minimise School Violence

In this section, we present a series of recommendations that were made by various participants in this inquiry. Following the conclusion, we will present our recommendations to develop policies that could eventually curb the scale of violence on learning environments.

24.12.1 Recommendations by Stakeholders on Reducing Violence

Almost all the teachers felt that the best action that needs to be taken is to educate teachers, learners, parents and different community structures regarding the negative impact of school violence. They further stated that they also needed campaigns on bullying and violence in schools. Teachers also felt that any solution to resolve violence should start in families. Parents should be empowered with parenting skills. Problem-solving skills should be taught at home. Teachers at echoed the fact that "LO must be taught by teachers who are qualified to do so. Learning programmes are rigid"; "as teachers, we don't have enough time to engage about life skills, and this is such an important aspect in learners' lives" (Teacher interview, 22 May 2019). Positive role models (religious and sports) must be allowed to visit the school frequently to encourage learners. Learners also need to be encouraged to spend more time in sports.

Teachers felt that strict laws must be applied to schools, and security measures must be taken, such as having a security guard at school and the presence of the SAPS. One of the teachers mentioned that "random checks must be done, and learners must be checked whether they are still mentally functional because some of them have been using drugs for a very long time, and this could have affected their state of mind" (Teacher interview, 2 May 2019). The school must be fenced to keep thugs and outsiders away from the school environment. One of the teachers argued that "I think we need things like Imbizo and invite professionals to talk about the effects of violence" (Teacher interview, 30 Apr 2019). One of the stakeholders recommended that:

> We need to revive recreational activities in our communities to keep children and young people busy. Sport associations have strategies to identify talents in learners, let them go to schools and start identifying these talents in learners while they are still in high schools. It must be compulsory for learners to be engaged in sport activities, and there must be competitions organised seasonally (Stakeholder interview, 21 May 2019).

He further stated that if learners are engaged in sports, they would learn to understand others, become accustomed to working in teams, and be able to focus; their stress levels would be reduced, and they would have less time to think about their misery and, instead, become disciplined.

24.12.2 Collaboration and Communication Between Stakeholders

Collaboration among stakeholders was one of the most common recommendations by the stakeholders. School violence can be reduced if there is communication between stakeholders, that is, the community, teachers, parents and learners. One traditional leader's advice was that "let's sit down and talk and see what we can do. We cannot expect everything from the government. Let us stop destroying the little that we have" (Stakeholder interview, 14 June 2019). Stakeholders such as the SGB, CPF, teachers, parents, and various government departments must come together to work on the solution and cascade information onto the learners. The learners should be allowed to talk as peers about the dangers of violence.

24.12.3 Police Intervention

The common response from the community, teachers and parents was to strengthen security in schools. The policing of schools was a permanent need of the community. Some teachers recommended workshops for teachers and parents regarding the code of conduct and workshops on the SASA regarding what is expected from them. Teachers recommended that "morning assemblies must be held daily to cultivate a culture of discipline so that teaching and learning can be conducive" (Teacher interview, 5 Jul 2019).

24.12.4 Department of Basic Education

One of the teachers echoed that "the Department of Basic Education has much to do in terms of arresting violence". In concurring with this teacher, one of the traditional leaders voiced his frustration that "children have negative attitudes, and we are not allowed to use corporal punishment because they have rights. But if we could be allowed to use corporal punishment, these children can change for the better" (Stakeholder interview, 14 June 2019). Another teacher opined that:

> Love is lacking in our schools and communities. Before teachers teach learners, they must love them. If we want to plant love in our children, we must do a thorough spadework for the seed to have enough ventilation to grow. We first affirm an individual before reprimanding them (Teacher interview, 2 May 2019).

One stakeholders suggested what follows:

> We need to build forums where people are able to talk about issues affecting them in communities. We have to revive formations in communities [and] talk about our values and principles. If we do not have families in our communities, there would not be any values in schools (Stakeholder interview, 21 May 2019).

Another recommendation by one of the teachers was: "the government must introduce training programmes for parents and teachers. We need seminars for parents that are both entertaining and educational. We need to talk to artistic learners to write poems and drama that addresses social ills" (Teacher interview, 17 May 2019). Another recommendation by one of the teachers was that "the government must bring back the in-service training for teachers so as to empower them in their leadership role" (Teacher interview, 29 Apr 2019).

Almost all the parents recommended that the Department of Basic Education revisit the prohibition of corporal punishment. "It seemed to work because it worked during our school days" (Parent interview, 14 May 2019). Teachers and parents highly recommended that "the government ... promote sponsorships for recreational activities in schools" (Teacher interview, 29 Apr 2019).

One stakeholder felt that "we need community forums where we can talk about issues of violence" (Stakeholder interview, 14 June 2019). The stakeholder further stated that "the Department of Basic Education must implement more stringent disciplinary measures for learners because right now parents and teachers are helpless when it comes to disciplining learners." A recommendation by one of the teachers was to "let the government revisit the curriculum [to see] whether it is up to date with today's standards, like agriculture and technology subjects" (Teacher interview, 29 Apr 2019).

24.12.5 Psychosocial Support from Department of Social Development

One parent stated that "learners need counselling to cope with all the challenges they face in their homes" (Parent interview, 3 Jul 2019). Another recommendation by teachers was to "let us have more sport activities to keep learners busy". In agreement with the above recommendation, one of the teachers alluded to the fact that:

> Learners are dealing with a lot of issues, and they bottle them up because they don't have anyone to talk to. If DSD can provide our school with a social worker at least once a week, the level of stress in learners can be reduced, leading to reduction of school violence. The teachers are not qualified to deal with psychosocial issues, and when the teachers shift their focus from their core function to deal with other issues, it affects the teachers' work schedule (Teacher interview, 5 Jul 2019).

Almost all teachers recommended that "psychosocial support is needed from the Department of Social Development (DSD) to assist learners [to] cope with whatever issues they are going through or have gone through in their past experiences" (Teacher interview, 5 Jul 2019).

24.12.6 Parents and Community to be Part of the Running of the School

Almost all teachers agreed that they needed support from the parents and the community because they could not improve the learners' behaviour by themselves. Teachers further stated that some parents need parenting skills, and there is a great need for parents to be educated on how to groom their children for the future. Some teachers and parents were in agreement that "we need to set up structures where we can engage parents, teachers, and learners" (Teacher interview, 2 May 2019). "We must have forums for engagements, motivational speakers to encourage parents to understand that it is important to speak the same language" (Teacher interview, 29 Apr 2019). Parents should not distance themselves from school, and they must attend parents' meetings so that they can speak in one voice. Parents must have an interest in visiting the school more often to give support to the teachers and to show that they care about their children's future. One of the SGB members stated that parents' visibility in schools could make a big difference in terms of the learners' behaviour.

24.12.7 Motivational Speakers to be Brought to School to Motivate Learners

Some teachers felt that since learners lacked role models, they needed to be motivated by people whom they admired or those who were successful in life, such as well-known sports men and women and actors. They also believed that learners could benefit substantially from having life coaches and mentorship. The teachers felt that some of the male learners who were neglected by their fathers could benefit greatly from the mentorship programme. Life coaches were needed in communities to engage learners in sports and other activities that could help them to deal with the stress and traumas of their lives. Getting young people engaged in sports and other activities could also reduce the level of drug abuse in the community. "Motivational speakers will steer up learners' attitude because most of the learners are demotivated" (Teachers' focus group discussion, 5 July 2019) – this was the sentiment echoed by almost all the teachers.

24.12.8 Sports and Entertainment

"Let us have more recreational activities" – this was a recommendation by one of the stakeholders (Stakeholder interview, 14 June 2019). All stakeholders agreed that sports, arts, cultural activities and entertainment were missing in MHS and the communities. Therefore, these need to be introduced in schools in order to redirect learners' focus into doing something good. Some teachers echoed the sentiment that learners were stressed and traumatised by their socioeconomic conditions. There was no outlet for the learners' stress, anger, and frustration and they took it out on other people, such as the teachers and other learners. Some learners are heading households, as they do not have parents; some are from single-parent families who are unable to satisfy all their needs; and other learners have parents, but the parents are not part of their lives. Engaging learners in recreational activities could encourage them to forget about their situation for a short while and refocus on something positive. This could strengthen the learners' relationships with each other, their teachers and their parents as well. One stakeholder had an experience when he was the provincial manager in arts, culture and sports and employed by the Department of Basic Education; he observed that "extramural activities, like sports, arts, and cultural activities, enforce discipline in learners." He further stated that:

> In almost all Afrikaans schools within the province, like Kuswag, sports are compulsory in learners, and this is the reason why there is so much discipline in learners from these schools because learners are used to discipline through their engagement in sport (Stakeholder interview, 14 June 2019).

Further information was provided by various stakeholders that were regarded as valuable for novel interventions.

24.13 Discussion

This study set out to empower the school community, that is, the teachers, learners and parents with knowledge and skills on conflict transformation. Different participants had different understanding of what school violence is. Most of the participants understood school violence as learners' disrespect of teachers. The kind of violence they experienced was gender-based violence, bullying, rebellious learners, violence perpetuated by the SGB and the murder of teachers and learners.

It emerged from this inquiry that violence was endemic in schools and affected the wellbeing of many learners and the learning process. Furthermore, the main reasons for violence were identified as social ills like high rate of poverty, unemployment, drug and alcohol abuse, child-headed households and lack of parenting skills. Teachers, parents, learners and stakeholders mentioned bullying, gang violence, faction fights, drug and substance abuse, gender-based violence, the carrying of weapons, corporal punishment, verbal abuse, physical fighting, unruly or ill-disciplined learners, showing signs of aggressiveness toward the teachers, uncommitted teachers, violent parents, vandalism, burglary and unsupportive communities as some of the prevalent forms of school violence.

One of the ways in which the school dealt with violence was through corporal punishment. In some cases, traditional leadership was called in to scare learners, while in some instances police would be called in to intervene and sometimes the police would be stationed around the school premises for long periods.

It came out that the root causes of violence were intertwined, interrelated and interconnected. One issue would lead to another, and if the issues were not resolved or addressed at the appropriate time, it would become difficult to resolve them later in life. An example was the issue of a girl who was found with marijuana at school. Apparently, her father was selling drugs in order to provide food for his family, but he was also involved in using drugs. The girl started using the drugs because it was freely available at home; sometimes, she would sell them to customers when her father was not there. When she was caught with drugs at school, the police were called in. Unfortunately, she did not see anything wrong with what she was doing because she had to do what she was doing in order to provide food for the family. However, this had then put her at risk of being arrested or expelled from school, and her only recourse would be to continue with her father's business of selling drugs.

How they dealt with school violence at the time of the study was through corporal punishment, suspension and expulsion. Sometimes, parents were called in to explain about their children's unbecoming behaviour. On every visit to the school during the data collection phase for this study, there was a parent or parents called in because their children had done something wrong. In some cases, the police were called in to intervene. For example, during one of the visits to the school, the police were there to intervene on drug and substance abuse problems caused by some of the female learners. This presented an interesting finding for the study as, in most cases, drug and substance abuse is associated with the male gender. The principal revealed that, in his experience, female learners' introduction to substance use did not come from peers, but rather from the learner's fathers and father figures.

24.13.1 Reflecting on the Interventions

The outcomes of RJ suggest that schools can be transformed into safer spaces so that learners can benefit from good education and thus, become agents of change, especially if they are provided with peace education and learn addressing their unpleasant and antisocial experiences within peace circles contexts.

This study raised awareness about the extent to which learners desired alternatives to violence or the means for the peaceful resolution of conflict, in the case of the learner who came from a family of warlords, for example. During the peace circles, one learner narrated how much he hated being part of his family and that he could see himself becoming involved in the same actions as his father and uncles. The principal expressed how he felt excited about the ambitions of learners who showed a burning desire to end violence in the school environment.

A learner in Grade 9 shared his ambition to become the head boy in order to fix issues and help the principal put an end to violence. Due to the implementation of peace circles and continued interaction with students, the learner began mobilising fellow students in 2019 to support him as a candidate for the Learner Representative Council as he expressed a desire to transform the school environment into a peaceful space. The goal was to elect this student as the school president. When he was elected, he did precisely what he had said during peace circles that he would do. He had a very close relationship with the principal and other teachers who had much trust in him. He was always very helpful during the study sessions and would encourage others to act in correct ways. The learner shared his ambition in 2019 when the main researcher started collecting data at the school. When we started peace circles, he kept on repeating the same statement and in 2020 and 2021, he became the deputy president of the Learner Representative Council and, in 2022, he became the president.

One of the learners who was his (the president's) classmate in Grade 9 and one of the unruly ones had to repeat Grade 9 because of his conduct. However, through the president's behaviour, he was motivated to change, and had chosen the president as his role model. He also started canvassing to be elected to be the deputy president, and, to the surprise of all, he was voted in; this is because his conduct started changing for the better. The student was convinced that many younger learners would look up to him as the deputy president of the Learner Representative Council and he did not want to disappoint them because of his behaviour. His goal was to become the president when the current president left the school after completing matric. Because he had to repeat Grade 9 in 2020, he is still in Grade 10 and not part of the study, but we organised a special meeting with him because he fervently wanted to update the main researcher about his behavioural change. This experience made us realise that there was much potential in young people to become peacebuilders.

What was notable was the learners' responses during the baseline survey regarding the type of learners who were most unlikely to become involved in violence. The response was that those who were part of the Learner Representative Council were the most unlikely to become involved in violent actions because they knew that others would be looking up to them. It was impressive and uplifting to see how learners started referring to themselves as "peacebuilders" before the research team had not stated anything regarding peacebuilders during the sessions. It became evident that learners were capable of resolving their own issues with the correct type of support and correct mentality, So, learners were not as ill-disciplined as they were perceived to be – they could be very supportive, kind and gentle toward each other.

24.13.2 Restorative Justice and Ubuntu Humanistic Intervention

In this study, it was observed that 'peace circles' formed an essential part of violence intervention methods at schools. Peace circles foster connections through empathy and self-reflection while creating a sense of belonging/community through shared values, storytelling, and active listening (Herrera/McNair, 2020). Teachers can begin to share their stories with the learners and the learners follow the same strategy which humanises the teaching process. Humanising pedagogy promotes respect, trusting relationships between teachers and learners, academic rigour and learning contexts where power is shared by teachers and students (Kajee, 2019). Circles provide everyone with an opportunity to speak and listen to each other, and connect to dialogue, peace and conflict resolution as part of a proactive educative process that builds classroom community, strengthens relationships and infuses equity across curriculum (Parker/Bickmore, 2020).

The feelings developed by several learners and teachers who were exposed to peace circles were imbued with humanism. One of the best conceptualisation of humanism is found in the philosophy of Intersubjectivity of Emmanuel Levinas, the Jewish-French philosopher who was deeply affected by the Jewish Holocaust. In response to the millions and millions of those murdered side-by-side by the National Socialists:

> [...] Levinas returns love for hate, the "wisdom of love", the "humanism of the other," against the "hatred of the other"—a hard love, yes, a "difficult freedom," no doubt, but the subtleties no less than the brutalities of contemporary anti-humanism have never more sophisticated. (Levinas, 2003: viii)

Amidst and against violence that confronts us, humans still can look at the adversities of life with humanity. Violent confrontations and retaliation in schools and the larger society are driven by the lack of forgiveness and humility to tolerate and transcend the ills that affect person-to-person and different social groups' relationships. Restorative justice processes are led to empathy, forgiveness and reconciliation (Kiyala, 2018; 2022). The link between peace, reconciliation and forgiveness is well conveyed by the spirit of Ubuntu (Tutu, 1999). For Archbishop Tutu, "a person with ubuntu" describes the being, not possession of humanness; it is synonymous with generosity, hospitality, friendliness, care and compassion (Tutu, 1999: 34). Ubuntu bonds the person and other humans. Archbishop Tutu expresses this in the following terms: "We belong in a bundle of life. We say, 'a person is a person through other people'. It is not "I think therefore I am'. It articulates rather: 'I am human because I belong.'. I participate, I share" (Tutu, 1999: 35). So, school environment should be ubuntu-pervaded space—an environment where leaners and teachers care for one another, where forgiveness prevails over revenge, peace over violence, justice over injustice and unfairness, reconciliation and restoration over retribution.

Peace is achieved via reconciliation, forgiveness and justice that is imbued with humanism as observed in Levinas (2003). We experienced such feelings in our engagement with learners. Adding to the learning benefits of RJ, Karp and Sacks (2013: 5) argued that learning processes may be greater with restorative practices because student development involved "a holistic enterprise focused on moral concern, citizenship, and emotional intelligence". Peace circles appeared to have shown that.

We noted how peace-making circles touched learners and made them rethink their circumstances and adopt humane attitudes and response to issues that affected them. These involved issues such as family violence; parents not being well; learners' academic performance, such as disappointing test marks; learners failing to achieve the goals that they had set for themselves; and relationships with their peers that needed to be corrected. There were also positive aspects to talk about, such as their academic performance and excitement about how they were able to assist those facing challenges in their academic work. A positive factor was that the teachers had also felt the need to share their feelings, both good and bad. The mere interest of the class teacher in sharing her feelings and emotions humanised the teaching process. It made the learners realise that teachers are also human beings, that they also have feelings, and that learners are not the only ones who are going through unpleasant situations.

Some learners expressed their feelings about not being comfortable sharing their feelings with others. The main researcher clarified that in peace circles, participants should respect each other's feelings. Restorative practice aims at nurturing respect, responsibility and empathy among members of the school community and contributes to the building of supportive communities and school discipline. One of the ways in which people support each other is to respect others' feelings. In respecting people's feelings, we explained about drawing the line for oneself. There are issues above the line and those below the line. Issues above the line are those that individuals feel comfortable to share, and issues below the line are those that individuals are not comfortable to share (Bartsch, 1996: 59). The decision on whether the issue is below or above the line is left to the individual, and each individual has to draw their own line. No one must be forced to talk about something that they do not want to talk about. However, together with the class teachers, the research team followed up on the learners who had issues that they did not feel comfortable to share. During the circles, some serious issues emerged and the class teachers conducted some follow-ups on those learners with serious cases.

Some learners still felt uncomfortable sharing issues with their class teachers and kept their issues below the line for longer periods. This caused the research team to believe that there was a greater need for psychosocial support. There was a greater need for professional help for the learners who were hurting to avoid their psychological wounds from becoming septic. The danger of a septic psychological wound is that it ultimately affects other people negatively. If the undesirable feelings are not dealt with at the time that they arise, they can spill over into rebellious behaviour inside and outside of the classroom.

During the circle sessions, many emotions were shared as learners felt unable to cope with their daily life challenges. It was during one of the sessions that one of the class teachers acknowledged the mistakes that "we sometimes make as adults of comparing siblings". One of the learners who felt uncomfortable in sharing her problem had a sibling in Grade 12. The sibling was one of the highly intelligent learners at school, and the Grade 10 learner was not performing well; notably, she was repeating Grade 10. Sometimes, the teachers would be frustrated with the learner's performance so much so that they would remind the Grade 10 learner about her sibling's performance in Grade 12. The class teacher began to realise that this might have caused distress to the Grade 10 learner. The class teacher felt there was a need for her to talk to the Grade 10 learner in order to ease the pain.

Unfortunately, the learner flatly refused to open up to the class teacher about what was bothering her. The most unfortunate part of the inquiry was when the main researcher returned to MHS in March 2021 to finish the programme, at which time the learner's class teacher informed the main researcher that she had fallen pregnant and dropped out of school. The sibling who was in Grade 12 received an award for obtaining the best final examination results from all other Grade 12 learners, and she is currently at university pursuing her studies in medicine. According to the teacher, what concerned her was the fact that a gap between the two siblings had already been created.

Applying Ubuntu principles in education systems in areas of conflict resolution, disciplinary measures for learners' antisocial conduct, in transforming interrelations in school setting – inter-learners' relationships, leaders-teachers' relationships, learners-community relationships, schools-community's relations – can yield significant results in terms of making learners agents of change. As argued by Oviawe (2016:9), the Ubuntu philosophy or ethics has important implications for transforming the dominant "market-driven ethos of education into one that is more moralistic, emancipatory, cultured and relevant to the needs and aspirations of the people", because Ubuntu "helps to bridge the divide between the classroom and the community in order to foster a relationship that is reciprocal and symbiotic, and thus reflective of the Ubuntu ideal".

24.13.3 Significance of the Findings

What became evident was that traditional methods of dealing with violence do not produce positive results. Therefore, there is a need for alternative solutions to school violence. The conflict experienced at MHS can be transformed to peace-building. The application of conflict transformation to school violence must be designed in such a way that it is proactive and reactive at the same time. The parents and community members can be invited on a regular basis to have a discussion on the issues related to peaceful conflict resolution and management and how it can be transformed into something positive. Whatever the parents learn during these sessions can be transferred to their families and to the community.

According to Wahyudin (2018: 26), the practice of peace education is an opportunity to promote the total peace aspect to learners in classroom which becomes a building block for peace-building in the school and community. Emotions and social relationships affect learning, positive relationships, including trust in the teacher, and positive emotions, such as interest and excitement, open up the mind to learning (Darling-Hammond/Cook-Harvey, 2018). The approach is a whole classroom methodology that includes teaching learners the foundation abilities of conflict resolution, principles of conflict resolution and one or more aspects of problem-solving like negotiation, mediation and consensus decision-making (Saiden, 2017). Only positive affirmation is encouraged because this is a character-building exercise. In reflecting on the exercise focus must be more on the message that is being portrayed by the group rather than their acting skills. Introductions will be done on the first day of school for academic year by introducing their name, surname, expectations (from teachers, classmates and what can be expected from you), purpose (for yourself), guidelines for classroom processes and the vision for the class. The introduction of learners is done by telling others who they are and the meaning behind their names. This is part of positive affirmation and taking pride in who we are.

Introductions can be done during the LO period. The teachers introduce themselves first and the learners follow suit. In order to respect and protect each individual's dignity, it is important to emphasise that whatever is discussed during this exercise is confidential and it must therefore remain within the classroom.

24.14 Conclusion

Schools and institutions of higher education develop disciplinary mechanisms to respond to anti-social behaviour. This explains why some educational institutions apply coercive and punitive sanctions, and some others have structures to deal with violence that are based on the prosecutorial model, which unfortunately proves not to be working.

The outcomes of peace-making circles at MHS show positive results such as empathy. Transformation of relationships and healing were also identified as outcomes of the restorative justice (RJ) processes. We also discussed the benefits of RJ based on the document issued by the South African Department of Justice and Constitutional Development and Ubuntu.

The findings of the research revealed that different participants had different understandings of what school violence is. Most of the participants understood school violence as learners' disrespect for and physical attacks on teachers. The kind of violence they experienced was gender-based violence, bullying, rebellious learners, violence perpetuated by the School Governing Body (SGB) and murder of learners by other learners and murder of teachers by other teachers (fighting for positions). The schools deal with violence through violence, by instilling fear in learners and involving police. In this study, we support that peace circles can form an essential part of violence intervention methods at schools. Trust can be built in circles where learners can feel safe to share with their teachers and peers. Two similar reasons came out of this research and our previous research on the main reason why there is violence. Firstly, the main reasons for violence are a high rate of poverty, unemployment and high level of inequality. Secondly, it was revealed that trauma and violence is intergenerational.

The use of restorative justice at MHS yielded promising results that are corroborated by previous studies on the significance of RJ to inspire peaceful attitudes and behaviours among students (Kaman/Harris, 2000). Such efforts to make schools agents of change aim to stop violence in educational environment. We argued that building and maintaining a peaceful educational space that is free from violence and where learners and students can be provided with education and human development are essential to minimise violence when children become adults. At the same time, actions against violence should not be limited to normative instruments. RJ approaches need to be parts of educational, disciplinary and violence-prevention processes.

Based on the outcomes of previous studies that credit restorative justice with the ability to address justice by holding offenders accountable, leading them to acknowledge the consequences of their actions and help them make reparation, RJ clearly becomes a preferred strategy to tackle actual violence and transform society – to make it a better place.

Four most popular types of restorative justice were presented. We put emphasis on restorative justice conferences and circles because they are the most implemented in schools. Records reveal that they are very effective in tackling violence, transforming relationships and easing reintegration of offenders into school and university communities. We supported that transferring problematic students or education personnel to other communities without assisting them to resolve their difficulties is a weak option. However, such exclusion or removal is seen as the last resort, crossing the Rubicon, an irrevocable decision for unrepentant and extremely stubborn offenders because schools and universities have no correctional facilities.

Restorative justice as an emerging paradigm of justice may not be the ultimate solution against violence, nor a panacea to anti-social behaviour, but implementing it whenever the need arises could bring about substantial transformation to educational environments. A combination of RJ and education for peace are crucial to make our society a safe and empathic place to study and live in. Transformation through peacebuilding strategies should embrace ethics of mutual responsibility, ensure respect for the dignity of the human person created in the image of God and uphold the sanctity of life. When wounded relationships are mended, both the offender and the party that sustained hurt are healed together with the community. I contend that maintaining healthy relationships in schools and university campuses could lessen the brutality that affects many learners and students. The approach that can achieve these ends is RJ and the ubuntu spirit.

24.15 Recommendations

Given the overwhelming psychosocial and emotional needs evidenced throughout South Africa and at MHS in particular, there is a growing demand for counsellors to provide supportive counselling services and preventative mental health initiatives, such as family strengthening, life skills training and social support services. We felt that there was a need for psychosocial support because the teachers were also psychologically wounded and, therefore, not in a position to attend to the psychological wounds of learners. In their woundedness, the teachers had been persistently wounding learners. Even though the learners had indicated that they trusted teachers more than their parents, there were issues that they were not comfortable to share with their teachers.

The number of learners in other classrooms was too big, and the seating arrangements were uncomfortable. An uncontrollable number of learners in each class causes chaos and a lack of discipline; learners are unable to focus, and those at the back of the classroom are unable to hear what the teacher is saying (we had first-hand experience when they administered the questionnaires). Thus, the teachers have to continuously shout for the entire period. It is highly recommended that the Department of Basic Education address the problem by building more schools to reduce the overcrowding in schools.

When implementing policies, the government needs to adopt the bottom-up approach as opposed to the top-down approach that they are currently using. The people at the top, such as the government, are not aware of what is occurring at the school or community level. Policies must be established in accordance with what people want. The top-down approach is more dictatorial, which, therefore, causes people at the grassroots to resist such policies. This causes conflict, and much time and resources are wasted arguing about what needs to be done. The bottom-up approach is empowering in the sense that people feel respected and valued; being part of designing policies makes them feel good about themselves, and they have a sense of ownership and belonging. If people feel that they belong to and own something, they can take care of it and protect it.

Most adult South Africans are still carrying the emotional scars of the previous regimes, leading to the term "wounded nation" being used by the current South African president. This, therefore, suggests that there is a greater need for adults to be given an opportunity to share their feelings as a step toward healing and as a way to fast-track the transformative process in learners, families and communities.

We finally recommend:

1. Insertion of RJ in schools' disciplinary processes instead employing the "model code" with its retributive logic.
2. Introduction of peace education in school curricula.
3. Incorporation of the Alternative to Violence Project (AVP) in secondary education.
4. Creation of an office for learners' conflict resolution to involve them in peacemaking processes.
5. Exhortation of schools to pursue the goals of the DFA.
6. Development of module on Ubuntu education as it is done with religious education.

These key recommendations will empower learners to become both peacebuilders and agents of social change. This perspective starts at school and evolves further in their respective communities. If all schools embark on the same peace circles scheme and implement the above-mentioned recommendation, the larger South African society would be transformed, and violence could decrease drastically.

Appendix

Summary of the interventions

Activity	Purpose	Date(s)	Duration	Response
Presentation of findings to all stakeholders	To ensure that the stakeholders and the researcher have the same understanding	1 Feb 2020	1 day	We agreed on the suggested intervention strategy
Non-violence curriculum development	To ensure training content, methodology, and learning objectives were planned in advance	3–21 Feb 2020	19 days	Intervention tool was designed
Sharing the intervention programme with supervisor	To seek feedback and clarity and ensure that the method to be used was relevant to impart the skills and knowledge on non-violence	21–27 Mar 2020	6 days	Feedback received from the supervisor
Purchase of stationery and material and inspecting the venue and handling other necessary logistics	To ensure that the training material and venue were secured in time for the training	2–4 Mar 2020	3 days	Training material was secured for the training
Negotiating with the principal and deputy principal about the implementation of the intervention strategy	To ensure that we agreed on time slots and days for the implementation of the programme	10–19 Mar 2020	12 days	92 grade 10 learners were part of the training. We had put the training on hold until the following year
Negotiating with the principal and deputy principal about the implementation of the intervention strategy	To ensure that we agreed on the time slots and days for the implementation of the programme	3 Mar to 27 May 2021	3 months	81 learners were received in non-violence training
Holding the evaluation of the programme	Report back and share how participants have practiced non-violent methods in their day-to-day engagement with other people	6 Jul 2021	1 day	All grade 11 learners at MHS were invited to be part of the evaluation

References

Akinola, A.O., & Uzodike, U.O. (2018). Ubuntu and the quest for conflict resolution in Africa. *Journal of Black Studies, 49*(2), 91–113. https://doi.org/10.1177/0021934717736186

Almalki, S. (2016). Integrating quantitative and qualitative data in mixed methods research challenges and benefits. *Journal of Education and Learning*, 5(3), 288–296.

Anderson, A. (2003). *Restorative justice, the African philosophy of Ubuntu and the diversion of criminal prosecution.* Paper presented at the 17th international conference of the International Society for the Reform of Criminal Law, The Hague, Netherlands. https://www.readkong.com/page/restorative-justice-the-african-philosophy-of-ubuntu-and-4649952?p=1. Accessed 28 February 2019.

Asiamah, N., Kofi, H., Mensah, H.K., & Oteng-Abayie, E.F. (2017). General, target, and accessible population: demystifying the concepts for effective sampling. *The Qualitative Report, 22*(6), 1608–1621.

Bakalar, B. (2019). Book Review: Justice on both sides: Transforming education through restorative justice. *American Journal of Qualitative Research, 2*(2), 145–149.

Barton, B.A. (2023). Peacemaking circles: a new tool for juvenile court? http://law.gsu.edu/dyarn/fall06/law7062/Peacemaking%20Circles%20-%20ABA%20Presentation%20Materials.htm. 3 March 2023.

Bartsch, K., & Bartsch, E. (1996). *Stress and trauma healing: a manual for caregivers.* Durban: Artswork.

Battle, M. (2000). A theology of community: The ubuntu theology of Desmond Tutu. *Interpretation, 54*(2), 173–182. https://doi.org/10.1177/002096430005400206

Belle, J.L. (2017). Factors that influence student behaviour in secondary schools. *European Journal of Educational and Development Psychology, 5*(5), 27–36.

Bless, C., Higson-Smith, C., & Sithole, S. L. (2013). *Fundamentals of social research methods: an African perspective* (5th ed.). Cape Town: Juta & Company.

Bojesen, J.A. (2018). Communication in conflict and peace Reviewing peace theory in the frames of a network society. *Communication for Development*, Malmo University, Thesis.

Braithwaite, J. (2003). Principles of restorative justice. In A. Von Hirsch, J. Roberts, A. E. Bottoms, K. Roach & M. Schiff (Eds.), *Restorative justice & criminal justice: competing or reconcilable paradigms?* (pp. 1–20). Oxford: Hart Publishing.

Braithwaite, J. (2013). Does restorative justice work? In G. Johnstone (Ed.), *A restorative justice reader* (2nd ed., pp. 254–291). London: Routlege.

Centre for Justice and Reconciliation. (2008). What is Restorative Justice? http://www.d.umn.edu/~jmaahs/Correctional%20Assessment/rj%20brief.pdf. Accessed 19 November 2015.

Chiramba, E.N., & Harris, G. (2020). Restorative discipline practices: an action research project in Zimbabwean primary schools. *Educational Research for Social Change*, 9 (2): 118–122.

Christians, C.G. (2004). Ubuntu and communitarianism in media ethics. *Ecquid Novi: African Journalism Studies, 25*(2), 235–256.

Clarke-Habibi, S. (2018). Teachers' perspectives on educating for peace in Bosnia and Herzegovina. *Journal of Peace Education.* https://doi.org/10.1080/17400201.2018.1463209

Cremin, H., & Bevington, T. (2017). *Positive peace in schools - tackling conflict and creating a culture of peace in the classroom.* London: Routledge, Taylor and Francis Group.

Creswell, J. W. (2014). *Research design. Qualitative, quantitative and mixed methods approaches.* (Fourth Edition ed.). California: SAGE Publications, Inc.

Crowe, M., Inder, M., & Porter, R. (2015). Conducting qualitative research in mental health: Thematic and content analyses. *Australian & New Zealand Journal of Psychiatry, 49*(7): 616-623.

Darling-Hammond, L., & C.M. Cook-Harvey. (2018). Educating the whole child: improving school climate to support student success. Learning Policy Institute.

Department of Justice and Constitutional Development. (2023). Restorative Justice. https://www.justice.gov.za/rj/rj.html. Accessed 4 March 2023.

Dolamo, R. (2013). Botho/Ubuntu: the heart of African ethics. *Scriptura: International Journal of Bible, Religion and Theology in Southern Africa, 112*(1), 1–10.

Etikan, I., Musa, S.A., & Alkassim, R.S. (2016). Comparison of convenience sampling and purposive sampling. *American Journal of Theoretical and Applied Statistics, 5*(1), 1–4.

Ferrara, P., Franceschini, G., Villani, A., and Corsello, G. (2019). Physical, psychological and social impact of school violence on children. *Italian Journal of Pediatrics 45,* 76. https://doi.org/10.1186/s13052-019-0669-z

Finkelhor, D., Shattuck, A., Turner, H.A., & Hamby, S.L. (2014). Trends in children's exposure to violence, 2003 to 2011. *JAMA pediatrics, 168*(6), 540–546.

Flick, U. (2014). *An introduction to qualitative research* (5 ed.). London: Sage Publications, Inc.

Fusch, P. I. and Ness, L. R. (2015). Are we there yet? Data saturation in qualitative research. *The Qualitative Report*, 20(9), 1408–1416.

Gaylard, R. (2004). Welcome to the world of our humanity: (African) humanism, Ubuntu and black South African writing. *Journal of Literary Studies, 20*(3_4), 265–282. https://doi.org/10.1080/02564710408530357

Gershoff, E.T. (2017). School corporal punishment in global perspective: prevalence, outcomes, and efforts at intervention. *Psychology, Health & Medicine, 22,* 224–239. https://doi.org/10.1080/13548506.2016.1271955

Grobler, R. (2018). *Violence and killing at SA schools.* These stories shocked us in 2018. 22 Nov 2018. https://www.news24.com/news24/violence-and-killing-at-sa-schools-these-stories-shocked-us-in-2018-20181122. Accessed 7 August 2023.

Hemphill, M. A., Janke, E. M., Gordon, B., & Farrar, H. (2018). Restorative Youth Sports: An Applied Model for Resolving Conflicts and Building Positive Relationships. *Journal of Youth Development.* 13(3). https://doi.org/10.5195/jyd.2018.603

Hennigan, K.M., & Sloane, D. (2013). Improving civil gang injunctions: How implementation can affect gang dynamics, crime, and violence. *Criminology & Public Policy, 12*(1), 7-41.

Herrera, L.J.V., & McNair, R.L. (2020). Restorative and community-building practices as social justice for English learners. *TESOL International Association.* https://doi.org/10.1002/tesj.523

High, A.J. (2017). McGill University. Using restorative practices to teach and uphold dignity in an American school district. *McGill Journal of Education 52*(2), 525–534.

https://learningpolicyinstitute.org/product/educating-whole-child-report. Accessed 4 March 2023.

John, V.M. (2018). Peace education in post-apartheid South Africa: needs, responses, and constraints. *Asian Journal Peacebuilding*, 6(1), 55–74.

Kajee, L. (2019). A (de) humanising pedagogy: let the teachers speak. *Education and New Developments.* https://doi.org/10.36315/2019v2end052

Kaman, J., & Harris, G. (2000). Does studying peace make a difference?: an experiment at the University of Papua New Guinea. *Australian Journal of Adult Learning, 40*(1), 83.

Karp, D., & Dagle, E. (2013). Campus restorative justice. https://www.skidmore.edu/campusrj/. Accessed 25 June 2013.

Karp, D.R., & Sacks, C. (2013). Student Conduct, Restorative Justice, and Student Learning: Findings from the STARR Project (Student Accountability and Restorative Research Project). *Association for Student Conduct Administration Law and Policy Report, 28.* https://www.skidmore.edu/campuslife/karp/Misc/STARR_for_LPR.pdf. Accessed 8 August 2023.

Karp, D.R., & Sacks, C. (2014). Student conduct, restorative justice, and student development: findings from the STARR project: a student accountability and restorative research project. *Contemporary Justice Review, 17*(2), 154–172.

Karp, D.R. (2004). Introducing restorative justice to the campus community. In D.R. Karp & A.Thom (Eds.), *Restorative justice on the college campus: Promoting student growth and responsibility, and reawakening the spirit of campus community* (pp. 5–15). Springfield: Charles C. Thomas.

Kgosana, R. (2023). TUT student found stabbed to death—boyfriend taken in. *Sunday Times.* https://www.timeslive.co.za/news/south-africa/2023-02-03-tut-student-found-stabbed-to-death-boyfriend-taken-in/. Accessed 3 Mar 2023.

Kidde, J., & Alfred, R. (2011). Restorative justice: A working guide for our schools. Alameda Country Health Care Services Agency: San Leandro, CA.

Kiyala, J.C. (2016). Utilising a traditional approach to restorative justice in the reintegration of former child soldiers in the North Kivu Province, Democratic Republic of Congo. *Africa Insight,* 46(3), 33–50.

Kiyala, J.C.K. (2018). *Child Soldiers and Restorative Justice: Participatory Action Research in the Eastern Democratic Republic of Congo*. Cham ZG: Springer International Publishing. https://doi.org/10.1007/978-3-319-90071-1

Kiyala, J.C.K. (2021). Responding to Epistemic Injustice against Child Soldiers. *The International Journal of Children's Rights, 29*(3), 662-700. https://doi.org/10.1163/15718182-29030006

Kiyala, J.C.K. (2022). Indigenous Approaches of peacebuilding and Accountability of Former Child Soldiers: African Case Studies. In J.C.K. Kiyala & G.T. Harris (Eds.), *Civil Society and Peacebuilding in Sub-Saharan Africa in the Anthropocene: An overview* (pp. 423–464). Cham: Springer Nature Switzerland. https://doi.org/10.1007/978-3-030-95179-5_18

Kostaris, C., Sergis, S., Sampson, D.G., Giannakos, M.N., & Pelliccione, L. (2017). Investigating the Potential of the Flipped Classroom Model in K-12 ICT Teaching and Learning: An Action Research Study. *Educational Technology & Society, 20*(1), 261–273.

Lanni, A. (2005). Future of community justice. *Harvard Civil Rights-Civil Liberties Law Review*, 40(Suppl.), 359–405.

Latimer, J., Dowden, C., & Muise, D. (2005). The effectiveness of restorative justice practices: A meta-analysis. *The Prison Journal, 85*(2), 127–144.

Lester, S., Lawrence, C., & Ward, C.L. (2017). What do we know about preventing school violence? A systematic review of systematic reviews. *Psychology, Health & Medicine*, 22 (sup1), 187-223.

Levinas, E. (2003). *Humanism of the Other*. Chicago: University of Illinois Press.

Lopes Cardozo, M. T., Affiat, R. A., Zaman, F., Irawani, M., & Srimulyani, E. (2022). Silent struggles: women education leaders' agency for peacebuilding in Islamic schools in post-conflict Aceh. *Journal of Peace Education, 19*(2), 158-181. https://doi.org/10.1080/17400201.2022.2052826.

Louw, D.J. (2008). The African concept of ubuntu and restorative justice. In D. Sullivan & L. Tifft (Eds.), *Handbook of restorative justice* (pp. 161–173). London: Routledge.

Lutz, D.W. (2009). African Ubuntu philosophy and global management. *Journal of Business Ethics, 84*(3), 313–328.

Lyubansky, M., & Barter, D. (2019). Restorative justice in schools: Theory, implementation, and realistic expectations. In M.G.C. Njoku, L.A. Jason, & R.B. Johnson (Eds.), *The psychology of peace promotion: Global perspectives on personal peace, children and adolescents, and social justice* (pp. 309–328). Cham: Springer. https://doi.org/10.1007/978-3-030-14943-7_19

Mathikithela, M.S. (2020). Vulnerable youth as agents of change: A YPAR approach to making schools enabling spaces for learners. PhD, North-West University.

Mathis, S.M. (2013). From warlords to freedom fighters: Political violence and state formation in Umbumbulu, South Africa. *African Affairs*, 112(448), 421–439. https://doi.org/10.1093/afraf/adt041

Metz, T. (2011). Ubuntu as a moral theory and human rights in South Africa. *African Human Rights Law Journal, 11*(2), 532–559.

Miers, D. (2001). *An international review of restorative justice*. London: Home Office.

Morse, J.M., & Niehaus, L. (2016). *Mixed method design: principles and procedures*. New York: Routledge.

New International Version. (2023). The Meaning of Shalom in the Bible. https://www.thenivbible.com/blog/meaning-shalom-bible/. Accessed 4 Mar 2023.

Newenham-Kahindi, A. (2009). The transfer of Ubuntu and Indaba business models abroad: A case of South African multinational banks and telecommunication services in Tanzania. *International Journal of Cross Cultural Management*, 9(1), 87–108.

Noll, D. (2003). *Peacemaking: practicing at the intersection of law and human conflict*. Telford: Cascadia Pub House.

Obadire, O.T., & Sinthumule, D.A. (2021). Learner discipline in the post-corporal punishment era: What an experience! *South African Journal of Education, 41*(2), 1–8. https://doi.org/10.15700/saje.v41n2a1862

Okoro Okoro, E.U., Onodugo, U., Chris, I., Chukwudi, A.B., & Okeke, C. (2018). Perception of teachers on the influence of peace education in the South-East Nigerian secondary schools.

International Journal of Operational Research in Management, Social Sciences and Education 4(1), 222–236.

Oviawe, J.O. (2016). How to rediscover the ubuntu paradigm in education. *International Review of Education,* 62, 1-10. https://doi.org/10.1007/s11159-016-9545-x

Parker, C., & Bickmore, K. (2020). Classroom peace circles: Teachers' professional learning and implementation of restorative dialogue. *Teaching and Teacher education, 95*, October 2020, 103129. https://doi.org/10.1016/j.tate.2020.103129

Pranis, K. (1997). Peacemaking circles. *Corrections Today*, 59, 72–77.

Pranis, K. (2013). Peacemaking circles. In G. Johnstone (Ed.), *A restorative justice reader* (2nd ed., pp. 117–121). London: Routledge.

Rautenbach, C. (2015). Legal reform of traditional courts in South Africa: Exploring the links between Ubuntu, restorative justice and therapeutic jurisprudence. *J. Int'l & Comp. L., 2*, 275.

Republic of South Africa. (2011). Restorative justice: the road to healing. http://www.justice.gov.za/rj/2011rj-booklet-a5-eng.pdf. Accessed 26 June 2013.

Saiden, T. (2017). Towards Conflict Transformation and Management skills' training in Zimbabwean Primary School Teacher Education for effective Teaching, Learning and Peace in Zimbabwean Primary School: A Case Study of Mkoba Teachers' Training College. *Advances in Social Sciences Research Journal, 4*(16), 1–22.

Schirch, L. (2004). *The little book of strategic peacebuilding*. Pennsylvania: Good Books.

Scorgie, F., Baron, D., Stadler, J., Venables, E., Brahmbhatt, H., Mmari, K., & Delany-Moretlwe, S. (2017). From fear to resilience: adolescents' experiences of violence in inner-city Johannesburg, South Africa. *BMC Public Health*, 17(3), 51–64. https://doi.org/10.1186/s12889-017-4349-x

South Africa Government Communication and Information System. (2023). Grade 10 Geluksdal Secondary learner fatally stabbed. https://www.sanews.gov.za/south-africa/grade-10-geluksdal-secondary-learner-fatally-stabbed. Accessed 12 March 2023.

Stefanie, K. (2017). Positive peace. *The SAGE encyclopedia of political behaviour*. 640–641.

Supriyanto, A., & Wahyudi, A. (2018). Group guidance services based on Folklore for students junior high school. *International Journal of Indonesian Education and Teaching, 2*(1), http://e-journal.usd.ac.id/index.php/IJIET. Accessed January 2022.

Taylor, L.K. (2020). The Developmental Peacebuilding Model (DPM) of children's prosocial behaviour in settings of intergroup conflict. *Child Development Perspectives, 14*(3), 127–134, https://doi.org/10.1111/cdep.12377

Tuncel, G. (2017). Improving the Cultural Responsiveness of Prospective Social Studies Teachers: An Action Research. *Educational Sciences: Theory and Practice, 17*(4), 1317–1344. https://doi.org/10.12738/estp.2017.4.0269

Tutu, D.M. (1999). *No future without forgiveness*. London: Rider.

Umbreit, M.S., Coates, R.B., & Vos, B. (2002). *The impact of restorative justice conferencing: A review of 63 empirical studies in 5 countries*. Minnesota: University of Minnesota Center for Restorative Justice and Peacemaking, School of Social Work, University of Minnesota.

UNESCO. (2000). The Dakar Framework for Action. https://sustainabledevelopment.un.org/content/documents/1681Dakar%20Framework%20for%20Action.pdf. Accessed 4 March 2023.

United Nations. (2006). A Handbook of restorative justice programmes. Criminal justice handbook series. https://www.un.org/ruleoflaw/files/Handbook%20on%20Restorative%20Justice%20Programmes.pdf. Accessed 17 June 2018.

Uzunboylu, H., Baglama, B., Kucuktamer, T., & Kuimova, M. V. (2017). Opinions of school counselors about bullying in Turkish high schools. *Social Behavior and Personality: An International Journal, 45*(6), 1043–1055. https://doi.org/10.2224/sbp.6632

Van Ness, D., Morris, A., & Maxwell, G. (2003). Introducing restorative justice. In A. Morris & G. Maxwell (Eds.), *Restorative justice for juveniles: conferencing, mediation & circles* (pp. 3–16). Oxford: Hart Publishing.

van Zyl, M. (2022). Crime state/stats reveal children are not safe at schools. https://www.da.org.za/2022/11/crime-state-stats-reveal-children-are-not-safe-at-schools. Accessed 4 March 2023.

Vaughan-Eden, V., Holden, G.W., & LeBlanc, S.S. (2018). Commentary: Changing the Social Norm about Corporal Punishment. *Child and Adolescent Social Work Journal. 36*, 43–48. https://doi.org/10.1007/s10560-018-0592-y

Wahyudin, D. (2018). Peace Education Curriculum In the Context of Education Sustainable Development. *Journal of Sustainable Development Education and Research*, 2(1), 21–32.

Walker, L. (2012). *Restorative justice today: Practical applications*: SAGE Publications.

Wessel-Powell, C., Buchholz, B. A., & Brownell, C.J. (2019). Polic(y)ing time and curriculum: how teachers critically negotiate restrictive policies. *English Teaching: Practice & Critique.* https://doi.org/10.1108/ETPC-12-2018-0116

Wilson, M.E. (2006). Restorative Justice on the College Campus: Promoting Student Growth and Responsibility, and Reawakening the Spirit of Campus Community (review). *Journal of College Student Development, 47*(3), 355–359.

Winn, M.T. (2020). *Justice on both sides: Transforming education through restorative justice.* Harvard Education Press.

Zehr, H. (2002). *The little book of restorative justice.* PA: Good Books Intercourse.

Zehr, H., & Mika, H. (2003). Fundamental concepts of restorative justice. In E. McLaughlin, R. Fergusson, G. Hugues & L. Westmarland (Eds.), *Restorative justice: critical issues* (pp. 40-43). London: Sage Publications.

Chapter 25
Indaba and Violence Prevention in Educational Institutions: Perspectives from South Africa

Jean Chrysostome K. Kiyala, Primrose Sibusisiwe Mlambo, and Crispin M. C. Hemson

Abstract Violence in educational environments is a widespread phenomenon which often results in injury, loss of life, and sustained negative impact on the mental, psychological and emotional health of survivors. While schools, universities and colleges globally rely upon punitive justice and restorative sanctions as a means of deterrence, little is known about locally-based approaches of conflict resolution and peacebuilding to prevent violence in educational institutions. This article consists of a desktop study combined with evidence from an action-research study of violence prevention in a high school in a poverty-stricken area of South Africa. It explores the potential of a compounded alternative dispute resolution to prevent violence in educational institutions while forging students' accountability. That comprises restorative justice approaches, peace education and *Indaba* ('dialogic structure to resolve disputes'), inasmuch as these methods are imbued with *Ubuntu* spirit ('a quality that includes the essential human virtues; compassion and humanity'). The chapter sets out some key elements that should inform a modern student *Indaba*. It is argued that a synergic humanistic and communitarian relationship between *Indaba*, restorative justice, peace education and *Ubuntu* can leverage peaceful resolution of conflicts in educational institutions, promote reconciliation and prevent recurring violence.

Dr Jean Chrysostome K. Kiyala holds a PhD in Public Administration: Peace Studies from DUT; he is a Senior Lecturer at the International Centre of Nonviolence (DUT), Associate Professor and visiting lecturer at the Université de Bandundu and the Université Evangélique en Afrique (Democratic Republic of Congo); Emails: JeanK@dut.ac.za; kljeanchrysostome@gmail.com; ORCID: https://orcid.org/0000-0002-9587-0230

Dr Primrose Sibusisiwe Mlambo (PhD) is a researcher, trainer and facilitator with over 25 years' experience in the development sector, peace, violence and HIV and AIDS related programmes. She holds a master's degree in Community Development and PhD in Public Administration: Peace Studies. Email: feladies@gmail.com; ORCID: https://orcid.org/0009-0000-2042-7544

Mr Crispin C Hemson is the former Director of the International Centre of Nonviolence at Durban University of Technology. He holds Master's degree in Education; he is currently completing his PhD on educating educators for peace. Email: crispinh@dut.ac.za

J. C. K. Kiyala and N. Chivasa (eds.), *Climate Change and Socio-political Violence in Sub-Saharan Africa in the Anthropocene*, The Anthropocene: Politik—Economics—Society—Science 37,
https://doi.org/10.1007/978-3-031-48375-2_25

Keywords Educational institutions · *Indaba* · Peace education · Peacebuilding · Restorative justice · Violence prevention · *Ubuntu*

25.1 Introduction

Violence in its various forms has always been a concern of human communities. South Africa is no exception. The level of brutality experienced in the South African educational environment is reflective of the same patterns of violence in the larger South African society. For instance, the South African Police Service's (SAPS) release of the crime statistics for the first quarter of 2023 demonstrates the scale of violence. The homicide rate was 43.3 per 100,000 of population in 2023, against a global average of 6.1 (South African Police Service, 2023), while reported rapes in 2019–2020 were extremely high at 72.1 per 100,000 (World Population Review, 2023). Violence in educational environments is of the same nature as that of the violence occurring in the general population as reported by the SAPS, though not necessarily of the same extent.

To further illustrate these statistics, a few cases of violence in educational institutions need to be mentioned. These include the case of Khensani Maseko, a 23-year-old third-year BA Law student at Rhodes University who took her own life at her home in Johannesburg, in protest to the rape culture on campus of which she herself was a victim (Nomahlubi, 2018); the murder of Zollie Khumalo; and the murder of a 21-year-old female student who was shot dead at a campus residence of the Mangosuthu University of Technology in 2018. Students continue to fall prey to sexual-related violence which occurs in society, such as the murder of Uyinene Mrwetyana, a 19-year-old student at the University of Cape Town who was raped and killed by a man employed at a post office in Cape Town. Her brutal murder triggered mass protests against gender-based violence (GBV) throughout the country. Two major studies of violence in schools (Burton/Leoschut, 2013; Mncube/Harber, 2013) reveal the depth and extent of all forms of violence within schooling. In areas like Ilovu Township, in the south coast of KwaZulu-Natal, a primary school learner killed another during school hours (Mlambo, 2022) and a high school student was killed when they had interschool cultural activities. Student protests have been violent at times; in conflicts at Durban University of Technology in 2019, a junior administrator had her face smashed in by a brick and a student was shot dead by a security guard. This is a clear indication of how violence has become endemic in schools and tertiary education. Silence is no longer sustainable in the face of such inhumanities.

Both the statistical evidence and media reports support the view that South Africa is a country caught in what Pearce (2007) defines as chronic violence:

> Chronic violence is present where rates of violent death are at least twice the average for high and low income countries respectively; where these levels are sustained for five years or more and where frequent acts of violence not necessarily resulting in death, are recorded across several socialisation spaces, including the household, the neighbourhood, the school, inter

community and the nation state public space (which brings in disproportionate, sanctioned and non-sanctioned acts of violence attributed to state security forces)' (Pearce, 2007: 54).

Children and young people do not exist in a vacuum, and their characters, social and emotional beings develop in homes, educational environments and in communities from where they come. Children and young people are socialised and shaped in their environment to behave in a particular way (Goldweber et al., 2013: 469). Burton/Leoschut (2013: 1) explain that violence within schools, colleges and universities will not be effectively addressed without dealing with the various environments and contexts in which students live and develop. Mlambo (2022) argues that violence is learnt and that there is always the possibility that teachers, learners and parents in contexts of extreme violence unlearn what has been learned in the past and relearn new ways of resolving conflict.

25.2 Purpose and Nature of the Study

The purpose of this chapter is to explore the complexities of violence in the learning environment, focusing on the South African case, and to examine current approaches used in responding to violence on campus colleges, schools and universities. It further explores the potential of the *Indaba* dialogic institution to forge a path of violence prevention through peace education and holding perpetrators accountable via restorative justice approaches. This is done through review of the relevant literature and by drawing also on a doctoral study (Mlambo, 2022). This was an action-research study; the researcher used mixed-methods to understand how learners, teachers and parents understand the nature, cause and impacts of violence in two schools. Mlambo (2022) then worked with learners at the one school to develop greater commitment to processes of nonviolence in addressing conflicts.

The paradigm within which this chapter is written is interpretivist; the emphasis is on how people frame and speak about the relevant issues and the significance of the meanings they give. The theoretical understanding underlying the analysis is Galtung's distinction between direct, structural and cultural violence (Galtung, 1967, 1990), an approach that directs attention to the links between structures of society, cultural forces and the immediate forms of interpersonal violence.

The central argument is that a major strategy for preventing violence is the involvement of students in forms of restorative justice through *Indaba*. *Indaba* is a widespread community practice across central and southern Africa that can be used to involve participants in discussion and reflection on major problems and conflicts. Through such a process, the root causes of violence, its consequences, and accountability processes are agreed upon. Such a strategy will lessen the negative impact of top-down mechanisms where students' participation and role in decision-making are sidelined.

The chapter proceeds by reviewing the phenomenon of violence in education and then exploring the tradition and elements of *Indaba*. It then relates the Indaba process to peace education and concludes with specific recommendations.

25.3 Violence and Disciplinary Approaches in Education Institutions

Violence in learning environments manifests in various ways such as corporal punishment, bullying, cyberbullying, gang violence, gender violence, violence against teachers, and school shootings (Mlambo, 2022). Mncube/Harber (2013: 26) report that in 90 out of 197 countries monitored by the Global Initiative to End Corporal Punishment of Children, corporal punishment has remained legal despite evidence of its harmful effects on students. Dogutas (2013: 85) states that it is a "common feature for students to be carrying weapons to school and to witness robberies and fights taking place inside the school premises." Students with a "lower level of school connectedness" are significantly more likely to be involved in victimisation and gang violence (Espelage et al., 2014: 289).

Schirch (2004) notes that humans harm each other in various ways as a result of their choices regarding what they decide to consume and how they interact with one another, as well as when they deem the use of force against others as necessary to satisfy their needs. This pursuit of individuals' unfettered whims at the expense of others' rights constitutes violence, which the World Health Organization (WHO) defines as follows:

The intentional use of physical force or power threatened or actual, against oneself, another person, or against a group or community, that either results in or has a high likelihood of resulting in injury, death, psychological harm, maldevelopment or deprivation (Krug et al., 2002: 5).

Such violence is experienced throughout the world, and the South African educational environment is no exception. The patterns of brutalities which occur in schools, campus colleges and universities are reminiscent of the violence seen in society in general. In higher learning spaces, violence is multifaceted and is manifest in bullying, sexual violence and GBV, and external violence (gangs, weapons, beatings, stabbings and shootings) which are caused by substance abuse, namely the presence of drugs in schools and the illicit trafficking of narcotic substances (UNESCO, 2009). Such violence is generally strongly gendered, with very high levels of assault against male students. The prevailing violence against female students is sexual assault, date-rape, intimate partner violence, domestic violence and stalking (UNESCO, 2009).

In order to tackle violence in learning environments, many institutions have largely relied on retributive sanctions or the referral of offenders to the police service and the criminal justice system. Karp (2004: 6) asserts that school and college administrations intensify surveillance and amplify sanctions because they cannot simply trust students' "internal controls". Increasing incidents of school and campus violence raise the question of the effectiveness of punitive sanctions in making learning environments safer spaces. Indeed, scholars such as Noll (2003) and Zehr (2013) point out that the mainstream criminal justice system has limited effects in deterring crimes and preventing re-offending. This justifies the movement towards 'revolutionising' the criminal justice system and the perception of crime (Zehr, 1990) or to 'civilise' civil justice (Skelton, 2013), and promoting "community justice" (Lanni, 2005). These revolutionary waves to reform the justice system are gaining ground and have been considered by universities in the United States, Canada, Europe and Australia as a model for student offenders' accountability and for peacebuilding on schools and campus colleges. Unfortunately, South African schools and campus rules and disciplinary sanctions remain substantively reliant upon a punitive justice stream that focuses on offenders who are expected to be dealt with harshly in order to prevent further misconduct. An alternative deterrence is conceivable only after grasping substantively the main drivers of school and campus violence, their consequences and appraising the effectiveness of existing preventive measures. This procedure is useful as it allows us to identify the weaknesses of ongoing violence prevention strategies in order to adopt a more productive approach.

25.4 Factors Contributing to Violence in Learning Environments

Low levels of social cohesion among teachers and learners often lead to antisocial and violent interactions (Le Mottee/Kelly, 2018: 53). With low levels of social cohesion, there is a lack of trust, no sense of belonging, and lack of focus toward achieving the same goal. Edberg (2017: 55) observed that there is "insufficient dialogue" in low- and middle-income country settings to harmonise people's efforts to address violence. Negative peer influences, poor relationships with classmates, poor parent–child communication, a lack of parental monitoring, and living in an unsafe environment also increase the risk of being a bully (Shetgiri et al., 2013: 90). Hinduja/Patchin (2013: 711) argue that during adolescence, many "behavioural choices are influenced and conditioned by the role of major socializing agents," including friends, families, and adults in the learning environment. According to Ramarola/Joyce (2014: 12), drug abuse in learning environments is one of the causes of violence – it is also a global problem that poses a severe threat to the goal of establishing peace and democracy. Data collected in South Africa between 1991 and 2001 showed that a growing proportion of adolescents in South Africa use alcohol and drugs, with cannabis being the most common drug, followed by the combinations of cannabis and methaqualone ('Mandrax') and other drugs (Parry et al., 2004a, b).

In the case of South Africa, there is also long history of violent suppression of people. Mamdani (2021: 4) asserts that "The violence of postcolonial modernity mirrors the violence of European modernity and colonial direct rule." The violence of the past is reproduced both through social learning (Bandura, 1971), where young people learn to imitate the violence that they learn, and through the mechanism of trauma. Eagle/Kaminer (2013: 85), writing of the years under colonialism and apartheid, use the term *continuous traumatic stress* (CTS) 'to describe the experience and impact of living in contexts of realistic current and ongoing danger, such as protracted political or civil conflict or pervasive community violence.' There is ample evidence of both exposure to extreme situations of violence, leading to post-traumatic stress disorder (Herman, 1992) and of CTS in the narratives of school violence in the two schools surveyed by Mlambo (2022).

These forms of violence are described by Schirch as secondary violence that emanate from structural violence and which take the form of 'self-destruction' (alcohol and drug abuse, suicide, depression), "community destruction" (crime, interpersonal violence, domestic brutalities, rape), and "national and international destruction" (Schirch, 2004: 24). It is important to understand these complexities in view of putting in place adequate strategies that can help curb inhumanities in society and in the educational environment.

25.5 Consequences of Violence in Learning Environments

In 2009, UNESCO reported that all forms of violence are detrimental to the mental and physical health of offended students (UNESCO, 2009). Violence affects social skills, leads to depression, loneliness, a loss of self-esteem, humiliation and anger; triggers retaliatory feelings; sustains stigmas; causes drop-out; and causes sustained anxiety and brutal behaviour, particularly when empathy or feelings of care for others have dissipated. The most destructive and perilous consequences of sexual and gender-based violence are increasing levels of infection with sexually transmitted diseases, unwanted pregnancy, low self-esteem and a deterioration of performance in educational undertakings (UNESCO, 2009).

Lester, Lawrence and Ward (2017: 187) observed that the long-term consequences of violence in learning institutions can be "very costly to the whole society." According to UNESCO (2017: 10), violence in learning environments "impacts both victim and perpetrator," and the consequences could include being unable to effectively socialise, having relationship difficulties, and displaying criminal behaviour. The effects of violence in learning environment can be "devastating to the economy" of the country because it drains the economy and consumes resources that could be harnessed for poverty eradication and development (John, 2018: 60). Kudenga (2017: 186) states that even though many people regard "corporal punishment positively as a customary and necessary technique of child rearing," it increases aggressive behaviour, especially toward other people.

Violence in learning environment has "systematic adverse effects on students' learning behaviour, as well as school and college personnel functioning and teachers/lecturers and parents' partnerships" (McMahon et al., 2014: 754). Students exposed to violence are at risk of developing health problems later in life such as depression, alcohol abuse, and lower educational achievement (Merrill, 2018: 1). According to UNESCO (2017: 10), violence in learning environment harms the physical health and emotional wellbeing of students and adolescents, and sexual violence increases the rates of teenage pregnancy and HIV and other sexually transmitted diseases.

According to Zulu et al. (2006), violence hinders the culture of teaching and learning in South Africa, in the same way as crime affects victims. In a similar vein, discussing Gender-based Violence (GBV), Gordon and Collins (2013) note that GBV in learning environment creates an atmosphere of terror, with female students living in fear. That constitutes an obstacle to the educational development of such students and those who may be indirectly affected for being exposed to atrocities experienced by their peers, either as witnesses or sharing with survivors about their sexually-related ordeal.

25.6 Tackling Violence in Learning Environments

Davids and Waghid (2016: 28) observe that among the largest challenges facing leaders of education institutions concerning violence is that they have "not acquired sufficient training to deal with violent encounters, and often respond in equally violent and violating ways." Society in general and educational institutions in particular are grappling with violence prevention measures, and both schools and campus colleges are striving to find the best strategies to counter aggressive behaviours, whether physical or verbal, in their space (Mayhew et al., 2011). Educational institutions have largely adopted either a punitive approach or the referral of offenders to the criminal justice system and police service. A law enforcement and policing strategy is used to coerce students to comply with school and college policies and the criminal law, since schools, universities and colleges do not have many options (Karp/Conrad, 2005).

A survey was conducted by Karp and Sacks (2014) based on the analyses of 659 cases of student conduct involving mainly student offenders, student conduct officers, and other participants in the contact processes. They found that restorative justice practices had a significant impact on students' learning processes as compared to the disciplinary hearing model (Karp/Sacks, 2014). Another study, conducted by Gevers et al. (2013), established that strategies to counter school and campus violence should include: strengthening gender equality and confronting dominant masculinity tendencies and widespread callousness in the face of violence; ameliorating dispute resolution and communication abilities; empowering students to build constructive relations; controlling abuse of substances; and improving the gun control system. All these findings are pertinent to making the educational environment conducive to quality learning and teaching, in view of the difficulty faced by college management structures in dealing with students who are incapable of 'self-regulation'. Such is the conundrum noted in the United States, which similarly reflects the global context of schools and colleges which are dependent on coercion to make students comply with rules and the criminal law, and since students are not able to "internally control" themselves, colleges have to intensify surveillance and punitive sanctions (Karp/Conrad, 2005).

In relating research to existing initiatives, UNESCO advocates a holistic strategy which requires the participation of all stakeholders, namely students, parents, school staff and the community (UNESCO, 2009). This collective action intends to make students partners in preventing violence; promotes a constructive discipline technique against bullying; stimulates students' resilience and helps them to respond to life's challenges constructively; encourages them to be positive role models by speaking out against sexual violence and GBV, and to be advocates for school safety mechanisms; supports efforts for making the educational environment safer and a more welcoming space for students; endows students with learning violence prevention and conflict resolution skills; and finally, discourages violence and discrimination against students with disabilities and those from indigenous, minority and other marginalised communities (UNESCO, 2009).

It is important to observe that while universities around the world are moving towards restorative justice approaches in tackling violence (Rinker/Jonason, 2014; Wilson. 2006), South African institutions of learning continue to rely largely on law enforcement and criminal-justice-oriented models for deterrence and holding students accountable. This observation is supported by the disciplinary system in South African schools and universities where a range of punitive sanctions are exerted on students in classrooms, campus and in campus residences. These include exclusion, suspension, conduct probation and conduct review, etc. (Durban University of Technology, 2014; University of KwaZulu-Natal, 2008). Despite the presence of tight security, surveillance and school and campus patrols, violence and crimes are not eradicated in learning environments. Moumakwa (2010) advocates the inclusion of restorative justice principles within educational systems and other forums constituted for conflict management. This is consistent with the South African legislation on the "Extension of diversion and compulsory diversion in South Africa" which encourages the use of restorative mechanisms such as "victim-offender mediation programs and performance of community service as an alternative to prosecution" (Anderson, 2003: 6).

Restorative justice procedures incorporate apology, restitution, enlightened community service, changed behaviour and generosity (Karp/Conrad, 2005). Restorative justice processes encourage offenders to voluntarily participate in a dialogue that reminds offenders of their wrongdoings and gives them the chance to take responsibility for repairing the harm and showing evidence that they are apt to become "good citizens" and will refrain from committing the same reprehensible acts which have caused hurt to others (Karp/Conrad, 2005). This practice of conflict resolution and peacebuilding needs to be incorporated into the processes for holding student offenders accountable and settling scores among students in order to prevent the escalation of violence and further atrocities.. However, the use of indigenous jurisprudence and dialogic institutions to advance the social agenda of violence prevention in learning environment remains understudied. The following section examines the functionality and potentiality of *Indaba* to promote societal harmony and build peace by preventing violence.

25.7 Perspectives on an Indaba Dialogic Institution

Africa has a long tradition of conflict resolution which is rooted in a humanistic and communitarian conception of the human person and their relation to society. In pre-colonial Africa, justice existed prior to the inception of the Western criminal justice system. Conflict resolution and holding offenders accountable for violations of ancestral and societal norms, which included civil and criminal lawsuits, were handled by a council of elders at various levels of the administrative organisation of society, namely at the locality and chieftaincy levels.

Regarding untapped African sources of conflict resolution, Mahlubi (2018) points out that traditional mechanisms of community dispute settlement such as *Lekgotla* (a gathering of villagers for court cases, and also a meeting place for village assemblies and meetings of village leaders among the Batswana people of Botswana and South Africa), *Imbizo* (gathering addressed by the King) or *Indaba* (gathering of elders). These traditional mechanisms, critically over-ridden or displaced during the colonial and (in the case of South Africa) apartheid era, may provide pivotal elements to procedures which prove to be more adequate and effective and in which parties with a stake in a conflict could feel more comfortable (Lee, 2009). When *Indaba* is used as an alternative dispute resolution, it can help avert the negative effects of procedural liquidations in courts while at the same time help manage and resolve disputes (Kiyala, 2022b). Court proceedings end in settlements that see parties separating in bitterness (Lee, 2009). It is worthwhile therefore to deepen our understanding of *Indaba* in the context of Southern Africa.

25.7.1 Traditional Indaba and Conflict Resolution

The concept of *Indaba* is versatile. In the Zulu traditional society, its multifunctional dimension was recognised. It is used as a platform where local traditional chiefs debate issues pertaining to social welfare, development, societal harmony; it also serves as an institution that helps to maintain communal stability and social cohesion in rural settings (Kiyala, 2022b). *Indaba* has derived from *ndaba* (a problem, matter, case, situation or issue, in the IsiZulu language of South Africa) in Nguni languages such as Isixhosa, IsiZulu and IsiSwati. It has evolved as a public forum and processes during which critical matters that affect a particular community are addressed by the community elders and chiefs (Kiyala, 2022b). Byerly (1998: 1) describes *Indaba* as "a Zulu word meaning both an update on topical matters and a conference, symposium or meeting-place at which members of a group contribute equally towards the solution of a problem".

What makes an *Indaba* a more desirable forum is the fact that anyone in a community leadership position can call for it; that could be an Induna (headsman) or ward councillor in rural settings, of a mayor, in urban setting (Kiyala, 2022b). The *Induna* of an *isigodi* (a region, a district, and given area in IsiZulu) has the prerogative to convoke an *Indaba* upon being notified of a complaint that needs to be handled to preserve peace; the Induna then calls an *Indaba* during which all the people of the *Isigodi* have to participate (Kiyala, 2022b). The expected outcomes of an *Indaba* gathering are contributed by a variety of people who participate in the discussions and deliberations of matters that led to the convocation of this public gathering; often the participants in *Indaba* seek to reach a consensus over the final ways forward to be followed by the community (Kiyala, 2022b). Very often, most of the attendees leave the *Indaba* meeting feeling heard and happy to have contributed towards the social good and common good.

Indaba follows the patterns of the African palaver during which stakeholders deliberate issues in a participatory manner and all intervening participants hold equal rights to give their views over what is at stake, and subsequently, propose ways of resolving them. Such ways of handling contentious community issues and resolving disputes are common in several communities of Sub-Saharan Africa, namely the DRC, Burundi, Rwanda, Uganda, Mozambique, Sierra Leone, etc. (Kiyala, 2022a). It uses the tenets of the palaver, which demands inclusivity; it rejects Marginalisation and seeks to attain and preserve the common good (Kiyala, 2022b). It has an ethical dimension, because community life requires abiding by the community social ethos to respect the dignity of others and foster communal and societal harmony (Kiyala, 2022b).

The inspiration we gain from the African communitarian way of dealing with conflicts, violence, and development via institutions such as *Indaba* and palaver (Kiyala, 2022b) focuses on the social and common good, deepens a sense of communal mutual responsibility for peace and welfare and fosters solidarity. Collective action for peace and peacebuilding are negotiated and attained via the Palaver, rituals and analogous ceremonies and practice to achieve durable peace in the community (Kiyala, 2022b). Even in the context of large-scale conflicts characterised by human atrocities, such indigenous approaches have proved effective to maintain social cohesion and pave the way of post-conflict community reconstruction (Kiyala, 2022a). Indaba conveys the notion of dialogue and problem-solving, as noted by Byerly (1998: 1), it "suggests combined efforts at working towards consensus through discussion of matters concerning the group as a whole". Indaba can also be referred to as a story and conveys several interpretations, not limited to public gathering, open discussion, dialogue, Palaver, etc.

Indaba knowledge has multiple functions and can be exploited to achieve educational objectives and shape the discourse about higher learning environments. This assertion is sustained by several studies that affirm the effectiveness of indaba indigenous concept, its credibility and its use as "an original group approach for expounding indigenous, educational psychological concepts" (Edwards et al., 2020). The *Indaba* is also perceived as a useful process in framing campus and university values of freedom and equality (Waghid et al., 2023). Adding to that, Indaba is constantly emerging as a rational alternative to the increasing impasse in South Africa; it is also considered as an option for "educational planning for social change" (Karodia, 1988; Moulder, 1991; Karodia, 1988; Moulder, 1991). The *Indaba* has been utilised to frame SDGs' agenda in higher learning institutions such as UNISA (Nhamo, 2021).

In this chapter, it denotes both an institution and a communal problem-solving space that aims at bringing social transformation and attaining communal welfare.

The effectiveness of indaba and the use of the palaver are sustained by the search and pursuit of collective wellbeing, individual rights, the dignity, and the sacred character of life—these principles lie at the heart of *Ubuntu* (Kiyala, 2022b). Violence, war and devious behaviours that violate the well-being of the community and of an individual are antithetical to the spirit of *Ubuntu*. Since schools and universities are macrocosms of society as are families, the *Indaba* forum can be utilised to instil and foster the values for peace and encourage children, learners and students to live up to the humanism conveyed by *Ubuntu* philosophy. These philosophical and ethical postulates determine the morality of human action in an African context, where the Baraza and *Indaba*, and other dialogic institutions are predominantly employed to pursue the common good and the welfare of the global community, achieve post-conflict reconciliation, nation-building, and national-healing. *Ubuntu*, Ujamaa and the Palaver are juxtaposed with perennial conflict, enmity, social exclusion, marginalisation, indifference, war, and quarrels. These moral propositions are essentially the principles that guide Baraza and *Indaba*. For this reason, it is obvious to conclude that Baraza and *Indaba* stand as a form of justification of the African social ethos and the humanist and communitarian African worldview.

Resorting to traditional methods for the resolution of conflicts may, however, have limitations. Mlambo (2022) observes that in one of the schools at Ilovu, in the south coast of Kwa-Zulu Natal, stakeholders (parents, teachers, traditional leadership, learners, ministers of religion, and ward councillors) were invited to meetings that were not fruitful until traditional leadership threatened to beat learners if they ever become violent. This is the conventional South African approach that approves of violence as a means of 'discipline' and is applied in a majority of households in South Africa, mainly in townships and rural settings. According to the parents and traditional leadership, this is what they are used to. They stated that they grew up with intimidation, it's part of who they are and as one of the cultural practices, it works, but the problem is that it is not sustainable. This points to two key features of South African responses to violence that in turn reproduce violence.

The first is that what is defined as violence often comes to be 'the things that bad people do'. What, typically, 'we', the 'good people' do, is not violence or it is necessary, even essential violence. This is explored fully in Collins (2013: 3):

> Violence is… widely seen as an appropriate and effective way of regulating interpersonal relationships. It is understood as an essential tool for raising children, a useful disciplinary technique in educational institutions, an acceptable strategy in pursuing sexual encounters, an indispensable resource in intimate relationships, and an effective way of establishing social status. [There is an assumed] …imagined clear boundary between the non-violence of decent law-abiding citizens and the violence of antisocial criminals.

Similarly, Collins and Plüg (2020) report how mobilisation of a group around what was wrongly assumed to be a case of child abduction by a foreigner led to the brutal deaths of two innocent men. They make the point that this must have been understood by the perpetrators as justifiable action against criminals, given the discourses around violence that shape people's understandings and actions. Similarly, in Mlambo (2022), her analysis of stakeholders statements about school violence reveals that the understandings are often very partial. It is often assumed to be carried out only by learners in the school (and not teachers); there is no overall recognition of what constitutes violence and thus no coherence in responding to violence.

This points to the need to involve those caught up in conflict in processes that are fully nonviolent and fully inclusive of those affected, in which all can speak freely. In a context in which those responsible for justice may themselves enact violence; it is critical to develop an experience of a truly alternative process that can open up an imagination of what is possible.

Here, thus, we envisage the use of *Indaba* as a dialogical institution of that nature. We set out what we propose as its features, possibilities and limitations. Taking into consideration that campus violence is a social evil that needs to be tackled by collective efforts, we rely on the potential of bottom-up approaches of addressing violence and building peace, particularly in the context of higher learning spaces where violence and inhumanity have become endemic.

25.7.2 Modern Applications of Indaba

Indaba constitutes a versatile concept that is used in the business sphere as a strategy to stimulate economic growth (Newenham-Kahindi, 2009); in politics and law, it carries the connotation of an administrative structure and a transformative framework which allows the rethinking of the legislation, and furthering a judiciary system which lines up with social and political changes, such as the context of South Africa emerging from the legacy of past injustices and extreme human rights violations (Hund, 1989; Lynch, 1987). The *Indaba* institution features within dispute resolution discourses at various levels of society including in faith-based organisations (Bridger/Goddard, 2014; Lee, 2009). Willms/Werner (2009) describe *Indaba* as a meeting procedure for valuable debates, and it is also viewed as an important indigenous paradigm of conflict resolution and a model of conversation where critical matters, such as sexuality in a faith organisation, are debated, such as in the Anglican church (Bridger/Goddard, 2014).

The concept of *Indaba* has permeated the field of jurisprudence as a non-adjudicative dispute resolution institution, and it has also incentivised the framework of the KwaZulu-Natal Bill of Rights, as noted by Hund (1989). This was done through working out a legal framework in which various social groups and individual rights are recognised in absolute respect of what inspires values, in both the Western and African worldview (Hund 1989). *Indaba* is singled out as a viable model of conflict transformation and 'problem-solving'. Lee (2009: 148) describes *Indaba* as "a meeting or a talk-shop; it can also mean a story, a legal case or an item of business".

For the purposes of this paper, the *Indaba* institution is discussed from the perspective of peacebuilding, conflict prevention and resolution, and violence prevention.

25.7.3 Indaba, Ubuntu and Violence Prevention

Several studies have established the relations between *Indaba* and conflict management at university level, specifically in issues pertaining to deliberations concerned with freedom of expression, communality and equality; however, the concern is that the import of concepts such as *Indaba* to university context has either undermined or left out its core values (Waghid et al., 2023). This observation is the more correct because campus violence often stems from students' perception of injustices and unfair treatment by university management (James et al., 2015). These scholars note: "General strain theory (GST) and procedural justice theory suggests that when students perceive unfairness in school rules or treatment from teachers, they will be more likely to participate in violence" (p. 1). They further argue" "GST proposes that the strength of these relationships may be reduced by social support" (p. 1). We support the view that, as youths become aware of social discrimination, unfair treatment may develop violent behaviours. This argument is explained by positive youth development and social justice theory (Daiute/Fine, 2003). Student's victimisation is an apparent cause of violence (DeVoe/Bauer, 2011; James et al. 2015). To illustrate this fact, James, Bunch and Clay-Warner (2015, p. 1) comment:

> Data from the 2009 School Crime Supplement of the National Crime Victimization Survey show that students who perceive unfair treatment from teachers are more likely to bring a weapon to school and fight at school than are students who believe that their teachers are fair. Students who perceive that rules are unfair are more likely to bring a weapon to school than are students who believe rules are fair.

This evidence is not limited to the US, but it is worldwide; this was evident in 2011 in the case of Ntsako Mogobe, the chairperson of the Congress of South African Students (COSAS), when he called upon learners "to fight fire with fire and hit back at teachers who use corporal punishment" on them (Le Mottee/Kelly, 2017: 56). It is explained by the general theory of strain (Agnew/White, 1992) and the general theory of crime (Pratt/Cullen 2000; Jones/Quisenberry, 2004). According to Agnew and White (1992: 474), the GTS contends that strain occurs under three circumstances, when others: "(1) prevent or threaten to prevent you from achieving positively valued goals, (2) remove or threaten to remove positively valued stimuli that you possess, or (3) present or threaten to present you with noxious or negatively valued stimuli" (p. 475). One of the reactions to such hostile constraints by others may be resorting to violence. But the general theory of crime emanating from Gottfredson and Hirschi explains the origin of some deviant behaviours (Hirschi/Gottfredson, 1993; Pratt/Cullen, 2000; Jones/Quisenberry, 2004). This theory claims that self-control and opportunity and analogous causes to self-control such as age may explain criminal and other deviant behaviours (Hirschi/Gottfredson, 1993): "In view of the theory, opportunities to commit one or another crime or analogous acts are limitless [...] self-control and opportunity may therefore interact for specific crimes, but are in the general case independent" (p. 50). We agree with the theory that the associations for events, factor, opportunities and self-control can lead to the commission of a crime or to analogous deviant behaviours. In one way or another, such circumstances can be prevented when any causes of deviant conducts originate from the negation of the alterity and humanness or *Ubuntu* of the other by the self.

Thus, we link campus violence to *Ubuntu* and *Indaba* on the premises that ubuntu is essentially about fairness, and *Indaba* is the forum or institution that inspires and pursues justice and fairness in interrelation (Kiyala, 2022b). This is the rationale for exploring the viability of *Indaba* and *Ubuntu* to prevent campus violence. For instance, campus racial injustice has been identified as a factor that can trigger youth violence (Arellano/Vue, 2019). The use of *Ubuntu* as fairness is applicable to students' interactions and interrelations on campus (Letseka, 2014). Violence which causes learners' and students' drop-outs threatens the future of South Africa (Letseka/Maile, 2008). Justice as fairness can be achieved by *Ubuntu* (Letseka, 2014; Ngubane/Makua, 2021). Letseka (2014: 544) notes: "Ubuntu is not only a moral theory concerned with infusing humane dispositions. It also embodies values, morals, and notions of traditional African communal justice. Indeed, in Southern Africa Justice is perceived as Ubuntu fairness". *Ubuntu* is the source of community strength and vitality when such community lives up to the communitarian values of social justice, procedural justice and distributive justice, where there is mutual respect of each other's dignity. In a similar vein Ngubane and Makua (2021: 1) point out:

> At the intersection of Ubuntu pedagogy and social justice, classroom practices are designed to respond to students' cultural competencies and to embrace all linguistic repertoires that students bring to the classroom for learning. Students are treated equally and with dignity and respect regardless of their social backgrounds.

We contend, from an African cosmological and social ethical point of view that the scale of violence we experience on campus and university environments stems from the erosion of the African fundamental value of *Ubuntu*. Violence constitutes the negation of *Ubuntu* of the other and, thus, the failure to recognise one's own *Ubuntu* in an intersubjective deontological ethics.

A "student *Ubuntu-Indaba*" forms a nexus of violence prevention and conflict management in which *Ubuntu* serves as the psychological and ethical underlying principle and *Indaba* as the material space for the pedagogical space where *Ubuntu* is learned, taught and popularised to pervade the whole society. When anti-social acts are committed by students against each other, or by campus management against students or vice-versa, *Indaba* is a viable structure to deal with any contentious matters whether that is related to discipline, violations of campus rules, or settling scores. Furthermore, a student *Indaba* can serve as the channel of communication among students, a conflict resolution space, and student accountability institution. Adopting this approach will lessen victimisation, unfairness, injustice, and low self-esteem often experienced by students. Adding to that, student *Indaba* will minimise the impact of teacher/rule unfair practices and the lack of collaboration among stakeholders like the civil society, parents, community, teachers, the business community and different government departments, the lack of which increases the level of school violence in learning environments.

UNESCO (2017) similarly comments on the absence of students' participation in the decision-making processes, limited involvement of learners when planning interventions, lack of counselling and other supportive services and structures, a situation that, we argue, can be resolved through the use of *Indaba* and *Ubuntu* and their adoption on campus structure and practices.

Goldweber et al. (2013: 470) postulate that the learning environment climate is a "multidimensional construct" consisting of shared values, beliefs, and attitudes that shape student–student and student–teacher dynamics and "set the tone" for behaviours that are acceptable and normative. The non-existence of violence prevention programmes negatively impacts classroom and campus management teachers. Teachers and lecturers have reported "class/lecture room management" to be one of the greatest concerns in their work environment, often leading to burnout, job dissatisfaction, and early exit from the profession (Sullivan et al., 2014: 43). At the school in Mlambo's (2022) study, teachers have had to risk their lives trying to protect learners from outsiders who come to school to attack learners for various reasons. There was a dispute between a student from a notorious family of warlords and another student; members of this family came fully armed and ready to kill. Teachers had to risk their lives by telling the family members that the student they were looking for had ran away. Because of such incidents staff members in institutions of learning experience anxiety, depression, somatic symptoms, lower professional functioning, lower efficacy in classrooms, and lower overall wellbeing (Kanrich, 2015: 5).

25.7.4 Rethinking Indaba for Students' Accountability

Indaba embraces the African humanistic and communitarian values which promote both individual and collective rights and foster social harmony. The spirit of *Ubuntu* prevails during this forum. With the *Ubuntu* spirit, *Indaba* forges the path to reconciliation and peaceful coexistence among former antagonists. This is the agenda pursued by restorative justice which, like *Indaba*, emerges from the indigenous Maori culture of dealing with human violations and the harm caused within a family that also affects a family community. *Indaba* is founded on the same philosophical assumptions and ethical values such as *Ubuntu*, which underscores restorative justice and the African Palaver. Just as the *Indaba* conflict resolution approach helped to resolve the contentious problem of "Anglican communion" (Bridger/Goddard, 2014), *Indaba's* widespread usefulness as jurisprudence (Hund, 1989), its reconciliation teleological effects (Lee, 2009), and its progressive development and business strategy (Newenham-Kahindi, 2009), demonstrate its undeniable potential to redress the erosion of African communitarian values such as *Ubuntu* and the state of living together harmoniously.

From the perspective of an African indigenous jurisprudence, justice is restorative and sustained by the spirit of *Ubuntu*, which can inspire diversional measures in holding offenders accountable. The South African Department of Justice and Constitutional Development underlines the positive impact of restorative justice by contrasting the adversarial consequences of prosecutorial trials with the participatory nature of restorative justice, which enhances partnership between the offender and the offended (Republic of South Africa, 2011). This argument further sustains the necessity of considering and adapting the *Indaba* forum as a tool for conflict resolution and violence prevention in spaces of higher learning.

Notwithstanding the prospects of successful engagement through *Indaba*, namely leading offending students towards reforming their behaviour, and fostering reconciliation with victims, there is a need to consider the following: firstly, "schools and colleges are not correctional facilities" (Karp/Conrad, 2005: 318), and a student who poses a serious threat to the safety of others must definitely be expelled; secondly, a refusal by a student to respond to "judicial proceedings" and failure to duly accomplish the "sanctioning tasks" should make it a necessity that he/she be removed (Karp/Conrad, 2005: 318). Due to the longstanding practice of restorative justice on college campuses and its compatibility with the ethics underlying the *Indaba* dialogue, the synergies of these two mechanisms of redressing after a harm has been caused deserve more consideration in the campaign to eradicate violence in educational environments.

Giving warnings, restricting student offenders' privileges and restorative sanctions should be prioritised. In cases where all these avenues aiming at bringing the transformation of student offenders prove ineffective, coercion may be utilised to gain the compliance of unruly students in executing sanctions. Karp (2004: 9) argues that suspension and expulsion, though they can be maintained, should be minimised whenever possible because they are "anti-communitarian devices". He argues that taking a student away from school or campus implies transferring the problem to another community without settling it. There is an absolute necessity for peace education on campus. This points to the need for an *Indaba* to be more than a disciplinary process; it needs to explore issues from a peacebuilding perspective.

25.7.5 Student Indaba and Peace Education

The World Report on Violence and Health acknowledges the key role of culture and traditions in preventing violence, and the same report suggests that cultural norms need to be respected, because people look to their culture as a "source of protection against violence" (Krug et al., 2002: 16). The need for educational campaigns to reduce violence through school and university programmes developed by departments of education and culture emerges in this report (Krug et al., 2002). Peace education here refers to processes which aim to maintain societal harmony and peaceful cohesion among individuals, and which offer a nonviolent alternative response to violence and its consequences. Student *Indaba*, restorative justice, and peace education could build a strong peace infrastructure in the fight against interpersonal conflict before it degenerates into violent confrontation, and between students and university management before unresolved conflict turns into vandalism.

Efforts to promote peace education in schools and universities reflect the vision of UNESCO that states that, although some students may unfortunately experience violence in their homes, teachers can provide them with alternative ways of being by modelling constructive, non-violent behaviour and by fostering empathy and peaceful conflict resolution skills (UNESCO, 2009: 7). For this reason, a bottom-up peace education and alternative dispute resolution model, the ownership of which rests on students' shoulders, could stimulate a positive response and yield more promising outcomes, namely violence prevention and the increase in individual safety.

Harris (1999) notes that peace education can be achieved by infusing the concepts of justice and peace into the existing curricula of schools, colleges and universities and should include, to some extent, the techniques of conflict resolution and ways to make learning environment safer spaces and should emphasise peacebuilding interventions. In schools, peace education can be incorporated in the Life Orientation (LO) period. This can be done through role plays and students can have an opportunity of reflecting on what they have observed. Students can be given opportunities to engage in a topic via participation in role plays, writing posters with peace messages, and writing poems or composing songs on issues relating to peace. The approach is fun, entertaining and it gives students an opportunity to release stress since the approach is different from the traditional way of learning. Peace education can be introduced to parents and stakeholders by inviting them to be spectators and observers of this activity in schools. At the end of the class, observers can be encouraged to comment on the negative or positive message of the activities and develop solutions on how to respond to such situations.

Mlambo (2022) observed that in grade 10 students begin to exhibit tendencies where they develop an "I know it all or I don't care attitude". At this stage they begin to be rebellious against any form of authority and have difficulty in taking instructions from adults. One needs to be proactive; in the case of her study, Mlambo (2022) introduced programmes that developed ways of tackling the relevant issues through peaceful interactions as a peacebuilding programme. By the time students start at university and colleges they should have already mastered strategies to resolve conflict peacefully.

Children learn through observation, hearing, and sometimes through experiences, and the learned behaviour becomes a norm in human interactions. According to Mncube and Harber (2017: 9), when children are trained, they learn to train others. Children who are lectured to, learn to lecture; if they are admonished, they learn how to admonish; if scolded, they learn how to scold; if ridiculed, they learn how to ridicule; if humiliated, they learn how to humiliate; if their psyche is killed, they learn to kill.

25.7.6 Necessary Elements of a Student Indaba

It is essential that from the outset an *Indaba* proceeds with a process and a set of values that are distinct from those of conventional formal meetings and formal disciplinary hearings. The first we would point to is that it is a space that does not reproduce the violence of the society – or of the institution. To develop a sense of student accountability and ownership, it is critical that the ground rules of such a process are developed and agreed upon by those in the *Indaba*, and not imposed by school or university authorities. In practice we have found this a simple task that participants readily understand, but one that is nonetheless very useful as a point of reference.

In our context, the most frequently articulated principle of that of respect (*ukuhlonipha* in IsiZulu). What this means may need some discussion, because the term is generally used for the respect that children must give adults; it is not always seen as reciprocal. Other typical proposals include the principle that one may disagree without seeking to undermine the other person, that people should collaborate to ensure success and the need to ensure confidentiality on certain sensitive issues. Such ways of working are not rigid rules; they should be seen as subject to revision and strengthening.

The way an *Indaba* is constituted needs to address how it is facilitated. In formal institutions, there is a chairperson who is typically also the most senior manager; in the case of an SRC, the President. However, the three roles of leadership, management and facilitation differ; we would argue that an *Indaba* should consider the possibility of selecting a skilled facilitator for the plenary discussion and attend also to the facilitation roles of smaller groupings as needed. Even if the decision is to give that task to the formal leader, there is value in making the choice explicit, as the role of facilitator of an inclusive dialogue needs to be different from that of formal decision-making.

There is no need for all *Indaba* sessions to follow the same format. The principles of collaborative enquiry, of inclusivity and of *Ubuntu* can sometimes best be served in group discussions, role plays or listening dyads (Weissglass, 1990). Similarly, Mlambo (2022) reports on the range of peace education methods she used in the intervention with learners in one school.

25.8 Conclusion

The prospect of a decent society, freed from violence, requires initiating and putting in place transformation strategies which should permeate the whole fabric of society. For schools, universities and campus colleges, this involves a holistic approach which exploits the synergies of *Indaba*, restorative justice, *Ubuntu* and peace education. The main challenge remains the lack of experimental studies on *Indaba* and *Imbizo* as dispute resolution alternatives and violence prevention mechanisms in the environments of institutions of learning. This is a field for further study which needs to be pursued in order to finally establish, on the basis of evidence, that the threefold perspectives of campus violence prevention and peacebuilding, namely the *Indaba* forum, peace education and restorative justice, are viable alternatives to the Western-oriented criminal justice model which relies substantively on retribution, which shows limited effects in deterring aggressive behaviour and the violation of rules and policies in schools, colleges and universities. Instilling principles of nonviolence in students through a bottom-up structure such as *Indaba* built on ubuntu – African humanism – used by top student leaders to provide direction and guidance, would potentially avert the detrimental consequences of violence in learning environments and students' unruly behaviour.

References

Agnew, R., & White, H.R. (1992). An empirical test of general strain theory. *Criminology, 30*(4): 475–500.

Anderson, A.M. (2003). "Restorative justice, the African philosophy of Ubuntu and the diversion of criminal prosecution." 17th international conference of the International Society for the Reform of Criminal Law, The Hague, Netherlands. https://www.readkong.com/page/restorative-justice-the-african-philosophy-of-ubuntu-and-4649952?p=1. Accessed 28 February 2019.

Arellano, L., & Vue, R. (2019). Transforming campus racial climates: Examining discourses around student experiences of racial violence and institutional (in) action. *Journal of Diversity in Higher Education, 12*(4): 351.

Bridger, F., & Goddard, A. (2014). Learning from Indaba: Some lessons for post-Pilling conversations. *Anvil, 30*(1), 41–56.

Burton, P., & Leoschut, L. (2013). *School Violence in South Africa: Results of the 2012 National School Violence Study*. Monograph Series (12). Cape Town. Centre for Justice and Crime Prevention.

Byerly, I.B. (1998). Mirror, mediator, and prophet: The music indaba of late-apartheid South Africa. *Ethnomusicology, 42*(1), 1–44.

Daiute, C., & Fine, M. (2003). Youth perspectives on violence and injustice. *Journal of Social Issues, 59*(1), 1–14.

DeVoe, J.F., & Bauer, L. (2011). Student victimization in US schools: Results from the 2009 School Crime Supplement to the National Crime Victimization Survey. NCES 2012-314. National Center for Education Statistics, https://nces.ed.gov/pubs2012/2012314.pdf. Accessed 7 August 2023.

Durban University of Technology. (2014). *Handbook for 2014: General handbook for students*. Durban University of Technology.

Edwards, S.D., Holdstock, T.L., Nzima, D. R., & Edwards, D.J. (2020). Indaba: Towards an African psychology of education founded on indigenous knowledge systems. *Indilinga African Journal of Indigenous Knowledge Systems, 19*(2), 155–164.

Eshowe.com. (2001–2019). Zulu hierarchy and political systems. https://eshowe.com/zulu-hierarchy-political-systems/. Accessed 5 November 2019.

Gade, C.B.N. (2013). Ubuntu and restorative justice: addressing the strife and divisions of the past in post-apartheid South Africa. PhD, Aarhus University. https://konfliktloesning.dk/wp-content/uploads/2017/03/Christian-B-N-Gade-phd-marts-2013.pdf. Accessed 7 August 2023.

Gevers, A., Nwabisa, J-S., & Yandisa, S. (2013). Gender-based violence and the need for evidence-based primary prevention in South Africa. *African Safety Promotion: A Journal of Injury and Violence, 11*(2), 14–20.

Goldweber, A., Waasdorp, T. E., & Bradshaw, C. P. (2013). Examining associations between race, urbanicity, and patterns of bullying involvement. *Journal of Youth and Adolescence*, 42, 206–219. https://doi.org/10.1007/s10964-012-9843-y

Gordon, S.F., & Collins, A. (2013). "We face rape. We face all things": Understandings of gender-based violence amongst female students at a South African university. *African Safety Promotion: A Journal of Injury and Violence Prevention, 11*(2), 93–106.
Harris, I. (1999). Peace Education: Colleges and Universities. *Encyclopedia of Violence, Peace, and Conflict, 2,* 679–689.
Hirschi, T., & Gottfredson, M. (1993). Commentary: Testing the general theory of crime. *Journal of Research in Crime and Delinquency, 30*(1), 47–54.
Hund, J. (1989). A bill of rights for South Africa. *American Journal of Jurisprudence,* 34, 45–64.
James, K., Bunch, J., & Clay-Warner, J. (2015). Perceived injustice and school violence: An application of general strain theory. *Youth Violence and Juvenile Justice, 13*(2), 169–189.
Jones, S., & Quisenberry, N. (2004). The general theory of crime: How general is it? *Deviant Behavior, 25*(5), 401–426.
Karodia, S. (1988). The Indaba education proposals: Mass schooling for a divided society. *Critical Arts, 5*(1), 146–153.
Karp, D.R. (2004). Introducing restorative justice to the campus community. In D.R. Karp & A. Thom (Eds.), *Restorative justice on the college campus: Promoting student growth and responsibility, and reawakening the spirit of campus community* (pp. 5–15). Springfield: Charles C. Thomas.
Karp, D., & Conrad, S. (2005). Restorative justice and college student misconduct. *Public Organization Review, 5*(4), 315-333.
Karp, D.R., & Sacks, C. (2014). Student conduct, restorative justice, and student development: findings from the STARR project: a student accountability and restorative research project. *Contemporary Justice Review, 17*(2), 154–172.
Krug, E.G., Dahlberg, L. L., Mercy, J. A., Zwi, A. B., & Lozano, R. (2002). World report on violence and health. https://apps.who.int/iris/bitstream/handle/10665/42495/9241545615_eng.pdf. Accessed 7 August 2023.
Kiyala, J.C.K. (2022a). Indigenous approaches of peacebuilding and accountability of former child soldiers: African case studies. In J.C.K. Kiyala & G.T. Harris (Eds.), *Civil society and peacebuilding in Sub-Saharan Africa in the Anthropocene: An overview* (pp. 423–464). Springer Nature Switzerland. https://doi.org/10.1007/978-3-030-95179-5_18
Kiyala, J.C.K. (2022b). Underlying moral justification of Indaba and Baraza dialogic institutions in African social ethics and philosophy. In L.C. Rodrigues & J.O. Chimakonam (Eds.), *African Ethics: A guide to key ideas* (pp. 159–184). London: Bloomsbury Publishing.
Lanni, A. (2005). Future of community justice. *Harvard Civil Rights-Civil Liberties Law Review,* 40 (Suppl.), 359–405.
Lee, P.J. (2009). Indaba as obedience: A Post Lambeth 2008 assessment 'If someone offends you, talk to him'. *Journal of Anglican Studies* 7(2), 147–161.
Letseka, M. (2014). Ubuntu and justice as fairness. *Mediterranean Journal of Social Sciences,* 5(9), 544.
Letseka, M., & Maile, S. (2008). *High university drop-out rates: A threat to South Africa's future.* Human Sciences Research Council Pretoria.
Louw, D.J. (2008). The African concept of ubuntu and restorative justice. In D. Sullivan & L. Tifft (Eds.) *Handbook of restorative justice* (161–173). London: Routledge.
Lynch, E.A. (1987). The KwaZulu/Natal Indaba: A federalist proposal for South Africa. *Publius: The Journal of Federalism,* 17(3), 231–248.
Mayhew, M.J., Cardwel, R.J., & Goldman, E.G. (2011). Defining campus violence: A phenomenological analysis of community stakeholder perspectives. *Journal of College Student Development, 52*(3), 253–269.
Metz, T. (2011). Ubuntu as a moral theory and human rights in South Africa. *African Human Rights Law Journal, 11*(2), 532–559.
Mlambo, P.S. (2022). Interventions strategies to reduce school violence in Bhekulwandle rural community. Doctoral thesis. Durban University of Technology.
Mncube, V., & Harber, C. (2013). *Dynamics of violence in South African schools.* Pretoria: UNISA,

Moulder, J. 1991. Karodia's working paper on Indaba's Education Report: A guide to some neglected policy questions. *Critical Arts, 5*(4), 1–12.

Moumakwa, P.C. (2010). *The Botswana Kgotla System: A mechanism for Traditional Conflict Resolution in modern Botswana. Case study of the Kanye Kgotla* (Master's Thesis in Philosophy of Peace and Conflict Transformation Faculty of Humanities online). University of Tromsø. https://munin.uit.no/bitstream/handle/10037/3211/thesis.pdf?sequence=1. Accessed 7 August 2023.

Nabudere, D.W. (2005). Ubuntu philosophy: memory and reconciliation. https://repositories.lib.utexas.edu/bitstream/handle/2152/4521/3621.pdf. Accessed 20 June 2020.

Newenham-Kahindi, A. (2009). The transfer of Ubuntu and Indaba business models abroad: A case of South African multinational banks and telecommunication services in Tanzania. *International Journal of Cross Cultural Management, 9*(1), 87–108.

Ngubane, N.I. & Makua, M. (2021), Intersection of Ubuntu pedagogy and social justice: Transforming South African higher education, *Transformation in Higher Education, 6*, 1–8. https://doi.org/10.4102/the.v6i0.113

Nhamo, G. (2021). Localisation of SDGs in Higher Education: Unisa's whole institution, all goals and entire sector approach. *Southern African Journal of Environmental Education, 37*(1), 63–85. https://doi.org/10.4314/sajee.v37i1.5

Noll, D. (2003). *Peacemaking: practicing at the intersection of law and human conflict*. Telford: Cascadia Pub House.

Nomahlubi, J. (2018). Rhodes and NMU on 'shutdown' after alleged student rapes." Tiso Blackstar Group - August 6, 2018. https://www.talkofthetown.co.za/2018/08/06/rhodes-and-nmu-on-shutdown-after-alleged-student-rapes/ Accessed 6 August 2018.

Parry, C.D.H., Plüddemann, A., Louw, A., & Leggett, T. (2004a). The 3-metros study of drugs and crime in South Africa: Findings and policy implications. *The American Journal of Drug and Alcohol Abuse, 30*(1), 167–185.

Parry, C.D.H., Myers, B., Morojele, N.K., Flisher, A.J., Bhana, A., Donson, H., & Plüddemann, A. (2004b). Trends in adolescent alcohol and other drug use: findings from three sentinel sites in South Africa (1997–2001). *Journal of Adolescence*, 27(4), 429–440.

Pratt, T.C., & Cullen, F.T. (2000). The empirical status of Gottfredson and Hirschi's general theory of crime: A meta-analysis. *Criminology, 38*(3), 931–964.

Republic of South Africa. (2011). *Restorative justice: the road to healing*. Available via the Department of Justice and Constitutional Development. http://www.justice.gov.za/rj/2011rj-booklet-a5-eng.pdf. Accessed 26 June 2013.

Rinker, J.A., & Chelsey, J. (2014). Restorative justice as reflective practice and applied pedagogy on college campuses. *Journal of Peace Education, 11*(2), 162–180.

Scheid, A.F. (2011). Under the palaver tree: Community ethics for truth-telling and reconciliation. *Journal of the Society of Christian Ethics, 31*(1), 17–36.

Schirch, L. (2004). *The little book of strategic peacebuilding*. Pennsylvania: Good Books.

Seedat, M., Van Niekerk, A., Jewkes, R., Suffla, S., & Ratele, K. (2009). Violence and injuries in South Africa: prioritising an agenda for prevention. *The Lancet, 374*(9694), 1011–1022.

Setlalentoa, B., Pisa, P., Thekisho, G., Ryke, E., & Loots Du, T. (2010). The social aspects of alcohol misuse/abuse in South Africa. *South African Journal of Clinical Nutrition, 23,* 11–15.

Skelton, A. (2013). The South African Constitutional Court's restorative justice jurisprudence. *Restorative Justice, 1*(1), 122–145.

South African Government News Agency. (2015). Zulu King calls for protection of foreign nationals https://www.sanews.gov.za/south-africa/zulu-king-calls-protection-foreign-nationals. Accessed 7 August 2023.

South African Police Service. (2023). Republic of South Africa: Crime situation. https://www.saps.gov.za/services/crimestats.php. Accessed 7 Augut 2023.

UNESCO. (2009). *Stopping violence in schools: A guide for teachers*. https://unesdoc.unesco.org/ark:/48223/pf0000184162. Accessed 7 August 2023.

University of KwaZulu-Natal. (2008). *University of KwaZulu-Natal: Rules for Students (and University Traffic Rules) 2008*. www.ukzn.ac.za. Accessed 10 October 2015.

Waghid, Y., Terblanche, J., Shawa, L. B., Hungwe, J. P., Waghid, F., & Waghid, Z. (2023). (Re)-imagining the Indaba Concept: In quest for a communal African university of deliberation, freedom of expression and equality. In Y. Waghid, J. Terblanche, L. B. Shawa, J. P. Hungwe, F. Waghid & Z. Waghid (Eds.), *Towards an Ubuntu university: African higher education reimagined* (pp. 39–57). Cham: Palgrave Macmillan. https://doi.org/10.1007/978-3-031-06454-8_6

Willms, D.G., & Karolina, W. (2009). Climate change and conflict in Africa. In H. Besada & N.K. Sewankambo (Eds.), *Climate change and conflict in Africa: Adaptation, mitigation and governance challenges* (pp. 26–30). Waterloo: The Centre for International Governance Innovation.

Wilson, M.E. (2006). Restorative justice on the college campus: Promoting student growth and responsibility, and reawakening the spirit of campus community (review). *Journal of College Student Development, 47*(3), 355–359.

World Population Review. (2023). Murder rate by country 2023. https://worldpopulationreview.com/country-rankings/murder-rate-by-country. Accessed 4 August 2023.

Writer, S. (2019). *South Africa crime stats 2019: everything you need to know*. https://businesstech.co.za/news/government/340513/south-africa-crime-stats-2019-everything-you-need-to-know/. Accessed 14 October 2019.

Zehr, H. (1990). *Changing lenses: a new focus for crime and justice*. Scottsdale: Herald Press

Zehr, H. (2013). Retributive justice, restorative justice. In G. Johnstone (Ed.), *A restorative justice reader* (pp. 23–35). London: Routledge.

Zulu, B.M., Urbani, G., Van der Merwe, A., & Van der Walt, J.L. (2006). Violence as an impediment to a culture of teaching and learning in some South African schools. *South African Journal of Education, 24*(2), 170–175.

Chapter 26
Peace-building Processes to Prevent and Manage Xenophobia in South Africa

Clayton Hazvinei Vhumbunu

Abstract The history of xenophobia in South Africa dates back to pre-1994 period where many political and economic refugees who had flocked into the country from other Sothern African countries and beyond for jobs and other economic opportunities were subjected to different forms of violence, hostilities and discrimination by locals. However, the post-1994 era witnessed several episodes of intensified and more brutal xenophobia against immigrants in South Africa as immigrants continued to be blamed by locals for crime, unemployment and spreading diseases. For instance, the 2008 xenophobic attacks resulted in the death of over 60 people across South Africa. The Government of South Africa, in partnership with various stakeholders, has been implementing different peace-building activities that have been aimed at preventing the recurrence, escalation and continuation of xenophobic attacks. In March 2019, the Government launched the National Action Plan (NAP) to Combat Racism, Racial Discrimination, Xenophobia and Related Intolerance in addition to the already existing National Strategy for Developing an Inclusive and a Cohesive South African Society. This paper reflects on the successes, challenges and prospects of addressing xenophobia in South Africa through peace-building interventions. Methodologically, the paper utilises secondary data from various sources for analysis whilst the concept of *peace-building* provides a conceptual framework of analysis for the study. Findings from the study – in the form of gaps, limitations, complexities, and success stories in addressing xenophobia in South Africa – are salient in

Dr Clayton Hazvinei Vhumbunu is a Research Fellow in the Faculty of Humanities (Centre for the Advancement of Scholarship) at the University of Pretoria; South Africa. He holds a PhD in International Relations and his research interests are in the areas of International Relations, Regional Integration, Conflict Transformation, Public Governance, Public Policy and China-Africa Relations.University of Pretoria, Faculty of Humanities (Centre for the Advancement of Scholarship), Hatfield; Pretoria; South Africa. Email: cvhumbunu@gmail.com

C. H. Vhumbunu (✉)
Faculty of Humanities (Centre for the Advancement of Scholarship), University of Pretoria, Pretoria, South Africa
e-mail: cvhumbunu@gmail.com

J. C. K. Kiyala and N. Chivasa (eds.), *Climate Change and Socio-political Violence in Sub-Saharan Africa in the Anthropocene*, The Anthropocene: Politik—Economics—Society—Science 37,
https://doi.org/10.1007/978-3-031-48375-2_26

proposing new interventions and adjusting existing peace-building initiatives in order to reduce the incidences of xenophobia in the country.

Keywords Xenophobia · Peacebuilding · South Africa

26.1 Introduction

The history of xenophobia in South Africa dates back to the late 19th century when political and economic refugees who had flocked into the country from other Sothern African countries and beyond for jobs and other economic opportunities were subjected to different forms of violence, hostilities and discrimination by locals. Migrant labour, mostly from Southern Africa, who provided cheap labour largely in white-owned mines and farms, and later in established factories in South Africa were subjected to discrimination and violence (see Klotz, 2016; Harris et al., 2018; Keer et al., 2019; Phakathi, 2019; Misago, 2016) although these negative anti-immigrant sentiments and hostilities were not as intense, brutal and widespread as has been the case in post-Apartheid South Africa. Since 1994, there has been several incidents of xenophobia across South Africa, with most of them concentrated in the provinces of Gauteng, Western Cape and KwaZulu-Natal (KZN). From 1994 to 30 June 2023, the African Centre for Migration & Society (ACMS, 2023)'s Xenowatch[1] reports that there have been 1017 incidents of xenophobia in South Africa across the country's nine provinces, and these incidents have resulted in the displacement of over 124,706 immigrants and deaths of 659 people whilst 5111 shops have been looted during these xenophobic incidents. This Chapter reflects on the experiences, successes and challenges faced in implementing the several peace-building interventions by the South African Government in collaboration with various partners and stakeholders as they collectively seek to prevent the recurrence, escalation and continuation of xenophobic attacks.

26.2 Methodology and Materials

This analysis relied on secondary data sources in the form of reports, policy documents, books, journal articles and other relevant sources. Key considerations were made to ensure that the utilised secondary data sources and materials are authentic, credible and reputable. The consulted sources were thoroughly examined to ensure that the applied data collection methods were appropriate whilst other factors such

[1] Xenowatch is an open-source tool used to collect, share and monitor xenophobic discrimination information in South Africa. It was developed by the ACMS, Witswatersrand University; South Africa. *See* the Xenowatch Dashboard at https://www.xenowatch.ac.za/statistics-dashboard/.

as periodicity of collected data and purpose for which the data was collected were also considered (*see* Johnston, 2017; Smith, 2008; Whiteside et al., 2012).

26.3 Conceptual Understanding of Xenophobia and Peacebuilding

In general terms, xenophobia[2] is understood as irrational fear, dislike or hatred of foreigners, outsiders or strangers (in-group versus out-group), which is often manifested in negative and discriminatory attitudes and behaviour. The United Nations High Commissioner for Refugees (UNHCR, 2020:16)'s *Guidance on Racism and Xenophobia* defines xenophobia a attitudes, prejudices and behaviour that reject, exclude and often vilify persons, based on the perception that they are outsiders or foreigners to the community, society or national identity.

The UNHCR definition is not very distinct or dissimilar from other definitions of xenophobia (*for instance see definitions in* Bordeau, 2009; Wimmer, 1997; Ndlovu-Gatsheni, 2010; Reilly, 2001). UNESCO (2017)'s Glossary of Migration Related Terms defines xenophobia as attitudinal orientation of hostility against; and rejection, exclusion and vilification of; non-nationals based on the perception that they are outsiders or foreigners in a given population, community, or society. Authors such as Crush et al (2008) make a submission that the xenophobia as a term is interchangeable with related terms such as *'xeno-racism', 'immigration-phobia', 'ethno-exclusionism', 'anti-immigrant prejudice', 'autochthony',* and *'nativism'*. In all the definitions, the common denominator is that those considered to be 'foreigners', 'outsiders', or 'strangers' are denied equal rights on the basis of their geographic origins, physical identity characteristics or shared ancestry (whether real or perceived). Whilst the factors that may instigate the fear can vary depending on circumstances and context, xenophobia manifest itself in both violent forms (such as physical attacks with intention to cause bodily harm) and/or non-violent forms (such as prejudice, stereotype, name-calling, harassment and discrimination).

Due to the inherent discrimination, prejudices, stereotypes, harassment, violence, human rights abuses, intolerance and unequal treatment of human beings that is often associated with xenophobia; all national and international conventions prohibit xenophobia and often equates it to racism. The Universal Declaration of Human Rights of 1948; the International Convention on the Elimination of All Forms of Racial Discrimination of 1965; the Declaration on the Human Rights of Individuals Who Are Not Nationals of the Country in which They Live of 1985; the Vienna Declaration and Programme of Action adopted by the World Conference on Human Rights in 1993; the Durban Declaration and Programme of Action (DDPA) adopted by the World Conference against Racism, Racial Discrimination, Xenophobia and Related Intolerance in 2001; the International Covenant on Civil and Political Rights

---

[2] The term itself is etymologically traced to the Greek words: '*Xenos*' meaning 'stranger' or 'foreigner', and '*phobos*' meaning 'fear'.

of 1966; the UN Convention on the Protection of the Rights of All Migrant Workers and Members of their Families (CMW) of 1990; Convention Relating to the Status of Refugees of 1951; and other several Resolutions of the United Nations (UN) General Assembly and those of the UN Commission on Human Rights (such as Resolutions 7/34, 16/33, 25/32 and 76/1 of 2008, 2011 and 2014 respectively) all prohibit xenophobia in all sectors of private and public life; and call upon all member states to eliminate such practices.

At continental and regional level, the African Union Convention Governing the Specific Aspects of Refugee Problems in Africa of 1969; the African Charter on Human and People's Rights of 1981; the Revised SADC Treaty of 1992 and the SADC Protocol on the Facilitation and Movement of Persons of 2005. The South African Constitution of 1996 protects the rights of all people in the country (including non-nationals). In the Preamble of the Constitution, a declaration is made that "[w]e the people of South Africa…[b]elieve that South Africa belongs to all who live in it, united in our diversity". Section 9 of the Constitution further prohibits discrimination against anyone any grounds including race, ethnic or social origin, and birth. These rights apply to immigrants, refugees and asylum seekers, and they are also enshrined in the Bill of Rights of the Constitution and other South Africa laws such as the Refugees Act of 1998.

Despite the existence of various legal and policy frameworks that prohibit xenophobia, there is an international trend in xenophobic incidences which have also been given rise to by the ever-increasing number of international migrants, refugees and displaced people. The International Organization for Migration (IOM, 2022) estimates that there were at least 281 million international migrants in the world in 2020, which is three times the 1970 estimate. On the other hand, the UNHCR (2022) states that there are at least 82.4 million displaced people in the world, with nearly 26.4 million being refugees. Whilst a plethora of factors and theories have tended to be conceptually and practically explain xenophobia, economic and political factors are predominant (see Figure 26.1).

26.4 Contextualising Xenophobia in South Africa within the Theories of Xenophobia

The Xenophobic incidents that have been experienced on South Africa since pre-Apartheid era can be well understood within the context of theories that explain the phenomenon. All this can be better appreciated within the context of the migration phenomenon in the SADC region, specifically the influx of migrants in South Africa during the colonial era (Bantu migration which predated the arrival of white settlers in 1652), the country's post-1994 immigration regime, the general retardation and/or deterioration of economies within the Southern Africa and beyond, and the refugee factor driven by conflict developments in selected countries in the Great Lakes Region and Horn of Africa such as the Democratic Republic of Congo (DRC), Ethiopia,

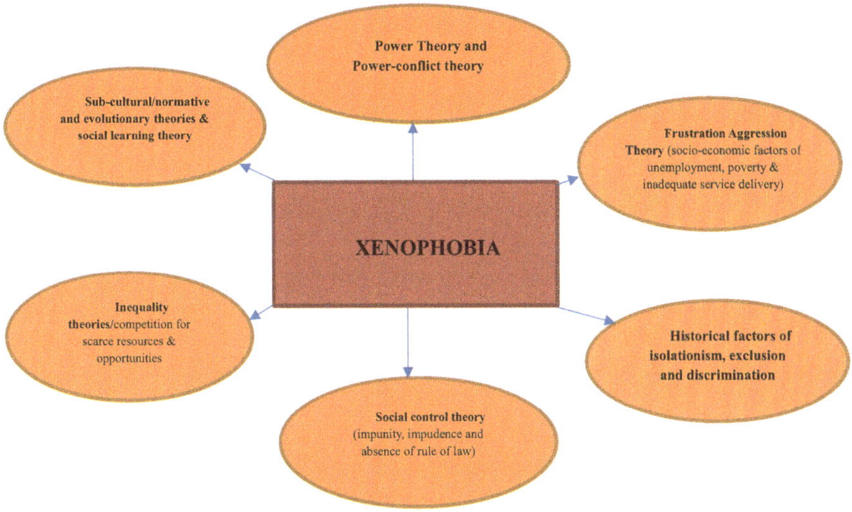

Fig. 26.1. Drivers and theories of Xenophobia. *Source* Author's construction based on factors from different sources and analysis

Somalia, and Burundi; as well as South Africa's steady economic growth since 1994 which created a number of economic opportunities whose 'seductive' power was felt from as far as the Far East in China, Bangladesh, Pakistan, and India.

The influx of black African migrants into South Africa is usually traced back to the Witwatersrand Gold Rush that followed the discovery of gold around the 1880s in the now Gauteng Province. With limited labour supply to work in the mine industry, an association of employers established by the South African Chamber of Mines – the Rand Native Labour Association – engaged labour initially from Mozambique. Later, the incorporation of the Witwatersrand Native Labour Association (WNLA) in the early 1990s, saw the recruitment, transportation, training, deployment and repatriation of labour from Basutoland, Bechuanaland, Belgian Congo (DRC), Germany West Africa (Namibia), Mauritius, Northern Rhodesia, Nyasaland, Portuguese West Africa (Angola), Portuguese East Africa (Mozambique), Southern Rhodesia, Swaziland, Tanganyika, Kenya and as far as Uganda, Egypt, and Liberia (*see* Gewald, 2015; Forrest, 2015). Bilateral labour agreements between South Africa and the labour supplying countries ensured constant supply and migrant labour came in huge numbers. For instance, in 1973 it was reported that 140,000 Malawians were working in South African mines whilst in 1975, Mozambican workers totalled 127,000 (Wentzel/Tlabela in Kok et al., 2006: 74–75).

The post-Apartheid witnessed changes in the Migration Policy of South Africa. These comprised the adoption of the White Paper on International Migration in Mar 1999 and 2017; amendment of the Aliens Control Act of 1991 in 1995; promulgation of the Refugee Act No. 130 of 1998; enactment of the Immigration Act No. 13 in 2002 and its amendment in 2014; and the gazetting of Immigration Regulations in the same year. All these changes were in response to the increasing inflows of migrants into South Africa from the region and beyond. As of 2021, South Africa was host to around 2.9 million immigrants (Migration Policy Institute, 2021), which is almost 5% of the national population of 60 million. The majority of the migrants are from Zimbabwe (24%), Mozambique (12%), Lesotho (7%) and Malawi (3%), whilst the other countries of United Kingdom, DRC, Somalia, Botswana, Angola and Eswatini all account for 2% of total immigrants each (United Nations Population Division, 2020).

As more and more immigrants flocked into South Africa, there have been community tensions across all South African provinces between locals and foreign nations, which have been manifested in xenophobic violence, discrimination and harassments. As cited above from Xenowatch, 889 incidents of xenophobia have been reported in South Africa since 1994, and these incidents have resulted in the displacement of over 122 immigrants, death of 621 people and looting of 4831 shops.

There was reported xenophobic violence from 1994 in a number of provinces of Free State, Gauteng, KwaZulu-Natal, Limpopo, and Western Cape. For example, the Human Rights Watch Report on the Abuse of Un-Documented Migrants, Asylum Seekers, and Refugees in South Africa of March 1998, reported widespread xenophobia and abuse of foreigners in Cape Town, Johannesburg, Pretoria, Krugersdorp, Pollsmoor, and Modderbee prisons. The report indicated that migrants from countries such as Burundi, Malawi, Mozambique, Nigeria, Senegal, Uganda, and Zimbabwe were physically and verbally assaulted by the public and blamed for being "responsible for the current crime wave, rising unemployment, or even spread of diseases" and other social ills (Human Rights Watch, 1998: n.p). There were xenophobic attacks from Dec 1994 to Jan 1995 as Operation *Buyelekhaya* ('Go Home') in Alexandra, and the call by the Greater Johannesburg Hawkers Association for South Africans to boycott goods sold by *amakwerekere* (black African migrants) and those from Asia and North Africans in 1996 and Nov 1997 (Human Rights Watch, 1998: n.p). Further, xenophobic attacks were also experienced in the years 1999 to 2001, then every year from 2004 to 2021 (*see* Cibane, 2015; KwaZulu-Natal Christian Council, 2016; Monson/Misago, 2016; Simelane, 2021; ACMS, 2022). Recently, at the beginning of 2022, there has been Operation Dudula[3], which is raising tensions and fears of xenophobia (Simelane, 2021).

In explaining the underlying causes, drivers and triggers of xenophobia in South Africa, a number of theories are instructive. What has to be acknowledged from the onset, however, is the irrefutable fact that human action and human behaviour is influenced by a multitude and multiplicity of factors such as personal attitudes, emotions, ethics, culture, values, authority, hypnosis, genetics, persuasion, and all these vary from one person/group to the other (*see* Hewstone et al, 2008; Mishra, 2008; Zander, 1979). It is even made more complex by the fact that group or community action is

[3] Operation Dudula is a protest movement led by South African citizens and community activists which is targeting undocumented foreign nationals in South Africa, claiming that they are engaged in illegal drug trade, prostitution, sale of expired products and taking away jobs meant for locals. Operation protesters are working in conjunction with the *Put South Africa First Movement*.

also further influenced by group dynamics and related social psychological principles of conformity, common motives, belongingness, perception, re-adjustment, change and power, et cetera (*see* Zander, 1979; Hewstone et al., 2008; Mishra, 2008). The theories presented in Fig. 27.1, albeit their limitations and constraints in explaining some human behaviour dimensions are therefore complementary and collectively assist to explain the root and immediate causes of xenophobia.

Historical factors of isolation, exclusion and discrimination are often cited by a number of scholars as contributing factors to xenophobia (*see* Choane et al., 2011; Zondi, 2008; Ngcamu/Mantzaris, 2021). The common and recurrent argument is that South Africa's long history of isolation during Apartheid plays an influential role in perpetuating xenophobia. Apartheid separated South Africa from significant parts of the continent and the rest of the world. The implementation of various sanctionary measures by the international community from the 1960s following the UN General Assembly Resolution 1761 of November 1962 which condemned South African Apartheid policies, which were only lifted by November 1993, meant that there was very minimal interaction between South Africa and the world hence the hostility towards foreign nationals, including immediate neighbouring citizens, as 'strangers' and 'outsiders'. Even South Africa's National Action Plan (NAP) to Combat Racism, Racial Discrimination, Xenophobia and Related Intolerance (2019: 5), states in its Preamble:

> The many years of a racist and isolationist policy of apartheid have planted seeds of xenophobia, particularly towards Africans, undoing centuries of brotherhood and sisterhood among Africans in South Africa and those from other parts of the continent. This is how Africans have come to be the worst victims of xenophobia in contemporary South Africa.

Moreover, some of these 'outsiders' become established businessmen and entrepreneurs as evidenced by booming small medium enterprises in South Africa's townships (*see* Lamb et al., 2019; Moiloa, 2015; Liedeman et al., 2013). This success weighs on outsider-insider power dynamics as suggested by the Power Theorists and this may breed conflicts as locals attempt to neutralise the dominant foreign nationals with growing economic power and influence as explained by Power-Conflict Theorists.

One may also not escape the role of inequalities as explained by the inequality theories and augmented by the Frustration Aggression Theory. The inequalities are worsened by rising unemployment, poverty and deterioration of social services. South Africa is considered the most unequal society in the world with 90% of the country's wealth being owned by only 10% of the population whilst the UN Human Development Report of 2020 mentions that 55.5% of South Africans are living below the poverty line (*see* Vhumbunu, 2021). In terms of unemployment, the third quarter of 2021, the number of employed people declined from 600,000 to 14,3 million people compared to the second quarter, with official unemployment rate at 34,9% (StatsSA, 2021) compared to 20.5% in 1994. The level of dissatisfaction with service delivery in South Africa is evidenced by the numerous service delivery protests since 1994. For instance, in the six months between Aug2020 and

Jan 2021, the South African Police Service (SAPS) Incident Registration Information System (IRIS) recorded a total of over 900 service delivery protests across the country (Martin, 2021). The fact that South Africa's economic growth and Gross Domestic Product (GDP) is still way higher than other countries in the region and second highest after Nigeria does not in any way dismiss the socio-economic challenges facing ordinary citizens.[4] All these conditions affect the basic aspirations of South African citizens to live a comfortable life, and frustrate their efforts to achieve their life goals and enjoy their socio-economic rights (*see* Frustration-Aggression Hypothesis by authors such as John Dallard, Leonard Doob, Neal Miller, Robert Sears, and others). With intensifying competition for ever-shrinking opportunities, it is unfortunate that foreigners become the scapegoat for all the socio-economic ills in South Africa as some portions of the society, including rising criminal activities.[5]

Sub-cultural/normative, evolutionary and social learning theories play a role in sustaining a culture of xenophobia. Sub-cultural and normative theories denote that individuals have a tendency to conform, and are often compelled to think or act according to, social situations and norms they find themselves in (*see* Tiberius/Hall, 2010; Woodward, 2021). In some instances, as explained by Bandura (1977) in social learning theory, citizens often (consciously or sub-consciously) identify, observe, assess, learn, model, imitate, adopt, retain and adapt to the behavours, attitudes, emotions, fundamental beliefs and views of those within their social contexts and environments as part of human and cognitive learning processes of interaction. Xenophobic tendencies, just as the case with racism, prejudices, discrimination and stereotyping in communities, are often perpetuated, inspired and reproduced through social learning.

How xenophobic tendencies often recur in communities as is the case in South Africa since pre-1994 period may be explained by the social control theory (attributed to classical social psychologists and authors such as Albert Reiss, Jackson Toby, Ivan Nye, Walter Reckless, among others). The social control theory – as related to functionalist theories of criminology – according to these authors, assumes that people's norms, values, beliefs and relationships within a society encourage them to abide by binding laws through some form of self-control, external control and voluntary commitment as they feel attached to societal norms, values and beliefs. However, the external means self-control, which forces individuals to conform due to the threat of sanctions or consequences of law in criminal or anti-social behaviour, may be diminished in societies where citizens consistently engage in illegalities and violate national laws with impunity, impudence and immunity from prosecution. Except for a few individuals, justice in South Africa has not been administered on most citizens involved in xenophobic attacks and the accompanying looting and wanton destruction

[4] By 2019, South Africa's GDP was US$388 billion, which was way over than the other countries in the region namely Botswana (US$17 billion), Eswatini (US$5 billion), Lesotho (US$2 billion), Mozambique (US$15 billion), Zimbabwe (US$19 billion), and Zambia (US$23 billion); *see* https://data.worldbank.org/country

[5] South Africa is ranked among the top three (3) countries with the highest crime rates in the world, behind Venezuela and Papua New Guinea, *see* https://worldpopulationreview.com/country-rankings/crime-rate-by-country

of private property (*see* Human Rights Watch, 2008; Shaban, 2019; Human Rights Watch, 2020). This is despite overwhelming incriminating evidence and ubiquitous footages of identifiable perpetrators of xenophobic violence committing premeditated crimes of looting, arson, malicious damage to property, murder, attempted murder, aggravated assault and battery, unlawful possession of firearms, hate crimes, xenophobia-related *crimen injuria,* and other violence-related criminal acts which should ordinarily attract severe punishment and long custodial sentences under South African criminal law for the purposes of deterrence and prevention of recidivism by xenophobic violence convicts.

All these theories assist to understand xenophobic causal factors and manifestation of xenophobia. They lay a firm foundation to understand how xenophobia preventive and management measures in South Africa can be well-thought out, structured and implemented whilst also providing valuable lenses to assess the efficacy of such measures. Peacebuilding initiatives may be helpful in both preventing future recurrence of xenophobia and also resolving and managing xenophobia undercurrents. The UN Peacekeeping (2021) provides a comprehensive, lexical, partitive and functional definition of peace-building as follows;

> Peacebuilding aims to reduce the risk of lapsing or relapsing into conflict by strengthening national capacities at all levels for conflict management, and to lay the foundation for sustainable peace and development. It is a complex, long-term process of creating the necessary conditions for sustainable peace.

Prominent and classical theorists of peace-building such as Johan Galtung, John Paul Lederach and others consider peace-building as key in addressing the root causes of conflicts and preventing future conflicts (preventive and post-conflict peace-building) through conflict management and conflict transformation measures. In the quest for peace, Galtung distinguishes, in a binary fashion, between positive peace (absence of direct and structural violence) which requires positive peace-building that emphasises on engaging institutions, policies and political economic conditions versus negative peace (absence of violence or war) which requires negative peace-building measures that focus on addressing direct conflict drivers or factors to reduce violence (Galtung/Fischer, 2013).

Whilst peace-building, as a concept and practice, has often included components and elements of conflict resolution, conflict management, and conflict transformation; the underlying salient characteristic of any successful peace-building process remains the meaningful inclusion and involvement of various actors and stakeholders playing different roles at different levels in any polity from the community grass-roots level to the top leadership (*see* McCann, 2015), hence the multidimensionality and complexity of the process. Different authors present different components of the peace-building process, depending on their own conceptualiation. For instance, Barnett et al (2007), in *Peacebuilding: What is in a Name?,* identify five components of peace-building as conflict prevention, conflict management, conflict resolution, post-conflict reconciliation and conflict transformation.

All in all, it takes holistic approaches (in terms of scope of measures and actors involved) in order nature positive relations, attitudes, behaviour and structures within a society. Lederach and Mansfield (2008)'s strategic peace-building pathways 'wheel' assists to comprehend how peace-building comprises, or ought to comprise, a multiplicity of inter-related engagements, components and sub-components, elements and sub-elements, dimensions and sub-dimensions, categories and sub-categories. These include structural and institutional change; transitional justice and healing; violence prevention, conflict management and transformation as well as initiatives for development, education and community dialogue (*see* Lederach/Mansfield, 2008).

For such a process to deliver results and have impact on the society; it requires substantial and substantive resources, political will, creativity and innovation (in both the design process and implementation process), community trust and confidence, and effective participation of all stakeholders. An analysis of South Africa experience in addressing xenophobia through peace-building processes will assist to identify the related contextual challenges and complexities faced in executing peace-building measures.

26.5 Xenophobia-Prevention and Management Interventions: Successes and Challenges

Having understood that xenophobia in both pre-1994 and post-1994 South Africa has often manifested in violence, discrimination and stereotype in both ordinary and professional life; the South African Government has managed to design and implement peace-building measures aimed at constructively building personal, group and community relationships (across race, class, religion, nationalities and other social boundaries) in order to peacefully resolve xenophobic injustices and transform the structural conditions that drive xenophobic conflicts within societies. The five main peace-building measures discussed in this analysis are as follows:

- substantive policy action and implementation;
- establishment of structures to deal with xenophobia-related matters;
- legal interventions and law-related institutional infrastructure for addressing xenophobia;
- grassroots peace-building interventions; and
- structural interventions to address socio-economic development challenges, poverty, unemployment and inequality.

26.6 Substantive Policy Action and Implementation

In terms of substantive policy action, the National Action Plan (NAP) to Combat Racism, Racial Discrimination, Xenophobia and Related Intolerance was launched in Mar 2019 by the South African government as a framework to eliminate xenophobia, racism and related intolerance. This was developed in consistent with the national undertaking made by member states in the Durban Declaration and Programme of Action adopted in 2001 at the UN World Conference against Racism, Racial Discrimination, Xenophobia and related intolerance. NAP (2019:14) seeks to address xenophobia through the promoting and protecting human rights; raising awareness of xenophobia among public officials, civil society and general public; encouraging the collection of data regarding xenophobia and allow for comprehensive assessment of the needs so as to effectively combat it; increasing the effectiveness and coherence of measures against xenophobia; strengthening programmes for individuals and groups encountering xenophobia across all social services sectors and access to justice where necessary; facilitating the identification of legislation that needs to be amended and/ or adopted so as to improve the protection of xenophobic victims; and building more equal society and strengthen the rule of law and democracy. The Government had also put in place a National Strategy for Developing an Inclusive and a Cohesive South African Society in 2012[6] with a view to create a socially cohesive, harmonious, peaceful and inclusive society.

At Presidium and Ministerial level, the Government engages with diplomatic missions in South Africa from time to time and discusses migration-related issues. For instance, through Bi-National Commissions, the Department of International Relations and Cooperation (DIRCO) engages their foreign counterparts which usually result in migration policy initiatives that assist to combat xenophobia. The Government has also put in place a White Paper on International Migration in 2017 and a Draft National Labour Migration Policy is now in place. The former seeks to provide a blue print for the country to strategically manage migration. However, the extent to which these policy frameworks prevent and address xenophobia may be debatable as the policy frameworks have often been followed by tightening of the visa regimes and shrinking of employment opportunities for foreign nationals on the labour market due to revision of foreign quotas and stringent employment requirements for foreign nationals (*see* Business Tech, 2022b) as the Government seeks to balance between critical skills shortages and rising unemployment amongst locals. At the same time, the tough and uncompromising political positions (which are oftentimes pernicious and toxic) adopted by some politicians and political parties towards migrants, especially during the run-up to national elections as they seek electoral sympathy has a high risk and likelihood of sowing seeds of xenophobia. Although most of the politicians are clear that they abhor criminal activities committed by foreign nationals and express their resentment of undocumented migrants as was during the South African Municipal Elections in Nov 2021 (*see* Isilow, 2021; Hlatshaneni, 2021), it is the

[6] *See* http://www.dac.gov.za/sites/default/files/approved-national-social-cohesion-strategy-a5-booklet_v2-6.pdf.

perceptions created by such politicking and electioneering that negatively influence community attitudes towards foreign nationals and stoke tensions, especially in the townships where some sections of the population that maybe swayable and more susceptible to political manipulation.

26.6.1 Establishment of Structures to Deal With Xenophobia-Related Matters

Through NAP, the South African Government has to be credited for establishing functional structures meant to deal with xenophobia issues and underlying causes of the tensions between hosting communities and foreign nationals in the form of three Inter-Ministerial Committees (IMCs). These comprise the Inter-Ministerial Committees on Migration, Inter-Ministerial Committees on Social Cohesion, and the Inter-Ministerial Committees on Population Policy. In August 2020, the Government also established the IMC on the Employment of Foreign Nationals (convened by the Minister of Employment and Labour, and co-chaired by the Minister of Home Affairs) to provide guidance on employment matters and labour relations issues relating to foreign nationals.

These IMCs have been working on the review of migration and xenophobia-related policy and legal frameworks. For instance, the IMC on the Employment of Foreign Nationals has contributed to the review of decisions on Zimbabwe Exemption Permits (ZEP), and amendment of the Immigration Act and Employment Services Act 4 of 2014 (SA News, 2020). The IMC also managed to establish the Technical Committee of Director Generals of all departments to support Ministers on migration and xenophobia issues, by initiating and implementing measures meant to address the root causes of xenophobic tensions and also complement the work of IMCs. The same IMC has also continued to engage stakeholders to address xenophobic tensions. For instance, the IMC on Migration and Labour established National and Provincial Trucking, Logistics and Road Freight Forums comprising association representatives covering all provinces (Department of Employment and Labour, 2021). Whenever there are tension in the sector between local drivers and foreign nationals, the IMC has been convening meetings to address grievances from all the concerned parties (Department of Employment and Labour, 2021). All the IMCs have been involved in coordinating public responses and stakeholder comments on reviewing the Draft National Labour Migration Policy and Draft Employment Services Amendment Bill (ESAB) both approved by the Cabinet in mid-Feb 2022 as frameworks to regulate the employment of foreign nationals in various economic sectors, labour relations policies and also to protect migrant rights (*see* Parliamentary Monitoring Group, 2021; Business Tech, 2022b). The role of the National Joint Operational and Intelligence Structure (NATJOINTS) – together with SAPS and community police forums – has been vital in coordinating government's responses to xenophobia outbreak emergencies and restore peace and order in communities. However, NATJOINTS has often been criticised in some instances for late detection of, and strategic response to, xenophobic violence due to ineffective tactical coordination and formal liaison with other stakeholders such as the National and Provincial Disaster Management Centres as well as Provincial Joint Operational and Intelligence Structures (PROVJOINTS) (Edwards/Freeman, 2021: 18–19).

26.6.2 Legal Interventions and Law-Related Institutional Infrastructure for Addressing Xenophobia

Since the primary responsibility and role of Government is to promote and protect human rights of all people and prevent (investigate and prosecute) human rights violations as provided for under Section 26.7(2) of the Constitution and also under international law; legislative efforts have been invested by the South African Government to provide a legal framework to address xenophobia. The Department of Justice and Constitutional Development (DoJCD), SAPS and the National Prosecuting Authority (NPA) have been working together to ensure justice for reported xenophobic criminal cases and also ensure that an early warning system is established linked to a Rapid Response Mechanism, all complemented and supported by the work of the Justice, Crime Preventions and Security Cluster (JCPS).

A number of legislative frameworks have been enacted to address xenophobia as well as racism and other hate crimes on the cyber space (social media and other worldwide web platforms). These include the Promotion of Equality and Prevention of Unfair Discrimination Act (PEPUDA) of 2000, Protection from Harassment Act 17 of 2011, National Cyber Security Policy Framework (NCPF) in 2012, the Cybercrimes Act 19 of 2020, and the Prevention and Combating of Hate Crimes and Hate Speech Bill of 2018 (which is now at the public consultation stage as of Feb 2022) which all discourage unfair discrimination, harassment and hate speech that often ignite xenophobic violence. Equality Courts created by PEPUDA and opened in 2003 to hear matters related hate speech, harassment and discrimination; have been critical in addressing some of the manifestations of xenophobia by providing avenues for justice. Their work has also been complemented by the South African Human Rights Commission (SAHRC) which allows the public to lodge human rights violation complaints. The SAHRC, for example, has provided an early warning mechanism on potentially triggering events or developments that usually stoke and incite xenophobic violence flames (*see* SAHRC, 2021) whilst at the same time carrying out *in loco* assessments and inspections within communities to identify and verify any reported xenophobia-related human rights abuses (*see* SAHRC, 2019a). The SAHRC, and other organisations such as Lawyers for Human Rights (LHR), Legal Resources Centre (LRC), Legal Aid South Africa, and others continue to assist those who may not afford legal services or qualify for legal aid so that they access justice at Equality Courts. In addition, foreign nationals access to Courts have also been assisted by the development of a database of foreign language interpreters by DoJCD to facilitate the reporting and trial of cases/matters relating to foreigners.

Whilst the NPA has committed to make contingency plans to prioritise, manage and deal with xenophobia related cases efficiently as a way of administering justice to victims of xenophobia and punishing perpetrators; one of the most disturbing trend has been the absence of arrests and prosecution of several perpetrators of xenophobic violence, especially those who engage in arson and looting during xenophobic violence outbreaks (*see* Human Rights Watch, 2008; Shaban, 2019; Human Rights Watch, 2020).

26.6.3 Grassroots Peacebuilding Interventions

The South African Government has also initiated anti-xenophobia programmes and social cohesion campaigns in communities, schools, colleges, universities and societies. Community-based anti-xenophobic campaigns continue to be conducted to promote human rights and educate citizens on the vitality of social cohesion, integration, tolerance and peaceful co-existence. For instance, the SAHRC has been arranging facilitated community conversations and dialogues on xenophobia to educate people on such issues that lead to foreign nationals leaving their respective countries and their contribution to the South African economy (*see for instance* SAHRC, 2019b). For example, the International Organization for Migration (IOM) launched, in 2009, the ONE Movement Campaign and funded by the US Department of State Bureau for Population, Refugees and Migration (PRM) in partnership with METRO FM (a national radio station owned by the South African Broadcasting Corporation) as a media partner and several goodwill ambassadors from across Africa. The ONE Movement Campaign has been hosting public awareness, education campaigns across the country as well as outreach programmes, community dialogues and training programmes for service providers and public officials on migrant rights, social cohesion, tolerance, unity in diversity and Ubuntu (Muli, 2011) which are critical in preventing xenophobia. The SAHRC and other civil society organisations (CSOs) such as the Moral Regeneration Movement (MRM), Nelson Mandela Foundation and the Ahmed Kathrada Foundation have been hosting community dialogues with various stakeholders on xenophobia (Nelson Mandela Foundation, 2015). Africa Day Celebrations by Government Departments as well as Universities on the 25th of May have also been utilised to host seminars and webinars that promote African unity messages and peaceful inter-continental relations. At provincial level, for instance, the KZN Province rolled out the 'Communities-in-Dialogue Programme' to promote social cohesion, tolerance, peace and harmony in communities.

Despite the grassroots peace-building interventions by the various stakeholders across all the nine provinces in South Africa, xenophobic attitudes are still engrained in substantial sections of the society. For instance, the Socio-Economic Justice for All (SEJA) Baseline Survey Report by the Foundation for Human Rights (2018:23) found out that a significant portion of South Africans (44%) agreed that "foreigners should not be allowed to live in South Africa because they take jobs and benefits away from South Africans". Of those who shared this view, almost half are white South African respondents (48%), and coloured people respondents (48%), black Africans (43%) and Indians/Asians (36%) (*see* SEJA Baseline Survey Report, 2018:23). The Report (2021:24) also exposed the high levels of mistrust towards foreign nationals, with approximately two fifths of all adult South Africa indicating that they had no trust at all in people from other African countries (38%) or those from other non-African countries (39%). This reinforces xenophobic attitudes.

Similarly, the latest empirical survey (South African Reconciliation Barometer Survey of 2021) carried out by the Institute for Justice and Reconciliation (2021: 27) reveals that a sizeable share of South Africans are likely to prevent foreign nationals from accessing jobs (35%), and government services (35%) whilst 37% South Africans are likely to prevent foreign nationals from operating a business (37%) or moving into their neighborhood (36%). In as far as social cohesion and integration is concerned, the Report (2021: 29) summarizes:

> of particular concern for horizontal trust is the lack of trust in foreign nationals, especially from other African countries. Xenophobic attitudes are not uncommon among respondents, with more than a third reporting that they would be likely to prevent foreign nationals from accessing certain services.

This empirical reality requires more efforts to be dedicated towards community and grassroots-focused holistic interventions that do not only aim at changing the attitudes of locals towards foreign nationals, but also target at structurally transforming the socio-economic circumstances and conditions of all people.

26.6.4 Structural Interventions to Address Socio-Economic Development Challenges, Poverty, Unemployment and Inequality

With the understanding that some of the underlying causes of xenophobia are socio-economic in nature, the South African Government has been implementing an array of economic development plans and programmes to spur inclusive economic growth, create jobs, alleviate poverty, and improve access to quality social services for all. The National Development Plan 2030, drafted in Aug2012 provide an overall framework for eliminating poverty, reducing inequalities and transforming the South African society. To stir economic growth and create employment, the Government has been implementing programmes such as the New Growth Path, Revised Industrial Policy Action Plan (IPAP2), Black Economic Empowerment (BEE), Black Business Supplier Development Programme (BBSDP), Expanded Public Works Programme (EPWP), Community Work Programme, Local Economic Development (LED), Youth Build Programme, National Skills Development Strategy, and the Mine Rehabilitation Programme, among others whilst rural development has been championed through the Comprehensive Rural Development Programme, Recapitalization and Development Programme (RADP), Comprehensive Agriculture Support Programme (CASP), Land Reform Programme, National Rural Youth Service Corps Programme. In order to improve service delivery and infrastructure development, efforts are being invested through the National Infrastructure Plan (NIP), Infrastructure Investment Plan, Municipal Infrastructure Grant, Human Settlements Programmes and Projects, and Strategic Integrated Projects (SIPs) across the infrastructure sectors of energy, social infrastructure, water and sanitation, transport and IT.

Notwithstanding the existence of several cross-sector initiatives, strategies, plans, programmes and projects that are meant to address socio-economic development challenges in South Africa, which are all complemented by an avalanche of social protection programmes administered by the Department of Social Development and various NGOs; South Africa continues to face binding constraints in the form of lack of sufficient investments, insufficient skills, inequalities in the ownership of land and other productive assets of the economy, limited integration into the regional and global economy, and the impact of climate shocks and recent effects of COVID-19 (*see* World Bank, 2021a). Inequality still persist with 90% of the country's wealth being owned by only 10% of the population whilst the unemployment rate among youths aged between 15 and 24 standing at 64% (*see* World Bank, 2021a). COVID-19 worsened the socio-economic situation, having resulted in a 6.4% contraction of the South African economy in 2020, at a time when economic growth was already slowing down, having recorded a 0.1% growth in 2019 due to power outages (*see* World Bank, 2021a). Even the social protection schemes are also insufficient in terms of reach and adequacy considering the increasing unemployment rates, care dependents and old persons in need of social grants. This is despite the fact that South Africa's social assistance system – which comprise social welfare grants (such as Child Support Grants, State Old Age Pension Grants/Older Persons Grants, Care Dependency Grants, Disability Grants, War Veterans Grants, Grant in Aid, and COVID-19 Special Relief of Distress Grants), social insurance, disaster management interventions and food security interventions – remains relatively high compared to most countries in the region and across the world, accounting for 3.3% of the GDP and 15.4% of total government spending (World Bank, 2021b).

26.7 Conclusion and Recommendations

All in all, the various peace-building initiatives – namely substantive policy action and implementation; establishment of structures to deal with xenophobia-related matters; legal interventions and law-related institutional infrastructure for addressing xenophobia; grassroots peace-building interventions; and structural interventions to address socio-economic development challenges, poverty, unemployment and inequality – have recorded measures of success and challenges. The successes and challenges are detailed in most survey reports conducted countrywide. The changing community attitudes towards foreign nationals; increasing awareness of human rights; the improving levels of tolerance and social cohesion in communities; the establishment and utilisation of equality courts; the utility of IMCs, NATJOINTS, PROVJOINTS; and the notable efforts towards socio-economic transformation are some of the achievements recorded so far, whilst challenges relating to inadequacy of resources; lack of proper coordination of interventions; sustainability issues; continued existence of portions of intolerance towards migrants; tightening immigration laws and visa regimes; as well as delayed justice for victims of xenophobia remain challenges that have manifested in the continued outbreak of xenophobia in South Africa.

The following five recommendations are salient strengthening peace-building initiatives in order to reduce the incidences of xenophobia in the country;

1. Greater efforts are needed to address deep-rooted socio-economic development challenges, poverty, unemployment and inequality in South Africa as these remain some of the key underlying causes of xenophobia in the country as they intensify competition for scarce resources, survival and economic opportunities. Addressing structural, institutional and societal drivers of xenophobia calls for more comprehensive and inclusive interventions that are socio-economic, political, legal, diplomatic, and cultural in nature. Thus, eeconomic development initiatives and the country's socio-economic ideology may need to be re-adjusted to be pro-poor, broad-based and more inclusive such that no one is left behind.
2. Since several laws have been enacted to address xenophobia and other hate crimes, the Government, law enforcement agencies and justice delivery system may need to step up collective efforts in enforcing such laws in such a way that perpetrators of xenophobic violence are prosecuted expeditiously and sentenced so that there are limited cases of recidivism and recurrence of xenophobic violence.
3. Whilst substantive anti-xenophobia policies have been developed and adopted (such as NAP National Strategy for Developing an Inclusive and a Cohesive South African Society, and others) as well as the legal interventions and law-related institutional infrastructure for addressing xenophobia; it is the coordination of the programmes in provinces and districts that appears to be lacking among Government and non-governmental actors at all spheres of government (national, provincial and local government). A more coordinated and structured approach may assist to ensure that there is more seamless coverage of the country, especially xenophobia hotspots. Campaign messages may need to be spread more through the media whilst also ensuring that there is pro-activity rather than reactivity as most NGOs and Government Departments tend to engage more in anti-xenophobia activities post-xenophobic violence.
4. Although the South African Government is faced with a complex migration phenomenon, against competing and conflicting interests where it has to protect its citizens whilst also protecting the human rights of migrants; there may be need for broader consultations when reviewing the country's migration regimes and labour laws in light of the root causes of xenophobia such that the pertinent issues raised by migrants are reflected in the laws from time to time.

5. The South African Government may need to work with relevant CSOs and NGOs to disseminate findings gathered from empirical surveys and research on xenophobia, its causes and recommendations on creating socially cohesive, united, integrated and peaceful societies characterised by positive relationships between locals and foreign nationals.

References

African Centre for Migration & Society. (2023). Xenowatch Dashboard: Incidents of Xenophobic Violence in South Africa: 1994 – June 2023. ACMS, Witswatersrand University. https://www.xenowatch.ac.za/statistics-dashboard/. Accessed 30 June 2023.

Bandura, A. (1977). *Social Learning Theory*. Englewood Cliffs, NJ: Prentice Hall.

Barnett, M., Kim, H., O'Donnell, M., & Sitea, L. (2007). Peacebuilding: What is in a name? *Global Governance, 13*, 35.

Bordeau, J. (2009). *Xenophobia*. New York: The Rosen Publishing Group, Inc.

Business Tech. (2022a). South Africa announces 'quota system' for foreigners – what you should know. https://businesstech.co.za/news/business/563364/south-africa-announces-quota-system-for-foreigners-what-you-should-know/. Accessed 28 February 2022.

Business Tech. (2022b). New job hiring laws planned for South Africa – what you should know. 14 Feb 2022. https://businesstech.co.za/news/business/557872/new-job-hiring-laws-planned-for-south-africa-what-you-should-know/. Accessed 17 February 2022.

Choane, M., Shulika, L.S., & Mthombeni, M. (2011). An analysis of the causes, effects and ramifications of xenophobia in South Africa. *Insight on Africa, 3*(2), 129–142.

Cibane, B. (2015). Xenophobic attacks: South Africa can no longer hide behind the shame. *The Daily Maverick*, 20 Apr 2015. https://www.dailymaverick.co.za/opinionista/2015-04-20-xenophobic-attacks-south-africa-can-no-longer-hide-behind-the-shame/#.WG6Ouxt96Uk. Accessed 14 November 2021.

Crush, J. (2008). The perfect storm: The realities of xenophobia in contemporary South Africa. *Africa Portal*, 1–67. https://www.africaportal.org/publications/the-perfect-storm-the-realities-of-xenophobia-in-contemporary-south-africa/. Accessed 4 November 2021.

Department of Employment and Labour. (2021). Government commits to expedite policy changes in the road freight industry. 05 Nov 2021. https://www.labour.gov.za/government-commits-to-expedite-policy-changes-in-the-road-freight-industry. Accessed 26 January 2022.

Edwards, L., & Freeman, L. (2021). Policing and non-nationals: Analysis of police prevention, detection and investigation of xenophobic violence in South Africa. African Policing Civilian Oversight Forum. Mowbray: Cape Town. https://apcof.org/wp-content/uploads/policing-and-non-nationals-report.pdf. Accessed on 20 Feb 2022.

Forrest, K. (2015). Rustenburg's labour recruitment regime: shifts and new meanings. *Review of African Political Economy, 42*(146), 508–525.

Foundation for Human Rights. (2018). Socio-Economic Justice for All (SEJA) Baseline Survey Report. Jan 2018. https://www.fhr.org.za/2018/01/20/seja-baseline-survey-report/. Accessed 23 February 2022.

Galtung, J., & Fischer, D. (2013). Positive and Negative Peace. In J. Galtung. *SpringerBriefs on Pioneers in Science and Practice* (pp. 173–178). vol 5. Berlin, Heidelberg: Springer. https://doi.org/10.1007/978-3-642-32481-9_17

Gewald, J.B. (2015). Wenela, Katima Mulilo, a Zone of Transit in Barotseland: The Development of a Holding Zone for Migrants on the Extreme Frontier of the South African Empire. In A. Akinyoade & J.B. Gewald (Eds.), *African Roads to Prosperity African Roads to Prosperity*

People en Route to Socio-Cultural and Economic Transformations (pp. 45–59). E-Book (PDF): Brill.

Government of South Africa. (2019). National Action Plan (NAP) to Combat Racism, Racial Discrimination, Xenophobia and Related Intolerance. https://www.gov.za/sites/default/files/gcis_document/201903/national-action-plan.pdf. Accessed 15 January 2022.

Harris, A.S., Findley, M.G., Nielson, D.L., & Noyes, K.L. (2018). The economic roots of anti-immigrant prejudice in the global south: Evidence from South Africa. *Political Research Quarterly, 71*(1), 228–241.

Hewstone, M.E., Stroebe, W.E. & Jonas, K.E. (2008). *Introduction to Social Psychology*. New Jersey: Blackwell Publishing.

Hlatshaneni, S. (2021). SA political parties' anti-immigrant election promises nothing but hot air. *The Citizen,* 12 Oct 2021. https://www.citizen.co.za/premium/2719092/sa-political-parties-anti-immigrant-election-promises-nothing-but-hot-air/. Accessed 25 February 2022.

Human Rights Watch. (1998). Report on the Abuse of Un-Documented Migrants, Asylum Seekers, and Refugees in South Africa, Mar 1998. New York. https://www.hrw.org/legacy/reports98/sareport/. Accessed 13 November 2021.

Human Rights Watch. (2008). South Africa: Punish Attacker in Xenophobic Violence. May 22, 2008. https://www.hrw.org/news/2008/05/22/south-africa-punish-attackers-xenophobic-violence. Accessed 22 December 2021.

Human Rights Watch. (2020). They Have Robbed Me of My Life': Xenophobic Violence Against Non-Nationals in South Africa. Sep 17, 2020. https://www.hrw.org/report/2020/09/17/they-have-robbed-me-my-life/xenophobic-violence-against-non-nationals-south. Accessed 22 December 2021.

Institute for Justice and Reconciliation. (2021). South African Reconciliation Barometer Survey: 2021 Report. https://www.ijr.org.za/home/wp-content/uploads/2021/12/IJR_SA-Reconciliation-Barometer-2021.pdf. Accessed 22 February 2022.

International Organization for Migration (IOM). (2022). World Migration Report 2022. IOM. https://worldmigrationreport.iom.int/wmr-2022-interactive/. Accessed 7 January 2022.

Isilow, H. (2021). South African politicians play up anti-foreigner sentiments to win votes. *Africa,* 26 Oct 2021. https://www.aa.com.tr/en/africa/south-african-politicians-play-up-anti-foreigner-sentiments-to-win-votes/2403115. Accessed 25 February 2022.

Johnston, M.P. (2017). Secondary data analysis: A method of which the time has come. *Qualitative and Quantitative Methods in Libraries, 3*(3), 619–626.

Kerr, P., Durrheim, K., & Dixon, J. (2019). Xenophobic violence and struggle discourse in South Africa. *Journal of Asian and African Studies*, 54(7), 995–1011.

Klotz, A., (2016). Borders and the roots of xenophobia in South Africa. *South African Historical Journal, 68*(2), 180–194.

KwaZulu-Natal Christian Council. (2016). Draft of Xenophobia Timeline. http://kzncc.org.za/wp-content/uploads/2016/12/xenophobia-time-line.docx. Accessed 15 November 2021.

Lamb, B., Kunene, L.N., & Dyili, N.F. (2019). Lessons from foreign owned spaza shops in South African townships. *Journal of Reviews on Global Economics, 8*, 1351–1362.

Liedeman, R., Charman, A., Piper, L., & Petersen, L. (2013). Why are foreign-run spaza shops more successful? The rapidly changing spaza sector in South Africa. *Sustainable Livelihoods Foundation.* https://www.econ3x3.org/article/why-are-foreign-run-spaza-shops-more-successful-rapidly-changing-spaza-sector-south-africa Accessed 4 August 2023.

Martin, G. (2021). 900 service delivery protests in South Africa over six months. *Defenceweb,* 20 Apr 2021. https://www.defenceweb.co.za/featured/900-service-delivery-protests-in-south-africa-over-six-months Accessed 19 December 2021.

McCann, J. (2015). Local ownership - an imperative for inclusive peace-building. *Development Dialogue, 63*, 16–27.

Migration Policy Institute. (2021). South Africa Reckons with Its Status as a Top Migration Destination, Apartheid, and Economic Challenges. Nov 18, 2021. https://www.migrationpolicy.org/article/south-africa-immigration-destination-history. Accessed on 24 Nov 2021.

Misago, J.P. (2016). Responding to xenophobic violence in post-apartheid South Africa: Barking up the wrong tree?. *African Human Mobility Review, 2*(2), 443–467.
Mishra, B.K. (2008). *Psychology: The study of human behaviour*. PHI Learning Pvt. Ltd: New Delhi.
Moiloa, K. (2015). The real reason foreign-owned shops are thriving in South African townships. News24, 07 Feb 2015. https://www.news24.com/news24/xArchive/Voices/the-real-reason-foreign-owned-shops-are-thriving-in-south-african-townships-20180719. Accessed on 4 January 2022.
Monson, T., & Misago, J.P. (2016). Why history has repeated itself: the security risks of structural xenophobia. *South African Crime Quarterly, 9*(29), 25–34.
Muli, N., 2011. ONE Movement Campaign. *The Communication Initiative Network*, 21 Mar 2011. https://www.comminit.com/global/content/one-movement-campaign. Accessed 27 February 2022.
Ndlovu-Gatsheni, S.J. (2010). Do 'Africans' exist? Genealogies and paradoxes of African identities and the discourses of nativism and xenophobia. *African Identities, 8*(3), 281–295.
Nelson Mandela Foundation. (2015). The Nelson Mandela Foundation's view on xenophobia. 7 Jul 2015. https://www.nelsonmandela.org/news/entry/nelson-mandela-foundation-view-on-xenophobia. Accessed 28 February 2022.
Ngcamu, B.S., & Mantzaris, E. (2021). Localizing and Locating the Hidden Causes of Xenophobia and its Ramifications. In *Addressing Xenophobia in South Africa: Drivers, Responses and Lessons from the Durban Untold Stories*. Emerald Publishing Limited.
Parliamentary Monitoring Group. (2021). Question NW1525 to the Minister of Employment and Labour. 24 June 2021. https://pmg.org.za/committee-question/16561/. Accessed 19 February 2022.
Phakathi, M. (2019). The role of music in combating xenophobia in South Africa. *African Renaissance, 16*(3), 125–147.
Reilly, J.E. (2001). Nation building and the construction of identity: Xenophobia in South Africa. *Refuge: Canada's Journal on Refugees, 19*(6), 4-11.
SA News. (2020). Cabinet establishes IMC on employment of foreign nationals. Aug28, 2020. https://www.sanews.gov.za/south-africa/cabinet-establishes-imc-employment-foreign-nationals. Accessed 25 January 2022.
Shaban, A.R.A. (2019). South Africa must prosecute perpetrators of xenophobic attacks. 15 Apr 2019. https://www.africanews.com/2019/04/15/south-africa-must-arrest-prosecute-perpetrators-of-xenophobic-attacks//. Accessed 22 December 2021.
Simelane, B. C. (2021). Xenophobia-accused Operation Dudula returns to Hillbrow. *Daily & Maverick*, 21 Feb 2022. https://www.dailymaverick.co.za/article/2022-02-21-xenophobia-accused-operation-dudula-returns-to-hillbrow/. Accessed 17 November 2021.
Smith, E. (2008). *Using secondary data in educational and social research*. McGraw-Hill Education (UK): Open University Press.
South African Human Rights Commission. (2019a). Media Statement: SAHRC Condemns attacks on Non-Nationals. Apr 1, 2019. https://www.sahrc.org.za/index.php/sahrc-media/news-2/item/1827-media-statement-sahrc-condemns-attacks-on-non-nationals. Accessed 21 February 2022.
South African Human Rights Commission. (2019b). Media Invite: SAHRC Free State Provincial Office holds a human rights dialogue. 2 Sep 2019. https://www.sahrc.org.za/index.php/sahrc-media/news-2/item/2076-media-invite-sahrc-free-state-provincial-office-holds-a-human-rights-dialogue. Accessed 21 February 2022.
South African Human Rights Commission. (2021). Media Statement: SAHRC Concerned by a Video Which Could Lead to Incitement Against Non-Nationals. 13 Aug2021. https://www.sahrc.org.za/index.php/sahrc-media/news-2/item/2768-media-statement-sahrc-concerned-by-a-video-which-could-lead-to-incitement-against-non-nationals. Accessed 21 February 2022.
StatsSA. (2021). Quarterly Labour Force Survey (QLFS) – Q3:2021. 30 Nov 2021. http://www.statssa.gov.za/?p=14957. Accessed 3 December 2021.

Tiberius, V., & Hall, A. (2010). Normative theory and psychological research: Hedonism, eudaimonism, and why it matters. *The Journal of Positive Psychology*, 5(3), 212–225.

UNESCO. (2017). Glossary of Migration Related Terms. https://wayback.archive-it.org/10611/20180705022135/ http://www.unesco.org/new/en/social-and-human-sciences/themes/international-migration/glossary/. Accessed 15 January 2022.

United Nations High Commissioner for Refugees (UNHCR). (2020). *Guidance on Racism and Xenophobia: How UNHCR can address and respond to situations of racism and xenophobia affecting persons under its mandate.* UNHCR, Geneva: Switzerland. https://www.unhcr.org/5f7c860f4.pdf. Accessed 5 January 2022.

United Nations High Commissioner for Refugees (UNHCR). (2022). How many refugees are there in the world? UNHCR. https://www.unhcr.org/figures-at-a-glance.html. Accessed 7 January 2022.

United Nations Peacekeeping. (2021). Terminology. https://peacekeeping.un.org/en/terminology. Accessed 4 January 2022.

United Nations Population Division, 2020. International Migrant Stock 2020: Destination and Origin. https://www.un.org/development/desa/pd/content/international-migrant-stock. Accessed 22 November 2021.

Vhumbunu, C.H. (2021). The Jul 2021 Protests and Socio-political Unrest in South Africa. *Conflict Trends, 2021/3.* https://www.accord.org.za/conflict-trends/the-Jul-2021-protests-and-socio-political-unrest-in-south-africa/. Accessed 13 December 2021.

Wentzel, M., & Tlabela, K. (2006). Historical Background to South African Migration. In P. Kok, D. Gelderblom, J. O. Oucho & J. van Zyl (Eds.) *Migration in South and Southern Africa: dynamics and determinants* (pp. 71–96). Cape Town: HSRC Press.

Whiteside, M., Mills, J., & McCalman, J. (2012). Using secondary data for grounded theory analysis. *Australian Social Work*, *65*(4), 504-516.

Wimmer, A. (1997). Explaining xenophobia and racism: A critical review of current research approaches. *Ethnic and Racial Studies*, *20*(1), 17-41.

Woodward, J. (2021). *Causation with a human face: Normative theory and descriptive psychology.* London: Oxford University Press.

World Bank. (2021a). The World Bank in South Africa. 5 Oct 2021. https://www.worldbank.org/en/country/southafrica/overview#1. Accessed 28 February 2022.

World Bank. (2021b). South Africa: A New Social Assistance Assessment Aims to Help Strengthen Policies and Programs for the Poor. 7 Oct 2021. https://www.worldbank.org/en/news/press-release/2021/10/07/south-africa-a-new-social-assistance-assessment-aims-to-help-strengthen-policies-and-programs-for-the-poor. Accessed 28 February 2022.

Zander, A. (1979). The psychology of group processes. *Annual Review of Psychology,* 30(1), 417–451.

Zondi, S. (2008). Xenophobic attacks: towards an understanding of violence against African immigrants in South Africa. *Africa Insight, 38*(2), 26–35.

Chapter 27
State and Civil Society-led Peace Committees in Zimbabwe

David Makwerere

Abstract This study focused on the value of local peace committees (LPCs) in Zimbabwe. Local peace committees were created by the National Peace and Reconciliation Commission (NPRC), which operates at the provincial level in all the administrative provinces of the country. Several other committees that are operational at different levels were formed by various civil society organisations in the country. Notable civil society organisations that have set up local peace committees in Zimbabwe include the Zimbabwe Council of Churches (ZCC), the Zimbabwe Civic Education Trust (ZIMCET) and the Institute for Young Women Development (IYWD). The study found that the LPCs that were set up by the NPRC are operational at the provincial level, thus making them less visible at the district and ward levels. On the other hand, the LPCs set up by the ZCC are visible from the national level right down to the ward level. The other organisations operate LPCs that are largely localised at the ward level. The chapter argued that state-led LPCs have not yielded the much-needed transformative effect because of the way they are set up. They are dominated by senior government officials and traditional leaders, most of whom are accused of being partisan. On the other hand, the CSO-led LPCs have also failed to deliver the goods because their working relationship with the government is very constrained as they are often accused of having sinister agendas. Although there is greater room for collaboration between state-led and non-state civil society organisations (CSOs), these actors have failed to effectively work together for the good of the people. The chapter recommends that they seriously consider working together to enhance the effectiveness of the LPCs.

Dr David Makwerere is a Lecturer in the Department of Peace and Governance at the Bindura University of Science Education in Mashonaland Central Province, Zimbabwe. Email: david.makwerere@gmail.com

---

D. Makwerere (✉)
Department of Peace and Governance, Bindura University of Science Education, Mashonaland Central Province, Zimbabwe
e-mail: david.makwerere@gmail.com

J. C. K. Kiyala and N. Chivasa (eds.), *Climate Change and Socio-political Violence in Sub-Saharan Africa in the Anthropocene*, The Anthropocene: Politik—Economics—Society—Science 37,
https://doi.org/10.1007/978-3-031-48375-2_27

Keywords Civil society · Local peace committees · Collaboration · Transformation · Grassroots · Dialogue

27.1 Introduction

Zimbabwe has been experiencing intermittent conflicts since independence in 1980. There are concerns that the violence that has continued to plague the country is evidently organised and has manifested itself through repeated cycles of violence (World Bank, 2011). Zimbabwe has to deal with victims of conflict, who include internally displaced persons (IDPs), the widowed, orphaned, maimed and traumatised (Dube/Makwerere, 2012). The conflict typology in Zimbabwe is largely defined by political identity. Lack of political tolerance has seen several violent episodes of political violence, especially towards and during national plebiscites (Mukuhlani, 2014; Vambe, 2014), and the culture is threatening to destroy the social fabric in Zimbabwe. Machakanja/Mungure (2013) observed that the emergence of a strong Movement for Democratic Change (MDC) party in 1999 saw a gradual undermining of government institutions, leading to their manipulation for partisan interests. The events have left the country in dire need of structural and social transformations. As a result, politically motivated violence has increased since 2000, and the contest for political dominance has heightened. Consequently, inter-party and intra-party violence have characterised Zimbabwe's political terrain (Feltoe, 2004).

State efforts to address issues of political violence in Zimbabwe included a largely informal call for reconciliation by the then prime minister of the Republic of Zimbabwe, Robert Gabriel Mugabe, who called for forgiveness and reconciliation in 1980 (Muchemwa et al., 2013). However, his (Mugabe's) government failed to provide a clear road map for such forgiveness and reconciliation. After the violent and disputed election of 2008, the political actors signed the Global Political Agreement (GPA) in September 2008 (Muchemwa et al., 2013). The GPA (2008) included provisions for the creation of the Organ on National Healing, Reconciliation and Integration (ONHRI). This provision marked the first serious attempt by the state to provide a framework for peacebuilding. Article VII of the GPA calls for the country to formally recognise and acknowledge the need for national healing and reconciliation. Thus, ONHRI was set up under the provisions of Article 7.1 (c) of the GPA (2008:7), which stated:

> The parties hereby agree that the new government …shall give consideration to the setting up of a mechanism to properly advise on what measures might be necessary and practicable to achieve national healing, cohesion and unity in respect of victims of pre- and post-independence political conflicts.

However, the ONHRI was soon hamstrung by political polarisation, a lack of a clear mandate, and other factors, including a lack of funding (Muchemwa et al., 2013). Other scholars, like Machakanja (2010) and Mbire (2011), expressed concern about the top-down approach that was used by the ONHRI. They argued that the

ONHRI missed out on an opportunity to correct and address past wrongs and to set the country on a peaceful trajectory by playing into the hands of the politicians. As a result, healing and reconciliation initiatives in Zimbabwe have instead been top-down, elitist and mere tokens meant to silence the voices of those calling for a holistic approach to peacebuilding that guarantees human security. Such failures to employ comprehensive approaches only guarantee piecemeal activities that do not serve their intended goals of bringing justice and redress to victims (United Nations, 2012). Observing this, Muchemwa et al. (2013) lamented that achieving reconciliation in Zimbabwe is a nightmare.

Having missed the opportunity, the country was soon presented with another opportunity: the new Constitution of Zimbabwe, which was adopted in July 2013. Chapter 12 of the constitution created five independent commissions, including the National Peace and Reconciliation Commission (NPRC). Section 251 of the Constitutional Amendment Act No. 20 of 2013 provides for the formulation of an independent commission charged with fostering reconciliation in Zimbabwe. The NPRC has the mandate to:

- Ensure post-conflict justice, healing, and reconciliation,
- Develop and implement programmes meant to promote healing and reconciliation,
- Encourage people to tell the truth for national healing,
- Receive and consider complaints on conflict issues,
- Conciliate and mediate disputes among communities, and
- Develop early detection mechanisms for potential conflicts and disputes.

A key issue bedevilling the NPRC is the lack of a defined peacebuilding model to address past wrongs and engage with current challenges. In 2019, the NPRC decided to create Provincial Peace Committees (PPC), with the intention to eventually decentralise the arrangement to the districts (NPRC, 2021). The NPRC decided that to effectively decentralise their operations, they needed to constitute PPCs. This development raised more questions than it could answer. For instance, do these provincial communities have the funding to carry out provincial peace-related activities? Who are the committee members? Can they be partial? While this development can be viewed as progressive, the composition of the committees is largely compromised. Firstly, the inclusion of traditional leaders and some senior government officials can be seen as counter-productive. The traditional chiefs are captured, as are the senior government bureaucrats, most of whom are often seen at political party functions organised by the ruling party (Alexander/Mcgragor, 2013; Rukuni et al., 2015; Fayayo, 2018; Kurebwa, 2020). In addition, traditional leaders are perceived to be aligned with the ruling ZANU-PF party, and in recent times, the president of the Chiefs' Council in Zimbabwe has on several occasions declared his allegiance to ZANU-PF (Makwerere/Mandoga, 2012). Although some traditional leaders have remained neutral, the majority have been compromised. Moreover, the institution of traditional leaders is supposed to be impartial, as stipulated for in the constitution of Zimbabwe.

Against this backdrop, it is very difficult to believe that the co-option of traditional leaders into these PPCs will ensure peace and reconciliation in the provinces. Apart from the traditional leaders, the impartiality of these PPCs appears compromised already because the NPRC recruited senior government officials, most of whom have been labelled political appointees. It remains to be seen whether or not the PPCs can deliver justice. There are genuine fears that the NPRC, just like the OHNRI, might never be able to address the past and present effects of violence in Zimbabwe.

On the other hand, the idea of local peace committees is not entirely new to Zimbabwe, since civil society organisations doing peace-related programming have experimented with the idea for example the Zimbabwe Council of Churches and the Institute for Young Women Development. Others have abandoned it for various reasons, while others still pursue the idea. However, there has been no empirical assessment to determine the effectiveness of the peace committees.

27.2 The Problem

The NPRC created a total of 10 PPCs in Zimbabwe between May and June 2019 (NPRC, 2021). Each committee comprises 2530 members from across local government authorities, civil servants, traditional leaders, churches and civil society. According to the NPRC (2021), the work of the Provincial Peace Committees will basically be to:

- Promote peace within the province, create and/or facilitate spaces for dialogue between groups and communities, and facilitate the exchange of ideas on issues that may threaten peace and stability within the community.
- Engage in confidence- and trust-building activities between groups and communities in conflict within the province.
- Promote tolerance building and provide strategy, advice and early warning on potential threats to the peace and stability of the communities.
- Engage in public education, sensation and awareness about conflict indicators within the province and how to ameliorate them.
- Facilitate sports, arts, cultural and economic activities that build friendships, promote trust and goodwill between communities in conflict.
- Organise training and other capacity building programmes for the district peace committee.
- Respond to the issues identified within the province or those emanating from the work of district peace committees.
- Facilitate other activities necessary for the prevention of violence and the promotion of peace.

However, it has come to light that some CSOs in Zimbabwe, particularly the ZCC, the ZIMCET, the IYWD and the CCMT, have a similar arrangement in some of the areas where the NPRC operates. Therefore, the key question is whether the reinvention of these peace committees in Zimbabwe will finally open the door for genuine reconciliation in the country and ensure preventive approaches to minimise or curb future violent conflicts.

In light of this, the chapter sought to specifically examine the relevance of peace committees (state-led and civil society-led) in Zimbabwe, analyse their specific strategies towards peacebuilding at the national level, assess the effectiveness of their strategies, and generate practical recommendations for sharing with local and international peace practitioners.

27.3 Conceptualising Conflict

The word 'conflict' is a significantly broad and often ubiquitous term. The definition is therefore relative and dependent on the focus and context of the discussion. Simply put, conflict is defined as an incompatibility of goals (Ramsbotham et al., 2005: 3). Conflict is an expressed struggle between at least two interdependent parties that perceive incompatible goals, scarce resources, and interference from the other party in achieving their goals.

Lederach (1999: 18) views conflict as natural, inevitable in human existential dynamism, and good as a motor of change. In his earlier publication, Lederach (1995: 17) defines social conflict as:

> A phenomenon of human creation, lodged naturally in relationships. It is a phenomenon that transforms events, the relationship in which conflict occurs, and indeed its very creators. It is a necessary element of transformative human construction and reconstruction of social organisation and realities.

The above definition of conflict takes a constructivist approach to understanding and transforming conflict. It is critical to understand the context (including cultural and religious aspects) so as to effectively deal with its manifestations. An important point raised by Lederach is that conflict can be constructed and deconstructed by the same actors. In other words, the power to transform a conflict rests entirely with the conflicting parties. The constructivist approach is supported by Conteh-Morgen (2005: 73), who submits that "constructivism as an approach is a useful theoretical lens in understanding the true nature of issues such as collective violence, class, gender, and racial issues, among others".

Another definition relevant to this study was provided by Cooper (2007: 85), who viewed conflict as "an ineradicable part of human condition and it occurs when the interests of one party come by accident or design into engagement with those of another or others."

Each conflict is unique and must thus be understood in its particular context. For instance, the drivers of conflict in Zimbabwe are different from those in the Great Lakes and the Horn of Africa regions; hence, approaches and strategies dealing with these conflicts must be informed by the realities on the ground. Moreover, the perceptions are different and the actors are different as well. In this vein, Harris (2011: 24) noted that "each of us sees, hears and experiences the world uniquely and we spend our lives bridging the differences between our perceptions and (all the needs and wishes they generate) and the perceptions of others."

Conflict that is handled in a constructive and positive manner is bound to have positive effects and positive changes in society. On the contrary, if conflict is handled negatively, the outcomes are bound to be retrogressive and destructive. This often leads to violence, loss of life and other negative consequences.

Clearly, the illustration shows that if conflict is expressed negatively, the results can be calamitous. Consequently, lives can be lost, the environment can be destroyed and property can be damaged. Feeding into this, conflicts experienced in sub-Saharan Africa are largely violent, while most regimes are oppressive and repressive. Examples of protracted violent conflicts in the region included the Mozambique-Angolan conflict, which raged for decades. Conflicts that are expressed positively usually lead to better lives, better social relationships, sound social and government structures, and greater social justice. This understanding of conflict influenced the researcher's understanding of the work of LPCs in Zimbabwe.

27.4 Evolution of the Infrastructure of Peace in Zimbabwe

There were no formal infrastructures of peace for a very long time in independent Zimbabwe, in spite of the country's long history of violent conflict pre-and post-independence. Baker and Scheye (2007: 503) note that this is a common trend in many post-conflict states since state-centred approaches are prioritised, as opposed to people-centred and locally owned reconciliation initiatives. It is against this background that Zimbabwe witnessed a phenomenal growth of civil society organisations at the beginning of the twenty-first century, in line with global developmental trends as the UN prioritised the peacebuilding agenda. As noted by Muchemwa et al. (2013), all the notable efforts towards peacebuilding in Zimbabwe were a top-down affair from the onset of independence. In 1980, the then prime minister, Mugabe, simply announced the policy of reconciliation and racial integration (Huyse, 2003), with no reference to any specific criteria for how these were to be achieved. As a way of ending the *Gukurahundi* conflict of 1982–1987, the two conflicting parties, the Zimbabwe African National Union (ZANU) and the Zimbabwe African People's Union (ZAPU), signed the Unity Agreement and merged to become one entity, the ZANU-PF, in 1987. This approach was heavily criticised with Mashingaidze (2005) arguing that the process benefitted the elite and not the ordinary citizens.

The trend continued after 2000, as political violence escalated but the government did not demonstrate any interest in effectively dealing with the conflicts. In addition, the formation of the ONHRI in 2009 was also criticised as being more political as opposed to reconciling the conflicting parties. Consequently, the ONHRI was largely ineffective as it did not address issues of political violence. Makoni (2012: 4) observed:

> Flighting adverts on a national broadcaster and the holding of meetings by a designated body which does not engage victims and people at personal and grassroots level is not enough to bring national healing to a highly polarised nation like Zimbabwe.

The ONHRI suffered operational challenges as the secretariat (comprising representatives from ZANU-PF, MDC-Tsvangirai and MDC-Ncube) could not agree on the strategy to be used. ZANU-PF wanted an umbrella amnesty for people who committed grave human rights abuses, while the MDC formations wanted a South African model of the Truth and Reconciliation Commission, where perpetrators come out in the open and confess (Muchemwa et al., 2013).

In the absence of a serious government effort, a high number of civil society organisations working on peacebuilding emerged in post-2000 Zimbabwe. Previously, the Catholic Commission for Justice and Peace (CCJP), Zimbabwe Council of Churches and very few others had led the way in peacebuilding efforts in Zimbabwe. A survey by the Centre for Conflict Management and Transformation (2013) indicates that there are more than 100 organisations engaged in peace-related activities throughout Zimbabwe. The survey also established that most of the employees in these organisations did not have training in peacebuilding. Peacebuilding requires specialised skills; hence, this research seeks to explore the nature and models of some of these peacebuilding organisations, as well as develop a training module for the development of effective skills among organisations working with different communities in the quest for sustainable peace.

The flourishing peacebuilding idea amongst CSOs in Zimbabwe, was also, to a large extent, the reflection of global developments. The neo-liberal era witnessed civil society gradually occupying important spaces in the socio-political and economic development of states, especially in developing countries. For a period, civil society organisations have received a great deal of international mention in post-conflict reconstruction and peacebuilding, since they are expected to play a crucial role, especially in societies emerging from violent conflicts.

27.5 Methodology

This qualitative study adopted a social constructivist philosophical approach. The philosophical helped the researcher with framing questions on how the efforts by the state and non-state actors can be harnessed in a complementary and constructive way for positive peace outcomes in the country. The central focus of the chapter emanated from excerpts from the researcher's Ph.D. thesis and complementary interviews and focus group discussions to capture contemporary developments in the performance of the state and non-state civil society in Zimbabwe. The study targeted members of the NPRC secretariat, selected members of the PPCs in the country's 10 provinces, traditional leaders, academics, ordinary citizens and civil society organisations doing peace-related programming. The study was a follow up on the issues that emerged during the researcher's Ph.D. journey which was primarily action-research.

A total of 10 participants were selected through judgemental sampling from members of the NPRC secretariat. These were relatively easy to target because their work is of public interest as clarified in the Constitution of Zimbabwe. Their responses were obtained through semi-formal interviews. Through convenient sampling, a further 18 participants were obtained from various Civil Society Organisations in Zimbabwe including the Institute for Young Women Development (10), Zimbabwe Council of Churches (8), and Zimbabwe Civic Education Trust (2). These organisations have or had Peace Committees at District levels in different parts of Zimbabwe. Their responses were obtained through focus group discussions and interviews. Two traditional leaders, Chief Masembura and Chief Nyamaropa in Mashonaland Central Province also participated in the study through convenient sampling. These two participated in the study because they belonged to the Ward Peace Committees that were set up by the Institute for Young Women Development in Bindura (Chief Masembura) and Shamva (Chief Nyamaropa) Districts and had an insight into the operations of the non-state-led peace committees. A total of 4 members from the academia (2 in Harare and 2 in Bindura) also participated in the study through interviews. Two FGDs comprising participants in beneficiary communities in Bindura and Shamva districts (7 in the FGD in Bindura and 6 in the FGD in Shamva) took part in the study. In all, a total of 47 participants took part in the study. Data were analysed manually through thematic analysis.

27.6 CSO-led Experiences with LPC

LPCs are growing in many conflict ridden areas. The concept is also currently being used by the organisations that took part in the action research study. The participating organisations in Zimbabwe indicated that they basically relied on two approaches to peacebuilding and these are the local peace committees and the workshop approach. They indicated that they use the concept of peace committees, which is some form of decentralisation for the various provinces and districts. One participant said:

> We realised that peace committees are best placed to deal with conflicts, as they understand the local dynamics much better than us outsiders. The arrangement is sustainable and cost-effective because the committee members are all local people and our duty is to facilitate their institutionalisation.[1]

The committees reflect a cross section of the communities in which they serve. The peace committees are made up of people who include traditional leaders, church leaders, women, and the elderly in society, as well as other important constituencies, including CSOs. The CSOs also indicated that they do capacity building workshops in areas like vegetable gardening, mushroom farming and general project planning and management. The interview participants indicated that the Zimbabwean conflict was too complex to understand and that they often face challenges navigating the political terrain and the bureaucratic structures in their attempts to address the issues affecting the communities. One of the interviewees noted that:

> Zimbabwe is facing multiple challenges, and it is very difficult to establish an entry point to solve these multiple challenges. Furthermore, there are a lot of bureaucratic challenges that

[1] Interview with the Director (Female) of a women-led Civil Society Organisation in Bindura, Mashonaland Central Province, Zimbabwe on 21 August 2021.

> we face when dealing with the district administrators and other institutions like the police. They often delay the granting of permission to work in certain districts and constituencies.[2]

Another challenge arose related to dealing with constituencies in the rural outposts of Zimbabwe. Participants indicated that some communities were still very conservative and valued their culture to the extent that they did not respect any initiative that did not include the traditional institutions of peace and security in their respective areas. They gave several examples of activities that had to be cancelled because the traditional authorities had not been invited or were invited at short notice and could not turn out for the event.

Some of the key functions of the peace committee are to act as watchdogs, provide early warning systems on conflicts within communities, deliberate on conflicts that have arisen in communities and follow up on resolutions arrived at during deliberations (NPRC, 2021).

27.7 Problems with the Peace Committees

While the peace committees are noble, they tend to pose more challenges in their operationalisation. For example, as noted by Makwerere/Mandoga (2012), the integrity of the traditional leaders is largely undermined by the politicisation of their roles in society. The political conflict in rural Zimbabwe has seen traditional leaders being used as agents of

violence in some cases. For example, it becomes difficult now for the same people with a contested history to suddenly become agents of transformation and players in the peace committees, and yet, it is equally difficult to ignore or sideline them because of their esteemed position in society. Where they are involved, the traditional leaders often declare themselves leaders of the peace committees and thus attempt to influence the direction and flow of things (Kurebwa, 2020).

This creates a difficult scenario because other members within the committee often find it difficult to freely express themselves as they will face the same leaders in other spheres of development within the community. There is often a conflict of interest as the traditional leaders attempt to maintain the status quo, whereas on the other side, the church leaders, who are also part of the committee, are determined to facilitate truth telling. The ordinary citizens, the women, youths and the elderly who also make up the committees are also determined to raise issues that they believe will heal the communities but again, they have their work cut out as they have to deal with the traditional leaders who believe that they have supreme authority.

Furthermore, the peace committees appear to be parallel structures, as the traditional leaders will still convene their traditional courts, where they will still deal with more or less the same issues. They appear to be duplicating as well as usurping the

[2] Interview with the Programmes Officer of a Civil Society Organisation based in Harare but operating in all provinces on 17 August 2021. *All civil society organizations intending to hold meetings/ workshops are compelled to get a police clearance.

functions of the traditional courts in the communal areas, and traditional leaders have appeared to resist them sometimes directly and, in some instances, indirectly. The more conservative chiefs have insisted that the committees report to their respective councils for finalisation of issues.

A critical issue here is that the peace committees are now treated as adjudication committees, which are supposed to bring finality to the conflicts emerging from the communities. This is and can never be the case because peacebuilding is itself a process that cannot be rushed into finality. The idea sounds good, but the implementation in the context of the Mazowe and Bindura districts has been problematic.

The organisations indicated that most of their efforts have been fruitful, but they have faced challenges emanating from the fact that the conflict in Zimbabwe is largely indirect. This implied that the conflict dynamics in Zimbabwe have shifted from direct violence to largely indirect (structural violence) as well as cultural violence, especially in relation to with the violation of women's rights. The nature of the conflict suits the conceptualisation of violence by Galtung (1990). The communities are seemingly peaceful, but there is a lot of indirect violence going on. Examples cited included the allocation of land, farming inputs, flea market stalls and other community schemes along political party affiliation lines. The great ZANU PF and MDC divide played out its ugly face in the communities with the supporters of ZANU PF apparently enjoying an upper hand over their MDC counterparts as most of their people are in occupation of the critical government offices in charge of resource distribution.

However, overt violence is still prevalent, especially during national plebiscites. Supporters of ZANU PF and MDC often clash, with most of the confrontations becoming bloody. The CSOs are often accused of being aligned with the opposition MDC party, which makes their intervention in the communities very difficult, as they are accused of bias. The fact that the MDC has for a long time been accused of being an agent of neo-imperialism has not helped matters either. These CSOs receive most of their funding from the Western Donor communities and the ZANU PF-led government has often found it convenient to label the CSOs as fronts in this neo-imperialist agenda. This has led to increased polarisation between the government and most of the CSOs working on peace projects in the communities.

Such a scenario has proved to be very difficult for the CSOs, as reported by the participants. In fact, one of the organisations indicated that they have had their memorandum of understanding (MoU) cancelled on several occasions on allegations of dabbling in the political affairs of the district in violation of their scope of operation.

27.8 Room for Collaboration

Although the NPRC strategic plan indicated that they would work with various stakeholders, including the CSOs, to promote peace, the reality is that there has been hardly any collaboration between the commission and civil society organisations that are working to promote peace using LPCs. The efforts of the state and those of the CSOs have so far been parallel, with no genuine effort to work together. This will provide an opportunity to pool resources together.

Collaboration can be done at the capacity building level, where the state-led and CSO-led LPCs can conduct trainings to build consensus on the skills and approaches needed to deal with conflicts in the country. They may also collaborate in terms of programming. CSOs usually have a budget dedicated to these initiatives, and the fact that the NPRC-led PPCs have no budget makes it necessary to work closely with organisations that can assist with the resources.

27.9 Conclusions and Recommendations

The chapter looked at the operationalisation of LPCs in Zimbabwe and pointed to the fact that there are basically two types of LPCs: those that are initiated by the NPRC and operate at the provincial level, and those that are initiated by various CSOs and operate at different levels, from the national level to the district and ward levels. The study concluded that the LPCs have not reaped significant achievements owing to a variety of factors. For the state-led LPCs, the primary challenge is that they lack the power and resources to maintain a meaningful presence in the provinces where they are operational. Apart from that, they have also failed to gain the confidence of ordinary citizens, as most of the people who were roped into the provincial committees are accused of being politically active and affiliated to certain political parties. These accusations also extend to traditional leaders, who are also accused of being aligned with the ruling ZANU-PF party. On the other hand, the CSO-led LPCs are often accused of working against the state and thus find it difficult to carry out their work freely. There is also very little evidence to suggest that the CSOs and the NPRC-initiated LPCs are working together, even though there is an apparent need for all actors to work together for the good of the nation at large.

The study recommends:

- The state-led LPCs and the CSO-led LPCs work in collaboration for the transformation of conflicts in Zimbabwean societies. Although there have been promises and pledges to work together from the various actors, the efforts are still not visible on the ground and therefore, there is a need to move from the rhetoric to actual action.
- There is a need to capacitate both the CSO-led and State-led LPCs so that they have the requisite skills to develop effective peacebuilding interventions. These capacity building initiatives may focus on conflict-sensitivity, gender-sensitivity, conflict analysis, and healing and reconciliation.

References

Alexander, J., & McGregor, J. (2013). Introduction: Politics, Patronage and Violence in Zimbabwe, *Journal of Southern African Studies, 39*(4), 749–763, https://doi.org/10.1080/03057070.2013.862100

Baker, B., & Scheye, E. (2007). Multi-layered justice and security delivery in post-conflict and fragile states. *Conflict, Security & Development, 7*(4), 503–528.

Centre for Conflict Management and Transformation. (2013). *Mapping Peace Initiatives in Zimbabwe,* https://www.ccmt.co.zw. Accessed 8 June 2023.

Constitution of Zimbabwe (2013) *Amendment Number 20 of 2013.* Harare: Government Printers.

Conteh-Morgen, E. 2005). Peacebuilding and Human Security in Africa: A constructivist Perspective, *International Journal of Peace Studies, 10*(1), 16–38.

Cooper, N. (2007). Review Article: On the crisis of the Liberal Peace. *Conflict, Security and Development,* 7(4), 605–616. https://doi.org/10.1080/14678800701693025

Dube, D., & Makwerere, D. (2012). Zimbabwe: Towards a Comprehensive Peace Infrastructure, *International Journal of Humanities and Social Sciences, 2*(18), 133–149.

Fayayo, R. (2018). *Zimbabwe's 2018 Elections: The Changing Footprints of Traditional Leaders,* https://www.boell.de/en/2018/07/26/zimbabwes-2018-elections-changing-footprints-traditional-leaders. Accessed 27 August 2021.

Feltoe, G. (2004) The courts, democracy and the rule of law in Harold-Barry (ed). *Zimbabwe: The Past is the Future* (pp. 221–236) Harare: Weaver Press.

Galtung, J. (1976). Three approaches to Peace: Peacekeeping, Peacemaking, and Peacebuilding. In J. Galtung (Ed.) *Peace, War and Defense: Essays in Peace Research II* (pp. 292–304). Coppenhagen: Christian Elders.

Galtung, J. (1990). Cultural Violence, *Journal of Peace Research, 27*(3), 291-305.

Galtung, J., & Ikeda, D., (1995). *Choose peace.* East Haven CT: Pluto Press.

Global Political Agreement. (2008) *Agreement between the Zimbabwe African National Union-Patriotic Front ((ZANU-PF) and the two Movement for Democratic Change (MDC) formations, on resolving the challenges facing Zimbabwe.* Harare: Government Printers.

Harris, G. (2011). Ministries of Peacebuilding in Sub-Saharan Africa: Rationale, Functions and Establishment. *African Security Review, 20*(1), 122–133.

Huyse, L. (2003). Zimbabwe: Why reconciliation failed. In D. Bloomfield, T. Barnes & L. Huyse (Eds.), *Reconciliation after Violent Conflict: A Handbook Series.* (pp. 34–39). Stockholm: International Institute for Democracy and Electoral Assistance.

Kurebwa, J. (2020). The capture of traditional leaders by political parties in Zimbabwe for political expediency. In S. Chhabra (Ed.), *Civic Engagement in Social and Political Constructs* (pp. 196–219). Pennsylvanie: IGI Global. https://doi.org/10.4018/978-1-7998-2364-3.ch009

Lederach, J.P. (1995). *Preparing for peace: Conflict transformation across cultures.* New York: Syracuse University Press.

Lederach, J.P. (1997). *Building Peace: Sustainable Reconciliation in Divided Societies.* Washington DC: USIP Press.

Lederach, J.P. (1999). *The Journey toward Reconciliation,* Waterloo Canada: Herald Press.

Lederach, J.P. (2000). Journey from Resolution to Transformative Peacebuilding. In C. Sampson & J.P. Lederach (Eds.) *From the Ground Up: Menonite Contributions to International Peacebuilding* (pp. 9–27). Oxford: Oxford University Press.

Lederach, J.P. (2003). *The Little Book of Conflict Transformation.* Delaware: Good Books.

Lederach, J.P. (2005). *Preparing for Peace: Conflict Transformation across Cultures.* New York: Syracuse University Press.

Lederach, J.P., & Maiese, M. (2009). Conflict transformation, A circular journey with a purpose. *A Journal of Peace Research and Action, 14*(2), 7-11.

Lee, R. (2011). *Civil Society's present and future role in Zimbabwe,* Open Society Initiative for Southern Africa available at http://www.osisa.org/openspace/zimbabwe/civil-society Accessed 07 March 2021.

Machakanja, P. (2010). *National Healing and Reconciliation in Zimbabwe: Challenges and Opportunities*. Wynberg: Institute of Peace Leadership and Governance, Africa University. https://africaportal.org/wp-content/uploads/2023/05/IJR_AP_Monograph_1_Zimbabwe_WEB.pdf. Accessed 7 August 2023.

Machakanja, P., & Mungure, S. (2013). *Mapping Transitional Justice Actors in Zimbabwe*. Harare: HansMak Designs.

Makoni, S. (2012). *Challenges Facing the Organ on National Healing and Reconciliation in Zimbabwe*. The Standard. 13 Apr 2012.

Makwerere, D., & Mandoga, E. (2012). Rethinking Traditional Institutions of Peace and Conflict Resolution in Post 2000 Zimbabwe, *International Journal of Social Sciences Tomorrow*, 1(4), 18-24.

Mashingaidze, T.M. (2005). The 1987 Zimbabwe National Unity Accord and its Aftermath: A case of Peace without Reconciliation. https://www.google.com/search?q=The+1987+Zimbabwe+National+Unity+Accord+and+its+Aftermath%3A+A+case+of+Peace+without+Reconciliation.&client. Accessed 22 August 2021.

Mbire, M. (2011). *Seeking Reconciliation and National Healing in Zimbabwe: Case of the Organ on National Healing, Reconciliation and Integration (ONHRI)*, Research paper submitted in partial fulfilment of a Masters of Arts Degreein Development Studies, The Hague, Netherlands.

Muchemwa, C., Ngwerume, E.T., & Hove, M. (2013). When will the Long Nightmare come to an End? Challenges to National Healing and Reconciliation in Post-Colonial Zimbabwe. *African Security Review, 22*(3), 145-159.

Mukuhlani, T. (2014). Zimbabwe's Government of National Unity: Successes and challenges of restoring peace and order, *Journal of Power, Politics and Governance*, 2(2), 169-180.

National Peace and Reconciliation Commission. (2021). *Provincial Peace Committee Meetings Chairperson's Statement*. https://www.nprc.org.zw/ Accessed 20 December 2021.

Ramsbotham, O., Woodhouse, T., & Miall, H. (2005). *Contemporary Conflict Resolution*. Cambridge: Polity Press.

Rukuni, T., Machingambi, Z., Musingafi, M. C., & Kaseke, K. E. (2015). The role of traditional leadership in conflict resolution and peace building in Zimbabwean rural communities: The case of Bikita District. *Public policy and administration research*, *5*(3), 75–77.

The World Bank. (2011). *World Development Report: Conflict, Security and Development*, Washington DC: World Bank.

United Nations, (2012). *Africa's youth and prospects for inclusive development*. Available at https://www.ohchr.org/sites/default/files/Documents/Issues/Youth/UNEconomicCommissionAfrica.pdf. Accessed 21 October 2022.

Vambe, M.T. (2014). Songs that won South Sudan's political independence: A Warchild perspective. *Journal of Music Research in Africa*, 11, 4–17.

Chapter 28
Personal Identity and Peacebuilding: A Critical Reflection

Daniel Erdmann

Abstract This chapter discusses the complex setting of a person's identity and how it interferes with the process of peace building. The wide range of potential character strains is highlighted and it is pointed out, in which way an individual's background supports the building of a specific personal's identity. The author places a crucial focus on how to deal with this wide-range of potential diversity during professional dialog facilitation that ought to lead to peace building in the long run. In this course, he challenges traditional, mostly academic approaches towards peace building, in order to address the urgent necessity to equip people on all levels of society with an understanding of conflict management that finds its place beyond strategies, methods, structures, and models.

Keywords Personal identity · Peace building · Deep listening · Critical thinking · Wise speaking

28.1 Introduction

Before we get started, I would like to describe briefly the nature of this article. Due to the fact that peace building in its most specific but also in its widest sense is definitely a practical art, the content of this part of the book, its impact, and the practical application of what I am going to introduce, is taken out of practice, was approved as valuable in real-life, and is presented to you in a form that allows you to introduce it to your professional and private environment. By saying this, I refer

---

Prof Daniel Erdmann is the Director of World Mediation Organisation. He worked for several years as a member of the managing board for humanitarian organizations. Nowadays he is the Director General and Founder of the World Mediation Organization based in Berlin–Germany. Email: mail@worldmediation.org

---

D. Erdmann (✉)
World Mediation Organization, Berlin, Germany
e-mail: mail@worldmediation.org

J. C. K. Kiyala and N. Chivasa (eds.), *Climate Change and Socio-political Violence in Sub-Saharan Africa in the Anthropocene*, The Anthropocene: Politik—Economics—Society—Science 37,
https://doi.org/10.1007/978-3-031-48375-2_28

to the point that my findings are based on practical experiences, accurate reflections, and many improving adjustments. This strong practical approach turns the article more likely into a practitioner's diary, introducing the reader to a unique and heavily personal perspective of the topic. Accordingly, life itself approved these findings in the most reliable way. I wish this journey may take us far.

28.2 A Balanced and Unbalanced Environment

Before we embark on our journey into the depth of an individual's identity, we need to create a common ground of understanding or more likely meet upon the level, and part upon the square. Hereby, I refer to the same level of awareness towards the topic that this book deals within the long run, specifically: building sustainable peace or simply peace building.

Prior to the examination of what ‚building peace' may mean, we may have a closer look at what peace itself actually means to you and me. More often than not, mankind demands far too fast a societal status of peace, without properly understanding its nature, its requested foundation, and its potential impact on fragmented parts of such societies, as well as its potential support regarding fragmentation and extremism per se.

Taking a superficial look at the term "peace", it may mean the same thing or at least something similar to all of us. A peaceful environment may be imagined by the majority of us by witnessing mutual respect between the citizens, who are enjoying social justice in their community, and a setting of coexistence without armed impact.

Leaving our urban communities for a moment and having a look at nature, we will see that nature itself does not offer peace, as we may understand it, but more likely a balancing environment that includes a wide variety of natural laws based on physical, chemical, and biological aspects. This indicates that such a balancing environment is set up by a multitude of variables. It is flexible, constantly changing, and stands for itself. The balancing process is not what we see and consider to be peaceful when we go hiking in a beautiful landscape. Environmental changes and balancing processes are moving slowly and by being visitors to such a landscape, we take a snapshot with us but are unable to look behind the scenes in order to obtain a true understanding of such processes. Save this indication and refer to it later on in this article.

Coming back to our social urban environments, we may wonder if all the personal diversity of locals and area-related citizens may identify with the term 'peace' in such a general and inclusive way. At first sight, we may think that what works for nature, also work for humans. But soon, we will need to learn more about the ambivalence of the term 'peace' and we will see that there are significant differences between nature and man-made society. While natural laws exist due to their very own necessity of establishing an environmental balance and ensuring the creation of a biotope, in order to enable the survival of the species, the laws that govern human co-existence, apart from the biological needs, are man-made. Such man-made laws are an attempt to refrain from natural laws based on the previously listed physical, chemical, and biological aspects. Man-made laws are built on a rational understanding of how to establish a social order, but first of all from the perspective of the person in charge of drafting the corresponding law.

While nature's laws give direct birth to the ongoing balancing process in a specific environment, man-made laws can be the first hindrances on our way to what we may consider peace. Potentially, a constitutional law shall provide a country with general equality, justice, and order. By having a look at our daily news, we will see that in more likely all countries, people and groups appear who see their personal rights being violated by such a constitutional law. While in nature, such separatists are often important impulses for change, people who see themselves being shortened in their civil liberty, may find themselves helpless to initiate changes at all or at least prompt changes regarding their constitutional law. Profound discontent and skepticism at a fundamental level towards their political system are born. The denial of one's own constitutional law turns out to give a person a basic instability in a field where confidence and acceptance potentially enable a person's identity to develop in a sound way. Such instability, discontent, and skepticism support the creation of an enemy image and reinforce a group's mutual belief system and identity (Sommer/Fuchs, 2004).

A more flexible and adjustable system in communities and countries is the political party that serves a fixed term of office. While being discontent with one's constitutional law, an individual or group can try to find a matching identification via a political camp that potentially serves one's fundamental interests in a short- and long-term. But what if mainstream separatists remain with the feeling of being violated and restrained by the ruling constitutional law, not being properly represented by one of the political parties, or even start or continue denying democracy and/or communism per se?

Obviously, it would be beyond the ruling constitutional law, beyond the established political system, and beyond the impact of the serving political party, where we could find a non-armed environment of mutual respect, tolerance, and social support, generally known as peace that properly includes all such perspectives. What peace may mean to one person may mean occupation to another. The fact that we find 'peace' on a fourth level, namely beyond the previously listed three instances, gives us a slight hint on how delicate and complex peacebuilding turns out to be in real life and in a sustainable way.

Finding oneself apart from the constitutional law, the political system, and the serving political camp, can have a multitude of reasons based on insufficient education, integration, tolerance, and sensitiveness on both sides. What we mainly see in the metropolises on our planet is that such persons and groups find themselves in the necessity of establishing their own doctrine and justice system, including honor killing, blood money, and arranged marriages; a parallel society is born, and the fragmentation in the way of tribalism is completed (Sommer/Fuchs, 2004).

A fragmented society or individual may live for decades for itself or as part of a parallel community or phenomenon in the shadow of the mainstream. While ignorance of such fragmented society and individuals is no acceptable option, the question is how such will ever raise awareness about their existence and necessities. From what can be learned by observing the media, the step towards extremist activities and terrorism seems to be a very tiny one and turns out to be widely practiced. Assuming extremist activities and even terrorism to be a manner of communication and awareness-raising, it is obvious to say that political camps ignored phenomena like insufficient education, integration, and results like the establishment of parallel societies and complete fragmentation for far too long.

28.3 A Person's Unpredictable Identity

Now that we have briefly discussed the complexity of the term 'peace' and we have a glimpse of how overwhelming sustainable peace building may be, we should start focusing on the kind of work that each of us can do on a private and/or professional level, namely, getting into dialogue with another individual.

What we can imagine from the previously discussed content, may be that learning about individuals and the impact of culture, religion, and politics in real life, cannot be learned by reading books and/or visiting an area. Being and remaining an alien person, we are more likely to experience all kinds of processes, visible and invisible ones, that take place.

We need to accept that studies and physical visits to certain areas may give us only a slight idea of what regional individuals may face. A person's identity remains bound to all kinds of environmental influences that accompanied the individual during the process of becoming an adult. This is a strong hint to the fact that foreigners perspectives may offer new triggers towards possible changes only, but more often than not, they hardly fail as all sorts of inner socio-cultural aspects are ignored in a conscious or unconscious way.

But still, claiming to be professionals in the field of international relations, intercultural exchange, etc., alien experts need to be aware of their limitations and should not be claiming to have an overall understanding. An overall understanding, if this can ever be achieved by such professionals, is not what is needed for change. We need local experts to work on local issues, meaning activating experts from a certain region, to work on contexts out of this proper area by making use of all kinds of insights and understanding that are not available to foreign experts. In order to improve a foreigner's impact, we need far deeper insights and one-on-one dialog that may guide us step by step to a more comprehensive level, what obviously would require time, turning out being a long-term process, and guiding us from confidence to relationship building (Hartley et al., 1989).

In order to better prepare for real-life circumstances and tasks such as interpersonal dialog and crisis exchange, it is widely common and broadly advised to study character and personality traits. By achieving a broader understanding of the human mindset and potential identities, practitioners, by studying these concepts – shall be empowered to detect or identify them in order to take accurate measures during potential high-tension exchanges. At this point, we shall do a reality check and investigate if the creators of these theories and models ever served or offered their services in armed contexts or high-risk environments. More often than not, we can identify a gap between theoretical content being communicated as a mass solution to issues, while such solutions where never approved to be of value in reality. Further, we may reflect on the true value of specific and truly approved actions in the field may have and if such would ever be communicated in a detailed way in a book. Hereby, I kindly ask you once again to consider the setting of this article: is it delivering answers to proper questions, or does it push you to find, identify, and create your own and proper understanding.

What makes it so difficult to approve such theoretical content in practice? Mainly, it is the fact that the creators of theories and models rely on academic sources and long-term surveys but have never had access to high-risk situations where intuitive and empathetic action is required, far beyond making use of theoretical content, and being bound to decision-making within very limited time frames. Even long-term surveys may have a questionable impact in a tremendously fast changing world. On top of this fast-changing world, we need to consider the fact that long-term surveys may not give sufficient credit, if any at all, to unconscious personality disorders based on personal traumatic experiences that could potentially end up finding and/or creating a new set of safe harbors or social groups per se (Schmitt/Altstötter-Gleich, 2010). Belonging to such a safe harbor or social group is a mainly positive experience of the individual and forms part of one's social identity (Sommer/Fuchs, 2004). Based on this positive experience, the individual will focus on strengthening the group's impact, power, and prosperity. Leaving such a positive environment turns out to be highly difficult act.

Another point to consider is that real-life environments offer a high level of diverse and influential social impact that makes each situation and each individual totally unique. There is no way to make use of any kind of theoretical background as real life will always be more complex than theories and models. The discrepancy is demonstrated by the fact that theoretical studies may offer us more likely an environment for meditative reflections, while practical reports always offer real-life insights and never give birth to theoretical concepts.

As previously outlined and confirmed by practice, several professional settings require prompt actions and do not allow any kind of analytical review of long-term studies in order to better prepare for a certain issue or task. Crisis dialoguing, crisis mediation, and crisis negotiation often take place in totally unpredictable settings, far from tranquility and non-distractive influences. But what is even more unpredictable is the personal identity of the individual that the exchange takes place.

By reviewing external sources, we may find proper indications that explain the broad variety and quantity of personality traits, and that it turns out being almost impossible to even learn them all by heart (Matthews et al., 2003). In order to make things easier to handle, the term of the 'Big 5' was created (Goldberg, 1990). Such Big 5 are meant to be the five major personality traits with which all of us may fit in, more or less. To fit in ,more or less looks to me as a very critical point, having in mind the kind of potential dialogue we are referring to and a highly demanding environment we may be facing. Trying to find a matching major group for the dialogue partner's personal identity may not turn out to be sufficiently matching and may be possibly cutting off crucial core points of his/her identity. Core points that may make the difference between advancing in the conversation or pushing it to collapse. Having the broad variety of possible character traits and personalities on my mind while working in such environments, proved to be totally unrealistic and more likely to be disturbing and distracting.

In order to slightly understand how unique all of us are, we need to sit back and reflect on the environmental impact that we face each day in our lives. Such impact includes every single aspect that we face, underlie, overcome, and take a lesson learned out of it. It is up to the reader to sharpen his/her mind and to unveil the complexity of the context described. Growing up remains a fragile process where more likely everything is absorbed and remains having an impact on the person's maturing process. The process continues in an unreflected way, until the person starts questioning, thinking for oneself, and challenging what was previously admitted to one's mindset. The process of questioning outlines a momentum that remains an optional turning point in one's further personal development. Heinz-Rolf Lückert shares the perspective that radical aspects of one's personality basically remain unconscious until they are identified by one's own proper reflections. Such radical aspects are not limited to one's own behavior but may also be copied from other person's acting and added to one's own key of behavior (Lückert, 1965).

28.4 Social Aspects and Social Dynamics

Excluding the above-mentioned potential turning point in one's further development process, we should look briefly at another important aspect, namely why a social identity is so crucial to a human being and what kind of dynamics it includes. Acting in a non-reflective way, makes a person project one's inner world of understanding onto the outer environment. This process is based on a social need that deals with security and feeling snug by building up a group of like-minded peers around oneself. Interesting enough, it turns out to be an educational and selective process at the same time. While the individual may more likely communicate the proper inner world towards the outer environment in an educative way, he/she also admits persons in an ad hoc method to enrich the like-minded fragmented parallel society or community in number. The series of events and steps taken by the group, creates a group's history that co-builds and / strengthens the group's identity. Such memories are as important to the group, as personal memories are to the individual (Bilali, 2012).

By seeing the number of like-minded peers increase, the initiator and group per se may note a rising level of conformity, stability, and power of impact. This confidence in one's own actions more often than not turns out to be a strong motivation towards action that may be understood being a fundamental necessity to the further achievement of the potential group's goals. Feeling oneself strong, grounded, and powerful opens the door to a possible agenda that was not originally part of the group's identity. Such achieved levels of strength and confidence generate, more often than not, new targets and/or even change the group's aims at all. Drifting off in such a way does not only create an uncertain destination for the individual's further identity developing process but includes the manipulation of the complete group in an unpredictable way.

Further, it should be noted that fragmentation per se includes very impactful dynamics, especially when a group achieves, as previously outlined, a certain level of self-confidence and/or power within society. History shows that fragmented groups in positions of power and impact tend to potentially fragment the rest of the society in order to destabilise it, and to identify one or more of the newly fragmented groups as potential enemies. Hereby, the recently created splinter groups or factions, more often than not, are subject to the aggressor in number and power. Such dynamics can find its climax in prosecution and ethnic cleansing (Zartman, 2010).

28.5 Identity Based Conversation

As previously outlined, progress requires dialogue. Thereby, we need to sum up the findings of this article and try to create and/or identify a meaningful pathway for our professional activities and impact. Working with or trying to remember an overwhelming quantity of theories and models turned out to be a non-supportive aspect of dialog related professional activities (Erdmann, 2018). Therefore, we need to identify the practicable takeaway that grants sufficient space to truly focus on the task, issue, and environment we work with or work in. Having a conversation on this topic, with a friend who works for one of the global intelligence agencies, both of us agreed that real-life crisis dialoguing requires an epic and brave step of the practitioner, namely, to reject potentially studied theories, concepts, and models – including the vast list of personality traits and even the minimised version of the Big 5 – and to truly follow one's intuition, empathy, and inner voice.

This finding seems alien to all academia-based researchers and lecturers, but this fact describes the previously mentioned gap between theory and practice. In my case, the focused minimisation of my professional tools towards the most essential core understandings turned out to be the door opened that granted space for my key tools of professional impact, namely: Deep Listening, Critical Thinking, Wise Speaking, or simply Living Empathy (Erdmann, 2021). Only such liberating step enabled me to focus mindfully on what presented itself to me during my professional tasks. It is not about thinking outside the box, or looking behind the curtain, but more likely to make a plumb role indicate the path to the proper center of a situation and to unveil the core necessities that such an environment requires, including the accurate handling of situations, high-tension dialoguing, and even the prevention of fragmentation. Such an applied dialogue may turn out being one of the very few meaningful acts to prevent fragmented groups of initiating violent actions (Zartman, 2010).

28.6 Conclusion

Trying to solve our professional tasks within tremendously limited time frames, with lots of content on our minds to sort out and make use of, turns us potentially into more likely stiff and inflexible individuals who ignore the character and the needs of conflict per se. Each conflict as well as each person we talk to has its own history, socio-cultural/religious environment, and impact (Erdmann, 2020). This indicates that a conflict and a person do have something in common, namely an outstanding uniqueness. Therefore, we need to work professionally with the previously listed tools in order to get to know the conflict or the person in depth and to handle our task without assumptions but only with our intuition, flexibility, and empathy. Finally, this article gives a strong hint that the art of dialoguing cannot be learned from a book but is a naturally born ability based on mindfulness and empathy.

References

Bilali, R. (2012). Remembering Intergroup Conflict. https://www.academia.edu/14628218/Remembering_intergroup_conflict. Accessed 15 October 2022.
Branthwaite, H. (1989). *The applied psychologist*. Philadelphia: Open University Press.
Erdmann, D. (2018). *Comprehensive Mediation - Towards an educational concept for the training of globally effective mediators*. Raleigh: Lulu Press Inc.
Erdmann, D. (2020). Escaping the Training Trap in Mediation and Conflict Management. https://pin-negotiation.org/en/publications/pinpoints/pinpoints-48-49/. Accessed 15 October 2022.
Erdmann, D. (2021). *Mindful mediation - The formless art*. Raleigh: Lulu Press Inc.
Goldberg, L. (1990). An Alternative "Description of Personality": The Big-Five Factor Structure. https://psychology.okstate.edu/faculty/jgrice/psyc4333/Goldberg_BigFive_1990.pdf Accessed 15 October 2022.
Lückert, H-R. (1965). Konflikt-Psychologie, Ernst Reinhardt Verlag, München/Basel.
Matthews, G., Deary, I.J., Whiteman, M. (2003). Personality Traits. https://catdir.loc.gov/catdir/samples/cam041/2003046259.pdf. Accessed 7 August 2023.
Schmitt, M., & Altstötter-Gleich, C. (2010). Differentielle Psychologie und Persönlichkeitspsychologie, Beltz Verlag, Weinheim/Basel.
Sommer, G., & Fuchs, A. (2004). Krieg und Frieden - Handbuch der Konflikt- und Friedenspsychologie, Beltz Verlag, Weinheim/Basel/Berlin.
Zartman, I.W. (2010). Preventing Identity Conflicts Leading to Genocide and Mass Killings. New York: International Peace Institute. https://www.ipinst.org/wp-content/uploads/publications/ipi_identity_conflicts_epub.pdf. Accessed 15 October 2022.

About the Editors

Jean Chrysostome K. Kiyala is a Senior Lecturer and researcher at the International Centre of Nonviolence based at the Durban University of Technology (DUT) in South Africa; Associate Professor and visiting lecturer at the University of Bandundu, The Evangelical University in Africa, the Catholic University of Bukavu, the Catholic University of Goma_La Sapientia in the Democratic Republic of Congo (DRC) and visiting lecturer at the University of Seychelles. He holds Higher Diplomas in Sacred Theology and in Mission from Tangaza University College, Nairobi (Kenya); Bachelor of Theology from the Pontifical Urbaniana University (Rome, Italy), BA Honours in Religious Studies from the Catholic University of Eastern Africa, Nairobi (Kenya), Higher Diploma in Philosophy from Saint Augustine University of Kinshasa (DRC), and Bachelor Philosophy from the Pontifical Urbaniana University. He holds a Master of Philosophy in Religious Education and Pastoral Theology from St.

J. C. K. Kiyala and N. Chivasa (eds.), *Climate Change and Socio-political Violence in Sub-Saharan Africa in the Anthropocene*, The Anthropocene: Politik—Economics—Society—Science 37,
https://doi.org/10.1007/978-3-031-48375-2

Augustine College of South Africa; he was conferred M Tech in Public Administration by DUT in 2012 and was awarded a Doctoral Degree in Philosophy specialising in Public Administration (Peace Studies) in 2016 from DUT for research on "Child soldiers and restorative justice". Chrys is the founder of the A Hipfunaneni [Let us help each other] Community Project that provides psychosocial assistance to Zimbabwean and Mozambican refugees in Limpopo Province (South Africa); the Youth Advocacy and Peacebuilding Organisation that trains young people in non-violent approaches of dispute resolution in KwaZulu-Natal Province (South Africa); and the Youth Fraternity for Peacebuilding and Reconciliation, based in North Kivu in the Democratic Republic of Congo, which aims to accompany former child soldiers and train them to become peacebuilders. His specialisations include child soldiers; restorative justice, post-conflict reconciliation and nation-building; civil society and peacebuilding; environmental studies; and peace studies. His publications include: Kiyala, J.C.K. (2018) *Child Soldiers and Restorative Justice: Participatory Action Research in the Eastern Democratic Republic of Congo*. Cham ZG: Springer International Publishing. ISBN 978-3-319-90070-4; ISBN 978-3-319-90071-1 (ebook). https://doi.org/10.1007/978-3-319-90071; Kiyala, J.C.K. & Harris, G.T. (Eds.) (2022) *Civil Society and Peacebuilding in Sub-Saharan Africa in the Anthropocene: An Overview* (Cham ZG: Springer Nature Switzerland), https://link.springer.com/book/10.1007/978-3-030-95179-5; Kiyala, J.C.K. (2021). Responding to Epistemic Injustice against Child Soldiers. *The International Journal of Children's Rights, 29*(3), pp. 662–700. https://doi.org/10.1163/15718182-29030006; Kiyala, J.C.K. (2020). Dynamics of Child soldiers' Psychosocial Rehabilitation and Well-being: Perspectives from Bioecological Systems Theory in the Democratic Republic of Congo. *Child Psychiatry & Human Development, 52*(3), pp. 376–388. https://doi.org/10.1007/s10578-020-01022-4; Kiyala, J.C.K. (2019) Combining restorative justice and social justice: Prospects of child soldiering transitional justice, *African Security Review*, 27(3–4), pp. 193–211. https://doi.org/10.1080/10246029.2018.1559868; Kiyala, J.C.K. (2018) The impact of globalisation on child soldiering: challenges and opportunities in the context of the Democratic Republic of Congo. *International Social Science Journal* 66(221–222), pp. 271–284. https://doi.org/10.1111/issj.12132; Kiyala, J.C.K. (2016) Utilising a traditional approach to restorative justice in the reintegration of former child soldiers in the North Kivu Province, Democratic Republic of Congo. *Africa Insight* 46 (3), pp. 33–50. ISSN (Print): 0256 2804; 1995-641X (online); Kiyala, J.C.K. (2015) Challenges of Reintegrating Self-Demobilised Child Soldiers in North Kivu Province: Prospects for Accountability and Reconciliation via Restorative Justice Peacemaking Circles. *Human Rights Review* 16(2), pp. 99–122. https://doi.org/10.1007/s12142-015-0361-7; Kiyala, J.C.K. (2022) Faith-based Organisations and Environmental Peace: Impact of Laudato Si' and its relevance in the Anthropocene in Africa. In J.C.K. Kiyala & G.T. Harris (Eds.) *Civil Society and Peacebuilding in Sub-Saharan Africa in the Anthropocene: An overview* (pp. 65–125). The Anthropocene: Politik—Economics—Society—Science, vol 34. Springer, Cham. https://doi.org/10.1007/978-3-030-95179-5_4; Kiyala, Jean Chrysostome K. (2022) Civil society, Climate Change, Security Risks Management and Peacebuilding in the Anthropocene. In

J.C.K. Kiyala & G.T. Harris (Eds.) *Civil Society and Peacebuilding in Sub-Saharan Africa in the Anthropocene: An overview* (pp. 67-93). The Anthropocene: Politik—Economics—Society—Science, vol 34. Springer, Cham. https://doi.org/10.1007/978-3-030-95179-5_3

http://www.theophilsapiens.co.za/
https://orcid.org/0000-0002-9587-0230
https://scholar.google.co.za/citations?user=7pQvqTQAAAAJ&hl=en
https://www.researchgate.net/profile/Jean-Chrysostome-Kiyala
https://www.webofscience.com/wos/author/record/HKF-0174-2023
https://www.scopus.com/results/authorNamesList.uri?sort=count-f&src=al&affilName=Durban+Univ

Address: Section 7, 13 Tilseth, 18 Bonamour Avenue, Glenwood, Durban, KwaZulu-Natal, South Africa
Email: JeanK@dut.ac.za; kljeanchrysostome@gmail.com

Norman Chivasa-PhD (Zimbabwean) is a Research Associate at the University of KwaZulu-Natal, South Africa. He holds a Master's and PhD in Conflict Resolution & Peace Studies from the University of KwaZulu-Natal, South Africa. Since 2010, he has worked as a part-time senior lecturer in the Department of Peace, Leadership, and Conflict Resolution at Zimbabwe Open University. He also lectured in the War and Strategic Studies department at the University of Zimbabwe. His research focuses on statelessness, hybrid conflict resolution, commodification of land and informal infrastructures for peace. He has long been involved in community peacebuilding efforts in Seke district, Zimbabwe, where he has been working closely with villagers in creating local peace committees. His consulting specialty involves designing, implementing and evaluating local infrastructures for peace at village level. His publications include:

Chivasa, N. (2022). An Exploration of gender and statelessness in Zimbabwe's new dispensation: Opportunities and Problems. *Alternation,* 29(2), 116–133, https://dol.org/10.29086/2519-5476/2022v29n2a7; Chivasa, N. (2022). Reflections on the everyday dimensions of peace building in Seke District, Zimbabwe. *Journal of Peacebuilding & Development*, 1–5. SAGE, https://doi.org/10.1177/15423166221114776; Chivasa, N. (2022). Reflections on Peace building Constructs in Seke District, Zimbabwe. SAGE Open, Jan-Mar, 1–13. SAGE, https://doi.org/10.1177/21582440221077246; Chivasa, N. (2021). Local Role-Players, Informal Peacebuilding Institutions and Participatory Peacebuilding in Zimbabwe. *African Journal of Peace and Conflict Studies*, 10(1), pp. 51–75. https://doi.org/10.31920/2634-3665/2021/v10n1a3; Chivasa, N. (2020). Instituting village savings and loan associations scheme through action research in Zimbabwe. *International Journal of Action Research,* 3, 184–201. https://doi.org/10.3224/ijar.v16i3.023; Chivasa, N. (2020). Anthropomorphisation of God, the 2018 presidential election outcome and political change in Zimbabwe, *Pharos Journal of Theology,* 101. ISSN2414-3324. https://www.pharosjot.com; Chivasa, N. (2020). *Kutanda botso* ritual among the Shona people of Zimbabwe: How gender is reflected, sustained and re-created? *Gender & Behaviour*, 18(2), June, 15746–15754. ISSN: 1596-9231; Chivasa, N. (2019). Action research by ordinary people: How communities are creating peace committees in Seke district, Zimbabwe, *African Journal of Social Work*, 9(2), 125–134. https://www.ajol.info/index.php/ajsw/article/view/192216/181329; Chivasa, N. (2019). A participatory approach to peacebuilding evaluation in Seke district, Zimbabwe. *International Journal of Action Research,* 15(3), 198–216, https://doi.org/10.3224/ijar.v15i3.03; Chivasa, N. (2018). Instituting dispute resolution procedures in the Apostolic Faith Mission in Zimbabwe church', *In die Skriflig*, 52(1), a228. https://doi.org/10.4102/ids.v52i1.2285. ISSN: (Online) 2305-0853, (Print) 1018-6441.

normanchivasa@gmail.com
https://orcid.org/0000-0001-5953-7855
https://publons.com/researcher/1326934/norman-chivasa
https://www.researchgate.net/profile/Norman_Chivasa3

Address: Domboshava, Borrowdale, Harare, Zimbabwe.

On Durban University of Technology

With approximately 23,000 students, the Durban University of Technology (DUT) is the first choice for higher education in KwaZulu-Natal (KZN). It is located in the beautiful cities of Durban and Pietermaritzburg (PMB). As a University of Technology, it prioritises the quality of teaching and learning by ensuring its academic staff possess the highest possible qualification that they can get.

The Durban University of Technology is a result of the merger in Apr 2002 of two prestigious technikons, ML Sultan and Technikon Natal. It was named the Durban Institute of Technology and later became the Durban University of Technology in line with the rest of the universities of technology.

DUT, a member of the International Association of Universities, is a multi-campus university of technology at the cutting edge of higher education, technological training and research. The university aspires to be a "preferred university for developing leadership in technology and productive citizenship", and to "making knowledge useful".

As a butterfly develops from a pupa, so have the students at our institution. From the moment they register as green freshers, to their capping at the hallowed graduation ceremony, our students undergo an intellectual evolution. Website: http://www.dut.ac.za/

J. C. K. Kiyala and N. Chivasa (eds.), *Climate Change and Socio-political Violence in Sub-Saharan Africa in the Anthropocene*, The Anthropocene: Politik—Economics—Society—Science 37,
https://doi.org/10.1007/978-3-031-48375-2

On International Centre of Nonviolence (ICON)

ICON is one of DUT's ten focus areas. Its mission is to make strategic interventions in education that challenge structural violence and help develop a culture of nonviolence.

ICON seeks to make nonviolence real through analysis of the local and global context, through research and through reflective practice. These inform its actions and strengthen its capacity in the field of nonviolence.

Accordingly, ICON is active in teaching nonviolence and leadership to undergraduates as part of in DUT's general education programme, using highly participatory learning methods. In addition, it has a large postgraduate programme in peacebuilding, with some 60 PhD and 15 master's students as at early 2017, drawn from many African countries. All postgraduate students follow an action research approach in their thesis work.

Website: http://www.icon.org.za/current.html

J. C. K. Kiyala and N. Chivasa (eds.), *Climate Change and Socio-political Violence in Sub-Saharan Africa in the Anthropocene*, The Anthropocene: Politik—Economics—Society—Science 37,
https://doi.org/10.1007/978-3-031-48375-2

On University of Bandundu

The **University of Bandundu (UNIBAND)** is a public university in the Democratic Republic of the Congo (DRC), located in Bandundu, the capital city of the province of Kwilu. At its inception, UNIBAND—then called *Centre Universitaire the Bandundu* (C.UB.) [Bandundu University Centre]—created on 1 Oct 2004, was an appendage of the University of Kinshasa (UNIKIN). By 2012, UNIBAND had 800 students spread through its six faculties. It became autonomous in 2010 following Ministerial Decree No. 157/MINESU/CABMIN/ EBK/PK/2010 27 Sep 2010. At present, UNIBAND hosts six faculties: Law, Medicine, Psychology and Sciences of Education, Economics and Agronomy. Based on Law No. 86/025 of 3 Oct 1981 and the Framework Law No. 89/005 of 22 Sep 1986 on the general

J. C. K. Kiyala and N. Chivasa (eds.), *Climate Change and Socio-political Violence in Sub-Saharan Africa in the Anthropocene*, The Anthropocene: Politik—Economics—Society—Science 37,
https://doi.org/10.1007/978-3-031-48375-2

organisation of higher learning institution and universities, which sets the objectives of universities and higher learning institutions in the DRC, UNIBAND, like any other university, has the following main mission: to ensure the training of intellectuals, researchers, academics, and managers in the diverse domains of life. As such, it provides training and lectures in various fields that encourage the emergence of new ideas and the development of professional skills: to organise basic and applied scientific research aimed at meeting the many challenges of social, scientific, economic, and political life. Because of its integrated academic scheme, UNIBAND's students enjoy ongoing formation provided by a qualified staff who pursues the goals of building a harmonious and integral development of the human person. UNIBAND's academic programmes bridge traditional training and innovative learning which promote research and community engagement, and train leaders capable of meeting the demanding criteria required to make society a better place to live in.

About the Contributors

Each contribution underwent a thorough peer-review process, initially by the editors and subsequently by two anonymous reviewers.

Alain Jean-Charles Tschudin (PhD) is Professor of Peace Studies and the Director of the International Centre of Nonviolence (ICON) based at Durban University of Technology. He is a former Executive Director of Good Governance Africa and former Visiting Professor at the WITS School of Governance. He is a registered psychologist with PhDs in Psychology and in Ethics. He was a Swiss Academy Postdoctoral Fellow at Cambridge and oversaw the Conflict Transformation & Peace Studies Programme at University of KwaZulu-Natal (UKZN) for several years. He has broad research and community engagement interests and has worked for various universities in Africa and Europe, with the European Commission, with local and international NGOs~~,~~ and as lead consultant for Save the Children, ~~and~~ UNICEF, and most recently The UN Office of the Special Adviser on Africa. Alain serves as President of the Association Montessori Internationale. Email**:** alaint@dut.ac.za

Aliye Fatma Mataracı (PhD) is Assoc. Prof. Dr., Political Science Program, Faculty of Arts and Social Sciences, International University of Sarajevo (IUS). She has been engaged with The World Social Capital Monitor, an open access survey tool assessing the Social Capital of countries around the globe in 50 languages, as the Researcher in Charge of the Western Balkans Social Capital Monitor since 2018. The monitor has been developed by the Basel Institute of Commons and Economics (http://commons.ch/) as a UN Sustainable Development partnership. She has delivered a number of lectures to share the results of the Monitor for the Western Balkans on yearly basis, in addition to publications. Email: amataraci@ius.edu.ba; ORCID ID: https://orcid.org/0000-0001-8289-8763

J. C. K. Kiyala and N. Chivasa (eds.), *Climate Change and Socio-political Violence in Sub-Saharan Africa in the Anthropocene*, The Anthropocene: Politik—Economics—Society—Science 37,
https://doi.org/10.1007/978-3-031-48375-2

Ayorinde Ogunyiola (PhD) is a Researcher in the Department of Forestry & Natural Resources, Purdue University, West Lafayette, Indiana, USA and an Assistant Professor in the Department of Political Science and Sociology, Murray State University, KY, USA. Email: aogunyiola@murraystate.edu

Bheki Magunga is a lecturer in the Department of Environmental Health, Some of the Module includes sustainable development and ecology as well as research project. He is in the final stages of his PhD in Environmental Health which he is pursuing with the Central University of Technology (CUT). His thesis is titled 'Thyme oil and Thyme oil Hydrosol as alternative fungicides against Phyllosticta citricarpa (causative agent of Citrus Black Spot)'. In 2018 he was selected for the award of young emerging researcher at the International Association of Food Protection (IAFP) in Salt Lake, Utah in the USA. Bheki is also contributing to research supervision, as he currently supervises Master's and undergraduate in the department of Environmental Health. He also supervises Master of Public Health (MPH) Student in the department. He is part of the undergraduate Research Ethics Committee (REC) in university of Johannesburg faculty of health science. Email address: bhekim@uj.ac.za

Charlotte Mokoatle holds a Master's degree in Environmental Health and currently in the final year of her doctoral studies in Environmental Health. She has 20 years' experience in the profession of Environmental Health (two years at provincial health, six years at a metropolitan municipality and twelve years in academia up to date). Her research focuses on Environmental Health Promotion and Management, Food Control and Food Safety. She has been a lecturer at the University of Johannesburg and served as the Head of Department of Environmental Health at UJ for four years (2017–2021). Email address: charlottem@uj.ac.za; ORCID ID: https://orcid.org/0000-0002-6884-9225; Scopus Author ID: 57202952718

Claris Siyamayambo is a mathematical statistician and a researcher (PhD in Statistics (2021), M.A. in Statistics (2012), PGDE (2004), BSc General in Mathematics and Statistics, (2001) trained at the Northwest University, South Africa, University of Botswana and University of Zimbabwe. Dr Claris is currently a Postdoctoral Research Fellow in Public Health at the SAMRC/UJ Pan African Centre for Epidemics Research (PACER) Extramural Unit, Faculty of Health Sciences, University of Johannesburg. Her research interests are in the areas of sexually transmitted infections (STIs), time series analysis, Covid-19, Key populations and public policies in Africa including in Sub-Saharan Africa. Email address: clariss@uj.ac.za; ORCID profile: https://orcid.org/0000-0002-4884-681X

Clayton Hazvinei Vhumbunu (PhD) is a Senior Lecturer in International Relations and Governance within the Department of Political Studies and Governance at the University of Free State (UFS), Bloemfontein; South Africa. Previously, he has held academic and non-academic appointments at the University of Namibia (UNAM) and at the National Institute for Humanities and Social Sciences (NIHSS), University of Pretoria; Rhodes University, and University of KwaZulu-Natal (UKZN) in South Africa. He has also been a Research Fellow at the United Nations Economic

Commission for Africa (UNECA) in Ethiopia and has been an Asia Global Institute (AGI) Research Fellow at the University of Hong Kong (HKU), Hong Kong since 2020. He researches in the areas of International Relations, Regional Integration, Conflict Transformation, Peace and Conflict, and China-Africa Relations. Email: cvhumbunu@gmail.com VhumbunuCH@ufs.ac.za; https://orcid.org/0000-0002-4065-1073

Clementine Maramura Tafadzwa holds a PhD in Public Management and Governance from the North-West University, a Master of Public Policy, Honours Public Policy (Cum Laude) and a B. Soc. Sci (Cum Laude) from the University, of Fort Hare. Her research interests are in Governance, Socio-Economic Development and Public Policies spanning across Africa. She is a Brightest Young Minds (BYM) in Africa alumnus and she is currently a Senior Lecturer at the University of the Free State. Email: tmaramura@gmail.com; http://orcid.org/0000-0002-4419-1887; https://www.researchgate.net/profile/Tafadzwac-Maramura; https://scholar.google.co.za/citations?user=YwAKLXQAAAAJ&hl=en

Crispin Hemson is the former Director of the International Centre of Nonviolence at Durban University of Technology. He holds Master's degree in Education; he is currently completing his PhD on educating educators for peace. He has a background as an adult educator and was formerly Head of the School of Education at the University of KwaZulu-Natal. His current research interests are in addressing gender-based violence and in developing pedagogies that promote nonviolence. Apart from his teaching in the General Education programme of DUT, he has led the Durban Innovative Leadership Programme, which develops DUT students and community activists as leaders. As an environmental activist, he heads the Friends of Pigeon Valley, which supports a nature reserve in Glenwood, Durban. Email: crispinh@dut.ac.za

Daniel Erdmann (Prof) worked for several years as a member of the managing board for humanitarian organizations. Nowadays he is the Chairman and Founder of the World Mediation Organization. He holds a PhD and serves as a Professor in Mediation and Conflict Management at Euclid University. He dedicates his time to initiating positive and productive change of a sustainable character within society on a global level. His publication titled "Mindful Mediation – The Formless Art" is one of many tools that he uses to bridge cultural diversity and the wide range of understanding peace. Email: email@worldmediation.org

Daniel Nkosinathi Mlambo (PhD) holds a Postgraduate Diploma in Teacher Education from the Haaga-Helia University of Applied Sciences School of Vocational Teacher Education (Finland), a PhD and Master's degree in Public Administration, an honours in International Relations, and a junior degree in Development Studies all from the University of Zululand. His research focuses on African Political Economy, Regional Integration, Governance and Democracy, Migration, and Security Studies. He is currently a Post-Doctoral Research Fellow in the Office of the Executive Dean (Faculty of Humanities) at the Tshwane University of Technology (TUT) under the sponsorship of the National Institute for the Humanities and

Social Sciences (NIHSS). Email: mlambo1@ymail.com; ORCID: https://orcid.org/0000-0001-9898-1414; URL: https://linktr.ee/danielnmlambo

David Makwerere is a Lecturer in the Department of Peace and Governance at the Bindura University of Science Education in Mashonaland Central Province, Zimbabwe. His research focus is primarily on social justice, community development and social cohesion, Peacebulding and the agency of non-state actors in peace processes. He is also particularly interested in the potential role of sport in Peacebulding and has previously published in these areas of interest. Email: david.makwerere@gmail.com

Diaku Dianzenza Kunsikila (PhD) holds a Master of Commerce Conflict Resolution and Peace Studies (CRPS) from the University of KwaZulu Natal (UKZN). He lectured at the Faculté Maria Malkia in D.R. Congo from 2008 to 2010 and lectured in the Cornerstone Module from 2015 to 2016 at the Durban University of Technology (DUT). He works as a Freelancer and is a Private Tutor in empowering tertiary students with their academic work and is involved in Humanitarian work since August 2017 until this day. His ORCID profile: https://orcid.org/0000-0001-8365-4955; Email: diakudianzenza4@gmail.com

Dorothy Moyo (PhD) holds a BSc Sociology degree and an MSc in Peace Leadership and Conflict Resolution and a PhD in Public Administration: Peace Studies. She is a Commissioner and Deputy Chairperson in the Zimbabwe Human Rights Commission. She is also a Facilitator of the Alternatives to Violence Project Zimbabwe (AVPZ). Email: dorothyhopemoyo@gmail.com

Esther Rehema Matendo (Mrs) in completing her doctoral degree at the Department of Food Science and Nutrition, Jomo Kenyatta University of Agriculture and Technology, Nairobi, Kenya; she is a lecturer at the Faculty of Agronomic Sciences and Environment of the Université Evangélique en Afrique, and at the Faculty of Sciences, Department of Agronomy and Environmental Sciences at the Université Officielle de Bukavu, the, D.R. Congo. Her PhD research focuses on the utilization of local plant-based treatment to mitigate the mycotoxin contamination in some staple food in South Kivu, DRC. Email: vecordcongoesther@gmail.com

Jackson Ishara (Mr) is completing his doctoral degree at the Department of Food Science and Technology, Jomo Kenyatta University of Agriculture and Technology. He is a lecturer at the Department of Food Science and Technology, Université Evangélique en Afrique, DR Congo. His other activities include research in biochemistry of natural bioactive metabolites, the molecular mechanisms underlying their contribution in preventing and controlling life threatening conditions and chronic diseases through diverse cellular and physiological pathways modulation by functional and nutraceuticals foods mainly edible mushrooms. Email: jackishara17@uea.ac.cd; https://orcid.org/0000-0003-4789-4040; https://www.researchgate.net/profile/Jackson-Ishara

Jean Chrysostome K. Kiyala (PhD) is a Senior Lecturer at the International Centre of Nonviolence at Durban University of Technology (South Africa), Associate

Professor and visiting lecturer at the University of Bandundu, Evangelical University in Africa, and the Catholic University of Bukavu (Democratic Republic of Congo); and visiting lecturer at the University of Seychelles (Seychelles); emails: JeanK@dut.ac.za; kljeanchrysostome@gmail.com; ORCID: https://orcid.org/0000-0002-9587-0230 ResearchGate: https://www.researchgate.net/profile/Jean_Kiyala; URL: www.theophilsapiens.co.za

Jean Mubalama Mondo (PhD) is a lecturer and researcher in Sustainable Agriculture, Food Security, Seed Systems, and Agricultural Innovation Adoption from the Université Evangélique en Afrique, Bukavu, eastern D.R. Congo. Besides, he works with local and international organizations to improve smallholder farmers' resilience to climate shocks by promoting climate-smart agriculture, with particular emphasis on breeding and disseminating multiple stress-resistant crop varieties and crop diversification using indigenous crops. He is currently the country representative of the International Institute of Tropical Agriculture (IITA)'s AfricaYam Project. Email: mondojean@gmail.com; ORCID ID: http://orcid.org/0000-0001-7197-6551; Researchgate: https://www.researchgate.net/profile/Jean-Mondo-Mubalama

Josephine Kimanu Mauwa holds a PhD in Public Administration: Peace Studies at Durban University of Technology (International Centre of Nonviolence) in Peacebuilding programme. She is a lecturer at the Faculty of Social Sciences, in the Department of Peace and Conflict Transformation at the Evangelical University in Africa. Her field of studies is gender and peace-building. Email: mauwakimanu@yahoo.fr

Katcho Karume (Prof.) holds a PhD in Environmental Physics; he is currently serving as the Director General of DRC Geothermal Research Center after being the Dean of Agricultural and Environmental sciences at the Université Evangélique en Afrique (for the last 10 years) where lectures. He researches energy alternatives to biomass, mainly hydropower, wind, geothermal and solar energy. His other activities include identifying sites for energy production, the determination of their potentials, taking into account the environmental impact and Cost Benefit Analysis using Remote Sensing and Geographical Information Systems (RS-GIS) tool. He also works on the dynamics of exploited systems, essentially the dynamics of biomass for a given region, on climatic data collection and analysis for early warning. His additional fields of interest include Natural Resource Management in the conflict and post conflict context. Email: kkatcho@yahoo.com

Lawrence Mhandara (PhD) is a senior lecturer in Political Science, Department of Governance and Public Management, University of Zimbabwe, Harare, Zimbabwe. He received his BSc in Politics and Administration and MSc in International Relations from the University of Zimbabwe. He obtained his PhD from the Durban University of Technology, South Africa. Lawrence has also authored book chapters and articles in his areas of specialty and research interests, namely: International Relations, International Security, Governance & Politics in Zimbabwe, Conflict issues and Comparative Peacebuilding. Email: lmhandara@gmail.com

Lufuluabo Ebenezer Maweja is a medical doctor from the University of Lubumbashi (2008) and hold a Master's degree in Public Health (Occupational and Environmental health) from the University of Johannesburg (2022). Trained on emergency response, Dr. Maweja has focused his practice on Environmental and occupational accident prevention. His current interest is in training health and mining workers on the use of PPE (full name), and prevention of occupational and Environmental accident. Email: ebenmaweja78@yahoo.com

Maleni Thakur (PhD) is a Postdoctoral Fellow at the International Centre of Nonviolence, Durban University of Technology. She hols a PhD in Public Administration: Peace Sudies from DUT. Her areas of research includes water sustainability and conservation through behavioural change. Email: maleni.thakur@gmail.com; ORCID: https://orcid.org/0000-0003-0102-3151

Martha Chadyiwa is a registered EHP with the Health Professions Council of South Africa. She is a holder of a Bachelor of Science Degree in Environmental Health. She has an MBA in Environmental and Energy Management from Twente University, completed Honors in Leadership at the University of Johannesburg (UJ). She is finalising a PhD in Public Health at Witwatersrand University focusing on Occupational Health and Safety. Her niche area of research is occupational and environmental health. She was the 2021 Winner of Inspiring Fifty ladies in South Africa. Email: mchadyiwa@uj.ac.za; ORCID ID: https://orcid.org/0000-0001-6263-4402**;** Scopus profile ID: 57211097070; Google Scholar: https://scholar.google.com/citations?user=cwbLDbgAAAAJ&hl=en&inst=5937892379707079942&oi=ao

Moses Monday John (PhD) is a civil society activist, associate lecturer and researcher in conflict management at the School of Public Service, University of Juba, South Sudan. He holds a PhD in Public Administration -Peace Studies from Durban University of Technology, South Africa. He co-founded the Organisation for Nonviolence and Development (ONAD) based in Juba. He has more than 20 years' experience in planning and management of NGO development projects, designing and facilitating nonviolent action and peacebuilding training, organisational development and Do No Harm programming. ORCID ID: https://orcid.org/0009-0009-7083-838X. Email: mosesjoa@gmail.com

Muesiri O. Ashe (PhD) is a postdoctoral research fellow in the School of Social Sciences, University of Kwazulu-Natal, Durban, South Africa. He holds a Doctor of Philosophy degree in History and Strategic Studies from the University of Lagos, Nigeria, in 2018. He is an erudite scholar with a track record of academic excellence, earning him academic prizes at both the undergraduate and postgraduate levels. His research focus embraces history, migration and gender. He has published various articles in journals of international status and was a part-time lecturer at the University of Lagos and Lagos State University before completing his doctorate. Emails: muesiri@yahoo.com; Ashem@ukzn.ac.za; ORCID: https://orcid.org/0000-0001-9314-2101

Norman Chivasa holds a PhD in Public Administration: Peace Studies from Durban University of Technology; he served as a research fellow at the university of KwaZulu

Natal between 2016 and 2018 as well as an NRF Postdoctoral Fellow at the International Centre of Nonviolence, Durban University of Technology between 2020 and 2022. During that period, Norman also served as a visiting researcher at the University of Konstanz, German where he collaborated on a project entitled: Hybrid conflict resolution. He is currently a lecturer in the department of peace, security and society at the university of Zimbabwe. His consulting specialty involves designing, implementing and evaluating local infrastructures for peace at village level. Email: normanchivasa@gmail.com. ORCID: https://orcid.org/0000-0001-5953-7855

Ntombizakhe Moyo-Nyoni (PhD) is a lecture in Peace Studies at the Midlands State University in Gweru, Zimbabwe. She hols a PhD in Management Sciences Specialising in Public Administration (Peacebuilding Programme) from Durban University of Technology, South Africa. Besides teaching, she is a peacebuilding practitioner working with communities as a facilitator of Alternative to Violence Project and a consultant in community trauma processing session. Her research interest is including peace education, violent prevention and restorative justice. Email: zakhemoyo@gmail.com

Ozge Buyukdagli (PhD) is currently working in International University of Sarajevo (IUS) as Assistant Professor and Program Coordinator in Software Engineering. Her research areas are performance analysis and optimal control of stochastic systems, system modeling and quantitative analysis. She is an active member of Sustainable Development Goals working group in IUS for last 2 years. Email: obuyukdagli@ius.edu.ba; Orcid ID: https://orcid.org/0000-0001-5758-4607

Primrose Sibusisiwe Mlambo (PhD) hods a Master's degree in Community Development and PhD in Public Administration – Peacebuilding. She is an independent researcher, trainer and facilitator with over 25 years' experience in the development sector and working with individuals and communities affected by violence and HIV and AIDS with a specific focus on rehabilitative programmes, economic empowerment, stress and trauma healing and peacebuilding (preventative interventions, conflict transformation, mediation, management and promoting social cohesion). She is also the founder of Sisisipho Human Development Agency an organisation that seeks to address poverty and unemployment in the province of Kwa-Zulu Natal, co-founder of Agape Child Care and Community Health Programme for children who are vulnerable and in distress. Email: Feladies@gmail.com; ORCID: https://orcid.org/0009-0000-2042-7544

Peter Makaye (PhD) is a Zimbabwean academic who holds a PhD in Development Studies from Nelson Mandela University in South Africa, Master of Arts Degree in Economic History and Bachelor of Arts Honours Degree in Economic History. He has a combined university working experience of twenty-one years. His primary area of research interests are African political economy and development studies. Email: pmakaye2002@gmail.com

Shalin Bidassey-Manilal is a lecturer in the Environmental Health Department (Faculty of Health Sciences, University of Johannesburg) and is currently completing her Doctoral Degree in Environmental Health. Her main research interests are climate change, air pollution, and environmental health. One of the key objectives of The Council for Scientific and Industrial Research's major project to determine the impact of climate change on human health was addressed through Shalin's Master's Degree, which was undertaken to investigate learners' perception of thermal comfort and its heat-related health impacts as well as to examine the prevailing thermal conditions in classrooms. To supplement the findings of her master' study she undertook her doctoral studies. Shalin has published in both National and International Journals as well as presented at conferences. Email address: shalinb@uj.ac.za; ORCID ID: https://orcid.org/0000-0002-3007-3142

Sunday Israel Oyebamiji (PhD) is a researcher and teacher at the University of KwaZulu-Natal, where he received his Doctorate in International Relations. As a passionate and intellectually inclined person, he is conscious of developing new knowledge paradigms within the national and transnational African context. His research span across Strategic Studies, History, International Relations and Gender Studies; more specifically, he works in the niche of women in migration. Dr Oyebamiji has taught courses in Economic History, Political Science and IR at the undergraduate level focused on globalisation, political economy, capitalism and development studies. Email: Oyebamijis@ukzn.ac.za; Tel. +27 78 050 0804, ORCID: https://orcid.org/0000-0001-5949-6563

Thokozani P. Mbonane (PhD) is currently appointed as a Head of Department and Senior Lecturer in the Department of Health, University of Johannesburg. His highest qualification is a Doctoral degree in Environmental Health, University of Johannesburg. He lectures environmental epidemiology to undergraduate students where his research interest is. His recent research project focused on human lead levels and antisocial behaviour (violent behaviour and aggression). His current interests are on the health and social impact amongst vulnerable groups from environmental exposure to heavy metals (lead, mercury, cadmium). Email address: tmbonane@uj.ac.za;https://scholar.google.com/citations?user=OojwH98AAAAJ&hl=en; https://orcid.org/0000-0001-5802-4265; https://www.scopus.com/authid/detail.uri?authorId=57202956041

Torque Mude holds a PhD in International Politics (University of South Africa), MA in International Affairs (Midlands State University), BA (Honours) in History and Development Studies (Midlands State University) and Bachelor of Laws Degree (University of South Africa). He serves as a Postdoctoral Research Fellow at the Centre for Africa-China Studies of the University of Johannesburg. His research engages African politics, international relations, governance, peacebuilding, international law and development. Email: mudetorque@gmail.com; ORCID: https://orcid.org/0000-0002-9885-6637.

Washington Mazorodze (PhD) is currently a temporary full time lecturer at the University of Zimbabwe in the Department of Peace, Security and Society. He lectures in the fields of peacebuilding, leadership, public policy and disaster management. He holds a Bsc Honours degree in Politics and Administration, Masters in International Relations from the University of Zimbabwe and a PhD in Public Administration: Peace Studies fron Durban University of Technology. He also holds a Master Certificate in Grant Proposal writing. He has authored book chapters and journal articles; he is is also a field researcher involved in monitoring and evaluation projects such as Stopping Abuse and Female Exploitation (SAFE), Livelihoods and Food Security Programme (LFSP), Improving Access to Girls Educations (IGATE), among others in Zimbabwe. Email: washvvvone@gmail.com

Wilkister M. Shanyisa (PhD) is currently a lecturer and HoD -Justice and Peace Centre for Social Justice and Ethics at The Catholic University of Eastern Africa – Kenya. She is also a gender consultant. She holds a PhD in Public Administration – Peace Studies (ICON) from Durban University of Technology, South Africa, a Master's degree in Gender and Peacebuilding from the UN Mandated University for Peace – Costa Rica and a bachelor's degree in social sciences – Social Work from The Catholic University of Eastern Africa- Kenya. She is an author and co-author to several publications including Peacebuilding. She has varied skills; in Gender-based Violence Mitigation Strategy and peacebuilding. She has published journal articles, etc. She is passionate about gender and peace related issues. Shanyisa hopes to write more on the same and to help empower especially girls and women in the community to realise positive change where human rights remain paramount. Email: wmilimu@yahoo.com; ORCID ID: https://orcid.org/0000-0002-5476-4271

Index

J. C. K. Kiyala and N. Chivasa (eds.), *Climate Change and Socio-political Violence in Sub-Saharan Africa in the Anthropocene*, The Anthropocene: Politik—Economics—Society—Science 37,
https://doi.org/10.1007/978-3-031-48375-2

D

R

Z

SPRINGER NATURE

GPSR Compliance

The European Union's (EU) General Product Safety Regulation (GPSR) is a set of rules that requires consumer products to be safe and our obligations to ensure this.

If you have any concerns about our products, you can contact us on ProductSafety@springernature.com

In case Publisher is established outside the EU, the EU authorized representative is:

Springer Nature Customer Service Center GmbH
Europaplatz 3
69115 Heidelberg, Germany

The manufacturer's authorised representative in the EU is Springer Nature Customer Service Centre GmbH, Europaplatz 3, 69115 Heidelberg, Germany. If you have any concerns regarding our products, please contact ProductSafety@springernature.com

Printed and bound by CPI Group (UK) Ltd, Croydon, CR0 4YY

30/09/2026

02225301-0018